118   181 $41.35

178,57

D1748926

The Uncertainty Principle
and Foundations of
Quantum Mechanics

The Uncertainty Principle and Foundations of Quantum Mechanics
A Fifty Years' Survey

Edited by
William C. Price, F.R.S.
*Wheatstone Professor of Physics,
University of London King's College*

Seymour S. Chissick
*Lecturer in Chemistry,
University of London King's College*

A Wiley–Interscience Publication

JOHN WILEY & SONS
London · New York · Sydney · Toronto

Copyright © 1977, by John Wiley & Sons, Ltd.

All rights reserved.

No part of this book may be reproduced by any means, nor translated, nor transmitted into a machine language without the written permission of the publisher.

Library of Congress Cataloging in Publication Data:
Main entry under title:

The Uncertainty principle and foundations of quantum mechanics.

   'A Wiley–Interscience publication.'
   'A tribute to Professor Werner Heisenberg to commemorate the fiftieth anniversary of the formulation of quantum mechanics.'
   1. Quantum theory—Addresses, essays, lectures. 2. Heisenberg uncertainty principle—Addresses, essays, lectures, I. Price, William Charles, 1909–   . II. Chissick, Seymour S. III. Heisenberg, Werner, 1901–1976.
QC174.125.U5   530.1'2   76-18213

ISBN 0 471 99414 6

Set on Linotron Filmsetter and Printed in Great Britain by J. W. Arrowsmith Ltd., Bristol

A tribute to
Professor Werner Heisenberg to commemorate
the fiftieth anniversary of the formulation of
quantum mechanics.

List of Contributors

Bohm, David

Clarke, Christopher J. S.

Detrich, John, H.

Feldman, Gordon

Gudder, Stanley P.

Heisenberg, Werner

Hodgson, Peter E.

Kraus, K.

Kuryshkin, Vassili V.

Lanz, Ludovico

Ludwig, Guenther

Mignani, Roberto

Papp, Erhardt W. R.

Ratner, Mark A.

Rayski, Jacek M.

Rayski, Jerzy

Recami, Erasmo

Reece, Gordon

Roothaan, Clemens C. J.

Rühl, W.

Rylov, Yuri A.

Sabin, John R.

Santhanam, Thalanayar S.

Sławianowski, Jan J.

Stenger, William

Streit, Ludwig

Tassie, Lindsay J.

Trickey, Samuel B.

Van Horn, Hugh M.

Yunn, B. C.

Dedication

PROFESSOR SIR HERMANN BONDI, K.C.B.

A remarkable factor in the progress of science is the temporary concentration of interest on particular topics. Science is above all a social activity, the picture of the lonely scientist being largely a figment of an untutored imagination. Scientists hunt in packs and follow scents. Sometimes the scent is material: when the progress of technology has opened up a whole new method of experimental work, and the ways of using this newly available technique, the assimilation of the results obtained, the formulation of novel hypotheses and their means of being tested all attract a large pack of experimenters and theorists that in full cry produces astonishingly rapidly a large and novel output. On other occasions the scent is intellectual, when an awkward question has been asked and many try to find at least partial answers to it, answers that can often lead to fruitful insights and new and vital problems.

One characteristic of this pack hunting is that if a new theory leads in rapid succession to numerous and varied experimental tests, each passed with honours, each leading to yet newer applications, then hardly anybody will stop to examine critically and logically the philosophy and internal consistency of the theory. Nobody will want to do so, because if he finds no flaw, his work will be regarded as insignificant, while if he does find a flaw, his papers will be brushed aside with the comment: 'The theory works, so there must be some fault in his argument. Why waste time to sort it out when there are so many more fascinating things to be done?' Thus foundations cannot be coolly examined until well after the main part of the pack has passed the site of the excavations, many years later.

This volume brings together many illuminating phases of one of the most exciting and successful hunts in history, the formulation of quantum theory. Not only was this hunt outstanding in the range and wealth of experimental data it covered (including the extension of the applicability of a theory founded on the spectroscopy of atoms to nuclear physics), but also in its philosophical implications. Some of them appeared early, some were later grossly misunderstood and indeed exaggerated, but many are only now starting to be fully explored and begin to come into focus because only now is the site of the excavations sufficiently unencumbered to allow for deep investigations into problems of foundations.

x Dedication

Nothing could be more appropriate for such a hunt than to start with Heisenberg's own description of the origin of his celebrated uncertainty relations. Now that he is—alas—no longer with us, particular value attaches to this recent recollection of the most formative phase of modern physics by one of its foremost figures.

It is splendid to observe from the many contributions in the first two parts of the volume how lively and active the subjects of foundations and of measurement theory now are in so many parts of the world. The final two parts deal with novel aspects of formal theory and of applications, where again we live in a vigorous period of activity.

I trust that many will find this book thought provoking, enjoyable and indeed fascinating.

Foreword

HANS MATTHÖFER
Bundesminister für Forschung und Technologie, F.D.R.

Great achievements on the part of researchers are often the result of their having had the courage to leave familiar ground and to explore genuinely unknown fields. The discoverer of the quantum theory and the uncertainty principle was required to leave the solid ground of classical physics. One of the most significant changes in our comprehension of the universe—a change which is reflected in fields far removed from physics—was wrought by the departure from the determinacy of physical phenomena and by far deeper-reaching relativization of the law of causality. The quantum theory and the uncertainty principle are discoveries which have changed the basis of our way of thinking. We still cannot foresee their ultimate consequences.

Werner Heisenberg, whose passing we mourned during the preparation of this book, displayed not only the courage to leave the familiar terrain of classical physics. He also possessed the spirit to defend that which has been established as true in his field of science against nationalism and racism, even in the face of the most bitter political oppression. Both during and after the Second World War, he was therefore a guarantor of another Germany which desired peace and reconciliation among the peoples of the world.

Preface

The first thirty years of the twentieth century saw an explosive development in the physical sciences, the like of which it is improbable we shall see again. Many of the discoveries were made by comparatively young men and this has provided opportunities for the international scientific community to commemorate the fiftieth anniversary of some of the more fundamental discoveries during the lifetimes of their discoverers. This book, dedicated to Professor Werner Heisenberg, is one in a series of books, each designed as a tribute to one of the founders of modern physics. While the book was organized with the cooperation of Professor Heisenberg, it is with deep regret that we learned of his death on 15th February 1976, at the age of 74, just before going to press.

This book commemorates the formulation by Heisenberg in the Spring of 1925 of the system of mechanics known as quantum (or matrix) mechanics. The subsequent development of quantum mechanics by Heisenberg with Max Born and Pascual Jordan provided the basis for modern physics. One of Heisenberg's best known and far reaching contributions to the understanding of quantum mechanics was his Uncertainty Principle, which limits the precision of measurement of the dynamic variables of a system.

While Heisenberg's decisive contribution to physics, for which he received the Nobel Prize in 1932, was made at the age of 24, he continued to advance knowledge over a wide range of subjects: nuclear and sub-nuclear physics, S-matrix theory, solid state theory, plasma and thermonuclear physics, unified field theory, etc.

In compiling this volume, the editors have again been fortunate in securing the help and cooperation of scientists throughout the world. The aims were essentially similar to those of *Wave Mechanics, the First Fifty Years* (a tribute to Professor Louis de Broglie on the fiftieth anniversary of the discovery of the wave nature of the electron); to review aspects of the philosophical implications, past and current thinking and potential future developments in physics stemming from the fundamental discoveries associated with, in this case, Werner Heisenberg.

The Editors wish to record their thanks to the University of London King's College, for the facilities provided and to Professor David Bohm, Dr. R. J. Griffiths and Dr M. P. Melrose for reading various sections of the manuscript and for making helpful comments.

February 1976   William C. Price, F.R.S.
                                       Seymour S. Chissick

                                             University of London King's College

Contents

PART 1 QUANTUM UNCERTAINTY DESCRIPTION 1

1 Remarks on the Origin of the Relations of Uncertainty 3
Werner Heisenberg

2 In Praise of Uncertainty 7
Gordon Reece

3 On the Meaning of the Time–Energy Uncertainty Relation 13
Jerzy Rayski and Jacek M. Rayski, Jr.

4 A Time Operator and the Time–Energy Uncertainty Relation 21
Erasmo Recami

5 Quantum Theory of the Natural Space–Time Units 29
Erhardt W. R. Papp

6 Uncertain Cosmology 51
Christopher J. S. Clarke

7 Uncertainty Principle and the Problems of Joint Coordinate–Momentum Probability Density in Quantum Mechanics 61
Vassili V. Kuryshkin

PART 2 MEASUREMENT THEORY 85

8 The Problem of Measurement in Quantum Mechanics 87
Ludovico Lanz

9 The Correspondence Principle and Measurability of Physical Quantities in Quantum Mechanics 109
Yuri A. Rylov

10 Uncertainty, Correspondence and Quasiclassical Compatability 147
Jan J. Sławianowski

Contents

11	**A Theoretical Description of Single Microsystems** Guenther Ludwig	189
12	**Quantum Mechanics of Bounded Operators** Thalanayar S. Santhanam	227

PART 3 FORMAL QUANTUM THEORY — 245

13	**Four Approaches to Axiomatic Quantum Mechanics** Stanley P. Gudder	247
14	**Intermediate Problems for Eigenvalues in Quantum Theory** William Stenger	277
15	**Position Observables of the Photon** K. Kraus	293
16	**A New Approach and Experimental Outlook on Magnetic Monopoles** Erasmo Recami and Roberto Mignani	321
17	**Problems in Conformally Covariant Quantum Field Theory** W. Rühl and B. C. Yunn	325
18	**The Construction of Quantum Field Theories** Ludwig Streit	349
19	**Classical Electromagnetic and Gravitational Field Theories as Limits of Massive Quantum Theories** Gordon Feldman	365
20	**Relativistic Electromagnetic Interaction Without Quantum Electrodynamics** Clemens C. J. Roothaan and John H. Detrich	395

PART 4 APPLIED QUANTUM MECHANICS — 439

21	**The Uncertainty Principle and the Structure of White Dwarfs** Hugh M. Van Horn	441
22	**Applications of Model Hamiltonians to the Electron Dynamics of Organic Charge Transfer Salts** Mark A. Ratner, John R. Sabin and Samuel B. Trickey	461

23 Alpha-Clustering in Nuclei Peter E. Hodgson	485
24 Commutation Relations, Hydrodynamics and Inelastic Scattering by Atomic Nuclei Lindsay J. Tassie	543
25 Heisenberg's Contribution to Physics David Bohm	559
Author Index	565
Subject Index	567

PART 1

Quantum Uncertainty Description

1

Remarks on the Origin of the Relations of Uncertainty

The late Professor WERNER HEISENBERG
Director Emeritus of the Max Planck Institut für Physik und Astrophysik, Munich, Germany

The situation of quantum theory in the summer of 1926 can be characterized by two statements. The mathematical equivalence of matrix mechanics and wave mechanics had been demonstrated by Schrödinger, the consistency of the mathematical scheme could scarcely be doubted; but the physical interpretation of this formalism was still quite controversial. Schrödinger, following the original ideas of de Broglie, tried to compare the 'matter waves' with electromagnetic waves, to consider them as real, measurable waves in three-dimensional space. Therefore he preferred to discuss those cases where the configuration space had only three dimensions (one-particle systems), and he hoped, that the 'irrational' features of quantum theory, especially quantum 'jumps', could be completely avoided in wave mechanics. The stationary states of a system were defined as standing waves, their energy was really the frequency of the waves. Born on the other hand had used the configuration space of Schrödinger's theory to describe collision processes and he took the square of the wave amplitude in configuration space as the probability of finding a particle. So he emphasized the statistical character of quantum theory without attempting to describe what 'really happens' in space and time.

Schrödinger's attempt appealed to many physicists who were not willing to accept the paradoxes of quantum theory; but the discussions with him in July 1926 in Munich and in September in Copenhagen demonstrated very soon, that such a 'continuous' interpretation of wave mechanics could not even explain Planck's law of heat radiation. Since Schrödinger was not quite convinced it seemed to me extremely important to decide beyond any doubt whether or not quantum 'jumps' were an unavoidable consequence, if one accepted that part of the interpretation of matrix mechanics, which already at that time was *not* controversial, namely the assumption that the diagonal element of a matrix represents the time average of the corresponding physical variable in the stationary state considered. Therefore I discussed a system consisting of two atoms in resonance. The energy difference between two specified consecutive stationary states was assumed to be equal in the two

4 Uncertainty Principle and Foundations of Quantum Mechanics

atoms so that for the same total energy the first atom could be in the upper and the second in the lower state or vice versa. If the interaction between the two atoms is very small one should expect that the energy goes slowly forth and back between the two atoms. In this case it can easily be decided whether the energy of one of the atoms goes continuously from the upper to the lower state and back again or discontinuously by means of sudden quantum jumps. If E is the energy of this one atom then the mean square of fluctuations $\overline{\Delta E^2}$ is quite different in the two cases [equation (1)]. The calculation does not require more than the *non*-controversial assumption of matrix mechanics mentioned above. The result decided clearly in favour of the quantum jumps and against the continuous change.

$$\overline{\Delta E^2} = \overline{(E - \bar{E})^2} = \overline{E^2} - \bar{E}^2 \tag{1}$$

The success of this calculation seemed to indicate, that the non-controversial part of the interpretation of quantum mechanics should already determine uniquely the complete interpretation of the mathematical scheme, and I was convinced that there was no room left for any new assumptions in the interpretation. In fact, in the example mentioned above the square of the elements of that matrix, which transformed from the state where the total energy of the system was diagonal to the state where the energy of the one atom was diagonal, had to be considered as the corresponding probability. In the autumn of 1926 Dirac and Jordan formulated the theory of those general linear transformations which corresponded to the canonical transformations of classical mechanics and which nowadays are called the unitary transformations in Hilbert space. These authors correctly interpreted the square of the elements of the transformation matrix as the corresponding probability; this was in line with Born's older assumptions concerning the square of Schrödinger's wave function in configuration space and with the example of the resonating atoms. It was in fact the only assumption which was compatible with the old non-controversial part of the interpretation of quantum mechanics; so it seemed that the correct interpretation of the mathematical theory had finally been given.

But was it really an interpretation, was the mathematical scheme a theory of the phenomena? In physics we observe phenomena in space and time; the theory should enable us, starting from the present observation, to predict the further development of the phenomenon concerned. But at this point the real difficulties started. We observe phenomena in space and time, not in configuration space or in Hilbert space. How can we translate the result of an observation into the mathematical scheme? E.g. we observe an electron in a cloud chamber moving in a certain direction with a certain velocity; how should this fact be expressed in the mathematical language of quantum mechanics? The answer to this question was not known at the end of 1926.

For some time Schrödinger had discussed the possibility, that a wave packet obeying his wave equation could represent an electron. But as a rule a wave packet spreads out so that after some time it may be extended over a volume much bigger than that of the electron. In nature, however, an electron remains an electron; so this interpretation would not do. Schrödinger pointed out, that in one special case, the harmonic oscillator, the wave packet did not spread; but this property had to do with the special fact, that for the harmonic oscillator the frequency does not depend on the amplitude.

On the other hand there could be no doubt that de Broglie's and Schrödinger's picture of the three-dimensional matter waves did contain some truth. In the many discussions we had in Copenhagen during the months after Schrödinger's visit it was primarily Bohr who emphasized this point again and again. But what does this term 'some truth' mean? We had already too many statements which contained 'some truth'. We could, for example, compare the statements: 'The electron moves in an orbit around the nucleus.' 'The electron moves on a visible path through the cloud chamber.' 'The electron source emits a matter wave which can produce interferences in crystals like a light wave.' Each of these statements seemed to be partly true and partly not true, and certainly they did not fit together. We got the definite impression that the language we used for the description of the phenomena was not quite adequate. At the same time we saw that at least in some experiments such concepts as position or velocity of the electron, wavelength, energy had a precise meaning, their counterpart in nature could be measured very accurately. It turned out that for a well defined experimental situation we finally always arrived at the same prediction, though Bohr preferred to play between the particle- and wave-picture while I tried to use the mathematical scheme and its probabilistic interpretation. Still we were not able to get complete clarity; but we understood that the 'well defined experimental situation' somehow played an important rôle in the prediction.

In the beginning of 1927 I was for some weeks alone in Copenhagen, Bohr had gone to Norway for a skiing holiday. In this time I concentrated all my efforts on the question: How can the path of an electron in a cloud chamber be represented in the mathematical scheme of quantum mechanics? In the despair about the futility of my attempts I remembered a discussion with Einstein and his remark: 'it is the theory which decides what can be observed'. Therefore I tried to turn around the question. Is it perhaps true that only such situations occur in nature or in the experiments which can be represented in the mathematical scheme of quantum mechanics? That meant: there was not a real path of the electron in the cloud chamber. There was a sequence of water droplets. Each droplet determined inaccurately the position of the electron, and the velocity could be deduced inaccurately from the sequence of droplets. Such a situation could actually be represented in the mathematical scheme; the calculation gave a lower limit for the product of the inaccuracies of position and momentum.

It remained to be demonstrated that the result of any well defined observation would obey this relation of uncertainty. Many experiments were discussed, and Bohr again used successfully the two pictures, wave- and particle-picture, in the analysis. The results confirmed the validity of the relations of

6 Uncertainty Principle and Foundations of Quantum Mechanics

uncertainty; but in some way this outcome could be considered as trivial. Because if the process of observation itself is subject to the laws of quantum theory, it must be possible to represent its result in the mathematical scheme of this theory. But these discussions demonstrated at least that the way in which quantum theory was used in the analysis of the observations, was completely compatible with the mathematical scheme.

The main point in this new interpretation of quantum theory was the limitation in the applicability of the classical concepts. This limitation is in fact general and well defined; it applies to concepts of the particle picture, like position, velocity, energy, as well as to concepts of the wave picture like amplitude, wave length, density. In this connection it was very satisfactory that somewhat later Jordan, Klein and Wigner were able to show that Schrödinger's three-dimensional wave picture could also be subject to the process of quantization and was then—and only then—mathematically equivalent to quantum mechanics. The flexibility of the mathematical scheme illustrated Bohr's concept of complementarity. By this term 'complementarity' Bohr intended to characterize the fact that the same phenomenon can sometimes be described by very different, possibly even contradictory pictures, which are complementary in the sense that both pictures are necessary if the 'quantum' character of the phenomenon shall be made visible. The contradictions disappear when the limitation in the concepts are taken properly into account. So we spoke about the complementarity between wave picture and particle picture, or between the concepts of position and velocity. In later literature, there have been attempts to give a very precise meaning to this concept of complementarity. But it is at least not in the spirit of our discussions in the Copenhagen of 1927 if the unavoidable lack of precision in our language shall be described with extreme precision.

There have been other attempts to replace the traditional language of physics with its classical concepts for the description of the phenomena, by a new language which should be better adapted to the mathematical formalism of quantum theory. But the development of language is a historical process, and artificial languages like Esperanto have never been very successful. Actually, during the past 50 years, physicists have preferred to use the traditional language in describing their experiments with the precaution that the limitations given by the relations of uncertainty should always be kept in mind. A more precise language has not been developed, and it is in fact not needed, since there seems to be general agreement about the conclusions and predictions drawn from any given experiment in this field.

2

In Praise of Uncertainty

GORDON REECE
Imperial College, London

1. THE PSYCHOLOGICAL BASIS OF OUR NEED FOR CERTAINTY

The first post-natal experiences of a human being are necessarily associated with learning about the world in which he or she lives. Ideally, his emotional needs will be satisfied in much the same way as his physical requirements. Indeed, these various aspects are inextricably intertwined, centring on the mother's breast, which supplies at once food, warmth, reassurance and companionship.

From the point of view of a very young baby, the idea of contentment cannot be separated from his confidence in the consistency and reliability of the world as he sees it. For him, happiness means the certainty that his food will arrive when he needs it, at the correct temperature and of a reliable composition.

Later he becomes aware of non-animate objects, some of which fail to interact with him (passive objects like floors and walls), while others (like mattresses, blankets and rattles) respond when pushed or shaken. Gradually, a baby builds up a library of objects in which he can have confidence. Floors can safely be crawled on; thin air cannot. Walls can be bumped without apparent damage (to the walls) while balls and bottles roll away when pushed. He learns to categorize the objects around him. Fine gradations are learned from the varying degrees of, for example, softness of floor coverings, and intensities of light, noise and warmth. None of these distinctions, however, rivals the fundamental importance of simple 'yes/no' questions such as 'Am I hungry?' or 'Am I wet?' It is not until a baby is much older—say a year, when his feelings about the world will already have begun to gel—that he begins to confuse the issue with questions like 'Am I very hungry?'

The real source of the baby's confidence in the external world is the certainty that if something is wrong it will be remedied. Uncertainty ('Where is Mummy?', 'Where am I?' or 'Why am I still hungry?') represents insecurity, a loss of confidence in the external world and consequent unhappiness. The baby's confidence relies also on a belief in causality: 'If I cry, then Mummy will come', 'If I get milk, then I shall no longer be hungry', and the action of crying represents this reliance.

8 Uncertainty Principle and Foundations of Quantum Mechanics

Eventually the child acquires the verbal skills to express his feelings, and to extend them by asking questions. No-one who has lived under the ceaseless questioning of a normal four-year-old child can have failed to detect consistent trends in the style of interrogation. Typically, one is asked questions like: 'Why can't we live upside down?' . . . followed by 'How long can you stand on your head?', 'Why can you stand on your head longer than me? Because you're older than me?' Such questioning is aimed at imposing a simple logical structure on seemingly haphazard phenomena. The 'simplest' structures, for this purpose, are perfect correlations amounting to causal relationships (age correlated with ability to stand on one's head). Each set of phenomena is dealt with more or less in isolation. Thus it is unusual for a child to follow a set of questions like these with, say, the question 'Why does being older make you better at standing on your head?' Some causal mechanism is taken for granted, and the exact details are not necessarily of interest. Far more likely is the catch question: 'Then why can't Granny stand on her head longest? She's older still.' Already, however, the child has sufficient confidence in causality that he will tend to dismiss odd exceptions to general rules, whether these rules are ones that he has thought up for himself or generally accepted truths, such as 'The older you are the wiser and taller you are'. It takes a lot of dwarves and imbeciles to convince a child that this is not always so. And the realization that Granny, despite her age, can no longer stand on her head does not appear to have even the briefest effect on the child's confident quest for definite causal connections.

It is of little or no relevance to what extent the child's desire to impose a logical structure on the external world is in some sense innate, and to what extent it is a function of his upbringing. The only point of real significance is the *universality* of this desire, and its intensity. If it is a consequence of the direction of the child's thinking by the external world, and in particular by the adults in that world, this is a remarkable tribute to our ability to mould children in our image. There is, however, little evidence for such a view: it seems much more likely that we are born with a hefty predisposition towards a belief in causality and a desire for certainty. The most compelling evidence for this latter view is the fact that we can interpret the behaviour of animals in much the same way as we interpret the behaviour of human beings. We do not find any lack of 'logic' in the behaviour of chimpanzees, snakes or even amoebae. We do not need a special vocabulary to describe the intelligence of animals: indeed it is standard practice to use the behaviour of animals to help us understand ourselves. We *assume* that the same analysis as we know to be valid for human behaviour will give correct results when applied to other creatures: we are of course imposing our own preconceptions on their behaviour. Such methods have so far justified themselves by producing self-consistent results.

As time passes, the child grows up, matures and begins to 'think for himself'. It is an attribute of intelligent adolescents that they tend to question accepted values 'for the sake of it'. Unfortunately they have by this time lost their desire to challenge *really* fundamental 'truths' and their doubting has begun to take place within a well-defined framework of accepted authority and standard techniques. Above all, they have the confidence that all questions have answers, and that most questions have exact answers.

It is a mark of true maturity to be able to function in the absence of certainty: for example, the ability to be 'good' without the certainty of ultimate retribution for one's wickedness. It is much easier to search for minor deviations from an accepted truth, and to suggest the appropriate minor modifications, than to search freely for correlations and to discover one's own truths. In practice, it is also a good deal slower and less efficient: hence the popularity of ready-made orthodoxies of every kind.

2. THE HISTORY OF CERTAINTY

It is clear that the desire for certainty and the belief in causality are not restrictions upon human thought imposed by the peculiar requirements of the external world but vice versa. In other words we can answer Eddington's disturbing question 'How much do our theories tell us about Nature, and how much do we contribute ourselves?' as follows: the very notion of causality and the desire for certainty are imposed by us on Nature. To pursue a metaphor due to Eddington, we are inclined to trawl the data of physics with a causal net. Small wonder, then, that we turn up just what we hope to find. For example, we tend still to use the vocabulary and methodology of classical physics when dealing with the phenomena of a submicroscopic world. We search for macroscopic analogues, such as Bohr's atom, and set them up with such plausibility that they inevitably become obstacles to the further understanding of the very phenomena they purport to illuminate. Precisely because they are easy to understand in themselves the analogues tend to take on a life of their own. The first job of each succeeding generation of physicists is then to demolish the simplifications of their predecessors. The lay world, its representatives in the scientific establishment, and in times past even the Church, have all naturally thrown their weight behind conventional wisdom. The consequent emphasis on the destruction of bad old theories rather than the untrammelled construction of good new ones has hindered the development of science. In particular it has slowed down the process of acceptance of new theories by making them seem far more revolutionary than they really are.

As with so many of the implicit 'values' of science, that of precision can be attributed to the Ancient Greeks. Their preoccupation with, for example, the problem of 'commensurability' is best explained in terms of a feeling on their part that commensurable quantities (rational numbers) were 'good' and irrationals were 'bad'. The alternative possibility, that they found irrationals too difficult to handle, is not particularly plausible, since most of the theorems proved in Greek mathematics for rationals hold equally for irrationals. Thus Archimedes established the rule for balancing weights on a lever for commensurable ratios of weights, though the proof he gave did not of course require such an assumption. When the Pythagoreans proved the existence of irrationals, the notion of approximation was drawn into the vocabulary of physics.

10 Uncertainty Principle and Foundations of Quantum Mechanics

Once established, the Greek attitude to accuracy remained unchallenged for 2,000 years. During that time, Christian civilization had imposed religious standards of 'truth' on science. By associating scientific truth with religious dogma, the Church unwittingly gave science a new importance. That importance is still with us today: it stems from the need to establish new scientific theories beyond reasonable doubt before they could safely be taught. Once accepted, though, a theory wore a 'seal of approval', and could not easily be dislodged.

One dogma of science in which religion has a more than usually large stake is the idea of causality. If we do not need to seek a cause for apparently inexplicable events—such as the existence of the Universe—we do not need to turn to religion for the explanation. Moreover, once it is admitted that there are questions which not only need not but actually cannot be answered—such as 'Where is that electron and how fast is it going?', 'What is the opposite of giraffe?' or 'Why did God create the world?'—it soon becomes evident there are whole realms of human experience which will continue to defy a simple, precise causal analysis.

Plato's system of ideals, those 'absolute objects which cannot be seen other than by thought,' still underlies our attitude to mathematics and physics. We tend to think of real objects as imperfect ideal ones. Though no-one has ever found a perfectly smooth plane or an inviscid fluid, the theory of motion of solids and liquids treats friction and viscosity as unfortunate aberrations and irritations. It is more than 90 years since Rayleigh showed (Rayleigh, 1892) that the theory of viscous flow did *not* in general reduce to inviscid theory as the viscosity tended to zero: modern undergraduate mathematics has yet to acknowledge Rayleigh's discovery.

It was not in fact until the late eighteenth and early nineteenth centuries that scientists felt free to challenge orthodox theories of precision. The most difficult to swallow of all the assertions of the Greeks was the parallel axiom of Euclid. In an age of rationalism it seemed only natural to put it to the test. Gauss took his instruments and set out to establish the truth of the parallel axiom by the only method he knew—that of direct measurement. The inconclusiveness of his results opened up the road to non-Euclidean geometry. The notion that the angles of a triangle might add up to *about* 180 degrees rather than *exactly* 180 degrees was capable of overturning the whole edifice of certainty on which physics seemed to be built. The irrational numbers and the transcendentals could be treated as exceptional oddities, but a non-Euclidean world would make uncertainty pervade every aspect of physics.

Laplace is commonly credited with having conjured up a demon capable of predicting the course of all subsequent events, given complete information on every particle in the universe at any one instant. Conventionally, this is regarded as the embodiment of rationalist overconfidence. But this can be read in precisely the opposite way: even if the world were purely causal it would require an impossibly well-informed demon to make proper use of the fact. Consequently, the world will necessarily seem unpredictable to us. Because we cannot hope to obtain the necessary information, we must resign ourselves to an inability to predict the vast majority of phenomena. Thus, although this may not be a theoretical limitation, in practice it introduces a new level of uncertainty.

The laws of thermodynamics, likewise first properly formulated at the beginning of the nineteenth century, tell us more about what we can not know or do. In particular, they rule out the possibility of a perfect machine, and hence of perpetual motion. Thus another aspect of perfection had to be abandoned.

3. THE ACCEPTANCE OF UNCERTAINTY

Gradually, therefore, imperfection, inaccuracy, unpredictability, uncertainty and randomness were accepted into physics. It is reasonable to relate this increasing tolerance with the growing maturity of science. By analogy with the growth of sophistication in the human being, we see that the history of science is the story of the realization that the world is not so simple as we should like it to be, that we cannot hope to achieve absolute certainty, and that we cannot hope to know or understand everything. Nor is it necessary to 'explain' everything that we do not understand, as the manifestation of some supernatural force, simply because we do not understand it. Belief in such dogmas is arrogance thinly disguised as humility. *True* humility in science consists in knowing that we do not know.

The culmination of the acceptance of uncertainty came in the decade between 1925 and 1935. 1926–7 saw publications by Heisenberg identifying the inherent uncertainty associated with certain measurements. In 1931 Goedel published his *Ueber formal unentscheidbare Saetze der Principia Mathematica*, showing that the axiomatic method itself had inherent limitations. In 1934 Popper published the *Logik der Forschung*, which showed that the nature of a scientific hypothesis required its falsifiability, and which finally demolished the notion of absolute scientific proof (or disproof). It was thus doubt and scepticism that distinguished the scientist, and not confidence and certainty.

It is clear that the awareness of the fallibility of the tools they use has made scientists much more careful in the way they derive and present their results. For example, the interaction between observer and phenomenon is recognized as a crucial factor in any sociological investigation. Indeed, in all experimental work involving living creatures, the effect of the experiment itself—the presence of experimenters and their measuring instruments—on the outcome is now recognized.

Logically, the next step should be to review our approach to the publication of the results of scientific investigations. Success is essentially trivial: it is *failure* to detect a satisfactory, simple causal explanation of a phenomenon that stimulates speculation. Currently, scientific journals concentrate on the essential work of cataloguing success. How much more exciting would be the

12   Uncertainty Principle and Foundations of Quantum Mechanics

publication of phenomena that defy correlation. It is 'anomalies' (such as that of the motion of the planet Mercury) that point the way out of inadequate theories and into the excitement of new fields.

It is easy to point to analogies between the position of quantum mechanics in 1926, and that of fundamental particle physics in 1976. It may well be that in order to resolve their current dilemma physicists may once again have to think the unthinkable, and challenge the very foundations of their subject.

REFERENCE

Strutt, J. W. [3rd Baron Rayleigh] (1892) 'On the question of the stability of the flow of fluids', *Phil. Mag.*, **34,** 59–70.

On the Meaning of the Time–Energy Uncertainty Relation

JERZY RAYSKI and JACEK M. RAYSKI, JR.
Jagiellononian University, Cracow, Poland

Soon after the formulation of the usual uncertainty relations between Cartesian coordinates of particles and their momenta

$$\Delta x \cdot \Delta p_x \simeq \Delta y \cdot \Delta p_y \simeq \Delta z \cdot \Delta p_z \simeq \hbar \qquad (1)$$

there appeared the problem of the existence and meaning of a similar relation between time and energy

$$\Delta t \cdot \Delta E \simeq \hbar \qquad (2)$$

constituting a natural extension of the three relations [given in equation (1)] of Heisenberg from the point of view of the special theory of relativity. References to the energy–time uncertainty problems may be found discussed, for example, by Carruthers and Nieto (1968).

The relation (2) is not derivable from the formalism of quantum mechanics in the same way as relations (1) were derived, neither can the interpretation of relation (2) be quite analogous to the ordinary interpretation of (1).

First of all, in contradistinction to the coordinates x, y, z, the time variable t is not an operator associated with an observable characterizing the particle but is a universal parameter. Moreover, energy is not a generalized momentum canonically conjugate to the time variable t in the usual sense of this word. In consequence of the fact that t is not an observable but a parameter, the opinion of most physicists is not favourable towards the possibility of regarding the operator $i\hbar(\partial/\partial t)$ as the operator of energy, and is against the suggestion of relating the non-commutability of $i\hbar(\partial/\partial t)$ and t with an impossibility of their 'simultaneous' determination and, consequently, of an appearance of the uncertainty relation (2). Incidentally, it is not at all clear what is meant by a 'simultaneous determination' of t and of any other physical quantity. Any measurement of a physical quantity at a given instant of time means a simultaneous determination of both this quantity and the time at that instant.

For the sake of completeness it should be mentioned that time and energy do form a pair: a generalized coordinate and its canonically conjugated momentum within the framework of the so-called homogeneous canonical formalism, but this formalism does not constitute a basis and starting point for quantization. The latter is performed with the help of the ordinary canonical formalism where the energy, i.e. the Hamiltonian, is not to be regarded as one more of the generalized coordinates or momenta.

It was argued that the Hamiltonian H rather than the operator $i\hbar(\partial/\partial t)$ plays the role of the operator of energy and—in order to formulate the fourth uncertainty relation—one should look for an operator which would play the role of a generalized coordinate, canonically conjugate to the Hamiltonian (taken to be a generalized momentum) or, vice versa, to look for a generalized momentum canonically conjugate to the energy (the latter regarded to be a generalized coordinate). But there are serious difficulties with defining such an operator. To show some of them let us limit our considerations to one space-like dimension x. The Hamiltonian for a free particle in the non-relativistic quantum mechanics is $p^2/2m$, and the (formal!) operator $\underline{t}$ satisfying the relation

$$[\underline{H}, \underline{t}] = -i\hbar \tag{3}$$

is

$$\underline{t} = \frac{m}{2}\left(x\frac{1}{p} + \frac{1}{p}x\right) + f(x) \tag{4}$$

where $f(x)$ is an arbitrary function of x. (There is a correspondence between the operator (4) and the classical time if one puts $f(x) = 0$ and recalls the fact that in classical physics $p = mv$ and for a free particle $v = x/t$.) However, the trouble is that the operator (4) is not well defined because the inverse of the operator p does not exist inasmuch as the spectrum of p includes the value zero. Thus, the domains of definition of the operators H and t given by (4) are not the same. Consequently, the operator H cannot play the role of a generalized coordinate correlated with a canonically conjugate momentum and it is impossible to derive the fourth uncertainty relation in an analogous way to that which was used to prove the usual three relations of Heisenberg.

Recently Eberly and Singh (1973) claimed to have circumvented this difficulty by constructing a reciprocal time-operator. However, their determination of the fourth uncertainty relation has been achieved in a very round-about way so that we shall not present it here. In what follows, we will present another, very straightforward and direct derivation of this relation.

Not only a derivation of relation (2) but also its interpretation in a way which is closely analogous to the interpretation of the usual relations (1) seems to be impossible. In fact, according to quantum mechanics, if the energy spectrum is discrete we may construct a stationary solution of the Schrödinger equation describing the system in an eigenstate of energy. In this case energy is exactly known for any time instant determined with an arbitrarily high precision $\Delta t < \varepsilon$. But also in the case of a continuous energy spectrum it is possible to construct a solution so that the energy is determined up to an arbitrarily small uncertainty $\Delta E < \varepsilon$, and this solution remains almost stationary for a very long time interval, i.e. energy is known almost exactly for any instant (determined with an arbitrarily high precision) within a long time interval. This contradicts sharply a naive interpretation of formula (2) according to which Δt means uncertainty of the time instant at which the particle possessed an energy E known within the limits of inaccuracy ΔE.

One could look for an excuse and explanation of the appearance of the above-mentioned difficulties with the uncertainty relation (2) in the fact that the ordinary quantum mechanics is a non-relativistic theory. Being in disaccord with the requirements of relativity quantum mechanics is treating the time coordinate on a different footing as compared with the three space-like coordinates of the particles. This may constitute a reason why relation (2) does not hold true in the ordinary quantum-mechanical description of physical phenomena. On the other hand, relation (2) is known to be satisfied in quite another context, viz. as a relation between the uncertainty of energy and the mean lifetime of unstable particles. But unstable particles are not described satisfactorily within the framework of quantum mechanics. They may be described consistently only within the framework of quantum field theory where the number of particles is observable which does not need to be a constant of motion. As is well known, it is quantum field theory but not quantum mechanics that may be truly reconciled with relativity and so the accord of relation (2) with quantum field theory as well as the disaccord with quantum mechanics seem to be explicable.

The above excuse for the appearance of serious difficulties with the problem of relation (2) in quantum mechanics is not convincing. The ordinary quantum mechanics of a single particle may be regarded as a limiting case of quantum field theory in the low-energy region and in the subspace of one-particle states (as the number of massive particles becomes constant in the low-energy limit). Thus, if relation (2) holds true (in a certain sense) in quantum field theory, it should also remain valid in the above-mentioned limit. Indeed, relation (2) does not involve the magnitude of the mean value of energy and should be valid also in the low-energy limit.

We may present still another argument against the view that the non-relativistic form of the ordinary quantum mechanics is to be blamed for the difficulties appearing in connection with the problem of relation (2). It is a common feature of relativistic theories that 'fourth' relations (completing some three-dimensional relations known from the pre-relativistic physics) are often only formally analogous to their three-dimensional counterparts whereas their meaning and interpretation are different. Let us illustrate this statement with an example: In the relativistic extension of Newtonian mechanics there appears a fourth equation of motion of a point particle, formally quite similar to the ordinary three equations. But its physical and even mathematical sense is quite different: it does not introduce any new degree of freedom, does not increase the number of independent equations of motion because it is dependent upon the usual three equations of motion and relates the energy change to the work, i.e. expresses the law of conservation of energy. In the non-relativistic limit this fourth relation does not disappear because energy is conserved also in the non-relativistic dynamics. Similarly, if in a relativistic theory the existence of a fourth relation (2) is to be expected, this relation should appear also in the limiting case of a non-relativistic theory, although its motivation and its physical meaning do not need to be similar to those of the usual relations (1).

In order to show that the uncertainty relations (1) as well as (2) must hold true it is not necessary to invest the whole machinery of quantum mechanics but one may limit oneself to a consideration of de Broglie wave packets. Let us consider a function $f(x)$ in R_1 and define the dispersion (uncertainty) of x with respect to $f(x)$ in the usual way

$$\Delta x = (f, [x - (f, xf)]^2 f)^{1/2} \tag{5}$$

where

$$(f, A \cdot f) = \int dx f^* A f \tag{6}$$

Let us consider the Fourier transform of the function $f(x)$

$$g(k) = \frac{1}{(2\pi)^{1/2}} \int dx f(x) e^{ikx} \tag{7}$$

A well-known mathematical theorem (see e.g. Heisenberg, 1930) says that the minimum of the product of dispersions $\Delta x \cdot \Delta k$ is obtained if $f(x)$ is of a Gaussian shape

$$f(x) = \frac{1}{(2\pi)^{1/2}} \exp(-x^2/(\Delta x)^2) \tag{8}$$

Then also $g(k)$ is of a Gaussian form

$$g(k) = \frac{1}{(2\pi)^{1/2}} \exp(-k^2/(\Delta k)^2) \tag{9}$$

where Δx and Δk appearing in (8) and (9) are identical with the dispersions defined according to (5). Moreover, their product is shown to be

$$\Delta x \cdot \Delta k = \tfrac{1}{2} \tag{10}$$

This is a mathematical fact, quite independent of the meaning of the variables x and k. By identifying x with a Cartesian coordinate of a particle in the ordinary space and k with the inverse of de Broglie's wave length divided by 2π, so that $k = 2\pi \lambda^{-1}$ one gets the usual uncertainty relation between the Cartesian coordinate and momentum (expressed in such units that $\hbar = 1$). But we may as well replace x by t and k by the frequency ω which yields

$$\Delta t \cdot \Delta \omega = \tfrac{1}{2} \quad \text{or} \quad \Delta t \cdot \Delta E = \frac{\hbar}{2} \tag{11}$$

Thus, the fourth uncertainty relation (2), or more precisely (11), is a direct consequence of the wave aspect of matter.

While the existence of relation (11) is beyond any doubt the problem of its interpretation still remains open. Let us stress once more that the following interpretation: 'the information about the value E of the energy of a particle and the information about the instant t at which it possessed this amount of energy are incompatible unless both informations are subject to uncertainties ΔE and Δt whose product is not smaller than '$\hbar/2$' is incorrect. In order to find out a correct interpretation of the fourth uncertainty relation let us come back once more to a discussion of the ordinary uncertainty relations between position and momentum. In this case it is also incorrect to say simply that these relations mean an impossibility of surpassing the exactitude of information about momentum and position of a particle beyond the limits imposed by the formulae (1). This last statement is not correct because one can measure the position of a particle first (say at t_1) with an arbitrary exactitude and afterwards (say at t_2) measure its momentum also with an arbitrarily high precision so that in the interval (t_1, t_2) bounded by the two instants of measurements the exactitude of our information about position and momentum surpasses, indeed, the limits imposed by the relations (1).

Heisenberg's uncertainty relations, if correctly understood, mean something else, namely the following two facts: (*a*) A *simultaneous* direct measurement of coordinate and momentum of a particle with an exactitude surpassing the limits (1) is impossible. (*b*) If the two measurements were performed consecutively then only the result of the latter may be used for probabilistic predictions of the future while the result of the former measurement becomes completely disactualized* and invalidated due to the uncontrollable disturbance of the particle by the latter measurement.

Thus, the point of utmost importance as regards the correct interpretation of (1) is that it determines the limits for an accuracy of *simultaneous* (i.e. at a fixed instant t_0) measurements of x and p and, consequently, for a maximal precision of prescribing the *initial values* of the parameters of the system that are necessary for the computation of its temporal development.

Substituting x by t and p by E we also must not forget to perform *suitable substitutions* in the interpretational comments: Exactly as (1) is valid for a fixed value $t = t_0$, the relation (2) *must be valid for a fixed value $x = x_0$*, otherwise the analogy of the two uncertainty relations (in a two-dimensional space–time) would be incomplete and would lead us astray. But what does it mean that the relation (2) applies to a fixed point $x = x_0$? Obviously, it means that if one is observing the particle (represented by a wave packet with a given ΔE, to pass the point $x = x_0$ during its propagation along the x-axis then one is unable to say *when* it will pass the point $x = x_0$ with an exactitude greater than $\Delta t \simeq \hbar/\Delta E$. The more exact is the knowledge of the particle energy the less exact is the time instant of passing (of this particle) by a fixed point on the x-axis and vice versa: the more exactly we know the instant at which a particle passed by an arbitrary but fixed point (on the x axis) the less exact must be our knowledge about its

---

*It remains valid for probabilistic retrodictions of the past. (See J. Rayski, 1973).

energy. Such is the *proper sense of the time–energy uncertainty relation* for a free particle in a two-dimensional space–time. To our knowledge, such interpretation has not been stated explicitly in any of the extremely numerous scientific articles and textbooks on quantum theory.

Going over from two- to a three-dimensional space–time the fixed point turns over into a fixed line, and going over to a four-dimensional space–time it becomes a surface. In this case the fourth uncertainty relation may be interpreted as follows: Δt means the uncertainty of the instant when the particle will cross this surface. The product of this uncertainty Δt and the uncertainty of energy ΔE cannot be smaller than $\frac{1}{2}\hbar$.

The above-mentioned surface may be a closed surface constituting the boundary of a three-dimensional domain whose volume V may be assumed to be finite, and we may ask about the instant when a particle will cross this surface and enter the domain in question. Again, the knowledge of the instant of crossing this boundary by an ingoing particle cannot be made certain beyond the exactitude imposed by the uncertainty relation (2).

The question about ingoing particles which enter a given domain by passing from its exterior into its interior across its surface is a problem of boundary conditions. In quantum mechanics one usually considers either wave functions in the whole space or in a finite domain but with non-penetrable walls. In neither case does the problem of how many particles and when, enter or leave the domain in question appear. But it is a very natural problem to consider a finite domain in space and to ask for a solution of the Schrödinger equation in this domain under given initial conditions (say at $t = 0$) and under some boundary conditions determining the ingoing waves, i.e. the waves crossing the surface of the domain into its interior (for $t > 0$). Such mixed boundary-initial conditions determine uniquely the solution in this domain for $t > 0$ and enable one to compute the outgoing waves crossing the surface of the domain from its interior to its exterior. This is the most natural approach to a description of scattering phenomena occurring in a finite domain.

Now, whereas the initial conditions at $t = 0$ have to be consistent with the ordinary uncertainty relations between coordinates and momenta, the *boundary conditions* for $t > 0$ *must be consistent with the time–energy uncertainty relation*: The knowledge of *when* an ingoing particle enters the domain in question and the knowledge of its *energy* are subject to uncertainties satisfying the relation (11).

In conclusion it may be stated that the ordinary uncertainty relations are related to the initial value problems at a space-like hypersurface whereas the energy–time uncertainty relation is connected with the boundary problems on closed time-like hypersurfaces, e.g. abstract (i.e. freely penetrable) walls restricting a finite domain during a finite or infinite time interval.

Hitherto in our discussion we tacitly assumed wave packets describing free particles. This fact reminds us of an objection raised by Eberly and coworkers (1973) in a footnote to their article. We quote: 'The conventional understanding is essentially dichotomous. That is, the uncertainty times associated with wave packet spreading and with excited-state decay are regarded as unrelated consequences of the uncertainty principle. This point of view is apparent in every quantum mechanics text known to the authors.' The question arises whether this objection applies also to our understanding of the energy–time uncertainty relation.

The first reason for the appearance of this dichotomy is simply the fact that quantum mechanics is principally unable to describe unstable systems. Therefore the relation (2) applied to the lifetimes of unstable systems and the uncertainty of their rest masses can be applied only to quantum field theory where the numbers of particles are not constants of motion. But assuming quantum field theory we may ask the following question:

How can we know that an unstable particle has undergone a decay? Obviously by surrounding a macroscopic domain D in the interior of which the particle is situated by detectors in order to register the decay fragments outgoing from the domain through its boundary (equipped with detectors). But this is just a particular case of the above described boundary-initial problems:

At the initial instant t_0 we assume the presence in the domain D of an unstable system characterized by an uncertainty of energy ΔE. As for the boundary condition we assume that no particles will penetrate into the domain from its exterior at $t > t_0$. We look for outgoing waves of the decay fragments. According to our previous discussion the time of crossing the boundary by the decay fragments must remain uncertain within $\Delta t \simeq \hbar/\Delta E$. But the uncertainty of the instant of escaping from the domain is related to a similar uncertainty Δt of the decay instant. Thus, the time of decay counted from an arbitrary initial time instant t_0 (when the system was known to be still a bound state) could be anything within the interval $(t_0, t_0 + \Delta t)$. Consequently, the mean lifetime of the system is something like one half of Δt and the product $\Delta t \cdot \Delta E$ is, indeed, of the order of magnitude of Planck's constant.

Let us remark that the boundary condition consisting of an assumption that no particles enter the domain for $t > t_0$ was necessary because otherwise we would have to deal with an induced decay which might affect considerably the mean lifetime of the unstable system.

From the above discussion it is obvious that the two uncertainties: one connected with the spreading out of wave packets representing free stable particles and the other connected with the problem of the lifetime of unstable systems are not dichotomous and the objection of Eberly and Singh does not apply to our interpretation of the energy–time uncertainty relation.

Our explanation of the fourth uncertainty relation may be summarized as follows: if there existed a well defined 'time-operator' canonically conjugate to the Hamiltonian then the fourth uncertainty relation would be independent of the usual ones. But it is not the case. Time does not need to be and, in fact, is not an operator but a mere parameter. However, similarly as the fourth equation of Newton in relativistic mechanics is not independent from the remaining three equations but is a consequence of them, also the fourth uncertainty relation exists and is a straightforward consequence of the remaining three relations and the wave character of particles. The proper interpretation of the energy–time relation is connected with the problem of boundary conditions in quite an analogous way to that in which the usual uncertainty relations are connected with the problem of initial conditions. In particular, the uncertainty Δt is related to the problem of *when* a particle is penetrating across a given surface.

REFERENCES

Carruthers, P. and Nieto, N. M. (1968) *Rev. Mod. Phys.*, **40,** 411.
Eberly, J. and Singh, L. P. S. (1973) 'Time operators, partial stationarity, and the energy', *Phys. Rev. D*, **7,** 359–362.
Heisenberg, W. (1930) *Die Physikalischen Prinzipien der Quantentheorie*, S. Hirzel, Leipzig.
Rayski, J. (1973) 'The possibility of a more realistic interpretation of quantum mechanics', *Foundations of Phys.*, **3,** 89–100.

A Time Operator and the Time–Energy Uncertainty Relation

ERASMO RECAMI
University of Catania, Italy

1. INTRODUCTION

In nuclear physics and in elementary particle physics (at low energies) it is usual to have recourse only to monochromatic plane waves and to the *time-independent* formulation of quantum mechanics.

With the aim of making quantum mechanics as 'realistic' as possible, let us on the contrary adopt a *space–time description* of the collision phenomena, by introducing wave packets. Notice that, even when dealing with many wave packets, it is not necessary at all to have recourse to unphysical, multidimensional spaces. On the contrary, if we want to preserve the individuality of the considered packets, we must just supply a temporal (*realistic* and *physical*) description of them within the ordinary, three-dimensional space.

As soon as a space–time description of interactions has been accepted, one can immediately realize, even in the framework of the usual wave-packet formalism, (Olkhovsky and Recami, 1968, 1969) that a quantum operator for the observable time is operating. Namely, it is implicitly used for calculating the packet time-coordinate, the flight-times, the interaction-durations, the mean-lifetimes of metastable states and so on (Recami, 1970; Olkhovsky and Recami, 1970; Baldo and Recami, 1969; Olkhovsky, 1967). A preliminary, heuristic inspection of the formalism (Olkhovsky and Recami, 1968, 1970) suggests the adoption of the following 'operators' (Olkhovsky and Recami, 1970; Baldo and Recami, 1969)

$$\hat{t}_1 = -i\frac{\partial}{\partial E}; \quad \hat{t}_2 = -\frac{i}{2}\frac{\overleftrightarrow{\partial}}{\partial E}, \quad [E \equiv E_{\text{tot}}] \tag{1}$$

acting on a wave-packet space which we must carefully define [because of the differential character of the 'operators' (1)].

2. MATHEMATICAL INTRODUCTION

Let us first consider, for simplicity, a *free* particle in the *one-dimensional* case, i.e. the packet:

$$F(t, x) = \int_0^\infty dp \cdot \tilde{F}(E, p) \cdot \exp[i(px - Et)] \quad (2)$$

where $\hbar = 1$ and $E \equiv p^2/m_0$. The integral runs only over the positive values owing to the 'boundary' conditions imposed by the initial (*source*) and final (*detector*) experimental devices. Notice that, in so doing, we chose as the frame of reference that one in which source and detector are at rest: i.e. the laboratory reference frame. In particular, notice that we are considering for simplicity the case of source and detector *at rest one with respect to the other*.

Let us now observe that the packet (average) position is always to be calculated at a fixed time $t = \bar{t}$; analogously, the packet time-coordinate is always to be calculated (by suitably averaging over the packet) for a position $x = \bar{x}$ along a particular packet-propagation-ray. Therefore, in our case we can fix a particular $x = \bar{x}$, and restrict ourselves to considering, instead the packets (2), the functions:

$$F(t, \bar{x}) = \int_0^\infty dp \cdot f'(p, \bar{x}) \cdot \exp[-iEt] = \int_{(+)} dE \cdot f(E, \bar{x}) \cdot \exp[-iEt] \quad (3)$$

where $E \equiv E_{\text{tot}} \equiv E_{\text{kin}} \equiv p^2/m_0$; and $f' = f \cdot dE/d|p|$. Functions $F(t, \bar{x})$ and $f(E, \bar{x})$, being *only* functions either of t or of E, respectively, are neither wave functions (that satisfy any Schrödinger equation), nor do they represent states in the chronotopic or four-momentum spaces. Let us briefly set:

$$F \equiv F(t) \equiv F(t, \bar{x}); \quad f \equiv f(E) \equiv f(E, \bar{x}) \quad (4a)$$

It is easy to go from functions F, or f, back to the 'physical' wave packets, so that one gets a one-to-one correspondence between our functions and the 'physical states'. We shall respectively call 《space t》 and 《space E》 the functional spaces of the F's and of the transformed functions f, with the mathematical conditions that we are going to specify. In those spaces, for example, the norms will be:

$$\|F\| = \int_{-\infty}^\infty |F|^2 \, dt; \quad \|f\| = \int_0^\infty |f|^2 \, dE \quad (4b)$$

In any case, due to equations (3), the space t and the space E are representations of the same abstract space P, where we indicate

$$F \to |F\rangle; \quad f \to |f\rangle \quad (4c)$$

where $|F\rangle \equiv |f\rangle$. For reasons which we shall see later, let us now *specify* what has previously been said by assuming that space P is the space of the continuous, differentiable, square-integrable functions f that satisfy the conditions:

$$\int_0^\infty |f|^2 \, dE < \infty; \quad \int_0^\infty \left|\frac{\partial f}{\partial E}\right|^2 dE < \infty; \quad \int_0^\infty |f|^2 E^2 \, dE < \infty \tag{5}$$

Such a space is *dense* (von Neumann, 1932) in the Hilbert space of L^2 functions defined over the interval $0 \leq E < \infty$.

3. DEFINITION OF THE TIME OPERATOR

Still within the framework of the usual quantum mechanics with wave packets, let us define in the most natural way:

$$\boxed{\langle t(\bar{x}) \rangle \equiv \frac{\int_{-\infty}^\infty \rho(t, \bar{x}) t \, dt}{\int_{-\infty}^\infty \rho(t, \bar{x}) \, dt}} \quad ; \quad \rho \equiv |F|^2 \tag{6}$$

Then we can immediately calculate that

$$\langle t(\bar{x}) \rangle \equiv \langle F|\hat{t}|F \rangle = \frac{1}{N} \int_{-\infty}^\infty [F^* t F] \, dt \tag{7}$$

where N is the normalization factor, and verify that

$$\frac{1}{N} \int_0^\infty \left[f^* \left(-\frac{i}{2} \frac{\overleftrightarrow{\partial}}{\partial E} \right) f \right] dE = \langle t(\bar{x}) \rangle \tag{8}$$

whence:

$$\boxed{\langle F|t|F \rangle = \langle f| -\frac{i}{2} \frac{\overleftrightarrow{\partial}}{\partial E} |f \rangle} \tag{9}$$

This would suggest adopting as the time 'operator' the bilinear *derivation*

$$\boxed{\hat{t} \equiv \hat{t}_2 \equiv -\frac{i}{2} \frac{\overleftrightarrow{\partial}}{\partial E}} \tag{10a}$$

By easy calculations, one realizes that we can also adopt the (standard) *operator*

$$\hat{t} \equiv t_1 \equiv -i \frac{\partial}{\partial E} \tag{10b}$$

even if at the price of imposing on space-P functions the *subsidiary condition* $f(0, \bar{x}) = 0$, which is not fully desirable from a physical viewpoint. Since for using bilinear derivation (10a) as a (bilinear) operator a new formalism should be introduced (Olkhovsky and Recami, 1970), let us prefer here the time operator (10b).

4. TIME-OPERATOR PROPERTIES

Our operator (10b) has many good properties as listed below.

(1). Equation (9) shows that, in the space t, it reduces—as is very natural—to the mere multiplication by t.

(2). Relations such as equation (8) become physically clear when written:

$$\langle t(\bar{x}) \rangle = \frac{1}{N} \int_0^\infty dE |f|^2 \left(\frac{\bar{x}}{v} + \frac{\partial \arg f}{\partial E} \right) \tag{8a}$$

whence, in accordance with the *Ehrenfest principle*, it follows that:

$$\langle \hat{t} \rangle = t_0 + \bar{x}/\langle v \rangle \tag{11}$$

(3). When we pass to a new frame of reference, source and detector will no more be at rest: However, only the packet properties *relative* to the detector (and to the source) will still be essential. This is enough to secure the Galilean *invariance* of our operator.

(4). In the impulse representation, one meets the interesting correspondence ($\hbar = 1$):

$$\frac{i}{2} \frac{\vec{\partial}}{\partial E} \leftrightarrow \frac{m_0}{2} \left[\hat{x} \cdot \frac{1}{p} + \frac{1}{p} \cdot \hat{x} \right] + \frac{im_0}{2p^2} \tag{12}$$

where the last addendum vanishes when $\hbar \to 0$.

(5). We have seen already that the space of the (continuous, differentiable) functions satisfying conditions (5) is *dense* in the Hilbert space of L^2 functions *defined over the interval* $0 \le E < \infty$. Firstly equation (5) is the condition for square integrability. Secondly equation (5) requires that our operator (10) transform Hilbert-space vectors into Hilbert-space vectors. Thirdly equation (5) requires that in our space a 'good' energy-operator can be defined.

It is easy to verify that our operator (10b) is *canonically conjugated* (Heisenberg, 1944) *to the (total) energy*:

$$\boxed{[\hat{t}, \hat{E}] = -i\hbar} \tag{13}$$

(6). Under conditions (5), one gets that:

$$\boxed{\int_0^\infty f_1^* \hat{t} f_2 \, dE = \int_0^\infty (\hat{t} f_1)^* f_2 \, dE} \tag{14}$$

i.e. that our time operator is not only *Hermitian*, but also *symmetric*, according to the usual mathematical terminology (Akhieser and Gladsman, 1954).

(7). Having now the time operator (10b) at our disposal, we can immediately obtain—through the standard procedure (see, for example, Caldirola, 1966)—the uncertainty correlation:

$$\Delta E \cdot \Delta t \geq \frac{\hbar}{2} \qquad (15)$$

In our opinion, equation (15) means that in general the uncertainty ΔE that one meets when *measuring* the energy E of a particle is tied to the duration of the actual *measurement interaction* by relation (15). For example, let us suppose that we are *measuring* the energy of a particle by observing its track in a bubble chamber. If we examine (by means of a photograph) a *long* track segment, we will be able to have good 'statistics' in counting bubbles, and therefore a good determination of the (average) energy of the particle while producing that track; but the time instant at which the particle possessed that energy will be known with a large uncertainty. Vice versa, if we examine a *short* track segment, then we shall get a good time measure, but at the price of poor bubble-statistics (see Figure 1). In this example, the experiment—or better the *measurement*—is the track-segment *examination*.

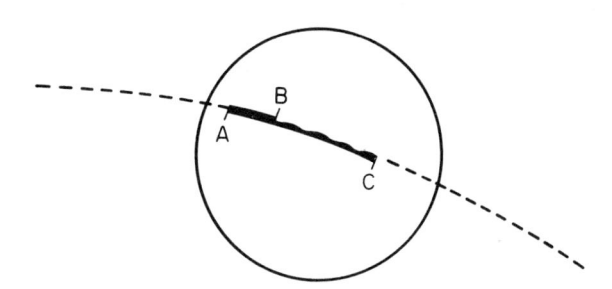

Figure 1 Track of particle in a bubble chamber

5. CASE OF POTENTIAL SCATTERING

When passing to the non-free case, things do not essentially change. Let us consider, for example, the case of the scattering of a (spin free) particle by a central potential $V(r)$. Inside the potential region, we have packets of partial l-waves, distorted by the potential (Calogero, 1967). By the introduction of certain functions $F_l^{(in)}(t, \bar{r})$, $F_l^{(out)}(t, \bar{r})$, and of the transformed ones $B_l^{(in)}(p, \bar{r})$, $B_l^{(out)}(p, \bar{r})$ (Olkhovsky, Recami and Gerasimchuk, 1974), the time durations are still got by using operator (10); and one will still write:

$$\langle F_l | t | F_l \rangle_{in,out} = \left\langle B_l \left| -\frac{i}{2} \frac{\overleftrightarrow{\partial}}{\partial E} \right| B_l \right\rangle_{in,out} \qquad (16)$$

Analogously, also equation (13) is still valid, and so on.

In the particular case of *metastable states* (Olkhovsky and Recami, 1968; Olkhovsky, 1968; Recami, 1970), let us admit that $V(\bar{r}) \equiv 0$ for $\bar{r} > R$, quantity R being the potential radius (see Figure 2). Let us analyse the process: free initial flight; unstable state formation; and decay with subsequent free final flight. Let us calculate the time τ_l spent by the particle (or better by its l partial wave) inside a sphere with centre in the potential centre and with radius $\bar{r} > R$.

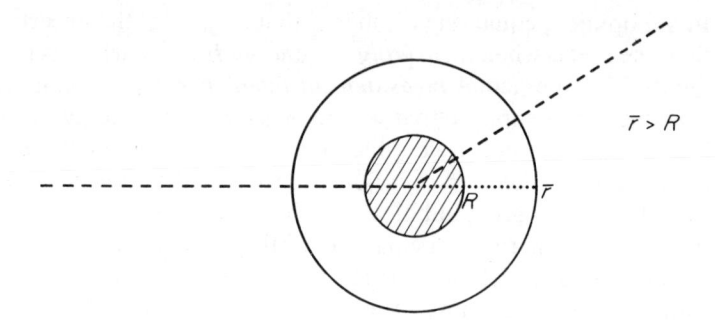

Figure 2 The scattering of a particle by a central potential

When in the presence of a *resonant* elastic scattering, we have:

$$S_l = \tilde{S}_l \frac{E - E_0 - i\Gamma}{E - E_0 + i\Gamma}; \quad \delta_l = \tilde{\delta}_l - \arctan \frac{\Gamma}{E - E_0} \tag{17}$$

where $\tilde{S}_l$ and $\tilde{\delta}_l$ are smoothly varying functions in the 'resonance' region. In the *narrow resonance* approximation, for sufficiently large values of $\bar{r}$ one obtains:

$$\tau_l \simeq 2\bar{r}\langle v^{-1} \rangle + \frac{\hbar}{\Gamma} \tag{18}$$

Analogously, one can calculate the duration of the *interaction* (Olkhovsky and Recami, 1968)—or of partial interactions (Olkhovsky and Recami, 1969)—in a two-wave packet collision (Olkhovsky, Sokolov and Zaychenko, 1969).

In particular, it seems useful to calculate the interaction duration, $\langle \Delta t \rangle_{\text{int}}$, corresponding to the cross-section enhancements: the necessary condition for the peak to be associated with a true resonance will be that $\langle \Delta t \rangle_{\text{int}}$ also has a maximum at the considered energy.

6. WHY A TIME OPERATOR WAS NOT INTRODUCED IN STANDARD QUANTUM MECHANICS

After what we have seen of the good behaviour of our operator (10), we can ask ourselves why a time-operator was not introduced in standard quantum mechanics, even if quantum mechanics is typically built up by associating an operator to every observable. The reason is that operator (10), defined as acting on the space P, does *not* become hypermaximal (von Neumann, 1932), because of the fact that P is a space of functions defined only over the interval $0 \le E < \infty$ and not over the whole E-axis. It follows that $\hat{t}$, while being *Hermitian* and *symmetric*, is however not self-adjoint, and does not allow identity resolution. Essentially because of these reasons, Pauli (1958) objected to the use of a time-operator, and this had the effect of practically stopping studies on the subject.

Von Neumann himself, however, had claimed—followed by other authors (e.g. Engelman and Fick, 1963, 1964, 1959; Razavy, 1969, 1967; Landau and Lifshitz, 1963; Aharonov and Bohm, 1961; Papp, 1971, 1972; Rosenbaum, 1969)—that considering in quantum mechanics only self-adjoint operators could be too restrictive. This is our conviction: In fact, even if operator $\hat{t}$ does not admit true eigenfunctions, nevertheless we succeeded in calculating the *average values* of $\hat{t}$ over our functions (and over the physical 'packets' corresponding to them). And that is enough for us. That is also the reason why, after equations (10), we have often written the bilinear form (10a) instead of the standard *operator* (10b).

To clarify the problem, we shall quote an explanatory example (von Neumann, 1932): Let us consider a particle Q, free to move in a semispace bounded by a rigid wall (see Figure 3). We shall then have $0 \le x < \infty$. Consequently, the impulse x-component of Q, which reads

$$\hat{p}_x = -i\frac{\partial}{\partial x} \qquad (19)$$

will be a non-hypermaximal, non-self-adjoint (but only Hermitian, symmetric) operator, even if it is an observable and has a simple physical meaning.

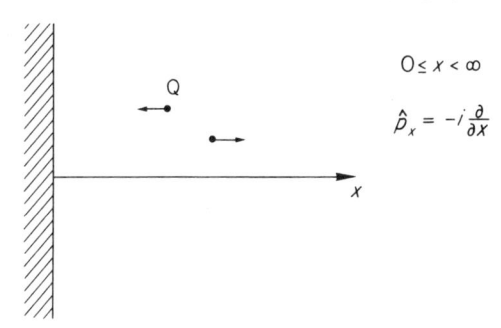

Figure 3   A particle free to move in a semispace bounded by a rigid wall

ACKNOWLEDGEMENTS

The author acknowledges that the core of the present matter was essentially developed in collaboration with Professor V. S. Olkhovsky, and he is also grateful to Professor M. Toller for very useful criticism. His thanks are due to Dr. S. S. Chissick, Dr. A. I. Gerasimchuk and Dr. E. Papp for their very kind interest.

REFERENCES

Aharonov, Y. and Bohm, D. (1961) *Phys. Rev.*, **122,** 1649.
Akhieser, N. I. and Gladsman, I. M. (1954) *Theorie der Linearen Operatoren in Hilbert Raum*, Akademie Verlag, Berlin.
Baldo, M. and Recami, E. (1969) *Lett. Nuovo Cimento*, **2,** 643.
Caldirola, A. (1966) 'Istituzioni di Fisica Teorica', Ambrosiana, Milano.
Calogero, F. (1967) *Variable Phase Approach to Potential Scattering*, Academic Press, New York.
Engelman, F. and Fick, E. (1959) *Supplem. Nuovo Cimento*, **12,** 63.
Engelman, F. and Fick, E. (1963) *Z. Phys.*, **175,** 271.
Engelman, F. and Fick. E. (1964) *Z. Phys.*, **178,** 551.
Heisenberg, W. (1944) *Die Physikalischen Prinzipien der Quantumtheorie*, 4th ed., Hirzel, Leipzig.
Landau, L. D. and Lifshitz, E. M. (1963) *Kvantovaya Mekhanika*, Nauka, Moscow.
Olkhovsky, V. S. (1967) *Nuovo Cimento*, **48 B,** 170.
Olkhovsky, V. S. (1968) *Ukr. Fis. Zh.*, **13,** 143.
Olkhovsky, V. S. and Recami, E. (1968) *Nuovo Cimento*, **53 A,** 610.
Olkhovsky, V. S. and Recami, E. (1969) *Nuovo Cimento*, **63 A,** 814.
Olkhovsky, V. S. and Recami, E. (1970) *Lett. Nuovo Cimento*, **4,** 1165.
Olkhovsky, V. S., Recami, E. and Gerasimchuk, A. I. (1974) *Nuovo Cimento*, **22 A,** 263.
Olkhovsky, V. S., Sokolov, L. S. and Zaychenko, A. K. (1969) *Soviet J. Nucl. Phys.*, **9,** 114.
Papp, E. (1971) *Nuovo Cimento*, **5 B,** 119.
Papp, E. (1972) *Nuovo Cimento*, **10 B,** 69, 471.
Pauli, W. (1958) *Handbuch der Physik*, Flügge, S. Ed., Vol. 5/1, p. 60 last ed., Springer-Verlag, Berlin.
Razavy, M. (1967) *Am. Journ. Phys.*, **35,** 955.
Razavy, M. (1969) *Nuovo Cimento*, **63 B.** 271.
Recami, E. (1970) *Acc. Naz. Lincei, Rendic. Sc.*, **49,** 77 (Rome).
Rosenbaum, D. M. (1969) *J. Math. Phys.*, **10,** 1127.
Von Neumann, J. (1932) *Matematischen Grunladen der Quantum Mechanik*, Hirzel, Leipzig.

Quantum Theory of the Natural Space–Time Units

ERHARDT W. R. PAPP
Polytechnic Institute of Cluj, Romania

1. INTRODUCTION

For more than 50 years the quantum-mechanical space–time description problem has aroused justified interest and has given rise to great power for insight. Overcoming difficulties, phycisists have investigated this subject initially from certain points of view, and reinvestigated it subsequently with respect to a relatively more evolved context. The history of the space–time quantization represents in fact the most significant and profound aspect of the history of quantum theory itself. Throughout the years attempts have been made to analyse, though only provisionally, the peculiarities of a common quantum-mechanical description of space–time and matter, and space–time quantization has come to be regarded as one of the fundamental problems in the scientific understanding of nature.

The conceptual new content of quantum mechanics is expressed by the explicit recognition that measurements cannot be objectively performed with indefinitely increasing accuracy. In these conditions we have to consider the existence of the ultimate (non-zero) accuracy of the space–time measurements. This ultimate accuracy principally results from the new role of the measuring apparatus as a physical object which is itself constituted from the really existing microparticles. Generally, the microparticles have to be considered neither as points, nor with a rigorously spatial extension. This assumption is supported by a certain structure of the physical microparticle and vice versa. Considering the microparticle coincidences as the elementary acts of the space–time measurements, there results the existence of an intrinsic space–time allowance (March, 1941). This allowance is able to offer by itself the possibility of defining—now in a natural way (Bohm and coworkers, 1970)—the existence of the natural space–time units. In this respect the quantum-mechanical space–time measuring process can be considered as the counting process of the successive elementary coincidences. Moreover, the structure of quantum mechanics as a proper physical theory, with a well-established form, can be generally deduced from the laws of the measurements (Ludwig, 1972).

It now becomes necessary for the mathematical formalism of quantum mechanics to be explicitly in agreement with the existence of the non-zero space–time imprecisions. For this purpose a suitably extended quantum-mechanical formalism is needed which has to contain the space–time imprecisions as fundamental entities. Such a formalism has also to permit the consistent definition of the natural space–time units as certain lower bounds of the space–time imprecisions. In this sense account has to be taken of the existence of certain profoundness levels in the quantum-mechanical description of the microparticles: atoms, nuclei and elementary particles. One would then expect to obtain the Bohr radius, the Compton wavelength and the electron radius as the natural units for atoms, for free elementary particles and for (interacting) electrons, respectively. There is also the Planck radius which has to be the natural space constant of the gravitational field.

Such a quantization programme can be materialized by a suitable use of the binary description formalism. In this sense the complex numbers $(\alpha - i\beta)$—and not exclusively the real ones—are allowed to describe the results of the measurements (Kálnay and Toledo, 1967). The result of the measurement is now expressed—in a relatively more complete manner—by the pair (α, β) of real numbers, and alternatively by the interval $[\alpha - \beta, \alpha + \beta]$ (or $(\alpha - \beta, \alpha + \beta)$) on the real α axis. This segment is non-equivocally defined by the complex number. In this respect the space imprecision approach proposed by Flint (1948) is relatively 'incomplete' as he uses not an interval on the real axis, but only a translated point. The space–time imprecisions become now inner elements of the theory, as they are defined as the imaginary parts of the binary (non-Hermitian) operator averages (Papp, 1972a, b; 1973; 1974a, b, c). In such conditions Neumann's axiom which legitimizes the description of the physical observables only by hypermaximal operators is in fact not rejected, but extended (Fick and Engelmann, 1964; Olkhovsky, Recami and Gerasimchuk, 1974).

A short history of the space–time quantization problem will be presented in Section 2. The meaning of the binary space–time description will be analysed—in terms of the collision-time evaluations—in Section 3. In this way it is proved that the space–time imprecisions are able to express certain limitations on the accuracy of the space–time measurements. Section 4 is devoted to the definition of the binary space–time operators and of the corresponding space and time imprecisions. There it is proved that the binary description of the space–time is mutually connected with the one of the action. In Section 5 the uncertainties of the binary space and time operators are evaluated. The high-energy approach to the space–time imprecisions will be performed in Section 6. The physical meaning of the electron radius, of the Compton wavelength, of the Bohr radius and of the Planck radius as natural units are also analysed.

Except in self-evident cases, units will be chosen so that $\hbar = c = 1$.

2. SHORT HISTORY OF THE SPACE–TIME QUANTIZATION

Some opinions about the existence of the space–time quanta were expressed before the development of quantum mechanics (Poincarè, 1913; Proca, 1928;

Kaluza, 1921). During the period of the main development of quantum mechanics the idea of an atomistic structure of space–time had been explicitly formulated by Thomson (1926), Lévi (1926), Pokrowski (1928), Latzin (1927), Beck (1929), Schames (1933) and others. In this respect a preferential meaning had been attributed to the electron and/or nuclear radius. Attempts were proposed to define a theory of the physical constants as a consequence of the existence of the space–time quanta and of the upper value of the elementary particle rest-mass (Beck, 1929; Schames, 1933). An essential step to take for assuming the existence of the ultimate accuracy of the space–time measurements in agreement with the mathematical formalism of quantum mechanics has been stimulated and supported by the Heisenberg (1927) uncertainty relations. In these conditions a conceptually more general approach to the space–time quantization has been formulated on the basis of the existence of the elementary space (h/m_0c) and time (h/m_0c^2) uncertainties by Ruark (1928), Flint and Richardson (1928), Fürth (1929), Wataghin (1930), Landau and Peierls (1931), Glaser and Sitte (1934) and Flint (1937). It is also the amplitude of the Zitterbewegung (Schrödinger, 1930) which has been interpreted as the result of the existence of the individual space imprecision (Iwanenko, 1931). During this period fundamental problems concerning the connection between the structure of the elementary particles and the existence of the space–time quanta (Fürth, 1929; Glaser and Sitte, 1934), the necessity of the synthesis between gravity and quantum theory (Fock and Iwanenko, 1929; Wataghin, 1932; Glaser and Sitte, 1934; Flint, 1935, 1937), the necessity of a more deep connection of electromagnetism and physical space–time description (Flint, 1935; Möglich and Rompe, 1939) were analysed and discussed.

Further progress in the analysis of the time–energy uncertainty relations are due to Madelstamm and Tamm (1945), Fock (1962), Fujiwara (1970) and Olkhovsky and Recami (1970), whereas certain objections concerning the meaning of the time–energy uncertainty relations were raised by Aharonov and Bohm (1964) and Bunge (1970). The space uncertainties have been evaluated for bound states by Remak (1931), and subsequently calculated for the interacting particles by Griffith (1974). A discrete space–time method has been used to evaluate the space–time uncertainties for the relativistic particles (Henning, 1956). Relativistic space uncertainties were calculated for fermions by Blokhintsev (1973). The uncertainty relations have been applied to the gravitational field by Peres and Rosen (1966) and Wheeler (1957) and to the electromagnetic field by Jordan and Fock (1930), Landau and Peierls (1931) and others. There is also the uncertainty-time operator which has been explicitly proposed for bound states by Eberly and Singh (1973).

To overcome the divergence difficulties of the present quantum field theory, to suitably define the high-energy production processes and also to favour the development of the theory for predicting the elementary particle rest-masses attempts were proposed to introduce a fundamental length in the quantum theory by Heisenberg (1936, 1938a, 1938b, 1942), March (1936, 1937a, b, c), Ambarzumian and Iwanenko (1930), Markov (1940) and others. Generally this fundamental length has to take the value of the particle size. In agreement also with the preferential meaning of the weak interactions (Heisenberg, 1938a) there are Kadyshevsky (1961) and Kim (1973) who have analysed explicitly the space constant of the weak interactions. In the papers cited above March advocates the necessity of a suitable redefinition of the short-distance geometry. In this sense certain contributions were also given by Wheeler (1957, 1962), Coish (1959), Takano (1961), Blokhintsev (1960, 1973), de Witt (1960) and others. To support the existence of the fundamental length, space-quantization approaches were performed by Snyder (1947) and Hellund and Tanaka (1954). In their approaches the space quantization is the result of a discrete space eigenvalue problem. Curved space approaches to the space quantization were also proposed (Yang, 1947; Flint, 1948). The compatibility between Lorentz invariance and the existence of the discrete space–time quanta has been analysed by Schild (1948) and Hill (1955). An alternative approach to the space quantization has been proposed by Darling (1950) who considers the irreducible volume character of events. Concerning the (cellular) discrete-space approach we have to mention the contributions given by Das (1960) and Peters (1974). It is significant that the present day non-linear, non-local, indefinite metric and higher derivatives field theories support in one way or another the existence of the fundamental length (see e.g. Vialtzew, 1965). There is also evidence for considering that the predictions formulated earlier by Heisenberg (1938a) concerning high-energy explosions are qualitatively in agreement with the present day multiparticle production processes.

Meaningful results were obtained by March (1941) in the description of the quantum-mechanical space–time measuring process. In this sense we have to consider that the quantum-mechanical measuring apparatus is essentially more complex and rather distinct from the one for relativity (March, 1937a). In this respect a first step in order to conceptually join relativity and quantum mechanics is to consider the reference frame as a component of the quantum-mechanical measuring apparatus (Wataghin, 1930). Certain difficulties concerning the co-existence of the fundamental length with the standard Lorentz invariance (Pavlopoulos, 1967) can be, at least qualitatively, overcome e.g. within the extended Lorentz invariance condition used by Schild (1948) and Hill (1955). However, there is evidence to conclude that the general theory of relativity is essentially more suitable for describing the extended particles than special relativity. These latter aspects were analysed by Motz (1962, 1972), Markov (1965, 1966), Penrose and MacCallum (1973), Sivaram and Sinha (1974), Lord and coworkers (1974) and others. In this connection the quantization of the gravitational field is of a special interest (see e.g. Wheeler, 1957; Treder, 1963; Brill and Gowdy, 1970). We may thus conclude that the problems raised by the space–time quantization have in fact not lost interest and opportunity since the appearance of quantum mechanics.

Progress was also obtained in the definition of the space–time operators as in the performing of the collision time evaluations (see e.g. Kálnay, 1971; Almond, 1973). It has been proved that further developments of the quantum-mechanical space–time description needs the extension of the standard quantum-mechanical formalism (Fick and Engelmann, 1964; Kálnay and Toledo, 1967; Broyles, 1970; Olkhovsky, Recami and Gerasimchuk, 1974 and others). There is also an increasing interest in analysing more deeply certain aspects of the very quantum-mechanical ('shell-pulsating') free-particle description (see e.g. Dirac, 1972). All the facts presented above permit us to assume that the quantum-mechanical space–time description—which is far from being completely resolved—is a fundamental problem characterizing all the steps in the evolution of the quantum theory.

3. THE BINARY CONTENT OF THE COLLISION-TIME DESCRIPTION

The time spent by the outgoing (reduced) particle in the interaction region is given for $l = 0$ by (Smith, 1960)

$$\tau_0(a, p) = \frac{2a}{v} + 2\frac{d}{d\omega}\delta_0(p) - \frac{1}{2\omega}\sin 2[pa + \delta_0(p)] \quad (1)$$

so that the collision time-shift is

$$\tau_0^{(\text{coll})}(a, p) \equiv \tau_0(a, p) - \frac{2a}{v} = 2\frac{d}{d\omega}\delta_0(p) - \frac{1}{2\omega}\sin 2[pa + \delta_0(p)] \quad (2)$$

where a is the interaction radius, $\omega = p^2/2m_0$ and $\delta_0(p)$ the phase-shift for $l = 0$. In agreement with the formal scattering theory we shall consider that the phase-shift does not explicitly depend on the interaction radius. In order to eliminate by all means the presence of the oscillating term a supplementary, outside of the theory, averaging device with respect to the interaction radius has been imposed by Smith (1960), Jauch and Marchand (1967) and Gien (1965). This averaging device is not only artificial but also physically meaningless. On the contrary, the presence of the oscillating term has to be maintained in order to preserve the macroscopic causality condition (Wigner, 1955) and especially the causal positivity of the interaction-time evaluation (Papp, 1972a; Baz, 1966; Peres, 1966). In this respect we can already suppose that the presence of the oscillating term presents a fundamental theoretical meaning.

Indeed, the interaction radius is not uniquely defined. More exactly, if a is the interaction radius, there is the larger value $a' > a$, too. In these conditions we can cause, at least formally, the last term of the time-shift (2) to oscillate, so that the punctual collision-time evaluation $2(d/d\omega)\delta_0(p)$ is in fact replaced by the interval

$$\left[2\frac{d}{d\omega}\delta_0(p) - \frac{1}{2\omega},\ 2\frac{d}{d\omega}\delta_0(p) + \frac{1}{2\omega}\right] \quad (3)$$

The width of this interval is independent from the dynamical peculiarities of the collision system. We can thus conclude that the actual observable meaning of the average $2\langle(d/d\omega)\delta_0(p)\rangle$ can be suitably defined only within a certain range, whose largest value is given by $\langle 1/2\omega \rangle$. In these conditions we have to consider that the real purpose of the quantum-mechanical description is a double one: to perform the observable time-shift evaluations and to state theoretically the existence of an objective degree of accuracy of the time-shift measurement. The above results express the essential step in the definition of the binary description formalism. In this sense the binary interval (3) describes the measurement in which the observable evaluation $2\langle (d/d\omega)\, \delta_0(p) \rangle$ is obtained within the imprecision $\langle 1/2\omega \rangle$. The fact that the above imprecision is twice larger than the one of $\langle 1/4\omega \rangle$ previously calculated (Papp, 1974a) can be explained noticing that the present imprecision does not refer to a single binary time-shift variable, but to the difference of two binary variables. We can also remark that the binary description formalism is consistent with the starting conditions concerning the necessity to assure the fulfilment of the macroscopic causality condition expressed by the positivity requirement of the interaction-time evaluation. Indeed, the inequality $|\alpha| > \beta$ expresses both the macroscopic causality condition and the necessary condition the binary variable $\alpha - i\beta$ to possess measurable meaning.

The above discussions preserve their meaning in the relativistic case, too. Thus, the outgoing time-shift of the elastically scattered Klein–Gordon particle is given, in the one-dimensional case, by

$$\tau_0^{(\text{coll})}(a, p) = 2 \frac{d}{dp_0} \delta_0(p) + \frac{p_0}{\mathbf{p}^2} \sin 2[pa + \delta_0(p)] \qquad (4)$$

where $p_0 = \sqrt{\mathbf{p}^2 + m_0^2}$ (Gien, 1965). Similarly to the non-relativistic case we are now able to define the existence of the relativistic binary time-shift description with the imprecision given by $\langle p_0/2\mathbf{p}^2 \rangle$. For the Dirac-particle one obtains the result

$$\tau_0^{(\text{coll})}(a, p) = 2 \frac{d}{dp_0} \delta_0(p) - \frac{m_0}{\mathbf{p}^2} \sin 2[pa + \delta_0(p)] \qquad (5)$$

thus stating the existence of the particular time imprecision $\langle m_0/2\mathbf{p}^2 \rangle$. In agreement with point (f) of the binary description formalism (Papp, 1973) we can see that the above obtained time imprecisions are binarily 'equivalent':

$$\frac{p_0}{2\mathbf{p}^2} \leq 2 \frac{m_0}{2\mathbf{p}^2} \qquad (6)$$

up to the threshold velocity $(\sqrt{3}/2)c$.

We have to mention that a similar concourse of events to the above one arises when comparing the results obtained for the lower bound of the phase-shift derivative by Wigner (1955) and by Goebel, Karplus and Ruderman (1955), respectively. The results so obtained allow us to conclude that the space–time imprecisions are essentially inner elements of the quantum-mechanical description.

4. THE BINARY SPACE-TIME DESCRIPTION

In the application of the correspondence principle there are cases when the resulting operators are not directly hermitian ones. To avoid the introduction of the non-hermitian operators, subsequent symmetrization devices were used. In line with Section 3 we shall consider that such symmetrization devices are in fact outside the proper theory. Consequently, the symmetrized operators cannot be principally used without also allowing the existence of the initial non-hermitian operators as physically meaningful. In such conditions we have to consider the initial non-hermitian operators as the binary operators which are able to originate the standard hermitian ones.

Thus the classical expressions for the projection of the position vector on the momentum direction and of the free evolution time corresponding to that direction are

$$r_p = \frac{1}{p}\mathbf{p}\cdot\mathbf{r} \quad \text{and} \quad t_p = \frac{m_0}{p^2}\mathbf{p}\cdot\mathbf{r} \tag{7}$$

respectively, where $p = |\mathbf{p}|$. Applying directly the correspondence principle we obtain the pair of the mutually conjugated binary space operators

$$\hat{r}_p = \mathbf{r}\cdot\widehat{\left(\frac{\mathbf{p}}{p}\right)}, \quad \hat{r}'_p = \hat{r}_p^* = \widehat{\left(\frac{\mathbf{p}}{p}\right)}\cdot\mathbf{r} \tag{8}$$

and the associated pair of the binary time-operators

$$\hat{t}_p = m_0\mathbf{r}\cdot\widehat{\left(\frac{\mathbf{p}}{p^2}\right)}, \quad \hat{t}'_p = \hat{t}_p^* = m_0\widehat{\left(\frac{\mathbf{p}}{p^2}\right)}\cdot\mathbf{r} \tag{9}$$

respectively. We then easily obtain, in agreement also with Lippmann (1966), the hermitian space-time operators as

$$\hat{r}_p^{(*)} = \tfrac{1}{2}(\hat{r}_p + \hat{r}'_p) \quad \text{and} \quad \hat{t}_p^{(*)} = \tfrac{1}{2}(\hat{t}_p + \hat{t}'_p) \tag{10}$$

respectively. Averaging the binary operators with respect to the non-relativistic wave packet

$$\varphi(\mathbf{r}, t) = (2\pi)^{-1/2} \int d\mathbf{p}\, a(\mathbf{p}) \exp i(\mathbf{p}\cdot\mathbf{r} - \omega t) \tag{11}$$

where j is the spatial dimension, there results

$$\langle\hat{r}_p\rangle = -\left\langle\frac{\mathbf{p}}{p}\cdot\frac{\partial}{\partial\mathbf{p}}\arg a(\mathbf{p})\right\rangle + i(j-1)\left\langle\frac{1}{2p}\right\rangle \tag{12}$$

and

$$\langle\hat{t}_p\rangle = -m_0\left\langle\frac{\mathbf{p}}{p^2}\cdot\frac{\partial}{\partial\mathbf{p}}\arg a(\mathbf{p})\right\rangle + i(j-2)\left\langle\frac{1}{4\omega}\right\rangle \tag{13}$$

thus allowing the definition of the space and time imprecisions as the imaginary parts of the above averages. In order to define the time operator for $j = 1$, the boundary condition

$$\lim_{p_1 \to 0} p_1^{-1/2} a(p_1) < \infty \tag{14}$$

is needed whereas for $j=2$ and $j=3$ the wave-packet form-factor has to be only bounded at the origin (Papp, 1974c). However, for $j=1$, appreciable limitations are not implied as the subspace defined by the well-behaved condition (14) is dense in the whole Hilbert space. The above boundary conditions maintain their meaning also in the relativistic case.

There is also a mutual connection between the binary description of the action and the one of space–time (Papp, 1974a). Indeed, the $\mathbf{r} \cdot \hat{\mathbf{p}}$-action average is given by

$$\langle \mathbf{r} \cdot \hat{\mathbf{p}} \rangle = -\left\langle \mathbf{p} \cdot \frac{\partial}{\partial \mathbf{p}} \arg a(\mathbf{p}) \right\rangle + t \left\langle \frac{p^2}{m_0} \right\rangle + j \frac{\hbar}{2} \tag{15}$$

so that the binary description of the action with the imprecision $j\langle \hbar/2 \rangle$ implies the existence of the binary description of the time with the imprecision given by $j\langle \hbar/4\omega \rangle$ and vice versa. Alternatively, there is implied a binary description of the space shift

$$\left\langle \frac{\mathbf{p}}{p} \cdot \frac{\partial}{\partial \mathbf{p}} \arg a(\mathbf{p}) \right\rangle$$

or of the time shift

$$m_0 \left\langle \frac{\mathbf{p}}{p^2} \cdot \frac{\partial}{\partial \mathbf{p}} \arg a(\mathbf{p}) \right\rangle$$

with the imprecisions given by $j\langle \hbar/2p \rangle$ and $j\langle \hbar/4\omega \rangle$, respectively. We can thus conclude that the purpose of the quantum-mechanical binary space–time description is indeed a double one: to perform the observable space–time (shifts) evaluations and to define the imprecisions of the space–time (shifts) measurements.

In the relativistic case the binary space, time and action ($p_0 t - \mathbf{p} \cdot \mathbf{r}$) operators are given by

$$\hat{r}_p = \mathbf{r} \cdot \widehat{\left(\frac{\mathbf{p}}{p}\right)}, \quad \hat{t}_p = \mathbf{r} \cdot \widehat{\left(\mathbf{p}\frac{p_0}{p^2}\right)} \quad \text{and} \quad \hat{a} = m_0^2 \mathbf{r} \cdot \widehat{\left(\frac{\mathbf{p}}{p^2}\right)} \tag{16}$$

respectively. Averaging the action operator $\hat{a}$ with respect to the Klein–Gordon wave packet

$$\Phi^{(+)}(\mathbf{r}, t) = (2\pi)^{-j/2} \int d\mathbf{p} \frac{1}{\sqrt{2p_0}} a^{(+)}(\mathbf{p}) \exp(-ipx) \tag{17}$$

where $px = p_0 t - \mathbf{p} \cdot \mathbf{r}$, one obtains

$$\langle \hat{a} \rangle = -m_0^2 \left\langle \frac{\mathbf{p}}{p^2} \cdot \frac{\partial}{\partial \mathbf{p}} \arg a^{(+)}(\mathbf{p}) \right\rangle + t \left\langle \frac{m_0^2}{p_0} \right\rangle + i(j-2) \left\langle \frac{m_0^2}{2p^2} \right\rangle \tag{18}$$

so that the implied space-shift and time (time-shift) imprecisions are

$$\delta^{(j)}s = (j-2)\left\langle \frac{1}{2p} \right\rangle \tag{19}$$

and

$$\delta^{(j)}t = (j-2)\left\langle \frac{p_0}{2p^2} \right\rangle \tag{20}$$

respectively. Requiring now the action imprecision to be larger than $\hbar/2$, it results that the relativistic binary description of the action maintains $\hbar/2$ as the natural unit of the action only when

$$\frac{m_0^2 c^2 \hbar}{2p^2} \geq \frac{\hbar}{2} \tag{21}$$

i.e. when the existence of the threshold velocity $(\sqrt{2}/2)c$ is quantum-mechanically allowed. In this respect we have to consider that for velocities larger than $(\sqrt{2}/2)c$, the elastically scattered particle (in the centre-of-mass system) ceases to preserve the initial single free-particle individuality. As a consequence we can no longer ignore the structure of the particle, so that—in the high-energy region $(v > (\sqrt{2}/2)c)$—the particle is in fact replaced by the system of its constituents. In this respect we shall attribute a fundamental theoretical meaning to the threshold velocity $(\sqrt{2}/2)c$, in agreement also with the fact that there is the same threshold velocity value which has been obtained within general relativity theory (Jaffe and Shapiro, 1972).

We can also remark that the existence of the natural unit of the action implies the existence of the natural space unit

$$\delta_c s = \frac{\hbar}{2m_0 c} \tag{22}$$

or, alternatively, of the natural time unit

$$\delta_c t = \sqrt{2} \frac{\hbar}{2m_0 c^2} \tag{23}$$

which is binarily 'equivalent' to the constant $\hbar/2m_0 c^2$. In such conditions the Compton wavelength $\hbar/2m_0 c$ can also be interpreted as the extent of the spatial localization region of the free particle. This interpretation is in agreement with the fact that the spatial localization (overlapping) of the free-particle field operators at two points in space is of the same order as the Compton wavelength (see e.g. Schröder, 1964; Griffith, 1974). However, the existence of factor $\sqrt{2}$ in the expression (23) needs some additional explanations. Firstly we have to mention that generally the threshold velocities implied by the space imprecisions are not identical with the ones corresponding to the time imprecisions (Papp, 1973). On the other hand a special juncture arises when we compare the results of Schild (1948) and Hill (1955). Allowing the existence of the space–time quanta and imposing the (extended) requirement of Lorentz invariance it is implied that there is only a certain set of allowed velocities which corresponds to the existence of the integral-number co-ordinates (Schild) and of the rational-number co-ordinates (Hill), respectively. But in the particular case when the rational number is also an integral one, the above sets are not—as one would expect—identical. In such a situation it would be justifiable to conclude that—in the high-energy region—the existence of a certain velocity allowance cannot be overcome (when also imposing the Lorentz invariance condition within approaches supporting the existence of the space–time quanta).

Averaging the binary time operator one obtains the imprecision

$$\delta^{(j)}t = (j-2)\left\langle \frac{p_0}{2p^2} \right\rangle + \left\langle \frac{1}{2p_0} \right\rangle \qquad (24)$$

which is larger than the expression (20) by the imprecision amount $\langle 1/2p_0 \rangle$. As

$$\frac{p_0}{2p^2} \leq 2\frac{1}{2p_0} \qquad (25)$$

for $v \geq (\sqrt{2}/2)c$, we may conclude—by virtue of the above binary 'equivalence'—that the average $\langle 1/2p_0 \rangle$ expresses in fact the time imprecision in the high-energy region. The binary space operator leads to the same space imprecision of $(j-1)\langle 1/2p \rangle$.

Analysing the binary meaning of the action operator $t\hat{p}_0 - \mathbf{r} \cdot \hat{\mathbf{p}}$, where t is now the time parameter, one obtains the action imprecision $(j-1)\hbar/2$, the time-shift imprecision

$$\delta^{(j)}t = (j-1)\left\langle \frac{p_0}{2p^2} \right\rangle \qquad (26)$$

the time-imprecision

$$\delta_0^{(j)}t = (j-1)\left\langle \frac{p_0}{2m_0^2} \right\rangle \qquad (27)$$

and the space-shift imprecision

$$\delta^{(j)}s = (j-1)\left\langle \frac{1}{2p} \right\rangle \qquad (28)$$

Excepting the dimensional factor $(j-1)$, it can be easily remarked that, at the threshold velocity $(\sqrt{2}/2)c$, the action and both the time imprecisions become identical with the ones of (18) and (22), respectively. Similarly, the $\mathbf{r} \cdot \hat{\mathbf{p}}$-action leads to the imprecisions

$$\delta^{(j)}a = j\frac{\hbar}{2}, \quad \delta^{(j)}t = j\left\langle \frac{p_0}{2p^2} \right\rangle, \quad \delta^{(j)}s = j\left\langle \frac{1}{2p} \right\rangle \qquad (29)$$

thus confirming again, at least 'binarily', the non-equivocal values of the space–time imprecisions.

The action operator $\widehat{(p_0, t_p)}$ possesses the imprecision

$$\delta^{(j)}a = \left|\left\langle \frac{jp_0^2 - 2m_0^2}{2p^2} \right\rangle\right| \tag{30}$$

which implies, now in a separate direct way, the existence of the threshold velocity $(\sqrt{2}/2)c$ for $j = 1$. It is worthwhile mentioning that the action operator $t\hat{p}_0$ possesses the imprecision $\hbar/2$. This operator is a binary one only for the Klein–Gordon particle. For the Dirac particle the binary action and space–time quantizations can be similarly performed.

The imprecision of the binary proper-time operator is essentially given—now in a manifestly Lorentz invariant way—by the Compton wavelength (Papp, 1972b). In this respect further evidence is also given concerning the meaning of this length as the natural space constant of the free particle. In this case the following space- and time-shift imprecisions are implied

$$\delta^{(j)}t = j\left\langle \frac{p_0}{2p^2} \right\rangle, \quad \delta^{(j)}s = j\left\langle \frac{1}{2p} \right\rangle \tag{31}$$

for the Dirac-particle, whereas

$$\delta^{(j)}t = (j-1)\left\langle \frac{p_0}{2p^2} \right\rangle, \quad \delta^{(j)}s = (j-1)\left\langle \frac{1}{2p} \right\rangle \tag{32}$$

are the space- and time-shift imprecisions for the Klein–Gordon particle. The implied time (t) imprecisions are given by

$$j\left\langle \frac{p_0}{2m_0^2} \right\rangle \quad \text{and} \quad (j-1)\left\langle \frac{p_0}{2m_0^2} \right\rangle \tag{33}$$

respectively. The imprecision $\langle p_0/2m_0^2 \rangle$ is binarily compatible with the one of $\langle p_0/2p^2 \rangle$ in the high-energy region:

$$\frac{p_0}{2p^2} \leq 2\frac{p_0}{2m_0^2} \tag{34}$$

for $v \geq (\sqrt{3}/3)c$. This latter threshold velocity agrees numerically with the one which has been defined in the general relativity theory, too (Treder, 1974). We may thus conclude that the existence of certain common threshold velocities allow us to assume that in fact some premises needed by a properly unified theory of space–time and matter have already been fulfilled.

5. THE UNCERTAINTIES OF THE BINARY SPACE–TIME OPERATORS

There is a certain formal analogy between the binary space–time description and the one of the space and time uncertainties. Indeed in both cases the existence of a certain interval associated with the measuring process is considered. But whereas in the first case the interval is the primary element of a theoretical description which has to express by itself the 'objective' imprecision of the measurement, the uncertainty interval (centred around the mean value, too) expresses in a rather conventional manner the general statistical accuracy limits of the measurement. In spite of these essential distinctions, we shall prove that certain space and time uncertainty contributions—the so-called uncertainty units—can be placed on the same footing as the space and time imprecisions (Papp, 1975). This fact is valid not only for the binary space–time operators, but also for the hermitian one-dimensional space operators.

Let us begin with the evaluations of the space uncertainty for the Dirac particle. Using the relations

$$\left(\frac{\mathbf{p}}{p}\cdot\frac{\partial}{\partial\mathbf{p}}\right)\left(\frac{\mathbf{p}}{p}\cdot\frac{\partial}{\partial\mathbf{p}}\right)\exp i\mathbf{p}\cdot\mathbf{r} = -\frac{1}{p^2}(\mathbf{p}\cdot\mathbf{r})^2 \exp i\mathbf{p}\cdot\mathbf{r} \tag{35}$$

$$u^*(\mathbf{p}, s')\mathbf{p}\cdot\frac{\partial}{\partial\mathbf{p}}u(\mathbf{p}, s) = \frac{\mathbf{p}^2}{2m_0 p_0}\delta_{s's} \tag{36}$$

$$\mathbf{p}\cdot\frac{\partial}{\partial\mathbf{p}}u^*(\mathbf{p}, s')\mathbf{p}\cdot\frac{\partial}{\partial\mathbf{p}}u(\mathbf{p}, s) = \frac{\mathbf{p}^2}{4m_0 p_0}\delta_{s's} \tag{37}$$

where $u(\mathbf{p}, s)$ is the positive energy spinor, there results

$$\langle \hat{r}_p \rangle = t\left\langle\frac{p}{p_0}\right\rangle - \left\langle\frac{\mathbf{p}}{p}\cdot\frac{\partial}{\partial\mathbf{p}}\arg b(\mathbf{p}, s)\right\rangle + i(j-1)\left\langle\frac{1}{2p}\right\rangle \tag{38}$$

and

$$\langle \hat{r}_p^2 \rangle = t^2\left\langle\frac{p^2}{p_0^2}\right\rangle + \left\langle\left(\frac{\mathbf{p}}{p}\cdot\frac{\partial}{\partial\mathbf{p}}\ln|b(\mathbf{p}, s)|\right)^2\right\rangle + \left\langle\frac{m_0^2}{4p_0^4}\right\rangle$$

$$+ \left\langle\left(\frac{\mathbf{p}}{p}\cdot\frac{\partial}{\partial\mathbf{p}}\arg b(\mathbf{p}, s)\right)^2\right\rangle - 2t\left\langle\frac{\mathbf{p}}{p_0}\cdot\frac{\partial}{\partial\mathbf{p}}\arg b(\mathbf{p}, s)\right\rangle$$

$$+ i(j-1)t\left\langle\frac{1}{p_0}\right\rangle - i(j-1)\left\langle\frac{\mathbf{p}}{p^2}\cdot\frac{\partial}{\partial\mathbf{p}}\arg b(\mathbf{p}, s)\right\rangle$$

$$- (j-1)(j-2)\left\langle\frac{1}{2p^2}\right\rangle \tag{39}$$

where the j-dimensional Dirac particle wave packet

$$\psi^{(+)}(\mathbf{r}, s, t) = (2\pi)^{-j/2}\int d\mathbf{p}\sqrt{\frac{m_0}{p_0}}u(\mathbf{p}, s)\exp(-ipx) \tag{40}$$

has been used. Allowing for simplicity the approximations

$$\left\langle\left(\frac{p_i}{p}\frac{\partial}{\partial p_i}\arg b(\mathbf{p}, s)\right)^2\right\rangle \simeq \left\langle\frac{p_i}{p}\frac{\partial}{\partial p_i}\arg b(\mathbf{p}, s)\right\rangle^2 \tag{41}$$

and

$$\left\langle \frac{\mathbf{p}}{p_0} \cdot \frac{\partial}{\partial \mathbf{p}} \arg b(\mathbf{p}, s) \right\rangle \simeq \left\langle \frac{p}{p_0} \right\rangle \left\langle \frac{\mathbf{p}}{p} \cdot \frac{\partial}{\partial \mathbf{p}} \arg b(\mathbf{p}, s) \right\rangle \quad (42)$$

there results

$$\Delta r_{p,j}^2 \equiv \langle \hat{r}_p^2 \rangle - \langle \hat{r}_p \rangle^2 \simeq \Delta v^2 t^2 + \Delta r_{p,j}^{(\min)2} + \Delta r_{p,j}^{(\text{unit})2} \quad (43)$$

where

$$\Delta r_{p,j}^{(\min)2} = \left\langle \left(\frac{\mathbf{p}}{p} \cdot \frac{\partial}{\partial \mathbf{p}} \ln |b(\mathbf{p}, s)| \right)^2 \right\rangle \quad (44)$$

is the square of the minimum space uncertainty,

$$\Delta r_{p,j}^{(\text{unit})2} = \left\langle \frac{m_0^2}{4p_0^4} \right\rangle - (j-1)(j-3) \left\langle \frac{1}{4p^2} \right\rangle \quad (45)$$

is the square of the space uncertainty unit and where Δv^2 is the square of the velocity uncertainty. The space uncertainty contribution Δvt can be neglected taking formally $t = 0$. Using the Kronecker symbols, the expression (45) takes the form

$$\Delta r_{p,j}^{(\text{unit})2} = (\delta_{j1} + \delta_{j3}) \left\langle \frac{m_0^2}{4p_0^4} \right\rangle + \delta_{j2} \left\langle \frac{m_0^2 p^2 + p_0^4}{4p_0^4 p^2} \right\rangle \quad (46)$$

We can easily remark that in the two-dimensional case there arises an additional contribution due to the space imprecision $\langle 1/2p \rangle$. For $j = 1$ and $j = 3$, one obtains

$$\Delta r_{p,j}^{(\text{unit})} \simeq \frac{\hbar}{2m_0 c} \left(1 - \left\langle \frac{v^2}{c^2} \right\rangle \right) \geq \frac{\hbar}{4m_0 c}, \quad j = 1, 3 \quad (47)$$

for $\langle v^2 \rangle \leq c^2/2$. The above results confirm our assumption that the space-uncertainty unit possesses the physical meaning of the space imprecision.

The calculations can be similarly performed for the Klein–Gordon particle, thus obtaining the results

$$\Delta r_{p,j}^{(\text{unit})2} = (3-j)(j-1) \left\langle \frac{1}{4p^2} \right\rangle - \left\langle \frac{p^2}{4p_0^4} \right\rangle \quad (48)$$

so that

$$\Delta r_{p,j}^{(\text{unit})2} = -(\delta_{j1} + \delta_{j3}) \left\langle \frac{p^2}{4p_0^4} \right\rangle + \delta_{j2} \left\langle \frac{m_0^2(p^2 + p_0^2)}{4p_0^4 p^2} \right\rangle \quad (49)$$

We can see that in the one- and three-dimensional cases the space uncertainty unit of the Klein–Gordon particle takes the maximum imaginary value at the threshold velocity $(\sqrt{2}/2)c$:

$$\Delta_c r_{p,j}^{(\text{unit})} = i \frac{\hbar}{4m_0 c}, \quad j = 1, 3 \quad (50)$$

42 Uncertainty Principle and Foundations of Quantum Mechanics

In these conditions the minimum space uncertainty of the Klein–Gordon particle has to be generally smaller than half of the Compton wavelength. The above requirement would also signify that the (elastically scattered) Klein–Gordon particle is not necessarily a free one in the high-energy region. Indeed, the one-dimensional space uncertainty evaluation has to take positive values as the space operator is hermitian. On the other hand, if we require the minimum space uncertainty $\hbar/2\Delta p$ to be larger than $\hbar/4m_0c$, it follows that $\Delta p < 2m_0c$. In these conditions, in order to assure—irrespective of any particular cases—the general validity of the quantum-mechanical description, we have also to consider the condition $\langle p \rangle < 2m_0c$, as there exist cases when $\Delta p^2 = \langle p^2 \rangle$. Contrarily, there would exist cases for which $\Delta p > 2m_0c$. Consequently the Compton wavelength is able to express the size of the Klein–Gordon particle only in the free case of a not too large energy (in order to also preserve the initial single particle individuality).

In the non-relativistic case the square of the space-uncertainty unit is

$$\Delta r_{p,j}^{(\text{unit})} = (j-1)\left\langle \frac{1}{2p^2} \right\rangle \tag{51}$$

so that the space-uncertainty unit is 'binarily' identical to the space imprecision $\langle 1/2p \rangle$.

The time-uncertainty units can be similarly calculated, thus obtaining the results (Papp, 1975)

$$\Delta t_{p,j}^{(\text{unit})2} = -5\left\langle \frac{m_0^2}{4p^4} \right\rangle \delta_{j1} + \left\langle \frac{1}{4p^2} \right\rangle \delta_{j2} + 3\left\langle \frac{m_0^2}{4p^4} \right\rangle \delta_{j3} \tag{52}$$

$$\Delta t_{p,j}^{(\text{unit})2} = -\left\langle \frac{5m_0^2 + p^2}{4p^4} \right\rangle \delta_{j1} + \left\langle \frac{3m_0^2 - p^2}{4p^4} \right\rangle \delta_{j3} \tag{53}$$

and

$$\Delta t_{p,j}^{(\text{unit})2} = -5\left\langle \frac{m_0^2}{4p^4} \right\rangle \delta_{j1} + 3\left\langle \frac{m_0^2}{4p^4} \right\rangle \delta_{j3} \tag{54}$$

for the Dirac, Klein–Gordon and non-relativistic particles, respectively. Besides the space imprecision $\langle 1/2p \rangle$, there is now the previously encountered collision time-shift imprecision $\langle m_0/2p^2 \rangle$ which is also implied. It can now be easily shown that the squares of the space and time uncertainty units of the Dirac particle are larger than the corresponding squares of the Klein–Gordon particle by the amounts $\langle 1/4p_0^2 \rangle$ and $\langle 1/4p^2 \rangle$, respectively. These results are in fact in agreement with the expressions (31) and (32) thus proving, in this way also, the general inner consistency of the binary description. There is also a mutual compatibility of the space–time imprecisions with the space–time uncertainty units.

We can show that there is not an irreconcilable difference between the space and time imprecisions. Thus the average $\langle 1/2p \rangle$ is not only the space imprecision, but it possesses the meaning of a time imprecision, too. Similarly, the average $\langle 1/2p_0 \rangle$ possesses also the meaning of a space imprecision.

6. THE HIGH-ENERGY SPACE–TIME IMPRECISION DESCRIPTION

Up to now the meaning and relevance of the Compton wavelength with respect to the (non-large energy) free (scattered) particle has been analysed. Proofs were given that the space–time imprecisions possess a well-defined physical significance within the quantum-mechanical collision time shift, binary space–time and space–time uncertainty descriptions. We shall now perform an approach which is able to define—in a unitary and direct manner—the Compton wavelength, the electron radius and (by extrapolation) the Bohr radius as natural space–time units.

For this purpose let us assume that there is a 'spectrum' of the relevant space shift evaluations which correspond to the various levels of the quantum-mechanical description of matter. Such a space-shift 'spectrum' has to be described by the generalized 'eigenvalue' equation

$$\frac{d}{dp}\delta_l(p) = N\frac{1}{2p} \tag{55}$$

where N is the space-multiplicity parameter ('eigenvalue') and where, for convenience, a well-defined value of the angular momentum has been chosen. We shall also subsequently consider that the production processes which are expected to arise in the high-energy region can be qualitatively supported by a resonance emission approximation. The equation (55) has to define by itself the physical meaning of the space imprecision in the high-energy region. In this sense the equation (55) has to establish a close connection between the existence of the natural space–time units and the high-energy structural effects raised by the validity of the resonance emission approximation. Consequently, there is also implied a high-energy interacting-particle approach, as the collision interaction can be qualitatively supported by the formation and subsequent decay of a resonance state (Peres, 1966). In such conditions the above space shift has to be also considered as an interaction space shift (Papp, 1972a). One would from the very beginning expect that among the physically relevant N-values there are the ones of $N = 1$ and $N = 3$ too. Indeed, in agreement with points (b) and (d) of the previous paper (Papp, 1973) the necessary condition for the binary variable to possess measurable meaning is $N \geq 1$, whereas $N \geq 3$ has to be considered the sufficient condition for the binary variable to possess (the well-defined) measurable meaning.

Besides the above-formulated approach to the high-energy space-shift description, we can define another variant, by using the high-energy time imprecision $\langle 1/2p_0 \rangle$. Starting from the (space-) time imprecision behaviour of the time shift

$$\frac{d}{dp_0}\delta_l(p_0) = N\frac{1}{2p_0} \tag{56}$$

one obtains the phase shift

$$\delta_l(p_0) = \frac{N}{2} \ln \frac{p_0}{m_0} \tag{57}$$

where the present N-parameter is not necessarily identical with the one of equation (55). We shall consider for convenience—in agreement with the previous remarks—the relation (56) as a space-shift relation, i.e. we shall take (dimensionally) $p_0 = \sqrt{p^2 + m_0^2 c^2}$. The scattered state function corresponding to the above phase shift is given (in the energy representation) by

$$\varphi_l^{(\text{scatt})}(p_0) = g_l(p_0) \sin \delta_l(p_0) \exp i\delta_l(p_0) \tag{58}$$

Neglecting the influence of the wave-packet preparation, we shall take the form factor $g_l(p_0)$ to be a constant. Consequently, the interaction time imprecision takes the form

$$\left\langle \frac{1}{2p_0} \right\rangle = \left[\int_{m_0}^{p_0^{(\max)}} dp_0 \sin^2 \delta_l(p_0) \right]^{-1} \int_{m_0}^{p_0^{(\max)}} dp_0 \frac{1}{2p_0} \sin^2 \delta_l(p_0) \tag{59}$$

so that

$$\left\langle \frac{1}{2p_0} \right\rangle = \frac{\hbar}{m_0 c} \frac{\pi(N^2+1)}{N^3} \left(\exp \frac{2\pi}{N} - 1 \right)^{-1} \tag{60}$$

where we have assumed that

$$\delta_l(p_0) \in [0, \pi], \quad p_0^{(\max)} = m_0 \exp \frac{2\pi}{N} \tag{61}$$

It can now be easily verified that

$$\left\langle \frac{d}{dp_0} \delta_l(p_0) \right\rangle \gtrsim \frac{e^2}{m_0 c^2} \tag{62}$$

for $N \gtrsim 1$, where instead of e^2/m_0c^2 we can also consider—without appreciably affecting the above approximation—the twice smaller value $e^2/2m_0c^2$. Similarly it results in

$$\left\langle \frac{d}{dp_0} \delta_l(p_0) \right\rangle \gtrsim \frac{\hbar}{2m_0 c} \tag{63}$$

for $N \gtrsim 3$. The existence of the inequality

$$\left\langle \frac{d}{dp_0} \delta_l(p_0) \right\rangle \gtrsim \frac{\hbar}{m_0 c}, \quad N \gtrsim 4 \tag{64}$$

agrees with the binary description formalism. Indeed, the time-shift evaluations for $N = 3$ and $N = 4$ are binarily equivalent ones, as there are the space constants $\hbar/2m_0c$ and $\hbar/m_0c$, too. We may thus conclude that the existence of the Compton wavelength and of the electron radius as the natural space units is a direct result of the binary description formalism of the space and time.

In order to perform the resonance emission approximation we have to impose the maximum-value condition of the scattered state function (Kilian and Petzold, 1970)

$$\delta_l(p_0^{(r)}) = \frac{\pi}{2} \tag{65}$$

so that the resonance energies are given by

$$p_0^{(r)} \equiv p_0^{(r,N)} = m_0 \exp \frac{\pi}{N} \tag{66}$$

On the other hand the energy average is given by

$$\langle p_0 \rangle^{(N)} = \frac{m_0 c}{2} \frac{N^2+1}{N^2+4} \left(\exp \frac{4\pi}{N} - 1 \right) \left(\exp \frac{2\pi}{N} - 1 \right)^{-1} \tag{67}$$

so that the meaning of the narrow resonance approximation can be analysed in this way also comparing, for the same N-values the expressions (66) and (67). Thus

$$p_0^{(r,1)} \simeq 23 m_0, \quad p_0^{(r,3)} \simeq 2.87 m_0, \quad p_0^{(r,4)} \simeq 2.26 m_0 \tag{68}$$

whereas

$$\langle p_0 \rangle^{(1)} \simeq 107 m_0, \quad \langle p_0 \rangle^{(3)} \simeq 3.12 m_0, \quad \langle p_0 \rangle^{(4)} \simeq 2.46 m_0 \tag{69}$$

Consequently, the above resonance emission approximation is mathematically consistent for the relatively larger N-values, but it could be qualitatively accepted, in a larger sense, even for $N \simeq 1$. Around the value $N = 1$, a very small variation of the N-parameter implies large variations of the $\langle p_0 \rangle$ and $\langle 1/p_0 \rangle$ averages. In these conditions the inequality

$$\langle p_0 \rangle \lesssim 107 m_0 \tag{70}$$

which is valid for $N \gtrsim 1$, can be practically replaced by the inequality

$$\langle p_0 \rangle \lesssim 137 m_0, \quad N \gtrsim 1 \tag{71}$$

Consequently, the electron radius is able to fulfil its role both as natural space unit and as intrinsic size of the interacting electron only up to the 'electromagnetic' threshold velocity $v^{(em)}$ (Papp, 1974b).

As a consequence of the inequality (62), a lower bound of the N-parameter values can be defined. Indeed

$$\left\langle \frac{1}{2p_0} \right\rangle < \frac{\hbar}{2m_0 c} \tag{72}$$

so that

$$N \frac{\hbar}{2m_0 c} > \left\langle \frac{d}{dp_0} \delta_l(p_0) \right\rangle \gtrsim \frac{e^2}{2m_0 c^2} \tag{73}$$

Consequently

$$N > N^{(min)} \equiv \frac{e^2}{\hbar c} = \alpha \simeq \frac{1}{137} \tag{74}$$

On the other hand, from the condition (61), one obtains

$$p_0 < m_0 \exp\frac{2\pi}{N} < p_0^{(\text{max})} \equiv m_0 \exp\frac{2\pi}{\alpha} \tag{75}$$

The so-defined upper energy bound agrees qualitatively with the upper cutoff momentum defined by Greenman and Rohrlich (1973). The extreme extrapolation $N \to \alpha$ implies not only the breakdown of the high-energy space–time (interaction) imprecision description, but also the breakdown of the linear quantum electrodynamics, as that extrapolation leads to the appearance of the deep non-linear effects.

Another extrapolation can be performed towards the large N-values. Thus requiring the interaction shift to equal the Bohr radius:

$$N\left\langle\frac{1}{2p_0}\right\rangle = \frac{\hbar^2}{2m_0 e^2} \tag{76}$$

it results that

$$N \simeq \frac{1}{\alpha} \simeq 137 \tag{77}$$

But the value $1/\alpha$ is in fact an upper bound of the N-parameter, as the energy average becomes in this case practically identical to the rest-mass energy. In the present case the Bohr radius is also a natural unit, but now with respect to another quantum-mechanical stratum of the bound states of the atomic electron. Indeed, the space-imprecision average performed with respect to the **p**-momentum representation state function of the hydrogen atom is given, for $l = 0$, by

$$\left\langle\frac{1}{2p}\right\rangle = \frac{\hbar^2}{m_0 e^2}\frac{4n}{\pi}\left[\sum_{k=1}^{n}\frac{1}{2k-1} - \frac{n^2}{4n^2-1}\right] \tag{78}$$

where n is the main quantum number. The smallest space imprecision is now given by $8\hbar^2/3\pi m_0 e^2$. This result confirms the above assumption concerning the role of the Bohr radius as a natural space unit.

There is formal analogy between the electrostatic and gravitational interactions of two point particles. Thus, whereas the Compton wavelength maintains unchanged its role as natural space unit, there is the Schwarzschild radius $g(m_0/2c^2)$—where g is the gravitational constant—which corresponds to the classical electron radius. In these conditions the Planck radius $(1/2c)\sqrt{\hbar g/c}$ (Planck, 1913) is even the geometrical average of the so defined gravitational space units. It is justifiable to consider that the Planck radius is of a fundamental theoretical significance (see e.g. Wheeler, 1957; Treder, 1963; Markov, 1966; Motz, 1972). In this sense this radius depends explicitly only on the universal constants and it is also the space constant which takes the smallest value. From the quantum-mechanical point of view we can consider—in a strong analogy with the results obtained for the electrostatic field by Heisenberg and Euler (1936)—that the Planck radius is mutually connected with the existence of the maximum observable value of the gravitational field strength. The meaning of the Planck radius as the critical distance value of the quantum gravity is in agreement with this latter result.

We may thus conclude that the above-analysed natural space–time units are in fact various aspects of the same space–time imprecision ($\langle 1/2p \rangle$ or $\langle 1/2p_0 \rangle$). The existence of the natural space–time units is actually required even by the mathematical binary description formalism, thus also proving both the relevance and the consistency of that formalism. In these conditions certain steps which are needed by a mathematically suitable description of the natural space–time units have been established, at least qualitatively.

7. CONCLUSIONS

Throughout this paper certain evidence concerning the existence of the space–time imprecisions as inner elements of an extended quantum-mechanical description has been analysed and discussed. Proofs have also been given that the existence of the natural space–time units is mathematically consistent with the binary description formalism. It turns out that the binary formalism expresses essential aspects of the quantum-mechanical description of space–time and matter. The space–time imprecisions—which were removed from the standard quantum-mechanical description—imply a quite natural extension of quantum mechanics. The so extended formalism fulfils—at least for the moment—the requirements needed to define a quantum theory of the natural space–time units. One would also assume the present binary description formalism to be not complete, so that further developments need subsequent extensions and refinements. Thus the binary description formalism cannot be adequately applied to the coulombian or to the static gravitational interactions (between two point particles) without additionally assuming the existence of a certain discrete-space model (Papp, 1974b). Indeed, we can assume that the experimental conditions to perform measurements on the short distance behaviour are more restrictive than the ones required by the usual large distance measurements. In this sense we have to consider that the space discretization methods imply certain additional restrictions needed to perform adequately the short distance measurements. The possibility exists to define the existence of the maximum observable value of the electrostatic field strength and of the upper bounds of the particle rest-mass and electric charge, too. All these facts led us once again to conclude on the relevance and the deep physical significance of the space–time binary description formalism.

REFERENCES

Aharonov, Y. and Bohm, D. (1964) 'Answer to Fock concerning the time–energy indeterminacy relation', *Phys. Rev.*, **134 B,** 1417–1418.

48 Uncertainty Principle and Foundations of Quantum Mechanics

Almond, D. (1973) 'Time operators, position operators, dilatation transformations and virtual particles in relativistic and nonrelativistic quantum mechanics', *Ann. Inst. Henri Poincaré*, **19 A**, 105–170.

Ambarzumian, V. and Iwanenko, D. (1930) 'Zur Frage nach Vermeidung der unendlichen Selbstrüskwirkung des Elektrons', *Z. Phys.*, **64**, 563–567.

Baz, A. I. (1966) 'Life-time of intermediate states', *Yader. Fiz.*, **4**, 252–260.

Beck, G. (1929) 'Die zeitliche quantelung der Bewegung', *Z. Phys.*, **51**, 737–739.

Blokhintsev, D. I. (1960) 'Fluctuations of space-time metric', *Nuovo Cimento*, **16**, 382–387.

Blokhintsev, D. I. (1973) *Space and Time in the Microworld*, Dordrecht, Boston.

Bohm, D., Hilley, B. J. and Stewart, A. E. G. (1970) 'On a new model of description in physics', *Int. J. Theor. Phys.*, **3**, 171–183.

Brill, D. R. and Gowdy, R. H. (1970) 'Quantization of general relativity', *Rep. Progr. Phys.*, **33**, 413–488.

Broyles, A. A. (1970) 'Space–time position operators', *Phys. Rev.*, **1 D**, 979–988.

Bunge, M. (1970) 'The so-called fourth indeterminacy relation', *Can. J. Phys.*, **48**, 1410–1411.

Coish, H. R. (1959) 'Elementary particles in a finite world geometry', *Phys. Rev.*, **114**, 383–388.

Darling, B. T. (1950) 'The irreducible volume character of events. A theory of the elementary particles and of fundamental length', *Phys. Rev.*, **80**, 460–466.

Das, A. (1960) 'Cellular space–time and quantum field theory', *Nuovo Cimento*, **18**, 482–504.

Dirac, P. A. M. (1972) 'A positive energy relativistic wave equation', *Proc. Roy. Soc. London*, **328 A**, 1–7.

Eberly, J. H. and Singh, L. P. S. (1973) 'Time operators, partial stationarity and the energy–time uncertainty relation', *Phys. Rev.*, **7 D**, 359–362.

Fick, E. and Engelmann, F. (1964) 'Quantentheorie der Zeitmessung', *Z. Phys.*, **178**, 551–562.

Flint, H. T. (1935) 'A relativistic basis of the quantum theory', *Proc. Roy. Soc. London*, **150 A**, 421–441.

Flint, H. T. (1937) 'Ultimate measurements of space and time', *Proc. Roy. Soc. London*, **159 A**, 45–56.

Flint, H. T. (1948) 'The quantization of space and time', *Phys. Rev.*, **74**, 209–210.

Flint, H. T. and Richardson, O. W. (1928) 'On a minimum proper time and its applications (1) to the number of chemical elements, (2) to some uncertainty relations', *Proc. Roy. Soc. London*, **117 A**, 637–649.

Fock, V. A. (1962) 'Criticism of an attempt to disprove the uncertainty relation between time and energy', *Zh. Eksp. Teor. Fiz.*, **42**, 1135–1139.

Fujiwara, I. (1970) 'Time-energy indeterminacy relationship', *Prog. Theoret. Phys.*, **44**, 1701–1703.

Fürth, R. (1929) 'Uber einen Zusammenhang zwischen quantenmechanischer Unschärfe und Struktur der Elementarteilchen und eine hierauf begründete Berechnung der Massen von Proton und Elektron', *Z. Phys.*, **57**, 429–446.

Gien, T. T. (1965) 'Relativistic formulation of the lifetime matrix in the potential theory of collision', *J. Math. Phys.*, **6**, 671–676.

Glaser, W. and Sitte, K. (1934) 'Elementare Unschärfe, Grenze des periodischen Systems und Massenverhältniss von Elektron und Proton', *Z. Phys.*, **87**, 674–686.

Goebel, G. J., Karplus, R. and Ruderman, M. A. (1955) 'Momentum dependence of phase shifts', *Phys. Rev.*, **100**, 240–241.

Greenman, M. and Rohrlich, F. (1973) 'Is there a maximal electrostatic field strength?', *Phys. Rev.*, **8 D**, 1103–1109.

Griffith, R. W. (1974) 'Explicit formula from field theory for the average intrinsic size of a real or virtual photon', *Nuovo Cimento*, **21 A**, 435–470.

Heisenberg, W. (1927) 'Uber den anschaulichen Inhalt der Quantentheoretischen Kinematik und Mechanik', *Z. Phys.*, **43**, 172–198.

Heisenberg, W. (1936) 'Die selbstenergie des Elektrons'. *Z. Phys.*, **65**, 4–13.

Heisenberg, W. (1938a) 'Uber die in der Theorie der Elementarteilchen auftretende universelle Länge', *Ann. Phys. Lpz.*, **32**, 20–33.

Heisenberg, W. (1938b) 'Die Grenzen der Anwendbarkeit der bisherigen Quantentheorie', *Z. Phys.*, **110**, 251–266.

Heisenberg, W. (1942) 'Die "beobachtbaren Grössen" in der Theorie der Elementarteilchen', *Z. Phys.* **120**, 513–538.

Heisenberg, W. and Euler, H. (1936) 'Folgerungen aus der Diracschen Theorie des Positrons', *Z. Phys.*, **98**, 714–732.

Hellund, E. J. and Tanaka, K. (1954) 'Quantized space–time', *Phys. Rev.*, **94**, 192–195.
Henning, H. (1956) 'Die Unschärferelation in der Dirac–Gleichungen und in der relativistischen Schrödinger–Gleichung', *Z. Naturforsch.*, **11 A**, 101–118.
Hill, E. L. (1955) 'Relativistic theory of discrete momentum space and discrete space–time', *Phys. Rev.*, **100**, 1780–1783.
Iwanenko, D. (1931) 'Die Beobachtbarkeit in der Diracschen Theorie', *Z. Phys.*, **72**, 621–624.
Jaffe, J. and Shapiro, I. I. (1972) 'Lightlike behaviour of particles in a Schwarzschild field', *Phys. Rev.*, **6 D**, 405–406.
Jauch, J. M. and Marchand, J. P. (1967) 'The delay time operator for simple scattering systems', *Helv. Phys. Acta*, **40**, 217–229.
Jordan, P. and Fock, V. (1930) 'Neue Unbestimmtheitseigenschaften des elektromagnetischen Feldes', *Z. Phys.*, **66**, 206–209.
Kadyshevsky, V. G. (1961) 'On the theory of quantization of space–time', *Zh. Ekspenm. Teor. Fiz.*, **41**, 1885–1894.
Kálnay, A. J. (1971) *The Localization Problem, Studies in the Foundations Methodology and Philosophy of Science*, Springer-Verlag, Berlin, **4**, 93–100.
Kálnay, A. J. and Toledo, B. P. (1967) 'A reinterpretation of the notion of localization', *Nuovo Cimento*, **48**, 997–1007.
Kaluza, T. (1921) *S. B. Akad. Wiss. Berlin*, 966.
Kilian, H. and Petzold, J. (1970) 'Zur Begründung der Gamowschen Zerfallsthorie', *Ann. Phys. Lpz.*, **24**, 335–355.
Kim, D. Y. (1973) 'A possible role of universal length in the theory of weak interactions', *Can. J. Phys.*, **51**, 1577–1581.
Landau, L. and Peierls, R. (1931) 'Erweiterung des Unbestimmtheitsprinzip für die relativistische Quantentheorie', *Z. Phys.*, **69**, 56–69.
Latzin, H. (1927) 'Quantentheorie und Realität', *Naturwissenschaften*, **15**, 161.
Lévi, R. (1926) 'L'atome dans la thèorie de l'action universelle et discontinue', *C.R. Acad. Sci. Paris*, **183**, 1026–1028.
Lippmann, B. A. (1966) 'Operator for time delay induced by scattering', *Phys. Rev.*, **151**, 1023–1024.
Lord, E. A., Sinha, K. P. and Sivaram, C. (1974) '"Cosmological" constant and scalar gravitons', *Progr. Theoret. Phys.*, **52**, 161–169.
Ludwig, G. (1972) 'An improved formulation of some theorems and axioms in the axiomatic foundation of the Hilbert space structure of quantum mechanics', *Commun. Math. Phys.*, **26**, 78–86.
Mandelstamm, L. and Tamm, I. (1945) *J. Phys. U.S.S.R.*, **9**, 249.
March, A. (1936) 'Die Geometrie kleinster Räume', *Z. Phys.*, **104**, 93–99, 161–168.
March, A. (1937a) 'Zur Grundlegung einer statistischen Metrik', *Z. Phys.*, **105**, 620–632.
March, A. (1937b) 'Statistische Metrik und Quantenelektrodynamik', *Z. Phys.*, **106**, 49–69.
March, A. (1937c) 'Die Frage nach der Existenz einer kleinsten Wellenlänge', *Z. Phys.*, **108**, 128–136.
March, A. (1941) 'Raum, Zeit und Naturgesetze', *Z. Phys.*, **117**, 413–436.
Markov, M. (1940) 'On the four "dimensionally" stretched electron in a relative quantum region', *Zh. Eksperim. Teor. Fiz.*, **10**, 1311–1338.
Markov, M. A. (1965) 'Can the gravitational field prove essential for the theory of elementary particles?', *Suppl. Progr. Theoret. Phys.* (extra number), 85–95.
Markov, M. A. (1966) 'Elementary particles with largest possible masses (quarks and maximons)', *Zh. Eksperim. Teor. Fiz.*, **51**, 878–890.
Möglich, F. and Rompe, R. (1939) 'Über einige Folgerungen aus der Existenz eines kleinsten Zeitintervalles', *Z. Phys.*, 740–750.
Motz, L. (1962) 'Gauge invariance and the structure of charged particles', *Nuovo Cimento*, **26**, 672–697.
Motz, L. (1972) 'Gauge invariance and the quantization of mass (of gravitational charge)', *Nuovo Cimento*, **12 B**, 239–255.
Olkhovsky, V. S. and Recami, E. (1970) 'About a space–time operator in collision description', *Lett. Nuovo Cimento*, **4**, 1165–1173.
Olkhovsky, V. S., Recami, E. and Gerasimchuk, A. J. (1974) 'Time operator in quantum mechanics (I. Nonrelativistic case)', *Nuovo Cimento*, **22 A**, 263–278.
Papp, E. (1972a) 'Interaction time measurement and causality', *Nuovo Cimento*, **10 B**, 69–78.

Papp, E. (1972b) 'The non-relativistic limit of a dynamical proper time', *Nuovo Cimento*, **10 B**, 471–482.
Papp, E. (1973) 'Peculiarities of the quantum-mechanical space–time description', *Int. J. Theoret. Phys.*, **8**, 429–441.
Papp, E. (1974a) 'Field theoretical space–time quantization', *Int. J. Theoret. Phys.*, **9**, 101–115.
Papp, E. (1974b) 'Imprecision description of the high-energy annihilation and production processes', *Int. J. Theoret. Phys.*, **10**, 123–143.
Papp, E. (1974c) 'An extended approach to the field theoretical time operators', *Int. J. Theoret. Phys.*, **10**, 385–389.
Papp, E. (1975) 'Meaning and bounds for the space and time uncertainty contributions', *Ann. Phys. Lpz.*, **32**, 285–296.
Pavlopoulos, T. G. (1967) 'Breakdown of Lorentz invariance', *Phys. Rev.*, **159**, 1106–1110.
Penrose, R. and MacCallum, M. A. H. (1973) 'Twistor theory: An approach to the quantisation of fields and space-time', *Phys. Rep.*, **6 C**, 243–315.
Peres, A. (1966) 'Causality in S-matrix theory', *Ann. Phys. (N.Y.)*, **37**, 179–208.
Peres, A. and Rosen, N. (1966) 'Quantum limitations on the measurement of gravitational fields', *Phys. Rev.*, **118**, 335–336.
Peters, P. C. (1974) 'Propagation in a space–time lattice', *Phys. Rev.*, **9 D**, 3223–3228.
Planck, M. (1913) *Vorlesungen über die Theorie der Wärmestrahlung*, Johann Ambrosius Barth, Leipzig, 167–169.
Poincarè, H. (1913) *Dernières pensées*, Flammarion, Paris.
Pokrowski, G. I. (1928) 'Zur Frage nach der Struktur der Zeit', *Z. Phys.*, **51**, 737–739.
Proca, A. (1928) *Sur la Théorie des Quanta de Lumière*, Blanchard, Paris.
Remak, B. (1931) 'Zwei Beispiele zur Heisenberg-schen Unsicherheitsrelation bei gebundenen Teilchen', *Z. Phys.*, **69**, 332–345.
Ruark, A. E. (1928) 'The limits of accuracy in physical measurements', *Proc. Nat. Acad. Sci. Wash.*, **14**, 322–328.
Schames, L. (1933) 'Atomistische Auffassung von Raum und Zeit', *Z. Phys.*, **81**, 270–282.
Schild, A. (1948) 'Discrete space–time and integral Lorentz transformations', *Phys. Rev.*, **73**, 414–415.
Schrödinger, E. (1930) 'Uber die kräftefreie Bewegung in der relativistischen Quantenmechanik', *Berliner Berichte*, 418–428.
Schröder, U. E. (1964) 'Lokalisierte Zustände und Teilchenbild bei relativistischen Feldtheorien', *Ann. Phys. Lpz.*, **14**, 91–112.
Sivaram, C. and Sinha, K. P. (1974) 'Gravitational charges and the quantization of mass, *Lett. Nuovo Cimento*, **10**, 227–230.
Smith, F. T. (1960) 'Lifetime matrix in collision theory', *Phys. Rev.*, **118**, 349–356.
Snyder, H. S. (1947) 'Quantized space-time', *Phys. Rev.*, **71**, 38–41.
Takano, Y. (1961) 'The singularity of propagators in field theory and the structure of space–time', *Progr. Theoret. Phys.*, **26**, 304–314.
Thomson, J. J. (1925/1926) 'The intermittence of electric force', *Proc. R. Soc. Edinb.*, **46**, 90–115.
Treder, H. (1963) 'Gravitonen', *Fort. Phys.*, **11**, 81–108.
Treder, H. (1974) 'Gravitationskollaps und Lichtgeschwindigkeit im Gravitationsfeld', *Ann. Phys. Lpz.*, **31**, 325–334.
Vialtzew, A. N. (1963) 'Discrete space and time' (in Russian), Isd. 'Nauka', Moskwa.
Wataghin, G. (1930) 'Uber die Unbestimmtheitsrelationen der Quantentheorie', *Z. Phys.*, **65**, 285–288.
Wataghin, G. (1932) 'Zur relativistischen Quantenmechanik', *Z. Phys.*, **73**, 121–129.
Wheeler, J. A. (1957) 'On the nature of quantum geometrodynamics', *Ann. Phys. (N.Y.)*, **2**, 604–614.
Wheeler, J. A. (1962) *Geometrodynamics*, Academic Press, New York.
Wigner, E. P. (1955) 'Lower limit for the energy derivative of the scattering phase shift', *Phys. Rev.*, **98**, 145–147.
De Witt, B. S. (1960) *The Quantization of Geometry*', Institute Field Physics, University of North Carolina.
Yang, C. N. (1947) 'On quantized space-time', *Phys. Rev.*, **72**, 874.

6

Uncertain Cosmology

CHRISTOPHER J. S. CLARKE
University of York, England

1. THE VIEWPOINT OF QUANTUM COSMOLOGY

The particularity of one's presuppositions should never be underestimated. I write from a set of assumptions which to the quantum physicist may seem peculiar: that one can and should provide a coherent mathematical scheme for the entire universe; that the scheme should admit a model that represents the universe as we perceive it to be, with ourselves as observers in it; that the occurrence of such a model within the scheme can, in some sense, explain both our own coming-into-being and also the nature of what we now observe.

Most interpreters of quantum theory, being concerned only with certain delineated subsystems of the universe, can with propriety *assume* the prior existence of the macroscopic world of daily experience as a background given in advance, in terms of which the subsystem must be explained. By rejecting this in favour of a cosmological view,* I am forced into the contentious position of using a quantum theory which includes the observer in the formalism. More than this: I wish to do it in a way consistent with a general relativistic† treatment of the universe, since such a treatment uniquely combines logical elegance with observational consistency. This raises three particular problems.

(1). If Newtonian space–time be abandoned, there is no reason, either physical or philosophical, to assume that the global properties of space–time should be the same as the local properties which our short-range observations reveal. In particular (as I have argued in detail elsewhere (Clarke, 1976)) the universe of general relativity may not admit any global time coordinate: if its existence (i.e. stable causality) is

---

*Certain cosmologists (Dicke, 1961; Collins and Hawking, 1973) have used our own existence as a constraint on cosmological parameters. But this only gives a non-tautologous explanation of the universe if the physical schema used is able to generate a reasonably restricted class of cosmological models before such a constraint is imposed; it does not absolve us from developing models which are potentially independent of our own existence.

†Within the term 'general relativity' I include all theories where space–time is represented as a C^1 manifold on which local inertial frames are specified by a metric of Lorentz signature, determined along with other physical entities by field equations which may not necessarily be (either of) those of Einstein.

assumed then it must be recognized as an additional postulate made either in anticipation of future experimental evidence or as a temporary expedient to ease the calculations. But without the assumption of global time the traditional quantum-mechanical picture of a system evolving in time is untenable.

(2). In a general space–time there is no symmetry group which will enable one to Fourier analyse a quantum field to give it an unambiguous particle interpretation. The definitions of particle number, particle creation rate, etc., are, in this situation, a matter of great controversy (Unruh, 1974).

(3). Since the structure of space–time is itself a dynamical variable, and not merely a fixed arena for other events, it must itself be quantized: firstly, because its source is composed of quantized fields; and, secondly, because it seems likely that there are regions of the universe where the space–time curvature is characterized by a length-scale small enough to be in the quantum domain. But the quantization of space–time is not only technically difficult. In addition, the removal of both a fixed background space–time and a reliable particle-representation leaves very little structure on which to hang an interpretation of any formalism proposed.

This third point leads one to the central difficulty of quantum cosmology: if everything is quantized—space, time, particles, observers—then everything dissolves into a structureless haze from which it is impossible to extract any semblance of concrete reality.

Such a viewpoint is intimately related to the place afforded to the uncertainty relations. For, as conceived by Heisenberg (1930) 'the statistical character of the relation (between values of dynamical quantities) depends on the fact that the influence of the measuring device is treated in a different manner than the interaction of the various parts of the system on one another ... The chain of (determinate) cause and effect could be quantitatively verified only if the whole universe were considered as a single system—but then physics has vanished, and only a mathematical scheme remains' (p. 58). The more cosmology has developed, the more this observation of Heisenberg has been confirmed, that the simple extension of quantum theory to the cosmological domain yields a mere 'mathematical scheme' that stands in need of something else before physics can emerge. For him this addition comes through alternative descriptions which (Heisenberg, 1974) are 'complementary' to quantum theory in being compatible with it, but not deducible from it.

The course of this chapter is the pursuit of this 'something else' in a context beyond the usual laboratory one: the context of a cosmological picture containing physically extreme regions where no distinction, however arbitrary, can be made between the observer and the observed, the 'measuring device' and the 'parts of the system', in Heisenberg's terms. Here the use of a complementary description in the original restricted sense of Bohr (1928) is impossible, and the uncertainty relations will not, as in the laboratory case, appear as a limitation on the applicability of a classical corpuscular description (Heisenberg, 1930).

2. THE WHEELER–EVERETT THEORY AND ADDITIONAL STRUCTURE

First we must examine an unusual cosmological view which seems to offer the hope of dispensing with any addition to quantum theory. Having given a detailed philosophical critique elsewhere (Clarke, 1974), I shall here summarize the conclusions and develop further mathematical points.

The theory (Wheeler, 1957; Everett, 1957; de Witt and Graham, 1973) regards the universe as a single quantum mechanical system whose state vector Ψ undergoes a determinate Hamiltonian evolution in a Hilbert space $\mathcal{H}$. When an observation is taking place $\mathcal{H}$ decomposes into the tensor product $\mathcal{H}_M \otimes \mathcal{H}_S$, where $\mathcal{H}_S$ describes the observed microsystem and $\mathcal{H}_M$ describes everything else. Correspondingly, $\Psi = \Psi_M \otimes \Psi_S$.

Suppose that $\Psi_S = \sum_i a^i \psi_i$, where $\psi_1, \psi_2, \ldots$ are eigenstates in which the quantity being measured has a definite value. The usual theory of measurement is followed, according to which the state $\Psi_M^0 \otimes \psi_i$ evolves (approximately) into a state $\Psi_M^{(i)} \otimes \psi_i$, where Ψ_M^0 represents the state of affairs before the measurement is made while $\Psi_M^{(i)}$ represents the state after a definite value, corresponding to ψ_i, has been found. Thus the measurement as a whole is described by an evolution from $\Psi_M^0 \otimes \Psi_S = \Psi_M^0 \otimes (\sum a^i \psi_i)$ into $\sum_i a^i \Psi_M^{(i)} \otimes \psi_i$.

The Wheeler–Everett theory interprets this last state as the simultaneous existence of many copies of the universe, one for each index i. This means that at each measurement the universe splits into many branches, each branch corresponding to a separate definite value for the measurement. Every person in the universe is thus split into many copies which henceforth evolve independently of each other (by linearity). Each copy is, by construction, aware only of the (definite) outcome of the measurement in his branch so that to him it appears as if the state-vector has 'collapsed' onto an eigenstate.

All the branches are equivalent: there is no need to try to attach meaning to one branch being 'more likely' than another. The statistical interpretation of quantum theory is derived, not from such an additional postulate, but from a consideration of long sequences of experiments. From this it is shown that, in the limit as the lengths of the sequences tend to infinity, in all branches of the universe except for a set of measure zero the relative frequencies of the various possible outcomes for the experiment accord with the usual quantum mechanical probabilistic interpretation of the a^i.

Now, it could well be argued that this splitting provides a mental picture that is simpler and more economical of hypothesis than that provided by the 'collapse of the wave packet' description of von Neumann (1955) and others. But in terms of physical verification the two approaches are equivalent, and both stand in need of definite criteria which determine when it is that a measurement takes place and how the Hilbert space is to be decomposed; information that is assumed to be given *ab extra* in both approaches to quantum theory. But in cosmology, by definition, nothing is extra to the dynamical formalism; there are no external fields—not even the geometrical structure of space–time if, as in general relativity, this geometry is itself a quantized dynamical variable. Thus I propose the following thesis: a system of quantum cosmology, as it is usually understood, cannot contain enough intrinsic structure to allow one to use criteria which might characterize the occurrence of measurement situations.

Let me clarify this by a comparison with a classical case. There the dynamical system might be represented by N functions $\mathbb{R} \to \mathbb{R}^3$ (specifying the positions of the N particles of the system as functions of time) determined by N coupled ordinary differential equations. In interpreting such a system, it suffices to take an existing structure within the mathematics (the relative positions of the particles) and match it with a corresponding observed structure. In quantum mechanics, however, the system is more usually represented mathematically by one function $\mathbb{R} \to \mathcal{H}$ (the vector in Hilbert space as a function of time), again governed by a differential equation. In interpreting this it is not enough to match a structure intrinsic to the mathematics with something observable: the interpretation itself sets up further implicit mathematical structure by distinguishing between different vectors in the Hilbert space.

Note that there is in this respect no essential difference between classical and quantum systems, only a difference in the presentation. The classical system could be presented as one function $\mathbb{R} \to \mathbb{R}^{3N}$, while the quantum system could be presented as an infinite set of equations for the coefficients of the state in a particular basis, and both are often found. But in quantum cosmology most of the structure beyond the specification of the Hamiltonian is ambiguous, and yet it contains virtually all the physical information. It is therefore vital to be self-conscious about the amount of structure already in the Hamiltonian, and the extent and role of the additional structure that is required.

This can be illustrated in three cases.

2(a) Field Theory

Let us first suppose that the quantum system is arrived at by using a background (flat) space–time. The customary procedure would be to set up a Fock-space or equivalent representation of the free fields, and then introduce some interaction. For a free field the pair $(\mathcal{H}, H)$, where $\mathcal{H}$ is the Hilbert space and H the Hamiltonian, conveys almost no information at all. The only intrinsic structure for such a pair is given by the spectral type of H (Plesner, 1969), and a free-field Hamiltonian has merely a homogeneous spectral type covering the positive reals with multiplicity d (the cardinality of the integers). The system only acquires some physical content when a particular Fock-space decomposition of $\mathcal{H}$ is specified.

When it comes to interacting fields, the main problem is one of establishing what $\mathcal{H}$ is by some renormalization procedure. In the absence of any rigorous treatment of this, one can only speculate as to the structure of $(\mathcal{H}, H)$; but it would be surprising if it differed at all from the free-field case.

In actual practice very little interest is paid to the nature of $\mathcal{H}$, the stress being entirely on the operator algebras derived from the fields. By virtue of their interpretation, these have a very rich implicit structure. But in cosmology it is the *states* which must be related to the universe we observe, since there are no external measurements to correspond to 'observables'; and so I shall concentrate on structure as it is manifested in the set of states.

2(b) Infinite Fock Space

The foregoing must be modified for cosmological application, in view of the likelihood (Schramm and Wagoner, 1974) of the universe being spatially infinite. Then it is more realistic to allow the space of 'states' to include descriptions of infinitely many particles (and not merely unboundedly many, as in Section 2(a)). A complication then arises when Fermi or Bose–Einstein statistics are needed because the appropriate 'state'-space $\mathcal{H}^*$ is not a subspace of the infinite tensor product $\mathcal{H}^\infty$ of one-particle states

$$\mathcal{H}^\infty = \bigotimes_{i=1}^{\infty} (\mathcal{H}_0)^{(i)}$$

but a subspace* of its algebraic dual. $\mathcal{H}^*$ is in fact not a Hilbert space; but a Hamiltonian is defined in it as the dual of an appropriately defined operator on the Hilbert space $\mathcal{H}^\infty$, and the arguments of Section 2(a) apply to this.

Because $\mathcal{H}^*$ is not a Hilbert space, its elements cannot be regarded as interpretable states. It has, however, a concept of orthogonality and hence of orthogonal projection. A (mixed) state can then be defined as a subadditive and orthogonally additive function of the projectors into [0, 1]: there are no pure states. As it does not seem to be known whether there is a representation theorem for this situation (of the type given, for instance, by Langerholc (1965)) one cannot pursue the matter much further. I shall assume that it will ultimately be possible to deal with these states in the same way as mixed states in the separable-Hilbert-space, to which I now turn.

2(c) Mixed States

Suppose one has the situation in Section 2(a), except that now the triple $(\mathcal{H}, H, \Phi_0)$ is given, where Φ_0 is a mixed state (a self-adjoint operator of trace

---

*Explicitly $\mathcal{H}^*$ is the annihilator of the subspace of $\mathcal{H}^\infty$ consisting of all finite combinations of vectors of the form

$$e_1^{(1)} \otimes \ldots \otimes e_i^{(i)} \otimes \ldots \otimes e_j^{(j)} \otimes \ldots \pm e_1^{(1)} \otimes \ldots \otimes e_j^{(i)} \otimes \ldots \otimes e_i^{(j)} \otimes \ldots$$

where $f^{(k)}$ denotes f as a member of $(\mathcal{H}_0)^{(k)}$, the kth copy of $\mathcal{H}_0$ and the + (resp. −) sign gives Fermi (resp. Bose–Einstein) statistics.

class) representing the universe at some initial time t_0. There is now room for considerable complexity since H and Φ will not in general commute; but I shall argue that interpreting quantum theory by using only this information, while not impossible, is highly unsatisfactory: the available structure cannot provide an acceptable basis for decomposing a vector into a superposition of macroscopic states, as is required by any interpretation, including that of Wheeler and Everett.

At first glance an obvious procedure presents itself. Suppose that we want to give an intrinsic characterization of the measurement process described in Wheeler–Everett terms at the start of this section, where $\mathcal{H}$ splits* into $\mathcal{H}_1 \oplus \mathcal{H}_2$ and Φ decomposes accordingly into

$$\Phi = (\Phi_{11} + \Phi_{12}) \circ P_1 + (\Phi_{21} + \Phi_{22}) \circ P_2$$

(where $\Phi_{ij} : \mathcal{H}_i \to \mathcal{H}_j \subset \mathcal{H}$, $P_i : \mathcal{H} \to \mathcal{H}_i$). It is a characteristic of the measurement situation (Daneri, Loinger and Prosperi, 1962) that in these circumstances statistical processes ensure that $\|\Phi_{12}\| = \|\Phi_{21}\| \to 0$, while the remaining components are H-stable in that $[\Phi_{11}, H]$ maps (approximately) from $\mathcal{H}_1$ to $\mathcal{H}_1$ and similarly for $\mathcal{H}_2$. Thus we can try to identify $\mathcal{H}_1$ and $\mathcal{H}_2$ by diagonalizing Φ and then looking for partitions of the set of eigenvectors into subsets which each span subspaces stable under H.

However, this procedure cannot separate out the subspaces $\mathcal{H}_1$ and $\mathcal{H}_2$ if the probabilities of the two corresponding experimental results are almost equal— as may happen if there is an exact symmetry between the two microscopic end states of the system being observed. In that case the nature of the diagonalization will be heavily influenced by the residual off-diagonal terms Φ_{12}, which would cause the procedure to select as possible outcomes states which should in fact be regarded as unacceptable superpositions of macroscopically distinct states.

The difference between the values of the various probabilities involved (the eigenvalues of Φ) is not the essential criterion in distinguishing possible outcomes. That criterion is the macroscopic dissimilarity of the various possibilities, and the diagonalization of Φ is linked with it only in certain cases. In general one must recognize the existence of some fundamental structure which corresponds to this dissimilarity, and is not equivalent to any intrinsic property of $(\mathcal{H}, H, \Phi)$.

3. COSMOLOGY WITH STRUCTURE

Let us suppose that as well as a Hilbert space and a Hamiltonian† we have also as additional structure a *particle decomposition*, in the form of a Fock-space representation of $\mathcal{H}$ (supposing, for simplicity, a finite but unbounded number

---
*Here $\mathcal{H}_1$ and $\mathcal{H}_2$ correspond to two different outcomes to the measurement.
†For the sake of definiteness, and to remain with familiar territory, I have phrased my account in terms of a conventional $(\mathcal{H}, H)$ formalism; but this is by no means essential, and is probably undesirable.

of particles). This might seem very drastic, but it is hard to see how one can get away with any less. One consequence of this is that the problem of defining particles in curved space in order to study particle creation in cosmology (Parker and Fulling, 1972) is now reversed: particles are supposed given at the outset, and the problem is now to define the space in which they are situated. Instead of particle-creation one has space-annihilation.

Although particles may be part of the *a priori* structure, they cannot directly give the sort of information needed to define macroscopic states and measurements. The universe cannot always be split into apparatus and microsystem along particle lines, since particle number could be one of the dynamical quantities being measured. The process of defining macroscopic states must be two-stage: first, the particle structure has to define a spatial structure; then criteria for macroscopic interpretability have to be formulated in terms of spatial structure.

The basic ideas for passing from particles to space are fairly simple, and limited progress has been made in their formal articulation. Consider an N-particle state of the form $\mathfrak{S}(\psi_1 \otimes \psi_2 \otimes \ldots \otimes \psi_N)$ where $\mathfrak{S}$ denotes (anti-) symmetrization. We may try to associate a 'distance' between ψ_1 and ψ_2 by examining $|\langle\psi_1|H|\psi_2\rangle|^2$; the greater this quantity the greater the probability of interaction between ψ_1 and ψ_2 and so the closer their 'distance'. The development of this idea (Penrose, 1972) has actually been in terms of models where there is no explicit Hamiltonian, but only quantities which may be thought of as scattering amplitudes. In the simplest case a geometry can be defined in terms of these quantities, which is a geometry of Euclidean directions. An important feature is that if one first sets up the amplitudes by using a conventional description, based on particle states which do not define precise directions, then the geometry which is deduced is still a normal Euclidean direction-geometry, but one that is not related in any simple way to the space with which one started. On this approach particles come first, and the 'real' space is the one which they define.

One can speculatively indicate the form which a cosmology based on this might take. An N-particle state, as above, could be examined to establish a rough criterion of 'nearness', or locality. Then a more detailed geometry could be defined which held locally, in the sense that Minkowski geometry holds locally in general relativity. Still working in terms of a group of nearby particles, states would certainly be regarded as macroscopically distinct if there was no isomorphism of the geometry which made a suitably smoothed-out particle-density for the two states coincide, even approximately. (More precisely, this would give a criterion for the distinguishability of two mixed states, each corresponding to specific states for a local group of k of the N particles, the remainder being unspecified.) Finally, transition probabilities would have to be specified between states which were macroscopically interpretable (i.e. which were not superpositions of distinguishable states). This would be done using either a Hamiltonian formalism, or combinatorial laws which arise more naturally in the twistor theory development already cited (Penrose, 1972).

Note one advantage of this emphasis on mixed states: one can pass with little modification to the infinite inverse described in Section 2(b), where there are no pure states.

If the theory were to progress along lines like these, one can discern three aspects which would prove especially interesting.

3(a) Renormalization

Infinities arise both from the unboundedly large momenta that occur in loops, and from the unboundedly large number of intermediate particles that can appear in the perturbation expansion for an interaction. The first divergence, with which renormalization concerns itself, is an essential part of the usual space–time descriptions used and of the field-theory approach of adding interactions onto a basically free-field structure. The theory I am envisaging must start with a system of finite matrix elements, i.e. it must already be renormalized. This is another point in favour of twistor theory, which automatically yields finite answers to scattering problems (Penrose, 1975).

The second infinity (caused by non-convergence of perturbation theory expansions) is unlikely to arise, because particle density is automatically limited in a theory which places particles before space. As more particles appear, so more space appears, since space is simply the numerical relations between the particles.

3(b) Local and Global Aspects

In general relativity it is assumed that the local aspect of the universe to which we have immediate access is similar to all other local aspects, and that these local glimpses can be pieced together into a global model. In the quantum case there seems to be no reason why this should be so. That is to say, while the *theory* refers to the universe as a whole, there is no reason to suppose that there should be globally defined macroscopic states and transition probabilities. The desire for this, which haunts much of current work in quantum cosmology, stems from the mistaken attitude of viewing the universe from the outside, as we would observe an atom in a crystal. In reality we are part of the universe and the states with which the scientist is concerned are states relative to himself. All our observations are local, at least in the sense that the domain of galaxies over which they extend is one in which the geometry departs only modestly from the Euclidean; and, on a purist view, they are very local, in that we are concerned directly only with photons and particles which arrive here on earth, and arguments as to their source are a matter of indirect inference.

So we should not be disappointed if we fail to obtain a god-like view of the universe as a whole. What we can demand is that which is scientifically testable: a theory which enables us to predict and understand future observations in terms of present ones. In practice, of course, one would make use of the global understanding of the universe which we think we already have: a mixed state in which a local group of particles only is specified can be regarded as a mixture of globally specified states, each of which can be analysed by analogy with a conventional cosmology. But since our present observations do not single out any unique cosmological model, many global models will be compatible with them; some having a global time, some being acausal and so on. All these participate in the mixed state which is defined by our observations, and define the states into which it may, probabilistically, turn.

The basic structures of the theory are global and comprehend the entire universe. But their translation into observation requires the selection of some particular viewpoint, in the form of a local group of particles small enough to enable a spatio-temporal structure to be defined. The 'additional structure' in the theory specifies the totality of possible viewpoints, one of which is ours.

3(c) Uncertainty

The probabilistic nature of the predictions which emerge could, if one wished, be ascribed to some kind of complementarity between the Hilbert space description and the macroscopic-state description. I would see this as unnecessarily dualistic, preferring to regard the Hilbert space structure as a kind of scaffolding on which to hang and manoeuvre the macroscopic states, and to use to calculate the transition probabilities. The physics is a physics of the macroscopic states, and the relations between them are by their nature probabilistic. This enables us to return to the uncertainty relations in a cosmological setting, when they become relations constraining an intrinsically probabilistic scheme. They *entail* limitations on our measuring abilities, rather than being consequences of limitations, because the probabilistic structures of which these are an instance are there constraining the physics of the universe even if nothing is happening which could conceivably be called a measurement.

In short, the indeterminate physics which is uncovered in our laboratories is simply an aspect of the entire uncertain cosmology of which it is a tiny part.

REFERENCES

Bohr, N. (1928) 'The Quantum Postulate and the Recent Development of Atomic Theory', *Nature*, **121**, 580–590.
Clarke, C. J. S. (1974) 'Quantum theory and cosmology', *Phil. Sci.*, **41**, 317–332.
Clarke, C. J. S. (1976) 'Time in general relativity', To appear in *Minnesota Studies in the Philosophy of Science*.
Collins, C. B. and Hawking, S. W. (1973) 'Why is the universe isotropic?' *Astrophys. J.*, **180**, 317–334.
Daneri, A., Loinger, A. and Prosperi, G. M. (1962) 'Quantum theory of measurement and ergodicity conditions', *Nucl. Phys.*, **33**, 297–319.
De Witt, B. S. and Graham, N. (Eds.) (1973) *The Many-Worlds Interpretation of Quantum Mechanics*, Princeton University Press, Princeton.

60 Uncertainty Principle and Foundations of Quantum Mechanics

Dicke, R. H. (1961) 'Dirac's cosmology and Mach's principle', *Nature*, **192,** 440–441.

Everett III, H. (1957) '"Relative state" formulation of quantum mechanics', *Rev. Mod. Phys.*, **29,** 454–462.

Heisenberg, W. (1930) *The Physical Properties of the Quantum Theory* (trans. Eckart, C. and Hoyt, F. C.), University of Chicago Press, Chicago.

Heisenberg, W. (1974) 'Double dialogue', *Theoria to Theory*, **8,** 11–34.

Langerholc, J. (1965) 'The trace formalism for quantum mechanical expectation values', *J. Math. Phys.*, **6,** 1210–1218.

Penrose, R. (1972) 'On the nature of quantum geometry', *Magic without Magic: John Archibald Wheeler*, Klauder, J. R. Ed., W. H. Freeman and Co., San Francisco.

Penrose, R. (1975) *Quantum Gravity*, Isham C. J., Penrose, R. and Sciama, D. W. (Eds.), Oxford University Press, Oxford.

Parker, L. and Fulling, S. A. (1972) 'Quantized matter fields and the avoidance of singularities in general relativity', *Phys. Rev. D*, **7,** 2357–2374.

Plesner, A. I. (1969) *Spectral Theory of Linear Operators*, Vol. II, Ungar, New York.

Schramm, D. N. and Wagoner, R. V. (1974) 'What can deuterium tell us?' *Phys. Today*, **27,** (12), 40–47.

Unruh, W. G. (1974) 'Alternative Fock quantization of neutrinos in flat space-time', *Proc. Roy. Soc. London A*, **338,** 517–525.

von Neumann, J. (1955) *Mathematical Foundations of Quantum Mechanics*, trans. Beyer, R. T., Princeton University Press, Princeton.

Wheeler, J. A. (1957) 'Assessment of Everett's "Relative State" Formulation of Quantum Theory', *Rev. Mod. Phys.*, **29,** 463–465.

7

Uncertainty Principle and the Problems of Joint Coordinate–Momentum Probability Density in Quantum Mechanics

V. V. KURYSHKIN
Peoples' Friendship University, Moscow, U.S.S.R.

1. INTRODUCTION

The 50-year old history of the development of quantum mechanics has been extremely rich in attempts to reconsider its interpretation, to alter its mathematical formalism and finally to create a new theory that would provide a more complete description of physical reality than the one offered by quantum mechanics. Among the investigations conducted in this field are those devoted to the search for the singular solutions of the equations of quantum mechanics (De Broglie, 1956; De Broglie and Andrade e Silva, 1971); the search for particle-like solutions of the non-linear field theory (Finkelstein and coworkers, 1956; Glasko and coworkers, 1958; Rybakov, 1974); the attempts to introduce all kinds of 'hidden' parameters (Bohm, 1952; Pena-Auerbach and coworkers, 1972); the realization of various stochastic approaches to quantum mechanics (Fényes, 1952; Bess, 1973); the attempts to explain quantum phenomena by the existence of an 'imaginary' or 'hidden' thermostat, 'subquantum medium' (Bohm and Vigier, 1954; Terletsky, 1960; De Broglie, 1964).

The authors of such investigations usually proceed from the assumption that generally accepted quantum mechanics does not completely describe the physical reality and that it is possible to create a more profound theory which would treat all experimentally measurable quantities as simultaneously existing physical realities.

The incompleteness of the quantum-mechanical description was implicit in the earliest works of the founders of quantum mechanics (De Broglie, 1927), and since 1935 it has been a kind of an accusation against the fully established quantum mechanics (Einstein and coworkers, 1935; Schrödinger, 1935). However, the thesis of the incompleteness of the quantum-mechanical description remains unproved so far. This is because all the proofs of the incompleteness can easily be refuted on the grounds that quantum mechanics, owing to the well-known Heisenberg uncertainty principle,

$$\langle(\Delta q_j)^2\rangle\langle(\Delta p_{j'})^2\rangle \geq \frac{\hbar^2}{4}\delta_{jj'} \qquad (1)$$

rejects the concept of the coordinate and the momentum of a system existing simultaneously as physical realities. But the statement of the completeness of the quantum-mechanical description remains an assumption that has not been proved either. This stimulates a search for a new theory, more profound than that of quantum mechanics; the existence of such a theory has not been doubted by many outstanding physicists (Einstein, 1948; De Broglie, 1953; Schrödinger, 1955).

In the construction of the above-mentioned profound theories the principle of the correspondence between the sought-for theory and quantum mechanics plays a major part. In the opinion of most investigators this correspondence means that the new theory must explain the fundamental propositions of quantum mechanics as of certain statistical theory which appears when the completeness of the description of physical reality is partially sacrificed (when certain statistical averaging is undertaken). Thus, quantum mechanics lays quite definite claims to the sought-for profound theory. It is quite natural that the authors of different profound theories are anxious to satisfy these claims first and foremost.

In its turn the assumption that the sought-for profound theory exists lays certain claims on quantum mechanics itself. This circumstance is usually neglected by most investigators. Meanwhile, it is the main obstacle in the way of creating a profound theory. This can be illustrated by the following reasoning.

Quantum mechanics in spite of its obvious and generally acknowledged statistical character is not a theory of the consistent probability nature. It does not make use of any joint probability distributions for physical quantities, for example for coordinate and momentum, it defines no conditional probabilities. This fact leads to no contradictions within quantum mechanics since the quantum-mechanical description does not require that all physical quantities be considered as simultaneously existing realities.

Let us assume now that there exists a theory giving a more complete description of physical reality than quantum mechanics and treating all quantities as simultaneously existing physical realities. Let this theory with the completeness of the physical reality description partially sacrificed (certain statistical averaging employed) lead to a statistical theory, coinciding with quantum mechanics. But renouncing the completeness of the description and resorting to probabilities in this theory we shall inevitably arrive at a statistical theory, in which along with the probability of values of the physical quantity A_1 and the probability of values of the quantity A_2 there will exist a joint probability of the values of quantities A_1 and A_2 the probability of values of the physical quantity A provided that A_1 has a certain definite value A'_1, i.e. the conditional probability. In other words, the statistical theory thus obtained, and coinciding in accordance with the tentative assumption with quantum mechanics, will inevitably follow the conventional probability scheme.

Hence, if the sought-for profound theory exists, then the concepts of joint and conditional probabilities can be introduced into quantum mechanics, i.e. quantum mechanics may be reduced to a consistent probability scheme.

Attempts to introduce the concepts of the joint probability density for various physical quantities and in the first place for a certain joint coordinate–momentum distribution (quantum distribution function, QDF) have been made repeatedly. The earliest works in this field (Wigner, 1932; Terletsky, 1937; Blokhintsev, 1940) did not aim at introducing the QDF into quantum mechanics and considered the proposed phase space functions only as possible mixed representations of the density matrix, which later proved extremely useful in concrete quantum-mechanical problems (Klimantovitch and Silin, 1960; Imre and coworkers, 1967; Arinshtein and Guitman, 1967; Gorshenkov and coworkers, 1973). It was only in 1949, that an attempt to interpret Wigner's function as a QDF was made apparently for the first time (Moyal, 1949). However, Moyal's statistical interpretation of quantum mechanics did not gain much support, since the sign-variability of Wigner's function prevents it from being treated as the joint coordinate–momentum probability density. In subsequent years a few more concrete functions that might be considered as QDF were suggested (Bopp, 1956; Margenau and Hill, 1961; Mehta, 1964; Cohen, 1966a, b; Shankara, 1967; Kuryshkin, 1968; Ruggery, 1971; Zlatev, 1974). Except for Bopp's function they all turned out to be sign-variable. Besides, investigations showed (Mehta, 1964; Cohen, 1966; Kuryshkin, 1968) that the choice of any of these functions for part of the QDF requires a certain correspondence rule (the rule of constructing quantum operators), which does not coincide with the rule (Neumann, 1932) used in quantum mechanics. In other words, the proposed QDF should be treated as no more than phase-space representation of the density matrix (Imre and coworkers, 1967; Kuryshkin, 1969a, b; Ruggery, 1971; Gorshenkov and Kognkov, 1973). Finally, in 1966 it was proved (Cohen, 1966a, b) that in the generally accepted quantum mechanics, whose operators satisfy Neumann's requirements (Neumann, 1932; Shewell, 1959) the QDF was non-existent, not only the non-negative QDF but the sign-variable one as well. Thus, the concept of the joint coordinate–momentum probability density cannot be introduced into generally-accepted quantum mechanics, i.e. the generally accepted quantum mechanics cannot be reduced to a consistent probability scheme. This conclusion was also formulated and discussed in a number of other works (De Broglie, 1964; Andrade e Silva and Lochak, 1969; Kuryshkin, 1974).

Thus, the generally accepted quantum mechanics compels us: (a) either to reject the assumption of the existence of a theory that can provide a more complete description of physical reality or (b) assuming that such a theory does exist, to question the validity of the generally accepted quantum mechanics itself.

Therefore, while favouring the search for a profound theory, it is necessary in the first place to reconsider the generally accepted quantum mechanics altering it so as to introduce into it a non-negative QDF interpreted as the joint coordinate–momentum probability density. Such alterations, naturally, must not lead to the violation of those propositions that can be experimentally checked. Heisenberg's uncertainty principle, which is a fundamental and indispensable law of quantum theory begins to play a very important part in this case. This is because correlation (1) forbids the physical system states with coordinate and momentum strictly determined, while any attempt to introduce a joint coordinate–momentum probability into the quantum theory is equivalent to an implicit assumption that a physical system can possess a definite momentum with a quite definite coordinate.

The principal possibility of altering quantum mechanics with the view of introducing QDF into it was shown in the works of the author of this paper (Kuryshkin, 1971, 1972a). Such alteration was based on the fact that the problem of constructing operators $O(A)$ of physical values A in quantum mechanics has not been completely solved.

The generally accepted quantum mechanics makes use of operators, satisfying a set of requirements, called the Neumann rule (Neumann, 1932; Shewell, 1959). However, as far back as 1935 it was shown that this rule is not single-valued and the attempts to get rid of that disadvantage lead to inner contradictions (Temple, 1935a, b). Other known correspondence rules (Born and Jordan, 1925; Dirac, 1958; Weyl, 1950; Tolman, 1938; Rivier, 1951; Yvon, 1946; Kuryshkin, 1968; Kerner and Sutsliffe, 1970) also suffer from a number of drawbacks (Shewell, 1959; Groenewold, 1935; Kuryshkin, 1969b; Cohen, 1970). It must be noted that all known correspondence rules, generally speaking, agree only in the statement that:

$$O(q_j)O(p_{j'}) - O(p_{j'})O(q_j) = i\hbar\, \delta_{jj'} \qquad (2)$$

Commutator (2) in the long run results in correlation (1).

The works criticizing the well-known correspondence rules made it possible to formulate a number of requirements to the 'uncontradictory' rule and to construct it (Kuryshkin, 1971). The application of this rule to the construction of quantum operators has led to a theory, named 'quantum mechanics with a non-negative QDF' (Kuryshkin, 1972a). To date this theory has been studied fairly fully (Kuryshkin, 1972b, c, 1973; Zaparovany, 1974; Zaparovany and coworkers, 1975).

In this paper therefore, we will mostly concentrate on the principles of constructing theories of a quantum mechanical type possessing a non-negative QDF as well as on a brief analysis of certain concepts distinguishing these theories from the generally accepted quantum mechanics.

It should apparently be stressed once again, that our concern will not be to offer another interpretation of the generally accepted quantum mechanics but to construct some new statistical theory which will comprise a major part of the quantum mechanical mathematical formalism and which will have a consistent probability character, and can, therefore without contradiction be considered as a statistical theory for the would-be more profound theories.

In order to pay maximum attention to the physical sense and not to burden our paper with a lot of mathematical formulae, we will consider one-body physical systems only [coordinate $r = (r_1, r_2, r_3)$, momentum $p = (p_1, p_2, p_3)$] and pure quantum-mechanical states represented by vector ψ. The task of the generalization of everything that follows for the case of many-body systems as well as for mixtures represented by density matrix ρ, presents no difficulties (Kuryshkin, 1972a, 1973; Zaparovany and Kuryshkin, 1975).

2. INITIAL POSTULATES AND THEIR COMPATIBILITY

In order to construct the most general class of statistical theories resembling quantum mechanics by their mathematical formalism and containing the non-negative joint coordinate–momentum distribution, treated as a phase-space probability density, let us proceed from the assumptions.

1. Interpretation Postulate

The state of a physical system at any instant of time t is completely described by a joint coordinate–momentum probability density $F(z, p, t)$; the physical quantity A can be represented by coordinate-momentum-time functions $A(r, p, t)$ and the experimentally observable value $\langle A \rangle$ of the physical quantity A for a system in an F-state is defined as:

$$\langle A \rangle = \int A(r, p, t) F(r, p, t) \, dr \, dp \tag{3a}$$

From the physical meaning of F, defined by this postulate, follow its essential properties:

$$\int F(r, p, t) \, dr \, dp = 1 \tag{3b}$$

$$F(r, p, t) \geq 0 \tag{3c}$$

This postulate is only a common statement of the classical statistical theory and in the case of δ-like distributions, of classical mechanics as well. However, considering the coordinate and the momentum as simultaneously existing physical realities, we extend this statement to physical systems possessing quantum properties. The question of the compatibility of this postulate with the general property of quantum systems, expressed by Heisenberg's uncertainty principle, remains open so far.

2. Mathematical Formalism Postulate

The state of a physical system at any instant of time t is completely described by a normalized vector $|\psi(t)\rangle$ of some states' space $\mathscr{L}$; any function of phase space and time $A(z, p, t)$ owing to a certain linear rule can be represented by a linear operator $O(A)$ in $\mathscr{L}$; and the experimentally observable value $\langle A \rangle$ of the physical quantity A for a system in a $|\psi\rangle$ state is defined as:

$$\langle A \rangle = \langle \psi(t)|O(A)|\psi(t)\rangle \tag{4a}$$

where $\langle \psi_1|\psi_2\rangle$ is a scalar product of vectors $|\psi_1\rangle$ and $|\psi_2\rangle$. The normalization requirement for the vector state, the linearity of operators and the linearity of the correspondence rule is understood as usual:

$$\langle \psi(t)|\psi(t)\rangle = 1, \tag{4b}$$

$$\left. \begin{array}{l} O(A)\{\alpha|\psi\rangle\} = \alpha O(A)|\psi\rangle \\ O(A)\{|\psi_1\rangle + |\psi_2\rangle\} = O(A)|\psi_1\rangle - O(A)|\psi_2\rangle \end{array} \right\} \tag{4c}$$

$$\left. \begin{array}{l} O(1) = \hat{1} \\ O(\alpha A) = \alpha O(A) \\ O(A_1 + A_2) = O(A_1) + O(A_2) \end{array} \right\} \tag{4d}$$

where α is a numerical coefficient, $\hat{1}$ is the unit operator in $\mathscr{L}$. The second postulate is practically a slightly paraphrased basic postulate of quantum mechanics. But in contrast to the generally accepted quantum mechanics the distinct forms and properties of the operators (with the exception of linearities (4c) and (4d) remain undefined here.

It is essential in the first place to prove the compatibility of the above-formulated postulates, i.e. to show that equations (3) and (4) do not contradict each other. With this purpose in view let us introduce into consideration the characteristic function

$$\tilde{F}(u, v, t) = (2\pi)^{-6} \int F(r, p, t) e^{-i(ur + vp)} \, dr \, dp \tag{5}$$

Expanding the exponent into a series and using correlation (3a) the characteristic function can be rewritten in the form:

$$\tilde{F}(u, v, t) = (2\pi)^{-6} \sum_{n,m} \frac{(-iu_1)^{n_1} \cdot \ldots \cdot (-iv_3)^{m_3}}{n_1! \cdot \ldots \cdot m_3!} \langle r_1^{n_1} \cdot \ldots \cdot p_3^{m_3}\rangle \tag{6}$$

where $n = (n_1, n_2, n_3)$, $m = (m_1, m_2, m_3)$ are integer vectors. Reconstructing now the probability density F from the characteristic function $\tilde{F}$ by the reverse transformation of the integral (5) and using equations (6) and (4a), we obtain:

$$F(r, p, t) = (2\pi)^{-6} \sum_{n,m} \int du\, dv\, e^{i(ur+vp)}$$
$$\cdot \frac{(-iu_1)^{n_1} \cdot \ldots \cdot (-iv_3)^{m_3}}{n_1! \cdot \ldots \cdot m_3!} \langle \psi(t) | O(r_1^{n_1} \cdot \ldots \cdot p_3^{m_3}) | \psi(t) \rangle \tag{7}$$

Taking into account linearity properties (4c) and (4d), equation (7) can be rewritten as:

$$F(r, p, t) = \langle \psi(t) | \hat{F}(r, p, t) | \psi(t) \rangle \tag{8}$$

where $\hat{F}(z, p, t)$ is linear in the $\mathscr{L}$ operator and parametrically dependent on coordinate, momentum and time. The form of operator $\hat{F}$, or rather its relation to operators $O(r_1^{n_1} \cdot \ldots \cdot p_3^{m_3})$, is defined by equations (7) and (8). The dependence of $\hat{F}$ on time in a general case is caused by the fact that the mathematical formalism postulate does not rule out the possibility of the time-dependence of operator $O(A)$ even if the corresponding function $A(r, p)$ does not depend on time.

Let us turn now to the physical meaning of operator $\hat{F}$ and its properties.

Substituting relation (8) into equation (3a) and comparing it with (4a) we will see that operator $\hat{F}$ completely determines the correspondence rule:

$$O(A) = \int A(r, p, t) \hat{F}(r, p, t)\, dr\, dp \tag{9}$$

Integrating equation (8) over phase-space and taking into account the normalizations (3b) and (4b) we obtain:

$$\int \hat{F}(r, p, t)\, dr\, dp = \hat{1} \tag{10}$$

At last from property (3c) and relation (8) it follows that:

$$\langle \psi | \hat{F}(r, p, t) | \psi \rangle \geq 0 \tag{11}$$

i.e. $\hat{F}$ is an operator positively determined in $\mathscr{L}$.

Thus, a theory, satisfying the two initial postulates, contains a linear operator $\hat{F}(r, p, t)$, positively determined in $\mathscr{L}$, parametrically dependent on coordinate, momentum and time, and normalized by condition (10).

The physical meaning of operator $\hat{F}$ is defined by relation (4a) of postulate 2 and equation (8), i.e. $\hat{F}(r, p, t)$ is the operator of the probability density of coordinate r and momentum p at the instant of time t.

One can easily determine the phase-space function $f(r, p, t)$ corresponding to the operator $\hat{F}$ in agreement with the correspondence rule (9). Indeed, writing down the probability density operator as

$$\hat{F}(\xi, \eta, t) = O(f(\xi, \eta, r, p, t)) \tag{12}$$

where ξ and η are the parameters of operator $\hat{F}$ and the function $f(r, p, t)$, from relations (7) and (8), determining operator $\hat{F}$, we obtain:

$$f(\xi, \eta, r, p, t) = \delta(\xi - r)\,\delta(\eta - p) \tag{13}$$

Here $\delta(x)$ is Dirac's three-dimensional δ-function. Hence, the coordinate–momentum probability density operator corresponds to the phase–space δ-function. This conclusion is in full agreement with correspondence rule (9) and the physical meaning of operator $\hat{F}$.

Let us consider now the problem of the dynamics which are permissible in a theory which satisfies the initial postulates.

Let $|\psi(t)\rangle$ and $|\psi(t')\rangle$ determine the states of a physical system at times t and $t' \geq t$, respectively. Since, in conformity with postulate 2 both these vectors belong to the same space $\mathscr{L}$, they can always be related by the transformation:

$$|\psi(t')\rangle = \hat{S}(t', t)|\psi(t)\rangle, \quad \hat{S}(t, t) = \hat{1} \tag{14}$$

where $\hat{S}(t', t)$ is a linear operator, parametrically dependent on t' and t. Assuming in (14) that $t' = t + \delta t$, in the limiting case when $\delta t \to 0$, we obtain:

$$\frac{\partial |\psi(t)\rangle}{\partial t} = \hat{X}(t)|\psi(t)\rangle \tag{15}$$

where $\hat{X}(t)$ is the linear operator parametrically dependent on t and related to operator $\hat{S}$ by the following equation

$$\hat{X}(t) = \left.\frac{\partial \hat{S}(t', t)}{\partial t'}\right|_{t'=t}$$

The fact that the permissible dynamic equation (15) contains only the first derivative of the vector state with respect to time is an immediate consequence of postulate 2: vector $|\psi(t)\rangle$ completely determines the state of a system and the knowledge of it is a sufficient condition for finding the vector state $|\psi(t')\rangle$ at any instant of time $t' > t$.

Since the operator $\hat{X}$ determines the evolution of the system in time it must be related to certain physical quantities. And since $\hat{X}$ is linear [the consequence of postulate 2] and since any physical quantity can be represented by a coordinate–momentum–time function [the requirement of postulate 1], operator $\hat{X}$ by the correspondence rule (9), is related to a certain phase–space function, which can, in a general case, be complex i.e.:

$$\hat{X} = O(X), \quad X(r, p, t) = Q(r, p, t) - iR(r, p, t) \tag{16}$$

Here Q and R are real functions.

Let us take into account now the normalization condition (4b), Since the normalized vector $|\psi(t)\rangle$ due to equation (15) must automatically result in the normalized vector $|\psi(t')\rangle$, we have:

$$\frac{\partial \langle \psi(t)|\psi(t)\rangle}{\partial t} = \langle \psi|\hat{X}|\psi\rangle + \langle \psi|\hat{X}^+|\psi\rangle = 0$$

Since the initial state $|\psi(t)\rangle$ can be arbitrarily chosen, it follows that

$$O(X) = -O^+(X) \tag{17}$$

where $O^+(X)$ is an operator in $\mathscr{L}$ conjugated to operator $O(X)$. Since the correspondence rule (9) is linear and owing to the properties of operator $\hat{F}$ (i.e. it gives self-conjugated operators for real functions) by substituting (16) into (17) we obtain:

$$O(Q) = O, \quad -iO(R) = O(X) \tag{18}$$

The dynamic equation (15) will in this case take the form:

$$i\frac{\partial |\psi(t)\rangle}{\partial t} = O(R)|\psi(t)\rangle \tag{19}$$

where $R(r, p, t)$ is a real function.

Hence, a theory, satisfying the two initial postulates, contains a real coordinate–momentum–time function $R(r, p, t)$ (a dynamic function) which, with the help of the corresponding operator $O(R)$ and equation (19), defines the evolution of a physical system in time.

By differentiating (8) with respect to t and using equation (19) it is possible in principle to obtain an equation for the probability density $F(z, p, t)$ as well. It will also contain only the first time derivative which agrees with postulate 1. But in order to determine the distinct form of this equation one must be able to reconstruct the function $R(r, p, t)$ from the operator $O(R)$, i.e. to know the distinct form of the coordinate–momentum probability density operator $\hat{F}$.

3. GENERAL PRINCIPLES OF CONSTRUCTING QUANTUM MECHANIC-LIKE STATISTICAL THEORIES WITH CONSISTENT PROBABILITY INTERPRETATION

The results (9)–(19) of the investigation of the compatibility of the two postulates made in the previous section make it possible to formulate the general principle of constructing the theories in question.

In order to construct a quantum mechanic-like statistical theory with a consistent probability interpretation it is necessary and sufficient to use the following procedure:

(1). To represent the physical quantities as coordinate–momentum–time functions $A(r, p, t)$ (20)
(2). To choose a space $\mathscr{L}$ of the physical system's vector states $|\psi\rangle$ (21)
(3). To indicate a linear probability density operator $\hat{F}(z, p, t)$ positively determined in $\mathscr{L}$ and parametrically dependent on coordinate, momentum and time and normalized by the condition (22)

$$\int \hat{F}(r, p, t) \, dr \, dp = \hat{1} \tag{22a}$$

70   Uncertainty Principle and Foundations of Quantum Mechanics

(4). To indicate a real dynamic function $R(r, p, t)$ of the phase–space and time, responsible for the evolution of the system (23)

The necessity of some solution of the above-listed problems was shown in the previous section. Its sufficiency can easily be demonstrated in the following way.

Assume that the problems (20)–(23) are in a certain way solved. Then bringing into correspondence to any function $A(r, p, t)$ the linear in $\mathscr{L}$ operator

$$O(A) = \int A(r, p, t)\hat{F}(r, p, t)\,dr\,dp \qquad (24)$$

and representing the physical state of the system by a normalized vector $|\psi(t)\rangle \in \mathscr{L}$ satisfying the equation

$$i\frac{\partial|\psi(t)\rangle}{\partial t} = O(R)|\psi(t)\rangle \qquad (25)$$

let us determine the values $\langle A \rangle$ of the physical quantity A in the state $|\psi\rangle$ as the scalar product

$$\langle A \rangle = \langle \psi | O(A) | \psi \rangle \qquad (26)$$

Correlations (24)–(26) now represent a quite definite enclosed theory both from the point of view of statistics and of dynamics. The correspondence rule (24) and operators $O(A)$ of such a theory possess the linear properties (4c) and (4d). A mere substitution of operators (24) into equation (26) results in the redetermination (3a) of the physical quantity values; defines the function $F(r, p, t)$ in accordance with correlation (8) and its properties (3b)–(3c); and therefore the only possible in such theory interpretation, i.e. $F(r, p, t)$ is the coordinate–momentum probability density. Finally, knowing the coordinate–momentum probability density $F(r, p, t)$ and the function representations $A(r, p, t)$ of physical quantities in this theory it is possible to calculate joint and conditional probabilities for any physical quantities by the conventional methods of the probability theory. The theory so obtained is consequently of a consistent probability character.

It is quite natural then, that the properties of the statistical theory thus obtained, and its results in the first place, will depend on the concrete solution of the problems (20)–(23). The natural questions arising from it are: What is the difference between these theories and the generally accepted quantum mechanics? Can this theory with some concretization of the problems (20)–(23) describe physical reality? And if so, in what way can this concretization be found?

In the sections that follow, we shall try to discuss these questions, omitting for brevity's sake all mathematical calculation.

4. MAIN CONSEQUENCES OF THE STATISTICAL THEORY UNDER CONSIDERATION

The main advantage of the statistical theory, whose methematical apparatus was given in the previous section is that it is of consistent probability character and at the same time does not violate the basic postulate of the quantum theory. The concrete form of such statistical theory, its properties and its results depend on the *a priori* solutions of the problems (20)-(23). However, irrespective of this concretization a number of general theoretical consequences can be pointed out, amongst which are the following:

(1). No concretization of problems (20)–(23) makes this statistical theory coincide with the generally accepted quantum mechanics. This is quite obvious, since the generally accepted quantum mechanics would otherwise be reduced to the consistent probability scheme.

(2). In the statistical theory under consideration Neumann's requirement is in a general case violated, i.e.

$$O(f(A)) \neq f(O(A))$$

The operator of the square of a physical quantity is not equal to that quantity's operator squared which can be written as:

$$O(A^2) = O^2(A) + \mathscr{D}(A) \tag{27}$$

The linear operator $\mathscr{D}(A)$ defined by the relation (27) depends on the concrete form of the probability operator $\hat{F}$ and, in a general case, is not equal to zero for all physical quantities.

(3). In the theory proposed here, as in any statistical theory, the physical quantity's value $\langle A \rangle$ in the state $|\psi\rangle$ is characterized by the uncertainty (dispersion)

$$\langle (\Delta A)^2 \rangle = \langle A^2 \rangle - (\langle A \rangle)^2 = \langle (A - \langle A \rangle)^2 \rangle = \langle (O(A) - \langle A \rangle)^2 \rangle + \langle \mathscr{D}(A) \rangle \geq 0 \tag{28}$$

whose non-negativeness is guaranteed here by the consistent probability character of the theory. In contrast to the generally accepted quantum mechanics, however, the value $\langle A \rangle$ of the quantity A here in the states with an eigenvector of $O(A)$ in a general case is not strictly determined. Thus, if

$$O(A)|\psi_\alpha\rangle = \alpha|\psi_\alpha\rangle \tag{29}$$

where α is the eigenvalue, coinciding with $\langle A \rangle$, then from (28) it follows that

$$\langle (\Delta A)^2 \rangle_\alpha = \langle \psi_\alpha | \mathscr{D}(A) | \psi_\alpha \rangle \geq 0 \tag{30}$$

Hence, over the eigenvectors of operator $O(A)$ the operator $\mathscr{D}(A)$ is non-negative and has the sense of a dispersion operator.

(4). If in a certain state $|\psi\rangle$ the dispersion of quantity A reaches its minimum value in the sense that

$$\langle (\Delta A)^2 \rangle_{\psi+\delta\psi} \geq \langle (\Delta A)^2 \rangle_\psi$$

where $|\delta\psi\rangle$ is an arbitrary infinitesimal deviation of the vector state, then

$|\psi\rangle$ satisfies the non-linear equation

$$\{O(A^2)-2\alpha O(A)+\alpha^2\}|\psi\rangle = d^2|\psi\rangle \tag{31}$$

where $\alpha = \langle\psi|O(A)|\psi\rangle$.

(5). In the statistical theory under consideration the precision of determining the value of even a single physical quantity is limited by the inequality

$$\langle(\Delta A)^2\rangle \geq (\delta A)^2 \stackrel{\text{def}}{=} \min_{(n)} \{d_n^2\} \tag{32}$$

where d_n^2 are the eigenvalues of equation (31). The uncertinties δA are finally determined by the probability density operator and may change with time.

(6). For the uncertainties of two physical quantities A and B in any state $|\psi\rangle$ there exists a correlation

$$\{\langle(\Delta A)^2\rangle - \langle\mathcal{D}(A)\rangle\} \cdot \{\langle(\Delta B)^2\rangle - \langle\mathcal{D}(B)\rangle\} \geq \tfrac{1}{4}|\langle\hat{C}\rangle|^2 \tag{33}$$

where $\hat{C} = [O(A), O(B)]$. Inequality (33) represents the uncertainty principle in the proposed statistical theory.

(7). In the case when

$$[O(A^2), O(A)]_- = 0 \tag{34}$$

is valid for a quantity A, the eigenvectors of equations (29) and (31) coincide. Therefore, providing that equality (34) is fulfilled the states with the most precise values of a quantity A (minimum dispersion) are defined, as in the generally accepted quantum mechanics, by the operator $O(A)$ eigenvalue equation. If the commutation condition (34) is fulfilled both for quantity A and for quantity B, the uncertainties correlation (33) takes the form:

$$\langle(\Delta A)^2\rangle\langle(\Delta B)^2\rangle \geq \tfrac{1}{4}|\langle\hat{C}\rangle|^2 + (\delta A)^2(\delta B)^2 \tag{35}$$

(8). In a similar way to the generally accepted quantum mechanics all probability characteristics of a physical system in the theory investigated are determined by the state $|\psi(t)\rangle$, the probability density of any physical quantity in the state $|\psi(t)\rangle$ being given by the expression

$$W(A, t) = \langle\psi(t)| \int \delta(A - A(r, p, t))\hat{F}(r, p, t) \, dr \, dp |\psi(t)\rangle \tag{36}$$

Here, however, the vector $|\psi\rangle$ does not generally have a distinct physical sense and can be considered as only a mathematical image of the probability density $F(r, p, t)$ carrying all its probability information.

(9). The condition for conserving the value $\langle A\rangle$ of a physical quantity A in time (the conservation law for quantity A) is formally the same as in the generally accepted quantum mechanics:

$$\frac{d\langle A\rangle}{dt} = 0, \quad \text{if} \quad \frac{\partial O(A)}{\partial t} = i[O(R), O(A)]_- \tag{37}$$

However, the fulfillment of this condition with R and A fixed essentially depends on the distinct form of the operator $F(r, p, t)$.

(10). The proposed theory in a general case results in concepts, that have no analogue in the generally accepted quantum mechanics. They will be further named 'subquantum' concepts. Among them one could name the 'subquantum' uncertainty δA of quantity A, which limits the precision of determining the value $\langle A \rangle$ of this quantity.

5. CONCRETIZATION OF PHYSICAL QUANTITIES AS PHASE–SPACE FUNCTIONS

According to the principle of constructing the statistical theories in question, which was formulated in paragraph 2, it is necessary above all to solve problem (20), i.e. to define physical quantities as certain coordinate–momentum–time functions. There obviously exist only two methods of such definition which divide the multitude of the theories under investigation into two classes: (*a*) all $A(r, p, t)$ coincide with and (*b*) all $A(r, p, t)$, or at least some of them, differ from the classical ones.

The second method involves considerable difficulties (Tyapkin, 1968; Zaparovany and Kuryshkin, 1975) since a constructive approach to the choice of such functions, with the exception of Tyapkin's condition ($A(r, p, t)$ at $\hbar \to 0$ turn into classical ones) has not been found as yet.

We shall assume, therefore, at least in this paper, that the function dependence of all physical quantities A of r, p and t, is given by functions $A(r, p, t)$, representing the same quantities in the classical theory.

6. CONCRETIZATION OF THE STATES' SPACE

If, while finding a solution to the problem (20), we made use of the analogy with the classical theory, it would seem quite natural to use the analogy with the generally accepted quantum mechanics when choosing the states' space.

Restricting ourselves (in this paper) to the consideration of non-relativistic theories alone, let us define $\mathscr{L}$ as the space of scalar square integrable functions of coordinates, i.e.

$$|\psi(t)\rangle = \psi(r, t), \quad \langle \psi(t)| = \psi^*(r, t) \tag{38a}$$

$$\langle \psi_1(t)|\psi_2(t)\rangle = \int \psi_1^*(r, t)\psi_2(r, t)\,dr \tag{38b}$$

For the sake of convenience in further investigations let us represent each operator $O(A)$ by a generation function $A_G(r, p, t)$, related to $O(A)$ with the help of transformations:

$$A_G(r, p, t) = e^{-(i/a)rp} O(A) e^{(i/a)rp} \tag{39a}$$

$$O(A)U(r, t) = (2\pi a)^{-3} \int A_G(r, p, t) e^{(i/a)(r-r')p} U(r', t)\,dr'\,dp \tag{39b}$$

where a is constant. Equality (39a) defines the generation function of operator $O(A)$, while (39b) reconstructs the operator when the generation function $A_G(r, p, t)$ is known.

Correlations (39) also bring in correspondence to the probability density operator $\hat{F}(\xi, \eta, t)$ some generation function $f_G(\xi, \eta, r, p, t)$ where ξ and η are parameters of the probability operator and its generation function. Then, from the correspondence rule (24), follows a connection of the generation functions of operators $O(A)$ with the generation function of the probability density operator $\hat{F}$:

$$A_G(r, p, t) = \int A(\xi, \eta, t) f_G(\xi, \eta, r, p, t) \, d\xi \, d\eta \tag{40}$$

It can be shown that, owing to the positiveness of the probability density operator and its normalization (22a), the generation function f_G can always be written as

$$f_G(\xi, \eta, r, p, t) = \sum_k e^{-(i/a)rp} \mu_k(\xi, \eta, r, t) \int e^{(i/a)r'p} \mu_K^*(\xi, \eta, r', t) \, dr' \tag{41}$$

where $\mu_K(r, p, \xi, t)$ is a certain set of functions of the phase–space (r, p), an additional configuration space ξ and time t, satisfying the normalization:

$$\sum_K \int \mu_K(r, p, \xi, t) \mu_K^*(r, p, \xi', t) \, dr \, dp = \delta(\xi - \xi') \tag{42}$$

7. THE CONCEPT OF THE 'SUBQUANTUM SITUATION'

Accepting the above concretizations of the functions $A(r, p, t)$ and the states' space $\mathcal{L}$, the whole of the statistical part of the theory under investigation is defined by a set of functions $\mu_K(r, p, \xi, t)$ satisfying normalization (42). The set of functions μ_K determines the generation function f_G of the probability density operator [see expression (41)] and the operator $\hat{F}$ (39b). In point of fact the operator $\hat{F}$ itself can be dispensed with since, with the set of functions μ_K fixed, the operators of all physical quantities are singularly determined by relations (39a), (40) and (41).

It should be noted that the functions μ_K themselves have no analogue either in classical or in the generally accepted quantum mechanics, i.e. in accordance with the above-accepted terminology, are 'subquantum' notions. The values of all 'subquantum' quantities, appearing in the theory ('subquantum' uncertainties (32), for instance) are determined by a set of functions μ_K. It is therefore suggested that one should say that the set of functions μ_K represents a certain 'subquantum situation' in the theory under investigation. Thus, the same physical system can be considered at various 'subquantum situations' (various sets of μ_K) in the proposed statistical theory and vice versa, different physical systems can be considered at one and the same 'subquantum situation' (a fixed set of μ_K).

The choice of a 'subquantum situation' (a certain set of functions μ_K) gives a single-valued definition of all operators, and, determines, consequently, the results of the theory. A change of the 'subquantum situation' leads to a change of the whole set of results. Therefore, assuming the correctness of the statistical theory investigated here, we are compelled to acknowledge that a 'subquantum situation' reflects a certain physical reality, which has no analogue either in classical or in the generally accepted quantum theory.

In a general case, a 'subquantum situation', as a physical reality, can change both in time and in space. For instance, together with the unconditional 'subquantum' uncertainty of the coordinate δr depending in a general case on t, one can consider a conditional 'subquantum' uncertainty of the coordinate

$$\delta r(r_0, t) \stackrel{\text{def}}{=} \sqrt{\min_{(\psi_{r_0})} \{\langle (\Delta r)^2 \rangle \psi_{r_0}\}} \tag{43}$$

where ψ_{z_0} are all possible states with $\langle r \rangle = r_0$. The coordinate's 'subquantum uncertainty' (43) is also determined by the 'subquantum situation', but it may depend not only on time, but on the system's location r_0 in space as well. This means that there exists a possibility of the space heterogeneity and anisotropy of the 'subquantum situation'.

Note should be made, however, that the 'subquantum situation' in the statistical theories under investigation is given by a set of functions μ_K only with the above-accepted concretizations of the functions $A(r, p, t)$ and the states' space $\mathscr{L}$. In a general case the 'subquantum situation' is given by the solution of the whole set of problems (20)–(23). It is essential, that the concept of a 'subquantum situation' is an indispensable part of the quantum mechanics-like statistical theories, possessing the joint coordinate–momentum probability density.

8. THE SIMPLEST CONCRETIZATION OF THE PROBABILITY DENSITY OPERATOR

Since in the present paper our task is only to make a brief analysis of the possibilities of the statistical theories obtained, we will henceforward make use of the simplest 'subquantum situation', given by a set of functions

$$\mu_K(r, p, \xi, t) = (2\pi a)^{-3/2} \varphi_K(r - \xi, t) e^{(i/a)\xi p} \tag{44}$$

where $\varphi_K(r, t)$ is an arbitrary set of squarely integrable functions, satisfying the normalization

$$\sum_K \int |\varphi_K(r, t)|^2 \, dr = 1 \tag{45}$$

A mere substitution of functions (44) into integral (42), with equality (45) taken into account, shows that the set of functions (44) possesses the required normalization.

Now the coordinate–momentum probability density operator is determined with an accuracy to an arbitrary set of coordinate–time functions $\varphi_K(r, t)$ normalized by condition (45).

The concretization (44) of the probability density operator is all the more significant because here the commutator of operators $O(r_j)$ and $O(p_j)$ does not depend on the distinct form or the number of functions φ_K:

$$[O(r_j), O(p_{j'})]_- = ia\, \delta_{jj'} \tag{46}$$

The commutator (46) follows from relations (39b), (40), (41), (44) and (45).

Besides 'subquantum situation' (44), in a particular case, can be stationary, space-homogeneous and isotropic. For this it is enough to choose:

$$\varphi_K(r, t) = \varphi_K(|r|) \tag{47}$$

The set of 'subquantum' functions (47), owing to relations (39b), (40), (41) and (44), results in a theory which is invariant with respect to time-shifts, translations and rotations of the space.

9. CONCRETIZATION OF THE DYNAMIC FUNCTION

In the previous sections the correspondence principle of the statistical theory investigated here with the classical and generally accepted quantum mechanics was used for the concretization of the functions $A(r, p, t)$ and the states' space $\mathscr{L}$.

Since the 'subquantum situation' has no analogue in the indicated theories, the probability density operator $\hat{F}$ so far remains determined with the accuracy to a normalized set of functions $\varphi_K(|r|)$ and the quantity a, present in correlations (39b), (41) and (44).

However, in spite of this uncertainty of the theory, the problem of the choice of a dynamic function $R(r, p, t)$ (23), due to the correspondence principle, has been definitely solved.

With the accepted concretizations of $A(r, p, t)$ and $\mathscr{L}$ for the coordinate and momentum uncertainties in any state ψ we have:

$$\langle (\Delta r_j)^2 \rangle \geq (\delta r)^2, \quad \langle (\Delta p_j)^2 \rangle \geq (\delta p)^2 \tag{48a}$$

$$\langle (\Delta r_j)^2 \rangle \langle (\Delta p_{j'})^2 \rangle \geq \frac{a^2}{4} \delta_{jj'} + (\delta r)^2 (\delta p)^2 \tag{48b}$$

where the 'subquantum' uncertainties δr and δp are the functionals of a set of functions φ_K (Kuryshkin, 1972a, 1972b, 1973).

Correlations (48) determine the conditions of the transition of the statistical theory under investigation into the classical theory. Since the classical theory allows F-distributions with the coordinate and the momentum precisely determined the conditions for such a transition will be:

$$\delta r \to 0, \quad \delta p \to 0, \quad a \to 0 \tag{49}$$

Differentitating the probability density $F(r, p, t)$ (8) with respect to time and using the evolution equation (19) after performing the limit transformation we come to the following conclusion (Kuryshkin, 1972b): the statistical theory under consideration satisfies the correspondence principle when, and only when

$$aR(r, p, t) = H(r, p, t) \tag{50}$$

where $H(r, p, t)$ is the system's Hamiltonian.

Correlations (48) also determine the conditions for the possible transition of the statistical theory under investigation into the generally accepted quantum mechanics. These conditions obviously are:

$$\delta r \to 0, \quad \delta p \to 0 \tag{51}$$

Comparing now commutator (46), correlation (48b) and equation (25) under conditions (50) and (51) with commutator (2), used in the generally accepted quantum mechanics, Heisenberg's uncertainty principle and the Schrödinger equation we come to the conclusion:

$$a = \hbar \tag{52}$$

Thus, the principle of the correspondence of the statistical theory under consideration with the generally accepted quantum mechanics requires that quantity a which is present in correlations (39b), (41), (44) and (50), coincide with Planck's constant.

10. A PARTICULAR CASE OF THE THEORY AND SOME OF ITS APPLICATIONS

The concretization of the functions $A(r, p, t)$, the states' space $\mathscr{L}$, the probability density operator $\hat{F}(r, p, t)$ and the dynamic function $R(r, p, t)$ introduced in the previous sections, results in a particular case of the quantum mechanic-like statistical theory with the consistent probability interpretation. The 'subquantum situation' in this theory is given by a set of functions $\varphi_K(|r|)$, normalized by the condition:

$$\sum_K \int |\varphi_K(|r|)|^2 \, r = 1 \tag{53}$$

The operators of physical quantities are defined by the correspondence rule

$$O(A)U(r, t) = (2\pi\hbar)^{-3} \int \varphi(r - \xi, p - \eta) A(\xi, \eta, t)$$
$$\cdot e^{(i/\hbar)(r-r')p} U(r', t) \, d\xi \, d\eta \, dr' \, dp \tag{54}$$

where $U(r, t)$ is an arbitrary coordinate–time function, $A(r, p, t)$ is a phase–space and time function, corresponding to quantity A in the classical theory, $\hbar$

is Planck's constant, $\varphi(r, p)$ is an auxiliary function related to 'subquantum' functions $\varphi_K(|r|)$ by correlations:

$$\varphi(r, p) = (2\pi\hbar)^{-3/2} e^{-(i/\hbar)rp} \sum_K \varphi_K(|r|) \tilde{\varphi}_c^*(|p|) \tag{55}$$

$$\tilde{\varphi}_K(|p|) = (2\pi\hbar)^{-3/2} \int e^{-(i/\hbar)rp} \varphi_K(|r|) \, dr \tag{56}$$

The physical system's state in such a theory is described by the vector (wave function) $\psi(r, t)$, normalized by the condition

$$\int |\psi(r, t)|^2 \, dr = 1 \tag{57}$$

and satisfying an equation of the same type as the Schrödinger equation

$$i\hbar \frac{\partial \psi(r, t)}{\partial t} = O(H)\psi(r, t) \tag{58}$$

where $H(r, p, t)$ is the system's Hamiltonian function, $O(H)$ is the operator, corresponding to it in accordance with rule (54). The value $\langle A \rangle$ of the physical quantity A in the ψ-state is determined by the formula:

$$\langle A \rangle = \int \psi^*(r, t) O(A) \psi(r, t) \, dr \tag{59}$$

The mathematical formalism of this theory's particular case given by formulae (53)–(59), immediately follows from relations (24)–(26) with the accepted concretization of the function $A(r, p, t)$ and equalities (38), (39a)–41), (44)–(45), (47), (50), (52) taken into account.

A mere substitution of operators (54) into formula (59) involving the auxiliary functions (55), (56) and normalizations (53), (57) results in:

$$\langle A \rangle = \int (r, p, t) F(r, p, t) \, dr \, dp \tag{60a}$$

$$F(r, p, t) = (2\pi\hbar)^{-3} \sum_K \left| \int \varphi_K^*(|r - \xi|) e^{-(i/\hbar)\xi p} \psi(\xi, t) \, d\xi \right|^2 \geq 0 \tag{60b}$$

$$\int F(r, p, t) \, dr \, dp = 1 \tag{60c}$$

Correlations (60) determine the consistent probability interpretation of the theory. The equation for $F(r, p, t)$ can be obtained from equation (58) with the help of correlations (54), (55) and (60b) (Kuryshkin, 1972c).

As has been noted above, the generally accepted interpretation of the wave function ψ cannot be accepted here since

$$W(r, t) = \int F(r, p, t) \, dp = \int |\psi(\xi, t)|^2 \sum_K |\varphi_K(|r - \xi|)|^2 \, d\xi \tag{61}$$

i.e. the square of the modulus $\psi(r, t)$ determines, but by no means coincides with, the coordinate probability density $W(r, t)$.

The statistical theory as represented by equations (53)–(59) does coincide with the 'quantum mechanics with the non-negative QDF' (Kuryshkin, 1972c) in the case of the stationary, homogeneous and isotropic 'subquantum situation' (Kuryshkin, 1973). The general theoretical concepts of this theory are at present fairly well studied (Kuryshkin, 1971, 1972a, 1972c, 1973; Zaparovany, 1973; Zaparovany and coworkers, 1975). The solution of actual problems within the framework of the mathematical formalism (53)–(59) yields quite satisfactory results (Kuryshkin, 1972c, 1973; Kuryshkin and Zaparovany, 1974; Zaparovany and coworkers, 1975).

Thus, equation (58), for example, results in the energy spectrum of a one-dimensional harmonic oscillator

$$E_n = \hbar\omega(n + \tfrac{1}{2}) + \varepsilon, \quad n = 0, 1, \ldots \qquad (62a)$$

where $\varepsilon \geq 0$ is the 'subquantum' energy, related to the 'subquantum' uncertainties of coordinate δr and momentum δp. ε does not affect the level-difference and is not therefore experimentally observable. The calculation of an oscillator average energy in a thermostat results in Planck's formula with the vacuum energy increased by ε.

A similar problem for an electron in a hydrogen-like atom in a second-order approximation with respect to the coordinate 'subquantum' uncertainty δr gives the energy spectrum:

$$E_{nlm} \doteq -\frac{Z^2 e^2}{2an^2} + T_0 + \varepsilon_{nl} \qquad (62b)$$

$n = 1, 2, \ldots; l = 0, 1, \ldots, n-1; m = -l, \ldots, 0, \ldots, l$. Here $T_0 \geq 0$ is the 'subquantum' kinetic energy related to δp, and $\varepsilon_{nl} \geq 0$ is the 'subquantum' energy shift, connected with δr and stipulating the split of the levels over l, resembling Lamb's shift. The result (62b) agrees with the experimental data when

$$\delta r \doteq 4.247 \times 10^{-12} \text{ cm} \qquad (63)$$

Energy levels (62) in contrast to the generally accepted quantum mechanics are not strictly determined. The dispersions of the levels, however, can be calculated when some real functions φ_K are chosen.

11. FURTHER CONCRETIZATION OF THE PROBABILITY DENSITY OPERATOR

Even in the particular case of the theory, analyzed in the previous section and permitting the solution of some concrete problems, the concretization of $\hat{F}$ is determined with the accuracy to a set of 'subquantum' functions φ_K. It is quite clear, that a further concretization of operator $\hat{F}$ is out of the question in the absence of some kind of an assumption concerning the physical nature of the 'subquantum situation'.

The arbitrary choice of 'subquantum' functions φ_K, however, can be partially eliminated by comparing the results of the theory with the experimental data. Thus, for instance, condition (63) obtained as a result of such comparison considerably reduces the arbitrariness in the choice of the 'subquantum situation'. One more opportunity of reducing the arbitrariness in the choice of functions can be pointed out. Thus, experiment shows that the dispersion of energy levels is either zero at at least very small. One may demand, therefore, that the eigenfunctions of the operator $O(H)$ coincide with the eigenfunctions of the minimum uncertainty equation (31) when $A = H$. This is possible only with the commutation of the type (34), i.e.

$$[O(H^2), O(H)]_- = 0 \qquad (64)$$

Obviously, at H fixed, equality (64) limits the choice of functions φ_K.

Assuming that condition (64) is fulfilled, one can estimate (qualitatively at least) the energy level dispersions (62). Calculations show:

$$\langle (\Delta E)^2 \rangle_n = C + 2\varepsilon E_n, \qquad (65a)$$

$$\langle (\Delta E)^2 \rangle_{nlm} \doteq \mathscr{D} + 4T_0(E_{nlm} - T_0) \geq 0 \qquad (65b)$$

where 'subquantum' quantities C, ε, $\mathscr{D}$ and T_0 remain non-negative functionals of the set φ_K, i.e. the additional concretization (64) and (63) of the probability density operator is not sufficient for a single-valued calculation of dispersions. However, equations (65) allow us to see the qualitative picture for the dispersion change relative to the energy-level increase

$$\langle (\Delta E)^2 \rangle_0 = C + \varepsilon \omega \hbar, \quad \langle (\Delta E)^2 \rangle_{n \to \infty} \to \infty$$

$$\langle (\Delta E)^2 \rangle_{000} \doteq \mathscr{D} - \frac{2T_0 Z^2 e^2}{a} \geq 0, \quad \langle (\Delta E)^2 \rangle_{n \to \infty, lm} \to \mathscr{D}$$

Thus, the minimum uncertainty for both the oscillator and the electron in a hydrogen-like atom is inherent in the ground-state (minimum energy) level. Hence, the additional concretization of $\hat{F}$, established by equality (64) leads to quite satisfactory qualitative results of the theory.

12. CONCLUSION

The investigations, the results of which are set forth in the present Chapter, makes it possible to conclude as follows:

(1). There exist a multitude of theories satisfying the principal postulate of quantum mechanics and permitting a statistical interpretation on the basis of coordinate–momentum probability density. These theories differ from each other in the concretization of physical quantities as functions of phase–space and time, states' space, probability density operator and the dynamic function. The generally accepted quantum mechanics does not belong to their number.

(2). Any specific theory out of the multitude of theories under consideration is of a consistent probability character. The existence of the coordinate–momentum probability density $F(r, p, t)$ in such a theory and the functional relations $A(r, p, t)$ of the physical quantities A with coordinates and momenta permit the calculation of all sorts of joint and conditional probabilities by the conventional methods of probability theory.

(3). Irrespective of the concretization, the theories in question bring into existence certain concepts ('subquantum' uncertainty, 'subquantum situation', etc.), which have no analogue either in the classical or in the generally accepted quantum mechanics.

(4). There exists a theoretical concretization which leads to quite satisfactory results. In one particular case of this concretization the statistical theory in question turns into the classical and partly (in the realm of physical quantities containing no products of the similar components of coordinate and momentum) into the generally accepted quantum mechanics.

(5). Violating Neumann's requirement for quantum operators, the theory in question is not subject to his theorem on the impossibility of 'hidden' parameters. Moreover, such statistical theory requires the introduction of some new physical concepts for the explanation of the physical nature of the 'subquantum situation'.

(6). In the statistical theories under consideration the concept of the uncertainty of physical quantities acquires a more general character than in accepted quantum mechanics. The correlation of the coordinate and momentum uncertainties (48b) which is nothing but Heisenberg's uncertainty principle reinforced by the 'subquantum' uncertainties is true in the particular concretization of the theory.

(7). The utilization of the joint coordinate–momentum probability density in the theory under investigation is equivalent to the assumption that a physical system always possesses quite definite coordinate and momentum. The uncertainty principle of the type of the Heisenberg principle, therefore, is not in contradiction with the concept of the coordinate and the momentum existing simultaneously as physical realities.

(8). The existence of the uncertainty principle in the statistical theory under investigation and the fact that the coordinate and momentum can be considered as simultaneously existing physical realities signify that this theory does not pretend to be a complete description of physical reality. In other words, the proposed statistical theory assumes the existence of a more profound, more deterministic theory, capable of also explaining, among other things, the physical nature of the 'subquantum situation'.

ACKNOWLEDGEMENT

The author wishes to express his most sincere gratitude to Professor L. de Broglie, Professor Ya. P. Terletsky, Professor J. Lochak and the participants of the seminars at the Peoples' Friendship University (Moscow) and the Henry Poincare Institute (Paris) for numerous and helpful discussions of the problem investigated in this paper.

REFERENCES

Andrade e Silva, J. L. and Lochak, G. (1969) *Quanta, Grains et Champs*, L'Univers de connaissances, Hachette, Paris.
Arinshtein, E. A. and Guitman, D. M. (1967) *Izvest. Vusov U.S.S.R., Fiz.*, No. 5, 123.
Bess, L. (1973) *Progr. Theoret. Phys.*, **49**, 1889.
Blokhintsev, D. I. (1940) *J. Phys.*, **2**, 71.
Bohm, D. (1952) *Phys. Rev.*, **85**, 166.
Bohm, D. and Vigier, J. P. (1954) *Phys. Rev.*, **96**, 208.
Born, M. and Jordan, P. (1925) *Z. Physik*, **34**, 858.
Bopp, F. (1956) *Ann. Inst. Henri Poincare*, **XV**, 81.
Cohen, L. (1966a) *J. Math. Phys.*, **7**, 781.
Cohen, L. (1966b) *The Philosophy of Science*, **33**, 317.
Cohen, L. (1970) *J. Math. Phys.*, **11**, 3296.
De-Broglie, L. (1927) *J. Phys. Radium*, **8**, 225
De-Broglie, L. (1953) *La Physique Quantique Restera-t-elle Indéterministe?*, Gauthier–Villars, Paris.
De-Broglie, L. (1956) *Une Interprétation Causale et Non Linéaire de la Mécanique Ondulatoire: la Theory de la Double Solution*, Gauthier–Villars, Paris.
De-Broglie, L. (1964) *Thermodinamique de la Particule Isolée*, Gauthier–Villars, Paris.
De-Broglie, L. and Andrade e Silva, J. L. (1971) *La Réinterprétation de la Mécanique Ondulatoire*, Gauthier–Villars, Paris.
Dirac, P. A. M. (1958) *The Principles of Quantum Mechanics*, Oxford University Press, Oxford.
Einstein, A., Podolsky, B. and Rosen, N. (1935) *Phys. Rev.*, **47**, 777.
Einstein, A. (1948) *Dialectica*, **11**, 320.
Fényes, I. (1952) *Z. Physik*, **132**, 81.
Finkelstein, R. J., Fronsdal, C. and Kaus, P. (1956) *Phys. Rev.*, **103**, 1571.
Glasko, V. B., Lerust, F., Terletsky, Ya. P. and Shushurin, S. F. (1958) *Zh. Eksperim. Teor. Fiz., U.S.S.R.*, **35**, 452.
Gorshenkov, V. N. and Kognkov, V. L. (1973) *Izvest. Vusov U.S.S.R., Fiz.*, No. 7, 140.
Gorshenkov, V. N., Denisova, N. A., Kognkov, V. L. and Ryasanova, L. Z. (1973) *Teor. Mat. Fiz., U.S.S.R.*, **15**, 288.
Groenewold, H. J. (1935) *Physica*, **12**, 405.
Imre, K., Ozizmir, E., Rosenbaum, M. and Zweifel, P. F. (1967) *J. Math. Phys.*, **8**, 1097.
Kerner, E. H. and Sutsliffe, W. G. (1970) *J. Math. Phys.*, **11**, 391.
Klimantovich, Yu. L. and Silin, V. P. (1960) *Uspe. Fiz. Nauk, U.S.S.R.*, **70**, 247.
Kuryshkin, V. V. (1968) *Sb. Nauchn. Rabot Aspirantov*, Peoples' Friendship University, Moscow, No. 1, 243.
Kuryshkin, V. V. (1969a) *Isvest. Vusov U.S.S.R., Fiz.*, No. 4, 111.
Kuryshkin, V. V. (1969b) *Sb. Nauchn. Rabot Aspirantov*, Peoples' Friendship University, Moscow, No. 6, 198.
Kuryshkin, V. V. (1971) *Izvest. Vusov U.S.S.R., Fiz.*, No. 11, 103.
Kuryshkin, V. V. (1972a) *Compt. Rend.*, **274**, Série B, 1107.
Kuryshkin, V. V. (1972b) *Ann. Inst. Henri Poincaré*, XVII, 81.
Kuryshkin, V. V. (1972c) *Compt. Rend.*, **274**, Série B, 1163.
Kuryshkin, V. V. (1973) *Int. J. Theoret. Phys.*, **7**, 451.
Kuryshkin, V. V. (1974) *Teor. Fiz.*, Peoples' Friendship University, Moscow, 78.

Kuryshkin, V. V. and Zaparovany, Yu. I. (1974) *Compt. Rend.*, **279**, Série B, 17.
Margenau, H. and Hill, R. N. (1961) *Progr. Theoret. Phys.*, **26**, 722.
Mehta, C. L. (1964) *J. Math. Phys.*, **5**, 677.
Moyal, I. E. (1949) *Proc. Cambridge Phil. Soc.*, **45**, 99.
Neumann, J. (1932) *Mathematische Grundlagen der Quantenmechanik*, Springer, Berlin.
Pena-Auerbach, L., Cetto, A. M. and Brody, T. A. (1972) *Letters alla Redazione, Nuovo Cimento*, **5**, 177.
Rivier, D. C. (1951) *Phys. Rev.*, **83**, 862.
Ruggery, G. J. (1971) *Progr. Theoret. Phys.*, **46**, 1703.
Rybakov, Yu. P. (1974) *Foundations of Physics*, **4**, 149.
Schrödinger, E. (1935) *Naturwissenschaften*, **23**, 807, 823, 844.
Schrödinger, E. (1955) *Nuovo Cimento*, **1**, 5.
Shankara, T. S. (1967) *Progr. Theoret. Phys.*, **37**, 1335.
Shewell, J. R. (1959) *Am. J. Phys.*, **27**, 16.
Temple, G. (1935a) *Nature*, **135**, 957.
Temple, G. (1935b) *Nature*, **136**, 179.
Terletsky, Ya. P. (1937) *Zh. Eksperim. Teor. Fiz.*, U.S.S.R., **7**, 1290.
Terletsky, Ya. P. (1960) *J. Phys. Radium*, **21**, 771.
Tolman, R. S. (1938) *The Principles of Statistical Mechanics*, Clarendon Press, New York.
Tyapkin, A. A. (1968) *Development of Statistical Interpretation of Quantum Mechanics by Means of the Joint Coordinate–Momentum Representation*, U.S.S.R, Dubna.
Weyl, H. (1950) *The Theory of Groups and Quantum Mechanics*, Clarendon Press, New York.
Wigner, E. P. (1932) *Phys. Rev.*, **40**, 749.
Yvon, J. (1948) *Cahiers Phys.*, **33**, 25.
Zaparovany, Yu. I. (1974) *Izvest. Vusov U.S.S.R., Fiz.*, No. 6, 18.
Zaparovany, Yu. I. and Kuryshkin, V. V. (1975) The article is deposited in VINITI U.S.S.R., No. 2353-75, Dep.
Zaparovany, Yu. I., Kuryshkin, V. V. and Lyabis, I. A. (1975) *Sovremen. Zadachi y tochnikh naukakh*, Peoples' Friendship University, Moscow, No. 1, 89 and 94.
Zlatev, I. S. (1974) *Compt. Rend.*, **27**, 311.

PART 2

Measurement Theory

The Problem of Measurement in Quantum Mechanics

Ludovico Lanz
Instituto di Fisica dell'Universià, Milan, Italy

1. INTRODUCTION

Quantum mechanics is itself a statistical theory of measurements. It works very well for some measurements, for example wonderfully well in atomic spectroscopy. It may appear surprising that a theory of measurement should be basic to quantum mechanics since quantum mechanics is explained in most textbooks, as well as being applied and further developed, without particular reference to such a theory of measurement. The peculiarity of quantum mechanics is that one firstly has measurements and then subsequently one must worry about what has been measured.

There are two very different attitudes towards quantum mechanics; namely:

(1). Quantum mechanics is the fundamental theory of physics; any physical theory is essentially a theory of measurements.

(2). Quantum mechanics is the fundamental theory of microsystems; therefore the theory of microsystems is essentially a theory of measurements on microsystems by macrosystems. A primary objective of physics is therefore to describe the nature of macrosystems. To reach such an objective the physics of microsystems is an essential ingredient.

The first attitude is the point of view one learns in textbooks of quantum mechanics, e.g. Dirac's fundamental book. There is an interpretation of physics, first expressed by J. von Neumann (1955) and favoured by Wigner (1971), in which the observer has a fundamental role: physics describes the observations of the observer and his impressions are the basic entities. Since observations are made by measuring apparatuses the following consistency problem arises. Any observable of a system must be equivalent to another observable of a second suitable system interacting with the first one, if the latter is to be interpreted as a measuring apparatus. Von Neumann gives a schematic solution of this problem of measurement. However the impressions of an observer are real but absolutely private entities. An observer cannot point out the impressions he has received from an observation, therefore such impressions are outside the realm of any science. On the contrary one needs objects as basic entities of physics. One can identify objective aspects in quantum mechanics: there are systems and sets of measurements on them which are dispersionless, i.e. the outcome of these measurements is certain. Then corresponding to each measurement of this kind one can attribute an objective property to the system. The set of such properties is the 'state' of a single system. Quantum mechanics has the following general feature. At any time a statistical collection of systems can be decomposed into subcollections such that to each system in each subcollection an objective 'state' can be attributed. One can identify in such properties the basic entities which are measured. Consistently with this point of view Jauch (1968, 1969) and Piron (1964) were able to obtain quantum mechanics as a consequence of simple axioms about yes–no experiments, properties and states. The 'state' of a macroscopic system should embody all the typical macroscopic properties known, for example, in the case of a measuring apparatus, a certain position of the pointer is a macroscopic property.

Let A be an apparatus with a pointer. The pointer moves from λ_0 to λ_1 when A interacts with a system which has a property p; the pointer does not move when the system has the property p^* (p^* not to have the property p). Let $h^A(h^S)$ be the Hilbert space of the apparatus (of the system); $h_p \subset h^S$ the subspace of h^S associated with the property p, $h_\lambda \subset h^A$ the subspace associated with the position λ of the pointer; H is the Hamiltonian of the joint system, t the duration of measurement. One must have

$$e^{-iHt} P^A_{\varphi_{\lambda 0}} \otimes P^S_\psi e^{iHt} = P^A_{\varphi_{\lambda 1}} \otimes P^S_{\psi'}, \quad e^{-iHt} P^A_{\varphi_{\lambda 0}} \otimes P^S_{\tilde\psi} e^{iHt} = P^A_{\varphi_{\lambda 0}} \otimes P^S_{\tilde\psi'} \quad (1)$$

for all $\psi \in h_p$, $\tilde\psi \in h^S \ominus h_p$, p_ψ being the projection on ψ; $\varphi_\lambda, \varphi'_\lambda \in h_\lambda$.

Consider now a case in which system S has neither property p, nor p^*, but it has another property p' such that $h_{p'}$ is not contained in h_p nor in $h^S \ominus h_p$. Let $\eta \in h_{p'}$ be the state of S:

$$\eta = c_1 \psi + c_2 \tilde\psi; \quad \psi \in h_p, \quad \tilde\psi \in h^S \ominus h_p; \quad \|\psi\| = \|\tilde\psi\| = 1$$

One has by (1)

$$e^{-iHt} P^A_{\varphi_{\lambda 0}} \otimes P^S_\eta e^{iHt} = |c_1|^2 P^A_{\varphi_{\lambda 1}} \otimes P^S_{\psi'} + |c_2|^2 P^A_{\varphi_{\lambda 0}} \otimes P^S_{\tilde\psi'} + \mathrm{Re}\, c_1 c_2^*(\ldots) \quad (2)$$

This formula contains the whole problem of measurement in the first-mentioned attitude; if the last term were zero, no problem would arise. The collection of systems A+S can be separated into two subcollections, the first containing a fraction $|c_1|^2$ of systems A+S each of which has part A with the pointer at λ_1, the second subcollection containing a fraction $|c_2|^2$ of systems A+S with part A having the pointer at λ_0. This is in complete agreement with the physical meaning which in quantum mechanics is given to coefficients. c_1, c_2 in $\eta = c_1 \psi + c_2 \tilde\psi$. However the third term in equation (2) is non-zero if c_1 and $c_2 \neq 0$. It is the infamous interference term. The mathematical reason for its presence is the following: $P^A_{\varphi_{\lambda 0}} \otimes P^S_\eta$ is a pure state, i.e. an extreme element of the convex set of states. If time evolution is represented by linear, invertible mappings on the set of states, an extreme element is mapped into an extreme element; if the interference term is zero, the right-hand side of equation (2) is a mixture of two states and not an extreme state, which is impossible. Due to the interference term the pointer of A after the interaction no longer has a position. The notion of objectivity in quantum mechanics is too restrictive to give an account of the objective properties of part A of the system A+S, if A and S interact.

More generally if one considers a composite system S_1+S_2, due to the interaction, there are no properties of S_1+S_2 to be described as: S_1 has a property p_1 and S_2 has a property p_2. As long as S_1 and S_2 are microsystems this can look strange but is not a basic difficulty: it is essentially the Einstein–Rosen–Podolsky paradox; if one of the components is macroscopic one meets a big difficulty, as it has been particularly stressed by d'Espagnat (1971b). The difficulty lies in the objectivity criterion or in the time evolution law. It is difficult to see how to generalize the objectivity criterion; time evolution is a consequence of symmetry under time translations of an isolated system. Since the interference terms depends critically on small external perturbations, it may be that they are not really meaningful in the physically realizable conditions of isolation, as remarked by Zeh (1971). If one considers a system as a subsystem of a larger one, one has no strict condition about its time evolution. 'Non-Hamiltonian' mappings, which are largely used in the theory of open systems, are admissible. Such mappings are not invertible and extreme states can be mapped into mixtures. The idea that any observed system should be considered as an open system has lead Everett (1957) to claim that no system can be isolated from the rest of the universe, in which the observer must also be included. However there are many examples of successful phenomenological theories for isolated macrosystems. The fact that in the quantum-mechanical description of macrosystems elements enter which are highly unstable, or foreign to the system, as Everett's wave function of the whole universe, is an indication that quantum mechanics does not work very well for macrosystems. Therefore, in my opinion, due to the difficulty in the problem of measurement, attitude (1) should be dismissed.

The second-mentioned attitude, at least at a linguistic level, was that held by Bohr. He puts the objective character of apparatuses in the foreground and perhaps misleadingly, describes them as 'classical systems'. This was often interpreted in the sense that there are systems, to be used as measuring apparatuses, for which pre-quantum physics is the right theory; since on the contrary quantum effects are well known in macrophysics, the concept of 'classical systems' seems fictitious. Ludwig (1972) has recently formulated a new general theory of macrosystems in which their objective character is built in. He has also developed a theory for a composite system (Macrosystem + microsystem). (Ludwig, 1966b). The problem of measurement is formally solved; it is shown that the registration of how a macroscopic apparatus is after the interaction with a microsystem is equivalent to the measurement of an observable of the microsystem. In principle such an observable can be calculated in terms of mathematical elements referring to the description of the macroscopic apparatus. Not only a physical content of a statistical operator for a microsystem can be read off from measuring apparatus A, but also the initial statistical operator can be read off from the macro-source S of the microsystem. In conclusion quantum mechanics for a microsystem α produced by the source S and measured by the apparatus A, describes to a certain extent the interaction between S and A, the microsystem being the vehicle of such an interaction; in principle the macroscopic description of system S + A can completely replace the quantum mechanics of α. A final primacy of the objectivistic way of describing the world or classical way, is established, consistently with quantum mechanics of microsystems. Such a result does not mean that one has gained a classical insight into microsystems, but rather one has a classical 'outsight' from the microsystems; no classical hidden variables have been found, but a classical anchorage for the microsystems has been achieved. It may be that the search for such an anchorage, which is basic goal of a sane philosophy, was one of the motivations for looking for hidden variables.

An important consequence of this point of view is that one has no need for objectivity criteria inside quantum mechanics of a microsystem. Properties and states for single microsystems are interesting but not basic features of the formalism. Peculiarities concerning properties of composed microsystems, as in the ERP paradox are no longer a difficulty. Quantum mechanics should not be based on properties and states of microsystems, but should be a theory for a certain class of experiments which, by means of an interaction between a source S and a measuring apparatus A, give evidence for a microsystem; this is just the starting point of Ludwig's axiomatic approach to quantum mechanics (Ludwig 1970). At such a point most physicists would be disappointed since this theory apparently gives a secondary role to quantum mechanics. However let me stress that the theory of macrosystems proposed by Ludwig is nothing more than a formalization of statistics and 'state' space. Every type of time evolution deterministic or not, Markoffian or not can be placed into it. Obviously all known examples of theories for macrosystems fit into this scheme, which, however, does not help one to find such theories. The sole general and conclusive way to build a theory of a macrosystem is to rely on its atomistic structure and describe it by the mechanics of the microcomponents. So indeed microsystems entered into physics first as an hypothesis, then as objects that could be emitted and revealed by macrosystems. The basic role of quantum mechanics is to provide the concept of a particle and an insight into the interactions between the particles. Particles and their interactions are then the starting point for the atomistic theory of macrosystems, which should finally fill up the formal scheme proposed by Ludwig. The actual theory of atomic structure is quantum statistical mechanics, in brief N-body quantum theory. Therefore we can state the problem of measurement in the following form. N-body quantum theory must yield the concrete input for Ludwig's theory of macrosystems. Quantum theory of an N-body system interacting with a microsystem must yield the concrete input for the new theory of a macrosystem interacting with a microsystem. The important point is the following: we need not rely on an objectivity criterion in quantum mechanics and pretend that macroscopicity is a 'property' (in the technical sense) of the N-body structure, which leads to the difficulty with the interference term in equation (2). One must simply extract from the N-body theory in a sufficiently general and precise way what is relevant as an input for the new theory of macrosystems and throw away the rest as physically irrelevant. This has not yet been done in a satisfactory way.

Anyway the problem of measurement in quantum mechanics is not a philosophical quarrel about the interpretation of the world, nor a basic difficulty of quantum mechanics as in equation (2), but a technical problem in N-body theory. This important conclusion has been reached long before the recent theory of a macrosystem, by Ludwig (1953) and by Daneri, Loinger and Prosperi (1962), who identified in the ergodic behaviour of many-particle systems the technically relevant point. Since these early attempts other main approaches and ideas in quantum statistics such as the master equation theory (Lanz and coworkers, 1971), the independent subdynamics theory (George and coworkers, 1972), C* algebra formalism (Hepp, 1972), have been confronted with the problem of measurement; all these attempts suffer from the lack of a clear and general mathematical characterization of macroscopicity. A general very readable survey on the way from microphysics to macrophysics is given by Caldirola (1974). The somewhat utilitarian exploitation of N-body theory by which the problem of measurement should be solved, is justified if one takes into account that N-body theory is a formal extrapolation of the quantum mechanics of microsystems to the case of systems with an extremely large number of particles, in which the quite unobservable correlations between all the particles are described. Let us suppose we have on the one side the quantum mechanics for microsystems, on the other Ludwig's scheme for macrosystems filled with the aid of N-body quantum theory then one can then hope that a new unified theory can be revealed, ranging from microsystems to macrosystems, which could perhaps cover also intermediate systems such as macromolecules (Ludwig, 1972a,b).

I shall discuss the problem of measurement having the second attitude mentioned above. In Section 2 quantum mechanics as a theory of measurements on microsystems will be discussed, with particular reference to the ERP paradox, in Section 3 a sketchy, but I hope not too distorted, account of Ludwig's theory of macrosystems will be given. Finally in Section 4, the problem of measurement within N-body theory will be stated in a precise way.

2. QUANTUM MECHANICS AS A THEORY OF MEASUREMENT ON MICROSYSTEMS

There are interactions between two macrosystems which can be explained in terms of a microsystem which the first macrosystem emits and by which the second one is affected. We shall call the first macrosystem the source or preparation part. The typical feature of such an interaction is that it causes a perturbation which spreads out from one or more pointlike regions inside the affected part. In a space–time description of the affected part these perturbations involve space–time points inside one or more cones with axes parallel to the future time axis. The occurrence of such perturbations is very stochastic, i.e. repetition of the same experiment under the same conditions of preparation and of affected parts yields a different pattern of the afore-mentioned pointlike regions. The single experiments are not reproducible and it would be meaningless to formulate a theory for them. If a single experiment is repeated n times, the frequencies of occurrence of a certain type of perturbation can be measured; if n is large enough such frequencies are reproducible and it is worth while formulating a theory for them. In fact such frequencies have in a certain sense a universal character. They are completely independent from very many features of the preparation and the affected part. Such a situation is an obvious consequence of the atomistic structure of matter: microsystems effect only some atoms of the affected part by interactions that have a universal character. We shall mean by an experiment on a microsystem a statistical collection of a large number of single experiments in each of which a single microsystem is emitted and revealed. It consists in principle in the repetition of the following steps. (a) production of the microsystem by a source which has been prepared by taking into account a certain set α of macroscopic prescriptions, the same procedure to be taken in the n repetitions, referred to a frame R; (b) observation of whether a measuring apparatus, which has been prepared taking a certain set β of macroscopic prescriptions into account, is affected or not in a certain prescribed way γ (also β, γ are referred to R and are fixed in the n repetitions) (c) counting how many times n^+ the apparatus is affected. The frequency n^+/n is the quantitative result of the experiment. Let us call (R, α) a preparation procedure and (β, γ) a measuring procedure.

Ludwig has obtained the axiomatic structure of quantum mechanics, in a generalized form, starting from suitable axioms about a physical input consisting of preparation procedures, measuring procedures and frequencies (Ludwig, 1970). Let me summarize the result in the particular case of no superselection rules. The set M of experiments which give evidence for a microsystem is described by means of a Hilbert space h_M with the following interpretation. Each statistical operator W on h_M represents a class of equivalent preparation procedures, where two preparation procedures are equivalent if one has the same frequencies for any measuring procedure. Each operator F on h_M, such that $0 \leq F \leq I$, represents a class of equivalent measuring procedures, where two measuring procedures are equivalent if for any preparation procedure one has the same frequency.

The operators F are called 'effects' by Ludwig; in the conventional axiomatics of quantum mechanics measuring procedures are associated only with projections on h_M. The frequency of the effect F, when the preparation W is made, is given by $\text{Tr}(FW)$. The symmetry of the theory under time-translations implies that a semigroup $\mathcal{V}(t)$, $t \geq 0$, exists of mappings of the set L_M of effects into L_M, such that $\mathcal{V}(t)I = I$, I being the identity operator. $\mathcal{V}(t)$ is contractive on $\mathbb{B}(h_M)$, the Banach space of linear bounded operators on h_M (Comi and coworkers, 1975). In the standard axiomatics of quantum mechanics one requires that $\mathcal{V}(t)$ can be extended to a group; then projection operators are mapped into projection operators and $\mathcal{V}(t)$ must have the following structure:

$$\mathcal{V}(t)F = V(t)FV\dagger(t)$$

$V(t)$ being a unitary group on h_M. Let iH be the generator of $V(t)$, then H is the hamiltonian of the system.

Let $\mathcal{V}'(t)$ be the adjoint operator of $\mathcal{V}(t)$; $\mathcal{V}'(t)$ maps the set $K(h_M)$ of all statistical operators on h_M into itself. $\mathcal{V}(t)$ is a contractive operator on $\tau\mathbb{C}(h_M)$, where $\tau\mathbb{C}(h_M)$ is the Banach space of 'trace class' operators on h_M. Due to the definition of $V'(t)$ one has:

$$\text{Tr}((\mathcal{V}(t)F)W) = \text{Tr}(F\mathcal{V}'(t)W)$$

Let us consider a preparation procedure $W \equiv (\alpha, R)$ and the transformed preparation $\mathcal{V}'(t)W$; $\mathcal{V}'(t)W$ is the preparation procedure W shifted back in time by a time interval of length t. The relation between W and $\mathcal{V}'(t)W$ can be described as follows: $\mathcal{V}'(t)W$ consists of the preparation W and of waiting a time t after W is finished. In the time interval $[0, t]$ only the microsystem evolves, therefore $\mathcal{V}'(t)$ can be considered as the time evolution operator of the microsystem.

Space-time symmetry has further consequences: in the non-relativistic case (for simplicity we consider only this case) one has on h_M a unitary, projective representation of the group G_0 of space translations, accelerations and rotations. Let us assume that such a representation is irreducible and that the theory describes at least the effects linked to the most simple perturbations of the effect parts; such most simple perturbations are those spreading out from one point-like region. The representations of G_0 which satisfy these requirements are characterized by two indexes, a half-integer number s and a positive number m, to be interpreted as spin and mass of an elementary particle. By suitable choices of the parameters s and m certain sets M of experiments can be described: the corresponding microsystems are the most simple ones, they are 'one particle' systems.

Let us consider two experiments concerning two particles I and II and let us build a correlated experiment, in which the two preparation procedures and the two measuring procedures are performed together.

Such correlated experiments can obviously be described by preparations

$$W^\text{I} \otimes W^\text{II} \in K^\text{I} \otimes K^\text{II} \subset \tau\mathbb{C}(h^\text{I}) \otimes \tau\mathbb{C}(h^\text{II})$$

and by effects

$$F^\text{I} \otimes F^\text{II} \in L^\text{I} \otimes L^\text{II} \subset \mathbf{B}(h^\text{I}) \otimes \mathbf{B}(h^\text{II})$$

Since
$$\tau\mathbb{C}(h^{\mathrm{I}}) \otimes \tau\mathbb{C}(h^{\mathrm{II}}) = \tau\mathbb{C}(h^{\mathrm{I}} \otimes h^{\mathrm{II}})$$
and
$$\mathbf{B}(h^{\mathrm{I}}) \otimes \mathbf{B}(h^{\mathrm{II}}) = \mathbf{B}(h^{\mathrm{I}} \otimes h^{\mathrm{II}})$$

$W^{\mathrm{I}} \otimes W^{\mathrm{II}}$ and $F^{\mathrm{I}} \otimes F^{\mathrm{II}}$ are very particular statistical operators and affect the Hilbert space $h^{\mathrm{I}} \otimes h^{\mathrm{II}}$. The axiomatic structure of quantum mechanics leads us to assume the existence of a microsystem to be associated with the Hilbert space $h^{\mathrm{I}} \otimes h^{\mathrm{II}}$, i.e. each $W \in K(h^{\mathrm{I}} \otimes h^{\mathrm{II}})$, $F \in L(h^{\mathrm{I}} \otimes h^{\mathrm{II}})$ should in principle correspond to a preparation and a measuring procedure. On $h^{\mathrm{I}} \otimes h^{\mathrm{II}}$ one can place a unitary representation of $G_0 \times G_0$ and at least effects linked with two-point perturbations of the affected part can be described.

Then symmetry allows the following structure for $V(t)$; $V(t) = e^{iHt}$, $H = H^{\mathrm{I}} \otimes I^{\mathrm{II}} + I^{\mathrm{I}} \otimes H^{\mathrm{II}} + H^{\mathrm{int}}$, H^{int} describing the interaction between the two particles. Microsystems of this type with a statistical operator such that interaction plays a role, so that H^{int} can be tested, are prepared in all scattering experiments and are emitted from macrosystems in many spontaneous decay processes. In conclusion if I and II are particles, a microsystem (I, II) also exists with Hilbert-space $h^{\mathrm{I}} \otimes h^{\mathrm{II}}$ and one can explain experiments about (I, II) by a suitable choice of H^{int}, at least in the non-relativistic case. Let us investigate the structure of microsystem (I, II) and its relation with particles I and II. Consider effects of the form $F^{\mathrm{I}} \otimes I^{\mathrm{II}}$ and $I^{\mathrm{I}} \otimes F^{\mathrm{II}}$, one has

$$\mathrm{Tr}\,(F^{\mathrm{I}} \otimes I^{\mathrm{II}} W) = \mathrm{Tr}_{h^{\mathrm{I}}}(F^{\mathrm{I}} W^{\mathrm{I}}), \quad \mathrm{Tr}\,(I^{\mathrm{I}} \otimes F^{\mathrm{II}} W) = \mathrm{Tr}_{h^{\mathrm{II}}}(F^{\mathrm{II}} W^{\mathrm{II}})$$

where
$$W^{\mathrm{I}} = \mathrm{Tr}_{h^{\mathrm{II}}}(W) \in K^{\mathrm{I}}(h^{\mathrm{I}}) \quad \text{and} \quad W^{\mathrm{II}} = \mathrm{Tr}_{h^{\mathrm{I}}}(W) \in K^{\mathrm{II}}(h^{\mathrm{II}})$$

for physically meaningful H^{int} one has that for t large enough, $t > \bar{t}$, $\mathcal{V}(t) W$ can be replaced with $\mathcal{V}'(t) \otimes \mathcal{V}''(t) \mathcal{S}$; $\mathcal{S}$ being a 'collision' mapping of K onto K. Then for $t > \bar{t}$ one has

$$\mathrm{Tr}_{h^{\mathrm{I}} \otimes h^{\mathrm{II}}}(F^{\mathrm{I}} \otimes I^{\mathrm{II}} \mathcal{V}'(t) W) = \mathrm{Tr}_{h^{\mathrm{I}}}(F^{\mathrm{I}} \mathcal{V}'(t) \tilde{W}^{\mathrm{I}}), \quad \tilde{W}^{\mathrm{I}} = \mathrm{Tr}_{h^{\mathrm{II}}}(\mathcal{S} W)$$

and similarly if I $\leftrightarrows$ II. The interpretation of these results is straightforward: we have two particles I and II, which for $t > \bar{t}$, no longer interact and are described by the initial statistical operators $\tilde{W}^{\mathrm{I}}$ and $\tilde{W}^{\mathrm{II}}$.

We can look for the occurrence of the joint effect $F^{\mathrm{I}} \otimes I^{\mathrm{II}}$ and $I^{\mathrm{I}} \otimes F^{\mathrm{II}}$, to be represented by $F^{\mathrm{I}} \otimes F^{\mathrm{II}}$, which has a frequency $\mathrm{Tr}\,(F^{\mathrm{I}} \otimes F^{\mathrm{II}} W)$. In conclusion: the microsystem (I, II) is a pair of particles I, II and a correlation law for joint measurements. The problem of the description of a microsystem (I, II) has been raised by the well-known ERP paradox (Einstein and coworkers, 1935). To make the discussion more specific let us consider the following example given by Bohm (1951): two spin $\frac{1}{2}$ particles emitted in a singlet state by a suitable source move in opposite directions and the components of the spins $\mathbf{S}_1$ and $\mathbf{S}_2$ in two directions $\mathbf{n}_1$, $\mathbf{n}_2$ are measured, e.g. by means of two Stern–Gerlach magnets followed by two revelators. If $u_{\pm}^{I,II}$ are the normalized eigenvectors of $S_z^{I,II}$ corresponding to eigenvalues $\pm\frac{1}{2}$, one has, neglecting for simplicity the spatial coordinates:

$$W = P_\psi, \quad \psi = \frac{1}{\sqrt{2}}(u_+^I \otimes u_-^{II} - u_-^I \otimes u_+^{II}) \in C_2^I \otimes C_2^{II}, \quad W^I = \frac{I^I}{2}, \quad W^{II} = \frac{I^{II}}{2}$$

Let P_n be the eigenprojection of $S_n = \mathbf{S_n}$ corresponding to the eigenvalue $\frac{1}{2}$; for the effects

$$P_n^I \otimes I^{II}, \quad I^I \otimes P_n^{II}, \quad P_n^I \otimes P_{\pm n}^{II}, \quad P_z^I \otimes P_{\pm x}^{II}, \quad P_z^I \otimes P_{\pm y}^{II}$$

one has the probabilities

$$\text{Tr}(P_n^I \otimes I^{II} P_\psi) = \text{Tr}(I^I \otimes P_n^{II} P_\psi) = \tfrac{1}{2}, \quad \text{Tr}(P_n^I \otimes P_n^{II} P_\psi) = 0,$$
$$\text{Tr}(P_n^I \otimes P_{-n}^{II} P_\psi) = \tfrac{1}{2}, \quad \text{Tr}(P_z^I \otimes P_{\pm x}^{II} P_\psi) = \tfrac{1}{4},$$
$$\text{Tr}(P_z^I \otimes P_{\pm y}^{II} P_\psi) = \tfrac{1}{4}.$$

The microsystem consists of two particles with isotropic spin statistics; it has the following property: total spin = 0; no spin property can be attributed to particles I and II. The simple correlation law can be anticipated from rotational symmetry. Take a collection of microsystems (I, II) with preparation P_ψ, put an apparatus for measuring S_n in the path of one of the two particles. Theoretically the frequency of $S_n = \frac{1}{2}, (-\frac{1}{2})$ is $\frac{1}{2}$. Then put an apparatus for measuring $S_{n'}$ on the path of the second particle and again measure the frequency $S_n = \frac{1}{2}, (-\frac{1}{2})$. The two sets of apparatus can coexist since

$$[P_n^I \otimes I^{II}, \quad I^I \otimes P_{n'}^{II}] = 0$$

We have the same effect as before

$$P_n^I \otimes I^{II} = P_n^I \otimes P_{n'}^{II} + P_n^I \otimes P_{n'}^{II}$$

and the same result as before. Apparatus II does not influence in any way the physics of particle I. One has the identity

$$\text{Tr}(P_n^I \otimes I^{II} P_\psi) = \text{Tr}(P_n^I \otimes I^{II} \bar{W}_{n'})$$

with

$$\bar{W}_{n'} = \tfrac{1}{2} W_{n'} + \tfrac{1}{2} W_{n'}, \quad W_{n'} = 2 P_{n'}^I \otimes P_{n'}^{II} P_\psi P_{n'}^I \otimes P_{n'}^{II}, \quad \forall \mathbf{n'}$$

We see that *as far as particle I* is concerned, we can describe the collection as built from two equal subcollections of microsystems with the property $S_{n'}^{II} = \frac{1}{2}$ for the first one and the property $S_{-n'}^{II} = \frac{1}{2}$ for the second one; this for any direction $\mathbf{n'}$, obviously without any consequence for particle I. Let us use measurement II to build a new statistical collection: we select those microsystems for which the apparatus II has yielded the result $S_{n'}^{II} = \frac{1}{2}, (-\frac{1}{2})$ and look for the effects produced by the microsystems of this collection. The statistical operator for this new collection is $W_{n'}, (W_{-n'})$ and a single microsystem of it has the properties, $S_{n'}^{II} = \frac{1}{2}, (S_{-n'}^{II} = \frac{1}{2})$. If $S_{n'}^I$ is measured the probability of the result

$S_\mathbf{n}^I = \frac{1}{2}$ depends on $\mathbf{n} \cdot \mathbf{n}'$, e.g. in the case $\mathbf{n} = \mathbf{z}$ $\mathbf{n}' = \mathbf{x}(\mathbf{y})$ it is $\frac{1}{2}$ in agreement with the probability $1/4$ of $S_\mathbf{z}^I$, $S_{\mathbf{x},(\mathbf{y})}^{II}$ for the initial collection. The transition $\bar{W} \to W_{\mathbf{n}'}$ is not a consequence of the interaction of particle II with the apparatus II, but is a consequence of the repreparation procedure in which measurement II is used. We stress that such a repreparation cannot coexist with another apparatus that measures $S_{\mathbf{n}''}^{II}$, $\mathbf{n}'' \neq \mathbf{n}'$; therefore the two decompositions of $\bar{W}$ according to the measurements of $S_{\mathbf{n}'}^{II}$, $S_{\mathbf{n}''}^{II}$ cannot be made together. There is nothing peculiar or paradoxical in this description. However it is possible that one would like to consider the microsystem (I, II) as a system of two correlated particles I and II in which case difficulties would arise. The correlation should be a correlation between properties of the two particles. Unfortunately due to the interaction, W^I, W^{II} are not pure states even if W is a pure state, this means that if a property can be attributed to a microsystem (I, II), such a property is not expressible as a property of particles I and II. Thus one has nothing to correlate and no vector state can be associated to the components I and II of microsystem (I, II); d'Espagnat describes this peculiarity as 'non-separability' of (I, II) into I and II (d'Espagnat, 1971a). In my opinion this indicated that one should not ascribe a basic role to the concepts of property and of state for a single microsystem, as I have already stressed in Section 1 where attitudes (1) and (2) were discussed. If attitude (1) is chosen one can assume that not the whole set $K(h^I \otimes h^{II})$ is physically meaningful, but only statistical operators with the following structure:

$$W = \sum_{jl} \gamma_{jl} P_j^I \otimes P_l^{II}$$

where $P_j^I \otimes P_l^{II}$ are projections on states of the form

$$u_j^I \otimes u_l^{II}, \quad u_j^I \in h^I, \quad u_l^{II} \in h^{II}$$

Statistical operators of this kind are called by d'Espagnat mixtures of the first kind, while the other mixtures are called of the second kind. This can be looked upon as a mixture of pure states $u_j^I \otimes u_j^{II}$ in each of which a property j of particle I is correlated to a property l of particle II. Therefore in such a case microsystem (I, II) could be considered as a system of two correlated particles.

However any state P_ψ with

$$\psi = \alpha u_1^I \otimes v_1^{II} + \beta u_2^I \otimes v_2^{II}$$

must be excluded; if I and II are two identical fermions, usual quantum mechanics claims that all pure states have this structure. It seems difficult to eliminate such states which provide the energy levels for atoms in excellent agreement with experimental results. However one could expect that scattering states, with well separated particles should be described as two correlated particles, e.g. in the example we considered before, one could expect that the singlet state transforms into a mixture of the kind (Janch, 1971)

$$W = \frac{1}{4\pi} \int P_\mathbf{n}^I \otimes P_\mathbf{n}^{II} \, d\mathbf{n}$$

A behaviour of this type has been discussed by Bohm and Aharonov (1957). Then $\mathcal{V}''(t)$ must transform pure states into mixtures. Usually one assumes that $\mathcal{V}'(t)$ is a group, this excludes the afore-mentioned behaviour. However, in the framework of Ludwig's axiomatics, as it is shown by Comi and coworkers (1975), the assumption that $\mathcal{V}''(t)$ is a group, seems to be unnecessarily restrictive. $\mathcal{V}'(t)$ can be a semigroup of linear mappings of K into K. Then the required behaviour can arise at least asymptotically (Barchielli and Lanz, 1975).

A very important point is that measurable correlations are different for separable and for non-separable microsystems (I, II). Such differences are the same as those which discriminate between the existence or non-existence of local hidden variables. In fact a violation of Bell's inequality would prove the existence of non-separable microsystems. (Selleri, 1971; Kasday, 1971). Let me comment briefly on the relation of quantum mechanics to hidden variables theory. We started the discussion of experiments about microsystems observing that for practical reasons only a statistical theory is needed, since single experiments are not reproducible. However one cannot exclude that a more fundamental theory exists which could be applied to very hypothetical, perhaps non-realizable, preparations, which are so accurate that all effects are reproducible, i.e. for each effect theory tells us whether it occurs or not.

Since the basic concept in macrophysics is the concept of state space, as will be shown in section 3, it is appealing to associate with a microsystem a 'hidden' state space Z^μ, which we assume to be a measure space, with a suitable set $\tilde{K}$ of measures on a suitable σ-algebra of subsets of Z^μ. Each preparation part prepares a microsystem to be represented by an element of Z. To a measuring part one associates a $\tilde{K}$-measurable function $\eta(z)$ which assumes the value unity if the measuring perturbation occurs or zero if it does not occur. In a statistical experiment one can assume that a measure $\mu_W \in \tilde{K}$ corresponds to a preparation procedure and an 'average' function $\bar{\eta}(z)$ corresponds to a measuring procedure, with $0 \le \bar{\eta}(z) \le 1$ and $\bar{\eta}(z)$ $\tilde{K}$-measurable; then the probability of an effect F after a preparation W would be

$$\int_{Z^\mu} \bar{\eta}_F(z) \mathrm{d}\mu_W(z)$$

The very hypothesis of the existence of a 'hidden' state space Z^μ for a microsystem does not contradict the basic axioms of quantum mechanics, if the latter is intended as the statistical theory of a certain class of interactions between two macrosystems; this is not so obvious if quantum mechanics is the theory of properties of a microsystem.

A famous negative theorem about the existence of 'hidden' variables has been given by von Neumann (1955) and in a more sophisticated way by Jauch and Piron (1963). The physical relevance of these negative theorems has been criticized in an important paper by Bell (1966). While existence or not of hidden variables has little to do with quantum mechanics, the properties of the function $\bar{\eta}_F(z)$ which refers to a measuring procedure and not to a measuring part, must be confronted with quantum mechanics. Bell considers the class of local hidden variables. Let me translate Bell's definition of local hidden variables into the language of the present discussion. Consider the effect $F \equiv (F_1, F_2) \equiv (F_1 \text{ and } F_2)$ where F_1, F_2 refer to spatially well separated effects, then the locality condition is $\bar{\eta}_{(F_1,F_2)}(z) = \bar{\eta}_{F_1}(z)\bar{\eta}_{F_2}(z)$. It is just such a requirement which makes hidden variables useful to describe a microsystem (I, II) and to explain the ERP paradox. Let us consider effects such that the pairs F_1, F_2; F'_1, F'_2; F_1, F'_2; F'_1, F_2 are coexistent and satisfy the locality condition. Typically $F_1, F'_1, (F_2, F'_2)$ can correspond to two different orientations of the same apparatus, e.g. the symmetry axis of a photon linear polarization analyser can be oriented in different directions. Then for any preparation, the probabilities $P_{(F_1,F_2)}, P_{(F_1,F'_2)}, P_{(F'_1,F_2)}, P_{(F'_1,F'_2)}$ satisfy the inequality (Bell, 1971).

$$|P(F_1, F_2) - P(F_1, F'_2)| + P(F'_1, F'_2) + P(F'_1, F_2) \leq 2$$

In the mathematical theory of microsystems (I, II) provided by quantum mechanics there are effects and preparations (having the feature of 'non-separability') which do not satisfy Bell's inequality. Therefore local hidden variables do not complete quantum mechanics, but contradict certain of its statistical predictions. It is a very important experimental problem to gain any evidence of a violation of Bell's inequality; this would rule out local hidden variables and indicate the reality of 'non-separable' microsystems. Recently interesting results on the two-photon system have been obtained by Kasday (1971) and by Clauser and Freedman (1972), which indicate a violation of Bell's inequality.

3 THEORY OF MACROSYSTEMS

A macrosystem is such that at any time one can say 'how it is'; physics is supposed to give a mathematical description of how 'a macrosystem' is. In place of the phrase 'how it is' let us speak of the 'state' of the macrosystem at time t and represent such a state by an element $z(t)$ of a suitable space Z. More precisely, let us consider a macrosystem which for times $t > 0$ is isolated; Ludwig postulates that its objective qualities at any time $t > 0$ are represented by a point $z(t)$ in a state space Z. Such a description of a macrosystem can be called, realistic, objectivistic or somewhat misleadingly 'classical'. Classical refers to the fact that it was the sole attitude of physicists before the development of quantum mechanics, but does not mean at all that one pictures a macrosystem as an assembly of molecules described by classical mechanics or that one derives all electromagnetic phenomena from the Maxwell equation. To make clearer what is meant, consider a black body at equilibrium: its state can be specified by a description of its walls, the temperature T and the distribution $U(\nu)$ of electromagnetic energy density on the eigenfrequencies of the electromagnetic field; neglecting all aspects of the walls except the volume V of the hollow part, one has $z \equiv (V, T, \{u(\nu)\}, z \in Z$; the average value of the variable $U(\nu)$ is the well-known Planck radiation law, in which the 'quantum theoretical' constant h appears. All statements about a macrosystem finally refer to a suitable space Z, which depends on the kind of system and on the level of the description. Examples of this are as follows: macrosystems schematized by a set of mass points, having at any time a position $\mathbf{x}(t) \in \mathbf{R}^3$ and a momentum $\mathbf{p}_j(t) \in \mathbf{R}^3$, $j = 1, 2 \ldots k$, which can be represented by an element of $\mathbf{R}^{6k}$; a fluid in local equilibrium inside a region $\Omega \subset \mathbf{R}^3$, is described in hydrodynamics by a mass density function $\rho(\mathbf{x}) \in \mathscr{L}^1(\Omega)$ by an internal energy density $u(\mathbf{x}) \in \mathscr{L}^1(\Omega)$ and by a velocity field $v(\mathbf{x}) \in \mathscr{L}^\infty(\Omega)$; then the fluid can be represented by an element of $\mathscr{L}^1(\Omega) \times \mathscr{L}^1(\Omega) \times \mathscr{L}^\infty(\Omega)$. A dilute gas in Ω is almost completely described by the Boltzmann distribution function $f(\mathbf{x}, \mathbf{p}) \in \mathscr{L}^1(\Omega \times \mathbf{R}^3)$.

Fortunately enough, in many cases fluctuations of the state are very small so that statistics can be forgotten, the difficult problem of defining measures in a function space can be avoided and actual states identified with average states.

By many examples, e.g. the afore-mentioned ones, one is lead to assume that Z is a complete, metric space. The space

$$Y = C(\Theta_+, Z), \quad \Theta_+ = (0, +\infty),$$

of all continuous functions $z(t)$, $t > 0$, $z(t) \in Z$, is called trajectory space. On Y one can define in well-known way a topology, by which Y becomes a metric, complete space. However the corresponding metric $d(y, y')$ has not a direct physical meaning. If two trajectories y, y' are physically appreciated to be in a certain vicinity this is not well represented by a condition such as $d(y, y') < \zeta$ the latter criterion being too restrictive. Ludwig shows that a new metric can be defined, leading to physically meaningful vicinities, which induces in Y the same topology (but a coarser uniform structure) as the topology $C_c(\Theta_+, Z)$ of the uniform convergence on compact subsets of Θ_+; Y with such a new metric is not complete. Its completion $\hat{Y}$ is a compact Haussdorf space; on $\hat{Y}$ the set of all continuous functions $z(t)$, $t > 0$ is dense. On $\hat{Y}$ a continuous time translation operator $T(\tau)$ can be defined for $\tau \geq 0: T(\tau)y = y'$ where $y \equiv z(t) \Leftrightarrow y' \equiv z(t+\tau)$, $\forall y \in Y$.

Let us consider the Borel σ-algebra $\mathscr{B}(\hat{Y})$ and the set of all signed Radon measures on $\mathscr{B}(\hat{Y})$. To this set a Banach space structure can be given, it coincides with $C'(\hat{Y})$, which is the dual space of the Banach space of continuous function on $\hat{Y}$.

A preparation procedure of a statistical collection of macrosystems is represented by a positive, normalized Radon measure $u(\omega)$ on $\mathscr{B}(\hat{Y})$. Once such a measure is explicitly given the whole statistical dynamics of a macrosystem of the prepared collection is known. In fact $u(\omega)$ is the probability that the trajectory of the macrosystem belongs to $\omega \subset \hat{Y}$. The whole physics of a macrosystem of a given type (i.e. describable in a given space Z) is known if the convex set of all possible preparations $K_m \subset C(\hat{Y})$ is known. Suppose K_m is given, let us see how the physics of the macrosystem can be gained. Consider the weak closure $\bar{K}_m$ of K_m in $C'(\hat{Y})$; $\bar{K}_m$ is convex and compact in the $\sigma(C'(\hat{Y}), C(\hat{Y}))$ topology; the extreme points u_l of $\bar{K}_m$ are 'elementary preparations', each preparation being a mixture of them. Then consider $u_i \in \bar{K}_m$. The support of $u_i(\omega)$ is the set of functions $z(t)$ and also of limit points of Y in $\hat{Y}$, which are possible trajectories for the macrosystem; if the support of u_i reduces to a point $y_i \in Y$ one has a deterministic dynamics, Y_i being the trajectory of the macrosystem. If one assumes that the cylindrical sets

$$\omega_{\eta,t} = \{y : z(t) \in \eta, \eta \in \mathcal{B}(Z), t > 0\}$$

are u_i measurable, $u_l(\omega_{\eta,t})$ is the probability that the state of a macrosystem of the prepared collection belongs at time t to $\eta \subset Z$. Obviously in the deterministic case

$$u(\omega_{zi(t)t}) \neq 0$$

For any $t > 0$, $u \in K_m$, $u(\omega_{\eta t})$ is a positive, normalized measure on $\mathcal{B}(Z)$. If for any $u_1, u_2 \in K_m$ the equality

$$u_1(\omega_{\eta,t}) = u_2(\omega_{\eta,t}), \forall \eta \in \mathcal{B}(Z)$$

$t \leq \varepsilon$, ε arbitrary > 0 implies $u_1 = u_2$, the theory is Markoffian. More general cylindrical sets can be considered

$$\omega_{\eta_1 t_1, \eta_2 t_2 \ldots \eta_k t_k} = \{y : z(t_i) \in \eta_i, i = 1, 2 \ldots k, \eta_i \in \mathcal{B}(Z); t_i > 0\}$$

then $u_i(\omega_{\eta_1 t_1, \eta_2 t_2 \ldots \eta_k t_k})$ provides a full description of the time correlations. Symmetry under time translations implies that if $u \in K_m$ also $u^\tau \in K_m$, $u^\tau(\omega) = u(T_\tau^{-1}\omega)$, $\tau > 0$; the mapping $u \to u^\tau$, defines a semigroup $\mathcal{V}'(t)$ of endomorphisms of $C'(\hat{Y})$, $u^\tau = \mathcal{V}(\tau)u$, which maps K_m into K_m. The great advantage of this formulation is as follows: no assumption about the dynamics of a macrosystem enters into the mathematical structure of the theory which, however, is precise enough to solve formally the problem of measurement in quantum mechanics. The unusual concept of trajectory space can be avoided at the price of the following, perhaps wrong, assumption about the dynamics of a macrosystem: by a suitable choice of Z the dynamics of a macrosystem is Markoffian.

Classical mechanics in phase space, hydrodynamics, the Boltzmann description of a gas are examples of Markoffian theories. In this case the preparations of a collection of macrosystems are represented by a set $K_{\hat{z}}$ of positive normalized measures on $\mathcal{B}(\hat{Z})$, $\hat{Z}$ being a suitable compactification of Z, and a semigroup $\mathcal{V}'_Z(t)$ of endomorphisms of $C'(\hat{Z})$ exists which maps $K_{\hat{z}}$ into $K_{\hat{z}}$. If $u \in K_{\hat{z}}$ is a statistical preparation of a macrosystem $(\mathcal{V}'(t)u)(\eta)$, $\eta \in \mathcal{B}(\hat{Z})$ is the probability that the state z of the macrosystem belongs at time t to a set $\eta \subset \hat{Z}$. For $\sigma \in \mathcal{B}(\hat{Z})$ let us define the linear operator χ_σ on $C'(\hat{Z})$ as

$$(\chi'_\sigma u)(\eta) = u(\sigma \cap \eta)$$

then

$$(\chi_{\eta_k}\mathcal{V}(t_k - t_{k-1})\chi_{\eta_{k-1}} \ldots \chi_{\eta_1}\mathcal{V}'(t_1)u)(\hat{Z})$$

is the probability measure of the cylindrical set of $Y: \{y : z(t_i) \in \eta_i, i = 1, 2, \ldots k, \eta_i \in \mathcal{B}(z)\}$. Let us consider a preparation procedure $W^1 \in K(h^1)$ of a collection of microsystems and a preparation $u^{II} \in K_m \subset C'(\hat{Y})$ of a collection II of macrosystems and let us correlate effects of the microsystem with observations of the trajectories of the macrosystem. Then the preparation of the composite system can be described by a positive normalized measure defined on $\hat{Y}$ with values in $\tau\mathbb{C}(h^I)$:

$$u^{I,II}(\omega) = W^I u^{II}(\omega), \quad \omega \in \mathcal{B}(\hat{Y})$$

the probability that the microsystem produces the effect F and the trajectory of the macrosystem belonging to a set $\omega \subset \hat{Y}$ is given by

$$\text{Tr}\,(W^I F^I) u^{II}(\omega) = \text{Tr}\,(u^{I,II}(\omega) F^I)$$

Therefore one is led to describe in general the preparation of the system: microsystem I + macrosystem II, by a suitable set $K^{I,II}$ of positive normalized measure $u^{I,II}(\omega)$ on $\mathcal{B}(\hat{Y})$, with values in $\tau\mathbb{C}(h^I)$, where normalization means that $\text{Tr}_{h^I}(u^{I,II}(\hat{Y})) = 1$. $\text{Tr}\,(F^I u^{I,II}(\omega))$ is the probability that effect F^I occurs and the trajectories of the macrosystem belong to the set $\omega \in \mathcal{B}(\hat{Y})$. By symmetry under time translations a semigroup of affine applications $V'(t)$ of $K^{I,II}$ into $K^{I,II}$ must exist, representing a preparation consisting in preparing u and waiting a time t, i.e. the free evolution during a time t, after the preparation u.

A measurement procedure on a microsystem with statistical operator W can be described in the following way. One has a system composed of an affected part, prepared with the preparation procedure u^{II} and of a microsystem prepared with the preparation procedure W^I. After a time T chosen in such a way that the micro- and the macrosystem, have interacted, one looks at the trajectories of the macrosystem with no regard to the microsystem. The probability that the trajectories of the macrosystem belong to a set $\omega \in \mathcal{B}(\hat{Y})$ is given by:

$$p(\omega) = \text{Tr}\,((\mathcal{V}'(T)\mathcal{J}W^I \cdot u^{II})(\omega) I^I) = \text{Tr}\,((\mathcal{V}'(T)\mathcal{J}W^I \cdot u^{II})(\omega))$$

where $(W^I \cdot u^{II})(\omega) = W^I u(\omega)$, $\mathcal{J}$ being a suitable affine mapping of $K(h^I) \times K_m$ into $K^{I,II}$. Since $p(\omega)$ is an affine functional of W^I on $K(h^I)$ and $0 \le p(\omega) \le 1$, there exists a uniquely identifiable effect $F^I(\omega) \in L(h^I)$ such that $p(\omega) = \text{Tr}\,(F^I(\omega) W^I)$. The set of effects $F^I(\omega)$, $\omega \in \mathcal{B}(\hat{Y})$ is an 'effect'—valued measure on the σ-algebra of Borel subsets of $\hat{Y}$; it defines an observable of the microsystem. This notion of an observable is a straightforward generalization of the usual representation of a set of compatible observables, by a set of commuting self-adjoint operators $A_1, A_2 \ldots A_k$; In fact one has the correspondence $\{A_i, i = 1, 2 \ldots k\} \leftrightarrow P(E)$, $E \in \mathcal{B}(\mathbf{R}^k)$, $P(E)$ being the common spectral measure of $A_1, A_2 \ldots A_k$ such that

$$A_i = \int_{\mathbf{R}^k} \lambda_i \, dP(\lambda)$$

$P(E)$ is a projection valued measure on the σ-algebra of Borel subsets of $\mathbf{R}^k$. The generalization consists in replacing the projection valued measure with an effect valued one and $\mathbf{R}^k$ by $\hat{Y}$. Let the effect part have a pointer whose position is $x \in \mathbf{R}$; then one can write with obvious notations

$$z = (x, z'), \quad \omega(t_0, E) = \{y : x(t_0) \in E\} \quad E \in \mathcal{B}(\mathbf{R})$$

and consider the effects $F_{t_0}(E) = F(\omega(t_0, E))$. For fixed t_0 and $E \in \mathcal{B}(\mathbf{R})$ one has an effect valued measure on $\mathbf{R}$, if in particular the effects are idempotents, the operator

$$\int_{-\infty}^{+\infty} \lambda \, dF_{t_0}(\lambda)$$

would be an ordinary observable. Therefore we see that affected parts prepared by a preparation procedure u^{II} and left to interact with a microsystem in a fixed time interval T, identifies an observable of the microsystem, which could be explicitly calculated if u^{II}, $\mathcal{V}'(T)$ were explicitly known.

Observables corresponding to different u are in general not compatible. If one assumes that, apart from superselection rules, all elements of $L(h^I)$ are effects, one has a deep, yet unexplored, link between the structure of Hilbert spaces and superselection rules for microsystems, the structure of spaces Z for macrosystems and the interactions between micro- and macrosystems. In this treatment it has been assumed for simplicity that the microsystem is not absorbed by the macrosystem, i.e. the possible transition of system (I, II) to system II is not taken into account. Let us consider the microsystem after the interaction. The fact that after a suitable time $\bar{t}$ the interaction is negligible can be formalized by

$$\mathcal{V}'(t)\mathcal{J}W^I \cdot u = \mathcal{V}'^0(t)\varphi\mathcal{J}W^I \cdot u^{II}, \quad t > \bar{t}$$

where $\mathcal{V}'^0(t)$ is the no interaction time evolution mapping

$$(\mathcal{V}'^0(t)u)(\omega) = e^{-iH_1 t} u(T_t^{-1}\omega) e^{iH_1 t}$$

and φ is an affine mapping of $K^{I,II}$ into $K^{I,II}$, which describes the 'collision' of the microsystem with the macrosystem.

The probability of an effect F with no regard to the macrosystem after the preparation $W^I \cdot u^{II}$ and its evolution in a time t, is given by

$$\text{Tr}\,(e^{-iH_1 t}\varphi(\mathcal{J}(W^I \cdot u^{II}))(\hat{Y}) e^{iH_1 t} F^I) = \text{Tr}\,(W^I(t) F^I),$$

with

$$W^I(t) = e^{-iH_1 t}\varphi\mathcal{J}(W^I \cdot u^{II})(\hat{Y}) e^{iH_1 t}$$

Consider any covering of $\hat{Y}$ by a numerable set of disjoint Borel sets ω_j of $\hat{Y}$; correspondingly one has the following decomposition

$$t > \bar{t}, \quad W^I(t) = \sum_j p_j(t) W_j^I(t)$$

where

$$W_j^I(t) = \frac{e^{-iH_1 t}(\varphi\mathcal{J}(W^I \cdot u^{II}))(T_t^{-1}\omega_j) e^{iH_1 t}}{\text{Tr}\,(\varphi\mathcal{J}(W^I \cdot u^{II}))(T_t^{-1}\omega_j)}, \quad p_j(t) = \text{Tr}\,(\varphi\mathcal{J}(W^I \cdot u^{II}))(T_t^{-1}\omega_j);$$

each component of such a decomposition is correlated to a certain set $T_t^{-1}\omega_j$ of trajectories of the macrosystem. In such a way the measuring process is explained and an explicit link between F and the description of the affected parts has been obtained. It is also possible to link W with the preparation parts. Finally the probability of F with a preparation W, can be formally expressed as the probability that the trajectories of the composed system, preparation part + affected part, belong to a certain set of the trajectory space of such a composed system (Ludwig, 1972b). Notice that, in attitude (2) of Section 1, the question about the statistical operator of a microsystem after a measurement (by which it is not absorbed) cannot be solved within the axiomatics of quantum mechanics: one only knows that a statistical operator exists, which represents a preparation including the interaction with the apparatus. Concepts such as 'measurements of the first type of a complete set of observables' are artificial ingredients by which simple exercises for students in quantum mechanics can be given.

Let me remark that I have made statements as 'the probability that the trajectories of the macrosystem belong to a certain set $\omega \subset \hat{Y}$', skipping for simplicity the problem of how such an objective fact can be ascertained. Such a point is treated in the theory of Ludwig, who formalizes the concept of 'registration' of trajectories. A concrete registration procedure, e.g. a registration by our senses, which registers certain trajectories and discriminates other ones, is always affected by an uncertainty in the registration of the trajectories which are at the boundary between the accepted and the rejected ones. To a registration procedure of trajectories a continuous function $f(y)$ on $\hat{Y}$ corresponds such that $0 \leq f(y) \leq 1$; $f(y)$ is called a 'trajectory effect'. The set $[0, 1]$ of $C(\hat{Y})$ represents the set of all registration procedures. The probability of registration f for a macrosystem with preparation u is given by $\int_{\hat{Y}} f(y) \, du(y)$; i.e. by the value at f of the functional which represents u in $C'(\hat{Y})$. Idealized registration procedures which accept or reject trajectories without uncertainty are represented by characteristic functions $\chi_\omega(y)$ of the Borel subset of $\hat{Y}$; in such cases one has

$$\int_{\hat{Y}} \chi_\omega(y) \, du(y) = u(\omega)$$

which is the result we have used. Analogous considerations hold for the composite system: macrosystem + microsystem. Trajectory effects have an important formal role: the whole theory can be put in a mathematical form which exhibits the same linear and order structures as quantum mechanics. Such formal resemblance could be relevant for the problem of connecting the axiomatic theory of macrosystems with N-body quantum mechanics.

4. RELATION TO N-BODY THEORY

The main tool for the description of macrosystems is N-body quantum theory, which in many applications can be replaced by N-body classical statistical

104 Uncertainty Principle and Foundations of Quantum Mechanics

mechanics. The practical success of this theory is very great and difficulties can be attributed to excessive technical difficulties. The general pattern by which such success is achieved can be described as follows. Let h be the Hilbert space of the N-body structure, H its Hamiltonian, $L(h)$ the set of effects and $K(h)$ the set of statistical operators. In correspondence to a space Z, typically a space of n-tuples of functions $\varphi_j(\zeta)$, $\zeta \in \mathbf{R}^k$, $j = 1, 2 \ldots n$, one guesses for a set of fields $\hat{\varphi}_j(\zeta)$, $\zeta \in \mathbf{R}^k$ $j = 1, 2 \ldots n$ of self-adjoint operators in h and a set of statistical operators $\tilde{K}$. For $W \in \tilde{K}$,

$$\mathrm{Tr}\,(e^{iHt}\hat{\varphi}_j(\zeta)e^{-iHt}W) = \langle \varphi_j(\zeta) \rangle_t$$

is interpreted as the average value of $\varphi_j(\zeta)$ at time t for the statistical collection described by W at $t = 0$. Sometimes also expressions

$$\mathrm{Tr}\,(e^{iHt}\hat{\varphi}_j(\zeta)e^{-iHt} - \langle \varphi_j(\zeta) \rangle_t)^2 W)$$

are calculated and interpreted as dispersions. An example of this procedure is as follows: $Z =$ space of Boltzmann distribution V functions (distribution functions) for a gas

$$\hat{\varphi}(\mathbf{x}, \mathbf{p}) = \int \psi^+\left(\mathbf{x} + \frac{\xi}{2}\right)\psi\left(\mathbf{x} - \frac{\xi}{2}\right)\frac{e^{i\xi\cdot\mathbf{p}}}{(2\pi)^3}d\xi$$

$\psi(x)$ being the field operator in the second quantization formalism. In general such a procedure meets many purposes in macrophysics but does not yield the statistical distribution on the trajectory space that underlies such average values and dispersions. Using such a procedure one does not have a sufficient input for Ludwig's theory of macrosystems. The simplest way to provide such an input would be to find a measure $F(\omega)$ on $\mathscr{B}(\hat{Y})$ having values in the set $L(h)$ of effects of N-body theory, such that

$$e^{iHt}F(\omega)e^{-iHt} = F(T_t^{-1}\omega) \quad t \geq 0$$

the last requirement arising since $e^{iHt} \ldots e^{-iHt}$ is the time translation mapping on $L(h)$ and ω is a subset of a trajectory space. Then $u_W(\omega) = \mathrm{Tr}\,(F(\omega)W)$, for all $W \in \tilde{K}(h)$, would be an element of K_m. The family $F(\omega)$ is an observable by Ludwig's more general definition. It is the macroobservable of the N-body system, which corresponds to the family of idealized effects $\chi(\omega)$ in the theory of a macrosystem. One does not expect that such a strong solution of the problem exists. In fact, due to the macroscopic irreversibility of $u_W \in K_m$ it follows that u_{W^T}, where W^T is W transformed by a time inversion, should no longer be sensible. Therefore one must give a 'weak' form to the previous requirement. A possible form could be:

(α_1). On $\mathscr{B}(\hat{Y})$ a measure $F(\omega)$ with values in $L(h)$ must exist and a set $\tilde{K} \subset K(h)$ can be found such that for all $W \in \tilde{K}$

$$\mathrm{Tr}\,(e^{iHt}F(\omega)e^{-iHt}W) = \mathrm{Tr}\,(F(T_t^{-1}\omega)W)$$

i.e.

$$u_{e^{-iHt}We^{iHt}}(\omega) = u_W(T_t^{-1}\omega), \quad t \geq 0$$

The requirement that $F(\omega)$ is a measure on $\mathcal{B}(\hat{Y})$ is mathematically very restrictive; physically it means that the apparatuses A_ω which, by a measurement on the N-body structure, register the sets ω of trajectories, can all coexist; since the apparatuses which measure a structure of $\sim 10^{23}$ particles are very hypothetical objects, the physical meaning of the previous statement is questionable. Therefore one is led to give a 'weak' formulation also for the measure character of $F(\omega)$ and in place of (α_1) one requires:

(α_2). On $\mathcal{B}(\hat{Y})$ a function $F(\omega)$ with values in $L(h)$ must exist and a set $\tilde{K} \subset K(h)$ can be found such that

$$\forall W \in \tilde{K}, \quad u_W(\omega) = \mathrm{Tr}\,(F(\omega)W)$$

is a measure on $\mathcal{B}(\hat{Y})$,

$$u_{e^{-iHt}We^{iHt}}(\omega) = u_W(T_t^{-1}\omega), \quad t \geq 0$$

Since in our attitude N-body theory is only a provisional tool to find the right theory a less strong requirement is meaningful, such as the following one:

(α_3). On $\mathcal{B}(\hat{Y})$ a family of functions $F^\delta(\omega)$, $\delta \geq 0$ with values in $L(h)$ exists, and a family $\tilde{K}^\delta \subset K(h)$ such that for all

$$W^\delta \in \tilde{K}^\delta, \lim_{\delta \to 0} \mathrm{Tr}\,(F^\delta(\omega)W^\delta)$$

exists for all $\omega \in \mathcal{B}(\hat{Y})$ and defines a measure $u_{\{W^\delta\}}(\omega)$ on $\mathcal{B}(\hat{Y})$ (no existence of $\lim_{\delta \to 0} F^\delta(\omega)$ or of $\lim_{\delta \to 0} W^\delta$ is required) further

$$u_{\{e^{-iHt}W^\delta e^{iHt}\}}(\omega) = u_{\{W^\delta\}}(T_t^{-1}\omega), \quad t \geq 0$$

Analogous considerations can be made if one assumes that the macrodynamics are Markoffian; then essentially one has $\hat{Z}$ in place of $\hat{Y}$ and this is a simplification, but the second part of α becomes the following: A semigroup $V'(t)$ exists on K_m such that

$$U'(t)u_W = u_{e^{-iHt}We^{iHt}}, \quad t \geq 0, \quad \eta \in \mathcal{B}(\hat{Z})$$

and this is a complication. If assumption (α_1) is further restricted by the requirement that $F(\omega)$ is idempotent it leads to the well-known problem of the macroobservables as self-adjoint commuting operators on h. Such a problem has no solution of appreciable generality and indeed it seems to be too naive a formulation of the problem of macroscopicity. Anyway since there is not a logically compelling reason to assume that dynamics in state space is Markoffian the usual master equation approach seems to be not completely appropriate for solving the problem.

The main step to obtain $F(\omega)$, $\omega \in \mathcal{B}(\hat{Y})$ is to build $F(\omega)$ for ω being cylindrical sets $(\eta_1 t_1, \eta_2 t_2, \ldots \eta_k t_k) \subset Y$, $t_i \geq 0$, $\eta_i \in \mathcal{B}(Z)$ which is already sufficient for most physical applications. It is also the sufficient input for the rather technical problem to prove the existence of the function $F(\omega)$, $\omega \in \mathcal{B}(\hat{Y})$, which assumes assigned values on the cylindrical sets. A final assumption on $F(\omega)$ and on $\tilde{K}$ which possibly has some implication for the allowed interactions between a microsystem and a macrosystem is as follows:

(β). For each microsystem S with Hilbert space h^S the following family of elements of $K(h^S)$

$$u_{W^S,W,t}(\omega) = \text{Tr}_h(I^S \otimes F(\omega)e^{-iHt}W^S \otimes We^{iHt}), \quad t \geq 0$$

for all

$$W^S \in K(h^S),$$

and for all

$$W \in \tilde{K},$$

is still a measure on $\mathscr{B}(\hat{Y})$; H is the Hamiltonian of the system: N-body structure + microsystem S; more generally according to (α_3) one could substitute the right-hand side of this equation by

$$(\beta_3) \qquad \lim_{\delta \to 0} \text{Tr}_h(I^S \otimes F^\delta(\omega)e^{-iHt}W^S \otimes W^\delta e^{iHt})$$

If $F(\omega)$ and $\tilde{K}$ were known, $u_{W^S,W,t}(\omega)$ would be the required input for Ludwig's formal scheme and the problems of measurement would be solved. In conclusion the difficulty with measurement in quantum mechanics has been shifted from formula (2) to the problem of building $F(\omega)$ and identifying $\tilde{K}$ such that (α_3) and (β_3) hold. The limit $\delta \to 0$ in (α_3), (β_3) should introduce macroscopicity as a limit situation of N-body theory. One can hope that this is only a technical difficulty.

Acknowledgement

The treatment in this section is based on a paper I am preparing with Dr. G. C. Lupieri. I wish to thank Dr. Lupieri for useful discussions about this subject.

REFERENCES

Barchielli, A. and Lanz, L. (1975) 'Non Hamiltonian description of two particle systems,' I.F.U.M., **181**, F. J. Milano.
Bell, J. S. (1966) *Rev. Mod. Phys.*, **38**, 447.
Bell, J. S. (1971) *Foundations of Quantum Mechanics, Proceedings of the IL Enrico Fermi International Summer School*, B. d'Espagnat, Ed., Academic Press, New York.
Bohm, D. (1951) *Quantum Theory*, Prentice Hall, New Jersey.
Bohm, D. and Aharonov, Y. (1957) *Phys. Rev.*, **108**, 1070.
Caldirola, P. (1974) *Dalla Microfisica alla Macrofisica*, Mondadori; Milano.
Comi, M., Lanz, L., Lugiato, L. A. and Ramella, G. (1975) *J. Math. Phys.* **16**, 910 (1975).
Daneri, A., Loinger, A. and Prosperi, G. M. (1962) *Nuclear Phys.*, **33**, 297.
Einstein, A., Podolsky, B. and Rosen, N. (1935) *Phys. Rev.*, **47**, 777.
d'Espagnat, B. (1971a) *Conceptual Foundations of Quantum Mechanics*, Benjamin, New York.
d'Espagnat, B. (1971b) *Foundations of Quantum Mechanics, Proceedings of the IL Enrico Fermi International Summer School*, B. d'Espagnat, Ed., Academic Press, New York.
Everett III, H. (1957) *Rev. Mod. Phys.*, **29**, 454.
Freedman, S. J. and Clauser, J. F. (1972) *Phys. Rev. Letters*, **28**, 938.
George, G., Prigogine, I. and Rosenfeld, L. (1972) *Dansk. Mat. Fys. Medd.*, **38**.
Hepp, K. (1972) *Helv. Phys. Acta*, **45**, 234.

Jauch, J. M. (1968) *Foundations of Quantum Physics*, Addison Wesley, Reading, Mass.
Jauch, J. M. (1971) *Foundations of Quantum Mechanics, Proceedings of the IL Enrico Fermi International Summer School*, B. d'Espagnat, Ed., Academic Press, New York.
Jauch, J. M. and Piron, C. (1963) *Helv. Phys. Acta*, **36**, 827.
Jauch, J. M. and Piron, C. (1969) *Helv. Phys. Acta.*, **42**, 842.
Kasday, L. (1971) *Foundations of Quantum Mechanics, Proceedings of the IL Enrico Fermi International Summer School*, B. d'Espagnat, Ed. Academic Press, New York.
Lang, L., Prosperi, G. M. and Sabbadini, A. (1971) *Nuovo Cimento*, **2 B**, 184.
Ludwig, G. (1953), *Z. Phys.*, **135**, 483.
Ludwig, G. (1970) *Lecture Notes in Physics*, **4**, Springer, Berlin.
Ludwig, G. (1973a) *Lecture Notes in Physics* **29**, Springer, Berlin and 'Makroskopische Systeme und Quantenmechanik', *Notes Math. Phys. Marburg* (1972).
Ludwig, G. (1973b) *Lecture Notes in Physics* **29**, Springer, Berlin and 'Mess-und Präparierprozesse', *Notes Math. Phys. Marburg* (1972).
von Neumann, J. (1955) *Mathematical Foundations of Quantum Mechanics*, Princeton University Press, Princeton.
Piron, C. (1964) *Helv. Phys. Acta.* **37**, 439.
Prosperi, G. M. (1971) *Foundations of Quantum Mechanics, Proceedings of the IL Enrico Fermi International Summer School*, B. d'Espagnat, Ed., Academic Press, New York.
Selleri, F. (1971) *Foundations of Quantum Mechanics Proceedings of the IL Enrico International Summer School*, B. d'Espagnat, Ed., Academic Press, New York.
Wigner, E. (1971) *Foundations of Quantum Mechanics, Proceedings of the IL Enrico Fermi International Summer School*, B. d'Espagnat, Ed., Academic Press, New York.
Zeh, H. D. (1971) *Foundations of Quantum Mechanics, Proceedings of the IL Enrico Fermi International Summer School*, B. d'Espagnat, Ed., Academic Press, New York.

The Correspondence Principle and Measurability of Physical Quantities in Quantum Mechanics

YURI A. RYLOV
Institute of Space Research U.S.S.R. Academy of Sciences, Moscow

I. Introduction

Fifty years have passed since the discovery of the uncertainty principle by Heisenberg. Quantum theory has scored big successes in many regions of physics especially in atomic and nuclear physics and solid-body theory. All physicists are agreed upon the formalism of quantum mechanics. But there is no agreement on questions of the interpretation of quantum mechanics and measurement theory. Most physicists merely ignore these problems, correctly believing that they are negligible in calculations of different quantum systems. There are many shades of interpretation of quantum mechanics and problems of measurement even in orthodox quantum theory. Some different viewpoints are found in numerous hidden-variable theories. Bibliographies on problems of quantum mechanics measurement and interpretations can be found in surveys by Margenau (1963), Pearle (1967), Ballentine (1970) and Reece (1973).

In this paper I am going to consider only the question about constraints imposed upon the measureability of physical quantities by the time–energy uncertainty relation. I shall not consider other questions connected with the interpretation of quantum mechnaics and measurement theory and shall only make a few remarks about them. I shall confine myself to the statistical interpretation of quantum mechanics.* According to this interpretation a wave function provides a description of the statistical properties of an ensemble of similarly prepared systems. Ballentine (1970) ascribed this interpretation to Einstein (1949), Popper (1959) and Blokhintsev (1968). In my opinion, the same interpretation was upheld by Mandelstam (1950) in his brilliant lectures on the theory of indirect measurements, which were given in Moscow State University in 1939 and were printed only in 1950.

The statistical interpretation differs from the Copenhagen one (Heisenberg, 1955), which asserts that the wave function provides a complete and exhaustive

---
*This terminology is used by Ballentine (1970).

description of an *individual* system. Ballentine (1970) has shown that the stronger constraint, which is used in the Copenhagen interpretation, is of no importance in applications of quantum mechanics, but in some cases leads to paradoxes. All sensible quantum mechanics statements, being statistical ones, always concern no single dynamical system but an ensemble of systems. For instance, the statement that a measurement of spin component σ_x in the state with $\sigma_x = 1/2$ gives the result value $-1/2$ with probability 0.5 is pointless, if it concerns only one single system. Really, this statement means that, measuring the component σ_x many times in the similarly prepared state, one gets the value $\sigma_x = -1/2$ in half of the cases. But many measurements cannot be performed on one system, because its state is perturbed after the measurement. For this reason it is necessary to have an ensemble of systems and to perform measurements upon many of them. There is another possibility: preparing the same initial state of the same individual system, to measure repeatedly. But this is an ensemble also.

The proposition, that there are sensible statistical statements about indvidual systems in quantum mechanics, is an illusion. To make a statistical statement sensible, it is necessary to deal with an ensemble of systems. Hence, there is no reason to insist that a wave function describes a state of an element of an ensemble (a single system), but not the ensemble as a whole. However, if the Copenhagen terminology is understood not word for word but as a peculiar physical slang, used for brevity, then there is no objection to it.

Following von Neumann (1932), let us give now the main statements of quantum mechanics.

(A1). Any state of an ensemble consisting of many identical systems is represented by a definite self-adjoint operator $\hat{U}$ (called a statistical operator or state operator), which is defined in a Hilbert space $\mathcal{H}$. The operator $\hat{U}$ has a real non-negative eigenvalues and obeys the conditions

$$Sp\hat{U} = 1 \qquad (1)$$

(A2) An observable physical quantity R is represented by a self-adjoint operator $\hat{R}$ in the Hilbert space $\mathcal{H}$. A physical quantity $F(R)$ is represented by the operator $F(\hat{R})$. In particular, the canonical variables q^i, p_i ($i = 1, 2, \ldots n$) of a classical system are corresponded by operators $\hat{q}^i, \hat{p}_i$ ($i = 1, 2, \ldots n$) in the space $\mathcal{H}$, which obey the commutation relations

$$\hat{q}^l \hat{p}_k - \hat{p}_k \hat{q}^l = i\hbar \hat{I} \delta_k^l \qquad (2)$$

where $\hat{I}$ is the unit operator on $\mathcal{H}$ and $\hbar$ is Planck's constant. A function $F(q, p)$ of canonical variables q, p is represented by the self-adjoint operator $F(\hat{q}, \hat{p})$. The order of the operators is chosen in some way which is not fixed.

(A3). The mathematical expectation $\langle R \rangle$ of an observable R in the state $\hat{U}$ is described by the relation

$$\langle \hat{R} \rangle_{\hat{U}} = Sp\{\hat{U}\hat{R}\} \qquad (3)$$

(A4). The state $\hat{U}$ of the physical system ensemble evolves in such a way that after time t it turns into the state $\hat{U}_t$

$$\hat{U} \to \hat{U}_t = e^{-i\hat{H}t/\hbar} \hat{U} e^{i\hat{H}t/\hbar} \tag{4}$$

where $\hat{H}$ is a self-adjoint operator on $\mathcal{H}$. The $\hat{H}$, called the Hamiltonian, is a function $H(\hat{q}, \hat{p})$ of $\hat{q}$ and $\hat{p}$. For physical systems, which have a classical description, the form of this function coincides with that of the Hamiltonian function of coordinates and momenta.

(A5). Measuring a quantity R on the physical system ensemble which is in the state $\hat{U}$, turns the state of R after measurement into a state $\hat{U}'$, which is defined by

$$\hat{U} \to \hat{U}' = \sum_n (\hat{U}\chi_n, \chi_n) \hat{P}_{[\chi_n]} \tag{5}$$

where (ψ, χ) denotes the scalar product of vectors ψ and χ on $\mathcal{H}$. $\chi_1, \chi_2, \ldots, \chi_n \ldots$ is a complete set of orthonormal eigenvectors of the operator $R \cdot \hat{P}_{[\chi_n]}$ is a projection operator on vector χ_n. The projection operator on a unit vector φ is defined by the relation

$$\hat{P}_{[\varphi]} f = (\varphi, f)\varphi \tag{6}$$

where f is an arbitrary vector on $\mathcal{H}$.

The superselection rules (Wick and co-workers, 1952) must be added to these propositions. But I shall not do this, because I shall not use these rules, and the above propositions do not pretend to be an axiomatization of quantum theory.

In certain cases the physical system ensemble state can be determined by pointing a unit vector ψ on $\mathcal{H}$. Such states are called pure ones. Their statistical operator can be represented in the form

$$\hat{U} = \hat{P}_{[\psi]} \tag{7}$$

where $\hat{P}_{[\psi]}$ is the projection operator on the unit vector ψ. In this case due to (6) the statement (3) takes the form

$$\langle R \rangle_\psi = Sp\{\hat{R}\hat{P}_{[\psi]}\} = \sum_n (\chi_n, \hat{R}\hat{P}_{[\psi]}\chi_n)$$
$$= \sum_n (\psi, \chi_n)(\chi_n, \hat{R}\psi) = (\psi, \hat{R}\psi) \tag{8}$$

The relation (4) can be written as

$$\psi \to \psi_t = e^{-iHt/\hbar} \psi \tag{9}$$

The set of vectors $\psi = \chi_1, \chi_2, \ldots \chi_n \ldots$ represents an orthonormal basis in Hilbert space $\mathcal{H}$.

The vector ψ, describing a pure state of the physical system ensemble is called the vector state or wave function. All statistical statements of quantum mechanics can be derived from the propositions (A1)–(A5). The detailed analysis of these statements can be found in the monograph by von Neumann

112 Uncertainty Principle and Foundations of Quantum Mechanics

(1932). The formal measurement theory can be derived from the above statements. This was shown by von Neumann (1932).

The relation (5), describing the violation of the ensemble state under the action of measuring a quantity R, represents the special process, which differs from the evolution of the ensemble, described by relation (4).

The quantum mechanical process of measurement has two aspects. The informative side of the measurement of a quantity R represents a registration of some definite statistical distribution of values R in the given ensemble state. The perturbing side of the measurement describes the violation of the ensemble state after measurement. The last property can be used for the preparation of the system ensemble in a definite state. The two sides of measurement are independent to such an extent, that Margenau (1963) insisted on distinguishing between the measurement process (registration) and that of state-preparation.

Really, the two sides of measurement are independent to such an extent, that they can be realized in principle by means of two different devices. One of them only prepares the state but does not record it, and the other only records but does not disturb it. Let us consider such an idealized measurement process. Let an ensemble E_ψ be described by the wave function ψ and consist of N similar single systems ($N \to \infty$). Let the preparation part M of the device, measuring the quantity R act on the systems of the ensemble E_ψ. One can imagine the preparing part M as a black box with one inlet and many outlet slits $S_1, S_2, \ldots$. Let all outlet slits be closed except for the slit S_1. The N_1 systems ($1 \ll N_1 \ll N$) of the ensemble E_ψ find themselves in turn in the box M. A proportion of them is absorbed by the box and a proportion passes out through the open slit S_1.*
Let n_1 systems pass through the slit S_1 ($1 \ll n_1 \ll N_1$). The box M transforms a part of the ensemble in E_ψ into another ensemble E_1, consisting of n_1 systems. In other words the box M prepares the ensemble E_1 in a certain state, which is not known exactly. In order to visualise the recording part of the measurement let us turn l_1 systems of the ensemble E_1 ($1 \ll l_1, k_1 = n_1 - l_1 \gg 1$) into a recording macroscopic device $\mathscr{P}$ with a 'pointer'. Let us assume that the influence of the sytem upon the device $\mathscr{P}$ deflects the pointer. The magnitude of this deflection shows the measured value of the quantity R. Let us assume that the box M has such an arrangement that an analysis of l_1 systems of the ensemble E_1 by means of the measuring instrument $\mathscr{P}$ gives the result R_1 for all these systems. We shall not be interested in what happens to the systems in the instrument $\mathscr{P}$ and shall not consider them. Because the measurement has given a value R_1 of the quantity R for all l_1 systems ($l_1 \gg 1$), we have the right to conclude that the quantity R has the value R_1 for the rest of the ensemble E_1

---

*For concreteness one can imagine any physical system as a moving electron. The measuring device measures the electron momentum. The black box M represents a region with a magnetic field, which is orthogonal to the direction of the electron motion. Depending on the magnitude of the electron momentum the electron is deflected through a certain angle and hits the screen with the slits $S_1, S_2, \ldots$. If only slit S_1 is open, then such a device prepares an ensemble of electrons with momentum p_1.

(k_1 systems, $k_1 = n_1 - l_1 \gg 1$)†. This conclusion can be drawn without analysing these k_1 systems by measurement with instrument $\mathscr{P}$.

Thus by subjecting N_1 systems of ensemble E_ψ to dynamical interaction in the box M and analysing part of them in the instrument $\mathscr{P}$, we get the ensemble E_1, consisting of k_1 systems. The quantity R has the value R_1 in all systems of ensemble E_1. If the ensemble E_ψ is a pure one then the ensemble E_1 must be a pure one and be described by a certain wave function ψ_1, because the ensemble systems were subjected *only to dynamical interaction* with the box M.

But the measurement process on the ensemble E_ψ is not finished, because only the one possibility that some systems have value $R = R_1$ has been investigated. For the investigation of the possibility that among the systems of the E_ψ there are ones, which have the value R_2 of the quantity R, it is necessary to open only the slit S_2 and to allow to pass through M N_2 systems ($N_2 \gg 1$, $N_2 \ll N$) of the ensemble E_ψ. Passing through the slit S_2 n_2 systems ($n_2 \gg 1$) of the E_ψ, one forms an ensemble E_2. Analysing l_2 systems ($l_2 \gg 1$, $k_2 = n_2 - l_2 \gg 1$) of the E_2 and finding a value R_2 of the quantity R for all of them one concludes that the rest of the systems of E_2 have the value $R = R_2$ and are described by a wave function ψ_2. An analogous process has to be produced for all slits $S_1, S_2, \ldots$. Let us note that the distribution of the ensemble E_ψ systems according to values of R is produced independently of reading the pointer $\mathscr{P}$, i.e. in the absence of any information about the state of the ensemble E_ψ. Of course, the box M is supposed to be arranged in such a way that any slit S_m corresponds to a definite value $R = R_m$.

Let us suppose that all N_m ($m = 1, 2, \ldots$) are equal and that the box M is arranged in such a way, that by all open slits $S_1, S_2, \ldots$ a system of the ensemble E_ψ, finding itself in the M, is to go out through one of the slits. Supposing that $l_m = \alpha n_m$, where $0 < \alpha < 1$ and the α is the same for all $m = 1, 2, \ldots$, one concludes that the number k_m ($m = 1, 2, \ldots$) of systems in the ensemble E_m is proportional to the probability of measuring the value R as equal to R_m among the systems of the E_ψ.

Let us unite all ensembles $E_1, E_2, \ldots$ into one ensemble E'. The E' is a mixture of pure ensembles $E_1, E_2, \ldots$ and cannot be described by means of a wave function. It is conditioned in that the E' is obtained as a result of a *set* of different dynamical actions but not one dynamical action. Really, some systems are subjected to the action of the box M with the only open slit S_1, other ones with slit S_2 only open and so on. It is different dynamical actions, which lead to different results.

I shall call this action the statistical action keeping in mind that the statistical character of the action manifests itself in a different dynamical action of the measuring device upon systems with different values R.

Of course the objection can be made to the above that it is not necessary to open only one slit in the box M. One can open all slits or even not use the box M. One can merely take a small part of the ensemble E_ψ and investigate the

---
†The concept of ensemble was introduced into quantum mechanics in order to know the state without perturbing it (von Neumann, 1932, Chap. 4, Section 1).

distribution of the values R by means of the measurement instrument $\mathcal{P}$. The same statistical distribution of the values R will be found in the rest of the E_ψ. Thus, one can find the distribution of values in the E_ψ without disturbing it. This is a valid objection. But it takes into account only the informational side of measurement, neglecting the state-preparing side. Essentially it is equivalent to the perfectly correct statement, that our knowledge about the ensemble state does not have an influence upon the state of the ensemble. However the measurement is not reduced merely to a change of information about the state of the ensemble. The measurement influences the ensemble state. All physicists are agreed upon this question. There is discordance of opinion only upon the question of how it influences the state of the ensemble. Influence of measurement upon the state of a system being measured is the main difference between the quantum theory of measurement and that in classical theory.

Let an electron ensemble state be described by a wave function ψ. Then $|\psi(\mathbf{q})|^2 \, dV$ represents a probability of finding the electron in the volume dV. To measure the electron position means, that it is necessary not only to measure a distribution $|\psi(\mathbf{q})|^2$ for the ensemble, but to determine the action of this measurement upon the ensemble description. To measure the electron position and to find it in the volume dV means selecting all the electrons of the ensemble, which have been found in the dV, to constitute a new ensemble of them and to solve the problem of the description of this new ensemble. Just such a problem arises in quantum measurement theory. For the elucidation of the nature of this problem I have divided the united measurement process into two parts: an informational part and a state-preparation part. Such a division is an idealization, which is possible only if the state preparation and the recording process happen instantaneously, and if the change of the ensemble state due to the process (4) can be neglected.

Usually the measurement device cannot be divided into parts: the informational one and state-preparation one. Besides nobody measures in the way that has been described, i.e. firstly the systems having the value $R = R_1$ are selected, the rest of them being given up, secondly the systems having the value $R = R_2$ are selected, the rest of them being given up and so on. Such a selection, produced blindly, is ineffective. In practice the measurement is performed in the following way. One finds the value of the quantity R for a single system. Depending on the value obtained for R, the system is attributed to one of the ensembles $E_1, E_2, \ldots$. For this reason some physicists believe that the appearance of a mixed state of the ensemble is connected with a change of information to an observer. Other authors connect its appearance with the fact that the measuring device is a macroscopic one. Some physicists reject the reduction of the pure ensemble state to the mixed one, stating (quite correctly), that dynamical action cannot reduce the pure state into a mixture (Wigner, 1963). If in addition to considering that the wave function describes a state of a single system and to understand this word for word, then the measurement action upon the system to be measured assumes in general a mystical character.

Thus, the quantum measurement process is a *set of single measurements*. This term will be used later on just in this sense. To measure the quantity R and to obtain a value R' means performing a set of single measurements of R, the selection of those systems for which the measurement has yielded the result R' and the constitution of a new ensemble of the selected systems. The measurement is in the first place a *statistical action* upon the ensemble systems, which can be accompanied by a dynamical one. The statistical action of measurement is conditioned by the statistical character of its description in quantum mechanics.* The system selection is an attribute of a measurement. The means and manner of how this selection is produced is of no importance. In any case this selection is not a result of a change of observer information about the ensemble state, because, as we have seen, this selection can be produced blindly without any information about the ensemble state.

The relation (5) describes a result of the measurement action upon the ensemble in the state **U**. The measurement is supposed to be performed instantaneously and the state evolution, described by (4). can be neglected. One can have doubts, that the measurement action is described by a projection operator $\hat{P}_{[\chi_n]}$ upon eigenvectors $\{\chi_n\}$ of the measured quantity operator, and propose another way. I believe that this is not a very essential detail. I have chosen the measurement action in the form (5) for the reason that its properties have been investigated in detail by von Neumann (1932).

Unfortunately, the measurement problem is not exhausted by this consideration. Quantum mechanics always attributes a result of measurement to a state $\hat{\mathbf{U}}$ of the ensemble to be measured. For such an attribution to be possible, the measurement would have to be performed sufficiently quickly, in principle, instantaneously. This means the following. Let the measurement of the quantity R continue during the time T and a set $\{R_T\}$ of results be obtained. This set $\{R_T\}$ depends in general on the duration T of a single measurement. If a limit of distribution $\{R_T\}$ with $T \to 0$ exists, then by definition, the measurement of the quantity R can be performed instantaneously. For an instantaneous measurement the result can be attributed to the state $\hat{\mathbf{U}}$, in which the ensemble has been found directly before measurement, even if the ensemble state has changed during the measurement process.

However, it is possible that some quantity R cannot be measured instantaneously, i.e. no limit of distribution $\{R_T\}$ of measurement results exists for $T \to 0$. For instance the energy and momentum of a particle are such properties. In

---

*The action of measurement upon the ensemble state takes place in the theory of Brownian motion, where the dynamical action of measurement can be neglected certainly. For instance, let an ensemble E_W of Brownian particles be described by a function $W(\mathbf{q}, t)$ satisfying the Einstein–Fokker equation. To measure the position of the Brownian particle at the same time t_0 and to find it in a volume V means selecting from the ensemble E_W only those particles which have been found in the V at the time $t = t_0$, and to constitute a new ensemble E_{W_0}, described by the function $W_0(\mathbf{q}, t)$, which does not vanish only within V at $t = t_0$. Later at $t > t_0$ the ensemble E_{W_0} will evolve in a different way from the ensemble E_W. In other words, the measurement at the time t_0 changes the probability of detecting the Brownian particle at the point $\mathbf{q}$ at the time $t > t_0$ although no dynamical action has been made upon the particle, and only selection (i.e. statistical action) has been effective.

accord with the uncertainty principle the smaller the measurement time T is the greater is the inaccuracy of the measurement of energy and momentum (Bohr, 1928; Heisenberg, 1930; Landau and Peierls, 1931; Mandelstam and Tamm, 1945; Fock and Krylov, 1947; Aharonov and Bohm, 1961; Fock, 1962). For a non-vanishing time T the measurement result cannot be attributed to the ensemble state directly before the measurement. Really, if this were possible, then it would be unclear why the measurement results have no limit for the non-vanishing measurement time $T \to 0$. The measurement results can be attributed to the ensemble state $\hat{U}$ during the measurement process only if the $\hat{U}$ is unchanged (or is changed very slightly) during the measurement time. Thus, if the measurement requires a non-vanishing time, then its result can be attributed to the ensemble state only for those states for which the change according to the relation (4) is negligible during the measurement time.

Using the relations (4) and (5), let us produce a formal consideration of the measurement process, continuing the time T during which the measurement instrument is switched on. Let the measurement of the quantity R be performed within the period $[0, T]$ and the ensemble state be described at the time instant $t = 0$ by a statistical operator $\hat{U}_0$. Let the operator $\hat{U}$ evolve within the period of time $[0, t]$ according to (4), turning into $\hat{U}_t$ at the instant t. Let the measurement process (5) be performed at the moment t the $\hat{U}_t$ turning into $\mathbf{U}'_t$. Let the $\hat{U}'_t$ within the period $[t, T]$ evolve according to (4), turning into $\hat{U}'_t$ at the instant T. A simple calculation shows that during the time T the statistical operator $\hat{U}_0$ turns into $\hat{U}'_T$

$$\hat{U}_0 \to \hat{U}'_T = \sum_n (\hat{U}_0 e^{i\hat{H}t/\hbar} \chi_n, e^{i\hat{H}t/\hbar} \chi_n) \hat{P}e_{[e^{-i\hat{H}(T-t)/\hbar}\chi_n]} \tag{10}$$

where $\chi_1, \chi_2, \ldots$ is a complete orthonormal set of the operator $\hat{R}$ eigenvectors. The instant t, at which the measurement process is performed, is supposed to be indefinite but within the period $[0, T]$.

For the described process to be a real measurement of the quantity R it is necessary for the state $\hat{U}'_T$ depends only on the initial state $\hat{U}_0$ and the operator $\hat{R}$. In particular the $\hat{U}'_T$ has not to depend on the instant t, at which the measurement (5) has been performed. It follows from (10), that the last condition is fulfilled, if the vectors $\{\chi_n\}$ are eigenfunctions of the Hamiltonian $\hat{H}$ and, hence

$$[\hat{R}, \hat{H}]_- \equiv \hat{R}\hat{H} - \hat{H}\hat{R} = 0 \tag{11}$$

This result can be found in von Neumann's book (1932, Chap. 5, Section 1). It means, that the action of the measurement device upon the measured system during the measurement process is to be of such a kind, that the Hamiltonian $\hat{H}$ would begin to commute with the operator of the measured quantity.

Suppose the relation is fulfilled. The question arises of to which state the measured values should be attributed. The instant of measurement is indefinite, and the ensemble state $\hat{U}$ changes within the period $[0, t]$ according to (4). For the measured values of the R to be attributed to a definite state $\hat{U}$, it must be stationary within the period $[0, t]$.* This leads to a condition

$$[\hat{U}_0, \hat{H}]_- = 0 \tag{12}$$

which is a great restriction upon $\hat{U}_0$. In particular the measurement of the distribution over the momenta of a free particle is possible (in the one-dimensional case) only if the momentum value is quite definite.

The above formal consideration is not consistent because on the one hand it uses the instantaneous measurement process, but on the other hand the measurement is supposed to continue for a non-vanishing period of time. Nevertheless it indicates that a long duration of measurement of some quantities should give rise to obstacles to their measurability.

The necessity of a long measurement of physical quantities of the energy–momentum pattern is conditioned in the end by the time–energy uncertainty relation. Unlike the position–momentum uncertainty relation (Heisenberg, 1927; Robertson, 1929) it cannot be derived from the statements (1)–(5) of quantum mechanics. Really, the formalism of quantum mechanics contains the time t as a parameter, which commutes with the Hamiltonian $\hat{H}$ and does not conjugate to $\hat{H}$ in the sense (2), as $\hat{q}$ and $\hat{p}$ do. Thus, the time–energy uncertainty relation is an additional statement, which should be taken into account in the formalism of quantum mechanics. It does not permit the instantaneous measurement of the energy–momentum pattern quantities and leads to the restrictions (11) and (12). All this is in contradiction to the basic statements (1)–(5) of quantum mechanics, which supposes the instantaneous measureability of physical quantities in any state, and apparently is connected with the non-relativistic character of quantum mechanics.

The subsequent analysis shows, that in reality relation (12) cannot be fulfilled for quantities of the energy–momentum pattern. In other words, the measured values of energy and momentum can never be attributed to any definite ensemble state. In this sense energy and momentum are not measureable, and this is the corollary of the time–energy uncertainty relation.

Later on I shall use the coordinate representation of vector state. In this representation every vector ψ in the Hilbert space $\mathcal{H}$ is represented by a square-integrable function ψ of the coordinate $\mathbf{q}$. The scalar product (φ, ψ) of two vectors φ and ψ is represented by

$$(\varphi, \psi) = \int \varphi^*(\mathbf{q}) \psi(\mathbf{q}) \, d\mathbf{q} \tag{13}$$

* denotes a complex conjugate. Integration is produced over all coordinates $\mathbf{q}$. The position operator $\hat{\mathbf{q}}$ and the momentum operator $\hat{\mathbf{p}}$ are defined respectively as the operator of multiplication by $\mathbf{q}$ and as the differentiation operator

$$\hat{p}_\beta = -i\hbar \frac{\partial}{\partial q^\beta} \tag{14}$$

*Within the period $[t, T]$ the state $\hat{U}'_t$ is stationary due to (11).

2. The Possibility of an Experimental Test of the Statistical Statements of Quantum Mechanics

In the present section I shall investigate the possibility of an experimental test of the statistical statements, represented by (A2) and (A3). Because this problem is very complicated, I shall confine myself to the investigation of a measurability of the simplest physical quantities such as coordinate and momentum. The complication consists of the impossibility of producing a formal analysis. For instance in measuring a momentum on the one hand it is stated that the momentum operator is represented by (14) and this operator has to be used in (8) for the calculation of the corresponding mean values, on the other hand it is necessary to describe some measurement process which is by definition the momentum measurement. If the measurement results disagree with the statements (8) of quantum theory, then its proponent can always say that this measurement process is not a momentum measurement, and for momentum measurement another measurement process should be used.

To reduce the number of possible measurement processes I shall require that any mesurement satisfies the correspondence principle. This means, that a measurement process and its result must not depend on the model of the phenomena used in the measurement (classical, quantum or some other kind). In particular, the measurement process applied to a system which permits a classical description must give results, which agree with classical mechanics.

The term 'correspondence principle' was introduced by Bohr (1918) to establish a connection between the old (before 1925) quantum mechanics and the classical theory of radiation. Originally it meant that the radiation frequency emitted by transition from one quantum orbit to another approaches asymptotically one of the frequencies which are obtained from a Fourier series of functions describing the motion of the electron.

In the contemporary version of the quantum mechanics the correspondence principle describes some correspondence between the formalism of quantum mechanics and that of classical mechanics. For instance, the operator $\hat{p} = -i\hbar\partial/\partial\mathbf{q}$ which is unlike the classical momentum, is interpreted as momentum and in many cases (for instance in approximate estimations) is substituted by the classical momentum. This is the corollary of the correspondence principle.

The following consideration is a base for such a correspondence. Let a particle ensemble be described by the wave function

$$\psi = \psi(q) = \sqrt{\rho(q)} \exp\left\{\frac{iS(q)}{\hbar}\right\} \tag{15}$$

where ρ and S are real functions of q. For simplicity only the one-dimensional case is considered. If ρ and $p \equiv \partial S/\partial q$ change slightly within the distance of a wavelength $\lambda = \hbar/p$, i.e.

$$\left|\frac{1}{\rho}\frac{\partial \rho}{\partial q}\right| \ll \frac{1}{\hbar}\left|\frac{\partial S}{\partial q}\right|, \quad \left|\frac{1}{p}\frac{\partial p}{\partial q}\right| \ll \frac{1}{\hbar}\left|\frac{\partial S}{\partial q}\right| \tag{16}$$

then such an ensemble can be described classically with $S(q)$ as an action. The density state of such an ensemble in a phase space (q, p) is described by the distribution function

$$W(q, p) = \rho(q)\delta\left(p - \frac{\partial S}{\partial q}\right) \tag{17}$$

Indeed, one obtains from (8) and (15)–(17) for

$$\langle F(q, p)\rangle_\psi = \int \psi^*(q) F\left(q, -i\hbar \frac{\partial}{\partial q}\right) \psi(q) \, dq \cong$$

$$= \int \rho(q) F\left(q, \frac{\partial S}{\partial q}\right) dq = \int F(q, p) W(q, p) \, dq \, dp \tag{18}$$

For this reason one concludes, that the operator (14) corresponds to momentum. The momentum is interpreted in the sense of classical mechanics. The function $W(q, p)$ describes an ensemble of classical systems. The state of every system is described as a 'point' in the phase space. Every 'point' has a volume, which is greater than the characteristic volume $\hbar$ of the phase space. The essential dependence of distribution (17) on the only coordinate q is conditioned by the fact, that the ensemble state is pure, i.e. the ensemble state is described by a wave function, not by a mixture of them.

To avoid a measurement process description for every single quantity it is natural to use the rich experience of classical mechanics. The correspondence principle is used for this purpose. For instance, it follows from the correspondence principle that the quantity p described by the operator (14) is to be measured in the same way as a momentum is measured in classical mechanics. The measurement connects the quantum mechanics formalism symbols with phenomena of the real world and puts a content and a sense into these symbols. Referring to classical mechanics, the correspondence principle formalizes a relation between the formalism of quantum mechanics and measurement. The measurement procedures for different quantities are supposed to be worked out in classical mechanics. Because the correspondence principle is the least formalized part of the theory, I shall make it responsible for a possible disagreement between experiment and the quantum theory formalism. This means that the experimental test of the statistical statements of the quantum theory is considered as a test of the correspondence principle.

While relation (8) describes only expressions for mean values if it is valid for all self-adjoint operators, then, as von Neumann (1932, Chap. 4) has shown, it contains all the statistical statements of quantum mechanics and permits the calculation of the probability of measuring a given value of any quantity R. In particular it follows from (8), that in every single measurement one can obtain only that value R', which is an eigenvalue of the operator $\hat{R}$.

Suppose, for instance, that a self-adjoint operator R has a discrete spectrum of eigenvalues $R_1, R_2, \ldots$. Let for simplicity every eigenvalue R_i ($i = 1, 2, \ldots$) be related to only one eigenvector χ_i. The vectors corresponding to unlike eigenvalues are orthogonal. Let us normalize them in such a way, that

$$(\chi_i, \chi_k) = \int \chi_i^*(\mathbf{q})\chi_k(\mathbf{q}) \, d\mathbf{q} = \delta_{ik}$$
$$i, k = 1, 2, \ldots \quad (19)$$

An arbitrary wave function ψ can be represented in the form

$$\psi(\mathbf{q}) = \sum_i a_i \chi_i(\mathbf{q}) \quad (20)$$

It follows from (19), (20) and the normalization condition of the wave function ψ, that

$$\sum_i a_i^* a_i = \sum_i |a_i|^2 = 1 \quad (21)$$

Let us calculate the mean value $\langle F(R) \rangle_\psi$ of a function F of the quantity R. Because the $\{\chi_i\}$ are eigenvectors of the operator $\hat{R}$

$$\hat{R}\chi_i = R_i \chi_i, \quad i = 1, 2, \ldots$$

one obtains from (8) and (20):

$$\langle F(R) \rangle_\psi = \sum_i F(R_i) |a_i|^2 \quad (22)$$

Since the (22) is valid for every function F, then it follows from (21) and (22) that the quantity R can take only values R_i ($i = 1, 2, \ldots$). The $|a_i|^2$ is the probability that the quantity R has a value R_i in the state (20). Thus any single measurement of the quantity R has to yield one of the eigenvalues R_i of the operator $\hat{R}$. It is appropriate to point out, that the last statement follows from (A3) only in the case when the condition (A3) is fulfilled for all operators $\hat{R}$.

Let us consider the problem of particle-position measurement. For simplicity, the particle is considered to be charged (for example an electron). Let there be some macroscopic device (generator), which prepares an electron in some state. For instance, an electron gun can serve as a generator. Let us imagine many similar generators (an ensemble) which are in the same macroscopic state. Every generator prepares an electron upon which a single measurement is performed. Single measurements performed upon different electrons yield, in general different results. The complex of single measurement results permits the determination of the quantity distribution in a state ψ in which we are interested. Let a detector, capable of recording the time at which an electron passes through it, be spaced some distance from the generator. The detector can be represented by a Geiger counter or other similar device. Let τ be the operating time of the detector. Let a set of experiments be performed. During each experiment the generator is switched on, and, if the detector trips, then it records the time of its tripping. If the detector trips in a period t after switching the generator on, then this means, by definition, that during the period $(t - \tau, t)$ the electron was found in the volume of the detector. Thus the electron coordinates in this period coincide with the detector coordinates within the precision of the detector size.

Let a set of N experiments be performed at a fixed position of the detector. Let N_0 be the number of times the detector has not tripped, N_1 the number of times it has tripped during the period τ [i.e. in the period $(0, \tau)$], N_2 the number of times it has tripped in the period 2τ and so on. We have

$$N = N_0 + N_1 + N_2 + \ldots$$

the limit

$$\lim_{N \to \infty} N_s/N$$

represents the probability of detecting an electron in the time $s\tau$ within the space taken up by detector. For this reason one has

$$|\psi(\mathbf{q}, s\tau)|^2 \, dV = \frac{N_s}{N}, \quad s = 1, 2, \ldots$$

where ψ is the electron wave function, $\mathbf{q}$ are the detector coordinates and dV is its volume. By performing measurements with different dispositions of the detector with respect to the generator, the $|\psi(\mathbf{q}, s\tau)|^2$ can be calculated for different positions $\mathbf{q}$ and times $s\tau$ $(s = 1, 2, \ldots)$ within the detector size and operating time τ. The detector size and its operating time are reduced as far as possible in order to increase the accuracy of the calculation of $|\psi(\mathbf{q}, s\tau)|^2$. An optical or electron microscope can be used if necessary. I shall not describe the measurement of the particle position by means of the microscope but refer to the paper by Mandelstam (1950). It should only be noted that by using a photon (electron) beam of sufficiently high energy, the position and the registration moment of an electron can be determined in principle with any desirable accuracy. This means that the electron coordinates can be measured with arbitrary accuracy and instantaneously.

Of course, a statement of such a kind is an idealization of a real state of affairs. Nevertheless I shall adopt this thesis, remembering that within non-relativistical physics nothing hinders in principle the accurate and instantaneous measurement of the electron position, if the energy is not too high and pairs generation can be neglected (see, however, Landau and Peierls, 1931; Pauli, 1933, Section 2).

Let us consider the problem of the measurement of electron momentum. In classical mechanics it is supposed that the influence of measurement upon electron motion can be made infinitesimal. So to measure the electron momentum it is sufficient to measure two neighbouring positions $\mathbf{q}$ and $\mathbf{q} + \Delta\mathbf{q}$ of a single electron which are separated by a short period of time Δt and to calculate the electron velocity $\mathbf{v} = \Delta\mathbf{q}/\Delta t$. Thereafter the momentum $\mathbf{p}$ is defined by

$$\mathbf{p} = m\mathbf{v} = m\frac{\Delta\mathbf{q}}{\Delta t} \tag{23}$$

Formula (23) determines the mean momentum over a period of time Δt. Measuring momentum in reducing periods of time Δt one obtains in the limit

122 Uncertainty Principle and Foundations of Quantum Mechanics

the exact value of the momentum. In this case the limit $\Delta q/\Delta t$ with $\Delta t \to 0$ is supposed to exist.

In quantum mechanics such a method of measurement is also possible. As the position can be measured, in principle, instantaneously, it is possible to measure two positions of an electron at two instants separated by a short period of time and then to use formula (23). By definition, the measurement result is a momentum averaged over a time period Δt. Let us suppose, that a set of such measurements is performed upon an electron ensemble described by a wave function ψ. A number of momentum values, described by a spread $\Delta \mathbf{p}$, is obtained. Generally, the spread (uncertainty) of momentum depends on the wave function before the first position measurement, but in any case due to the uncertainty principle

$$|\Delta \mathbf{p}| \geq \frac{\hbar}{|\Delta \mathbf{q}|} \tag{24}$$

where $|\Delta \mathbf{q}|$ is the distance between the position of the electron in the first and second measurements of its position. As $|\Delta q| \leq c \Delta t$, where c is the speed of light then it follows from (24) that

$$|\Delta p| \geq \frac{\hbar}{c \Delta t}$$

Thus, unlike classical mechanics a reduction in the time of measuring moments m leads to an increase of momentum uncertainty independent of the form of the wave function, which has described the electron ensemble before measurement. Although nothing prevents a set of measurements of momentum being averaged over a short period of time Δt, one cannot assert that the resulting momentum values represent those of an electron ensemble described by some wave function. These momentum values cannot be attributed to the wave function, which described the ensemble directly before measurement, because the measurement result depends on the measurement duration Δt (with $\Delta t \to 0$). The momentum values cannot be attributed to the wave function arising during measurement, because, however short the period is, the wave function changes in this period essentially, and that moment, to which the measured values should be attributed, is unknown.

Thus, although a measurement can be made, the results cannot be connected with any wave function, and, hence with the statistical statements of quantum mechanics.

Let us consider the measurement of momentum averaged over a long period of time T. Let there be a generator localized in some region Ω with linear dimension of order Δq. The generator prepares an electron ensemble in a state described by a wave function ψ. Let the electron be detected at a distance q from the generator in time T after switching the generator on ($|q| \gg \Delta q$). Let us assume, by definition, that the average of the electron momentum over a period

T is measured. It is defined by

$$\mathbf{p} = \frac{m\mathbf{q}}{T} \tag{25}$$

Essentially the relation (25) is the relation (23) which is used for the case when T is much more than the characteristic time of evolutional change of the wave function. The inaccuracy $\Delta \mathbf{p}$ of momentum measurement is determined by the relation

$$\Delta p = \frac{m \Delta q}{T} \tag{26}$$

For a fixed generator size the inaccuracy is less the longer the measurement time. I shall call momentum defined by relation (25) q-momentum (from the word 'quantum') in contrast to the momentum defined by the relation (23) with $\Delta t \to 0$. I call the latter c-momentum (from the word 'classical').

In the case of the absence of an electromagnetic field the q-momentum distribution can be determined by performing a set of single measurements of q-momentum. This distribution is determined by the relation

$$W(\mathbf{p}) \, d\mathbf{p} = |\psi_\mathbf{p}|^2 \, d\mathbf{p} \tag{27}$$

where $W(\mathbf{p}) \, d\mathbf{p}$ is the probability of measuring a momentum $\mathbf{p}$ within the region $d\mathbf{p} = dp_1 \, dp_2 \, dp_3$ and

$$\psi_\mathbf{p} = \frac{1}{(2\pi\hbar)^{3/2}} \int e^{-i p q / \hbar} \psi(\mathbf{q}, t) \, d\mathbf{q} \tag{28}$$

is the Fourier-component of the wave function.

It should be noted that the measurement of q-momentum (25) is reduced to a position measurement at the moment $t = T$. The wave function of the free electron evolves in such a way that for t long enough the form of $|\psi(q, t)|^2$ determines $|\psi_\mathbf{p}(t)|^2$.

I shall show this in the simple example of one-dimensional motion. Let the wave function at the initial moment have the form

$$\psi(q) = \frac{1}{\sqrt{2\pi\hbar}} \int_{-\infty}^{\infty} e^{ipq/\hbar} \psi_p \, dp \tag{29}$$

$$\psi_p = \sqrt{\frac{\Delta}{\sqrt{\pi}\hbar}} \exp\left\{ -\frac{\Delta^2}{2} \frac{(p - p_0)^2}{\hbar^2} \right\} \tag{30}$$

where Δ is a constant representing an effective width of the wave packet, p_0 is the mean value of the q-momentum. According to (27) and (30) the probability of detecting q-momentum p within the range dp which is

$$W(p) \, dp = |\psi_p|^2 \, dp = \frac{\Delta}{\hbar\sqrt{\pi}} \exp\left\{ -\frac{\Delta^2 (p - p_0)^2}{\hbar^2} \right\} dp \tag{31}$$

has a normal distribution form with mean value p_0 and dispersion $\hbar^2/(2\Delta^2)$. At the initial moment the packet centre was found at the point $q = 0$. With the passage of time the wave function evolves according to Schrödinger's equation. It turns at the moment t into

$$\psi(q,t) = \frac{1}{\sqrt{2\pi\hbar}} \int e^{(ipq/\hbar)-(ip^2t/2m\hbar)} \psi_p \, dp \qquad (32)$$

Substitution of (29) onto (32) and calculation yields the result

$$\psi(q,t) = \sqrt{\frac{\Delta}{(\Delta^2+i\hbar t/m)\sqrt{\pi}}} \exp\left\{ -\frac{(q-p_0 t/m)^2}{2\Delta^2(1+i\hbar t/\Delta^2 m)} - \frac{ip_0 t}{2m\hbar} + \frac{ip_0 q}{\hbar} \right\} \qquad (33)$$

Hence, one gets for the probability dW of detecting the particle in the vicinity of the point q within range dq at the moment t

$$dW = |\psi(q,t)|^2 \, dq = \sqrt{\frac{\Delta^2}{\pi(\Delta^4 + \hbar^2 t^2/m^2)}} \exp\left\{ -\frac{\Delta^2(q-p_0 t/m)^2}{\Delta^4 + \hbar^2 t^2/m^2} \right\} dq \qquad (34)$$

For $t \gg m\Delta^2\hbar^{-1}$ the coordinate distribution reproduces the initial distribution (31) over q-momenta. Indeed in accordance with (25) assuming

$$p_t = \frac{mq}{t} \qquad (35)$$

and substituting q from (35) into (34), one gets

$$W(p_t) \, dp_t = \frac{\Delta}{\hbar\sqrt{\pi}B} \exp\left\{ -\frac{\Delta^2(p_t-p_0)^2}{\hbar^2 B^2} \right\} dp_t \qquad (36)$$

where

$$B = \sqrt{1 + \frac{m^2\Delta^4}{\hbar^2 t^2}} \qquad (37)$$

If $t \to \infty$, then $B \to 1$ and the q-momentum distribution, obtained from the coordinate distribution at the moment t, coincides with the distribution (31).

It follows from the example, that the q-momentum distribution subsequently turns into a coordinate distribution and can be measured.

It should be noted that measuring $|\psi_p|^2$ in such a way, we have no right to assert that the momentum distribution is measured in any definite state. The fact is that the wave function has changed essentially during measurement time. At first it has been localized in the vicinity of the generator and then it spreads over space. By measurement one obtains only time-independent characteristics of the wave function such as amplitudes $|\psi_p|^2$.

Thus the above manner of q-momentum measurement does not permit the measurement of momentum or momentum distribution in the state described by a wave function. The best that can be measured is the momentum distribution averaged in some way over states with different wave functions. In the end it is connected with the fact, that due to the uncertainty principle the precise measurement of momentum needs a long time in which the wave function changes essentially.

Let us consider momentum measurement based on the law of the conservation of momentum. For the measurement of the electron momentum a particle of mass M is placed in the electron path. By collision with the particle the electron is captured by the particle and passes its momentum to it. In measuring the particle momentum, by definition, one measures the electron momentum at the instant of impact. For the electron momentum to be measured with an accuracy Δp, it is necessary that the initial momentum of the particle is of the order Δp. According to the uncertainty principle this is possible only if before collision the particle is placed within a region with line size of the order Δq and

$$\Delta q \geq \frac{\hbar}{\Delta p} \tag{38}$$

For simplicity I shall consider only one dimension. Let the wave function describing an electron ensemble having the form of the wave packet (32) with spread L. The time uncertainty of hitting an electron with a particle is determined by the relation

$$\Delta t \geq \frac{m}{p}(L + \Delta q) \tag{39}$$

where m is the electron mass and p is its momentum. If one measures the momentum long enough it can be determined with great accuracy. Thus, the uncertainty of electron momentum measurement is conditioned only by the uncertainty of the initial momentum of the particle.

For the measured electron momentum to be attributed to any definite wave function, it is necessary that the wave function changes slightly during the measurement time Δt. In the optimum case the spread δp of ψ_p in the region of the variable p is determined by the uncertainty relation

$$\delta p \geq \frac{\hbar}{L} \tag{40}$$

During the time Δt the phases

$$\varphi_p = \left(pq - \frac{p^2 t}{2m}\right) / \hbar \tag{41}$$

corresponding to unlike unvanishing Fourier-components of ψ_p change. The greatest phase difference arising during Δt is

$$\Delta \varphi = \frac{p \delta p \Delta t}{m \hbar} \tag{42}$$

Substituting (39) and (40) into (42) one gets

$$\Delta\varphi \geq 1 \qquad (43)$$

This means that the wave function always changes essentially during the measurement time, and the measured values of momentum cannot be attributed to any definite wave function. They can be attributed only to an ensemble state averaging in some sense over the period Δt. In this sense the q-momentum distribution cannot be measured in this way. The reason preventing this can be formulated as the time–energy uncertainty relation. Indeed, the relations (42) and (43) can be formulated as

$$\Delta\varphi = \frac{\Delta E \, \Delta t}{\hbar} \geq 1 \qquad (44)$$

where ΔE is the uncertainty of the electron energy.

Let us consider momentum measurement based on the Doppler effect. Let an atom in an excited state radiate a photon with frequency ω_0. If the atom moves, then the photon, radiated into the direction of motion, has a frequency

$$\omega = \omega_0\left(1 + \frac{V}{c}\right) \qquad (45)$$

where V is the atom velocity. Measuring the photon frequency, one can determine the atom velocity from (45) and hence the atom momentum. The least error $\Delta\omega$ of the frequency determination is given by the relation

$$\Delta\omega \geq \frac{1}{T} \qquad (46)$$

where T is the measurement time, i.e. the period during which the objective of the spectrometer being used for the photon frequency measurement is open.

Let the atom ensemble be described by a wave function, having the form of a wave packet with spread L. Again for simplicity the one-dimensional case is considered. Supposing that for the atom speed $V \ll c$, one gets for the uncertainty Δt of the photon radiation time

$$\Delta t = T + \frac{L}{c} \qquad (47)$$

Caculating the atom velocity by means of (45), one obtains the following expression

$$\Delta p = M\Delta V = \frac{Mc\,\Delta\omega}{\omega_0} \geq \frac{Mc}{\omega_0 T} \qquad (48)$$

for atom momentum error measurement. Let us write the wave function of the atom in the form

$$\psi(q, t) = \frac{1}{\sqrt{2\pi\hbar}} \int e^{(ipq/\hbar) - (iE_p t/\hbar)} \psi_p \, dp \qquad (49)$$

where E_p is the energy of the atom with momentum p. At the moment of photon radiation the atom energy reduces by $\hbar\omega$. Photon radiation can be produced at any moment of the period Δt. This entails a phase uncertainty $\Delta\varphi_p$ of one of the Fourier-components ψ_p, which is determined by

$$\Delta\varphi_p = \frac{\hbar\omega\Delta t}{\hbar} \geq \omega T \geq \frac{\omega}{\Delta\omega} \geq 1 \tag{50}$$

Thus the phase difference between different ψ_p changes essentially in the period Δt. This means, that the wave function changes essentially during measurement time Δt.

Thus, the measured momentum cannot be attributed to any definite wave function, i.e. the momentum cannot be measured in the sense that it is customary to treat measurement in quantum mechanics. The best that can be done is to state that the measured momentum distribution is attributed to a set of wave functions having constant modules $|\psi_p|$ of Fourier-components and indefinite phases. Statements of such a kind are absent in conventional quantum mechanics. Apparently they do not contradict quantum mechanics, but one cannot say that they confirm it.

The distribution $|\psi_p|^2$ over q-momenta can be obtained, because it does not involve any wave function describing a free particle, but this is not the measurement of momentum distribution. The restriction of momentum measurement is born not from the impossibility of measuring momentum but from the impossibility of attributing measured values to any definite state.

All the ways considered of measuring momentum fit the case when the particle motion obeys the laws of classical mechanics. In other words the method of measurement does not depend on the model which is used for the explanation of physical phenomena. In classical mechanics also it is necessary to attribute the measured value to a definite state but in this case there are no wave functions and no uncertainty principle. It is values of coordinates and momenta that determine a state. For this reason in classical mechanics the problem of attributing measured values to a definite state does not exist.

Let us consider the measurement of the component p_1 of the momentum of a free particle at the point q. Let the free-particle ensemble be described by means of a wave function ψ. Let us measure the particle position at the moment $t+T$. In principle this is possible. Let us select only those single measurements which have given the particle position in a small vicinity of position q at the moment t. Let these single measurements constitute the ensemble E_q. Suppose further in these cases that the particle position measurement at moment $t+T$ gives a result $\mathbf{q}+\Delta\mathbf{q}$, where $\Delta\mathbf{q}$ is generally different for different single measurements. Let us suppose that each single measurement determines the component p_1 of the particle momentum at the point $\mathbf{q}$. It is determined by the relation

$$p_1 = m\frac{\Delta q^1}{T} \tag{51}$$

128 Uncertainty Principle and Foundations of Quantum Mechanics

Performing many single measurements, one gets a distribution of the momentum component p_1 at the point $\mathbf{q}$. This distribution depends essentially on the choice of the period T between two consecutive positions of the particle. The shorter the period T the more precise the determined position of the particle is and the more energetic the beam of sounding particles (electron or photon) to be used for the particle position determination. As a result the particle motion will be disturbed and the distribution over component p_1 will be distorted. The shorter the period, the greater is the momentum dispersion of the particle. This means that the distribution over component p_1 of the particle momentum cannot be measured. The criterion of such an impossibility is the dependence of the distribution on the period T as $T \to 0$.

Let a particle position measurement be realized by means of a beam of sounding particles (for instance, electrons or photons). Let us take the beam to be directed normally to the first axis along which the momentum component is to be measured. One can expect that in this case the sounding particle beam does not influence the value of momentum component p_1 on the average. This means that the mean value $\langle p_1 \rangle_\mathbf{q}$ of the momentum component at the point $\mathbf{q}$ does not depend on the sounding particle energy. The formal criterion of this is the existence of a limit

$$\lim_{T\to 0}\left\langle m\frac{\Delta q^1}{T}\right\rangle_\mathbf{q} = \lim_{T\to 0} \langle p_1 \rangle_\mathbf{q} \tag{52}$$

The problem of the existence of the limit (52) can be investigated by means of a quantum mechanics formalism. I shall not do this, but shall confine myself only to the optimistic supposition that such a limit exists. The existence of the limit (52) for all points $\mathbf{q}$ and a certain ensemble of free particles, described by the wave function ψ, means the possibility of measuring the mean value $\langle p_1 \rangle_\mathbf{q}$ of the momentum component at the point $\mathbf{q}$ at the moment t and to attribute it to the state descibed by the wave function ψ.

Using proper sounding beams one can measure mean values of other momentum components at the point $\mathbf{q}$. A measurement possibility of the mean value $\langle p_1 \rangle_\mathbf{q}$ of the momentum at the point $\mathbf{q}$ in the state ψ means a measurement possibility of mean angular momentum $\langle [\mathbf{q} \times \mathbf{p}] \rangle_\mathbf{q}$ and other mean values, which are linear over the momentum components of a free particle.

Measuring electron position one is able to calculate a distribution $|\psi(q)|^2$ over coordinates for all moments of time. This permits the calculation of all moments $\langle q^l \rangle$ ($l = 1, 2, \ldots$) of the electron coordinates and their time dependence (for simplicity the one-dimensional case is considered). Using Schrödinger's equation for the free particle one can show that the moments $\langle p^l \rangle$ ($l = 1, 2, \ldots$) of momentum are expressed by the relations

$$\langle p^l \rangle = \frac{m^l}{l!}\frac{d^l}{dt^l}\langle q^l \rangle, \quad l = 1, 2, \ldots \tag{53}$$

where m is the mass of a particle. As long as moments $\langle q^l \rangle$ can be measured at all moments of time then in principle one can calculate all momenta $\langle p^l \rangle$

($l = 1, 2, \ldots$) and establish momentum distribution for all moments of time. As long as free-particle energy is expressed through particle momentum, then the formula (53) permits the calculation of the energy distribution for each moment of time. All these distributions can be attributed to a definite moment of time and consequently to a definite wave function.

However the essential problem consists in whether or not the foregoing procedure is a measurement of momentum distribution. Although to a degree this is a question of terminology it is usually taken that momentum distribution measurement is a procedure such that a certain value of momentum is obtained as a result of each single measurement. A set of all measured values on momentum constitutes a moment distribution. The described procedure is not one of such a kind. Here the result of a single measurement is a certain value of the coordinate. For this reason I shall not take this procedure as a momentum measurement.

Let us consider the angular momentum measurement in the experiment of Gerlach and Stern (1924). The detailed analysis of this experiment can be found in any textbook on quantum mechanics (see, for instance, Pauli, 1933; Blokhintsev, 1963; Bohm, 1965). I shall confine myself to only the analysis of to what extent this experiment proves that an angular momentum projection upon a certain axis takes the $\hbar$-fold values. Let there be an atom with an electron shell with a non-zero momentum **M** resulting from the orbital motion of the electrons. The electron spins are supposed to be compensated. The magnetic moment μ is connected with the angular momentum **M** by means of the relation

$$\boldsymbol{\mu} = \frac{e\mathbf{M}}{2mc} \tag{54}$$

where e is the electron charge and m is the electron mass.

A beam of such atoms passes between the poles of a magnet with a very inhomogeneous field, with field magnitude gradient directed along the magnetic field. In the magnetic field the atom obtains an additional energy and is affected by the force

$$\mathbf{F} = -\nabla(\boldsymbol{\mu}\mathbf{H}) \tag{55}$$

Passing through the magnetic field the atoms move normally to the lines of force. Under the action of the force (55) during motion through the magnetic fields, the atoms obtain a momentum in the direction of the force **F**, i.e. in the direction of the magnetic field. This momentum is different for different values of magnetic moment projection μ_H upon magnetic field direction. As a result the atom beam splits up into several beams depending on the magnetic moment projection μ_H. After some time the beams are separated in space. Each beam can be recorded by its dropping into a definite place on the screen. If the place of dropping is known, then one can calculate the corresponding value of the magnetic moment projection μ_H and the value M_H of the angular momentum projection upon the magnetic field direction. Experiments show that measured in such a way the value of the angular momentum projection is $\hbar$-fold. Such is the conventional interpretation of the Stern–Gerlach experiment. The discreteness and the multiplicity of $\hbar$ of the angular momentum projection M_H are explained by those of eigenvalues of the angular momentum operator (Bohm, 1965, Chap. 14).

On the other hand, if the stationary states of the atom are discrete and each state obtains an additional energy

$$\Delta E_i = \Delta E_i(\mathbf{H}) \qquad (56)$$

in the magnetic field, then independent of the nature of this energy change of the stationary state the atom, placed in the inhomogeneous magnetic field, in the ith state is affected by the force

$$\mathbf{F}_i = -\nabla(\Delta E_i) \qquad (57)$$

If the force is different for unlike discrete stationary states, then under its action the atom beam is split into several beams according to different stationary states. Thus, splitting into discrete beams is connected with discreteness of the atomic states and the difference of their energies in the external magnetic field. Strictly speaking, it is the energy of the atom in the external magnetic field, that is measured in the Stern–Gerlach experiment. This was noted by Pauli (1933) and Blokhintsev (1968). The discreteness of the angular momentum projection M_H results from the fact that the operator $\hat{M}_H$ commutes with the Hamiltonian and, hence, its eigenvalues can serve as a label of stationary states.

One can see from analysis of the atom beam motion in a inhomogeneous electric field that it is the second interpretation that is correct. If the atom is placed in an electric field $\mathbf{E}$ then the energy of its stationary state changes a little. This is the so called Stark-effect. For simplicity let us consider hydrogen. It is known (see, for instance, Landau and Lifshits, 1963, section 77) that for not too large an electric field the change of the stationary state energy ΔE is linear with the electric field $\mathbf{E}$. As a first approximation of perturbation theory one gets for the ith undisturbed state

$$\Delta E_i = [\mathbf{D}]_i \mathbf{E} \qquad (58)$$

where $[\mathbf{D}]_i$ means some quantity which depends on the undisturbed Hamiltonian eigenvectors and electric dipole operator $\hat{\mathbf{D}}$, but not on electric field $\mathbf{E}$. For some stationary states of hydrogen the $[\hat{\mathbf{D}}]_i$ is non-zero. The hydrogen atom beam, moving in the inhomogeneous electric field $\mathbf{E}$, is affected by the force

$$\mathbf{F} = -\nabla([\mathbf{D}]\mathbf{E}) \qquad (59)$$

which is different for atoms in different stationary states. As a result of the motion in a proper electric field the beam is split up into a few discrete beams. From a measurement of the beam deflection one can calculate values of $[\mathbf{D}]_i \mathbf{E}$ which take a set of discrete values. At the same time the electric dipole operator $\hat{\mathbf{D}}$ has the form

$$\hat{\mathbf{D}} = e(\mathbf{q}_p - \mathbf{q}_e) \qquad (60)$$

where q_p and q_e are position operators of a proton and a electron respectively. Components of operator $\hat{D}$ commute with each other and have a continuous spectrum of eigenvalues.

Thus the interaction operator ($-\mu H$ or $-DE$) responsible for beam splitting has a discrete spectrum in one case and a continuous one in another case. But splitting into discrete beams is produced in both cases. This means that the discreteness of the beams is produced not by the spectrum discreteness of the interaction operator, but by that of the whole Hamiltonian, i.e. by discreteness of the stationary states of the atom. This means that the Stern–Gerlach experiment cannot be used for testing the quantum mechanics statement which asserts that by measuring the quantity M_H, the operator of which $\hat{M}_H$ has a discrete spectrum, only those values can be obtained which are equal to the eigenvalues of operator $\hat{M}_H$. Thus, in the Stern–Gerlach experiment a sorting of stationary states is produced. It can be considered as the measurement of angular momentum provided values M_H are labels of the states.

Let us analyse to what extent the measured result can be attributed to a definite wave function. Suppose in the Stern–Gerlach experiment that the wave packet describing the atom ensemble moves in the positive direction of the x-axis, as is shown in Figure 1. First of all the wave-packet spread in the

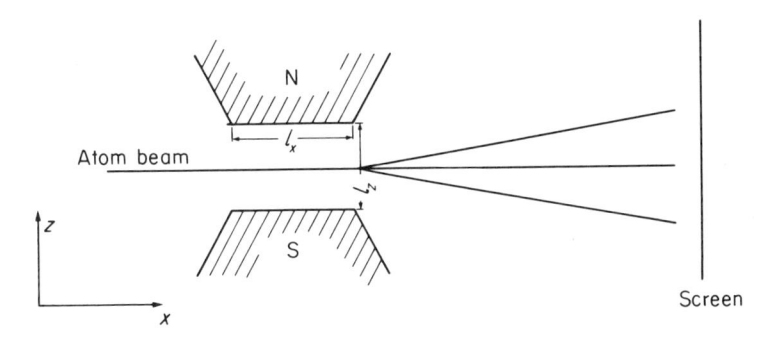

Figure 1 Stern–Gerlach experiment with the wave packet moving in the positive direction of the x-axis.

x-direction has to be essentially more than the size of the apparatus, otherwise during the time T of the wave packet passing through its spread, L_x, the phase difference (42) in the exponent of formula (49) will change within limits

$$\Delta\varphi = \frac{\Delta ET}{\hbar} = \frac{p_x \Delta p_x}{M\hbar} T \geq \frac{p_x \Delta p_x}{M\hbar} \frac{ML_x}{p_x} = \frac{L_x \Delta p_x}{\hbar} \geq 1 \qquad (61)$$

where p_x is an atom momentum x-component, i.e. the wave function has time to change during the experiment which cannot be shorter than T.

132 Uncertainty Principle and Foundations of Quantum Mechanics

Let the wave packet size L_x be much more than the apparatus size l_x, and the momentum uncertainty of the atom Δp_x along the x-axis be small. Let L_z be the wave-packet spread in the direction of the z-axis. During passage between the poles of the magnet the atom acquires a momentum

$$p_z = \mu_H \frac{\partial H}{\partial z} T \tag{62}$$

where

$$T = \frac{Ml_x}{p_x}$$

is the interaction time of the atom with the magnetic field, i.e. transit time. For beam separation it is necessary that

$$p_z > \Delta p_z > \frac{\hbar}{L_z} \tag{63}$$

where Δp_z is the uncertainty of the momentum of the atom along the z-axis. Atoms with different magnetic moment projection have different energies. This conditions the change of phase difference of the different Fourier-components of the wave function which correspond to different values of μ_H. At best during the time T the phase difference change is

$$\Delta\varphi = \frac{\Delta E T}{\hbar} \approx \frac{1}{\hbar} \mu_H \frac{\partial H}{\partial z} L_z T \tag{64}$$

Substituting T from (62) and using (63), one gets

$$\Delta\varphi \geq 1 \tag{65}$$

This means that during the measurement time the wave function changes essentially and the measurement result cannot be attributed to a definite wave function.

Thus, it follows from the analysis that quantities such as momentum, energy and angular momentum cannot be measured in the sense that *measured values cannot be attributed to definite ensemble states*. This means that quantum mechanics statements cannot be verified experimentally for all physical quantities R, because for this test it is necessary to attribute the measured value to a definite state. Statistical statements (8) can be verified for $R = F(q)$, and perhaps for $R = p$, but there are those that cannot be verified, for instance, for $R = p^3$.

Thus the statistical statements of quantum mechanics can be verified only in particular cases and cannot be verified in general. This is the corollary of the time–energy uncertainty principle. In principle, this is connected with the non-relativistic character of quantum mechanics, according to which a wave function is given at one moment of time. This is in contradiction to the time–energy uncertainty principle which requires that an esemble description is 'spread over time'. Indeed, in quantum mechanics the particle description by means of a wave function makes it 'spread over space', while relativistic symmetry requires it 'to spread it more over time'.

If statistical statements of quantum mechanics cannot be tested experimentally then their consequences remain doubtful: for instance, von Neumann's theorem on hidden variables or the statement that in the measurement of a quantity R only values equal to eigenvalues of the operator $\hat{R}$ can be obtained. For instance, as we have seen, the Stern–Gerlach experiment does not prove at all that angular momentum takes only $\hbar$-fold values. At the same time some consequences of formula (8) may be correct, even if it is not always valid. At least it is possible to consider formula (8) as correct because it has not been proved that it is incorrect, but only that it can not be proved experimentally.

At the same time I believe an alternative conception would be welcome which would explain the impossibility of making measurements of quantities having energy–momentum character and which would, in general, consider as observable only those quantities which could be measured experimentally.

3. IDEAS OF RELATIVISTIC STATISTICS

Let us try to have a look at quantum mechanics from another viewpoint. Let us imagine that the motion of a microscopic particle is not deterministic (for instance, because of its interaction with the surroundings), i.e. its behaviour is like a Brownian particle. This means that the particle's world-line appears to be strained in a random way. Let us suppose that a statistical description of such non-deterministic world-lines can lead to the same results as those obtained by quantum mechanics. Of course, it is hopeless to try to obtain all basic statements (1)–(5) of quantum mechanics because on account of the von Neumann theorem on hidden variables they are certainly inconsistent with the supposition that a particle is described by means of a definite world-line (see, Moyal, 1949). As we have seen, not all statements of quantum mechanics can be tested experimentally. For this reason some hope remains for the successful formulation of world-line statistics in such a way that disagreement with statements (1)–(5) will occur only within an unobservable field.

It should be noted that attempts to interpret quantum mechanics from a classical or quasi-classical point of view are numerous. They are known as hidden-variables theories. A review of different versions of such theories and their bibliography can be found in the survey by Kaliski (1970) and in a monograph by Belinfante (1973). In most cases such theories represent attempts to interpret the basic statements of quantum mechanics from different viewpoints. Attempts to obtain the results of quantum mechanics starting from classical statistics in their pure form have not succeeded as a rule, because non-relativistic statistics have been used. This is motivated by the fact that quantum mechanics is also non-relativistic.

As far as possible I shall confine myself consequently to the relativistic viewpoint. First of all it is necessary to differentiate between non-relativistic and relativistic notions of state. The non-relativistic state (n-state) of a system is a set of quantities given at a certain moment of time. For instance, the particle n-state is determined at a certain moment by coordinates q and momenta p, i.e. by a point in the phase space of coordinates and momenta. In non-relativistic physics the division of physical phenomena descriptions into state and equations of motion, is connected with the existence of absolute simultaneity and the existence of an invariant division of space–time into space and time (two invariants: time period and distance). Correspondingly, a particle is considered as a point in the three-dimensional space.

The relativistic state (r-state) is given over all space–time. For instance, the particle r-state is its world-line, described by the equation $q^i = q^i(\tau)$ ($i = 0, 1, 2, 3$). τ is a parameter along the world-line. Equations of motion play the part of restrictions imposed upon possible r-states. In relativistic physics the united description (without division into state and equations of motion) is connected with the absence of an invariant division of space–time into space and time (the only invariant: interval of space–time). Correspondingly, a particle is considered as a one-dimensional line in space–time (but not a point in space).

In non-relativistic physics a statistical method is used for the descriptions of the non-deterministic system (i.e. systems with uncertainty in equations of motion or systems with uncertainty in initial state). One considers a statistical ensemble, i.e. a set of many identical systems which are in different states. The dynamical systems constituting the ensemble are known as elements of the ensemble. The statistical ensemble is a deterministic dynamical system even if constituting systems are non-deterministic. This means that the ensemble n-state can be calculated at time t if its n-state at time t_0 ($t_0 < t$) is determined. For example, let a dynamical system A consist of a non-deterministic particle. The ensemble consists of N such independent particles ($N \to \infty$). The state of every particle is represented by a point in the phase space. Let $d\Omega$ be an element of volume of phase space, and let dN be the number of points in $d\Omega$. Then

$$dN = W\, d\Omega \tag{66}$$

where $W = W(\mathbf{q}, \mathbf{p})$ is a state density.

Although any individual system of an ensemble is indeterministic, it is found that the evolution of an esemble n-state W can be calculated because the W obeys some equation the form of which depends on the character of random forces, acting upon particles of the ensemble. In the case, when systems A constituting an ensemble are deterministic, the form of the equation which is obeyed by W is determined uniquely by means of the equation of motion of system A. Thus, the W is the n-state of the statistical ensemble as a dynamical system.

The equation which is obeyed by W is invariant with respect to the transformation $W \to CW$ when C is a constant. This is so because the ensemble behaviour does not depend on the number of systems constituting the ensemble, if this number is large enough. The constant C can be chosen in such a way that $W(\mathbf{q}, \mathbf{p})$ represents the probability of detecting the n-state of a particle in the volume $d\Omega$ of the phase space.

That the ensemble n-state is a probability density is connected with the representation of the state of the system by a point (but not a line or surface) in the phase space. As $W(\mathbf{q}, \mathbf{p})$ is the probability density of the detection of the physical system in the state $(\mathbf{q}, \mathbf{p})$, then calculating $W(\mathbf{q}, \mathbf{p})$ at any moment t by means of the equation of motion of the statistical ensemble, one can calculate the evolution of the mean value $\langle F(\mathbf{q}, \mathbf{p}) \rangle$ of any function F of the n-state $(\mathbf{q}, \mathbf{p})$.

I have described in general the conventional scheme of the statistical ensemble application for describing the behaviour of the non-deterministic systems. Three essential points in this scheme should be stressed.

(1) The transition from a physical system to the ensemble, i.e. the method of construction of the ensemble state.
(2) The determination of equations which are obeyed by the ensemble state, and the solution of these equations.
(3) The transition from an ensemble state to an individual system, i.e. the method of calculation of the statistical characteristics of the non-deterministic system proceeding from an ensemble state.

In the generalization of the statistical method to the relativistic case different variants are possible. I shall consider three of them.

(1) In the first variant the base statement asserts that in the relativistic case the ensemble state is represented by a probability density. For this it is necessary that the ensemble state is represented by a point in a phase space. To reach this the particle description by means of the r-state (world-line) is dropped and one considers the intersection points of the world-line with different three-dimensional surfaces or points marked by parameters such as proper time (Hakim, 1967a, b, 1968). Such points together with momenta represent a state of a particle, as in non-relativistic physics. Requirements of relativity are taken into account imposing Lorentz-invariance conditions upon corresponding equations. Such an approach is most widely accepted, but I shall not use it. I believe that using the world-line as the main subject of a theory is far more important than the application of the developed formalism of probability theory.
(2) The second method consists in considering that any particle r-state (world-line) is a point in a certain functional phase space M, the r-state being kept as the basic subject of the theory and the ensemble state being a probability density on M. However, if l is an intercept of the world-line located within a region Ω of space–time, then the question of which region on M corresponds to Ω will be solved depending on the behaviour of the world-line outside Ω. This means non-locality of description. Such a description seems to me unsatisfactory.

(3) The method, which I shall use, keeps the r-state as a basic subject of theory. It realizes a local description, but the ensemble state, i.e. the density state of systems constituting the ensemble, cannot be treated as a probability density. For instance, let there be an ensemble of systems consisting of one particle, i.e. an ensemble of world-lines. Let ds_i ($i = 0, 1, 2, 3$) be an infinitesimal area at the point q^i and dN be the number of world-lines crossing ds_i. Then

$$dN = \sum_{i=0}^{3} j^i \, ds_i \qquad (67)$$

where j^i is a factor which is, by definition, the density of r-states (world-lines) at the point q of space-time. The j^i considered at a certain moment of time is the ensemble n-state. The same j^i considered in the whole space–time (or in some region) is the r-state. In this sense the r-state of the ensemble coincides with the n-state.

In the case, when the ensemble elements are dynamical systems consisting of N particles, the r-state of such a system described by an N-dimensional surface on 4N-dimensional space $V_{12...N} = V_1 \otimes V_2 \otimes \ldots \otimes V_N$ which is a tensor product of spaces V_i ($i = 1, 2, \ldots N$) for each of the particles (see, for instance, Hakim, 1967a, b). The state density of such systems is described by the antisymmetrical over all indices pseudotensor $j^{a_1 a_2 \ldots a_N}$ on $V_{12...N}$ ($a_1, a_2, ,,, a_N = 1, 2, \ldots 4N$).

Later on for simplicity I shall confine myself to the case, when the ensemble element is a dynamical system consisting of one particle. The method of transition from the system r-state to the statistical ensemble is defined. It coincides with the conventional method: the ensemble state is the state density of systems constituting the ensemble. The reverse transition from the statistical ensemble to the properties of the system cannot coincide with the conventional one, because the conventional method is based on the fact that an ensemble state $W(\mathbf{q}, \mathbf{p})$ is the probability density of detecting a system in the state $(\mathbf{q}, \mathbf{p})$. Strictly speaking, neither j^i nor $j^{a_1 a_2 \ldots a_N}$ can be treated in such a way. For this reason the transition from the statistical ensemble to the non-deterministic system is based on the use of additive quantities.

DEFINITION. The quantity B is an additive one if the value of B for several independent systems is equal to the sum of values of B for every system.

Energy, momentum, angular momentum and their densities are examples of additive quantities. The statistical ensemble is a set of independent systems. For this reason any additive quantity attributed to the statistical ensemble as a dynamical system is a sum of values of this quantity for all systems constituting the ensemble. As the ensemble behaviour does not depend on the number of systems in the ensemble, the equations for the ensemble state j are invariant with respect to transformation $j \to Cj$, where C is a constant and j denotes any ensemble state: $W, j^i \ldots$. Hence, j can be normalized on one system. In this case a value of any additive quantity of the ensemble is equal to the mean value over the ensemble of this quantity for systems constituting an ensemble. In the non-relativistic approximation, when one of the components of j is non-negative and conserved (for instance, j^0 for a one-particle system), it is possible to treat this component as a corresponding probability density. In this case it is possible to obtain additional information.

Let us formulate the above in the axiomatic form.

THE STATISTICAL PRINCIPLE

A deterministic dynamical system, which is called a statistical ensemble corresponds to a non-deterministic* dynamical system A whose state is described by quantities ξ.

(1) A state j of the statistical ensemble is a density state of systems A.
(2) The equations for the ensemble state j are invariant with respect to transformation

$$j \to Cj, \quad C = \text{constant} \qquad (68)$$

(3) If the ensemble state j has a proper normalization (on one system), every additive quantity B attributed to the statistical ensemble as a dynamical system is the mean value of quantity B for system A.

The statistical principle fits either the relativistic or non-relativistic case. It settles the question about the determination of the ensemble state and about the determination of the non-deterministic system properties, but it does not determine which equations are obeyed by the ensemble state. However if the ensemble elements are deterministic systems, then the equation satisfied by an ensemble state j is determined by equations which a single ensemble element obeys. This fact can be used to simplify derivation of equations which the ensemble state j obeys.

I shall show this by a simple example of an ensemble E the elements of which are free particles of mass m, i.e. all possible timelike straight lines in space–time. Although it is possible to introduce a density state for such an ensemble by means of (67), it is still not possible to describe the ensemble completely. This means that if the state j^i is given at any moment t, then, in general, the j^i cannot be determined at another moment. The state of such an ensemble can be described by a distribution function $f(q, p)$, $q = \{q^0, q^1, q^2, q^3\}$, $p = \{p_0, p_1, p_2, p_3\}$ which satisfies equation (69)

$$\frac{\partial}{\partial q^i}(p^i f(q, p)) = 0 \qquad (69)$$

where p^i is 4-momentum of a particle. Here and later on summation is made on

---

*The statistical principle can be applied to the deterministic system, if its initial conditions are not exactly determined.

like arabic super- and subscripts from zero to three. The distribution function $f(q, p)$ vanishes except over the surface

$$p_i g^{ik} p_k = m^2 c^2 \tag{70}$$

where g^{ik} is the metric tensor

$$g_{ik} = \begin{Vmatrix} c^2 & 0 & 0 & 0 \\ 0 & -1 & 0 & 0 \\ 0 & 0 & -1 & 0 \\ 0 & 0 & 0 & -1 \end{Vmatrix}, \quad g^{ik} = \begin{Vmatrix} c^2 & 0 & 0 & 0 \\ 0 & -1 & 0 & 0 \\ 0 & 0 & -1 & 0 \\ 0 & 0 & 0 & -1 \end{Vmatrix} \tag{71}$$

amd m is the mass of a particle. The meaning of the distribution function is determined by the fact that the stream density j^i of world-lines at the point q is

$$j^i = \int (p^i/m) f(q, p) \, d^4 p, \quad d^4 p = dp_0 \, dp_1 \, dp_2 \, dp_3 \tag{72}$$

However, instead of the description by means of a distribution function one can use the following. Let us consider the ensemble E as consisting of elements E_p, where E_p is an ensemble whose elements are straight lines having the direction of the unit vector p_i/mc. The ensembles E_p are completely described by the pseudovector j^i. I shall call such an ensemble a pure one. As long as the E_p are dynamical systems, they can be elements of the ensemble E. It is reasonable to normalize the states j^i of the E_p in a similar way.

It is easy to verify that the equations which the $j^i(q)$ satisfy have a form

$$j^i \frac{\partial}{\partial q^i} \left(\frac{j^l}{\sqrt{j^s g_{sk} j^k}} \right) = 0, \quad l = 0, 1, 2, 3 \tag{73}$$

Let us normalize the j^i on one system by means of

$$\int_\Sigma j^i \, ds_i = 1 \tag{74}$$

where Σ is an infinite space-like hypersurface and ds_i is an element of the hypersurface. All physical quantities attributed to a pure ensemble as a dynamical system can be expressed through the state j^i.

In the given case, when all ensembles E_p can be labelled by parameters p_i, the state of ensemble E can be described by means of non-negative quantities $W(p)$ which represent a probability density of detecting a pure ensemble E_p in the ensemble E. Finally, here all can be reduced to a distribution function $f(q, p)$.

However, in the case, when non-deterministic one-particle systems are considered the equations which the pure ensemble state satisfy can be more complicated than equations (73), and the possibility of labelling all their solutions by means of the finite number of parameters is not evident.

For example suppose it is insufficient for the description of a single system to give coordinates and momenta, but it is necessary to give quantities $\dot{q}, \ddot{q}, \dddot{q}, \ldots$

where $q = \{q^0, q^1, q^2, q^3\}$ and the dot denotes differentiating in proper time τ. For example, such a situation arises in the consideration of Lorentz–Dirac equations. In this case the distibution function $f(q, \dot{q}, \ddot{q})$ is used (Hakim, 1967b). In the non-relativistic case the more general consideration can be found in the book by Vlasov (1966).

In place of the consideration of distribution functions of the type $f(q, \dot{q}, \ddot{q}, \ldots)$ one can consider an arbitrary ensemble E as an ensemble in which the elements are pure ensembles E_j described by means of j^i with j^i being a function of q only. The dependence of the ensemble E on distributions over $\dot{q}, \ddot{q}, \ldots$ manifests itself as the implicit dependence of E on E_j. This has been shown in a simple example of the distributive function $f(q, p)$. The consideration of the ensemble E as consisting of elements E_j has the advantage that, in general, one cannot be interested in how many derivatives there are like $\dot{q}, \ddot{q}, \ldots$ and in which way the distribution function depends on them. All that it is necessary to know are the equations which the state of the pure ensemble satisfies. The state of an arbitrary ensemble satisfies equations which can be obtained merely as a result of a formal transformation of the pure ensemble equations. Just this fact explains the mysterious circumstances that the quantum mechanics pure state (wave function) depends only on coordinates, while the usual classical distribution function depends on coordinates and momenta. The dependence on momenta (and not only on momenta, but, in general, on quantities like $\dot{q}, \ddot{q}, \ldots$) is taken into account in quantum mechanics in consideration of mixed states, i.e. ensembles whose elements are pure ensembles.

In the general case the state of an ensemble E is described by a function W of states j^i of pure ensembles E_j which are elements of the ensemble E. The state of each pure ensemble E_j is completely described by four functions j^i ($i = 0, 1, 2, 3$). This means that j^i satisfy certain equations, and that all physical quantities are expressed through j^i. In the case in which the state W of the ensemble E is considered as an r-state, the W should be considered as a function of four quantities j^i ($i = 0, 1, 2, 3$) which are considered as functions of space coordinates q and time t. As it is necessary to use the relation derived using non-relativistic statistics, it is convenient to consider the state W of the ensemble E as the n-state, determined at any moment of time t. In this case the n-state of ensemble W is a function of four quantities j^i ($i = 0, 1, 2, 3$), considered as the function of the space coordinates q. Besides the W is a function of time t. It is supposed that, if for a pure ensemble E_j the state j^i ($i = 0, 1, 2, 3$) is given at moment t, then it can be determined at all subsequent moments of time. The dependence of the n-state $W[j^i]$ on time can be determined from the relation

$$W_t[j_t] = W_0[j_{t=0}], \quad j = \{j^0, j^1, j^2, j^3\} \tag{75}$$

The index 't' of W_t and j_t shows the moment at which these values are taken. The relation (75) expresses the fact that the change of form of the functional $W[j]$ is conditioned only by a change of functions j^i which is described by the equations of motion. Let the equations of motion of the n-state of pure

140 Uncertainty Principle and Foundations of Quantum Mechanics

ensemble E_j have a form

$$\frac{\partial j^i}{\partial t} = G^i(j, \partial), \quad \partial = \left\{ \frac{\partial}{\partial q^1}, \frac{\partial}{\partial q^2}, \frac{\partial}{\partial q^3} \right\}, \quad i = 0, 1, 2, 3 \qquad (76)$$

G^i are some functions of j^i and their derivatives. Differentiating the (75) with respect to t and using the expressions (76) for $\partial j^i/\partial t$, one gets the following linear equation with variational derivatives.

$$\frac{\partial W_t}{\partial t} + \sum_{i=0}^{3} \int G^i(j(\mathbf{x}), \partial) \frac{\delta W_t}{\delta j^i(\mathbf{x})} d\mathbf{x} = 0 \qquad (77)$$

Equations of motion (76) of a pure ensemble are characteristics of the linear equation (77).

Let us apply the above consideration to an arbitrary ensemble of non-deterministic world-lines. In the paper (Rylov, 1971) in the non-relativistic approximation I considered a pure ensemble the elements of which are non-deterministic systems, each one consisting of one particle moving in a given electromagnetic field. It has been shown that it is possible to choose the equation of motion (76) of a pure ensemble in such a way that from statistical principles all basic results of one-particle quantum mechanics could be deduced. Namely, supposing that j^α/j^0 ($\alpha = 1, 2, 3$) has a potential and can be represented in the form

$$\frac{j^\alpha}{j^0} = \frac{1}{m} \frac{\partial \phi}{\partial q^\alpha}, \quad \alpha = 1, 2, 3 \qquad (78)$$

it was shown that the function

$$\psi = \sqrt{\rho} \exp\left\{ \frac{i\varphi}{\hbar} \right\}, \quad \rho = j^0 \qquad (79)$$

satisfies the Schrödinger equation

$$i\hbar \frac{\partial \psi}{\partial t} = \hat{H}\psi \qquad (80)$$

where

$$\hat{H} = \frac{1}{2m} \sum_{\alpha=1}^{3} \left(-i\hbar \frac{\partial}{\partial q^\alpha} - \frac{e}{c} A_\alpha \right)^2 - \frac{e}{c} A_0 \qquad (81)$$

A_i ($i = 0, 1, 2, 3$) is the 4-potential of the electromagnetic field, e and m are the charge and mass of the particle, respectively.

With proper normalization of the wave function (79) the rule for calculation of the mean values $\langle R \rangle$ of the quantity R has the form (8) for quantities representing momentum, energy, angular momentum and arbitrary functions of coordinates. The stationary states of an ensemble are described by a wave function which is an eigenfunction of the Hamiltonian (81).

Thus the difference of these results from the basic statements of quantum mechanics consists only in that (8) is fulfilled not for all physical quantities, but only for certain ones. There are differences just in those points that, as we have seen, cannot be tested experimentally.

Now, using the statistical principle, I am going to generalize the results obtained for a pure ensemble of one-particle systems on the case of an arbitrary ensemble of one-particle systems. In the case, when restriction (78) is fulfilled the j^i ($i = 0, 1, 2, 3$) determined by the wave function (79). In this case the n-state W of an arbitrary ensemble E can be considered as a function $W[\psi]$ of two independent quantities ψ and ψ^* (ψ^* is the complex conjugate of ψ). The relation (77) takes the form

$$\frac{\partial W}{\partial t} + \frac{1}{i\hbar} \int \left\{ \frac{\delta W}{\delta \psi(x)} \hat{H}\psi(x) - (H\psi(x))^* \frac{\delta W}{\delta \psi^*(x)} \right\} dx = 0 \qquad (81a)$$

* denotes complex conjugate, and $\hat{H}$ is the hamiltonian (81). As $W[\psi]$ represents the probability density of finding a pure ensemble described by the wave function ψ, then the mean value $\langle R \rangle_E$ of the quantity R over the ensemble E can be written in the form

$$\langle R \rangle_E = \int W[\psi] \langle R \rangle_\psi \, d[\psi] \qquad (82)$$

where the integral denotes integration over the whole functional space of the values of functions ψ and ψ^*, and $\langle R \rangle_\psi$ denotes the mean value of the quantity R over the pure ensemble described by the wave function ψ. All functions are supposed to be normalized by means of the relation

$$\int \psi^*(x)\psi(x) \, dx = 1 \qquad (83)$$

Let us choose an orthogonal basis $\{\chi_i\}$ in the Hilbert space of functions ψ. Let the function ψ be decomposed on this basis according to the relations (20) and (21), with decomposition coefficients $a_i = a_i[\psi]$ being functionals of ψ and ψ^*. Due to (8) and (20) one obtains

$$\langle R \rangle_\psi = \sum_{i,k} a_i^*[\psi] a_k[\psi] R_{ik} \qquad (84)$$

where R_{ik} is a matrix element of the operator R on the basis $\{\chi_i\}$

$$R_{ik} = \int \chi_i^*(x) \hat{R} \chi_k(x) \, dx \qquad (85)$$

Substituting (84) into (82) one obtains

$$\langle R \rangle_E = \sum_{i,k} R_{ik} U_{ki} = Sp\{\hat{R}\hat{U}\} \qquad (86)$$

where the following notation is used

$$U_{ik} = \int W[\psi] a_i[\psi] a_k^*[\psi] \, d[\psi] \qquad (87)$$

and $\hat{U}$ denotes an operator with matrix elements U_{ik} on the basis $\{\chi_i\}$. The operator $\hat{U}$ depends only on the state $W[\psi]$ of the ensemble E but does not depend on the quantity R. The relation (86) coincides with (3), which describes the rule for the calculation of the mean value for an arbitrary ensemble.

Multiplying (81a) by $a_i[\psi]a_k^*[\psi]$ and integrating over the functional space of functions ψ one obtains after transformations an equation describing the evolution of the matrix elements U_{ik} of state operator $\hat{U}$. It has the form

$$i\hbar\frac{\partial \hat{U}}{\partial t}+\hat{U}\hat{H}-\hat{H}\hat{U}=0 \qquad (88)$$

and is equivalent to (4).

Thus, in the non-relativistic approximation one succeeds in showing that for an arbitrary ensemble of one-particle systems the basic statements of quantum mechanics (1)–(4) can be obtained starting from the statistical principle with the restriction that the calculation rule (86) for mean values can only be applied for arbitrary functions of coordinates and additive quantities: energy, momentum and angular momentum. The last restriction arises because the expression (84) for the $\langle R \rangle_\psi$ has been obtained from (8) which is valid in this case only for those quantities R.

Using the statistical principle, the results derived for a pure ensemble of two-particle systems (Rylov, 1973a, b) can be generalized to the case of an arbitrary ensemble.

It is interesting to consider how the process of moment measurement described in the second section looks from the point of view of relativistic statistics (this I call the conception which used the r-state and statistical principle). Let us imagine an ensemble of usual Brownian particles (specks of dust), spaced in a moving gas. The gas velocity is supposed to be less than the thermal velocity of the gas molecules, and the mass of a speck of dust is supposed to be of the order of the mass of the molecule. The motion of the particles has a character of random wandering. If the gas motion is neglected then the mean square displacement of a particle during the time t with respect to its initial position has the form

$$\langle(\mathbf{q}-\mathbf{q}_0)^2\rangle = 2Dt \qquad (89)$$

where $\mathbf{q}_0$ is the initial position of a particle, and $\mathbf{q}$ is its position at time t. D is a diffusion coefficient. Defining the mean value v_t during the period t by means of

$$\mathbf{v}_t = \frac{\mathbf{q}-\mathbf{q}_0}{t} \qquad (90)$$

one finds that during the time t the mean velocity of the particle will be reduced as the period t is increased and it will tend to zero as $t \to \infty$. It is conditioned by the fact that the root-mean-square displacement is proportional to $\sqrt{t}$, as is seen from (89). If the gas motion is taken into account, then the contribution of the systematic motion of the gas to the root-mean-square displacement is proportional to ut, where u is the mean velocity of the gas. This contribution dominates if t is large enough. As a result one of the gas streams takes the speck of dust away. The mean velocity during the time t will be different for different specks of dust, but, in general, if $t \to \infty$ it will not tend to zero but will tend to some value which is equal to the velocity of the gas stream taking the speck of dust away. The velocity of each element of gas is supposed to tend to a constant as $t \to \infty$. The magnitude of the velocity of the speck of dust averaged over period t (with $t \to \infty$) depends only on which gas stream catches that speck.

Let us estimate a distribution of the dust specks over velocities using (90). If the time t is of the order of the mean free time then the velocity distribution will be a Maxwellian one with a temperature close to that of the gas. With increasing time t the distribution remains Maxwellian, but its temperature reduces until the root-mean-square velocity of the specks calculated by means of (89) and (90) becomes of the order of the gas speed. With further increasing of time t the velocity distribution of the specks is determined mainly by the gas-stream velocities.

The dependence of the velocity distribution on the measurement time t is rather like that of an electron in the state described by a wave function. The difference is only in the absence of an agent taking the electron away as the gas does with the dust specks. In quantum mechanics the statistical ensemble (dynamical system!) plays the part of such a carrying agent. The conservation laws of energy and momentum are fulfilled for the ensemble. This prevents the electron momentum being dissipated and provides a constant motion of the electron in a definite direction. Thus, from the viewpoint of relativistic statistics the q-momentum measurement represents the measurement of the momentum of an electron averaged over a long period.

Relativistic statistics permits one to obtain all the basic statements of quantum mechanics with the reservation mentioned above. It is curious that this conception does not contain anything typical for quantum physics. It is based completely upon principles of classical mechanics and classical statistics. Besides the motion of microscopic particles is assumed to be non-deterministic. Such an assumption in itself is not specific to quantum mechanics. It occurs occasionally in classical statistics, for instance, in the description of Brownian particles. The only essential supposition is that about a form of the term which is added to the Lagrangian of a pure ensemble of deterministic particles for the description of particle indeterminacy. This term has a universal form (it does not depend on particle characteristics) with Planck's constant being a coefficient before it. This is the only place, where Planck's constant $\hbar$ is introduced into the theory. The introduction of such a term cannot be treated as a principle because some supposition about the character of the particle motion indeterminacy should be made in any case.

Being in its form a classical (non-quantum) conception, the relativistic statistics means in no case a returning to classical mechanics. It gives a less detailed description of a dynamical system than quantum mechanics does and far less detail than in the classical case. It is a further step on the way to the restriction of detailed description in a microcosm.

Being essentially a relativistic theory, the relativistic statistics coincide with quantum mechanics only in a non-relativistic approximation. Linearity (equation linearity, linear operators, the linear superposition principle), which is raised into a principle by quantum mechanics, is one of the main reasons for the advance of quantum mechanics. All this is conditioned by the non-relativistic character of the approximation.

It is most curious that the conventional way of joining quantum theory with relativity, using wave functions, the linear superposition principle and so on, cannot be understood from the viewpoint of relativistic statistics. Accordingly to relativistic statistics nothing of that kind should be done. The non-relativistic approximation must not be used: the problem should be solved exactly. This claim of relativistic statistics does not seem a complete absurdity, if it is taken into account that in a consequent application the relativistic statistics contains a possibility of the generation of pairs i.e. particle–antiparticle (Rylov, 1975). The future will show how firmly the claims of relativistic statistics are founded.

REFERENCES

Aharonov Y. and Bohm D. (1961) 'Time in the quantum theory and the uncertainty relation for time and energy', *Phys. Rev.*, **122**, 1649–1658.
Ballentine I. L. (1970) 'The statistical interpretation of quantum mechanics', *Rev. Mod. Phys.*, **42**, 358–381.
Belinfante F. J. (1973) *A Survey of Hidden-Variables Theories*, Pergamon Press, Oxford.
Blokhintsev D. I. (1963) 'Foundation of quantum mechanics', Moscow, Leningrad (in Russian).
Blokhintsev D. I. (1968) *The Philosophy of Quantum Mechanics*, Reidel, Dordrecht, Holland. (Original Russian edition 1965).
Bohm D. (1965) *Quantum Theory*, Prentice-Hall, Englewood Cliffs, New Jersey.
Bohr N. (1918) 'On the quantum theory of line spectra', *Koniglige Danske Videnskabernes Selskabs skrifter Naturvidenskabelig og. mathematisk Aufdelung 8 Raekke*, Bd. **4**, I, 1–118.
Bohr N. (1928) 'The quantum postulate and the recent development of atomic theory', *Nature*, **121**, 580–590.
Einstein A. (1949) in *Albert Einstein: Philosopher–Scientist*, P. A. Schlipp, Ed., Library of the Living Philosophers, Evanston (reprinted by Harper and Row, New York, p. 665).
Fock V. and Krylov N. (1947) 'On the uncertainty relation between time and energy', *J. Phys.* (*U.S.S.R.*), **11**, 112–120.
Fock V. A. (1962) 'On the uncertainty relation between time and energy and one attempt to disprove it', *J. Experim. Theoret. Phys.*, **42**, 1135–1139 (in Russian).
Gerlach W. and Stern O. (1924) 'Über die Richtungsquantelung im Magnetfeld', *Ann. Physik*, **74**, 673–699.
Hakim, R. (1967a) 'Remarks on relativistic statistical mechanics. I', *J. Math. Phys.*, **8**, 1315–1344.
Hakim R. (1967b) 'Remarks on relativistic statistical mechanics. II. Hierarchies for reduced densities', *J. Math. Phys.*, **8**, 1379–1400.
Hakim, R. (1968) 'Relativistic stochastic processes', *J. Math. Phys.*, **9**, 1805–1818.
Heisenberg W. (1927) 'Über den anschaulichen Inhalt der quantentheoretischen Kinematik und Mechanik', *Z. Physik*, **43**, 172–198.
Heisenberg W. (1930) *Die Physikalischen Prinzipen der Quantetheorie*, Leipzig, Section II, 2d.
Heisenberg W. (1955) 'Quantum theory and its interpretation', in *Niels Bohr and the Development of Physics*, London.
Kaliski S. (1970) 'A tentative classical approach to quantum mechanics, *Proceedings of Vibration Problems*, **11**, 3–17.
Landau L. and Peierls R. (1931) 'Erweiterung des Unbestimmtheisprinzips für die relativistische Quantentheorie', *Z. Physik*, **69**, 56–69.

Landau L. D. and Lifshits E. M. (1963) *Quantum mechanics*, Moscow, Section 77 (in Russian).
Mandelstam L. and Tamm Ig. (1945) 'The uncertainty relation between energy and time in non-relativistic quantum mechanics,' *J. Phys.* (U.S.S.R.), **9**, 249–254.
Mandelstam L. I. (1950) 'Lectures on foundation of quantum mechanics (Theory of indirect measurements),' in *Complete collection of proceedings*, **5**, 345–415. (in Russian).
Margenau H. (1963) 'Measurement in quantum mechanics,' *Ann. Phys.*, **23**, 469–485.
Moyal J. E. (1949) 'Quantum mechanics as a statistical theory,' *Proc. Cambridge Phil. Soc.*, **45**, 99–124.
Neumann J. V. von (1932) 'Mathematische Grundlagen der Quantenmechanik,' Berlin.
Pauli W. (1933) 'Die allgemeinen Prinzipen der Wellenmechanik,' in H. Geiger and K. Scheel *Handbuch der Physik*, 2nd ed., Vol. 24/1, Springer. Berlin, Chap. 2, pp. 83–272.
Pearle P. (1967) 'Alternative to the orthodox interpretation of quantum theory', *Am. J. Phys.*, **35**, 742–753.
Popper K. R. (1959) *The Logic of Scientific Discovery*, Basic Books, New York.
Reece G. (1973) 'Theory of measurement in quantum mechanics,' *Intern. J. of Theoret. Phys.*, **7**, 81–117.
Robertson H. P. (1929) 'The uncertainty principle', *Phys. Rev.*, **34**, 163–164.
Rylov Yu. A. (1971) 'Quantum mechanics as a theory of relativistic Brownian motion,' *Ann. Physik*, **27**, 1–11.
Rylov Yu. A. (1973a) 'Quantum mechanics as relativistic statistics. I: the two-particle case,' *Intern. J. Theoret. Phys.*, **8**, 65–83.
Rylov Yu. A. (1973b) 'Quantum mechanics as relativistic statistics. II: the case of two interacting particles,' *Intern. J. Theoret. Phys.*, **8**, 123–139.
Rylov Yu. A. (1975) 'The problem of particle generation in classical mechanics,' in *Investigation of Cosmic rays*, Moscow (in Russian, pp. 171–177).
Vlasov A. A. (1966) *Statistical Distribution Functions*, Moscow (in Russian).
Wick G. C., Wightman A. S. and Wigner E. P. (1952) 'The intrinsic parity of elementary particles,' *Phys. Rev.*, **88**, 101–105.
Wigner E. (1963) 'The problem of measurement,' *Am. J. Phys.*, **31**, 6–15.

10

Uncertainty, Correspondence and Quasiclassical Compatibility

JAN J. SŁAWIANOWSKI
Polish Academy of Sciences, Warsaw

1. INTRODUCTION

There was no exaggeration in the famous metaphor of Sommerfeld comparing the correspondence principle with a magic wand. In fact, within the framework of the Old Quantum Theory this Principle was the only method which enabled physicists to evaluate such physical quantities as the intensities of spectral lines and the polarizations of atomic radiation. However the very nature and the internal logic of the correspondence itself remained unknown. Using the more fashionable terms of cybernetics we might say that the Bohr correspondence principle has worked like a black box.

There were many such 'magic wands' in quantum mechanics; some of them strongly influenced the development of the theory, especially in its early days. There is nothing strange in this: quantum theory aims to describe microscopic phenomena which are so far from our everyday experience that when analysing them, all Newtonian intuitions break down. Therefore, from the point of view of philosophy based on Newtonian mechanics, all the famous quantum postulates were incomprehensible and could only be justified at the stage of final results. As a matter of fact, the whole of the old quantum theory consisted of mysterious magic wands. For example the theoretical status of the Bohr–Sommerfeld quantum conditions was completely inconceivable. The same is true of the Planck–Einstein postulates about the quantum granular structure of electromagnetic radiation. Similarly, the material waves as postulated by de Broglie, were more a mathematical idea than a physical picture in a classical sense.

From the heuristic point of view two so-called 'principles' were of special importance namely the afore-mentioned correspondence principle and the Heisenberg uncertainty principle. Some historical comments are necessary here. The correspondence principle was formulated in the days of the old quantum theory. When formulating and developing the main ideas of quantum mechanics, Heisenberg, Born and Jordan referred explicitly to it (Heisenberg, 1925; Born and Jordan, 1925; Born, Heisenberg and Jordan, 1926). On the

148 Uncertainty Principle and Foundations of Quantum Mechanics

other hand, the uncertainty principle was derived by Heisenberg later in 1927, two years after he had discovered matrix mechanics (Heisenberg, 1927). Hence it could be said that the uncertainty idea being a consequence of fundamental quantum rules, has not played any heuristic role in the development of quantum mechanics. However this is not the case. The uncertainty principle is only a mathematical/quantitative comment on the qualitative postulate which enabled Heisenberg to discover the quantum rules (Heisenberg, 1925). In the following, I shall call it the uncertainty postulate or 'postulate of the phase nonlocalizability'.

This postulate and the correspondence principle were just the magic wands which enabled Heisenberg to create the matrix mechanics—the first correct formulation of quantum theory.

Now let us consider these postulates with special emphasis to their qualitative content and philosophical assumptions.

(A) The Correspondence Principle

According to this principle classical laws are asymptotic/approximate expansions of the quantum ones. There are two meanings of the classical limit: (a) the formal one based on the transition $h \to 0$ and (b) the physical asymptotics of large quantum numbers. It depends on the kind of problem, which of these two asymptotics is to be used. The oldest formulation of the correspondence principle, which is due to Bohr, concerned only the theory of atomic radiation. Bohr formulated some methodological guiding hints which enabled him to get some qualitative and even quantitative results concerning intensities and polarizations of spectral lines. The selection rules, for example, have been formulated in such a way. The semi-classical theory of atomic spectra based on the correspondence principle was a hybrid of the classical theory of electromagnetic radiation and the Bohr–Sommerfeld quantum conditions imposed upon a classical multiply-periodic system. Hence it joined in a mysterious magic way two contradictory pictures of radiation: the Bohr–Sommerfeld quantum jumps and the classical continuous Hertz–Maxwell radiation.

The very idea of the correspondence principle was based on the philosophical belief in the existence of some rigorous quantum theory of atomic phenomena which had remained undiscovered up to 1925. This belief motivated all efforts at guessing this unknown theory by an appropriate reformulating of its asymptotic form, i.e. classical mechanics and electrodynamics (Born, 1925).

(B) The Uncertainty Postulate

This postulate was the basic idea of the epoch-making paper of Heisenberg (1925). According to it, microscopic phenomena are essentially non-local in both configuration and phase space. Concepts such as a trajectory and a hodograph of an electron become essentially inadequate within the framework of atomic phenomena. They are incompatible with the Heisenberg matrix representation of physical quantities. Therefore, the resulting quantum theory breaks with classical ideas even on the elementary level of *kinematics*. Before Heisenberg's discovery it was expected rather that the classical *dynamics* (the equations of motion) was to be replaced by some quantum theory, but nobody supposed the classical notions of state, position, etc., to be essentially inadequate.

The uncertainty principle derived by Heisenberg in 1927 is a mathematical comment to his uncertainty postulate. It describes in quantitative terms the phase space non-localness of quantum phenomena. The modern general formulation of this principle predicts the relationship between statistical dispersions of arbitrary quantities when measured in the same quantum state. When ΔA, ΔB are dispersions of quantities described by operators $\hat{A}$, $\hat{B}$ on the quantum state ρ then

$$\Delta A \, \Delta B \geq \frac{\hbar}{2} |\langle [\hat{A}, \hat{B}] \rangle_\rho| \qquad (1)$$

In particular for positions and momenta

$$\Delta q^i \, \Delta p_j \geq \frac{\hbar}{2} \delta^i_j \qquad (2)$$

Therefore quantum phenomena are non-local in a classical phase space. The critical phase volume characteristic for this non-localness is of the order $(\hbar/2)^n$ where n is a number of degrees of freedom.

Within the finished modern framework of the quantum theory the uncertainty and the correspondence principles seem to be rather secondary results of the basic assumptions and automatically following from them in a purely logical way. In spite of such views we are going to show that there are still some doubtful questions and physical problems of interpretation connected with these principles. The correspondence of some classical and quantum concepts is a delicate matter. This concerns for example pure states. It appears that the purely logical formal approach based on the $h \to 0$ asymptotics is not sufficient. Our analysis leads to interesting physical consequences concerning the relationship between concepts of information and symmetry on the classical and the quantum level (Sławianowski, 1973). We start with the derivation of the Weyl–Wigner–Moyal phase-space formulation of quantum mechanics directly from the uncertainty postulate.

2. THE WEYL–MOYAL QUANTIZATION
(if Heisenberg Had Started with Statistical Mechanics)

The uncertainty principle does not preclude us from describing quantum phenomena in terms of classical phase space. It implies only that such a description shall be essentially non-local.

150 Uncertainty Principle and Foundations of Quantum Mechanics

The phase-space formulation of quantum mechanics is due to Weyl (1928, 1931), Wigner (1932) and Moyal (1949). In particular, Wigner functions describing quantum statistical ensembles have long been used in physical chemistry. The phase-space methods as developed by Wigner seem to be a secondary accidental consequence of the usual Hilbert space approach. Weyl and Moyal developed more systematic formulations in which the classical phase space and its geometry played an essential role prior to the Hilbert space techniques. Nevertheless the basic ideas, motives and techniques of Moyal and those of Weyl were completely different. Moyal aimed at replacing an abstract geometry of Hilbert spaces by more familar statistical methods in a classical phase space. On the other hand the group-theoretical Weyl approach is based completely on the mathematical *a priori*.

The formulation we present here joins and unifies the approaches of Weyl and Moyal. Starting with the uncertainty principle as a primary idea, we suggest reformulating the classical statistical mechanics so as to turn it into non-local theory. It appears that the Weyl–Wigner–Moyal formulation of quantum mechanics is the most natural result. To get it, we will appeal to some group-theoretical results of Bargmann (1954). Let us note that starting with the uncertainty relations as a primary basis for the 'deductive' construction of quantum mechanics is historically justified and free from tautology. In fact the famous derivation of this principle based on the idea of the Heisenberg microscope appealed explicitly to the semi-classical model of interaction between electromagnetic field and electrons. Hence, it was certainly possible to achieve this result before 1925, when the matrix mechanics was formulated.

Unfortunately, the Weyl–Wigner–Moyal approach is applicable only to systems with affine symmetry of degrees of freedom:

An affine phase space is a triplet (P, Π, Γ) where:

(1). (P, Π) is an affine space; P is its underlying set, i.e. a manifold of classical states and Π is a linear space of translations (free vectors) on P.

(2). Γ is a covariant skew-symmetric and non-degenerate tensor of the second order on Π: $\Gamma \in \Lambda^2 \Pi^*$.

The only translation carrying $a \in P$ over onto $b \in P$ will be denoted as $\vec{ab} \in \Pi$. We will use only affine coordinates on P: when $(a; \ldots e_i \ldots)$ is an affine frame, i.e. $a \in P$ and $\{e_i\}$ is a linear basis in Π, then the corresponding affine coordinates of $b \in P$ are components of $\vec{ab}$ with respect to the basis $\{e_i\}$

$$\vec{ab} = \xi^i(b) e_i$$

The reciprocal contravariant tensor of Γ will be denoted as $\tilde{\Gamma}$. Raising and lowering of indices is to be understood in the sense of skew-symmetric 'metric' Γ. Instead of $\tilde{\Gamma}^{ab}$, we will write shortly Γ^{ab} this does not lead to misunderstandings

$$\Gamma^{ac} \Gamma_{cb} = \delta^a_b$$

Γ gives rise to the skew-symmetric scalar product on Π and to the Poisson bracket operation.

The *Poisson bracket* of smooth functions F, G on P is given by

$$\{F, G\} = \langle dF \otimes dG, \tilde{\Gamma}\rangle = \Gamma^{ab}\frac{\partial F}{\partial \xi^a}\frac{\partial G}{\partial \xi^b} \tag{3}$$

where ξ^a are affine coordinates on P.

The algebraic non-singularity of Γ implies that $\dim P = 2n$, where n is a natural number, the so-called *number of degrees of freedom*.

One can choose the affine frame in P in such a way that

$$\|\Gamma_{ab}\| = \begin{Vmatrix} 0 & I \\ -I & 0 \end{Vmatrix} \tag{4}$$

where I is an $n \times n$-identity matrix (Kronecker matrix), 0 is a matrix with all elements vanishing and Γ_{ab} are components of Γ with respect to the choosen frame. Denoting the corresponding affine coordinates as $(\ldots q^i \ldots, \ldots p_i \ldots)$ we find

$$\{F, G\} = \frac{\partial F}{\partial p_i}\frac{\partial G}{\partial q^i} - \frac{\partial F}{\partial q^i}\frac{\partial G}{\partial p_i} \tag{5}$$

Such coordinates are called *canonical* ones.

Affine structure gives rise to the translationally-invariant Lebesgue measure on P. It is unique up to normalization. One can normalize it in such a way that the integration consists in

$$f \mapsto \int f \, d\Omega = \int f \, dq^1 \ldots dq^n \, dp_1 \ldots dp_n \tag{6}$$

for any system of canonical coordinates. The above definition is correct because it does not depend on the particular choice of *canonical* coordinates.

Such a normalization is inconvenient from the point of view of statistical mechanics, because the measure has a physical dimension of the nth power of action. So there is no unique definition of the entropy of statistical ensembles until some unit of action λ is chosen. Such a choice enables us to get rid of the physical dimension in the volume measure on P. Namely we put this measure μ_λ as follows

$$\int f(p) \, d\mu_\lambda(p) = \lambda^{-n}\int f \, dq^1 \ldots dq^n \, dp_1 \ldots dp_n \tag{7}$$

where $(q^1 \ldots p_n)$ are arbitrary canonical coordinates on P.

The *entropy* $S_\lambda(\rho)$ of the probabilistic measure on P, being absolutely continuous with respect to μ_λ and having density ρ, is given as

$$S_\lambda(\rho) = \int \rho \ln \rho \, d\mu_\lambda \tag{8}$$

Roughly speaking $\mu_\lambda(U)$ is the number of λ^n cells contained in the domain $U \subset P$.

152 Uncertainy Principle and Foundations of Quantum Mechanics

According to the Planck theory of black-body radiation we have at our disposal the quantum of action which is at the same time the universal constant of nature

$$h = 6.54 \times 10^{-27} \text{ erg sec} = 2\pi\hbar$$

Hence, we could put $\lambda = h$. *This is just the first place where physical quantum notions taken from experimental analysis are introduced into the geometrical framework of the classical phase space.* To retain the correspondence with some commonly used formulas we will rather put $\lambda = h/2$. Obviously, this does not alter any essential matter; such a choice only simplifies our notation. In the following we will often write μ, S simply instead of $\mu_{h/2}$, $S_{h/2}$. Hence our measure is given as follows:

$$\int f \, d\mu = (2/h)^n \int f \, dq_1 \ldots dq^n \, dp_1 \ldots dp_n \tag{9}$$

Besides the above entropy argument there is one more reason justifying (7) with $\lambda = h/2$. In fact, according to the uncertainty relations any essentially $2n$-dimensional physical situation in P must be smeared out onto a region in P the Ω volume of which exceeds $(h/2)^n$. States of quantum systems are related to phase-space regions the $\mu_{h/2}$ volumes of which are of order one at least. Therefore roughly speaking, $\mu_{h/2}(U)$ gives an account of the number of quantum states within the phase-space region $U \subset P$. Similarly, one can reasonably expect that quantum statistics attaches well-defined probabilities to those subsets only which are large when compared with critical h^n cells (which 'contain' many quantum states). *The uncertainty Principle suggests that we reformulate the classical statistical mechanics so as to turn it into non-local theory.*

Let us start with the basic notions of classical statistics in P: *Physical quantities* i.e. *random variables* are described by real analytic functions on P.

Statistical ensembles are described by probabilistic measures on P. We are especially interested in measures absolutely continuous with respect to $\mu_{h/2}$; they are described by positive statistical densities normalized in such a way that

$$\int \rho \, d\mu_{h/2} = 1$$

Obviously

$$\rho^* = \rho \tag{9a}$$

$$\int \rho A^* A \, d\mu \geq 0 \quad \text{for arbitrary } A \tag{10}$$

In the following even when talking about measures concentrated on subsets of μ-measure zero we will describe them shortly by 'densities' ρ keeping only in mind that they are then distributions rather than usual functions.

The operational statistical interpretation of the above notions is based on the following concepts:

(1). *Expectation value* of a physical quantity A in the ensemble ρ:

$$\langle A \rangle_\rho = \int A(p)\rho(p)\, d\mu_{h/2}(p) \tag{11}$$

(2). The non-normalized *probability of detecting* a system in a statistical state ρ_2 when it is known to be in a state ρ_1:

$$P(\rho_1, \rho_2) = \int \rho_1(p)\rho_2(p)\, d\mu_{h/2}(p) \tag{12}$$

(3). *'Proper-ensemble'* of physical quantity A with an eigenvalue $a \in A(P)$. Measurements of A on this ensemble give the result a with certainty; there is no statistical spread. Therefore, A becomes constant and equal to a when restricted to the support of ρ:

$$A | \text{Supp}\, \rho \equiv a \tag{13}$$

This is equivalent to the eigenequation:

$$A\rho = a\rho \tag{14}$$

When A is a non-constant analytic function, then:

$$\rho = F\delta(A - a) \tag{15}$$

where F is non-negative on $\text{Supp}\, \rho$. Hence, 'proper ensembles' of A with an eigenvalue a are measures concentrated on the a-value surface of A:

$$M_{(A,a)} = \{p \in P : A(p) = a\} \tag{16}$$

When $F = 1$, (15) becomes an (A, a)-*microcanonical ensemble*, i.e. statistical distribution $\delta(A - a)$.

The only essentially local notion used in (1), (2) and (3) is the usual, pointwise product of functions (distributions). This is especially apparent in (3), i.e. in the classical spectral analysis of statistical ensembles and physical quantities. The lack of statistical dispersion of A measurements performed on the ensemble ρ is evidently a *physical* notion, prior to any particular mathematical model. It is only the structure of the associative function-algebra over P under the pointwise product, that is responsible for the *essentially local* relationship between proper ensembles of A and value surfaces of A. Therefore, the most natural way to achieve the non-local description of measurements in P compatible with the uncertainty principle is to replace the afore-mentioned associative algebra by some other function-algebra based on the non-localized product $A \perp B$.

$$(A \perp B)(p) = \int \mathcal{K}(p : p_1, p_2) A(p_1) B(p_2)\, d\mu(p_1)\, d\mu(p_2) \tag{17}$$

Such a product gives rise to the non-local statistical theory in P: one should only replace the pointwise product in (10), (11) and (12) and especially (14) by the above product (17). Now, we will find the appropriate form of $\mathcal{H}$, starting with some natural postulates:

We assume $\perp$ to be translationally invariant:

$$(A \perp B)_\pi = A_\pi \perp B_\pi \tag{18}$$

for arbitrary functions A, B and $\pi \in \Pi$; C_π denotes the function obtained by a π-translation of C:

$$C_{\vec{pq}}(q) = C(p) \tag{19}$$

Translational invariance implies

$$\mathcal{H}(p; p_1, p_2) = K(\overrightarrow{pp_1}, \overrightarrow{pp_2}) \tag{20}$$

for some function K.

We assume the non-local product (17) to be associative:

$$(A \perp B) \perp C = A \perp (B \perp C) \tag{21}$$

This results in the functional equation for K

$$\int K(x_1, x) K(x_2 - x, x_3 - x) \, dx = \int K(x_1 - x, x_2 - x) K(x, x_3) \, dx \tag{22}$$

(Sławianowski, 1974) where dx is a translationally-invariant Lebesgue measure on Π. The convolution-like structure of (22) suggests that we search for its solution in the Fourier representation. Our functional equation becomes then purely algebraic:

$$\hat{K}(\xi_1, \xi_2 + \xi_3) \hat{K}(\xi_2, \xi_3) = \hat{K}(\xi_1, \xi_2) \hat{K}(\xi_1 + \xi_2, \xi_3) \tag{23}$$

(Sławianowski, 1974), where $\hat{K}: \Pi^* \times \Pi^* \to C^1$ is a Fourier transform of K:

$$\hat{K}(\xi, \eta) = \int K(x, y) \exp[-i(\langle \xi, x \rangle + \langle \eta, y \rangle)] \, dy \, dx \tag{24}$$

Equation (23) is easily recognized to coincide with the functional equation for *factors of projective representations* of the abelian additive group Π^* (Bargmann, 1954). Even without any appealing to the theory of projective representations, it is easy to show (in elementary terms) that the only *smooth* and *bounded* solutions of (23) are given by:

$$\hat{K}(\xi, \eta) = \exp[i\langle \xi \otimes \eta, B \rangle] = \exp[iB^{ab} \xi_a \eta_b] \tag{25}$$

where $B \in \Pi \otimes \Pi$ is an arbitrary real, contravariant tensor of the second order on Π (Bargmann, 1954).

The correspondence principle suggests that we put B proportional to $\tilde{\Gamma}: B = b\tilde{\Gamma}$, because of the fundamental role of Γ in the geometry of P. If we had chosen any other form of B, we would have broken the symplectic symmetry of the problem; there is no sufficient, non-arbitrary reason for any other choice of

B. One can easily show that when $B = b\tilde{\Gamma}$, then:

$$K(x, y) = \exp\left[\sigma \frac{i}{\hbar}\langle \Gamma, x \otimes y \rangle\right] = \exp\left[\sigma \frac{i}{\hbar}\Gamma_{ab}x^a y^b\right] \quad (26)$$

where σ is some real number depending on b, and using of $\hbar$ enables us to get rid of the physical dimension in the exponent, which is necessary if (26) is to be well-defined. It appears that the particular choice of σ does not matter— associative algebras corresponding to various values of σ are isomorphic with each other. To attain the correspondence with the currently used notations, we put $\sigma = 2$. (Obviously, there is no physics nor mathematics in any such choice.) Finally:

The non-local Weyl–Moyal product is defined as:

$$(A \perp B)(p) = \int \exp\left[\frac{2i}{\hbar}\langle \Gamma, \overrightarrow{p_1 p} \otimes \overrightarrow{p_2 p}\rangle\right] A(p_1) B(p_2) \, d\mu_{h/2}(p_1) \, d\mu_{h/2}(p_2) \quad (27)$$

Let us quote the following properties of $\perp$:

$$\int A \perp B = \int AB = \langle A^* | B \rangle \quad (28)$$

$$(A \perp B)^* = B^* \perp A^* \quad (29)$$

$$1 \perp A = A \perp 1 = A \quad (30)$$

moreover, the constant function 1 is the only function satisfying (30) for all A.

$$A^* \perp A \neq 0 \quad (31)$$

unless A is a function vanishing almost everywhere.

$$\langle C \perp A | B \rangle = \langle A | C^* \perp B \rangle = \langle C | B \perp A^* \rangle \quad (32)$$

If $C \perp A = 0 = A \perp D$ for all A, then $C = D = 0$ \quad (33)

Contrary to the pointwise product, the Weyl multiplication is non-commutative; its centre consists of constant functions only.

Besides, let us notice the following asymptotic formulas which give account of the correspondence principle:

$$\lim_{h \to 0} A \perp B = AB \quad (34)$$

$$\lim_{h \to 0} \frac{1}{\hbar i}(A \perp B - B \perp A) = \{A, B\} \quad (35)$$

provided A, B are smooth and $A \perp B$ is well-defined. Replacing in (10), (11), (12) and (14) the pointwise product AB by the Weyl product $A \perp B$ we get the non-local statistical theory in P. The property (28) implies that (11) and (12) do not change then at all. On the contrary (14) is replaced in a non-trivial way by the following eigenequation:

$$A \perp \rho = a\rho \qquad (36)$$

where both a and ρ are unknown. We are looking only for probabilistic solutions of (36), i.e. such functions (distributions) ρ which satisfy the normalization condition and are positively semi-definite in the non-local sense:

$$\int \rho \, d\mu = 1 \qquad (37)$$

$$\langle \rho | A^* \perp A \rangle = \int \rho(A^* \perp A) \, d\mu = \int \rho \perp A^* \perp A \, d\mu \geq 0 \qquad (38)$$

for all functions A. The subset of R composed of those values of a for which probalistic solutions of (36) exist, is in no direct way related to $A(P)$. The relationship between solutions of (36) and value surfaces of A is essentially non-local. Let us notice that contrary to the classical semi-definiteness condition: $\langle \rho | A^* A \rangle \geq 0$, (38) does not imply that $\rho \geq 0$. Quantum density functions are allowed to take negative values. They become positive in the usual, pointwise and local sense after coarse-graining over subsets of P which are large when compared with critical Heisenberg h^n cells. The quantity

$$P_U = \int_U \rho \, d\mu \qquad (39)$$

can be approximately interpreted as a non-negative probability of localization in $U \subset P$ only when $\mu_{h/2}(U) \gg 1$.

Quantum spectral analysis and quantum measurements, being based on (36) are *essentially* non-local in P. The very structure of (27) implies that this non-localness is of the Ω-order h^n, which agrees with the uncertainty principle.

The non-local Weyl–Moyal statistical mechanics is obviously isomorphic with the usual Hilbert space formulation of quantum mechanics. The corresponding isomorphism, the so-called *Weyl prescription* attaches to functions on P, operators in an appropriate Hilbert space. Denoting the operator corresponding to A as $\hat{A}$, we have:

$$\int A \, d\mu = \text{Tr} \, \hat{A}, \quad \widehat{aA + bB} = a\hat{A} + b\hat{B}, \quad \widehat{A \perp B} = \hat{A}\hat{B},$$

$$\widehat{A^*} = \hat{A}^+, \quad \langle A | B \rangle = \int A^* \perp B = \int A^* B = \text{Tr}(\hat{A}^+ \hat{B})$$

The $L^2(R^n)$ can be used as a Hilbert space of wave functions. In fact, let (q^i, p_i) be canonical affine coordinates on P and $A(q^i, p_i)$ some function on P (obviously A itself is a function on R^{2n} hence $A(q^i, p_i)$ is to be understood as a function on P resulting from superposing two mappings: $(q^i, p_i): P \to R^{2n}$ and $A: R^{2n} \to C^1$. Do not take $A(q^i, p_i)$ to be a value of A at a point of R^{2n}!). Then, denoting the corresponding operator by $\hat{A}$ and the kernel of its integral representation as $\langle x^i|\hat{A}|y^i\rangle$:

$$(\hat{A}\Psi)(x^i) = \int \langle x^i|\hat{A}|y^i\rangle\Psi(y^i)\,d_n y \tag{40}$$

we find:

$$\langle x^i|\hat{A}|y^i\rangle = \left(\frac{1}{2\pi\hbar}\right)^n \int \exp\left[\frac{i}{\hbar}p_i(x^i-y^i)\right]A\left(\frac{x^i+y^i}{2},p_i\right)d_n p \tag{41}$$

(41) is just the famous *Weyl prescription*.

A pure state corresponding to the wave function $\Psi(x^i)$ is described in the Weyl–Moyal language by the following *Wigner function* (Moyal, 1949; Wigner, 1932):

$$\rho = \left(\frac{1}{2\pi}\right)^n \int \Psi^*\left(q^i - \frac{\hbar\tau^i}{2}\right)\exp(-i\tau^i p_i)\Psi\left(q^i + \frac{\hbar\tau^i}{2}\right)d_n\tau \tag{42}$$

Obviously (42) implies then:

$$\rho \perp \rho = \rho \tag{43}$$

In contrast to the system of classical 'proper equations':

$$A_i\rho = a_i\rho, \quad i=1\ldots m \tag{44}$$

the corresponding quantum systems of eigenequations:

$$A_i \perp \rho = a_i\rho, \quad i=1\ldots m \tag{45}$$

need not be compatible. The compatibility condition has a form:

$$[A_i, A_j] = \frac{1}{\hbar i}(A_i \perp A_j - A_j \perp A_i) = C_{ij}^k \perp (A_k - a_k) \tag{46}$$

where C_{ij}^k are some functions (Dirac, 1964).

One can show that ρ describes a pure state, i.e. (42) or equivalently (43) is satisfied if and only if it satisfies a maximal (impossible to non-trivial extending) compatible system (45).

In more rigorous terms: let B denote the space of such functions A on P so that the corresponding operators $\hat{A}$ (via the Weyl prescription) are bounded. Let $\rho \in B$ describe a quantum statistical ensemble:

$$\int \rho = \text{Tr}\,\hat{\rho} = 1, \quad \langle\rho|A^*\perp A\rangle \geq 0$$

Now, we introduce the following space of functions:

$$E_\rho = \{A \in B : A \perp \rho = 0\} \tag{47}$$

Then, ρ is a *pure state* if and only if E_ρ is a *maximal left ideal* in the associative (although non-commutative algebra $(B, \perp)$. When functions $(A_i - a_i)$ occurring in (45) generate such an ideal, then any consistent system of eigenequations:

$$A_i \perp x = a_i x, \quad F \perp x = 0 \qquad (48)$$

is equivalent to (45), hence, there exist functions F_i such that:

$$F = \sum F_j \perp (A_j - a_j) \qquad (49)$$

For an arbitrary, not necessarily pure statistical ensemble ρ, a real subspace of E_ρ composed of real-valued functions is a Lie algebra under the quantum Poisson bracket (i.e. it is closed under this operation).

3. CORRESPONDENCE PRINCIPLE AND WKB-ASYMPTOTIC EXPANSION
(Information and Symmetry of Statistical Ensembles)

There are two kinds of asymptotics describing the correspondence principle: 'large quantum numbers' and 'small values of the Planck constant'. They are supposed to be essentially equivalent, however, up to now there is no rigorous and general proof of this equivalence. Besides, in spite of some current views neither of these methods leads automatically to classical laws when starting with quantum theory. Some kind of physical intuition and 'feeling' is necessary to avoid mistakes; there are some dangers and traps typical in either of these descriptions.

Roughly speaking, the asymptotics of large quantum numbers consist of the limit transition:

$$\Delta n \to \infty, \quad n_0/\Delta n \to \infty \qquad (50)$$

where n_0 is a mean quantum number of a physical situation and Δn is a spread of quantum numbers (a width of a quantum state is n representation). In technical terms:

$$n_0 \gg \Delta n \gg 1 \qquad (50a)$$

Quantum formulae should approach then those derived from the classical laws.

The typical danger of such a method is the neglect of the condition $\Delta n \gg 1$. In particular, let $\{\Psi_n\}$ be the wave functions of stationary states of a bounded system (e.g. an atom). In general, it is not true for large values of n that Ψ_n become quasiclassical throughout the whole configuration space. (By quasiclassical we mean here the following: interpretable in terms of geometric objects of the classical Hamilton–Jacobi theory.) However, when superposing such quantum states with coefficients slowly varying inside the interval $(n_0 - \Delta n/2, n_0 + \Delta n/2)$ and vanishing outside of it, the non-classical terms of the various Ψ_n approximately cancel each other, provided (50a) is satisfied. In such expressions the functions Ψ_n can be replaced by their classical counterparts $\Psi_{cl\,k}$ built from the Hamilton–Jacobi objects and described by continuous 'quantum numbers' k. Obviously summation over n is then to be replaced by integration over k.

Basing the asymptotics on large quantum numbers is a convincing physical procedure. Its main idea is the cancelling of interference in situations described by rapidly oscillating wave functions (just those with large quantum numbers or short de Broglie waves).

The asymptotics of 'small values' of the Planck constant is of a rather formal nature. The Planck constant h is treated then as a free parameter of the theory. All fundamental mathematical expressions of quantum mechanics are analytic functions of h on the positive real semi-axis: $0 < h < \infty$. Quasiclassical analysis is based on the expansion of these expressions in asymptotic series about the dangerous point $h = 0$. Passing over from finite to vanishing h is connected with some qualitative discontinuities. As a rule, the afore-mentioned asymptotic series are divergent. According to the correspondence principle, their lowest order terms are expected to coincide with the appropriate classical expressions. The structure of asymptotics is essentially the same as that in the 'large quantum numbers' approach: when $h \to 0$, quantum interference phenomena break down because we are then dealing with rapidly oscillating functions. In fact the basic quantum formulas involve expressions such as $\exp(iW/h)$ where W does not depend on h. Obviously they become rapidly oscillating when $h \to 0$. This is just the idea of the method of stationary phase (Born and Wolf, 1964; Erdélyi, 1956).

Technically, the $h \to 0$ approach is much easier and more 'automatic' than that based on $n \to \infty$. However, it is rather formal and one must not forget that the asymptotics of small h are only the convenient conceptual shorthand of the physical asymptotics based on large quantum numbers. In fact, the transition $h \to 0$ transforms formally the whole conceptual structure of the quantum theory into a classical one, but it has nothing to do with the real laboratory conditions of 'quasiclassicality'. Physics is interested rather in answering the question 'what are quasiclassical situations in the real world, when the Planck constant has a fixed value?' Of course, it is only the $n \to \infty$ asymptotics that are able to answer this question. It shows that quantum laws both kinematical and dynamical, when applied to states compatible with (50) asymptotically degenerate to classical laws. Conditions (50) justify even the possibility of an approximate description of a quantum system in terms of the classical notion of state.

Let us notice that in situations described by (50), h is small when compared with typical values of physical quantities, which are then of the order hn_0, $h \Delta n$. This is why the asymptotics $h \to 0$ is justified, but only as a shorthand for 'large quantum numbers'.

Typical dangers of the $h \to 0$ asymptotics are as follows: Let $\{\Psi_n\}$ again be the stationary states of a bounded system. As calculated by means of the h-dependent Schrödinger equation, Ψ_n depend in addition on h. Let us indicate this explicitly by using $\Psi_{(n,h)}$. It would be meaningless to calculate $\lim_{h \to 0} \Psi_{(n,h)}$ and expect any relationships with classical expressions; moreover, as a rule, such a limit does not exist at all. The reason is that it is only the fixed quantum number n, or equivalently the *number of nodes* of Ψ_n, that describes the quantum structure of a state and decides to what extent $\Psi_{(n,h)}$ is essentially quantum or quasiclassical. The varying of h in $\Psi_{(n,h)}$ when n is fixed, would be non-physical, because the scale, order of 'classicality' of Ψ_n depends only on n. To be supported by an extreme example: we could consider a ground state corresponding to the smallest possible value of n. It is obvious that there is neither classical limit nor anything reasonable in putting $h \to 0$ in a wave function of a ground state because n is then *fixed* and *small*, and $h \to 0$ is only the theoretical shorthand for $n \to \infty$. Similarly, it is meaningless to apply the $h \to 0$ asymptotics when studying spinning particles, because then some quantum numbers (describing internal angular momentum) are small.

In the following, we are using the methods of $h \to 0$ asymptotics, carefully comparing some results with the analysis of the order of quantum numbers. Of course, we remain within the Weyl–Wigner–Moyal framework. This formulation of quantum mechanics, being based on the geometry of classical phase spaces is especially convenient when studying quasiclassical phenomena and the correspondence principle.

As we have mentioned in the previous section, the non-local Weyl–Moyal operations reduce in the limit $h \to 0$ to the local, pointwise operations. More strictly:

Let A, B be smooth functions over P, for which the Weyl product $A \perp B$ is well-defined (the integral (27) converges). Expanding (27) into an asymptotic series about $h = 0$ by means of the method of stationary phase (Born and Wolf, 1964; Erdélyi, 1956), we get:

$$A \perp B = AB + \frac{\hbar i}{2}\{A, B\} + \ldots \qquad (51)$$

Therefore:

$$\lim_{h \to 0} A \perp B = AB \qquad (52)$$

$$\lim_{h \to 0} \frac{1}{\hbar i}(A \perp B - B \perp A) = \{A, B\} \qquad (52a)$$

Hence the pointwise product and the classical Poisson bracket are classical limits of the Weyl product and the quantum Poisson bracket (Moyal bracket), respectively.

On the quantum level, the structures of Lie algebra and associative algebra are directly, algebraically related to each other. In fact the quantum Poisson bracket $[A, B] = 1/\hbar i(A \perp B - B \perp A)$ is algebraically built from the associative product (via a commutator). It is not the case on the classical level: $\{A, B\}$ involves differentials of A, B and fails to be an algebraic function of A, B. This is the main qualitative discontinuity of the classical limit $h \to 0$. There are serious physical consequences of this fact. Namely, let us consider a left eigenequation:

$$A \perp \rho = a\rho \tag{53}$$

where A is an arbitrary physical quantity ($A^* = A$, $\hat{A}$ is Hermitian) and ρ is a statistical ensemble: $\int \rho \, d\mu = 1$, $\int \rho(B^* \perp B) \, d\mu \geq 0$, for arbitrary B, $\rho^* = \rho$.

Obviously (53) is the Weyl–Moyal counterpart of the left operator eigenequation for density operators:

$$\hat{A}\hat{\rho} = a\hat{\rho} \tag{53a}$$

Taking the complex conjugate of (53) and subtracting it from (53), we get via (29):

$$[A, \rho] = \frac{1}{\hbar i}(A \perp \rho - \rho \perp A) = 0 \tag{54}$$

or, in operator terms:

$$\hat{A}\hat{\rho} - \hat{\rho}\hat{A} = 0 \tag{54a}$$

But (54) means that the statistical ensemble ρ is invariant under a one-parameter unitary group generated infinitesimally by A.

EXAMPLE: let $A = L_i n^i$ be a component of the angular momentum along the axis given by the unit vector $\mathbf{n}$. Now, let ρ be a statistical ensemble with the sharply defined value m of $L^i n_i$:

$$L_i n^i \perp \rho = m\rho$$

Then, ρ is invariant under the one-parameter group of rotations about the n axis. The records of the detector in a scattering experiment remain unaffected when the source producing particles in such an eigenstate ρ, is subject to rotations about the n axis.

Therefore, the *informative property* (53) (no statistical spread when measuring A on ρ), *implies* the *invariance property* (54). This is not the case on the classical level, i.e. for vanishing $\hbar$, because the equation

$$A\rho = a\rho \tag{55}$$

possesses solutions for which the equation:

$$\{A, \rho\} = 0 \tag{56}$$

is not satisfied.

Roughly speaking: On the *quantum* level (finite $\hbar$) *information implies symmetry*. On the *classical* level (vanishing $\hbar$) this relationship *breaks down*. Therefore, the purely formal methods of the $\hbar \to 0$ asymptotics are unsatisfactory and insufficient when studying classical counterparts of eigenstates. In fact, on the quantum level, the eigenstates of A are uniquely defined by (53) or (53a). Therefore, the equation (52) suggests that we define classical eigenvalues and eigensembles of A as those satisfying (55). However, from the physical point of view, it is hard to accept such an approach, because the eigenstates would then lose their fundamental physical symmetries. The physical qualitative understanding of the correspondence principle rather suggests that we should define classical eigenstates of A as those satisfying both (55) and (56) i.e. the pair

$$A\rho = a\rho$$
$$\{A, \rho\} = 0 \tag{57}$$

where a is a regular value of A.

In fact, from the *physical laboratory* point of view, the notions of *information and symmetry* are *essentially independent* of each other in spite of their accidental relationship within the framework of Hamiltonian quantum mechanics. They involve two different kinds of physical operations:

(1). Measurements, mainly scattering experiments and statistical analysis of spreads of results, recorded by films or counters for example.

(2). Transformations, motions performed on the experimental set-up, e.g. on sources and detectors. The main example being rigid translations and rotations.

Therefore, to retain the physical structure of eigenstates, we have to define them on the classical level by means of (57). This is confirmed by the quasiclassical scattering experiments.

When A is an analytic function, then the only probabilistic solutions of (57) are distributions:

$$\rho = F\delta(A - a) \tag{58}$$

where:

$$\{F, A\} = (A - a)G \tag{59}$$

for some function G. The equation (59) is equivalent to:

$$\{F, A\}|M_{A,a} \equiv 0 \tag{59a}$$

where:

$$M_{(A,a)} = \{p \in P : A(p) = a\} \tag{59b}$$

REMARK: F in (58) need not be differentiable, we assume only that $\{F, A\}$ does exist, i.e. there exists the derivative of F in the direction of the vector field $\widetilde{dA}$ which is related to the Pfaff form dA via the Γ-lowering of indices:

$$(\widetilde{dA})^a = \Gamma^{ab}\frac{\partial A}{\partial \xi^b} \tag{59c}$$

In general, (57) or equivalently (59a) possesses many solutions. To restrict this arbitrariness it is necessary to perform additional measurements or in mathematical terms to add some similar conditions. Hence, let us consider the following system of classical eigenconditions:

$$A_i \rho = a_i \rho$$
$$\{A_i, \rho\} = 0 \qquad i = 1 \ldots m \qquad (60)$$

where $a = (a_1 \ldots a_m) \in R^m$ is a regular value of the mapping $A = (A_1, \ldots, A_m): P \to R^m$ (this mapping transforms $p \in P$ onto $(A_1(p) \ldots A_m(p)) \in R^m$).

Contrary to the system of proper equations:

$$A_i \rho = a_i \rho, \quad i = 1 \ldots m \qquad (61)$$

which is always compatible, the eigenconditions (60) need not be compatible and as a rule they are not.

To be more precise we give the following definitions. A classical system of eigenconditions (60) is said to be *compatible*, when:

(1). It is completely integrable in the sense of the theory of systems of differential equations;
(2). It is regular in the sense that $a = (a_1 \ldots a_m)$ is a regular value of the analytic mapping

$$A = (A_1 \ldots A_m): P \to R^m$$

One can easily show that the regular system (60) is compatible if and only if there exist functions C_{ij}^k such that:

$$\{A_i, A_j\} = \{A_i - a_i, A_j - a_j\} = C_{ij}^k (A_k - a_k) \qquad (62)$$

When $A_1 \ldots A_m$ are functionally independent (which ensures the regularity of (60) then the only probabilistic solutions of (60) have the form:

$$\rho = F \delta(A_1 - a_1) \ldots (A_m - a_m) \qquad (63)$$

where:

$$\{A_i, F\} = FC_{ij}^j + G_i^j (A_j - a_j) \qquad (64)$$

for some functions G_i^j.

Classical compatibility conditions (62) are obvious counterparts of the corresponding quantum equations (46). They possess a geometric interpretation which is interesting from the point of view of both physics and mathematics. In fact let

$$M_{(A,a)} = \{p \in P : A_i(p) = a_i, i = 1 \ldots m\} \qquad (64a)$$

then:

$$\{A_i, A_j\}|M_{A,a} = \{A_i - a_i, A_j - a_j\}|M_{A,a} = 0 \qquad (62a)$$

Equations (62) and (62a) imply that $M_{(A,a)}$ is what Dirac, Bergmann, Goldberg and others have called *I-class constraints* (Bergmann, 1966; Bergmann, 1970;

Bergmann and Goldberg, 1955; Dirac, 1950; Dirac, 1951; Dirac, 1955 a,b; Dirac, 1964; Sławianowski, 1971; Sławianowski, 1975; Tulczyjew, 1968).

Roughly speaking, the submanifold $M \subset P$ is said to be of *first class* if any vector Γ-orthogonal to M must be at the same time tangent to M. More strictly let $p \in M$ and let $T_p M$ denote the linear subspace tangent to M at p. Now let $k \in \Pi$ be such a vector that for arbitrary $u \in T_p M$:

$$\langle \Gamma, k \otimes u \rangle = \Gamma_{ab} k^a u^b = 0 \tag{65}$$

M is a *first-class submanifold*, if for arbitrary $p \in M$, (65) implies: $k \in T_p M$.

Let $V(M)$ denote the set of analytic functions vanishing on M for arbitrary subset $M \subset P$:

$$V(M) = \{f \in C^\omega(P) : f|M = 0\} \tag{66}$$

Obviously, $V(M)$ is an *ideal* in the *associative* algebra $C^\omega(P)$, i.e. it is closed under pointwise multiplications by arbitrary analytic functions. What is interesting is that $V(M)$ is at the same time a *Lie algebra* (i.e. it is closed under the Poisson bracket) if and only if M is a first-class submanifold. In particular let ρ be an arbitrary classical probability distribution and

$$\mathscr{E}_\rho \stackrel{\text{def}}{=} \{f \in C^\omega(P) : f\rho = 0\} \tag{67}$$

[compare with (47)]. Obviously, $\mathscr{E}_\rho = V(\text{Supp})$. For arbitrary ρ, $\mathscr{E}_\rho$ is an associative ideal. It is a Lie subalgebra if and only if the support of ρ, Supp ρ, is a I-class submanifold, i.e. if ρ satisfies some compatible system of classical eigenconditions. Let us remember the corresponding property of the quantum space E_ρ. The analogy is obvious.

In the following, the ideals $V(M)$ corresponding to the 1-class manifolds M will be called *self-consistent*. Any such ideal has a form $\mathscr{E}_\rho = V(\text{Supp } \rho)$ where ρ satisfies some compatible system (60). Probability distributions ρ such that $\mathscr{E}_\rho$ is not self-consistent (Supp ρ fails to be a 1-class submanifold) must not be looked on as classical counterparts of quantum states. They are non-interpretable from the point of view of the correspondence principle, because they violate the uncertainty principle on the quasiclassical level. A typical example of such a 'wrong' distribution is:

$$\rho = F \delta(q^1 - a^1) \delta(p_1 - b_1) \tag{68}$$

Both q^1 and its conjugate momentum p_1 are simultaneously dispersion-free on such a ρ, hence Supp ρ fails to be of 1-class because $\{q^1, p_1\} = 1 \neq 0$ and ρ is non-interpretable in terms of the quasiclassical theory.

On the quantum level special attention is paid to the *pure states* which carry the *maximal possible information* and could be defined as those satisfying a *maximal system of compatible eigenequations* (45), i.e. *answering a maximal number of operational 'questions'*. The corresponding left ideal E_ρ [cf. expression (47)] then becomes maximal. The question arises as to the classical counterparts. The correspondence principle suggests that we look for classical distributions ρ for which $\mathscr{E}_\rho$ is a *maximal self-consistent ideal*. Roughly speaking, such a distribution ρ satisfies some *maximal system of compatible eigenconditions* (60): 'maximal' means here that any extended system of eigenconditions

$$A_i x = a_i x \qquad Fx = 0$$
$$\{A_i, x\} = 0 \qquad \{F, x\} = 0 \tag{60a}$$

is consistent only if

$$F = \sum_j (A_j - a_j) F_j \tag{60b}$$

for some functions F_j. Hence (60a) is equivalent then to (60).

Obviously, $\mathscr{E}_\rho$ is a maximal self-consistent ideal of the type $V(M)$ if and only if Supp ρ is a minimal closed analytic submanifold of I-class in P. The lowest possible dimension of such a manifold equals the number of degrees of freedom: $n = 1/2 \dim P$. The I-class manifold becomes then what mathematicians call the *lagrangian manifold* i.e. maximal Γ-self-orthogonal manifold. In rigorous terms: $M \subset P$ is said to be an *isotropic manifold* if any vector tangent to it is at the same time Γ-orthogonal to M that is if for an arbitrary pair of vectors tangent to M at p : $u, v \in T_p M \subset \Pi$; the following orthogonality equations hold:

$$\langle \Gamma, u \otimes v \rangle = \Gamma_{ab} u^a v^b \tag{69}$$

One can easily show that $\dim M \leq n$.

An isotropic manifold is said to be *lagrangian* if there are no vectors transversal to it and Γ-orthogonal to it at the same time. Obviously an isotropic manifold is lagrangian if and only if its dimension equals n (Weinstein, 1971; Weinstein, 1973; Sławianowski, 1971; Arnold, 1974).

When Supp ρ is a closed and connected lagrangian manifold, then obviously $\mathscr{E}_\rho = V(\text{Supp }\rho)$ is a maximal self-consistent ideal in $C^\omega(P)$ and ρ is a *classical counterpart of a pure quantum state*. Typical and suggestive examples are as follows:

$$\rho = \delta(q^1 - a^1) \ldots \delta(q^n - a^n) \tag{70a}$$
$$\rho = \delta(p_1 - b_1) \ldots \delta(p_n - b_n) \tag{70b}$$

(where a and b are constants), or, in a six-dimensional phase space P:

$$\rho = \delta(L_i n^i - m)\, \delta(L^2 - l^2)\, \delta(\tfrac{1}{2} p^2 + V(r) - E) \tag{71}$$

where

$$L_i = \varepsilon_{ij}^{\;k} q^j p_k, \quad L^2 = \sum_i L_i^2, \quad p^2 = \sum_i p_i^2, \quad r^2 = \sum_i q^{i\,2}$$

n is a three-dimensional unit vector, and m, l, E are constants. Obviously equations (70) describe classical statistical ensembles with sharply defined values of q^i, p_i, respectively (on the contrary p_i and q^i then become completely undetermined according to the classical uncertainty relations). Expression (71) is an ensemble with defined values of the nth component of the angular momentum, the square of the angular momentum and energy. Hence, (71) corresponds to the classical partial wave analysis. Distributions (70) and (71) are especially convenient when analysing classical scattering experiments. They are directly related to the Hamilton–Jacobi theory. Moreover, it appears that properties of classical probability distributions concentrated on langrangian submanifolds are interpretable in terms of the classical Hamilton–Jacobi theory. There is nothing surprising in this: such distributions are classical counterparts of pure quantum states which admit the description in terms of wave functions. But it is known even from elementary textbooks that the $h \to 0$ asymptotics of the Schrödinger equation imposed on the wave function $\Psi = \sqrt{D} \exp(iS/\hbar)$, leads to the classical Hamilton–Jacobi equation imposed on the phase S and to the continuity equation for probabilistic fluids (with velocity-field given by the gradient of S) (Landau and Lifshitz, 1958; Messiah, 1965).

Let us investigate these problems in some details. We start with interpreting some of our results in terms of wave functions.

Let us consider a pure quantum state, the wave function of which is $\Psi(q^i)$ (q^i, p_i are canonical affine coordinates on P). The corresponding Wigner function ρ is given by (42)

$$\rho = \left(\frac{1}{2\pi}\right)^n \int \Psi^*\left(q^i - \frac{\hbar\tau^i}{2}\right) \exp(-i\tau^i p_i) \Psi\left(q^i + \frac{\hbar\tau^i}{2}\right) d_n\tau \qquad (42)$$

Let us put:

$$\Psi = \sqrt{D} \exp\left(\frac{i}{\hbar} S\right) \qquad (72)$$

We will write $\rho[D, S]$ to indicate explicitly the functional dependence of ρ on D, S. D is assumed to be continuous and S twice differentiable. We aim to find the classical limit of $\rho[D, S]$ expressed in terms of D and S. In the lowest order of WKB-approximation, D and S are assumed to be independent of h (Landau and Lifshitz, 1958; Messiah, 1965). Moreover, the h-independence of the physical interpretation of D and S is obvious even without any use of the WKB-approximation. In fact D is a probability distribution for positions and S is related to a spread of momentum:

$$\langle\Psi|\hat{Q}^i|\Psi\rangle = \int D(q^1 \ldots q^n) q^i \, dq^1 \ldots dq^n \qquad (73)$$

$$\langle\Psi|\hat{P}_i|\Psi\rangle = \int D(q^1 \ldots q^n) \partial_i S(q^1 \ldots q^n) \, dq^1 \ldots dq^n \qquad (74)$$

provided Ψ is well-behaved at infinity (Mackey, 1963). These formulas do not involve the Planck constant explicitly.

Let $\mathcal{M}_S$ be a maximal submanifold of P on which all functions $p_i - \partial_i S(q^j)$ vanish:

$$[p_i - \partial_i S(q^j)]|\mathcal{M}_S = 0 \qquad (75)$$

It is obvious that $\mathcal{M}_S$ is a lagrangian submanifold:

$$\{p_i - \partial_i S(q^k), p_j - \partial_j S(q^k)\} = (\partial_{ij}^2 - \partial_{ji}^2)S(q^k) = 0$$

Now, let us introduce the following probability distribution $\rho_{cl}[D, S]$ on P, concentrated on $\mathcal{M}_S$:

$$\rho_{cl}[D, S] = |\Psi(q)|^2 \, \delta(p_1 - \partial_1 S(q^k)) \ldots \delta(p_n - \partial_n S(q^k)) \tag{76}$$

Obviously, ρ_{cl} is a classical probability distribution, non-negative in the usual local sense. Nevertheless as far as we are interested in analysing measurements of first-order polynomials of canonical affine coordinates, $\rho_{cl}[D, S]$ is essentially equivalent to ρ on the rigorous quantum level (finite $\hbar$):

$$\begin{aligned}\langle \Psi | \alpha_i \hat{q}^i + \beta^i \hat{p}_i | \Psi \rangle &= \int (\alpha_i q^i + \beta^i p_i) \rho_{cl}[D, S] \\ &= \int (\alpha_i q^i + \beta^i p_i) \rho[D, S]\end{aligned} \tag{74a}$$

Of course for higher order polynomials this formula breaks down (excepting polynomials depending on q^i only). However, in the classical limit ρ becomes exactly equivalent to ρ_{cl}. In fact, making use of the afore-mentioned independence of D and S on $\hbar$ in the lowest order of WKB-expansion we find

$$\lim_{\hbar \to 0} \rho[D, S] = \rho_{cl}[D, S] \tag{77}$$

Obviously, the limit is to be understood in the sense of generalized functions (Schwartz, 1950–1951).

Now, let H be an arbitrary analytic function on R^{2n} and $\hat{H}$, the operator corresponding to $H(q^1 \ldots q^n, p_1 \ldots p_n): P \to R$ in the sense of the Weyl prescription [cf. (40) and (41)]. We assume $\hat{H}$ to be bounded in $L^2(R^n)$.

Let us consider the eigenequation

$$\hat{H}\Psi(q) = E\Psi(q) \tag{78}$$

where $\Psi \in L^2(R^n)$. We put $\Psi = \sqrt{D} \exp(iS/\hbar)$. Equation (78) implies that

$$H(q, p) \perp \rho[D, S] = E\rho[D, S] \tag{79}$$

$$H(q, p) \perp \rho[D, S] - \rho[D, S] \perp H(q, p) = 0 \tag{80}$$

In the classical limit, these equations become

$$H(q, p)\rho_{cl}[D, S] = E\rho_{cl}[D, S] \tag{81}$$

$$\{H(q, p), \rho_{cl}[D, S]\} = 0 \tag{82}$$

Expression (81) implies that

$$H(\ldots q^i \ldots, \ldots \partial_i S(q^k) \ldots) \equiv E \tag{83}$$

which is none other than the time-independent Hamilton–Jacobi equation imposed on S. In geometric terms it means that

$$H(q, p)|\mathcal{M}_S \equiv E \tag{84}$$

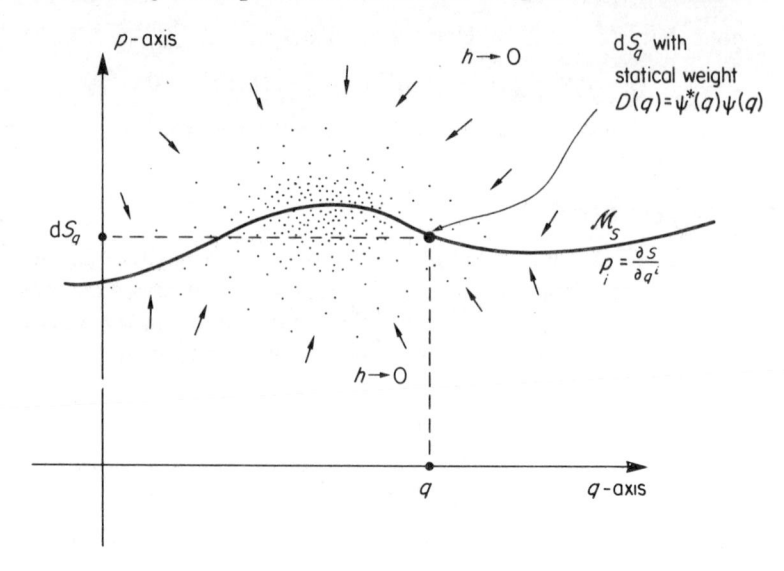

Uncertainty Principle and Foundations of Quantum Mechanics

Figure 1 Wave functions and phase–space distributions. Systems of dots represent an exact quantum Wigner function $\rho[\Psi] = \rho[D, S]$. The surface $\mathcal{M}_S$ given by equations $p_i = \partial S/\partial q^i$ is a support of the classical probability distribution:

$$\rho_{cl}[\Psi] = \rho_{cl}[D, S] = D(q)\,\delta\!\left(p_1 - \frac{\partial S}{\partial q^1}\right)\ldots\delta\!\left(p_n - \frac{\partial S}{\partial q^n}\right)$$

Both $\rho[D, S]$, $\rho_{cl}[D, S]$ lead to identical expectation values for canonical affine coordinates in the phase space. When $h \to 0$, ρ approaches ρ_{cl} and the above system of dots shrinks to $\mathcal{M}_S$.

Restricting the canonical vector field $\widetilde{dH}$ to $\mathcal{M}_S$ (this is possible, because $\widetilde{dH}$ is tangent to $\mathcal{M}_S$) and projecting it to the configuration space (the quotient space of P with respect to fibres on which all q^i are constant) we get the following S-velocity field $V[S]$

$$V[S]^i = \partial_{n+i}H(\ldots q^j \ldots, \ldots \partial_j S(q^k)\ldots) \qquad (85)$$

For example when

$$H = \frac{1}{2m} g^{ij} p_i p_j + f(q^j)$$

then:

$$V[S]^i = \frac{1}{m} g^{ij}\, \partial_j S(q) \qquad (85a)$$

and (82) is equivalent to:

$$\mathcal{L}_{V[S]} D(q) = 0 \qquad (86)$$

where $\mathcal{L}_{V[S]}D(q)$ denotes the Lie derivative of $D(q)$ with respect to $V[S]$. Expression (86) means that $D(q)$ is invariant during the motion, hence, it is a continuity equation for quasiclassical stationary states. As a geometric object, $D(q)$ is a density of weight one, hence (86) is a shorthand for

$$V[S]^i \partial_i D(q) + D(q) \frac{\partial V[S]^i}{\partial q^i} = 0 \qquad (86a)$$

Finally the quasiclassical counterpart of (78) is a couple of conditions:

$$H(\ldots q^i \ldots, \ldots \partial_i S(q^k) \ldots) \equiv E \qquad (87a)$$

$$V[S]^i \partial_i D(q^k) + D(q^k) \frac{\partial V[S]^i}{\partial q^i} = 0 \qquad (87b)$$

Conditions (87) could be derived directly from (78) after substitution of (40), (41) and (72) and expanding it up to the first asymptotic order in h. One should only make use of the method of stationary phase; conditions (87a) and (87b) result then as a real and imaginary part of the asymptotic formula respectively (Tulczyjew, 1968).

A system of equations (87) is completely integrable if and only if the conditions (62) are satisfied:

$$\{H_a, H_b\} = C_{ab}{}^c (H_c - E_c) \qquad (62a)$$

The maximum possible number of independent and compatible simultaneous eigenconditions (87) imposed on the same pair D, S equals the number of degrees of freedom n. Any system of n independent compatible eigenconditions (87) possesses a solution which is unique up to an additive constant in S and a multiplicative one in D. It is essentially equal to the quasiclassical solution found by Van Vleck (1928) and proved by him to coincide with the asymptotic WKB solution. In more detail let $A_i : R^{2n} \to R$, $i = 1 \ldots n$ be analytic functions such that the corresponding phase space functions $A_i(q, p)$ are in the involution

$$\{A_i(q, p), A_j(q, p)\} = 0 \qquad (88)$$

Let us consider the corresponding compatible system of eigenconditions imposed on the quasiclassical wave function $\sqrt{D} \exp(iS/\hbar)$:

$$A_i(\ldots q^j \ldots, \ldots \partial_j S(q^k) \ldots) \equiv a_i \qquad (89a)$$

$$V_i[S]^j \partial_j D(q^k) + D(q^k) \partial_j V_i[S]^j = 0 \qquad (89b)$$

The autonomous subsystem (89a) imposed on S possesses the unique (up to an additive constant) solution which depends in a parametric way on constants a_i. Let us insert this dependence explicitly into S by introducing a function $S : R^{2n} \to R$ such that $S(q^i, a^i)$ is a solution of (89a) for arbitrary values of a_i. Obviously, $S(q^i, a^i)$ is a *complete integral* for any of the Hamilton–Jacobi equations (89a). Substituting $S(q^i, a^i)$ into (89b) we get an autonomous system of first-order differential equations imposed on D. It is completely integrable and one can show that the only solution (up to normalization) of this system is given by the Van Vleck determinant:

$$D(q^i, a^i) = \text{Det} \left\| \frac{\partial^2 S}{\partial q^i \, \partial a^j} \right\| \tag{90}$$

In particular, S can be chosen to be an arbitrary complete integral of the stationary Hamilton–Jacobi equation

$$H(\ldots q^i \ldots, \ldots \partial_j S(q^k, a^k) \ldots) \equiv E \tag{91}$$

where H is a Hamiltonian function of the mechanical system.

The geometric interpretation of the Van Vleck solution in terms of symplectic geometry has been given in Sławianowski (1972). To some extent, the Van Vleck object can be guessed *a priori*, in terms of pure geometry, without any appeal to the quasiclassical approximation.

Obviously the quasiclassical Wigner–Moyal density in P corresponding to the Van Vleck solution of (89a) is given by

$$\rho_{cl}[D, S] = \delta(A_1 - a_1) \ldots \delta(A_n - a_n)$$

$$= \text{Det} \left\| \frac{\partial^2 S}{\partial q^i \, \partial a^j} \right\| \delta(p_1 - \partial_1 S(q, a)) \ldots \delta(p_k - \partial_k S) \tag{92}$$

REMARK: When ρ is a classical probability distribution the support of which is a closed connected lagrangian submanifold then $\mathscr{E}_\rho$ is a maximal self-consistent ideal in $C^\omega(P)$ and consequently ρ is a quasiclassical pure state. Contrary to what might be expected the converse statement is false. In fact let $H: P \to R$ be such a Hamiltonian that the corresponding dynamical system does not possess any non-trivial constants of motion (excepting the Hamiltonian itself, of course). This is the case for example when the system is *ergodic* (any classical trajectory is dense on some value-surface of H-). The quotient sets of value-surfaces of H with respect to the congruence of classical trajectories (integral curves of $d\widetilde{H}$) fail to be differential manifolds in any natural way. Then it is easy to see that the only solution up to a constant factor of the classical eigencondition

$$H\rho = E\rho \quad \{H, \rho\} = 0$$

is the microcanonical ensemble

$$\rho \approx \delta(H - E)$$

$\mathscr{E}_\rho$ is a maximal self-consistent ideal in $C^\omega(P)$, although Supp ρ fails to be a Lagrangian submanifold (it is only a 1-class submanifold of P).

We have shown above in what way and in what sense the asymptotics $h \to 0$ lead to the Hamilton–Jacobi theory of classical mechanics. The crucial point was that the purely formal asymptotic rules were inadequate to get physically reasonable and satisfactory results. They had to be completed by qualitative physical ideas such as information and symmetry properties. The necessity for such 'supplements' is a typical feature of the formal $h \to 0$ approach, in contrast to the more physical although less elegant asymptotics of large quantum numbers.

The physical analysis based on information and symmetry properties of statistical ensembles leads to results which do not agree with some current views concerning pure states. It is well-known that pure states carry the maximal information about the system. Hence it is reasonable to conjecture their classical counterparts to be distributions or measures the supports of which degenerate to the subsets of phase measure zero. One commonly believes that they are Dirac measures concentrated on single points in a phase space. Hence within the framework of such, currently used analogy classical pure states and points of the phase space are essentially identical notions. In spite of these views but in agreement with the old Hamiltonian ideas of Synge and Dirac (Dirac, 1964; Synge, 1953; Synge, 1954) we have shown that quasiclassical probability distributions corresponding to pure states are concentrated on lagrangian submanifolds the dimension of which equals the number of degrees of freedom. Hence, to any quasiclassical pure state there is attached some set of the usual, classical states—a lagrangian manifold in a phase space. Each of these classical states is taken with its own statistical weight. Hence classical states, that is points in a phase space, are 'hidden parameters' of quasiclassical ones. We have presented both physical and *a priori* geometrical arguments in support of our views. Let us notice that our approach agrees nicely with the Bohr–Sommerfeld conditions of the old quantum theory. In fact the objects upon which these conditions are imposed are none other than lagrangian manifolds in a phase space. Let $(\ldots q^i \ldots, \ldots p_i \ldots)$ be canonical affine coordinates on P; we introduce the Pfaff form ω

$$\omega = p_i \, dq^i \tag{93}$$

Obviously in arbitrary affine coordinates

$$\Gamma_{ab} = \tfrac{1}{2}(\omega_{a,b} - \omega_{b,a}) \tag{94}$$

Now, let $\mathcal{M}_a$ be a Lagrangian submanifold

$$\mathcal{M}_a = \{p \in P : A_i(p) = a_i, i = 1 \ldots n\} \tag{95}$$

where

$$\{A_i, A_j\} \equiv 0$$

The Bohr–Sommerfeld conditions imposed on $\mathcal{M}_a$ mean that the line integral of ω over any closed curve in $\mathcal{M}_a$ equals the Planck constant multiplied by some integer characterizing a loop of integration. When $\mathcal{M}_a$ are topologically equivalent to tori which is the case in the theory of multiply periodic systems, these conditions give rise to non-trivial restrictions on the admissible quantized (in the Bohr–Sommerfeld sense) values of physical quantities $A_1 \ldots A_n$.

The currently used analogy between pure quantum states and points of the phase space is based on the properties of Gaussian-shape Wigner functions. In spite of its formal correctness this argument is physically wrong. Let us investigate this in some detail. Let (q^i, p_i) be canonical affine coordinates and

$$H_{(a,b)} = \frac{1}{2}\sum_i (q^i - a^i)^2 + \frac{1}{2}\sum_i (p_i - b_i)^2 \tag{96}$$

Obviously, $H_{(a,b)}$ is a Hamiltonian of the harmonic oscillator the equilibrium of which is given by the point in P with coordinates (a^i, b_i). The corresponding Gaussian–Wigner function is given by

$$E_{(a,b)} = (\pi\hbar)^{-n} \exp\left[-\frac{2}{\hbar} H_{(a,b)}\right] \tag{97}$$

One can easily show that $E_{(a,b)}$ is a Wigner function describing some pure state

$$(2\pi\hbar)^{-n} \int E_{(a,b)} \, dq^1 \ldots dp_n = 1 \tag{98}$$

$$E_{(a,b)} = E^*_{(a,b)} \tag{99}$$

$$E_{(a,b)} \perp E_{(a,b)} = E_{(a,b)} \tag{100}$$

$$\int E_{(a,b)}(A^* \perp A) \, dq^1 \ldots dp_n \geq 0 \tag{101}$$

(As a matter of fact, $E_{(a,b)}$ is positive for arbitrary A in the local sense too: $E_{(a,b)} \geq 0$). One can easily show that

$$\lim_{\hbar \to 0} E_{(a,b)} = \delta(q^1 - a^1) \ldots (p_n - b_n) \tag{102}$$

This could be understood in such a way that at least some pure states possess classical counterparts concentrated on single points (point measures). However this is only a typical example of mistakes arising when no care is taken with regard to the scale of quantum numbers. In fact $E_{(a,b)}$ describes the physical situation with small quantum numbers. To see this let us notice that it is essentially concentrated in the region the phase volume of which is of the order $\hbar^n$. Moreover, $E_{(a,b)}$ is none other than the ground state of the harmonic oscillator:

$$H_{(a,b)} \perp E_{(a,b)} = \tfrac{1}{2}n\hbar E_{(a,b)} \tag{103}$$

where n is the number of degrees of freedom. Therefore according to what we have mentioned above the $\hbar \to 0$ asymptotics of $E_{(a,b)}$ is physically meaningless.

We finish this chapter with some remarks concerning quasiclassical problems of the superposition principle and interference of amplitudes. Let us consider the wave mechanics in R^n. An arbitrary wave function $\Psi = \sigma \exp(iS/\hbar)$ gives rise to some geometric figure in R^{n+1}, namely the graph of its phase:

$$P[\Psi] = P[\sigma, S] = \{(x, S(x)) : x \in R^n\} \tag{104}$$

Now let us consider an arbitrary m-parameter family of wave functions

$$\left\{\Psi_a = \sigma_a \exp\left(\frac{i}{\hbar}S_a\right), \quad a \in R^m\right\}$$

and the corresponding system of phase diagrams

$$P[\Psi_a] = P[\sigma_a, S_a]$$

Let $\phi = \mu \exp(it/\hbar)$ be some complex amplitude on R^m. It gives rise to the continuous superposition

$$\Psi = \sigma \exp\frac{i}{\hbar}S = \int \phi(a)\Psi_a \, da^1 \ldots da^m \tag{105}$$

which can be represented by its phase graph $P[\Psi]$.

As usual in the lowest order of the WKB approximation we assume σ_a, S_a, μ, t not be dependent on the Planck constant. Let $^0\Psi = {^0\sigma} \exp(i{^0S}/\hbar)$ denote the lowest order term of the asymptotic expansion of Ψ about $\hbar = 0$ renormalized so as to retain the probabilistic interpretation in the classical limit:

$$\langle {^0\Psi} | {^0\Psi} \rangle = \int \overline{^0\Psi}\,{^0\Psi} = 1$$

The corresponding phase diagram will be denoted as $P[^0\Psi] = P[^0\sigma, {^0S}]$.

Making use of the method of stationary phase when calculating (105) we get the result that essentially $P[^0\Psi]$ *is an envelope of the family*

$$\{P[\Psi_a \cdot \phi(a)]\} = \{P[\mu(a) \cdot \sigma_a, t(a) + S_a]\}$$

This is just the *quasiclassical Huyghens interference rule*. More rigorously

$$^0\Psi(x) = N\phi(a_0(x))\Psi_{a_0(x)}(x)\frac{\exp\left[\frac{\pi}{4}i \, \text{sign}\, \partial^2(S_{(\cdot)}(x) + \phi(\cdot))_{a_0(x)}\right]}{\sqrt{h^n}\,\text{hes}_{a_0(x)}(S_{(\cdot)}(x) + \phi(\cdot))}$$

$$= N\mu(a_0(x))\sigma_{a_0(x)}(x)\exp\left[\frac{i}{\hbar}S_{a_0(x)}(x) + \phi(a_0(x))\right]$$

$$\times \frac{\exp[\frac{1}{4}\pi i \, \text{sign}\, \partial^2(S_{(\cdot)}(x) + \phi(\cdot))_{a_0(x)}]}{\sqrt{h^n}\,\text{hes}_{a_0(x)}(S_{(\cdot)}(x) + \phi(\cdot))} \tag{106}$$

where $a_0(x)$ is the solution of equations:

$$\frac{\partial}{\partial a^j}[S_{(\cdot)}(x) + \phi(\cdot)] = 0 \tag{107}$$

N is an h-dependent normalizing factor ensuring that $^0\Psi$ retains its probabilistic interpretation and $\partial^2 f_y$ is the matrix of the second-order derivatives of f at y, $\text{hes}_y f$ is the determinant of this matrix (hessian of f at y) and $\text{sign}\, \partial^2 f_y$ is a signature of the symmetric matrix $\partial^2 f_y$, (i.e. a difference between the number of positive and negative eigenvalues).

174 Uncertainty Principle and Foundations of Quantum Mechanics

When the equation (107) possesses more than one solution, then, (106) becomes the sum of terms corresponding to all solutions $a_0(x)$.

The Huyghens–Fresnel interference rule (107) gives rise to the classical Huyghens superposition principle. In fact when all $S_a(q^i)$ are solutions of the same Hamilton–Jacobi equation (83) then $^0S(q^i)$ corresponding to the envelope of

$$\{P[\Psi_a \cdot \phi(a)] = P[\mu(a) \cdot \sigma_a, t(a) + S_a]\}$$

is a solution of this equation too (Caratheodory, 1956). The envelopewise superposition principle for the classical Hamiltonian–Jacobi equation appears to be the correspondence principle counterpart of the usual linear superposition principle for the Schrödinger equation (Sławianowski, 1971; Sławianowski, 1975). In particular one can solve initial (boundary) problems for the Hamilton–Jacobi equation by the envelopewise superposition of classi-

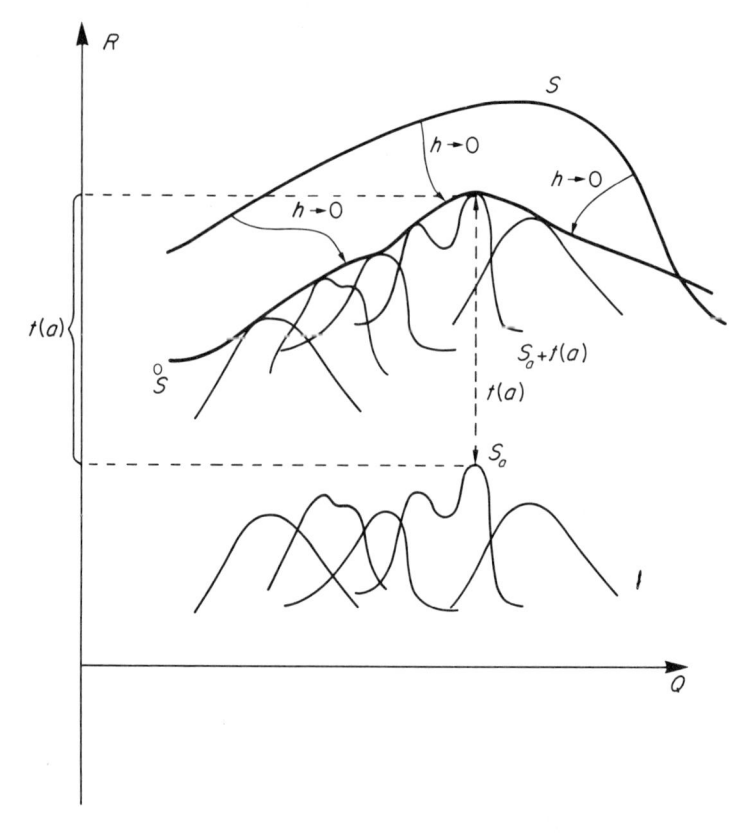

Figure 2   The classical Huyghens superposition rule. Continuous superposition of wave functions $\Psi_a = \sigma_a \exp(iS_a/\hbar)$ with coefficients $\varphi(a) = \mu(a) \exp[it(a)/\hbar]$, $\Psi = \sigma \exp(iS/\hbar) = \int \varphi(a) \psi_a \, da$. When $\hbar \to 0$, the diagram of S becomes an envelope of diagrams of $\{S_a + t(a)\}$ (i.e. the diagram of ^0S).

cal 'propagators' (Sławianowski, 1971; Sławianowski, 1975) with initial boundary conditions as 'coefficients' of 'superposition' (Sławianowski, 1975).

4. ALGEBRAIC AND PHYSICAL PROPERTIES OF EIGENSTATES AND PURE STATES. QUASICLASSICAL COMPATIBILITY

In the previous section we have derived some results concerning the correspondence principle for mechanical systems with affine symmetry of degrees of freedom. The Weyl–Wigner–Moyal formulation enabled us to achieve this in an almost deductive way. Unfortunately, this method does not work for general mechanical systems. This is why we have paid special attention to structures which did not depend explicitly on the particular properties of the space of states. Now starting with these structures we present some general statements concerning the analogy and correspondence between classical and quantum theory, with special attention to the quasiclassical compatibility problem. The geometric guidance from the theory of mechanical systems in affine spaces suggests some general analogies between classical and quantum concepts without explicit calculations and asymptotic expansions. To justify these analogies strictly one should appeal to the correspondence principle formulated in terms of large quantum numbers. However we will not do this here because we believe that the general information and symmetry properties of statistical ensembles do not depend on the particular structure of the manifold of states.

We start with the classical probability calculus of discrete sets. Let I be a countable set of elementary events. No additional structure in I is supposed; it is only a set.

Let $C(I)$ denote the linear space of complex-valued functions over I. Obviously, $C(I)$ is an associative algebra under the local, pointwise product. *Statistical ensembles* are described by probability distributions on I, i.e. real and normalized functions ρ on I:

$$\rho^*_{(i)} = \rho(i), \quad \rho(i) \geq 0, \quad \sum_i \rho(i) = 1$$

Physical quantities, i.e. random variables are described by real functions on I, $A: A^*(i) = A(i)$. Expectation values are given by the usual formula:

$$\langle A \rangle_\rho = \sum_i A(i) \rho(i)$$

Now let $\mathscr{E}_\rho$ be a linear subspace of $C(I)$ composed of all random variables which vanish on the support of ρ

$$\mathscr{E}_\rho = \{F \in C(I): F\rho = 0\} \tag{108}$$

Obviously $\mathscr{E}_\rho$ is an ideal in the associative algebra $C(I)$—it is closed under pointwise multiplications by all elements of $C(I)$. All random variables in $\mathscr{E}_\rho$

176 Uncertainty Principle and Foundations of Quantum Mechanics

are free from statistical dispersion on the ρ-ensemble; all measurements give a sharply defined result namely zero. In particular let $F \in \mathscr{E}_\rho$ and $F = (A - a)$ where a is some constant. Then A has a sharply defined value $a \in A(I)$ on ρ and the following eigenequation is satisfied:

$$A\rho = a\rho \qquad (109)$$

The greater $\mathscr{E}_\rho$ is, the more information is carried by ρ: $\mathscr{E}_{\rho_1} \subset \mathscr{E}_{\rho_2}$ implies:

$$\sum_i \rho_1(i) \ln \rho_1(i) \le \sum_i \rho_2(i) \ln \rho_2(i) \qquad (110)$$

Roughly speaking by imposing additional eigenequations (109) upon ρ we increase its informative content, because the number of measurements with completely predictable results is then increased: Now, let us assume that $\mathscr{E}_\rho$ is a maximal non-trivial ideal. Then, ρ answers the maximal number of non-trivial questions, i.e. the maximal number of measurements has unique, certain results (the maximal number of eigenconditions (109) is satisfied). Such a ρ is a *pure state* of the classical probability calculus. It is obvious that for an arbitrary pure state ρ there exists such a point $i \in I$ that: $\rho(j) = \delta_{ij}$, i.e. ρ is a point measure and:

$$\rho\rho = \rho \qquad (111)$$

Information (entropy) takes then its maximal (minimal) i.e. vanishing value:

$$\sum_i \rho(i) \ln \rho(i) = 0 \qquad (112)$$

All physical quantities A have then sharply defined values $A(i)$ and $\mathscr{E}_\rho$ consists of functions vanishing at $i \in I$. Then $\mathscr{E}_\rho$ is *maximal* if and only if *all* random variables are dispersion-free on ρ. This is just the main peculiarity of the classical probability calculus.

Now let us turn to *quantum statistics*. The *state space* of a quantum system is a separable Hilbert space H. Let $B(H)$ denote the associative but non-commutative algebra of linear bounded operators in H. Physical quantities are described by Hermitian elements of H: $A^+ = A$.

Statistical ensembles are described by density operators ρ i.e. Hermitian, positively definite and trace-class elements of $B(H)$:

$$\rho^+ = \rho, \quad \operatorname{Tr} \rho = 1, \quad \operatorname{Tr}(\rho A^+ A) \ge 0 \text{ for any } A \qquad (113)$$

An expectation value of $A \in B(H)$ on ρ is given as:

$$\langle A \rangle_\rho = \operatorname{Tr}(A\rho) \qquad (114)$$

Similarly, as in the classical probability calculus we define:

$$E_\rho = \{F \in B(H) : F\rho = 0\} \qquad (115)$$

Obviously, E_ρ is a *left ideal* in the associative algebra $B(H)$; it is closed under all transformations: $F \mapsto GF$. Algebraic and physical interpretation of E_ρ is analogous to some extent to that of $\mathscr{E}_\rho$ in classical statistics. In fact, E_ρ describes all measurements which, performed on the statistical ensemble ρ give sharp results without any statistical dispersion. Hermitian elements of E_ρ describe physical quantities which take a definite value, namely zero, when measured on ρ. When $A \in B(H)$ is a physical quantity ($A^+ = A$) which takes a value $a \in \mathrm{Sp}\, A$ on the ensemble ρ without any statistical spread, i.e.:

$$A\rho = \rho A = a\rho \qquad (116)$$

then obviously

$$F = (A - aI) \in E_\rho$$

The greater E_ρ is the more experimental questions are uniquely (spread-free) answered by ρ. Similarly as in the classical case $E_{\rho_1} \subset E_{\rho_2}$ implies the following inequality for information (entropy):

$$\mathrm{Tr}(\rho_1 \ln \rho_1) \leq \mathrm{Tr}(\rho_2 \ln \rho_2)$$

Pure states are defined as those answering a maximal number of experimental questions i.e. such as that E_ρ is a maximal (non-trivial) left ideal in $B(H)$. It is easy to show that for an arbitrary maximal left ideal E_ρ there exists a one-dimensional linear subspace $V \subset H$ such that E_ρ consists of operators vanishing on V:

$$E_\rho = \{F \in B(H) : F|V = 0\} \qquad (117)$$

This implies ρ to be a *projector* mapping H onto V and

$$\rho\rho = \rho \qquad (118)$$

Similarly as in classical statistics information (entropy) then takes its maximal (minimal) value

$$\mathrm{Tr}(\rho \ln \rho) = 0 \qquad (119)$$

All the above equations in $B(H)$ *are formally analogous to the classical ones in* $C(I)$. However, the non-commutativity of $B(H)$ involves us in serious physical differences. In fact when ρ is a classical probability distribution describing a pure state (point measure; $\mathscr{E}_\rho$-maximal) then obviously, the quotient space $C(I)/\mathscr{E}_\rho$ is isomorphic with the one-dimensional field of complex numbers C^1. Therefore the pure ensembles of classical statistics are characterized by definite values of *all* random variables; they satisfy eigenequations (109) for any $A \in C(I)$. This is not the case in quantum theory. Even if $\rho \in B(H)$ is a pure state (E_ρ is maximal) there exist physical quantities which take no definite value on ρ. A physical quantity A is spread-free on ρ if and only if the image V of ρ (i.e. $V = \rho(H)$) is an eigenspace of A:

$$A|V = a\mathrm{Id}_V$$

There exists one more peculiarity of quantum statistics which has no counterpart in the classical probability calculus. It is also strictly related to some formal algebraic differences between $C(I)$ and $B(H)$. In fact, the only algebraic structure carried naturally by $C(I)$ is that of commutative associative algebra over the complex field C^I. On the contrary, the non-commutative associative product in $B(H)$ (superposition) gives rise to another non-trivial algebraic structure namely that of *Lie algebra* under the quantum Poisson bracket:

$$[A, B] = \frac{1}{\hbar i}(AB - BA) \tag{120}$$

which is skew-symmetric and satisfies the Jacobi identity:

$$[[A, B], C] + [[B, C], A] + [[C, A], B] = 0 \tag{120a}$$

As we have mentioned above all informative essentially statistical properties of statistical ensembles were described in terms of the associative algebra structure (ideals, E_ρ, $\mathscr{E}_\rho$, eigenvalues, eigenequations and so forth). By contrast the structure of Lie algebra in $B(H)$ is strictly related to the symmetry properties of statistical states. In fact, any Hermitian element $A \in B(H)$ gives rise to two physically distinct kinds of physical operations:

(1). One can measure A on a statistical state ρ and perform a statistical analysis of the spread of the results. The corresponding mathematical expression describing the statistical dispersion is given as:

$$\sigma_2(A, \rho) = \text{Tr}(A^2\rho) - (\text{Tr}(A\rho))^2 \tag{121}$$

Spread-free eigenensembles satisfy the operator eigenequation:

$$A\rho = \rho A = a\rho \tag{122}$$

Hence, informative, statistical notions are related to the associative algebra structure.

(2). A gives rise to the one-parameter group of unitary automorphisms of the theory:

$$B(H) \ni B \mapsto_t B_t = \exp\left[\frac{i}{\hbar}tA\right] B \exp\left[-\frac{i}{\hbar}tA\right] \in B(H) \tag{123}$$

An infinitesimal description of this group is given in terms of the Lie algebra structure:

$$\frac{d}{dt}B_t = [B_t, A] \tag{124}$$

$B_0 = B$, hence:

$$\left.\frac{d}{dt}B_t\right|_{t=0} = [B, A]$$

Automorphisms (123) preserve all laboratory measurable quantities, i.e. expectation values and probabilities of detection, $[\text{Tr}(A\rho), \text{Tr}(\rho_1\rho_2)]$. This is because they preserve both the trace and the associative product that gives rise to the following compatibility conditions for the associative and Lie algebraic structures:

$$[A, BC] = [A, B]C + B[A, C] \tag{125}$$

As we have mentioned in the previous chapters from the laboratory point of view information and symmetry properties are essentially independent of each other. Nevertheless in Hamiltonian quantum mechanics they become interrelated namely information implies symmetry. In fact the operator eigenequation (122) implies that

$$[A, \rho] = 0$$

Therefore the lack of statistical dispersion of A on the ensemble ρ implies that ρ is invariant under the one-parameter unitary group generated by A. This implies that $\operatorname{Re} E_\rho$ that is the real linear subspace of E_ρ composed of the Hermitian operators is a Lie algebra over the field of real numbers. This is true for an arbitrary density operator $\rho \in B(H)$.

To summarize the above considerations we compare now the main algebraic and physical features of quantum statistics and classical probability calculus:

(1). If ρ is a classical probability distribution then $\mathscr{E}_\rho$ is an associative ideal in $C(I)$ (cf. (108)). This structure gives an account of the informative properties of ρ.
(2). If $\rho \in B(H)$ is a density operator then: (a) E_ρ defined in (115) is a left ideal in the associative algebra $B(H)$. This structure describes the informative properties of ρ. (b) $\operatorname{Re} E_\rho$ is a real Lie subalgebra in the Lie algebra $B(H)$ (under the quantum Poisson bracket). This structure gives an account of the symmetries of ρ implied by its informative properties.

Now let us investigate the corresponding structures in classical Hamiltonian mechanics. In contrast to the previous sections we will not assume the affine geometry of degrees of freedom.

Hence we must appeal to the general formulation of mechanics based on symplectic geometry (Abraham, 1967, Arnold, 1971; 1974; Sternberg, 1964; Hermann, 1970; Kostant, 1970; Souriau, 1970). Let us start with some mathematical preliminaries:

The *classical phase space* of a mechanical system is a pair (P, γ) where P is an analytic differential manifold (the set of classical states of a system) and γ is a non-degenerate and closed differential two-form on P of class $C^\infty(P)$: if $\langle \gamma, X \otimes Y \rangle = 0$ for an arbitrary vector field Y, then

$$X = 0 \tag{126}$$

$$d\gamma = 0 \tag{127}$$

Making use of local coordinates ξ^a on $U \subset P$, we have:

$$\gamma | U = \gamma_{ab} \, d\xi^a \wedge d\xi^b$$

where:

$$\det\|\gamma_{ab}\| \neq 0 \tag{126a}$$

$$\gamma_{ab,c} + \gamma_{bc,a} + \gamma_{ca,b} = 0 \tag{127a}$$

Hence dim $P = 2n$ where n is an integer, the so-called *number of degrees of freedom*.

The Darboux theorem (Abraham, 1967; Sternberg, 1964) implies the existence of canonical coordinates ξ^a such that:

$$\|\gamma_{ab}\| = \left\|\begin{matrix} 0 & I \\ -I & 0 \end{matrix}\right\| \tag{128}$$

[compare with expression (4)]. One uses then the historical notation: $(\xi^a) = (q^i, p_i)$ and:

$$\gamma|U = dp_i \wedge dq^i \tag{128a}$$

The contravariant skew-symmetric tensor reciprocal to γ will be denoted as $\tilde{\gamma}$:

$$\gamma_{ab}\tilde{\gamma}^{bc} = \delta_a^c \tag{129}$$

We will also write γ^{ab} simply instead of $\tilde{\gamma}^{ab}$.

The *Poisson bracket* of the differentiable functions F, G is defined as:

$$\{F, G\} = \langle dF \otimes dG, \tilde{\gamma}\rangle \tag{130}$$

In the coordinates:

$$\{F, G\}|U = \gamma^{ab}\frac{\partial F}{\partial \xi^a}\frac{\partial G}{\partial \xi^b} \tag{130a}$$

It is skew-symmetric and (127) implies that the Jacobi identity holds hence the Poisson bracket turns $C^\omega(P)$ (and $C^\infty(P)$) into Lie algebra.

Now, let F be a differentiable function of P and dF—its differential. 'Raising the index' of the Pfaff form dF by means of $\tilde{\gamma}$, one obtains a contravariant vector field on P, the so-called *Hamiltonian vector field generated by F*. One denotes it as $\widetilde{dF}$: $\langle \gamma, \widetilde{dF} \otimes Y\rangle = -\langle dF, Y\rangle$; in the coordinates:

$$(\widetilde{dF})^a = \gamma^{ac}\frac{\partial F}{\partial \xi^c}$$

The components of $\widetilde{dF}$ with respect to the canonical coordinates (q^i, p_i) are given by:

$$\left(\frac{\partial F}{\partial p_i}, -\frac{\partial F}{\partial q^i}\right)$$

The tensor γ gives rise to the skew-symmetric scalar product and skew-symmetric orthogonality of vectors. We say that two vectors attached at the same point $p \in P$ are γ-orthogonal if:

$$\langle \gamma_p, u \otimes v\rangle = \gamma_{p\,ab}u^a v^b = 0 \tag{131}$$

Obviously, all vectors are self-orthogonal (skew symmetry of γ). Hence similarly as in the previous section we can define I-class submanifolds and lagrangian submanifolds in P:

The submanifold $M \subset P$ is said to be of *I-class* if any vector γ-orthogonal to M is at the same time tangent to M.

$M \subset P$ is called *isotropic* if any vector tangent to M is at the same time γ-orthogonal to M.

$M \subset P$ is said to be *Lagrangian* when it is both I-class and isotropic. Obviously, the dimension of lagrangian submanifolds equals the number of degrees of freedom, $n = \frac{1}{2} \dim P$. Now, let:

$$M = \{p \in P = F_i(p) = 0, \quad i = 1 \ldots m\}$$

M is a I-class submanifold if and only if the Poisson brackets of F_i vanish weakly:

$$\{F_i, F_j\}|M = 0$$

i.e.:

$$\{F_i, F_j\} = C_{ij}^k F_k \tag{132}$$

for some functions C_{ij}^k (Dirac, 1964).

Any I-class submanifold M of the co-dimension m is foliated by a family $K(M)$ of m-dimensional isotropic submanifolds in such a way that all vectors tangent to this foliation are γ-orthogonal to M (Tulczyjew, 1968; Sławianowski, 1971, Dirac, 1950; Bergmann and Goldberg, 1955). $K(M)$ is called the *characteristic* or *singular foliation* of M. Some global properties of the characteristic foliation are important in quasiclassical problems. The I-class submanifold M is called *simple* when its quotient set $P(M) = M/K(M)$ carries the natural $2(n-m)$ dimensional differential structure of class C^ω projected from M. When M is simple, then any smooth function on $P(M)$ gives rise to some non-trivial function on M which becomes constant when restricted to any fibre of $K(M)$. For example when M is a value surface of the Hamiltonian H then the corresponding dynamical system is completely degenerate and admits $(2n-1)$ non-trivial independent and autonomous constants of motion (the Hamiltonian itself included among them). The quotient manifold $P(M)$ of a simple submanifold M carries natural phase space structure, because $\gamma|M$ is projectable from M to $P(M)$. The resulting phase space $(P'(M), \gamma')$ is called the *reduced phase space* of M; it describes gauge-free physical degrees of freedom of M.

Now let us describe an opposite case: I-class submanifold M is said to be *primitive* when the only smooth functions on M constant on fibres of $K(M)$ are those constant all over the whole M. For example the value surfaces of an *ergodic* Hamiltonian are primitive submanifolds; the only non-trivial constant of motion is the Hamiltonian (energy) itself.

Obviously the arbitrary lagrangian submanifold is both simple and primitive.

The skew-symmetric tensor γ gives rise to the nowhere vanishing differential form of maximal possible degree $2n$, namely:

$$\gamma^n = \underbrace{\gamma \wedge \ldots \wedge \gamma}_{n} \tag{133}$$

It is well-known that such a form gives rise to some *measure* on P (Abraham, 1967; Schouten, 1954; Sławianowski, 1975). It is convenient to divide it by $(2\pi\hbar)^n = h^n$ (cf. Section 2); the corresponding dimensionless measure on P will be denoted as μ_h or simply μ. Obviously when using canonical coordinates $(q^1 \ldots p_n)$, the measure μ consists in

$$\int f \, d\mu = \left(\frac{1}{2\pi\hbar}\right)^n \int f \, dq_1 \ldots dq^n \, dp_1 \ldots dp_n \tag{134}$$

for an arbitrary smooth function f the support of which is contained in the domain on which coordinates $(q^1 \ldots p_n)$ are defined. The existence of a canonical measure enables us to describe *statistical ensembles* in P by means of non-negative scalar functions or distributions ρ normalized to unity:

$$\int \rho A^* A \, d\mu \geq 0 \quad \text{for any } A \tag{135}$$

$$\int \rho \, d\mu = 1 \tag{136}$$

Physical quantities are described by analytic functions on P. The linear space of all analytic functions on P, $C^\omega(P)$ carries two natural algebraic structures: (*a*) $C^\omega(P)$ is an associative algebra (obviously a commutative one) under the pointwise product; (*b*) $C^\omega(P)$ is a Lie algebra under the Poisson bracket operation.

In contrast to the quantum case, the associative and Lie algebraic structures in $C^\omega(P)$ are algebraically independent which gives rise to the separation of information and symmetry properties of statistical ensembles. Nevertheless, they are compatible in the sense that the Lie structure gives rise to the derivations of associative structures, i.e. the Leibniz rule is satisfied:

$$\{A, BC\} = \{A, B\}C + B\{A, C\} \tag{137}$$

which is an obvious counterpart of (125).

Hence just as in quantum theory an arbitrary physical quantity $A \in C^\omega(P)$, $A^* = A$ can be related to two kinds of physical operations: measurements and transformations.

Statistical analysis of A is based on the well-known formulas of classical probabilistic calculus involving associative algebraic structure on $C^\omega(P)$:

$$\langle A \rangle_\rho = \int A\rho \, d\mu \tag{138}$$

$$\sigma_2(A, \rho) = \int A^2 \rho \, d\mu - \left(\int A\rho \, d\mu\right)^2 = \langle A^2 \rangle_\rho - \langle A \rangle_\rho^2 \tag{139}$$

Spread-free statistical ensembles of A satisfy the eigenequation

$$A\rho = a\rho \tag{140}$$

A one-parameter group of automorphisms of $C^\omega(P)$ (*canonical transformations*) generated by A consists of transformations $B \mapsto B_t$ such that

$$\frac{d}{dt} B_t = \{B_t, A\}, \quad B_0 = B \tag{141}$$

hence

$$\left.\frac{d}{dt} B_t\right|_{t=0} = \{B, A\}$$

Such transformations preserve both the associative product (due to expression (137) and its Poisson bracket (due to the Jacobi identity). Hence, they preserve all measurable quantities of the theory.

Now we are able to investigate classical counterparts of quantum ideals E_ρ and pure states in some detail. Let us start with some definitions.

An ideal $V \subset C^\omega(P)$ in the *associative* algebra $C^\omega(P)$ is said to be *probabilistic* if there exists such a subset $M \subset P$ that

$$V = \{F \in C^\omega(P) : F|M = 0\} \tag{142}$$

Such an ideal is denoted as $V(M)$. Let ρ be an arbitrary probability distribution and let us put

$$C^\omega(P) \supset \mathscr{E}_\rho = \{F \in C^\omega(P) : F\rho = 0\} \tag{143}$$

Obviously

$$\mathscr{E}_\rho = V(\text{Supp } \rho) \tag{144}$$

which justifies the name we have used above. Such an ideal describes uniquely the space of all physical quantities which are dispersion-free on a given statistical ensemble.

Any ideal in associative algebra $C^\omega(P)$ is contained in some probabilistic ideal. As an example let us mention an ideal of all functions which vanish on a given subset, together with their derivatives up to some fixed order.

An associative ideal $V \subset C^\omega(P)$ is called *self-consistent* when it is at the same time a Lie subalgebra of $C^\omega(P)$. A probabilistic ideal $\mathscr{E}_\rho$ is self-consistent if and only if Supp ρ is a first-class submanifold of P. When physical quantities F, $G \in C^\omega(P)$ are simultaneously spread-free on ρ and $\mathscr{E}_\rho$ is self-consistent, then

$$\{F, G\}|\text{Supp } \rho = 0, \quad \text{i.e. } \{F, G\} \in \mathscr{E}_\rho \tag{145}$$

These are just the *classical compatibility conditions*. In particular, when both (q^i, p_i) are dispersion-free on ρ, then $\mathscr{E}_\rho$ fails to be self-consistent. Hence, the geometric *a priori* of symplectic manifolds seems to anticipate on the purely classical level the Heisenberg uncertainty principle.

184 Uncertainty Principle and Foundations of Quantum Mechanics

In the classical probabilistic calculus of discrete sets we had no compatibility restrictions; all probability distributions and all ideals $\mathscr{E}_\rho$ were admissible and consistent. This is not the case in Hamiltonian mechanics. The correspondence-principle analysis based on information and symmetry suggests that the only physically justified probability distributions ρ on P are those for which ideals $\mathscr{E}_\rho$ are self-consistent. Hence, the classical probability distribution the support of which fails to be a I-class submanifold is only a technically convenient shorthand for probability distributions closely concentrated around the 'support' but essentially smeared out beyond it. Hence, the lowest possible dimension of the support of the probability distribution satisfying classical compatibility and the uncertainty restrictions, equals the number of degrees of freedom (the lowest dimension of an I-class submanifold). In particular, point measures are incompatible.

As we have mentioned in the previous section classical counterparts of pure states are probability distributions ρ for which $\mathscr{E}_\rho$ is a maximal self-consistent probabilistic ideal. In particular, any probability distribution the support of which is a connected closed lagrangian lagrangian submanifold is a quasiclassical pure state. Probability distribution $\rho[D, S]$ (76) is the most typical example because of its obvious relationship with the wave functions Ψ through their phases S and their moduli D.

However it is interesting that Supp ρ need not be a lagrangian submanifold to be able to ensure the maximality of the probabilistic ideal $\mathscr{E}_\rho$. In fact when ρ is an arbitrary distribution the support of which is a *primitive* submanifold (cf. definition above) then $\mathscr{E}_\rho$ is maximal provided Supp ρ is analytic and connected. Although such a distribution is pure in the sense of answering the maximal number of compatible questions (measurements), it is hard to relate it to any wave function. The problem of quasiclassical interpretation of such distributions from the point of view of the correspondence principle is still *open*. As a typical example we refer to the *microcanonical ensembles* of *ergodic* Hamiltonians. Hence one can hope that such distributions are in some sense related to the quantum ergodic theory (Ludwig, 1961).

REMARK: Let M be an arbitrary submanifold of P not necessarily a self-consistent one. There exist self-consistent ideals in $C^\omega(P)$ (associative ideals being at the same time Lie algebras) all elements of which vanish on M. However they are of a non-probabilistic type; any such ideal is a non-trivial proper subspace of $V(M)$. A typical example is

$$V(M) = \{f \in C^\omega(P) : f(p) = 0, \, df_p = 0, \, p \in M\} \qquad (146)$$

When ρ is a classical probability distribution and Supp ρ is a primitive analytic submanifold, then obviously $\mathscr{E}_\rho = V(\text{Supp } \rho)$ is a *maximal* self-consistent *probabilistic* ideal and ρ is a conceptual counterpart of a pure quantum state. However if we omitted the word 'probabilistic' in the above statement it would become false because there exist essentially larger self-consistent ideals of a non-probabilistic type. In fact, let $p \in P$ and let $U_p \subset T_p P$ be some n-dimensional isotropic subspace of the tangent space at p, i.e. n-dimensional linear space of pairwise γ-orthogonal vectors attached at $p \in P$:

$$\langle \gamma_p, w \otimes v \rangle = \gamma_{p\,ab} w^a v^b = 0 \tag{147}$$

provided $w, v \in U_p$. Now let us put

$$S(U_p) = \{f \in C^\omega(P) : f(p) = 0, \widetilde{df}_p \in U_p\} \tag{148}$$

One can easily show that $S(U_p)$ is a self-consistent ideal in $C^\omega(P)$ moreover it is a maximal self-consistent ideal. Let $M \subset P$ be an arbitrary lagrangian submanifold and let $p \in M$. Then obviously

$$V(M) \subset S(T_p M) \tag{149}$$

and this is a non-trivial proper inclusion.

Hence the point measures in P are related to maximal self-consistent ideals of a non-probabilistic type in $C^\omega(P)$; Let δ_p be a Dirac measure concentrated at $p \in M$. There exists a system of $2n$ generators of $S(U_p)$ namely $F_1 \ldots F_{2n}$ such that:

$$F_i \delta_p = 0$$

$$\{F_i, \delta_p\} = 0 \tag{150}$$

Nevertheless, $\mathscr{E}(\delta_p)$ is essentially larger than $S(U_p)$ and this is why the point measures violate the relationship of information and symmetry suggested by the correspondence principle, although the $2n$ invariance conditions $0 = \{F_i, \delta_p\}$ are satisfied.

Let us finish our chapter with some remarks concerning the quasiclassical description of *projectors* in terms of symplectic geometry. We present only general ideas; more detailed information is given in (Sławianowski, 1971; Sławianowski, 1975).

As we have mentioned above, there exists an exact relationship between lagrangian submanifolds in P and phases of quasiclassical wave functions. It is a well-known peculiarity of the quasiclassical WKB- approximation that all relations between phases become separated so as to satisfy some autonomous closed algebra quite independent of what happens to the moduli of the wave functions. Let $D(P)$ denote the set of all closed lagrangian submanifolds in P. Now let $M \subset P$ be a simple closed submanifold of P (cf. the definition above) and let $D(M) \subset D(P)$ denote the set of all closed lagrangian submanifolds of P contained in M (i.e. $\mathscr{M} \in D(M)$ if and only if $\mathscr{M} \subset M$). $D(M)$ is non-empty because M is a I-class submanifold (compatibility conditions satisfied by quasiclassical wave functions).

One can show that except for some special cases of singular intersections there exists for an arbitrary $\mathscr{M} \in D(P)$ only one $(\Lambda_M \mathscr{M}) \in D(M)$ such that

$$(\mathscr{M} \cap M) \subset (\Lambda_M \mathscr{M}) \tag{151}$$

The natural mapping $\Lambda_M : D(P) \to D(M)$ satisfies the following rules:

(1). It is a retraction of $D(M)$:

$$\Lambda_M | D(M) = id_{D(M)} \tag{152}$$

in particlar, it is idempotent:

$$\Lambda_M \circ \Lambda_M = \Lambda_M \tag{153}$$

(2). When $M, N, M \cap N$ are simple and closed, then:

$$\Lambda_M \circ \Lambda_N = \Lambda_N \circ \Lambda_M = \Lambda_{M \cap N} \tag{154}$$

(3). When

$$\Lambda_M \circ \Lambda_N = \Lambda_N \circ \Lambda_M \tag{155}$$

then $M \cap N$ is a simple submanifold and (154) is satisfied.

(4). When $f : P \to P$ is an arbitrary canonical mapping then:

$$\Lambda_{f(M)} = F \circ \Lambda_M \circ F^{-1} \tag{156}$$

Where $F : D(P) \to D(P)$ is a mapping in $D(P)$ induced in an obvious way by f. (By canonical mapping we mean here an analytic diffeomorphism of P, preserving γ.)

Λ_M describes quasiclassical projection on to a subspace of quasiclassical wave functions characterized by definite sharp values of those physical quantities which are described by smooth functions, constant on M.

EXAMPLE: Let us consider an affine phase space with canonical affine coordinates (q^i, p_i), obviously: $\gamma = dp_i \wedge dq^i$. Let us put:

$$M = \{p \in P : p_1(p) = b\} \tag{157}$$

$$D(P) \ni \mathcal{M} = \{p \in P : q^i(p) = a^i, i = 1 \ldots n\} \tag{158}$$

Then

$$\Lambda_M \mathcal{M} = \{p \in P : p_1(p) = b, q^2(p) = a^2, \ldots, q^n(p) = a^n\} \tag{159}$$

Hence fixation of the value of p_1 by means of Λ_M results in a complete indeterminacy of q^1 (quasiclassical uncertainty principle).

Quasiclassical theory becomes complete when besides the projectors Λ_M, classical Huyghens–Fresnel superpositions are taken into account (cf. the definition at the end of the previous section). The actual definition and properties of Λ_M are based on the geometric *a priori* of a symplectic manifold endowed with the second order Pfaff form γ with a local description $dp_i \wedge dq^i$. Similarly, the geometric structure of the Huyghens–Fresnel superposition can be deduced from the geometric *a priori* of the so-called *contact manifolds*, the geometry of which is based on the first-order Pfaff form Ω with a local representation: $-dz + p_i \, dq^i$ (Sławianowski, 1971). The contact manifold is a fibre bundle over symplectic manifold with a one-dimensional fibre. Roughly speaking it arises from the classical phase space when the action variable (i.e. the phase of a quasiclassical wave function) is taken into account as an additional dimension (Souriau, 1970; Arnold 1974).

REFERENCES

Abraham, R. (1967) *Foundations of Mechanics*, Benjamin, New York.
Arnold, V. I. (1971) *Obyknovennyie Differetsialnyie Uravneniya*, (Ordinary Differential Equations, in Russian), Nauka, Moscow.
Arnold, V. I. (1974) *Matematicheskie Metody Klassicheskoj Meckhaniki* (Mathematical Methods of Classical Mechanics, in Russian), Nauka, Moscow.
Arnold, V. I. and Avez, A. (1968) *Ergodic Problems of Classical Mechanics*, Benjamin, New York.
Bargmann, V. (1954) 'On unitary ray representations of continuous groups' *Ann. Math.*, **59**, 1–46.
Bergmann, P. G. (1966) 'Hamilton–Jacobi and Schrödinger theory in theories with first-class Hamiltonian Constraints', *Phys. Rev.*, **144**, 1078–1080.
Bergmann, P. G. (1970) *Quantisation of the Gravitational Field*, Aerospace Research Laboratories–Report ARL 70-0066.
Bergmann, P. G. and Goldberg I. (1955) 'Dirac bracket transformations in phase space', *Phys. Rev.*, **98**, 531–538:
Born, M. (1925) *Vorlesungen über Atommechanik*, Springer, Berlin.
Born, M., Heisenberg, W. and Jordan P. (1926) 'Zur Quantenmechanik. II', *Z. Phys.*, **35**, 557–615.
Born, M. and Jordan, P. (1925) 'Zur Quantenmechanik', *Z. Phys.*, **34**, 858–888.
Born, M. and Wolf, E. (1964) *Principles of Optics*, Pergamon Press, London.
Caratheodory, C. (1956) *Variationsrechnung und partielle Differentialgleichungen erster Ordnung*, B. G. Teubner, Leipzig.
Dirac, P. A. M. (1950) 'Generalized Hamiltonian dynamics', *Canad. J. Math.*, **2**, 129.
Dirac, P. A. M. (1951) 'The Hamiltonian form of field dynamics', *Canad. J. Math.*, **3**, 1.
Dirac, P. A. M. (1958a) 'Generalized Hamiltonian dynamics', *Proc. Roy. Soc. London*, **A 246**, 326–332.
Dirac, P. A. M. (1958b) 'The theory of gravitation in Hamiltonian form', *Proc. Roy. Soc. London*, **A 246**, 333.
Dirac, P. A. M. (1964) 'Hamiltonian methods and quantum mechanics', *Proc. Roy. Inst. Acad. Sect. A*, **63** 49–59.
Erdélyi, A. (1956) *Asymptotic Expansions*, Dover, New York.
Fröman, N. and Fröman, P. O. (1965) *JWKB Approximation*, North-Holland Publishing Co., Amsterdam.
Heisenberg, W. (1925) 'Über Quantentheoretische Umdeutung kinetischer und mechanischen Beziehungen', *Z. Phys.*, **33**, 879–893.
Heisenberg, W. (1927) 'Über den anschaulichen Inhalt der quantentheoretischen Kinematik und Mechanik', *Z. Phys.*, **43**, 172–198.
Hermann, R. (1970) *Vector Bundles in Mathematical Physics*, Benjamin, New York.
Kostant, B. (1970) *Lecture Notes in Mathematics*, Springer, New York.
Landau, L. D. and Lifschitz E. M. (1958) *Quantum Mechanics*, Pergamon Press, London.
Ludwig, G. (1961) 'Axiomatic quantum statistics of macroscopic systems (ergodic theory)', in *Ergodic Theories, Proceedings of the International School of Physics 'Enrico Fermi'*, XIV Course, P. Caldirola, Ed., Academic Press, New York.
Mackey, G. W. (1963) *The Mathematical Foundations of Quantum Mechanics*, Benjamin, New York.
Messiah, A. (1965) *Quantum Mechanics*, North-Holland Publishing Co., Amsterdam.
Moyal, J. E. (1949) 'Quantum mechanics as a statistical theory', *Proc. Cambridge Phil. Soc.*, **45**, 99.
Schouten, J. A. (1954) *Ricci Calculus*, Berlin.
Schouten, J. A. and Kulk, W. (1949) *Pfaffs Problem and its Generalizations*, Clarendon Press, Oxford.
Schrödinger, E. (1926a) 'Quantisierung als Eigenwertproblem, I', *Annln. Phys.*, **79**, 361–376.
Schrödinger, E. (1926b) 'Quantisierung als Eigenwertproblem, II', *Annln. Phys.*, **79**, 489–527.
Schrödinger, E. (1926c) 'Über das Verhältnis der Heisenberg–Born–Jordanschen Quantenmechanik zu der Meinen', *Annln. Phys.* **79**, 734–756.

188 Uncertainty Principle and Foundations of Quantum Mechanics

Schrödinger, E. (1926d) 'Quantisierung als Eigenwertproblem, III', *Annln. Phys.* **80**, 437–490.
Schrödinger, E. (1926e) 'Quantisierung als Eigenwertproblem. IV', *Annln. Phys.* **81**, 109–139.
Schwartz, L. (1950–1951) Theorie des distribution, Hermann, Paris.
Sławianowski, J. J. (1971) 'Quantum relations remaining valid on the classical level', *Rep. Math. Phys.*, **2**, 11–34.
Sławianowski, J. J. (1972) 'Geometry of Van Vleck Ensembles', *Rep. Math. Phys.*, **3**, 157–172.
Sławianowski, J. J. (1973) 'Classical pure states. Information and symmetry in statistical mechanics', *Int. J. Theoret. Phys.*, **8**, 451–462.
Sławianowski, J. J. (1974) 'Abelian groups and the Weyl approach to kinematics', *Rep. Math. Phys.*, **5**, 295–319.
Sławianowski, J. J. (1975) *Geometria Przestrzeni Fazowych* (Geometry of Phase Spaces, in Polish), Polish Scientific Publishers, Warsaw.
Souriau, J. M. (1970) *Structures des Systemes Dynamiques*, Dunod, Paris.
Sternberg. S. (1964) *Lectures on Differential Geometry*, Prentice Hall, New York.
Synge, J. L. (1953) 'Primitive quantization in the relativistic two-body problem', *Phys. Rev.*, **89**, 467.
Synge, J. L. (1954) *Geometrical Mechanics and de Broglie Waves*, Cambridge University Press, Cambridge.
Synge, J. L. (1960) *Classical Dynamics*, Springer, Berlin.
Śniatycki, J. and Tulczyjew, W. M. (1971) 'Canonical dynamics of relativistic charged particles', *Ann. Inst. Henri Poincare*, XV, 177–187.
Tulczyjew, W. M. (1968) Unpublished results.
Van Vleck, J. H. (1928) 'The correspondence principle in the statistical interpretation of quantum mechanics', *Proc. Nat. Acad. Sci.*, **14**, 178–188.
Weinstein, A. (1971) 'Symplectic manifolds and their lagrangian submanifolds', *Advan. Math.*, **6**, 329–346.
Weinstein, A. (1973) 'Lagrangian submanifolds and Hamiltonian systems' *Ann. Math.*, **98**, 377–410.
Weyl, H. (1928) 'Quantenmechanik und Gropentheorie', *Z. Phys.*, **46**, 1.
Weyl, H. (1931) *The Theory of Groups and Quantum Mechanics*, Dover, New York.
Wigner, E. (1932) 'On the quantum correction for thermodynamic equilibrium', *Phys. Rev.*, **40**, 749.

11

A Theoretical Description of Single Microsystems

GUENTHER LUDWIG
Universitaet Marburg, Germany

The fundamental relation for the interpretation of quantum mechanics is:

$$m = \operatorname{tr}(WE) \tag{1}$$

E being a projection operator in a Hilbert space $\mathcal{H}$ and W being a self-adjoint operator with $W \geq 0$ and $\operatorname{tr}(W) = 1$. As it is well known, the trace in (1) can be calculated by

$$\operatorname{tr}(WE) = \sum_\nu \langle \phi_\nu, WE\phi_\nu \rangle$$

ϕ_ν being any complete orthonormal set of vectors in $\mathcal{H}$; $\langle \ldots, \ldots \rangle$ denotes the inner product in $\mathcal{H}$. The real number m in (1) satisfies $0 \leq m \leq 1$. The fundamental statistical interpretation of quantum mechanics is as follows: m is the probability of E.

Better known than the general form (1) is the special case $W = P_\phi$, P_ϕ being the projection operator which projects onto the one-dimensional subspace of $\mathcal{H}$ spanned by the vector ϕ ($\|\phi\| = 1$). Then (1) takes on the form

$$m = \operatorname{tr}(P_\phi E) = \langle \phi, E\phi \rangle \tag{2}$$

Any experimental test of quantum mechanics employs the relation (1).

A description of quantum mechanics based on the general formula (1) is given in (Ludwig, 1975) and in a more consequent way in Ludwig (in preparation a).

More general, but reducible to (1), is the well-known interpretation that

$$\operatorname{tr}(WA) \tag{3}$$

(and $\operatorname{tr}(P_\phi A) = \langle \phi, A\phi \rangle$ in the case of $W = P_\phi$) is the expectation value of the observable A (in expression (3) A is a self-adjoint operator).

The Heisenberg uncertainty relation is nothing other than the physical interpretation of the mathematical theorem that Heisenberg's commutation relation $PQ - QP = (\hbar/i)\mathbf{1}$, holding for the position and momentum operators Q and P, yields

$$\Delta P \cdot \Delta Q \geq \frac{\hbar}{2} \tag{4}$$

with ΔP and ΔQ defined by

$$\Delta P^2 = \text{tr}\,(W(P-\alpha 1)^2)$$
$$\Delta Q^2 = \text{tr}\,(W(Q-\beta 1)^2)$$
$$\alpha = \text{tr}\,(WP),\ \beta = \text{tr}\,(WQ)$$

The physical interpretation of expression (4) depends on the physical interpretation of (3).

In order to reduce expression (3) to (1), it is necessary to introduce the conception of 'simultaneous measurability', i.e. of 'commensurability'. Usually it is assumed that commensurable E_λ's (E_λ being a projection operator in $\mathcal{H}$) commute, i.e. $E_\lambda \cdot E_\rho = E_\rho \cdot E_\lambda$. By means of a family of commuting projection operators and a measuring scale it is possible to deduce (3) from (1). We will give such a deduction in Section 5.

We may summarize: for the interpretation of quantum mechanics, we need the following concepts (the physical meaning of which must be exactly specified):

(1). Probability denoted in expression (1) by the real number m.

(2). A physical interpretation of the operator W in (1) (or the vector in ϕ in (2)). The following expressions are often used: W is the mathematical image of an 'ensemble' or of a 'state'. Sometimes the expression 'state' is used only in the case $W = P_\phi$, calling ϕ the 'state of the system'. W is also called the 'statistical operator' (or the 'statistical matrix', if given in matrix form).

(3). E in (1) is considered as the image of a 'yes–no observable', a 'yes–no measurement'. Also words like 'questions' or 'propositions' are used.

(4). A conception of 'commensurability'. Sometimes, instead of 'commensurability', one is speaking of 'measurability at the same time'.

All discussions concerning quantum mechanics ultimately depend on the concepts mentioned above under points (1) through (4). Many misunderstandings are based on the fact that various authors use the same words for different conceptions. All mistakes, all paradoxa are based on inadmissable fictions attached to the expressions noted under (1) through (4). It is impossible to give here a survey of all the discussions concerned with the various concepts.

Such discussions do not appear clear enough since, up to now, one has tried to clarify the meaning of the concepts under (1) through (4) using common language. This was necessary, because the description of single experiments and their results had to be given in common language. The gap between relation (1) and experiments is much too big to correlate immediately experiments and theory. One has to use common language to bridge this gap.

This necessity may be seen if we want to describe an experiment on one individual microsystem, e.g. an individual trace in a cloud chamber. There is no term in the mathematical framework of quantum theory which could be used as an image of this individual experiment; neither the 'probability' m, nor the 'statistical operator' W, nor the 'observable' A.

Thus it seems only natural to use common language in describing single experiments.

In the mathematical framework of quantum mechanics, we have the 'set of statistical operators', the 'set of projection operators', but there is no set the elements of which could be used as 'images' of individual microsystems. The concept of an 'individual microsystem' itself is not cleared up theoretically by quantum mechanics; it may be understood only intuitively.

Someone might object that quantum theory contains terms usable as images for individual systems in the same way as classical mechanics does: in classical mechanics every individual system is described by a point in the Γ-space, the 'state space' of classical mechanics; in quantum theory every individual system is described by a 'state' ϕ, ϕ being an element of the Hilbert space $\mathcal{H}$. The surface of the unit sphere in $\mathcal{H}$ has to be taken as the 'state space' of quantum mechanics. In my opinion, this is a false interpretation of quantum mechanics; in any case, in quantum theory, it is generally impossible to determine the state ϕ of an individual system in a single experiment.

All these problems sketched above motivate the development of a more comprehensible mathematical framework for quantum theory, which opens the possibility to interpret this framework *before* using relation (1) and the concepts given above under points (2) through (4). The experiments on individual microsystems and the statistics of these experiments should be described directly by their respective 'images' in this new mathematical framework. Then the physical *interpretation* of quantum mechanics does not depend on the 'statistical operators' W and the 'observable operator' A; the physical interpretation will depend only on two fundamental notions just as in all classical theories, namely on the notions of 'physical system' and of 'statistics'.

The fundamental concepts of quantum mechanics, in this new form, will not be essentially different from those of any other classical theory. The interpretation may be done in exactly the same way as in classical theories. The concepts mentioned above under points (2) through (4) will not be fundamental, but will be derived from the fundamental concepts of physical system and of statistics. Their interpretation will be given automatically by their deduction and by the interpretation of the two fundamental concepts. If only these two concepts are accepted as fundamental, all paradoxa and misunderstandings disappear; only mistakes will be possible but mistakes can be corrected.

1. THE CONCEPTION OF 'PHYSICAL SYSTEM'

The conception of physical system is based on the possibility of making 'experiments' on such systems. The first step in such experiments is to 'produce', to 'manufacture', to 'prepare' the systems. Examples of microsystems: an accelerator produces 'ions'; another accelerator produces 'electrons', a third 'pairs of particles', using colliding beams. The last examples shows that an individual system is not necessarily an elementary system (in the sense of not being composed). An individual system is only one of a large family produced by a preparing apparatus.

After their production, the various systems can be 'recorded' experimentally. This recording may also be described by saying that 'something has been measured on the system'. Such recording of a system may be realized, e.g. by a trace in a cloud chamber, by a signal of a counter, etc.

We will give examples for preparing and recording of microsystems: atoms can be produced (prepared) by a canal-ray tube and the photons emitted by these atoms may be recorded; electron–proton pairs can be prepared, and after the collision between electrons and protons the electrons may be recorded; nuclei can be prepared and (in the case of β-decay) the electrons emitted by the nuclei can be recorded.

To show that the usual experimental procedures concerning classical systems have the same structure, we will give just one example: a gun may 'prepare' projectiles and the trajectory of the projectiles can be recorded; e.g. the point of impact as a part of the trajectory may be recorded.

The examples show that the fundamental parts of 'experiments on physical systems' are the processes of preparing and recording.

It is impossible to give here a more detailed description of preparing and recording procedures composing the experiments on systems. However, we want to stress that preparing and recording procedures may be described without referring to the physical systems that are prepared and recorded. A gun, for instance, may be described as a procedure (i.e. its construction and instruction for use) without referring to a special projectile fired by this gun; also the impact may be described without explaining the cause of this event. Also, in the case of experiments on microsystems, the experimentator can describe the apparatus used to prepare the systems and the apparatus recording them. The 'evaluation' of such experiments giving values of quantities defined by a theory of the system follows *after* this description; cross section, wavelengths, etc., are calculated only after the description of the experimental procedures and the recording of the response of the measuring apparatus has been given (e.g. by a computer).

To give a mathematical picture of such experiments on individual systems, it is necessary to introduce a set M, the elements of which shall be 'images' of the systems. Given a special atom 'x', in an experiment, the relation $x \in M$ should be the mathematical form of the proposition: x is a physical system. However, the relation $x \in M$ reflects the proposition 'x is a physical system' only if the set M is *endowed with a structure* as an image of preparing and recording procedures. (A systematical description of this method of theoretical physics, employed here in a more intuitive way, is given in Ludwig (in preparation b).

We will not discuss the question concerning the reasons which allow us to speak of given microsystems and, in this sense, of *real* microsystems of which the elements of M are images. The reader interested in such questions will find a detailed discussion in Ludwig (1970, 1972a) and in a very short form in

(Ludwig, 1974a). In these references, a theoretical description is given how to 'recover' the microsystems, starting by the description of preparing and recording procedures only. However, to understand quantum mechanics it is not necessary to *justify* the existence of real microsystems, an existence which was more or less founded intuitively in the history.

(A) SELECTION PROCEDURES

According to the short sketches given above, the preparing and recording devices have a common structure, namely, that by these procedures physical systems can be *selected*. Therefore, it seems to be useful to begin with a mathematical description of this common structure.

Physical as well as mathematical reasons motivate the introduction of a more general species of structure henceforth called a 'selection procedure':

A subset $\mathscr{S} \subset \mathscr{P}(M)$ ($\mathscr{P}(M)$ the potential set, i.e. the set of all subsets of M) is called a *structure of selection procedures* or shortly a *selection structure* on M if the following axiom holds ($b \setminus a$ being the relative complement of a in b):

AS 1.1 $a, b \in \mathscr{S}$ and $a \subset b \Rightarrow b \setminus a \in \mathscr{S}$

2   $a, b \in \mathscr{S} \Rightarrow a \cap b \in \mathscr{S}$

A physicist would like to 'understand' why we have postulated AS 1.1, 2. However, it is more or less difficult to 'make plausible' this rather *general* conception of selection procedures. Therefore we can only give some hints:

If $M \in \mathscr{S}$, AS 1.1, 2 has the consequence that $\mathscr{S}$ would be a Boolean algebra of sets. We have *not* postulated $M \in \mathscr{S}$, nevertheless, the assumption $M \in \mathscr{S}$ would not lead to mathematical contradictions in the following. However, it seems to us that the postulate $M \in \mathscr{S}$ would be unrealistic on physical grounds. To see this, we will try to elucidate the physical significance of AS 1.1, 2.

The physical interpretation of '$x \in a$ and $a \in \mathscr{S}$' is as follows: the physical system x has been selected by the selection procedure a. In this sense, an element $a \in \mathscr{S}$ represents the method of selecting as well as the family of physical systems selected by this method.

If there are two selection procedures a and b given by special physical methods, it is not difficult to construct the following selection procedure: select all $x \in M$ which are selected by a as well as by b, i.e. the set $a \cap b$. This is the meaning of AS 1.2. Of course, $a \cap b = \emptyset$ is possible, namely, in the case when there are no systems which can be selected by a as well as by b, i.e. if the selection procedures a and b are incompatible.

If $a \subset b$, the selection procedure a is called finer than b. If a is finer than b, and if one selects by the method a all x of b, the remaining systems of b are those of $b \setminus a$; AS 1.1 says that the selection of these remaining systems is also a selection procedure.

194 Uncertainty Principle and Foundations of Quantum Mechanics

The following two examples show the unrealistic feature of the postulate: $M \in \mathscr{S}$.

Consider an apparatus producing steel balls. This apparatus is an example of a selection procedure for steel balls. Let us denote this selection procedure by 'a'. Then $a \subset M$ and $a \in \mathscr{S}$ hold, m being the 'set of all steel balls'. The set $M \setminus a$ is then characterized as the set of all those elements which have *not* been selected at all. The knowledge of the construction of the machine makes it possible, to specify various properties of the systems of a. However, there are not properties which can be ascribed to the systems of $M \setminus a$. Therefore, we have not included $M \setminus a$ in the set of selection procedures.

We have a similar situation in the case of an accelerator of electrons. The knowledge of the construction of the accelerator yields very essential information on the electrons produced by this apparatus. Such information is necessary for any physicist who wants to perform experiments on these electrons. However, what can be said about all those electrons which are not produced by this particular accelerator?

In the following, let $\mathscr{S}(a)$ be the abbreviation for the set:

$$\mathscr{S}(a) = \{b | b \in \mathscr{S} \text{ and } b \subset a\}$$

As a consequence of AS 1.1, 2 $\mathscr{S}(a)$ is a Boolean algebra of sets, a being the unit element of $\mathscr{S}(a)$.

To every set $\Sigma \subset \mathscr{P}(M)$ there exists a smallest set $\mathscr{S}$ of selection procedures with $\mathscr{S} \supset \Sigma$. $\mathscr{S}$ is called the set of selection procedures generated by Σ.

(B) STATISTICAL SELECTION PROCEDURES

In this section we want to give a mathematical formulation of the second fundamental concept of quantum mechanics, i.e. the concept of *statistics*, of *probability*. Many authors consider the familiar mathematical probability theory as sufficient for the foundation of quantum mechanics. Other authors state that the quantum mechanical probability given by (1) cannot be formulated in the framework of the familiar mathematical probability theory. Indeed, this last opinion will prove wrong as we will see in the following. There are two reasons why we will recall now some essential aspects of the familiar mathematical probability theory:

REASON 1: The axioms of mathematical statistics are so simple that we are able to list them in order to give a complete survey of all fundamental concepts of quantum mechanics.

REASON 2: We will formulate the axioms in a more 'physical' form, i.e. in a form more suitable for describing experiments with physical systems.

In the results of experiments, probability appears in the form of frequencies with which a selection procedure b finer than a selects systems of a.

In other words: if, in an experiment, N systems $x_1, \ldots x_N \in M$ are selected by the procedure a and if one selects out of these N systems those N' systems which fulfil also the conditions of b, the number N'/N is called the frequency with which b has selected systems of a. In many (not in all!) cases the experiments show a 'reproducibility' of this frequency, i.e. if one repeats experiments employing selection procedures a and b, one obtains nearly the same frequencies if N and N' are 'large' numbers. If such a reproducible frequency exists, we say that b depends statistically on a. To give a mathematical description of such a statistical dependence we introduce a mathematical structure called 'statistical selection procedures' or shortly a 'statistical selection structure' which is defined as follows:

A set $\mathscr{S} \subset \mathscr{P}(M)$ is called a structure of statistical selection procedures if AS 1 holds and if a mapping λ is given, mapping

$$\mathscr{T} = \{(a,b) | a, b \in \mathscr{S}; a \supset b \text{ and } a \neq \emptyset\}$$

into the interval [0, 1] of real numbers and if the following axioms hold:

AS 1.1 $a_1, a_2 \in \mathscr{S}, a_1 \cap a_2 = \emptyset, a_1 \cup a_2 \in \mathscr{S} \Rightarrow$
$\lambda(a_1 \cup a_2, a_1) + \lambda(a_1 \cup a_2, a_2) = 1$

2    $a_1, a_2, a_3 \in \mathscr{S}, a_1 \supset a_2 \supset a_3, a_2 \neq \emptyset \Rightarrow$
$\lambda(a_1, a_3) = \lambda(a_1, a_2) \lambda(a_2, a_3)$

3    $a_1, a_2 \in \mathscr{S}, a_1 \supset a_2, a_2 \neq \emptyset \Rightarrow \lambda(a_1, a_2) \neq 0$

$\lambda(a,b)$ is usually called the probability of b relative to a. $\lambda(a,b)$ is the mathematical picture of the frequency with which b selects relative to a as described above. With this 'interpretation' of $\lambda(a,b)$ at hand, the reader may easily check the 'physical' significance of axioms AS 2, 1 to 3 (see also Ludwig (1975) and in preparation a).

From AS 2.1 to 3 we obtain:

$$\lambda(a,a) = 1, \quad \lambda(a, \emptyset) = 0;$$

and for $a_1 \supset a_2$, $a_1 \supset a_3$ and $a_2 \cap a_3 = \emptyset$:

$$\lambda(a_1, a_2 \cup a_3) = \lambda(a_1, a_2) + \lambda(a_1, a_3)$$

By $\mu(b) = \lambda(a,b)$, an additive measure on the Boolean algebra $\mathscr{S}(a)$ is defined.

In the sequel, the following definition will be important:

DEFINITION: A decomposition $a = \bigcup_{i=1}^n b_i$ of an $a \in \mathscr{S}$ with $b_i \neq \emptyset$, $b_i \in \mathscr{S}$, $b_i \cap b_k = \emptyset$ if $i \neq k$ is called a *'demixture'* of a into the b_i's and a is called a 'mixture' of the b_i's. $\lambda(a, b_i)$ is called the 'weight' of b_i in a.

From AS 2.1 through 3 we obtain

$$\sum_i \lambda(a, b_i) = 1$$

(C) PREPARING PROCEDURES

As we have mentioned in the beginning, we want to give a mathematical picture of the procedures by which physical systems are prepared. To this end, we introduce a structure on M (M being the set of systems) by a set $\mathcal{Q} \subset \mathcal{P}(M)$ (the elements of $\mathcal{Q}$ shall be the pictures of the various preparing procedures), for which the following axiom holds:

APS 1 $\mathcal{Q}$ is a statistical selection structure.

The probability function defined by APS 1 will be denoted by $\lambda_\mathcal{Q}$. '$x \in a$' is the mathematical form of the proposition: the physical system x has been prepared by the procedure a.

(D) RECORDING PROCEDURES

It is a bit more complicated to give a mathematical picture of the procedures by which physical systems are recorded. The recording process is characterized by two steps:

(1). Construction and employment of the recording apparatus.

(2). Selection according to signals which appeared (or did not appear on the recording apparatus employed.

Accordingly, we define another mathematical structure on M by choosing two other subsets of $\mathcal{P}(M)$: $\mathcal{R}_0$ and $\mathcal{R}$. $\mathcal{R}_0$ and $\mathcal{R}$ satisfy the following axioms:

APS 2 $\mathcal{R}$ is a selection structure.
APS 3 $\mathcal{R}_0$ is a statistical selection structure.
APS 4.1 $\mathcal{R}_0 \subset \mathcal{R}$.
    2   $b \in \mathcal{R}, b \supset b_0 \in \mathcal{R}_0$ and $b \neq \varnothing \Rightarrow b \in \mathcal{R}_0$.
    3   To each $b \in \mathcal{R}$ there is a $b_0 \in \mathcal{R}_0$ with $b_0 \supset b$.

In order to describe the physical meaning of APS 2 through 4, we must say of what the elements of $\mathcal{R}$ and $\mathcal{R}_0$ are pictures.

An element $b_0 \in \mathcal{R}_0$ represents the *construction and employment* of a recording apparatus. We may clarify it by an example: The constructed apparatus may be a Geiger counter; then b_0 is the set of all those microsystems, to which this Geiger counter is employed. $x \in b_0$ is the mathematical form of the proposition: the Geiger counter b_0 has been used to record x. This does not imply that a recording signal has been produced by x. Therefore, we call $\mathcal{R}_0$ the set of *recording methods*.

The Geiger counter (mentioned above as an example), used to record x, can respond or not; b_+ may be the selection procedure for all those systems $x \in b_0$ to which the Geiger counter has responded; hence $b_+ \subset b_0$. Correspondingly b may be the set of all those $x \in b_0$ to which the counter has not responded; hence $b_- = b_0 \setminus b_+$. b_+ and b_- are elements of $\mathcal{R}$. Generally $\mathcal{R}$ is the set of all those selection procedures which are finer than the procedures of $\mathcal{R}_0$; finer by virtue of the influence of the microsystems on the apparatus, represented by the elements of $\mathcal{R}_0$. We express this briefly by saying: $\mathcal{R}$ is the set of all *recording procedures*. Concerning the axioms APS 2 through 4, we will make only short remarks; a more general discussion is given in Ludwig (1975 and in preparation a).

APS 3 means that the statistical dependence between the various recording methods has nothing to do with the microsystems. In contradistinction to the elements of $\mathcal{R}_0$, the selection procedures $b \in \mathcal{R}$ depend essentially on the influence of the microsystems. For this reason we did *not* state in APS 2 that $\mathcal{R}$ should be a *statistical* selection structure. We may illustrate this situation with the example of a counter: In nature there are no *reproducible* frequencies $\lambda(b_0, b_+)$ for the response of the counter as such; $\lambda(b_0, b_+)$ would describe frequencies independent of the surroundings of the counter. In reality, the frequency of the response of the counter depends essentially on its surroundings.

The probability function corresponding to $\mathcal{R}_0$ will be denoted by $\lambda_{\mathcal{R}_0}$.

(E) THE DEPENDENCE OF THE RECORDING ON THE PREPARING PROCESS

The first physical problem is raised by the question: which preparing procedures and measuring procedures may be combined together. Unfortunately, this problem is not trivial.

We define quite naturally: $a \in \mathcal{Q}$ and $b_0 \in \mathcal{R}_0$ are said to be *combinable* if $a \cap b_0 \neq \emptyset$. The combination problem amounts to finding axioms (as laws of nature in mathematical form) concerning the set

$$\mathcal{C} = \{(a, b_0) | a \in \mathcal{Q}, b_0 \in \mathcal{R}_0, a \cap b_0 \neq \emptyset\}$$

A discussion of this combination problem would be beyond the scope of this paper (see Ludwig, in preparation a). Here it seems sufficient to give a very simple axiom—though this axiom, in fact, is not very realistic. (This simple axiom can be replaced by another more realistic one with essentially the same mathematical consequences.)

If
$$\mathcal{Q}' = \{a | a \in \mathcal{Q}, a \neq \emptyset\}$$
$$\mathcal{R}_0' = \{b_0 | b_0 \in \mathcal{R}_0, b_0 \neq \emptyset\}$$

we formulate as an axiom:

$$\text{APS 5} \quad \mathcal{C} = \mathcal{Q}' \times \mathcal{R}_0'$$

The central problem of quantum mechanics is the description of the statistical dependence of the recording on the preparing process. To begin with, we define

$$\Theta = \{c | c = a \cap b \text{ and } a \in \mathcal{Q}, b \in \mathcal{R}\}$$

198 Uncertainty Principle and Foundations of Quantum Mechanics

An element $c = a \cap b$ of Θ is the set of all systems x prepared by the procedure a and recorded by the procedure b. Let $\mathscr{S}$ be the smallest set of selection procedures for which $\Theta \subset \mathscr{S}$.

In general, neither $\mathscr{Q} \subset \mathscr{S}$ nor $\mathscr{R} \subset \mathscr{S}$ holds!

Now we formulate the experience that the combination of preparing and recording procedures leads to reproducible frequencies:

APS 6 $\mathscr{S}$ is a statistical selection structure.

The probability function corresponding to $\mathscr{S}$ will be denoted by $\lambda_{\mathscr{S}}$. The three probability functions $\lambda_{\mathscr{Q}}, \lambda_{\mathscr{R}_0}, \lambda_{\mathscr{S}}$ cannot be independent for physical reasons. Physical experience suggests that there is no dependence of the recording *methods* $b_0 \in \mathscr{R}_0$ on the procedures $a \in \mathscr{Q}$. This fact is expressed by the following axiom:

APS 7 $a_1, a_2 \in \mathscr{Q}; a_2 \subset a_1; b_{01}, b_{02} \in \mathscr{R}_0$
$b_{02} \subset b_{01}$ and $a_1 \cap b_{01} \neq \emptyset$
implies:
1 $\lambda_{\mathscr{S}}(a_1 \cap b_{01}, a_2 \cap b_{01}) = \lambda_{\mathscr{Q}}(a_1, a_2)$;
2 $\lambda_{\mathscr{S}}(a_1 \cap b_{01}, a_1 \cap b_{02}) = \lambda_{\mathscr{R}_0}(b_{01}, b_{02})$

Axiom APS 7 implies the important theorem (for proof see Ludwig, in preparation a):

The function $\lambda_{\mathscr{S}}$ is determined uniquely by $\lambda_{\mathscr{Q}}$ and the special values

$$\lambda_{\mathscr{S}}(a \cap b_0, a \cap b)$$

for $a \in \mathscr{Q}', b_0 \in \mathscr{R}'_0, b \in \mathscr{R}$ and $b \subset b_0$.

If one looks at the various experiments, it is easy to see that only the values $\lambda_{\mathscr{S}}(a \cap b_0, a \cap b)$ are tested by experimental physicists. Only one example should be sketched: By a preparing procedure a pair of particles may be produced to perform a collision experiment. The particles are recorded after the collision by a recording method b_0. Let b (with $b \subset b_0$) be the recording procedure counting if b_0 has responded by a certain signal. $a \cap b_0$ are all those systems (i.e. *pairs* of particles!), which are prepared by a and for which the recording method b_0 is employed. $a \cap b_0$ characterizes the collision experiment, $a \cap b$ are all those systems to which the recording apparatus has responded.

We define

$$\mathscr{F} = \{(b_0, b) | b_0 \in \mathscr{R}'_0, b \in \mathscr{R}, b \subset b_0\}$$

and call $\mathscr{F}$ the set of all *effect procedures*.

The following real function is defined on $\mathscr{Q}' \times \mathscr{F}$:

$$\mu(a, f) = \mu(a, (b_0, b)) = \lambda_{\mathscr{S}}(a \cap b_0, a \cap b) \tag{5}$$

The function $\mu(a, f)$ defined by (5) plays a central rôle in the statistical description of physical systems, especially of microsystems. The axioms imply the following theorem (for proof see Ludwig, in preparation a).

The function $\mu(a, f)$ satisfies the following relations:

(1). $0 \leq \mu(a, f) \leq 1$.
(2). To every $a \in \mathcal{Q}'$, there is a $f_0 \in \mathcal{F}$ for which $\mu(a, f_0) = 0$.
(3). To every $a \in \mathcal{Q}'$, there is a $f_1 \in \mathcal{F}$ for which $\mu(a, f_1) = 1$.
(4). Every demixture $a = \bigcup_i a_i$ (see the end of Section 1(B)) implies

$$\mu\left(\bigcup_i a_i, f\right) = \sum_i \lambda_i \mu(a_i, f) \text{ and } 0 < \lambda_i = \lambda_\mathcal{Q}(a, a_i) \leq 1, \sum_i \lambda_i = 1$$

(5). $b_{01} \supset b_{02} \supset b$ ($b_{01}, b_{02} \in \mathcal{R}_0'$) and $f_1 = (b_{01}, b)$, $f_2 = (b_{02}, b)$ implies

$$\mu(a, f_1) = \lambda_{\mathcal{R}_0}(b_{01}, b_{02})\mu(a, f_2) \text{ for all } a \in \mathcal{Q}'$$

(6). Every demixture $b_0 = \bigcup_i b_i$ (i.e. $b_i \in \mathcal{R}$, $b_i \cap b_k = \emptyset$ if $i \neq k$) implies (with $f_i = (b_0, b_i)$)

$$\sum_{i=1}^n \mu(a, f_i) = 1, \quad \text{for all } a \in \mathcal{Q}'$$

(7). $\mu(a, (b_0, b)) = 0$ is equivalent to $a \cap b = \emptyset$.

According to the following theorem (proof by H. Neumann, to be published in Ludwig, in preparation a), the statistical structure of the theory is completely described by the function μ. Given $\lambda_\mathcal{Q}$, the conditions (1), (4), (5) and (7) for the function $\mu(a, f)$ imply the existence of a uniquely defined probability function $\lambda_\mathcal{S}$ with

$$\lambda_\mathcal{S}(a \cap b_0, a \cap b) = \mu(a, (b_0, b))$$

$\lambda_{\mathcal{R}_0}$ is determined by $\mu(a, f) = \mu(a, (b_0, b))$.

By the formulation of the axioms APS1 through APS 7, we have reached our first aim, namely, the definition of the concept of physical systems: those components of experiments which are represented by elements of a set M (according to the mapping principles of the physical theory) are called physical systems if the set M is endowed with a structure $\mathcal{Q}, \mathcal{R}, \mathcal{R}_0$ such that the axioms APS 1 through APS 7 hold, and if the elements of $\mathcal{Q}, \mathcal{R}, \mathcal{R}_0$ are pictures of preparing and recording procedures (according to the mapping principles of the theory). All probabilities concerning the outcome of experiments on a physical system are determined by the function μ.

However, this structure is not yet typical for microsystems, as may easily be seen in looking at the example of the gun (as an $a \in \mathcal{Q}'$) and the impacts of the projectiles (as a $b \in \mathcal{R}$).

2. ENSEMBLES AND EFFECTS

The next essential step in the development of the theory will be the introduction of the notions of ensembles and effects. We will introduce these notions on the basis of preparing and recording procedures. it is to be stressed that these notions do not agree in every respect with the customary intuitive usage of the words ensembles and effects.

Uncertainty Principle and Foundations of Quantum Mechanics

In the discussion of problems in the interpretation of quantum mechanics, many difficulties arise due to the fact that, in using 'common language' for a description of experiments, usually no difference is made between preparing procedures and ensembles (or 'states'): a family of microsystems, prepared by a procedure $a \in \mathcal{Q}'$, is often called an ensemble or a set of systems in a 'state', where the ensemble or state is described by a statistical operator W. It is impossible to give here a survey of all misunderstandings and mistakes caused by not distinguishing between preparing procedures and ensembles. Only in the case of the so-called Einstein–Podolski–Rosen paradox, we will demonstrate (see Section 8) how the situation will be clarified by the use of the concepts introduced here.

Also the notion of effects (also known as yes–no measurements or questions) is used in a different sense by various authors. It may be stressed here from the outset that in the Hilbert space representation of the theory the effects are *not* always represented by projection operators!

Since the function μ is defined by (5) on the whole of the set $\mathcal{Q}' \times \mathcal{F}$, the relation defined by

$$\mu(a_1, f) = \mu(a_2, f) \quad \text{for all } f \in \mathcal{F}$$

is an equivalence relation on $\mathcal{Q}'$: $a_1 \sim a_2$. The notion of ensembles is *defined* by:

DEFINITION 1: Let $\mathcal{K}$ be the set of all equivalence classes in $\mathcal{Q}'$. An element of $\mathcal{K}$ is called an *ensemble* (or a *state*); $\mathcal{K}$ is called the set of ensembles (or states).

The relation

$$\mu(a, f_1) = \mu(a, f_2) \quad \text{for all } a \in \mathcal{Q}'$$

defines an equivalence relation $f_1 \sim f_2$ on $\mathcal{F}$.

DEFINITION 2: Let $\mathcal{L}$ be the set of all equivalence classes in $\mathcal{F}$. An element of $\mathcal{L}$ is called an *effect*; $\mathcal{L}$ is called the set of effects.

The following theorem holds: For $w \in \mathcal{K}$, $g \in \mathcal{L}$ and $a \in w$, $f \in g$ the equation

$$\tilde{\mu}(w, g) = \mu(a, f)$$

defines a real function $\tilde{\mu}$ on $\mathcal{K} \times \mathcal{L}$. $\tilde{\mu}$ satisfies:

(1). $0 \leq \tilde{\mu}(w, g) \leq 1$.
(2). $\tilde{\mu}(w_1, g) = \tilde{\mu}(w_2, g)$ for all $g \in \mathcal{L} \Rightarrow w_1 = w_2$.
(3). $\tilde{\mu}(w, g_1) = \tilde{\mu}(w, g_2)$ for all $w \in \mathcal{K} \Rightarrow g_1 = g_2$.
(4). There is a $g_0 \in \mathcal{L}$ such that $\tilde{\mu}(w, g_0) = 0$ for all $w \in \mathcal{K}$.
(5). There is a $g_1 \in \mathcal{L}$ such that $\tilde{\mu}(w, g_1) = 1$ for all $w \in \mathcal{K}$.

Definition 1 gives a precise *definition* of the notion of ensemble (or of state). Nevertheless, we shall give an explanation of this definition since the intuitive usage of the notion of ensemble does not always coincide with the notion defined in Definition 1.

An ensemble w is *not* a set or a family of microsystems, since w is not a subset of M. w is a subset of $\mathcal{P}(M)$, an equivalence class the elements of which are subsets of M. It is an important feature of quantum mechanics that a class w has more than one element. We will demonstrate this in the case of the example given in Section 8.

It should be stressed that we do *not* use the notion of ensemble to formulate the connection between experiment and mathematical theory, i.e. to intepret quantum mechanics. The interpretation of quantum mechanics given here depends only on the notions of preparing and recording procedures. Also, the notion of ensemble is not necessary for the statistical description which is already given by the function $\lambda_{\mathscr{S}}$. The notion of ensemble is used only to analyse the structure which already has been founded in Section 1.

The following definition proves useful in the subsequent analysis of the structure of the theory.

DEFINITION 3: The canonical mapping which maps an element $a \in \mathscr{Q}'$ onto its corresponding equivalence class, $w \in \mathscr{K}$, will be denoted by ϕ; correspondingly, let ψ be the canonical mapping of $\mathscr{F}$ onto $\mathscr{L}$. For $f = (b_0, b)$, we also write $\psi(b_0, b)$ instead of $\psi(f)$.

In the following, we simplify the notation in writing μ instead of $\tilde{\mu}$; the arguments in the function will show whether μ is defined on $\mathscr{Q}' \times \mathscr{F}$ or $\mathscr{K} \times \mathscr{L}$. In this sense, the equality

$$\mu(a, f) = \mu(\phi(a), \psi(f))$$

holds.

The relation (4) of Section 1(E) implies: If $a = \bigcup_i a_i$ is a demixture of a preparing procedure a, we have for all $g \in \mathscr{L}$

$$\mu(\phi(a), g) = \sum_i \lambda_i \mu(\phi(a_i), g) \qquad (6)$$

with $\lambda_i = \lambda_{\mathscr{Q}}(a, a_i)$, $0 \le \lambda_i \le 1$ and $\sum_i \lambda_i = 1$.

We define, in analogy to equation (6):

DEFINITION 4: Let $w \in \mathscr{K}$ be an ensemble, let λ_i be a set of real numbers with $0 \le \lambda_i \le 1$ and $\sum_i \lambda_i = 1$ and let $w_i \in \mathscr{K}$ be a set of ensembles such that for all $g \in \mathscr{L}$

$$\mu(w, g) = \sum_i \lambda_i \mu(w_i, g) \qquad (7)$$

holds; then (7) is called a *demixture* of w with respect to the *components* w_i with weights λ_i.

It is as essential as in the case of ensembles to avoid a false interpretation in the case of effects. Similar to the case of the notion of ensembles it is to be stressed that effects are *classes* of effect procedures (b_0, b). The mapping ψ, defined in Definition 3, maps several effect procedures onto the same effect. By applying the mapping ψ, parts of the structure $(\mathscr{R}_0, \mathscr{R})$ may get lost in the image $\mathscr{L}$ of ψ. This is actually the case as we will see in discussing coexistent effects in

Section 5(B). An effect procedure is characterized by a recording method b_0 and by a recording procedure b. Two recording methods $b_0^{(1)}$ and $b_0^{(2)}$ differing in their technical specification together with different responses $b^{(1)}$ and $b^{(2)}$ can be representatives of the same effect $g \in \mathscr{L}$, i.e.

$$\psi(b_0^{(1)}, b^{(1)}) = \psi(b_0^{(2)}, b^{(2)}) = g$$

The frequently used expressions 'yes–no measurement', for the element (b_0, b) of $\mathscr{F}_1$ is due to the fact that b represents the response of the apparatus b_0. A more or less precise conception of yes–no measurement is used by many authors. Unfortunately, various authors attach a different meaning to the words 'yes–no measurement'. Some people do not use this word for the experimental situation we are describing mathematically by the elements (b_0, b) of $\mathscr{F}$. Our mathematical description has the advantage that all possible misunderstandings may be avoided. Also, the word *question* instead of yes–no measurement is used by some authors, but not always in the same sense. Our definition is: $(b_0, b) \in \mathscr{F}$ is a question, b the answer 'yes' and $b_0 \setminus b$ the answer 'no'.

In contradistinction to our definition, some authors use the words yes–no measurement or question to denote the elements of $\mathscr{L}$ which we called effects; some authors use these words only for the elements of a subset of $\mathscr{L}$, which we shall call 'decision effects', in the following (see Section 3).

In introducing the notions of ensemble (or state) and effect, we have taken a first step towards the 'usual' representation of quantum mechanics. The next step is the introduction of the concept of 'simultaneous measurement' and the concept of 'observable'. Again, these concepts will be *defined* with the use of the fundamental concepts introduced in Section 1. The interpretation of the theory has already been given in Section 1 and new concepts are not introduced for the interpretation, but rather for the sake of a structural analysis of the theory.

However, before introducing new concepts, we will analyse in the next chapter the connection between the mathematical symbols introduced so far, and relation (1) of Section 1.

3. LAWS FOR THE PREPARATION AND RECORDING OF MICROSYSTEMS

For a 'physical' approach to the structure represented by relation (1), it seems best to introduce axioms (i.e. physical laws in mathematical form) for the preparing and recording procedures and to deduce from these axioms the following theorem: the function $\mu(w, g)$ on $\mathscr{K} \times \mathscr{L}$ can be represented in the form $\mu(w, g) = \mathrm{tr}\,((\alpha w)(\beta g))$ with injective mappings α, β. We will formulate such axioms without a detailed physical discussion (cf., e.g., Ludwig, 1964, 1967a, b, c, d, 1968, 1970, 1971a, b) neither will we present the proofs of the theorems (see Ludwig, 1970, 1972b; Stolz, 1971 and Ludwig, in preparation b).

Since one always performs only a *finite* number of experiments (see the extensive discussions of the 'finiteness of physics' in Ludwig 1970, 1974b, and in preparation b), we assume that the sets $M, \mathcal{Q}, \mathcal{R}$ are denumerable (this is equivalent to the assumption that the completions of these sets are separable if endowed with a physically meaningful uniform structure—see Ludwig 1970, and in preparation b). If $M, \mathcal{Q}, \mathcal{R}$ are denumerable sets, so are $\mathcal{K}$ and $\mathcal{L}$.

The relations given in Section 2 imply the following theorem:
There is a pair of real Banach spaces $\mathcal{B}, \mathcal{B}'$ ($\mathcal{B}'$ denotes the dual Banach space of $\mathcal{B}$) and an embedding of $\mathcal{K}$ into $\mathcal{B}$ and of $\mathcal{L}$ into $\mathcal{B}'$ such that

(1). The canonical bilinear form (x, y) defined on $\mathcal{B} \times \mathcal{B}'$ coincides with $\mu(w, g)$ on $\mathcal{K} \times \mathcal{L}$, i.e.

$$\mu(w, g) = (w, g)|_{\mathcal{K} \times \mathcal{L}}$$

(2). $\mathcal{B}$ is a base norm space (see, for instance, Nagel, 1974), the base K being equal to $\overline{\text{co}}\,\mathcal{K}$ ($\overline{\text{co}}\,\mathcal{K}$ denotes the norm-closed convex set generated by $\mathcal{K}$); the cone, generated by K is norm-closed;

(3). The linear hull of $\mathcal{L}$ is $\sigma(\mathcal{B}', \mathcal{B})$-dense in $\mathcal{B}'$.

Points (1) through (3) determine $\mathcal{B}$, and $\mathcal{B}'$ uniquely up to isomorphism. $\mathcal{K}$ denumerable implies that $\mathcal{B}$ is separable.

In the following, we will denote the bilinear form (x, y) by $\mu(x, y)$ because of point (1) of the theorem.

$\mathcal{B}$ being a base norm space implies: $\mathcal{B}'$ is an order unit space. Because of $0 \leq \mu(w, g) \leq 1$ for $w \in \mathcal{K}$ and $g \in \mathcal{L}$, it follows $0 \leq \mu(w, g) \leq 1$ also for $w \in K$ and $g \in \mathcal{L}$, i.e. $\mathcal{L} \subset [0, 1]$, $[0, 1]$ being the order interval between the zero element and the order unit.

We denote by L the set $\overline{\text{co}}\,\mathcal{L}$, the closure of co $\mathcal{L}$ in the $\sigma(\mathcal{B}', \mathcal{B})$-topology. Let $\mathcal{D}$ be the norm closure of the linear hull of $\mathcal{L}$. We have $1 \in \mathcal{D}$. $\mathcal{D}$ is a separable Banach subspace of $\mathcal{B}'$ ($\mathcal{D}$ is also an order unit space). $\mathcal{D}$ is $\sigma(\mathcal{B}', \mathcal{B})$-dense in $\mathcal{B}'$. K is $\mathcal{S}(\mathcal{B}, \mathcal{D})$-precompact and $\sigma(\mathcal{B}, \mathcal{D})$-separable.

$\mathcal{B}$ may be identified with a subspace of $\mathcal{D}'$ ($\mathcal{D}'$ being the dual Banach space of $\mathcal{D}$, $\mathcal{D}'$ is a base norm space), and consequently K may be identified with a subset of $\mathcal{D}'$. Let $\bar{K}$ be the $\sigma(\mathcal{D}', \mathcal{D})$-closure of K in $\mathcal{D}'$. $\bar{K}$ is $\sigma(\mathcal{D}', \mathcal{D})$-compact and L is $(\mathcal{B}', \mathcal{B})$-compact. For the sets $\bar{K}$ and L the theorem of Krein–Milman holds.

The topologies $\sigma(\mathcal{B}', \mathcal{B})$ and $\sigma(\mathcal{D}', \mathcal{D})$ are of considerable physical significance: first, the topologies $\sigma(\mathcal{D}', \mathcal{D})$ and $\sigma(\mathcal{D}', \mathcal{L})$ are identical *on K and $\bar{K}$* and the topologies $\sigma(\mathcal{B}', \mathcal{B})$, $\sigma(\mathcal{B}', K)$ and $\sigma(\mathcal{B}', \mathcal{K})$ are identical *on L*. The topologies $\sigma(\mathcal{D}', L)$ on K (and $\mathcal{K}$) and $\sigma(\mathcal{B}'\,\mathcal{K})$ on L (and $\mathcal{L}$) are suited to distinguish *through experiment* between different ensembles and different effects, respectively. This should be clarified, at least to some extent, in the case of ensembles: $\mu(w_1, g) = \mu(w_2, g)$ for all $g \in \mathcal{L}$ implies $w_1 = w_2$. However, in experiments, an ensemble can only be tested by finitely many recording procedures, i.e. by *finitely* many $g \in \mathcal{L}$ and not by *all* $g \in \mathcal{L}$). Likewise, the probability $\mu(w, g)$ can be tested only with a finite error. Thus we see that the inequalities

$$|\mu(w_1, g_i) - \mu(w_2, g_i)| < \varepsilon \quad (i = 1, 2, \ldots n)$$

may be tested experimentally only for a finite number of effects $g_1, \ldots g_n$ and a finite error ε. These inequalities determine (for various ε, n, g_i) a neighborhood base of the topology $\sigma(\mathscr{D}', \mathscr{L})$.

To formulate the laws for preparing and recording we give some definitions:

DEFINITIONS:

$$K_0(B) = \{w \mid w \in K, \mu(w, g) = 0 \text{ for all } g \in B \subset L\}$$
$$K_1(B) = \{w \mid w \in K, \mu(w, g) = 1 \text{ for all } g \in B \subset L\}$$
$$L_0(A) = \{g \mid g \in L, \mu(w, g) = 0 \text{ for all } w \in A \subset K\}$$

$K_0(B)$ and $K_1(B)$ are closed faces of K, $L_0(A)$ is a closed face of L. If B has only one element g, we write intead of $K_0(B)$ simply $K_0(g)$ and similarly for K_1 and L_0.

It may be easily seen that the ordering $y_1 < y_2$ in $\mathscr{B}'$ is equivalent to the relation

$$\mu(w, y_1) \leq \mu(w, y_2) \quad \text{for all } w \in K$$

Let us formulate the first law (concerning recording procedures) by the axiom:

AV 1.1 To every pair $g_1, g_2 \in L$ there exists a $g_3 \in L$ such that $g_3 > g_1$, $g_3 > g_2$ and $K_0(g_1) \cap K_0(g_2) \subset K_0(g_3)$.

AV 1.1 is equivalent to the following statement: Every set $L_0(A)$ has a greatest element (see Ludwig, 1970, and in preparation b); this greatest element may be denoted $eL_0(A)$.

All elements of K (not only of $\mathscr{K}$) will be called *ensembles* (or states), all elements of L (not only of $\mathscr{L}$) will be called effects. The elements of the form $eL_0(A)$ will be called *decision effects*. The set of all *decision effects* will be denoted by G.

AV 1.1 implies $G \subset \partial_e L$ (for proof see Ludwig, 1970 and in preparation b), $\partial_e L$ being the set of extremal points of L.

Since, for a subset $\{A_\alpha\}$ of $\mathscr{P}(M)$, the relation $L_0(\bigcup_\alpha A_\alpha) = \bigcap_\alpha L_0(A_\alpha)$ holds, the set $\{L_0(A) \mid A \subset K\}$ is a complete lattice, the order relation being given by the set theoretical inclusion. The mapping $L_0(A) \to eL_0(A)$ is an order isomorphism of $\{L_0(A) \mid A \subset K\}$ onto G; hence also G is a complete lattice with respect to the ordering induced on G by the ordering of $\mathscr{B}'$.

Let $\hat{L}$ be the $\sigma(\mathscr{B}', \mathscr{B})$-closure of $\{y \mid y = \lambda g, \lambda \in \mathbf{R}, g \in L \text{ and } 0 \leq \lambda g \leq 1\}$. As a second law (concerning again recording procedures), we postulate the axiom:

AV 1.2 $g \in \hat{L}, e \in G$ and $K_0(e) \subset K_0(g)$ implies $K_1(e) \supset K_1(g)$.

Let $C(w)$ be the norm-closed face of K generated by w. Since $\mathscr{B}$ is separable, every norm-closed face of K is of the form $C(w)$ with a suitably chosen w.

The next law (concerning both preparing and recording procedures) is given by the axiom:

AV2 $w_1, w_2 \in K$ and $C(w_1) \supsetneq C(w_2)$ imply that there is a $g \in L$ such that $w_2 \in K_0(g)$ but $w_1 \notin K_0(g)$.

AV 2 is equivalent to the relation: $K_0 L_0(F) = F$ for every norm-closed face F of K.

The axioms AV 1.1, 2 and AV 2 imply

$$L = \hat{L} = [0, 1]$$

In the theory, we need the following axiom AVid which may be regarded as a mere mathematical idealization:

AV id $K_1(e) \neq \emptyset$ for all $e \in G$ with $e \neq 0$.

AV id cannot be tested by experiments! AV id is equivalent to: $(1-e) \in G$ for all $e \in G$.

The mapping $e \rightarrow 1-e$ is an orthocomplementation in the lattice G. The axioms AV 1.1 through AV id imply that G is an orthocomplemented, orthomodular lattice.

We define the following 'distance' between two closed faces of K:

$$\Delta(C(w_2), C(w_3)) = \tfrac{1}{2} \inf \{\mu(w, e_3) | w \in C(w_2)\}$$
$$+ \tfrac{1}{2} \inf \{\mu(w, e_2) | w \in C(w_3)\}$$

e_i being an abbreviation for $eL_0 C(w_i)$. Two faces $C(w_2)$ and $C(w_3)$ are called strictly separated if $\Delta(C(w_2), C(w_3)) \neq 0$.

The next law (concerning preparing procedures) is given by the axiom:

AV3 If $w_1, w_2, w_3 \in K$ with $C(w_1) \subset C(w_3) \subset C(\tfrac{1}{2} w_1 + \tfrac{1}{2} w_2)$ and if the faces $C(w_2), C(w_3)$ are strictly separated, we have $C(w_1) = C(w_3)$.

All of the axioms AV1. 1, AV1. 2, AV2, AV id and AV3 hold for all (!) known theories of physical systems, even for the so-called classical theories. The axiom distinguishing between 'classical' systems and *micro*systems is the following:

AV 4 For every face $C(w)$ of K there is a sequence $w_\nu \in K$ such that $C(w_\nu)$ is of finite dimension, $C(w_{\nu+1}) \supset C(w_\nu)$ and

$$C(w) = \bigvee_\nu C(w_\nu)$$

It should be stressed that $\bigvee_\nu C(w_\nu)$ is *not* the set-theoretical union of the sets $C(w_\nu)$, but the smallest closed face of K which includes all $C(w_\nu)$.

The concept of *micro*system may now be defined as follows:

DEFINITION 6: If the axioms AV 1, AV 2, AV 3 *and* AV 4 hold, the set M endowed with the structure $\mathcal{Q}, \mathcal{R}_0, \mathcal{R}$ is called a set of microsystems.

Since for the particular case of 'classical' systems the axiom AV 4 does not hold, we will call axiom AV 4 the 'law of microsystems'. 'Classical' physical systems can be defined as such systems for which all (!) $C(w)$ are infinite-dimensional *and* all decision effects are commensurable [for the concept of commensurability see Section 5(B); other equivalent forms of axioms for classical systems are given by H. Neumann (1972, 1974a, b) it has been demonstrated in these references how to regain the Γ-space from preparing and recording procedures.]

It can be proved (see Ludwig 1970, 1972b; Stolz 1971, and Ludwig, in preparation b), that the axioms AV 1, AV 2, AV id, AV 3 and AV 4 are equivalent to the following statement:

K and L can be identified with the base of the Banach space $\mathcal{B}(\mathcal{H}_1, \mathcal{H}_2, \ldots)$ and with the order interval $[0, 1]$ of the Banach space $\mathcal{B}'(\mathcal{H}_1, \mathcal{H}_2, \ldots)$ respectively, $\mathcal{B}(\mathcal{H}_1, \mathcal{H}_2, \ldots)$ being the space of all sequences $(W_1, W_2, \ldots)$, where every W_i is a self-adjoint operator of the trace class in the Hilbert space $\mathcal{H}_i$ such that $\sum_i \text{tr}((W_i^2)^{1/2}) < \infty$. The dual Banach space $\mathcal{B}'(\mathcal{H}_1, \ldots)$ can be identified with the space of all sequences $(A_1, A_2, \ldots)$, any A_i being a self-adjoint and bounded operator in $\mathcal{H}_i$ such that $\sup_i \|A_i\| < \infty$. The canonical bilinear form is given by:

$$((W_1, W_2, \ldots), (A_1, A_2, \ldots)) = \sum_i \text{tr}(W_i A_i)$$

K is the set of all sequences $(W_1, W_2, \ldots)$ such that $0 \leq W_i$ and $\sum_i \text{tr}(W_i) = 1$. L is the set of all sequences $(F_1, F_2, \ldots)$ such that $0 \leq F_i \leq 1$.

The axioms imply that the $\mathcal{H}_i$ are Hilbert spaces over the fields **R** (of real numbers) or **C** (of complex numbers) or **Q** (of quaternions). There are *physical* arguments to eliminate the cases **R** and **Q**.

4. ENSEMBLES AND EFFECTS IN QUANTUM THEORY

The identification (given in Section 3) of K with the base of $\mathcal{B}(\mathcal{H}_1, \ldots)$ and of L with the order interval $[0, 1]$ of $\mathcal{B}'(\mathcal{H}_1, \ldots)$ makes it possible to interpret the mappings ϕ and ψ (defined in Section 2, Definition 3) as mappings of $\mathcal{Q}'$ into $\mathcal{B}(\mathcal{H}_1, \ldots)$ and $\mathcal{F}$ into $\mathcal{B}'(\mathcal{H}_1, \ldots)$, respectively. $\phi\mathcal{Q}'$ is then norm-dense in the base K of $\mathcal{B}(\mathcal{H}_1, \ldots)$ and $\psi\mathcal{F}$ is $\sigma(\mathcal{B}', \mathcal{B})$-dense in $L = [0, 1]$.

The norm-closed subspace of $\mathcal{B}'(\mathcal{H}_1, \ldots)$ generated by $\psi\mathcal{F}$ was denoted by $\mathcal{D}$ (see Section 3). A more detailed characterization of $\mathcal{D}$ by axioms cannot be given here (see Ludwig, in preparation a). Nevertheless, it should be mentioned that one may formulate axioms in such a way that $\mathcal{D}$ becomes a set of sequences $(A_1, A_2, \ldots)$, A_i being a self-adjoint operator of a certain C^*-algebra $\mathcal{A}_i$ of operators in $\mathcal{H}_i$. Thus $\mathcal{D}$ becomes the real part of a C^*-algebra. We would like to call the attention of the reader to the fact that the 'set of states' in the theory of C^*-algebras has been denoted (in Section 3) by $\bar{K}$ (and not by K, as is usually done in the theory of C^*-algebras).

The set G of decision effects (as introduced in Section 3) may then be identified with the set of all sequences

$$e = (E_1, E_2, \ldots)$$

E_i being a projection operator in $\mathcal{H}_i$. The special decision effects

$$e_i = (0, 0, \ldots 1_i, \ldots)$$

define the superselection rules. We shall refrain here from a detailed discussion. However, it is not difficult to see that it is very practical to investigate theoretically as well as experimentally the various 'sorts' of microsystems (characterized by the e_i's) separately. Each sort is described in one particular Hilbert space. In this way, one obtains the 'usual' quantum mechanical formalism. We cannot do this here in detail. Instead we want to show that the sets $\mathcal{Q}$ of preparing procedures and $\mathcal{R}$ of recording procedures may serve to elucidate some of the conceptions of 'usual' quantum mechanics.

As a first example we treat a structure very similar to the famous Heisenberg uncertainty relation.

As mentioned above, it is sufficient to discuss the case of one Hilbert space only. The following theorem holds (see Ludwig, 1970): There are two decision effects E_1 and E_2 such that, for every ensemble W, at least one of the following inequality relations must be false:

$$\operatorname{tr}(W(E_1 - \alpha_1 1)^2) \leq \tfrac{1}{16}$$
$$\operatorname{tr}(W(E_2 - \alpha_2 1)^2) \leq \tfrac{1}{16}$$

where

$$\alpha_1 = \operatorname{tr}(WE_1), \qquad \alpha_2 = \operatorname{tr}(WE_2).$$

This theorem is analogous to Heisenberg's uncertainty relation, formulated for decision effects. This theorem shows precisely, that the Heisenberg uncertainty relation has nothing to do with experimental errors (at least not in principle; see also the discussion in Ludwig, 1975), since in measuring the decision effects E_1 and E_2 only the two values one or zero may be obtained.

We will simplify the following discussion by assuming $\phi \mathcal{Q}' = K$ and $\psi \mathcal{F} = L$ (which is not essential). Then we can express the physical content of the above theorem as follows:

There are two recording methods (i.e., it is possible to construct two recording apparatus) $b_0^{(1)}$ and $b_0^{(2)}$ with responses $b^{(1)}$ and $b^{(2)}$, respectively, such that $\psi(b_0^{(1)}, b_0^{(1)}) = E_1$ and $\psi(b_0^{(2)}, b_0^{(2)}) = E_2$ are decision effects and such that, for every preparing procedure a, at least one of the two probabilities

$$\lambda_{\mathscr{S}}(a \cap b_0^{(1)}, a \cap b^{(1)}), \qquad \lambda_{\mathscr{S}}(a \cap b_0^{(2)}, a \cap b^{(2)})$$

is essentially different from zero or one. One can make experimental efforts as strong as possible to construct preparing procedures, at least one of the recording procedures $b^{(1)}$, $b^{(2)}$ will respond indeterministically even in case $\psi(b_0^{(1)}, b^{(1)})$ and $\psi(b_0^{(2)}, b^{(2)})$ are decision effects.

Thus we may conclude that the Heisenberg uncertainty relation is a relation which concerns merely the possibility of constructing preparing apparatus. All the more Heisenberg's uncertainty relation does not tell us anything about the possibility of 'simultaneous measurement'. We will have to come back to this problem in the next chapters. The vagueness in some discussions of Heisenberg's uncertainty relation arises from the fact that in most cases there is no clearcut distinction between preparing and recording procedures, since only the so-called 'ideal' measuring processes of the 'first' kind are discussed. These particular measuring processes are recording and preparing procedures at the same time; therefore, the Heisenberg uncertainty relation for the preparing procedure forbids the simultaneous recording of the values of position and momentum.

However, the concept of simultaneous measuring can be defined in a very natural way without any recourse to the so-called ideal measuring processes of the first kind. We will do this in Section 5(A).

5. OBSERVABLES

The concept of an observable can be defined in a very natural way starting from the experimental situation of recording, described by the terms $\mathcal{R}_0$ and $\mathcal{R}$.

(A) Coexistent recording procedures

A pair $(b_0, b) \in \mathcal{F}$ was called an effect procedure. The mapping ψ connects with each effect procedure (b_0, b) an effect $\psi(b_0, b) \in L\mathcal{CB}'(\mathcal{H}_n, \dots)$. All discussions in Sections 3 and 4 were concerned with the mapping of one $f = (b_0, b)$ onto one $g = \psi(f) = \psi(b_0, b)$ only. In reality, b_0 represents in general an apparatus which has several possibilities of response, namely, all b with $b \subset b_0$. Let us denote by $\mathcal{R}(b_0)$ the set of all b with $b \subset b_0$. The elements of the set $\mathcal{R}(b_0)$ represent *all* recording procedures that are possible in applying the recording method b_0. In many experiments on microsystems, one uses methods b_0, where the set $\mathcal{R}(b_0)$ is so large that it is practically impossible to determine all $b \in \mathcal{R}(b_0)$.

This situation may be illustrated by the following two typical examples of recording apparatus.

1. A SYSTEM OF MANY COUPLED COUNTERS

Each microsystem can produce responses of some of these counters. A recording procedure b can for instance be characterized by the answer of three particular fixed counters. In this case the set $\mathcal{R}(b_0)$ is very well known to technicians: $\mathcal{R}(b_0$ is the Boolean algebra of switching between the various counters.

2. The recording method b_0 is a bubble chamber (or a cloud chamber)

In this case the set of all b's is immense: every possible bubble corresponds to one b, but also every connected trace of bubbles corresponds to one b.

The two examples demonstrate another feature of quantum mechanics: The various responses $b \in \mathscr{R}(b_0)$ of the apparatus b_0 do *not* necessarily appear at the *same* time. In general a response b need not be instantaneous but may have a finite duration, for instance in the case where b represents the simultaneous response of two coupled counters responding with a time delay. Another example is a trace in a bubble chamber.

These examples show that the 'simultaneous recording' of the various b's of $\mathscr{R}(b_0)$ has nothing in common with 'measuring at the *same* time'. The frequently used formulation: Some observables as, for instance, position and momentum are 'not measurable at the same time' but at 'different times' is at least incomprehensible, if not false (see Ludwig, 1975).

We define:

DEFINITION 7: The recording procedures $b \in \mathscr{R}(b_0)$ are called *coexistent* with respect to the recording method b_0. Several $(b_0, b) \in \mathscr{F}$ which have the *same* b_0 are called *coexistent effect procedures*.

(B) Coexistent effects

If (b_0, b) is a family of coexistent effect procedures, then the set of the effects $\psi(b_0, b)$ is a subset of $\psi\mathscr{R}(b_0)$. Therefore, we may define a mapping ψ_0 of $\mathscr{R}(b_0)$ into L by

$$\psi_0(b) = \psi(b_0, b).$$

It is not difficult to prove the following theorem: The mapping ψ_0 is an additive and effective measure on the Boolean algebra $\mathscr{R}(b_0)$, which maps the unit element of $\mathscr{R}(b_0)$ onto $1 \in L$.

As an idealization we define:

DEFINITION 8: A set $A \in L$ is called a set of *coexistent* effects if there is a Boolean algebra Σ endowed with an additive measure, $F: \Sigma \to L$ such that $A \subset F\Sigma$.

The essential conception of coexistent effects has been defined (though not yet in a clearcut way) in Ludwig (1964, 1967a, b, c, d). This conception is fundamental for the notion of observables.

DEFINITION 9: A set $A \subset G$ (i.e. a set of decision effects) is called a set of commensurable decision effects if there is a Boolean algebra Σ endowed with an additive measure $F: \Sigma \to G$ such that $A \subset F\Sigma$.

It may be proved that every set of coexistent decision effects is also a set of commensurable decision effects and that the decision effects are commensurable if and only if the projection operators (in $\mathcal{H}_i$) belonging to these decision effects commute. This last condition is very well known. However, the foundation of this condition is usually presented with much 'philosophy'.

(C) Observables

The notion of observable is nothing but an idealization of the correspondence $\mathcal{R}(b_0) \xrightarrow{w_0} L$. This idealization is obtained in a process of completion (see Ludwig 1970 and in preparation a): If a Boolean algebra Σ is endowed with an additive and effective measure $F: \Sigma \to L$, then a metric may be defined on Σ by

$$d(\sigma_1, \sigma_2) = \mu(w_0, F(\sigma_1 \wedge \sigma_2^*) + F(\sigma_2 \wedge \sigma_1^*))$$

σ^* being the complement of σ; w_0 is an effective ensemble, for instance, $w_0 = \Sigma_\nu \lambda_\nu w_\nu$, $\lambda_\nu > 0$, $\Sigma_\nu \lambda_\nu = 1$, where the set $\{w_\nu\}$ is dense in K. Σ can be completed with respect to this metric (see Ludwig 1970 and in preparation a).

DEFINITION 10: A Boolean algebra Σ endowed with an additive measure $F: \Sigma \to L$ is called an *observable* if Σ is complete and separable (with respect to the metric defined above).

This general concept of an observable has been introduced and analysed in (Ludwig, 1970); a more detailed analysis is contained in (Neumann, H., 1971) and in (Ludwig, in preparation a). It is impossible to give a structural analysis of the concept of an observable in this short article. We only wanted to stress that the concept of an observable is no more than an idealization of recording methods.

To show at least the connection between the notion of an observable defined in Definition 10 and the 'customary' notion, we add the following definitions:

DEFINITION 11: An observable is called a *decision observable* if the mapping F of definition is a mapping into G.

DEFINITION 12: A mapping $\mathbf{R} \xrightarrow{\sigma} \Sigma$ of the set $\mathbf{R}$ of real numbers into a Boolean algebra Σ is called a measuring scale of Σ if $\alpha_1 \geq \alpha_2$ implies $\sigma(\alpha_1) > \sigma(\alpha_2)$ and if $\sigma(-\infty) = 0$, $\sigma(+\infty) = \varepsilon$ (ε being the unit element of Σ) and if the set of all $\sigma(\alpha)$ generates the whole algebra Σ.

A decision observable endowed with a measuring scale is identical with what is 'usually' called an observable.

6. PREPARATORS

So far, in the discussions of quantum mechanics only the simultaneous measurability of decision observables has been of interest. However, the question of the possibilities of simultaneous preparation was neglected or, at the most, discussed as a partial aspect of simultaneous measurability, since one had in mind only 'ideal' measuring processes 'of the first kind'. These 'ideal' measuring processes are, indeed, connected with a certain idealized form of 'repreparing' processes (see Ludwig, 1972a, 1975). At any rate many fundamental questions of the interpretation of quantum mechanics will become much more transparent if the discussion of simultaneous preparation is separated from that of simultaneous recording. Within the scheme of quantum mechanics as outlined here, a quite natural question arises: what is the condition for simultaneous preparation.

Equation (6) combined with the identification of K with a subset of $\mathcal{B}(\mathcal{H}_1, \ldots)$ implies the theorem:

Let $a = \bigcup_{i=1}^{n} a_i$ be a demixture of the preparation procedure a, then

$$\phi(a) = \sum_{i=1}^{n} \lambda_i \phi(a_i) \qquad (8)$$

holds with $\lambda_i = \lambda_2(a, a_i)$, $0 < \lambda_i \leq 1$ and $\sum_i \lambda_i = 1$.

If there are two demixtures of the *same* preparing procedure

$$a = \bigcup_{i=1}^{n} a_i = \bigcup_{k=1}^{m} \tilde{a}_k$$

then we have

$$\phi(a) = \sum_{i=1}^{n} \lambda_i \phi(a_i) = \sum_{k=1}^{m} \tilde{\lambda}_k \phi(\tilde{a}_k)$$

In this case, also, these two demixtures generate a third one, namely:

$$a = \bigcup_{i,k}{}' (a_i \cap \tilde{a}_k), \qquad (9)$$

where the union $\bigcup'$ is taken over those pairs i, k for which $a_i \cap \tilde{a}_k = \varnothing$. Expression (9) implies

$$\phi(a) = \sum_{i,k}{}' \lambda_{ik} \phi(a_i \cap \tilde{a}_k) \qquad (10)$$

where $\lambda_{ik} = \lambda_2(a, a_i \cap \tilde{a}_k)$.

It is very useful to introduce the following mapping $\phi_a : \mathcal{S}(a) \to \check{K}$; ($\check{K}$ being the cap of the cone generated by K i.e.

$$\check{K} = \bigcup_{0 \leq \lambda \leq 1} \lambda K)$$

defined by the equation:

$$\phi_a(\tilde{a}) = \lambda_2(a, \tilde{a}) \phi(\tilde{a})$$

It is not difficult to see that $\phi_a : \mathcal{S}(a) \to \check{K}$ is an additive measure on the Boolean algebra $\mathcal{S}(a)$ with $\phi_a(a) = \phi(a) \in K$.

DEFINITION 13: An element $\tilde{w} \in \check{K}$ with $\tilde{w} < w \in K$ is called a mixture component, or shortly a component of w.

If $\tilde{w}$ is a component of w, then so is $w - \tilde{w}$; and $w = \tilde{w} + (w - \tilde{w})$ is a demixture of w.

Two demixtures of one and the same preparing procedure

$$a = \bigcup_{i=1}^{n} a_i = \bigcup_{k=1}^{m} \tilde{a}_k$$

yield two demixtures of the ensemble $\phi(a)$:

$$\phi(a) = \sum_{i=1}^{n} \phi_a(a_i) = \sum_{k=1}^{m} \phi_a(\tilde{a}_k)$$

for which the components $\phi_a(a_i)$ and $\phi_a(\tilde{a}_k)$ are elements of the image of $\mathscr{S}(a)$ under ϕ_a. That leads to the following definition:

DEFINITION 14: Two demixtures of an ensemble $w \in K$

$$w = \sum_{i=1}^{n} w_i = \sum_{k=1}^{m} \tilde{w}_k \qquad (w_i, \tilde{w}_k \in \check{K})$$

are called *coexistent* if there is a Boolean algebra Σ endowed with an additive measure $W: \Sigma \to \check{K}$ such that $W(\varepsilon) = w$ (ε being the unit element of Σ) and w_i, $\tilde{w}_k \in W\Sigma$.

Two demixtures of one and the same preparing procedure a give two coexistent demixtures of the ensemble $\phi(a)$.

DEFINITION 15: A set $A \subset \check{K}$ is called a set of coexistent components of w if there is a Boolean algebra Σ endowed with a measure $W: \Sigma \to \check{K}$ such that $W(\varepsilon) = w$ and $A \subset W\Sigma$.

The set

$$\phi_a \mathscr{S}(a) = \{\phi_a(\tilde{a}) | \tilde{a} \in Q, \tilde{a} \subset a\}$$

is a set of coexistent components of $\phi(a)$.

If $W: \Sigma \to \check{K}$ is an additive and effective measure on the Boolean algebra Σ, then

$$d(\sigma_1, \sigma_2) = \mu(W(\sigma_1 \wedge \sigma_2^*) + W(\sigma_2 \wedge \sigma_1^*), 1)$$

defines a metric in Σ. Σ may be completed in this metric and the measure W may be defined uniquely on this completion. This leads to the following idealization of $\phi_a: \mathscr{S}(a) \to \check{K}$:

DEFINITION 16: A Boolean algebra Σ endowed with an additive measure $W: \Sigma \to \check{K}$ such that $W(\varepsilon) \in K$ and Σ is complete and separable is called a *preparator*.

Let Σ be the completion of $\mathcal{S}(a)$ in the metric defined by the measure ϕ_a then $\Sigma \xrightarrow{\phi_a} \check{K}$ is a preparator. Many experiments (especially in elementary particle physics) may be described by a structure of the form $\mathcal{S}(a) \xrightarrow{\phi_a} \check{K}$. This structure is also of paramount significance for a discussion of fundamental problems of quantum mechanics. We must refrain from a general mathematical structure-analysis of the concept of preparators [see for instance Ludwig, (1975) and in preparation a]. rather, we will discuss some 'gedanken' experiments to elucidate the significance of preparators and coexistent and non-coexistent demixtures.

First, we shall show that there exist non-coexistent demixtures. It is sufficient to prove this in the case of one Hilbert space $\mathcal{H}$ only.

Let $\phi, \psi \in \mathcal{H}, \|\phi\| = \|\psi\| = 1$, $\phi \perp \psi$ and $0 < \lambda < 1$; then $w = \lambda P_\phi + (1-\lambda) P_\psi$ is an ensemble. $\lambda P_\phi + (1-\lambda) P_\psi$ is a demixture of w, but it is easy to find others: Let χ be another vector in the plane generated by ϕ and ψ. Then it is possible to choose a real number μ $(0 < \mu < 1)$ and a vector $\eta \in \mathcal{H}$ such that

$$w = \mu P_\chi + (1-\mu) P_\eta$$

is another demixture of the same w. Proof: χ has the form

$$\chi = \phi a + \psi b; \quad |a|^2 + |b|^2 = 1$$

The conclusion is obtained by putting:

$$\mu = \frac{\lambda(1-\lambda)}{\lambda |b|^2 + (1-\lambda)|a|^2}$$

$$\eta = \frac{1}{\sqrt{\lambda^2 |b|^2 + (1-\lambda^2)|q|^2}} [\varphi \lambda \bar{b} - \psi(1-\lambda) a]$$

If $a \neq 0$ and $b \neq 0$, the two demixtures

$$w = \lambda P_\phi + (1-\lambda) P_\psi = \mu P_\chi + (1-\mu) P_\eta$$

cannot be coexistent as we will prove immediately:

Let Σ be a Boolean algebra endowed with an additive measure W such that there are $\sigma_1, \sigma_2 \in \Sigma$ satisfying $W(\sigma_1) = \lambda P_\phi$, $W(\sigma_2) = \mu P_\chi$. That implies $W(\sigma_1^*) = W(\epsilon) - W(\sigma_1) = w - \lambda P_\phi = (1-\lambda) P_\psi$ and $W(\sigma_2^*) = (1-\mu) P_\eta$. We want to calculate $W(\sigma_1 \wedge \sigma_2)$. One has $W(\sigma_1 \wedge \sigma_2) \leq W(\sigma_1) = \lambda P_\phi$ and $W(\sigma_1 \wedge \sigma_2) \leq W(\sigma_2) = \mu P_\chi$. Since $v \in \check{K}$ and $0 \leq v \leq P_\phi$ $0 \leq v \leq P_\chi$ implies $v = 0$, we get $W(\sigma_1 \wedge \sigma_2) = 0$. In the same way, it follows that $W(\sigma_1 \wedge \sigma_2^*) = 0$, $W(\sigma_1^* \wedge \sigma_2) = 0$, and $W(\sigma_1^* \wedge \sigma_2^*) = 0$. Now $\epsilon = (\sigma_1 \wedge \sigma_2) \vee (\sigma_1 \wedge \sigma_2^*) \vee (\sigma_1^* \wedge \sigma_2) \vee (\sigma_1^* \wedge \sigma_2^*)$ implies $w = W(\epsilon) = W(\sigma_1 \wedge \sigma_2) + W(\sigma_1 \wedge \sigma_2^*) + W(\sigma_1^* \wedge \sigma_2) + W(\sigma_1^* \wedge \sigma_2^*) = 0$ in contradiction to $w = \lambda P_\phi + (1-\lambda) P_\psi \neq 0$.

7. PROPOSITIONS CONCERNING EXPERIMENTS ON INDIVIDUAL MICROSYSTEMS

So far, one is accustomed to speak about individual microsystems and about the results of experiments with individual microsystems using common (not mathematical) language. On the contrary, mathematical language is employed in classical mechanics of mass points, where every individual mass point is represented by a mathematical trajectory which is then compared with the 'real' trajectory obtained in measurement. Every proposition concerning an individual system may be formulated in mathematical form. The situation is quite different in customary quantum mechanics, where an individual experimental result, e.g. an individual trace in a cloud chamber, cannot be compared with the theory, since the usual quantum mechanical mathematical picture comprises only terms describing the statistics.

For instance, it is not possible to translate into mathematical language such propositions as 'the position of this individual electron has been measured in the region $\mathcal{V}$'. It is not possible to find a corresponding mathematical relation for this proposition. Only after a series of such measurements of the position have been carried out, the statistics of the results can be compared with the theory using the relation $m = \text{tr}\,(WE_{\mathcal{V}})$. Here, $E_{\mathcal{V}}$ is the following decision effect: the measured position is localized in $\mathcal{V}$.

One has tried out many ways of coping with this defect of quantum mechanics. Some approaches should be mentioned.

Some physicists call the projection operators (or, in our terminology, the decision effects) properties of the individual microsystems, and formulate propositions such as: The microsystem x has the property e; I know that x has the property e; x has the property 'not e'; x has not the property e; I do not know whether x has the property e; x has the property e with the probability α; the property e has been measured on x; etc. No wonder that one gets into difficulties in using this type of language. Some people have tried to avoid these difficulties in introducing a more precise form of this language and a new logic, too.

Another attempt is interpreting the projection operators as propositions. In this case, the authors have to say *which* propositions (formulated in common language) should be represented symbolically by projection operators. Correspondingly some authors interpret the lattice of projection operators as a proposition-logic, the so-called quantum logic. This conception of quantum logic is to be distinguished from others, where the lattice of decision effects is formally called 'quantum logic', however, without claiming that the word 'quantum logic' should have anything in common with a logic of propositions.

We want to show that every experimental result (also if concerning an individual microsystem) can be described by a mathematical relation if one accepts the foundations of quantum mechanics outlined in the first sections of this paper. This description is possible without employing any new form of logic and using only the usual logic of mathematics.

It is not possible to give here a complete survey of all possible ways of formulating propositions on individual microsystems in mathematical language, only some examples may briefly be sketched. To avoid mathematical difficulties connected with the fact that $\phi\mathcal{Q}'$ is only dense in K and $\psi\mathcal{F}$ only dense in L, we will sharpen our axioms in postulating: $\phi\mathcal{Q}' = K$ and $\psi\mathcal{F} = L$. This sharpening is without any fundamental physical relevance, since every comparison between theory and experiment can only be made with a finite inaccuracy (anyway, this sharpening may be avoided in using more sophisticated mathematics).

For the following discussions it is essential to realize that the physical interpretation of '$x \in a$, $a \in \mathcal{Q}'$' and of '$x \in b$, $b \in \mathcal{R}$' has already been given in Section 1. All other propositions may be deduced from these two fundamental relations. For instance, a proposition of the form 'the microsystem x has the property e' makes no sense until one has given this proposition a mathematical form which is reducible to relations of the form $x \in a$, $a \in \mathcal{Q}'$ and $x \in b$, $b \in \mathcal{R}$. No intuitive meaning of properties is introduced. Every interpretation must be in the last line reducible to the interpretation of preparing and recording procedures.

We define the following sets:

DEFINITION 17: For every $e \in G$ let

$$\mathcal{Q}(e) = \{a | a \in \mathcal{Q}', \text{ and } \phi(a) \in K_1(e)\}$$

$$M_p(e) = \bigcup_{a \in \mathcal{Q}(e)} a$$

We want to discuss the following relation:

$$x \in M_p(e) \qquad (11)$$

The physical interpretation of (11) is clear by Definition 17, since e is defined by preparing and recording procedures as has been done in Section 2, and $M_p(e)$ is the union of some preparing procedures. However, one usually wants to express relation (11) by a short formulation in common language. To this end, we introduce the following terminology: The elements $e \in G$ will be called *pseudo-properties* (not simply properties — to avoid misunderstandings). We express the relation (11) in the form: 'the microsystem x has been prepared with the pseudo-property e'.

Relation (11), formulated for microsystems, may also be formulated for macrosystems and, in this context, it represents a very interesting technical procedure. For instance, for steel balls manufactured by a machine, it reads: 'this individual steel ball x has the property e, i.e., that the radius lies in the interval $[r_1, r_2]$'.

Relation (11) is a proposition on a pseudo-property of the prepared microsystem x. No misunderstanding is possible if we use relation (11) for such propositions. The logical negation of (11) is $x \notin M_p(e)$. There is another relation: $x \in M_p(e^\perp)$, $e^\perp = 1 - e$. However, the two relations $x \notin M_p(e)$ and

216 Uncertainty Principle and Foundations of Quantum Mechanics

$x \in M_p(e^\perp)$ are not equivalent! This fact has nothing to do with logic, but only reflects the structure of the family of sets $\{M_p(e), e \in G\}$.

Another possible proposition on an individual microsystem may be obtained using the following definition:

DEFINITION 18: For every $e \in G$, we define:

$\mathcal{R}(e) = \{b | b \in \mathcal{R} \text{ and there is a } b_0 \in \mathcal{R}_0 \text{ such that } b_0 \supset b \text{ and } \psi(b_0, b) \leq e\}$;

$$M_r(e) = \bigcup_{b \in \mathcal{R}(e)} b$$

We want to discuss the relation

$$x \in M_r(e) \tag{12}$$

We express (12) in common language by the proposition: 'The pseudo-property e of the microsystem x has been recorded'.

We define:

DEFINITION 19: $M(e) = M_p(e) \cup M_r(e)$

The mathematical relation

$$x \in M(e) \tag{13}$$

is equivalent to

$$x \in M_r(e) \quad \text{or} \quad x \in M_p(e)$$

We express (13) in normal language by: 'The microsystem x has the pseudo-property e'. It seems again in order to stress that the two relations $x \notin M(e)$ and $x \in M(e^\perp)$ are *not* equivalent.

At the beginning of Section 7, we mentioned some of the propositions usually used in the interpretation of quantum mechanics. These propositions have the disadvantage that it is often unclear what they mean, and various authors may give them various meanings. This is not possible while using relations (11), (12) and (13).

What about a new logic to treat the relations (11), (12) and (13)!

At first, it is to be stressed that in 'handling' mathematically relations (11), (12) and (13), one has to use customary mathematical logic! What does this imply? We will demonstrate this only by examples.

Let, in the following, x be an individual microsystem, we are experimenting with. Let e be a well-defined element of G, for instance, the following decision effect: 'the position is in a certain space-region $\mathcal{V}$'. We will now discuss some possibilities of describing what we have done in experimenting with the microsystem x.

(1). We have shown in Ludwig (1970, 1975) that experimental results and procedures may be written down in mathematical form. These relations representing the experiments are, from the point of view of mathematical theory, new axioms denoted in Ludwig (1970, 1975) by (—)$_r$. We assume that these axioms (—)$_r$ are added to the mathematical theory and that, for instance, (13) is a theorem in this stronger mathematical theory. If this is the case, one says for instance: I know that x has the (pseudo-)property e. The words 'I know' could be misunderstood in the following way: someone could mean that my subjective knowledge is essential here. However, what we, indeed, mean is that the real experimental situation (expressed by the axioms (—)$_r$ together with the theory) make it possible to deduce (13) as a theorem. In the case where (13) is deducible as a theorem, also another sentence is used: The proposition (13) is 'true'. If one uses these words 'I know' or 'true' only as abbreviations for the fact that (13) is deducible as a theorem, there is no objection and no real mistakes are possible.

(2). For instance, relation (12) is no theorem, however, it may be added without contradiction as an axiom to the theory complemented by the axioms (—)$_r$. In this case, one may say: It is possible (but not necessary) to record the pseudo-property e on the microsystem x. Or: The pseudo-property e can be recorded on x. It is exactly this situation which, in the opinion of some authors, may be described only by a new logic. In fact, any mathematician knows that a relation like (12) may be added to a theory without any contradiction, although (12) is not a theorem of the theory. In this case, it is possible to add (without any contradiction) also the negation of the relation (12): $x \notin M_r(e)$. If one wants to interpret the word 'true' by 'deducible as a theorem' (see above), then, in this sense, mathematical logic is 'many-valued' since the very beginning of mathematics. In a mathematical theory a relation A can be a theorem, or the relation (not A) can be a theorem, or (as a *third* possibility) neither A nor (not A) are theorems, but A as well as (not A) can be added as axioms without any contradiction (naturally, not both of these relations!). However, no new logical axioms different from the usual mathematical axioms are necessary to discuss quantum mechanics.

(3). The negation of relation (12) is a theorem in the theory complemented by the experimental results (—)$_r$. One says: It is impossible that x could be recorded with the pseudo-property e. It may be stressed again that $x \notin M_r(e)$ does not imply $x \in M_r(e^\perp)$!

After having interpreted words such as 'I know', 'true', 'possible', 'impossible', we want to discuss in more detail example (2). We assume, in particular, that the microsystem x has been prepared experimentally by a procedure a (but not yet recorded) and that the well-known construction of the preparing apparatus makes it possible to determine the element $\phi(a)$ of K. Under this assumption, relation (12) may be added as an axiom without any contradiction if and only if $\mu(\phi(a), e) \neq 0$.

If this is the case, one sometimes states that relation (12) is true with probability $\mu(\phi(a), e)$. However, this statement could be misunderstood. In fact, the question as to the 'possibilities' of (12) is very complicated. Therefore, we ask: How is it possible to 'realize' the relation (12)? We ask for an experimental possibility of a recording procedure, such that after writing down the experimental results of recording (in the form of axioms (—)$_r$), relation (12) will appear as a theorem. However, this experimental possibility does not only depend on the probability, it also depends on a certain 'arbitrary choice'.

Arbitrary is here the selection of the recording method b_0. $x \in b_0$ can be realized for many different recording methods b_0. It is up to the experimentalist which method b_0 he will apply. If a certain apparatus b_0 has been installed to be used in the experiment, the relation $x \in b_0$ (as one of the axioms of (—)$_r$) is to be added to the theory. Now we have a situation very typical for microsystems:

By the selection of a certain b_0, i.e. by writing down $x \in b_0$ for a certain b_0, the 'possibilities' have changed essentially. By b_0 i.e. by $\psi_0 : \mathcal{R}(b_0) \to L$ an observable is defined. It may be that the decision effect e in (12) is *not* coexistent with all effects $\psi(b_0, b)$. In particular, b_0 can be chosen in such a manner that the following theorem holds:

There is no $b \in \mathcal{R}$ such that $b \neq \varnothing$, $b \subset b_0$, and $\psi(b_0, b) \leq e$.

If $x \in a$ and $x \in b_0$ is fixed by the experiment and if the theorem above holds for b, then relation (12) is false, i.e. $x \notin M_r(e)$ is a theorem in the theory with axioms $x \in a$, $x \in b_0$. We say: it is *impossible* that x could be recorded with the pseudo-property e.

However, if one has not yet fixed the recording method b_0, one is free to choose another one. For instance, it is possible to choose a recording method $\tilde{b}_0$ such that there exists a $b \in \mathcal{R}$ such that $b \subset \tilde{b}_0$ and $\psi(\tilde{b}_0, b) \leq e$. Now suppose we have chosen such a $\tilde{b}_0$ experimentally. What can we say then about the 'possibility' of relation (12)?

It is clear that (12) may be added as an axiom without contradicting the theory containing the axioms, $x \in a$ and $x \in \tilde{b}_0$, if there is a $b \in \mathcal{R}(\tilde{b}_0)$ such that $\psi(\tilde{b}_0, b) \leq e$ and $\mu(\phi(a), \psi(\tilde{b}_0, b)) \neq 0$. If there are $b_1, b_2 \in \mathcal{R}(\tilde{b}_0)$ such that $\psi(\tilde{b}_0, b_1) \leq e$, $\psi(\tilde{b}_0, b_2) \leq e$, then also the relation $\psi(\tilde{b}_0, b_1 \cup b_2) \leq e$ holds, which can be simply proved: $b_1 \cup b_2 = [b_1 \cap (\tilde{b}_0 \setminus b_2)] \cup b_2$ implies

$$\psi(\tilde{b}_0, b_1 \cup b_2) = \psi(\tilde{b}_0, b_1 \cap (\tilde{b}_0 \setminus b_2)) + \psi(\tilde{b}_0, b_2)$$

One has:

$$g_1 = \psi(\tilde{b}_0, b_1 \cap (\tilde{b}_0 \setminus b_2)) \leq \psi(\tilde{b}_0, b_1) \leq e$$

and $g_2 = \psi(\tilde{b}_0, b_2) \leq e$. If we write $g = \psi(\tilde{b}_0, b_1 \cup b_2)$, we get $g = g_1 + g_2$. This equation and the relations $g_1 \leq e$, $g_2 \leq e$ imply $K_0(g) \supset K_0(e)$, which is equivalent to $g \leq e$.

To avoid measure-theoretical arguments, we will assume that the union of *all* elements, b of $\mathcal{R}(\tilde{b}_0)$ such that $\psi(\tilde{b}_0, b) \leq e$, is an element $\tilde{b}$ of $\mathcal{R}(\tilde{b}_0)$. $\tilde{b}$ is the greatest element of $\mathcal{R}(\tilde{b}_0)$ satisfying $\psi(\tilde{b}_0, b) \leq e$. Therefore, if $\tilde{b}_0$ has been chosen as the recording method, we say: relation (12) is possible with probability $\mu(\phi(a), \psi(\tilde{b}_0, \tilde{b}))$. If $\tilde{b}_0$ has been chosen in such a way that $\psi(\tilde{b}_0, \tilde{b}) = e$, the probability is equal to the greatest possible value $\mu(\phi(a), e)$.

We now assume that the recording of the system x has been accomplished and that our $\tilde{b}$ has given a response. Then the relation $x \in \tilde{b}$ must be added to the theory as an axiom, i.e. as a mathematical formulation of the experimental result. In this more comprehensive theory, relation (12) is a theorem; we say: 'the microsystem x has been recorded with the pseudo-property e'. Since (13) is a theorem, too, we also say after this experiment has been performed: 'The microsystem x has the pseudo-property e'.

However, if $\tilde{b}$ has not responded, i.e. if $\tilde{b}_0 \setminus \tilde{b}$ has given a response, then $x \in \tilde{b}_0 \setminus \tilde{b}$ is to be added as an axiom. This implies that the relation $x \notin M_r(e)$ is a theorem; we say: 'it is impossible that the system x has been recorded with the pseudo-property e'. But $x \in M(e)$ can be a theorem if $\psi(a) \in K_1(e)$!

Some authors are using *intuitively* the proposition 'x has the property e' as equivalent to relation (13). To others, this proposition seems inadmissible. We will see why.

Let e_1, e_2 be two incommensurable decision effects such that $M(e_1) \cap M(e_2) \neq \emptyset$ (for instance, let e_1 be the decision effect: the position in a very small space-region $\mathcal{V}$, and e_2: the momentum is in a very small region Π of the momentum space).

$M(e_1) \cap M(e_2) \neq \emptyset$ implies that the relation

$$x \in M(e_1) \quad \text{and} \quad x \in M(e_2) \tag{14}$$

could be an allowed hypothesis (see Ludwig 1974b, and in preparation b), i.e. that in the case of a suitable experimental situation relation (14) may be added to the theory without any contradiction. It may also be that the experimental situation is such as to admit relation (14) as a theorem.

At first sight, relation (14) seems to contradict the well-known quantum-mechanical laws, for instance, the Heisenberg uncertainty relation, if the regions denoted above by $\mathcal{V}$ and Π are small enough. Relation (14) expressed in common language would read: The microsystem x has the pseudo-property e_1 (for instance, a position in $\mathcal{V}$) *and* the pseudo-property e_2 (for instance, a momentum in Π). Such a sentence in common language could, indeed, seem to be in contradiction to quantum mechanical laws. However, in reality, there is no contradiction. We may clear up this problem in formulating it in mathematical symbols. Relation (14) is, for instance, a theorem if the experiment is of such a type that $x \in a$, $\phi(a) \in K_1(e_1)$, $x \in b$, b being such that there is a $b_0 \supset b$ and $\psi(b_0, b) \leq e_2$.

Another objection to accepting the proposition 'x has the pseudo-property e' as a translation of (13) into common language is the following: Let $a \in \mathcal{Q}$ be such that $\phi(a) \in K_1(e)$ and $x \in a$. Then (13) is a theorem, even if the recording procedure b is such that $\psi(b_0, b)$ is not coexistent with e. After (!) the recording method b_0 had been employed, the microsystem x was influenced by the apparatus b_0 in such a way that the microsystem does not have the pseudo-property e any more. This objection is not correct. It is true that there was an influence of the recording apparatus on the microsystem. However, it is not essential for the interpretation of quantum mechanics to know what happens *after* the recording procedure. Only the correlation between preparing and recording are essential. Why do many physicists believe that a measurement (we called it a recording procedure) yields an information on the system after (!) measurement? Only because they tacitly assume the so-called 'ideal measurement of the first kind'! However, in general, the recording is much more complicated, considering the influence of the recording apparatus on the microsystem see Ludwig (1972a, 1975). In our interpretation of quantum mechanics, the recording response yields an information on a microsystem such as it was *before* the recording process has materialized. The essential relations $x \in a$ ($a \in \mathcal{Q}'$) and $x \in b$ ($b \in \mathcal{R}$) are relations which give an information on the system x *after preparing* and *before recording*. The probability $\mu(\phi(a), \psi(b_0, b))$ gives the correlation between preparing and recording processes, a correlation due to the so-called microsystem conceived as a system between preparing and recording.

Instead of our description of the basic quantum-mechanical processes, given by the structure $\mathcal{Q}, \mathcal{R}_0, \mathcal{R}$, a lot of different postulates for the measuring process are considered to be fundamental by other authors. The majority of these postulates have their origin in the famous postulate (M) in J. v. Neumann's book *Mathematical Foundation of Quantum Mechanics* (1955). This postulate (M) reads:

(M.) If the physical quantity, R, is measured twice in succession in a system, S, then we get the same value each time. This is the case even though R has a dispersion in the original state of S, and the R-measurement can change the state of S.

V. Neumann's postulate (M) was the source of the widespread opinion as if it were essential, in a measurement, to fix the state *after* the measurement. The existence of such measurements of the first kind, as they are postulated by (M), is not necessary for the interpretation of quantum mechanics; the structure $\mathcal{Q}, \mathcal{R}_0, \mathcal{R}$ is sufficient, moreover this structure $\mathcal{Q}, \mathcal{R}_0, \mathcal{R}$ seems to represent a more natural basis for quantum mechanics considered as a theory describing microsystems *between* preparing and recording.

The description of a microsystem as a real entity between preparing and recording should not be taken as a hint to endow such microsystems with additional intuitively motivated structures. Such additional (read: not deducible from $\mathcal{Q}, \mathcal{R}_0, \mathcal{R}$) structures could eventually prove to contradict quantum mechanics [see, for instance, the theories of hidden variables, see also Ludwig (1975)].

The examples (11) through (14) should suffice to demonstrate that, on the conceptual basis given in Section 2, it is possible to speak of individual microsystems without any contradiction.

8. THE EINSTEIN–PODOLSKI–ROSEN PARADOX

We will discuss only the spin example of the Einstein–Podolski–Rosen paradox (in the following abbreviated as the EPR paradox).

Pairs of particles may be prepared in such a way that the total spin is equal to zero; the spin of each single particle of the pair may be equal to 1/2.

To describe the EPR paradox, one needs three Hilbert spaces, $\mathcal{H}_i$: Let $\mathcal{H}_1$ be the Hilbert space of the particles of sort 1, $\mathcal{H}_2$ the Hilbert space of the particles of sort 2 and $\mathcal{H}_3$ the Hilbert space of the pairs (1, 2). To simplify the problem we shall consider only the spin spaces, i.e. $\mathcal{H}_1$ and $\mathcal{H}_2$ are both two-dimensional and

$$\mathcal{H}_3 = \mathcal{H}_1 \times \mathcal{H}_2$$

We install a preparing apparatus a which prepares only pairs of total spin zero. The ensemble $\phi(a)$ is described by a statistical operator in $\mathcal{H}_3 = \mathcal{H}_1 \times \mathcal{H}_2$. We have $\phi(a) = P_\chi$, χ being the vector

$$\chi = \frac{1}{\sqrt{2}}(u_+(1)u_-(2) - u_-(1)u_+(2))$$

We have denoted the eigenvectors of the 3-component of the spin of particle 1 by $u_+(1)$, $u_-(1)$, and the corresponding eigenvectors of particle 2 by $u_+(2)$, $u_-(2)$.

Let $b_0^{(2)}$ be a recording method that measures the 3-component of particles 2. Then, there is a recording procedure $b^{(2)} \subset b_0^{(2)}$ such that $\psi(b_0^{(2)}, b^{(2)}) = 1 \times P_{u_+}$, and $\psi(b_0^{(2)}, b_0^{(2)} \setminus b^{(2)}) = 1 \times P_{u_-}$ (as operators in $\mathcal{H}_3$).

Now we want to construct a *new* preparing apparatus composed of the apparatus a and the apparatus $b_0^{(2)}$. We get three new preparing procedures, preparing particles of sort 1.

We do this in the following way (see Figure 1): The pairs prepared by a are leaving the preparing apparatus in such a manner that the particles of sort 1 and

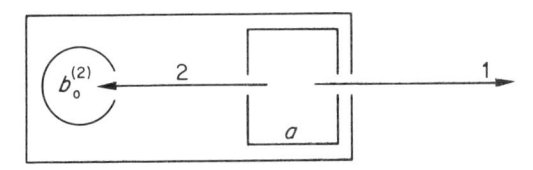

Figure 1 Preparing apparatus composed of apparatus a and $b_0^{(2)}$

2 leave the apparatus in opposite directions. Particles of sort 2 are entering the recording apparatus $b_0^{(2)}$. The particles of sort 1 are leaving the new apparatus composed of a and $b_0^{(2)}$. The apparatus composed of a and $b_0^{(2)}$ gives us three preparing procedures: The preparing procedure a_1^3 comprises all particles of sort 1, leaving a. The preparing procedure a_1^{3+} comprises all particles of sort 1 such that $b^{(2)}$ has given a response. The preparing procedure a_1^{3-} comprises all particles of sort 1 such that $b_0^{(2)} \setminus b^{(2)}$ has given a response.

Apparently, the following relations hold:
$$a_1^3 = a_1^{3+} \cup a_1^{3-}, \qquad a_1^{3+} \cap a_1^{3-} = \emptyset,$$
i.e. $a_1^3 = a_1^{3+} \cup a_1^{3-}$ is a demixture of a_1^3. $\phi(a_1^3)$, $\phi(a_1^{3+})$, $\phi(a_1^{3-})$ are operators in $\mathcal{H}_1$:
$$\phi(a_1^3) = W = \tfrac{1}{2}1, \qquad \phi(a_1^{3+}) = P_{u_-}, \qquad \phi(a_1^{3-}) = P_{u_+}.$$
Since
$$\lambda_2(a_1^3, a_1^{3+}) = \lambda_2(a_1^3, a_1^{3-}) = \tfrac{1}{2}$$
the demixture $a_1^3 = a_1^{3+} \cup a_1^{3-}$ leads (according to Section 6(A)) to the demixture of the ensemble W: $W = \phi(a_1^3) = \tfrac{1}{2}\phi(a_1^{3+}) + \tfrac{1}{2}\phi(a_1^{3-})$, i.e.
$$W = \tfrac{1}{2}P_{u_-} + \tfrac{1}{2}P_{u_+}$$

It is not difficult to construct many more preparing procedures in a similar way. We combine the preparing apparatus a with a recording apparatus $\tilde{b}_0^{(2)}$ that measures the 1-component of the spin of particles of sort 2. Then we get three preparing procedures a_1^1, a_1^{1+}, a_1^{1-} such that $\phi(a_1^{1+}) = P_{v_-}$, $\phi(a_1^{1-}) = P_{v_+}$, v_+ and v_- being the eigenvectors of the 1-component of the spin. $a_1^1 = a_1^{1+} \cup a_1^{1-}$ is a demixture of a_1^1, which implies
$$\phi(a_1^1) = W = \tfrac{1}{2}1 = \tfrac{1}{2}P_{v_-} + \tfrac{1}{2}P_{v_+}$$

Since $\phi(a_1^1) = \phi(a_1^3)$, we have obtained two different demixtures of the same W.

Since the two preparing procedures, a_1^1 and a_1^3, differ only in the measuring parts, one could mean that we should postulate: $a_1^1 = a_1^3$. However, although the postulate $a_1^1 = a_1^3$ seems very 'plausible', at first sight, it leads to a contradiction. Indeed, we will prove: $a_1^1 \cap a_1^3 = \emptyset$, which contradicts the postulate $a_1^1 = a_1^3$, since $a_1^1, a_1^3 \in \mathcal{Q}'$, i.e. $a_1^1 \neq \emptyset$, $a_1^3 \neq \emptyset$. Hence a paradox (i.e. a contradiction) arises if one attempts to add to the theory the (perhaps plausible) axiom: $a_1^1 = a_1^3$.

To prove $a_1^1 \cap a_1^3 = \emptyset$, we start with the relation
$$a_1^3 \cap a_1^1 = (a_1^{3+} \cap a_1^{1+}) \cup (a_1^{3+} \cap a_1^{1-}) \cup (a_1^{3-} \cap a_1^{1+}) \cup (a_1^{3-} \cap a_1^{1-})$$
Since $\mathcal{Q}$ is a selection structure, $a_1^{3+} \cap a_1^{1+} \in \mathcal{Q}$ holds. If $a_1^{3+} \cap a_1^{1+} \in \mathcal{Q}'$ (i.e. $a_1^{3+} \cap a_1^{1+} \neq \emptyset$ were true, $\phi(a_1^{3+} \cap a_1^{1+})$ would be an ensemble.

The two demixtures
$$a_1^{3+} = (a_1^{3+} \cap a_1^{1+}) \cup \tilde{a}_1^{3+} \quad \text{and} \quad a_1^{1+} = (a_1^{3+} \cap a_1^{1+}) \cup \tilde{a}_1^{1+}$$
where
$$\tilde{a}_1^{3+} = a_1^{3+} \setminus (a_1^{3+} \cap a_1^{1+}), \qquad \tilde{a}_1^{1+} = a_1^{1+} \setminus (a_1^{3+} \cap a_1^{1+})$$
imply the demixtures
$$\phi(a_1^{3+}) = \lambda\phi(a_1^{3+} \cap a_1^{1+}) + (1-\lambda)\phi(\tilde{a}_1^{3+})$$
$$\phi(a_1^{1+}) = \mu\phi(a_1^{3+} \cap a_1^{1+}) + (1-\mu)\phi(\tilde{a}_1^{1+})$$

Since $\phi(a_1^{3+}) = P_{u_-}$ and $\phi(a_1^{1+})P_{v_-}$, the demixtures take on the form:
$$P_{u_-} = \lambda W' + (1-\lambda)W'' \quad \text{and} \quad P_{v_-} = \mu W' + (1-\mu)W'''$$
$a_1^{3+} \cap a_1^{1+} \neq \emptyset$ would imply $\lambda \neq 0$ and $\mu \neq 0$. Since P_{u_-}, P_{v_-} are extreme points, $\lambda \neq 0$ and $\mu \neq 0$ would have the consequence: $W' = P_{u_-}$ and $W' = P_{v_-}$, which is a contradiction. Thus $a_1^{3+} \cap a_1^{1+} = \emptyset$ is proved.

In a similar way, $a_1^{3+} \cap a_1^{1-} = \emptyset$, etc, can be proved. Thus also $a_1^3 \cap a_1^1 = \emptyset$ is proved.

It is a consequence of $a_1^3 \cap a_1^1 = \emptyset$ that the two demixtures $W = \frac{1}{2}P_{u_-} + \frac{1}{2}P_{u_+} = \frac{1}{2}P_{v_-} + \frac{1}{2}P_{v_+}$ of $W = \frac{1}{2}1$ are not coexistent.

The example, $\phi(a_1^3) = \phi(a_1^1)$, shows an essential feature of quantum mechanics, namely, the fact that something of the structure of the preparing procedures gets lost by the mapping ϕ of $\mathscr{Q}'$ on to K. The introduction of the structure $\mathscr{Q}, \mathscr{R}_0, \mathscr{R}$ as a basis for the interpretation of quantum mechanics is essential to clear up all problems of the interpretation.

The example above is an 'extreme' case of two non-coexistent preparing procedures. We define:

DEFINITION 20: Two demixtures of an ensemble $w \in K$
$$w = \sum_i w_i = \sum_k \tilde{w}_k$$
$(w_i, \tilde{w}_k \in \check{K})$ are called *complementary* if the assumption $\varphi(a) = \varphi(\tilde{a}) = w$ together with two demixtures $a = \bigcup_i a_i$, $\tilde{a} = \bigcup_k \tilde{a}_k$, such that $\varphi_a(a_i) = w_i$, $\varphi_{\tilde{a}}(\tilde{a}_k) = \tilde{w}_k$, implies $a \cap \tilde{a} = \emptyset$.

Example: the two demixtures
$$W = \frac{1}{2}1 = \frac{1}{2}P_{u_-} + \frac{1}{2}P_{u_+} = \frac{1}{2}P_{v_-} + \frac{1}{2}P_{v_+}$$
are complementary.

Definition 20 may be generalized in such a form that only abstract Boolean algebras (instead of $\mathscr{Q}(a), \mathscr{Q}(\tilde{a})$) are used (see Ludwig, in preparation b).

Another possibility of formulating the intuitive idea that a_1^1 and a_1^3 should not 'essentially' differ would be the following:

It is true that not every preparing procedure $a \in \mathscr{Q}'$ can be demixed 'arbitrarily' by constructing a corresponding apparatus. However, one might imagine that, ideally, there should exist much more and finer 'preparing procedures' than those of $\mathscr{Q}'$. For instance, it is conceivable that procedure a could be demixed in such a way that the spin-3-components of the components of the demixture are well-determined.

According to this point of view, we shall now attempt to add new axioms to the theory: Let $\tilde{\mathscr{Q}}$ be the set of all 'imagined preparing procedures'. We postulate $\tilde{\mathscr{Q}} \supset \mathscr{Q}$ and all the axioms APS 1 through APS 7 also for the set $\tilde{\mathscr{Q}}$. naturally, it would make no sense to postulate the other axioms AV 1 through AV 4 also for $\tilde{\mathscr{Q}}$, since we assume that there are possibly more imagined preparing procedures than those of $\mathscr{Q}$. In the same way as $\mathscr{Q}$, $\tilde{\mathscr{Q}}$ also splits into equivalence classes. Since the effect procedures $f \in \mathscr{F}$ are 'tested' with the help of a larger set of preparing procedures, namely, those of $\tilde{\mathscr{Q}}$, the partition of $\mathscr{F}$ into equivalence classes with respect to $\tilde{\mathscr{Q}}$ will, in general, be finer than that of $\mathscr{F}$ with respect to $\mathscr{Q}$.

According to the idea that the elements of G are symbols for some *properties* which the microsystems are endowed with, we want to express that an $f \in \mathscr{F}$, such that $\psi(f) \in G$, characterizes only the fact that the microsystem has been found to 'have' the property $\psi(f)$. Therefore, we postulate the axiom:

$$\mu(a, f_1) = 1, \qquad a \in \tilde{\mathscr{Q}}$$

(not only $a \in \mathscr{Q}$!), and

$$\psi(f_2) = \psi(f_1) \in G \quad \text{implies} \quad \mu(a, f_2) = 1$$

In this axiom, we express the idea that the difference between the two effect procedures, f_1, f_2, is not essential if the microsystems recorded by these procedures 'have' the property $\psi(f_1) = \psi(f_2)$. If an 'imagined preparing procedure' involves only systems endowed with the property $e = \psi(f_1) = \psi(f_2)$, then the probability must be one for f_1, as well as for f_2.

All axioms introduced for the set $\tilde{\mathscr{Q}}$ are satisfied if one puts $\tilde{\mathscr{Q}} = \mathscr{Q}$. They cannot lead to any contradiction. The essential difference between $\tilde{\mathscr{Q}}$ and $\mathscr{Q}$ shall be a much wider possibility to demix preparing procedures. According to the experiments of the EPR-paradox the following axiom seems to be very 'plausible'.

To every demixture of a recording method, $b_0 \in \mathscr{R}_0$, in the form

$$b_0 = b_1 \cup b_2 \cup b_3$$

(i.e. $b_i \in \mathscr{R}$ and $b_i \cap b_k = \emptyset$ if $i \neq k$), such that

$$\psi(b_0, b_1) = e_1 \in G, \qquad \psi(b_0, b_2) = e_2 \in G, \qquad \psi(b_0, b_3) = e_3 \in G$$

hold there exists, to every $a \in \mathscr{Q}'$ (not necessarily $a \in \tilde{\mathscr{Q}}'$), a demixture $a = \tilde{a}_1 \cup \tilde{a}_2 \cup \tilde{a}_3$ such that $\tilde{a}_i \in \tilde{\mathscr{Q}}$ and $\mu(\tilde{a}_i, (b_0, b_i)) = 1$ (if $\tilde{a}_i \neq \emptyset$).

This 'demixing' axiom means the following: every set of a microsystems prepared in a 'normal' preparing procedure a may be thought to be demixed with respect to every triplet of decision effects e_1, e_2, e_3, satisfying $e_1 + e_2 + e_3 = 1$, in such a manner that the microsystems of the components $\tilde{a}_i$ 'have' the same property e_i.

The axioms introduced above allow us to define a mapping Φ of G into $\mathscr{P}(M)$ by:

$$\Phi(e) = \bigcup_{a \in \tilde{\mathscr{Q}}_1(e)} a$$

where $\tilde{\mathscr{Q}}_1(e) = \{a | a \in \tilde{\mathscr{Q}}' \text{ and } \mu(a, f) = 1 \text{ if } \psi(f) = e\}$. It may be proved (see Ludwig, 1975):

(1). $e_1 \leq e_2$ implies $\Phi(e_1) \subset \Phi(e_2)$;
(2). $\Phi(e) \cap \Phi(e^\perp) = \emptyset$;

(3). $e_i \in G$ and $e_1 + e_2 + e_3 = 1$ implies

$$\Phi(e_1) \cup \Phi(e_2) \cup \Phi(e_3) = M$$

These relations lead to a contradiction (as can be proved by the methods given in (Bell, 1966; see also Ludwig, 1975). Therefore, the idea of a set $\tilde{\mathcal{Q}}$ of 'imagined preparing procedures', such that the 'demixing axiom' holds, is forbidden, even though this idea might seem very plausible.

9. CONCLUSIONS

We have presented some examples to demonstrate the following:

(1). How one may work with the theory founded in Sections 1 and 2.

(2). How to make precise theoretical propositions concerning individual microsystems,

(3). How to formulate additional 'imagined' structures by axioms — with the intention to test these additional structures on their compatibility with the theory.

We have shown that there is *no* difference between classical systems and microsystems as to the formulation of propositions concerning individual systems. The reason for the difference between classical systems and microsystems is only a different form of the statistical laws, due to axiom AV 4.

ACKNOWLEDGEMENT

Thanks are due to Professor Jenč, Dr. Kanthack, Professor Melsheimer and Professor Neumann for critical reading of the manuscript and numerous improvements in the text.

REFERENCES

Bell, J. S. (1966) 'On the problem of hidden variables in quantum mechanics', *Rev. Mod. Phys.*, **38**, 3, 447–452.
Ludwig, G. (1964) 'Versuch einer axiomatischen Grundlegung der Quantenmechanik und allgemeinerer physikalischer Theorien', *Z. Physik*, **181**, 233–260.
Ludwig, G. (1967a) 'An axiomatic foundation of quantum mechanics on a nonsubjective basis', in *Quantum Theory and Reality*, Springer, Berlin, pp. 98–104.
Ludwig, G. (1967b) "Attempt of an axiomatic foundation of quantum mechanics and more general theories II', *Commun. Math. Phys.*, **4**, 331–348.
Ludwig, G. (1967c) 'Hauptsätze des Messens als Grundlage der Hilbert raumstruktur der Quantenmechanik', *Z. Naturforsch.*, **22a**, 1303–1323.
Ludwig, G. (1967d) 'Ein weiterer Hauptsatz des Messens als Grundlage der Hilbertraumstruktur der Quantenmechanik', *Z. Naturforsch.*, **22a**, 1324–1327.
Ludwig, G. (1968) 'Attempt of an axiomatic foundation of quantum mechanics and more general theories III', *Commun. Math. Phys.*, **9**, 1–12.

Ludwig, G. (1970) 'Deutung des Begriffs "Physikalische Theorie und axiomatische Grundlegung der Hilbertraumstruktur der Quantenmechanik durch Hauptsätze des Messens', *Lecture Notes in Physics*, **4**, Springer, Berlin.

Ludwig, G. (1971a) 'The measuring process and an axiomatic foundation of quantum mechanics', *Foundation of Quantum Mechanics*, B. D'Espagnat ed., Academic Press, New York, pp. 287–315.

Ludwig, G. (1971b) 'A physical interpretation of an axiom within an axiomatic approach to quantum mechanics and a new formulation of this axiom as a general covering condition', *Notes in Math. Phys.*, **1**, Marburg.

Ludwig, G. (1972a) 'Meß- und Präparierprozesse', *Notes in Math. Phys.*, **6**, Marburg.

Ludwig, G. (1972b) 'An improved formulation of some theorems and axioms in the axiomatic foundation of the Hilbert space structure of quantum mechanics', *Commun. Math. Phys.*, **26**, 78–86.

Ludwig, G. (1974a) 'Measuring and preparing processes', *Lecture Notes in Physics*, **29**, Springer, Berlin, pp. 122–162.

Ludwig, G. (1974b) *Einführung in die Grundlagen der Theoretischen Physik*, Vol. 1, Bertelsmann, Düsseldorf.

Ludwig, G. (1975) *Einführung in die Grundlagen der Theoretischen Physik*, Vol. 3, Bertelsmann-Vieweg, Düsseldorf-Wiesbaden.

Ludwig, G. (in preparation a) *Fundaments of Quantum Mechanics*, Springer, Berlin, 2nd ed. of *Die Grundlagen der Quantenmechanik* (1954).

Ludwig, G. (in preparation b) 2nd ed. of Ludwig (1970).

Nagel, R. J. (1974) 'Order unit and base norm spaces', *Lecture Notes in Physics*, Vol. **29**, Springer, Berlin, pp. 23–29.

Neumann, H. (1971) 'Classical systems and observables in quantum mechanics', *Commun. Math. Phys.*, **23**, 100–116.

Neumann, H. (1972) 'Classical systems in quantum mechanics and their representation in topological spaces', *Notes in Math. Phys.*, **10**, Marburg.

Neumann, H. (1974a) 'On the representation of classical systems', *Lecture Notes in Physics*, Vol. **29**, Springer, Berlin, pp. 316–321.

Neumann, H. (1974b) 'A new physical characterization of classical systems in quantum mechanics', *Int. J. Theoret. Phys.*, **9**, 225–228.

v. Neumann, J. (1955) *Mathematical Foundation of Quantum Mechanics*, Princeton University Press.

Stolz, P. (1971) 'Attempt of an axiomatic approach of quantum mechanics and more general theories IV', *Commun. Math. Phys.*, **23**, 117–126.

Quantum Mechanics of Bounded Operators

THALANAYAR S. SANTHANAM
Institute of Mathematical Sciences, Madras, India

1. INTRODUCTION

Heisenberg (1925) started the golden age of modern quantum mechanics. The essence of his discovery has been in the identification of physical observables in terms of Hermitian matrices (operators) leading to the fact that operators corresponding to canonical variables do not commute. The 'uncertainty' in the simultaneous measurement of canonical variables is then a simple manifestation of this non-commuting behaviour of the corresponding Hermitian operators. Weyl (1931) rewrote the Heisenberg commutation relation in an exponential form leading to a certain nilpotent Lie group. Von Neumann (1931) and Stone (1930) simultaneously solved the problem of uniqueness for Weyl commutation relations. The Weyl group has a single irreducible representation which is precisely the Schrödinger representation. The complete equivalence of the Heisenberg and the Weyl commutation relations has been established (Rellich, 1946; Dixmier, 1958; Nelson, 1959; Cartier, 1966). The canonical commutation relation (CCR) of Heisenberg implies that the Hilbert space on which these operators act is infinite dimensional and that one (or both) of these operators should be necessarily unbounded. A natural question arises whether one can write an *analogue* of CCR for operators with a discrete bounded spectrum and acting on a finite dimensional vector space. The answer is yes. To achieve this, we start with the Weyl form (Weyl, 1931) of the representations of the Abelian group of unitary rotations in ray space. We define the canonical generators (Hermitian) as the logarithms of the finite dimensional unitary rotations. We then compute explicitly the commutator of these generators. It is naturally trace free. Also, we demonstrate that in the limit of continuous spectrum, valid as the dimension goes to infinity, the new commutator reduces to the standard CCR. Thus, our relation is the correct discrete analogue of CCR and it is unique if one starts with the Weyl form. We elevate the commutator for bounded operators with discrete spectrum to what we call quantum mechanics on discrete space (QMDS) (Santhanam and Tekumalla, 1975).

We show that the angular momentum operators satisfy QMDS with the corresponding phases. To demonstrate this, we reformulate the theory of angular momentum by quantising the azimuthal direction instead of the zenithal angle (Levy-Leblond, 1973).

It turns out that there is no 'uncertainty' in the measurement of canonical variables (canonical in the sense of QMDS). The uncertainty in the measurement of usual canonical variables (CCR), in our formulation, is a manifestation of the continuous nature of the spectrum. QMDS is, however, distinct from the classical theory since the operators do not commute.

There is an approach due to Schwinger (1960) to discuss quantum mechanics in finite dimensions and eventually take suitable limits to get the usual one. We will make use of this technique.

Besides, the representation theory of generalized Clifford algebras has been studied by Ramakrishnan and coworkers (1969a, b). We shall make use of some of their results.

2. WEYL'S FORM OF HEISENBERG'S RELATION

Suppose A and B are two elements of the Abelian group of unitary rotations in ray space so that

$$AB = \varepsilon BA \tag{1}$$

where ε is the primitive nth root of unity. By iteration we have

$$A^k B^l = \varepsilon^{kl} B^l A^k, \qquad k, l = 0, 1, \ldots, n-1 \tag{2}$$

From this equation it follows that A^n commutes with B and B^n commutes with A and if the Abelian rotation group is irreducible it follows from Schur's lemma that

$$A^n = I, \qquad B^n = I \tag{3}$$

where I is the $(n \times n)$ unit matrix. The order of any element of an irreducible Abelian rotation group in n dimensions is consequently a factor of n. In the diagonal representation for B, i.e.

$$B = \mathrm{diag}\,(1, \varepsilon, \varepsilon^2, \ldots, \varepsilon^{n-1}) \tag{4}$$

the matrix A has the form of a cyclic permutation matrix

$$A = \begin{bmatrix} 0 & 1 & 0 & 0 & \cdots & 0 \\ 0 & 0 & 1 & 0 & \cdots & 0 \\ 0 & 0 & 0 & 1 & \cdots & 0 \\ & & \vdots & & & \\ 0 & 0 & 0 & 0 & \cdots & 1 \\ 1 & 0 & 0 & 0 & \cdots & 0 \end{bmatrix}. \tag{5}$$

The action of A and B on the components of an n-dimensional vector is then

$$A: x'_k = x_{k+1}$$
$$B: x'_k = \varepsilon^k x_k \tag{6}$$

More generally,

$$A^s \quad x'_k = x_{k+s}$$
$$B^t: x'_k = \varepsilon^{kt} x_k \tag{7}$$
$$s, t = \text{any integer}$$

The transition to continuous groups is now carried out. Following Weyl (1931) we set

$$A \equiv e^{i\xi P}$$
$$B \equiv e^{i\eta Q} \tag{8}$$

where ξ and η are real infinitesimal parameters and we pass to the limit $n \to \infty$. ε has therefore to be identified as

$$\varepsilon = e^{2\pi i/n} = e^{i\xi \eta} \tag{9}$$

which yields

$$n\xi\eta = 2\pi \tag{10}$$

We see that

$$A^s = e^{i\xi s P} \equiv e^{i\sigma P}, \qquad \sigma = \xi s$$
$$B^t = e^{i\eta t Q} \equiv e^{i\tau Q}, \qquad \tau = \eta t \tag{11}$$

Since

$$\varepsilon^{kt} = e^{i\xi \eta k t} = e^{i\xi k \tau} \tag{12}$$

the eigenvalues of Q are given by

$$q = \xi k \bmod n\xi \tag{13}$$

where k runs through all integral values. As $n\xi = 2\pi/\eta$, by choosing η infinitesimal, we see that q may assume all real numbers from $-\infty$ to $+\infty$. In identifying

$$x_k = \sqrt{\xi} \psi(\xi k), \qquad \xi k = q \tag{14}$$

where $\psi(q)$ is an arbitrary function satisfying the normalization

$$\int |\psi(q)|^2 \, dq = 1 \tag{15}$$

We find then that the quantity $e^{i\tau q}$ is represented by the linear operator

$$\psi(q) \to e^{i\tau q} \psi(q) \tag{16}$$

230 Uncertainty Principle and Foundations of Quantum Mechanics

and the operator representing $e^{i\sigma p}$

$$\psi(q) \to \psi(q+\sigma) \tag{17}$$

From equation (17) it is clear that $e^{i\sigma P}$ acts as the translation operator. If these linear operators are infinitesimal we have

$$\begin{aligned} q&: \delta\psi(q) = q\psi(q) \\ p&: \delta\psi(q) = \frac{1}{i}\frac{d}{dq} \end{aligned} \tag{18}$$

Thus, it follows that the Schrödinger representation (wave equation) is a necessary consequence of Weyl's commutation relation. To summarize we have *the theorem (of von Neumann)* as follows.

THEOREM. Let $U(\xi) = e^{i\xi P}$ and $v(\eta) = e^{i\eta Q}$ be one parameter continuous unitary groups on a separable Hilbert space $\mathcal{H}$ satisfying the Weyl relation

$$U(\xi)v(\eta) = e^{i\xi\eta}v(\eta)U(\xi) \tag{19}$$

Then there are closed subspaces $\mathcal{H}_l$ so that

(1). $\mathcal{H} = \bigoplus_{l=1}^{N} \mathcal{H}_l$ (N is a positive integer or ∞)

(2). $U(\xi): \mathcal{H}_l \to \mathcal{H}_l$
    $v(\eta): \mathcal{H}_l \to \mathcal{H}_l$ for all $\xi, \eta \in \mathbf{R}$

(3). For each l, there is a unitary operator $T_l: \mathcal{H}_l \to L^2(\mathbf{R})$ such that $T_l U(\xi) T_l^{-1}$ is translation to the left by ξ and $T_l v(\eta) T_l^{-1}$ is multiplication by $e^{in x}$. It also follows from the theorem that if P and Q denote the generators of $U(\xi)$ and $v(\eta)$, respectively then there is a dense domain $D \subset \mathcal{H}$ so that

(a) $P: D \to D$
    $Q: D \to D$
(b) $[Q, P]\phi = i\phi$, for all $\phi \in D$.

and

(c) P and Q are essentially self adjoint on D.

Thus the Schrödinger representation is the only representation of CCR. It is not difficult to see that a pair P, Q of self-adjoint operators satisfying the canonical commutation relation $[Q, P] = iI$ cannot both be bounded. If they were bounded then

$$PQ^n - Q^n P = -inQ^{n-1}$$

and thus

$$n\|Q^{n-1}\| \le 2\|P\|\|Q\|^n$$

and hence

$$2\|P\|\|Q\| \ge n$$

for all n which is a contradiction. Therefore, either P or Q or both must be unbounded. This can also be seen by simply taking the trace on both sides of CCR.

Mackey (1949) replaced $v(\eta) = e^{i\eta Q}$ by its spectral measure E such that

$$v(\eta) = \int e^{i\eta x} \, dE(x)$$

The measure E and $U(\xi)$ is an imprimitivity system for R based on R and the uniqueness theorem is then a consequence of the imprimitivity theorem.

3. A BASIS IN FINITE DIMENSION

In this, we essentially follow the method of Schwinger (1960) in the construction of a unitary operator A given by the cyclic permutation matrix. The action of A in Dirac's notation is

$$\langle a^k | A = \langle a^{k+1} |, \quad k = 1, 2, \ldots n \tag{20}$$

where we identify

$$\langle a^{n+1} | = \langle a^1 | \tag{21}$$

since

$$A^n = I \tag{22}$$

The eigenvalues of A are then the n roots of unity

$$v' = \varepsilon^k, \quad k = 0, 1, 2, \ldots n-1$$
$$\varepsilon = e^{2\pi i/n} \tag{23}$$

The sylvester matrix defined as

$$S = \frac{1}{\sqrt{n}} \begin{bmatrix} 1 & 1 & 1 & & 1 \\ 1 & \varepsilon & \varepsilon^2 & & \varepsilon^{n-1} \\ 1 & \varepsilon^2 & \varepsilon^4 & & \varepsilon^{2n-2} \\ \vdots & \vdots & \vdots & & \vdots \\ 1 & \varepsilon^{n-1} & \varepsilon^{n-2} & & \varepsilon \end{bmatrix} \tag{24}$$

$$SS^+ = S^+S = I$$

diagonalizes any circulant matrix and in particular the cyclic permutation matrix A. Hence

$$S^{-1}AS = B \tag{25}$$

where
$$B = \text{diag}(1, \varepsilon, \varepsilon^2, \ldots \varepsilon^{n-1}) \tag{26}$$

Thus any two operators connected by the transform equation (25) have as their eigenvalues the n roots of unity and if we denote their eigenvectors (of A and B) by $|\varepsilon^k\rangle$ and $|\varepsilon^l\rangle$, respectively, then

$$\langle \varepsilon^l | \varepsilon^k \rangle = S_{kl}$$
$$= \frac{1}{\sqrt{n}} e^{2\pi i/n \cdot kl} \tag{27}$$

In fact, we have the *Theorem* of Schwinger.

THEOREM. The basis of any finite dimensional vector space can be mapped by a unitary transformation (Sylvester transform) to a basis furnished by the roots of unity.

Suppose we have two unitary operators U and V satisfying

$$VU = \varepsilon UV$$
$$\varepsilon^n = 1 \tag{28}$$
$$U^n = V^n = I$$

Then the n^2 operators defined by

$$X_{kl} = \frac{1}{\sqrt{n}} U^k V^l, \quad k, l = 0, 1, 2, \ldots n-1 \tag{29}$$

are linearly independent. All X_{kl} except the unit operator are traceless and with the multiplication defined by equation (28) form an associative algebra c_2^n. We shall later discuss the representations of the generalized Clifford algebra c_m^n. Suffice it now to say that c_2^n is isomorphic to the matrix ring $M_{n \times n}$. Thus, the operators defined in equation (29) furnish a unitary operator basis and this fact has been particularly used by Schwinger (1960). Now consider an arbitrary Y. We notice that

$$\sum_{k,l} X_{kl} Y X_{kl}^\dagger = \frac{1}{n} \sum_{k,l} U^k V^l Y V^{-l} U^{-k}$$
$$= \frac{1}{n} \sum_{k,l} V^k U^l Y U^{-l} V^{-k} \tag{30}$$

It is easy to show that this operator commutes with U and V. Hence

$$\sum_{k,l=0}^{n-1} X_{kl} Y X_{kl}^\dagger = rI$$

where
$$r = \text{Tr } Y \tag{31}$$

We refer to U and V as a complementary pair of operators. From equation (31) it follows that

$$U^k V^l U V^{-l} U^{-k} = \varepsilon^l U$$
$$U^k V^l V V^{-l} U^l = \varepsilon^{-k} V \tag{32}$$

which exhibits the unitary transformations that produce only cyclic spectral translations. If Y is an arbitrary function of U and V we see from equation (31) that

$$\frac{1}{n^2} \sum_{k,l} F(\varepsilon^l U, \varepsilon^{-k} V) = \frac{1}{n} \operatorname{Tr} F \tag{33}$$

which is a kind of ergodic theorem.
The operators V^l and U^k, $l, k = 0, 1, 2, \ldots n-1$ will satisfy the same operator relation as V and U, viz.

$$V^l U^k = e^{2\pi i/n} U^k V^l \tag{34}$$

provided

$$kl = 1 \bmod n \tag{35}$$

with the unique solution given by the Fermat–Euler theorem

$$l = k^{\phi(n)-1} \bmod n \tag{36}$$

where $\phi(n)$ = the number of integers less than and relatively prime to n. The pair of operators U^k, V^l are also complementary. Suppose now we write

$$n = n_1 n_2 \tag{37}$$

where the integers are relatively prime.
Then we can rewrite

$$e^{2\pi i/n \cdot k} = e^{2\pi i k_1/n_1} e^{2\pi i k_2/n_2} \tag{38}$$

with

$$k = k_1 n_2 + k_2 n_1$$
$$k_1 = 0, 1, 2, \ldots n_1 - 1 \tag{39}$$
$$k_2 = 0, 1, 2, \ldots n_2 - 1$$

Thus a single basis defined by ε^k can be written as

$$|\varepsilon^k\rangle = |\varepsilon_1^{k_1}\rangle |\varepsilon_2^{k_2}\rangle \tag{40}$$

where ε_1 and ε_2 are of periods n_1 and n_2, respectively. Then the single pair of complementary operators can be replaced by two pairs satisfying

$$V_1 U_1 = \varepsilon_1 U_1 V_1$$
$$\varepsilon_1^{n_1} = 1, \quad V_1^{n_1} = U_1^{n_1} = I \tag{41}$$

234 Uncertainty Principle and Foundations of Quantum Mechanics

and
$$V_2 U_2 = \varepsilon_2 U_2 V_2$$
$$\varepsilon_2^{n_2} = 1, \quad V_2^{n_2} = U_2^{n_2} = I \tag{42}$$

where
$$U_1 = U^{n_2}, \quad V_1 = V^{l_1 n_2}$$
$$U_2 = U^{n_1}, \quad V_2 = V^{l_2 n_1} \tag{43}$$

with
$$l_1 = n_2^{\phi(n_1)-1} \bmod n_1$$
$$l_2 = n_1^{\phi(n_2)-1} \bmod n_2 \tag{44}$$

The two pairs commute with each other. The basis now becomes

$$X_{kl} \to X_{k_1 l_1, k_2 l_2} = \frac{1}{\sqrt{n}} U_1^{k_1} U_2^{k_2} V_1^{l_1} V_2^{l_2}$$

where
$$k_1, l_1 = 0, 1, 2, \ldots n_1 - 1$$
$$k_2, l_2 = 0, 1, 2, \ldots n_2 - 1 \tag{45}$$

Or
$$X_{kl} = \prod_{j=1}^{2} X_{k_j l_j} \tag{46}$$

with
$$X_{k_j l_j} = \frac{1}{\sqrt{n_j}} U_j^{k_j} V_j^{l_j}$$
$$k_j, l_j = 0, 1, 2, \ldots n_j - 1 \tag{47}$$

In general, since any integer can be written in terms of primes
$$n = \prod_{j=1}^{f} \nu_j \tag{48}$$

where f is the total number of primes including repetitions. The resulting factored basis is then
$$X(kl) = \prod_{j=1}^{f} X(k_j, l_j) \tag{49}$$

where
$$X(k_j, l_j) = \frac{1}{\sqrt{\nu_j}} U_j^{k_j} V_j^{l_j}$$
$$k_j, l_j = 0, 1, 2, \ldots, \nu_j - 1 \tag{50}$$

In the particular case of $\nu = 2$, the complementary pair of operators anticommute and the basis forms the Clifford algebra. In the next section we shall study the structure of generalized Clifford algebras.

4. GENERALIZED CLIFFORD ALGEBRAS

The problem that Dirac (1926) faced was to linearize

$$x_1^2 + x_2^2 + \ldots + x_m^2 = \left(\sum_{i=1}^{m} \alpha_i x_i\right)^2 \tag{51}$$

and consequently the α's satisfy

$$\begin{aligned} \alpha_i \alpha_j &= -\alpha_j \alpha_i, \quad i \neq j \\ \alpha_i^2 &= I \\ i, j &= 1, \ldots m \end{aligned} \tag{52}$$

The set of elements defined by

$$a = \prod_{i=1}^{m} \alpha_i^{\rho_i}, \quad \rho_i = 0, 1 \tag{53}$$

which are 2^m in number are linearly independent and with the product defined by equation (52) form an associative algebra which is the familiar Clifford algebra c_m^2. In this case, it is well-known Boerner (1963) that when $m = 2\nu =$ even, there is a single irreducible representation of dimension 2^ν and $c_{2\nu}^2 \approx M_{2^\nu \times 2^\nu}$. When $m = 2\nu + 1$, $C_{2\nu+1}^2 = C_{2\nu}^2 + C_{2\nu}^2$ where the elements of the second are simply the negatives of the first. The elements defined by

$$\begin{aligned} \alpha_{ij} &= [\alpha_i, \alpha_j] \\ \alpha_{0i} &= \alpha_i \end{aligned} \tag{54}$$

satisfy the algebra $O(m+1)$ of the orthogonal group in $(m+1)$ dimensions. In fact, they furnish the spinor representation of $O(m+1)$. A natural question arises whether one can solve the linearization

$$x_1^n + x_2^n + \ldots + x_m^n = \left(\sum_{i=1}^{m} e_i x_i\right)^n \tag{55}$$

The answer is yes if the e's satisfy the ordered commutation relation

$$\begin{aligned} e_i e_j &= \varepsilon e_j e_i, \quad i < j \\ i, j &= 1, 2, \ldots m \\ \varepsilon &= e^{2\pi i/n} \\ e^n &= 1 \end{aligned} \tag{56}$$

The basis defined by

$$b = \prod_{i=1}^{m} e_i^{\tau_i}$$
$$\tau_i = 0, 1, 2, \ldots n-1 \tag{57}$$

which are n^m in number are linearly independent and with the product given by equation (56) form an associative algebra called the generalized Clifford algebra C_m^n. Morinago and Nono (1952), Yamazaki (1964) and Morris (1967) have studied the algebra in detail and Ramakrishnan and co-workers (1969b) have studied exhaustively the particular realizations and their connections with physical problems. Sufficient for our purpose to state the following theorem.

THEOREM: When $m = 2\nu$, the algebra C_m^n has one irreducible representation of dimension n^ν and $C_{2\nu}^n \approx M_{n^\nu \times n^\nu}$, the matrix ring of dimension n^ν. When $m = 2\nu+1$, $C_{2\nu+1}^n = C_{2\nu}^n + C_{2\nu}^n + \ldots + C_{2\nu}^n$ (n copies). The elements of the second, third, etc., are obtained from the first by multiplication with ε^i, $i = 1, 2, \ldots n-1$.

The explicit realizations can be obtained either by a straight forward extension of the method of Brauer and Weyl (1935) or by using the method of Ramakrishnan (1967) or by Ramakrishnan and co-workers (1969) an extension of Rasevskii's method (1969b). It should have become obvious by now that the algebra satisfied by the canonical pair of unitary operators U and V of the last section is simply $C_2^n \approx M_{n \times n}$. Since in the basis X_{kl}, all except the unit element are traceless, suitable linear combinations furnish a Hermitian basis and thus give the self representation of the group $su(n)$. (Ramakrishnan and co-workers, 1969a).

5. COMMUTATOR IN FINITE DIMENSIONS

We now start with the Weyl algebra. We have seen that it has a *single* irreducible representation (which in the limit of infinite dimensions is in fact, the Schrödinger representation) in finite dimensions. We define then the generators as the 'formal' logarithm of the Weyl operators and we solve for them. Then we compute their commutator. We then show in the limit of infinite dimensions it reduces to the one of Heisenberg.

We start with the operators $V(\xi)$ and $U(\eta)$ satisfying Weyl algebra

$$VU = \varepsilon UV,$$

$$\varepsilon = \exp \frac{2\pi i}{n} \tag{58}$$

$$V^n = U^n = I$$

We have seen that the single irreducible representation of equation (58) in finite dimensions is given by

$$U = \text{diag}(1, \varepsilon, \varepsilon^2, \ldots \varepsilon^{n-1}) \qquad (59)$$

We define the Hermitian operators P and Q by

$$U \equiv e^{i\sqrt{2\pi/n}Q}$$
$$V \equiv e^{i\sqrt{2\pi/n}P} \qquad (60)$$

where we have chosen $\xi = \eta = \sqrt{2\pi/n}$.

Then, formally

$$Q = -i\sqrt{\frac{n}{2\pi}} \log U$$

$$P = -i\sqrt{\frac{n}{2\pi}} \log V \qquad (61)$$

The logarithms of U and V are well-defined since they are non-singular (Gantmacher, 1959). Further since

$$S^{-1}VS = U \qquad (62)$$

where S is the Sylvester matrix

$$S = \frac{1}{\sqrt{n}} \begin{bmatrix} 1 & 1 & 1 & \cdots & 1 \\ 1 & \varepsilon & \varepsilon^2 & & \varepsilon^{n-1} \\ 1 & \varepsilon^2 & \varepsilon^4 & & \varepsilon^{n-2} \\ \vdots & \vdots & \vdots & & \vdots \\ 1 & \varepsilon^{n-1} & \varepsilon^{n-2} & & \varepsilon \end{bmatrix}, \quad S^{-1} = S^\dagger \qquad (63)$$

From the definition of the logarithm of a matrix, in view of equation (62), we have

$$\log V = S \log U S^{-1} \qquad (64)$$

Therefore,

$$Q = -i\sqrt{\frac{n}{2\pi}} \log U$$

$$P = -i\sqrt{\frac{n}{2\pi}} S(\log U)S^{-1} \qquad (65)$$

where

$$\log U = (\log \varepsilon) \, \text{diag}(0, 1, 2, \ldots n-1) \qquad (66)$$

The commutator of Q and P is then

$$K_{rs} = [Q, P]_{rs} = -\frac{n}{2\pi}[\log U, S(\log U)s^{-1}]_{rs} \qquad (67)$$

where the matrix indices are labelled from $(0, 1, 2, \ldots n-1)$. Explicitly evaluating equation (67), we have

$$K_{rs} = -\frac{(\log \varepsilon)^2}{2\pi}(r-s)\sum_{u=0}^{n-1} u\varepsilon^{u(r-s)} \qquad (68)$$

If $\varepsilon^{r-s} = x = 1$, then

$$K_{rs} = -\frac{(\log \varepsilon)^2}{2\pi}(r-s)[\tfrac{1}{2}n(n-1)] \qquad (69)$$

If $x \neq 1$, since $x^n = 1$, there results

$$K_{rs} = -\frac{(\log \varepsilon)^2}{2\pi}(r-s)\left[\frac{n}{\varepsilon^{r-s}-1}\right] \qquad (70)$$

since

$$\sum_{u=0}^{n-1} ux^u = \frac{n}{x-1} \qquad (71)$$

Thus, we find

$$[Q, P]_{rs} = \frac{(\log \varepsilon)^2}{2\pi}(s-r)\begin{cases} \tfrac{1}{2}n(n-1) & \text{if } \varepsilon^{r-s} = 1 \\ \dfrac{n}{\varepsilon^{r-s}-1} & \text{if } \varepsilon^{r-s} \neq 1 \end{cases} \qquad (72)$$

We notice that this commutator is off-diagonal and hence trace-free, as it should be for bounded operators.

Alternately, we can directly sum the expression in equation (68) which yields

$$\begin{aligned}[Q, P]_{rs} &= 0, \quad r = s \\ &= -\frac{(\log \varepsilon)^2}{2\pi}(r-s)\cdot n\left[\frac{1}{2} - \frac{i}{2}\cot g\frac{\pi}{n}(s-r)\right], r \neq s\end{aligned} \qquad (73)$$

We call the commutation relations (72) or (73) quantum mechanics in discrete space (QMDS).

Let us now evaluate the following commutator which we shall use when we discuss the application of QMDS to the algebra of angular momentum operators later. Let

$$\Omega = [Q, V] \qquad (74)$$

From equations (65) and (66) it is clear that

$$\Omega = -i\sqrt{\frac{n}{2\pi}} \log \varepsilon [N, SUS^{-1}] \tag{75}$$

where the matrix

$$N = \text{diag}\{0, 1, 2, \ldots (n-1)\} \tag{76}$$

Explicit evaluation of equation (75) yields

$$\Omega = i\sqrt{\frac{n}{2\pi}} (\log \varepsilon) K \tag{77}$$

where the matrix

$$K = \begin{bmatrix} 0 & 1 & 0 & 0 & \cdots & 0 \\ 0 & 0 & 1 & 0 & \cdots & 0 \\ 0 & 0 & 0 & 1 & \cdots & 0 \\ & & \vdots & & & \\ 0 & 0 & 0 & 0 & \cdots & 1 \\ -(n-1) & 0 & 0 & 0 & \cdots & 0 \end{bmatrix}$$

$$= V - L \tag{78}$$

where

$$L = \begin{bmatrix} 0 & & & & & \\ n & 0 & 0 & 0 & \cdots & 0 \end{bmatrix} \tag{79}$$

6. LIMITING CASE

We shall now show that the commutator of the bounded operators given by equation (72) does reduce to the usual form of Heisenberg in the limit of continuous spectrum, valid as $n \to \infty$. Beginning with equation (68), we relabel the rows and columns from $-(n-1)/2$ to $(n-1)/2$ instead of from 0 to $n-1$ and replace the sum by the integral. In other words, we let the matrix indices take continuous values and take the limit $n \to \infty$. This is the method of Heisenberg (1931) and Dirac (1930) to pass from the discrete to the continuous case. Then we have,

$$[Q, P]_{rs} = -\frac{(\log \varepsilon)^2}{2\pi}(r-s) \cdot \int_{-\infty}^{\infty} u \exp\{2\pi i u(r-s)/n\} du$$

$$= -i(r-s)\frac{d}{d(r-s)} \int_{-\infty}^{\infty} \exp\{2\pi i u(r-s)/n\} d\left(\frac{u}{n}\right)$$

$$= -i(r-s)\delta'(r-s)$$

$$= i\delta(r-s)$$

where δ, δ' are the Dirac delta function and its derivative, respectively. We have used the fact that in the limit considered above $\log \varepsilon = 2\pi i/n$, i.e. we have retained only its principal part.

This is exactly what Weyl does, by choosing the parameter η infinitesimal. The same limiting method can be demonstrated by starting from equation (73). Thus, QMDS reduces to the usual theory in the limit mentioned. Since the representation of the Weyl group is unique, it is clear that QMDS is unique too. Since the commutator (QMDS) is off-diagonal, the diagonal measurements 'commute' and hence there is no 'uncertainty'. However, the commutator is not zero as in the classical case.

7. APPLICATION TO ANGULAR MOMENTUM

In this section, we apply the concept of QMDS studied in detail in the previous sections to the study of the angular momentum operators which provide an excellent example of bounded operators with a discrete spectrum. We shall reformulate the angular momentum algebra by quantizing the azimuthal direction instead of the zenithal angle (Levy-Leblond, 1973). We also briefly remark on the related problem of defining a phase canonically conjugate to the number operator which has a lower bound.

Denoting the generators of rotation by J_x, J_y and J_z, we know that they satisfy the commutation relation

$$[J_i, J_j] = i\varepsilon_{ijk}J_k, \quad i, jk \text{ cyclic} \tag{80}$$

or in the Cartan canonical form (choosing J_z diagonal)

$$[J_z, J_\pm] = \pm J_\pm$$
$$[J_+, J_-] = 2J_z \tag{81}$$

where

$$J_\pm = J_x \pm iJ_y \tag{82}$$

What is usually done is to choose a basis diagonal in J_z and $J^2 = J_x^2 + J_y^2 + J_z^2$ i.e.

$$J_z \psi_m^j = m \psi_m^j$$
$$J^2 \psi_m^j = j(j+1)\psi_m^j, \quad -j \leq m \leq j \tag{83}$$

By polar decomposing J_+ we have

$$J_+ = J_T Y \tag{84}$$

with J_T Hermitian and Y unitary. Taking the adjoint yields

$$J_- = (J_+)^\dagger = Y^{-1} J_T \tag{85}$$

It follows that

$$J_+ J_- = J_T^2 = J^2 - J_z^2 + J_z \tag{86}$$

We define the transverse component $J_\perp$ as

$$J_+ = YJ_\perp \tag{87}$$

and hence

$$J_\perp^2 = j_- J_+ = J^2 - J_z^2 - J_z \ldots \tag{88}$$

In the (J^2, J_z) diagonal basis given by equation (83) we find from equations (86) and (87) that (choosing a phase convention)

$$\langle jm|J_T|jn\rangle = \delta_{mn}[(j+m)(j-m+1)]^{1/2} \tag{89}$$

$$\langle jm|J_\perp|jn\rangle = \delta_{mn}[(j-m)(j+m+1)]^{1/2} \tag{90}$$

It follows from equations (84) and (89) and equations (87) and (90) that in this basis the operator Y is just the cyclic permutation matrix A defined in the last section. In fact,

$$Y \equiv A \tag{91}$$

and

$$N = -J_z + jI \tag{92}$$

where I is the unit matrix.

From equation (77) it can be seen that

$$[J_z, A] - A = -L \tag{93}$$

which has been derived by Levy-Leblond. It is also clear fom equation (72) that

$$[J_z, \phi]_{rs} = i(\log \varepsilon)\frac{r-s}{\varepsilon^{r-s}-1} \quad \text{for } \varepsilon^{r-s} \neq 1$$

$$= 0 \quad \text{for } \varepsilon^{r-s} = 1 \tag{94}$$

Since $Y(=A)$ acts as a cyclic permutation matrix for the basis in which J_z is diagonal it follows (Schwinger) that (J^2, Y) diagonal basis given by

$$\begin{aligned}J^2|j, \mu\rangle &= j(j+1)|j, \mu\rangle \\ Y|j, \mu\rangle &= \mu|j, \mu\rangle\end{aligned} \tag{95}$$

with

$$\mu = \varepsilon^\zeta, \quad \zeta = -j, \ldots +j,$$

is connected by the Sylvester transform to the (J^2, J_z) basis equation (83), i.e.

$$|j, \zeta\rangle = \sum_{m=-j}^{j} S_{\zeta m}|j, m\rangle$$

$$= \frac{1}{(2j+1)^{1/2}} \sum_{m=-j}^{j} \varepsilon^{m\zeta}|j, m\rangle \tag{96}$$

242 Uncertainty Principle and Foundations of Quantum Mechanics

The matrix elements of J_z in the new basis is given by

$$\langle j\zeta|J_z|j\lambda\rangle = \sum m S^{-1}_{\zeta m} S_{\lambda m}$$

$$= \frac{1}{2j+1}\sum_{m=-j}^{j} m\varepsilon^{(\lambda-\zeta)} \tag{97}$$

Of course, one knows that in a finite dimensional space we can go from one basis (jm) to another basis $(j\zeta)$ by a unitary transformation (s). But what is important has been the fact that the operator Y (not $J_\pm$) acts as a cyclic permutation matrix in the J_z diagonal basis and the commutation relation of Y with J_z is furnished by equation (93) which carries the essence of our QMDS. From equations (92) and (95) it follows that

$$[N,\phi]_{rs} = -i(\log \varepsilon)\frac{r-s}{\varepsilon^{r-s}-1} \quad \text{for } r-s \neq 1$$

$$= 0 \quad \text{for } \varepsilon^{r-s} = 1 \tag{98}$$

Thus the number operator N (with spectrum bounded below) and the phase operator are conjugate in the sense of equation (98). It is perhaps a great luxury to demand that they must be canonically conjugate.

8. CONCLUSIONS

We have discussed the quantum mechanics of bounded operators with a discrete spectrum acting on a finite dimensional Hilbert space. We have shown that in the limit of continuous spectrum with the dimension going to infinity one gets the usual theory. As an illustration we have studied the algebra of angular momentum operators reformulated by quantizing the azimuthal angle. We have briefly remarked about the phase operator canonically conjugate to the number operator. We believe that the QMDS can be applied to a system with periodicity like the cyclic lattice. Also, we may avoid many difficulties (divergences etc.) if we work with a finite number of states and eventually take suitable limits. It is to be seen how QMDS works with realities.

ACKNOWLEDGEMENTS

I thank Professors B. Gruber and W. Hink for their gracious hospitality at the University of Würzburg. The article was written during my stay at the International Centre for Theoretical Physics, Trieste. I am grateful to Professor Abdus Salam and the I.A.E.A. for their hospitality. A brief discussion with Professor C. N. Yang is gratefully acknowledged.

REFERENCES

Boerner, H. (1963) *Representations of Groups*, North-Holland, Publishing Co., Amsterdam, Chap. 8.
Brauer, R. and Weyl, H. (1935) 'Spinors in n dimensions', *Am. J. Math.*, **57**, 425-449.
Cartier, P. (1966) 'Quantum mechanical commutation relations and theta functions', *Proc. Symp. Pure Math.*, **9**, 361-383.
Dirac, P. A. M. (1926) *Proc. Roy. Soc.*, **109 A**, 642.
Dirac, P. A. M. (1930) *The Principles of Quantum Mechanics*, Oxford University Press, London.
Dixmier, J. (1958) 'Sur la relation $i(PQ - QP) = 1$', *Compositio Math.*, **13**, 263-269.
Gantmacher, F. R. (1959) *Matrix Theory*, Vol. I, Chelsea, New York, p. 239.
Heisenberg, W. (1925) *Zeit. Phys.*, **33**, 879.
Heisenberg, W (1931) *The Physical Principles of the Quantum Theory*, Dover, New York.
Levy-Leblond, J. M. (1973) 'Azimuthal quantization of angular momentum', *Rev. Mex. Fis.*, **22**, 15-23.
Mackey, G. W (1949) 'On a theorem of Stone and von Neumann', *Duke Math. J.*, **16**, 313-326.
Morinaga, K. and Nono, T. (1952) *J. Sci. Hiroshima Univ.*, **A6**, 13.
Moris, A. O. (1967) 'On a generalized Clifford algebra', *Quart. J. Math. Oxford* (2), **18**, 7-12.
Nelson, E. (1959) 'Analytic vectors', *Ann. Math.*, **70**, 572-615.
von Neumann, J. (1931) 'Die Eindeutigkeit der Schrödinger'schen Operatoren', *Math. Ann.*, **104**, 570-578.
Ramakrishnan, A. (1967) 'Dirac Hamiltonian as a member of a hierarchy', *J. Math. Anal. Appl.*, **20**, 9.
Ramakrishnan, A., Chandrasekaran, P. S., Ranganathan, N. R., Santhanam, T. S. and Vasudevan, R. (1969a) 'Generalized Clifford algebra and the unitary group, *J. Math. Anal. Appl.*, **27**, 164.
Ramakrishnan, A., Santhanam, T. S. and Chandrasekaran, P. S. (1969b) 'Representation theory of generalized Clifford algebras', *J. Math. Phys. Sci. (Madras)*, **3**, 307.
Rellich, F. (1946) 'Der Eindeutigkeitssatz für die Läsungen der quantum-mechanischen Vertauschungsrelationen', *Nachr. Akad. Wiss. Göttingen, Math. Physik.*, **K1**, 107-115.
Santhanam, T. S. and Tekumalla, A. R. (1976) *Quantum Mechanics in Finite Dimensions*, Foundations of Physics, **6**, 5, 583-587.
Schwinger, J. (1960) 'Unitary operator bases', *Proc. Nat. Acad. (USA)*, **46**, 570-579.
Stone, M. (1930) 'Linear transformations in Hilbert space III, operational methods in group theory', *Proc. Nat. Acad. Sci. (USA)*, **16**, 172-175.
Weyl H. (1931) *Theory of Groups and Quantum Mechanics*, Dover, New York, pp. 272-280.
Yamazaki, K. (1964) 'On projective representations and ring extensions of finite groups', *J. Fac. Sci. Univ. Tokyo, Sect.*, **T 10**, 147-195.

PART 3

Formal Quantum Theory

Four Approaches to Axiomatic Quantum Mechanics

STANLEY P. GUDDER
University of Denver, Denver, U.S.A.

1. INTRODUCTION

This is a survey article on contemporary approaches to axiomatic quantum mechanics. There are, at present, four main frameworks within which axiomatic quantum mechanics is being studied. These are the classical approach, the algebraic approach, the quantum logic approach and the convexity approach. Each of these approaches has its advocates and critics, its strengths and weaknesses, its history and literature. To do justice to any one of these approaches, an entire volume could easily be dedicated to each. Therefore, by necessity, this survey must be fairly superficial. I shall include an introduction to the framework of each approach, some of the interrelations between them and a few of the important results they encompass. I hope to give the reader a unifying viewpoint, expose him to results that are scattered throughout the literature and not previously compiled in one place, and finally to announce some little known and new results.

The importance of axiomatic quantum theories to mathematics and physics has perhaps not been sufficiently recognized. This field is not only important in its own right, but has had tremendous influence and spin-off to other areas. For example, in the physical sciences there are important applications of results, methods and concepts of these theories to statistical mechanics, thermodynamics, turbulence, solid-state physics and laser physics, among others. In mathematics, many of the most active areas of research owe their original conception and/or later development to axiomatic quantum mechanics. These include: Hilbert spaces, self-adjoint and symmetric operators, spectral theory, general operator theory, von Neumann algebras, Lie groups and algebras, group representations, Schwartz distributions, C*-algebras, Jordan algebras, modular lattices, orthomodular lattices, continuous geometries and functions of several complex variables. Axiomatic quantum mechanics is a prime example of the fertile interplay between mathematics and physics. Even if the present methods, concepts and results prove to be absolete and are eventually superseded, the applications and mathematics that they have inspired will justify their existence.

248   Uncertainty Principle and Foundations of Quantum Mechanics

It is generally recognized that the two most basic concepts in quantum mechanics are those of a state and an observable. These two concepts serve as the basic building blocks of most axiomatic theories and, in particular, the approaches considered here. Each of the four approaches of this article will have as its primitive axiomatic elements one of these entities and this is one of our unifying themes. More specifically, in the classical approach, the observables are assumed to be self-adjoint operators in a Hilbert space; in the algebraic approach, the observables are taken to be elements of a C*-algebra; in the quantum logic approach, the axiomatic elements are certain types of observables called propositions; and in the convexity approach, the states are taken to be elements of a convex structure.

2. THE CLASSICAL APPROACH

The classical approach to axiomatic quantum mechanics is not only the prototype for other approaches, it is the most widely used and probably the most popular among physicists. It was originated by Dirac (1930) and von Neumann (1932). There are three equivalent formulations for this approach.

(A) Formulation 1

The observables O of a physical system are described by self-adjoint linear operators acting on a complex Hilbert space H. Thus in this formulation, the observables are the basic axiomatic elements. We shall usually identify an observable with its corresponding operator.

A state of a physical system is a complete description of the preparation or condition of the system. In quantum mechanics the state is determined by the expectations or average values of the observables when the system is prepared according to that state. Thus, the states can be described by the set of expectation functionals $\mathcal{S}$ on O. For $s \in \mathcal{S}$ and $A \in O$, we define $s(A)$ to be the expectation of A in the state corresponding to s. Of course, for an unbounded observable A, the expectation $s(A)$ may not exist. It is usually convenient, therefore, to consider the set of bounded observables O_b. These are described by bounded self-adjoint operators. The set O_b is still large enough to determine a state.

What are the properties of the functionals $s(A)$ for $s \in \mathcal{S}$, $A \in O_b$? First, the identity operator I corresponds to the observable that always has the value one so $s(I) = 1$. Also, if A is a self-adjoint operator with a non-negative spectrum (this corresponds to an observable with non-negative values) then we should have $s(A) \geq 0$. Furthermore, one can argue that two observables whose self-adjoint operators commute are simultaneously measurable; that is, a measurement of one does not interfere with a measurement of the other. In this case, the expectation of their sum should be the sum of their expectations. In slightly more general form $s(\alpha A + \beta B) = \alpha s(A) + \beta s(B)$ for all $\alpha, \beta \in R$ (R is the set of real numbers) whenever A and B commute. Finally, a continuity condition is imposed. This is justified by the fact that if two observables are 'close', then their expectations should also be 'close'. But how do we define 'close' mathematically? Here it is defined in terms of strong convergence. That is, a sequence of bounded operators A_i *converges strongly* to a bounded operator A if $A_i\phi \to A\phi$ for every $\phi \in H$. The continuity condition is given as follows: if a sequence of bounded observables A_i converges strongly to A, then $s(A_i) \to s(A)$.

Using a truly amazing theorem due to Gleason (1957), the above four conditions give the following characterization of states. For every $s \in \mathcal{S}$ there is a positive trace class operator T_s (the *density operator*) of trace 1 such that $s(A) = \text{Tr}\,(T_s A)$ for every $A \in O_b$. Thus it follows that a state is not only a *linear* functional on O_b but it is given by a density operator.

There are many interesting consequences of these simple, far-reaching axioms. We now mention two of them and others will be seen later. Let ϕ be a unit vector and let P_ϕ be the one-dimensional projection onto the subspace determined by ϕ. Then P_ϕ is a self-adjoint operator and also a density operator. Thus P_ϕ can be interpreted as both an observable and a state. This dual nature of P_ϕ makes it possible to define transition probabilities in a succinct manner. The significance of P_ϕ as a state is that P_ϕ is a *pure* state since it cannot be written as a convex combination of other states. As an observable, P_ϕ can be interpreted as corresponding to the statement, 'the system is in the pure state P_ϕ'. If P_ϕ and P_ψ are two pure states, then $\text{Tr}\,(P_\phi P_\psi)$ is the expectation of the observable P_ψ in the state P_ϕ and is interpreted as the probability that the system will be found in the state P_ψ when we know it is in the state P_ϕ. This is the *transition probability* between the two states and is given by

$$\text{Tr}\,(P_\phi P_\psi) = \langle \phi, P_\psi \phi \rangle = |\langle \phi, \psi \rangle|^2$$

Notice that for a pure state P_ϕ, the expectation of an observable A is given by $\text{Tr}\,(P_\phi A) = \langle \phi, A\phi \rangle$. Another interesting consequence of the axioms is the fact that the product of the variances of two observables is given by

$$s([A - s(A)]^2)s([B - s(B)]^2) \geq 1/4[s([A, B])]^2$$

This inequality provides a lower bound for the simultaneous measurability of A and B and is a mathematical formulation of the Heisenberg uncertainty principle. It also shows that two commuting observables are simultaneously measurable which substantiates our earlier statement to that effect.

The dynamics of the system is also easily formulated within this framework. If an observable is given by the operator A at time $t = 0$, then this observable is given by an operator $W_t A$ at time t. This is the Heisenberg picture of the dynamics. There is an equivalent formulation called the Schrödinger picture in which the observables are kept constant and the states are assumed to evolve. If the system is in a pure state given by a unit vector ϕ at time $t = 0$, then at time t the system is in a pure state given by a unit vector $\hat{W}_t\phi$. It follows by a theorem of Mazur and Ulam (1932) (there is also a related theorem due to Wigner (1931)) that $\hat{W}_t$ is given by a unitary operator U_t. Furthermore, since

$$\langle \phi, W_t A \phi \rangle = \langle U_t \phi, A U_t \phi \rangle = \langle \phi, U_t^{-1} A U_t \phi \rangle$$

we see that $W_t A = U_t^{-1} A U_t$.

If the state at time t_1 is given by $U_{t_1} \phi$, then the state at time $t_2 + t_1$ is given by $U_{t_2+t_1} \phi = U_{t_2} U_{t_1} \phi$ and so we must have $U_{t_2+t_1} = U_{t_2} U_{t_1}$ for every $t_1, t_2 \in (-\infty, \infty)$. Furthermore, letting $t_2 = 0$ and $t_2 = -t_1$ in the above identity gives $U_0 = I$ and $U_{-t} = U_t^{-1}$ so U_t forms a one-parameter group of unitary transformations. It is also usually assumed that $t \mapsto U_t$ is strongly continuous. It follows from Stone's theorem (Stone, 1930) that there exists a unique self-adjoint operator H_0 such that $U_t \phi = e^{-iH_0 t} \phi$ for all $t \in (-\infty, \infty)$. The operator H_0 is identified with the *Hamiltonian* of the system. The differential form of the evolution laws become:

$$\frac{d}{dt} \phi_t = \frac{d}{dt} U_t \phi = -iH_0 U_t \phi = -iH_0 \phi_t$$

$$\frac{d}{dt} A_t = \frac{d}{dt} U_t^{-1} A U_t = i[H_0, U_t^{-1} A U_t] = i[H_0, A_t]$$

The first of these equations is called *Schrödinger's equation*.

(B) Formulation 2

Formulation 1 is the prototype of the algebraic approach to axiomatic quantum mechanics. We next briefly consider an equivalent formulation which is the prototype to the quantum logic approach. In this formulation the primitive axiomatic elements are the 'propositions' of a physical system. A proposition of a system represents a special type of observable that has at most two values 0 and 1 or true and false. For example, a counter which is either activated or unactivated is a proposition. A filter that passes only certain types of particles is a proposition since a particle either passes or does not pass. The basic postulate is that the propositions of a physical system are described by the set of closed subspaces $\mathcal{P}$ of a complex Hilbert space H. This is equivalent to describing the propositions by orthogonal projections on H.

The sets of states $\mathcal{S}'$ of the system now determine the probabilities that propositions are true. Thus, if $s \in \mathcal{S}'$, $P \in \mathcal{P}$ then $s(P) \in [0, 1]$. Since the identity projection I corresponds to the proposition that is always true, we must have $s(I) = 1$ for every $s \in \mathcal{S}'$. If P_i is a sequence of mutually orthogonal projections (i.e. $P_i P_j = 0$, $i \neq j$) then the probability that $\sum P_i$ is true should be the sum of the probabilities that each P_i is true. We therefore postulate that $s(\sum P_i) = \sum s(P_i)$ for every $s \in \mathcal{S}'$. In other words a state is described by a 'probability measure' on $\mathcal{P}$. Again by Gleason's theorem, if $s \in \mathcal{S}'$ there exists a unique density operator T_s such that $s(P) = \text{Tr}(T_s P)$ for every $P \in \mathcal{P}$.

The general observables come into the theory in the following way. If X is an observable and E is a Borel subset of the real line then the pair (X, E) corresponds to the proposition: 'X has a value in the set E'. Thus X can be thought of as a map from the Borel sets $B(R)$ into $\mathcal{P}$. It is easy to justify that $X: B(R) \to \mathcal{P}$ should satisfy

(1). $X(R) = I$;
(2). If $E \cap F = \varnothing$, then $X(E)X(F) = 0$;
(3). If E_i are mutually disjoint, then $X(UE_i) = \sum X(E_i)$.

Thus X can be thought of as a projection-valued measure. By the spectral theorem their exists a unique self-adjoint operator A such that $A = \int \lambda X(d\lambda)$. conversely, if A is a self-adjoint operator, then there is a projection-valued measure $E \mapsto P^A(E)$ such that $A = \int \lambda P^A(d\lambda)$. Thus there exists a one-to-one correspondence between observables and self-adjoint operators.

If $s \in \mathcal{S}'$ and A is an observable, then $s[P^A(E)]$ is the probability that A has a value in the set E when the system is in the state s. It is then clear that the expectation of A in the state s is

$$s(A) = \int \lambda s[P^A(d\lambda)] = \int \lambda \operatorname{Tr}[T_s P^A(d\lambda)] = \operatorname{Tr}\left[T_s \int \lambda P^A(d\lambda)\right]$$
$$= \operatorname{Tr}(T_s A)$$

We thus see that corresponding to a state $s \in \mathcal{S}'$ as defined in this formulation there is a state $s \in \mathcal{S}$ as defined in Formulation 1. Conversely, if $s \in \mathcal{S}$ as defined in Formulation 1 then in particular s can act on projections. If we restrict s to $\mathcal{P}$ then s is a state in $\mathcal{S}'$ as defined in Formulation 2. For this reason, Formulations 1 and 2 are equivalent.

(C) Formulation 3

This formulation of the classical approach to axiomatic quantum mechanics is the prototype of the convexity approach. In this formulation the states form the basic axiomatic elements. The basic postulate is that the states of a physical system can be described by density operators $\mathcal{S}$ on a complex Hilbert space H. Convexity comes into play since $\mathcal{S}$ is a (strong) convex set. That is, if $T_i \in \mathcal{S}$ and $\sum \lambda_i = 1$, $\lambda_i \geq 0$, then $\sum \lambda_i T_i \in \mathcal{S}$. We call $\sum \lambda_i T_i$ a *mixture* of the states T_i. The *extreme points* of $\mathcal{S}$ are the states that cannot be written as mixtures of other states. These states are also called *pure states* and are given by one-dimensional projections.

In Formulation 1, we mentioned that a state is determined by the expectation values it gives to observables. In that formulation the observables were the basic axiomatic elements and the states were derived from the observables. We now turn the situation around. Now the states are the axiomatic elements and we shall derive the observables from the states. Hence if A is a bounded observable (we are not now assuming that A corresponds to a self-adjoint operator, this will follow automatically) and $T \in \mathscr{S}$ a state, we define $A(T)$ to be the expectation of A in the state T. We now seek the properties of the function $T \mapsto A(T)$. First, it is natural to assume that this function preserves convex combinations. That is,

$$A\left(\sum_{i=1}^{n} \lambda_i T_i\right) = \sum_{i=1}^{n} \lambda_i A(T_i)$$

whenever $\lambda_i \geq 0$, $\sum \lambda_i = 1$. Second, since observables must have real expectations, we assume $A(T) \in R$ for every $T \in \mathscr{S}$. Finally, since states that are 'close' should give expectations that are 'close', a continuity condition is imposed. This continuity condition is usually given in terms of the trace norm. This norm is defined as follows. If $T_1, T_2 \in \mathscr{S}$ then there is a unique positive trace class operator T_3 such that $T_3^2 = (T_1 - T_2)^2$. This operator is denoted $|T_1 - T_2|$. The *trace norm* of $T_1 - T_2$ is $\|T_1 - T_2\|_1 = \text{Tr}\,|T_1 - T_2|$. The continuity condition becomes: if $T_i, T \in \mathscr{S}$ and $\|T_i - T\|_1 \to 0$ as $i \to \infty$, then $A(T_i) \to A(T)$ for every bounded observable A. If the above three conditions hold, it can be shown (Schatten, 1950) that for any bound observable A there exists a unique bounded self-adjoint operator A_0 on H such that $A(T) = \text{Tr}\,(TA_0)$. We thus see that this formulation is equivalent to Formulations 1 and 2.

(D) Strengths and Weaknesses

Now that we have seen a brief formulation of the classical approach, a natural question is, why look at other approaches? Is there something wrong with this approach that makes it necessary to abandon or modify it? A lucid discussion of this question can be found in (Emch, 1972). Here we shall limit ourselves to a few comments.

Let us begin with the strengths of the classical approach. First, and most important, this approach has been highly successful, especially for systems with a finite number of degrees of freedom. Second, it has the advantage of concreteness. The observables can be identified with self-adjoint operators, the states with density operators, the propositions with projection operators and the dynamics with unitary operators on a Hilbert space. Now all that is needed is to specify what Hilbert space is to be used and to give a prescription for the self-adjoint operator that corresponds to each observable. There is a very satisfactory way of doing this in the case of a finite number of degrees of freedom. Suppose the system has $3N$ degrees of freedom in a cartesian coordinate system $x_1, x_2, x_3, \ldots, x_{3N}$. Then the Hilbert space is taken to be the space $L^2(R^{3N}, d\lambda)$ of square integrable complex-valued functions on R^{3N}. Using the correspondence principle or other heuristic arguments, the position observables are prescribed as $Q_i f(x) = x_i f(x)$, $i = 1, 2, \ldots, 3N$, and the momentum observables as $P_i f(x) = -i\hbar(\partial/\partial x_i)f(x)$, $i = 1, 2, \ldots, 3N$. The other observables are now given in terms of these basic observables. The quantum mechanics is now completely described.

The above procedure is satisfactory for the following reason. The position and momentum operators satisfy the canonical commutation relations $[Q_j, P_k] = i\hbar \delta_{jk}$. (To be perfectly rigorous one should work with the Weyl form of the commutation relations but for simplicity we shall be a little imprecise here.) Now by a theorem of von Neumann (1931), if $Q_i^0, P_i^0, i = 1, 2, \ldots, 3N$, are an (irreducible) set of self-adjoint operators on a Hilbert space H which satisfy the above commutation relations, then H is unitarily equivalent to $L^2(R^{3N}, d\lambda)$ and Q_i^0, P_i^0 are equivalent to Q_i, P_i defined above, respectively. Thus, if the framework is to satisfy these basic commutation relations the Hilbert space and the observables (and hence the states, dynamics, etc.) are uniquely determined within a unitary equivalence.

Now for some of the weaknesses of the classical approach. Von Neumann's theorem does not extend to systems with an infinite number of degrees of freedom. In fact, one can show that in this case there are infinitely many (in fact, uncountably infinitely many) inequivalent representations of the canonical commutation relations. Each of these representations gives different results. How is one to choose the 'right' representation? In fact, if one chooses the most 'natural' representation, namely Fock space, the results are unsatisfactory when interactions are involved. This problem would be merely a mathematical curiosity if no important physical systm had an infinite number of degrees of freedom, but unfortunately all of quantum field theory lies within this range. Furthermore, this discussion would be unnecessary if the present quantum field theory were successful, but this is far from the case.

A second weakness of the classical approach is the basic axioms themselves. Where does the Hilbert space come from? Why describe observables by self-adjoint operators? There seems to be no really convincing reason. In other words, the axioms are *ad hoc*, devoid of empirical evidence. There are some who believe that the troubles encountered in quantum field theory may be due to the basic axioms. They feel that if these axioms were established on a firmer empirical foundation then many of the difficulties would dissolve.

Besides the basic axioms, there is another assumption, of a less important character, which is of a questionable nature. This is the continuity condition placed on the states. In Formulation 1, this was defined in terms of the strong convergence of operators. But there seems to be no physical significance for this type of convergence. In Formulation 2 this convergence is contained implicitly in the countable additivity of states. It is clear that states should be finitely additive but there is no physical reason for them to be countable additive.

3. THE ALGEBRAIC APPROACH

The algebraic approach was initiated by Jordan, von Neumann and Wigner (1934). It was later developed by Segal (1947), Haag and Kastler (1964) and many others. In this approach, the bounded observables are taken as the primitive axiomatic elements. We begin with a slight modification of Segal's formulation.

(A) Segal Algebras

A collection of objects $\mathcal{A}$ is called a *Segal algebra* if $\mathcal{A}$ satisfies the following postulates.

Axiom A. $\mathcal{A}$ is a linear space over the real numbers R.
Axiom B. There exists in $\mathcal{A}$ an identity I and for every $A \in \mathcal{A}$ and integer $n \geq 0$ an element $A^n \in \mathcal{A}$ which satisfies the following: If f, g and h are real polynomials, and $f(g(\lambda)) = h(\lambda)$ for every $\lambda \in R$, then $f(g(A)) = h(A)$; where

$$f(A) = \beta_0 I + \sum_{k=1}^{n} \beta_k A^k \quad \text{if} \quad f(\alpha) = \sum_{k=0}^{n} \beta_k \alpha^k$$

Axiom C. There is defined for each $A \in \mathcal{A}$ a real number $\|A\| \geq 0$ such that the pair $(, \|\cdot\|)$ is a real Banach space.
Axiom D. $\|A^2 - B^2\| \leq \max(\|A^2\|, \|B^2\|)$ and $\|A^2\| = \|A\|^2$.
Axiom E. A^2 is a continuous function of A.

Of course, a Segal algebra is supposed to describe axiomatically the set of bounded observables for a physical system. The underlying idea is that an observable is determined by its average values as given by laboratory experiments. Let A be an observable and $\lambda \in R$. If the average values of A as determined by a laboratory experiment are multiplied by λ, then this determines a new observable λA. If the average values of two observables A and B are added, then this determines a new observable $A + B$ whose average values are these sums. This argument justifies Axiom A. Unfortunately, this procedure cannot be used to define products since in general the expectation of a product is not the product of the expectations. For this reason, products of observables are not defined. Axiom B states that polynomials in a single observable exist and enjoy the usual properties. The norm in Axiom C can be thought of as the maximum absolute expectation value of an observable. The properties of a norm then easily follow. The completeness is included for mathematical convenience since if the system were not complete it could be completed in the usual way still preserving all the axioms. Axiom D can be justified in terms of the interpretation of the norm given above. Axiom E is a natural continuity condition. We henceforth call the elements of a Segal algebra *observables*.

An example of a Segal algebra is the set of bounded self-adjoint operators on a Hilbert space as given in Formulation 1 of the classical approach. Another example is the Banach space $\mathscr{C}(\Gamma)$ of all continuous real-valued functions on a compact Hausdorff space Γ under the supremum norm. This example corresponds to a classical mechanical system. In this case, Γ corresponds to a phase space and $C(\Gamma)$ is the set of observables (dynamical variables) which are necessarily compatible (or commuting or simultaneously measurable).

Let $\mathcal{A}$ be a Segal algebra. A *state* of $\mathcal{A}$ is a real-valued linear functional ω on $\mathcal{A}$ such that $\omega(A^2) \geq 0$ for all $A \in \mathcal{A}$ and $\omega(I) = 1$. The states are supposed to describe the expectation values of the observables for a particular preparation or condition of the physical system. With this interpretation, the above properties of a state are clear. A collection of states $\mathcal{S}$ on $\mathcal{A}$ is *full* if for any two distinct observables A, B there exists a state $\omega \in \mathcal{S}$ such that $\omega(A) \neq \omega(B)$. Segal (1947) has shown that any Segal algebra has a full set of states and that

$$\|A\| = \sup\{|\omega(A)| : \omega \in \mathcal{S}\}$$

for all $A \in \mathcal{A}$. This latter fact justifies our interpretation of the norm as the maximum absolute expectation value of an observable.

Although products of arbitrary observables are not defined, we can define a 'symmetrized product'. For $A, B \in \mathcal{A}$ the symmetrized product $A \circ B$ is defined by $A \circ B = \frac{1}{2}[(A+B)^2 - A^2 - B^2]$. The physical significance of $A \circ B$ is not clear, although it is a convenient mathematical construct. This product does not enjoy very many algebraic properties. It is clearly commutative and from Axiom B, $A \circ I = A$ for every $A \in \mathcal{A}$. However, in general it need not be homogeneous $((\lambda A) \circ B) = \lambda(A \circ B))$, distributive $(A \circ (B+C) = A \circ B + A \circ C)$, or associative $(A \circ (B \circ C) = (A \circ B) \circ C)$. In fact, the Segal algebra of all bounded self-adjoint operators on a Hilbert space is an example in which the associative law does not hold for the symmetrized product. We shall later give an example in which distributivity and homogeneity do not hold.

LEMMA 1. *The symmetrized product is homogeneous if and only if it is distributive.*

Proof. Suppose the symmetrized product is homogeneous. It follows that $-(A \circ B) = A \circ (-B)$. Writing this out in terms of the definition gives

$$(A+B)^2 + (A-B)^2 = 2A^2 + 2B^2 \tag{1}$$

It follows that

$$A \circ B = \tfrac{1}{4}[(A+B)^2 - (A-B)^2] \tag{2}$$

Now substitute $A+C$ and $A-C$ for A in (1) to get the following two equations.

$$[(A+B)+C]^2 + [(A-B)+C]^2 = 2(A+C)^2 + 2B^2$$
$$[(A+B)-C]^2 + [(A-B)-C]^2 = 2(A-C)^2 + 2B^2$$

Subtracting these last two equations and using (2) gives

$$(A+B) \circ C + (A-B) \circ C = 2(A \circ C) = (2A) \circ B$$

Replace $A+B$ by A and $A-B$ by B to get

$$A \circ C + B \circ C = (A+B) \circ C \tag{3}$$

Conversely, suppose the distributive law (3) holds. Then replacing A by B gives $2(B \circ C) = (2B) \circ C$. Now replacing B by $B/2$ gives $(\frac{1}{2}B) \circ C = \frac{1}{2}(B \circ C)$. Replacing A by $2B$ gives $3(B \circ C) = (3B) \circ C$, etc. Also, replacing A by $-B$ gives $(-B) \circ C = -(B \circ C)$. In this way $\lambda(B \circ C) = (\lambda B) \circ C$ for every rational λ. But since addition and squaring are continuous, so is the symmetrized product. It follows by continuity that the symmetrized product is homogeneous.

COROLLARY: If the symmetrized product is associative, then it is homogeneous and distributive.

Proof. If the symmetrized product is associative, then

$$(\lambda A) \circ B = (\lambda I \circ A) \circ B = \lambda I \circ (A \circ B) = \lambda(A \circ B)$$

The converse of this corollary does not hold as the Hilbert space example shows. We now give an example which shows that the symmetrized product need not be homogeneous (and hence not distributive or associative). This example is a simplified version of one due to Sherman (1956). Let $X = R^3$ and define addition and multiplication by scalars in the usual ways. Let $I = (1, 1, 1)$ and $(\alpha I)^n = \alpha^n I$ for $n \geq 0$ an integer. If $x - (x_1, x_2, x_3) \in X$, let $\bar{x} = \max x_i$, $\underline{x} = \min x_i$ and let $X_0 = \{x \in X : \bar{x} = 1, \underline{x} = -1\}$. If $x \in X_0$, define $x^n = x$ if n is an odd integer, and $x^n = I$ if n is an even integer. If $x \in X$, then it is easy to see that there exists an $x_0 \in X_0$ such that $x = \alpha x_0 + \beta I$, $\alpha, \beta \in R$. Define

$$x^n = (\alpha x_0 + \beta I)^n = \sum_{j=0}^{n} \binom{n}{j} \alpha^j \beta^{n-j} x_0^j$$

It is easy to see that x^n is well-defined. For $x \in X$ we define

$$\|x\| = |x| = \|\alpha x_0 + \beta I\| = \max\{|\beta - \alpha|, |\alpha + \beta|\}$$

Sherman (1956) has shown that with these operations X is a Segal algebra. Now let $a = (1, 1, 0)$ and $b = (1, 0, 1)$. We shall show that $2(a \circ b) \neq (2a) \circ b$. Indeed, $a = \frac{1}{2}(1, 1, -1)$ and $b = \frac{1}{2}(1, -1, 1) + \frac{1}{2}I$ and so $a + b = \frac{1}{2}(1, -1, -1) + \frac{3}{2}I$. Hence

$$2(a \circ b) = [(a+b)^2 - a^2 - b^2] = [(4, 1, 1) - (1, 1, 0) - (1, 0, 1)]$$
$$= (2, 0, 0)$$

Furthermore,

$$(2a) \circ b = \frac{1}{2}[(2a+b)^2 - (2a)^2 - b^2] = \frac{1}{2}[(9, 5, 1) - (4, 4, 0) - (1, 0, 1)]$$
$$= (2, \tfrac{1}{2}, 0)$$

A Segal algebra is *compatible* if the symmetrized product is associative. A collection of observables is *compatible* if the subalgebra generated by the collection is compatible. Segal (1947) has proved that a compatible Segal algebra is isomorphic (algebraically and metrically) with the algebra $\mathscr{C}(\Gamma)$ of all real-valued continuous functions on a compact Hausdorff space Γ considered earlier. It is well known that the states on $\mathscr{C}(\Gamma)$ consist of the regular Borel probability measures on Γ; that is, if ω is a state, there exists a regular Borel probability measure μ on Γ such that $\omega(f) = \int f \, d\mu$ for every $f \in \mathscr{C}(\Gamma)$. It follows from these results that a compatible Segal algebra can be thought of as the set of observables in a classical mechanical system. Furthermore, compatible observables can be thought of as being simultaneously measurable.

(B) C*-Algebras

The Segal algebras considered earlier were based upon axioms that had physical relevance. Unfortunately, their mathematical structure is so weak that not much further progress has been made in terms of using them for the study of quantum theory. To proceed further, additional axioms have been imposed which have not been given physical justification. One of these is to postulate the distributive law for the symmetrized product. In a mathematical sense, the distributive law is a rather mild required. In fact, by the proof of Lemma 1 this law is equivalent to requiring that $(-A) \circ B = -(A \circ B)$ and $(2A) \circ B = 2(A \circ B)$ for all $A, B \in \mathscr{A}$. However, the physical reasons for such a requirement are lacking. Furthermore, additional axioms have been imposed which are not nearly so mild. These are best stated in terms of C*-algebras.

We first review the terminology of C*-algebras. A *complex algebra* is a complex vector space equipped with an identity and a distributive, associative product AB. An *involution* on a complex algebra $\mathscr{B}$ is a map * of $\mathscr{B}$ into itself which satisfies $(A^*)^* = A$, $(A+B)^* = A^* + B^*$, $(\lambda A)^* = \lambda^* A^*$ and $(AB)^* = B^* A^*$ for every $A, B \in \mathscr{B}$ and complex λ with λ^* denoting the complex conjugate of λ. An *involution algebra* is a complex algebra equipped with an involution. A *Banach algebra* is an algebra equipped with a norm such that $\|AB\| \leq \|A\| \|B\|$ and which is complete in the norm topology. A C*-algebra is an involutive Banach algebra $\mathscr{B}$ satisfying $\|A^*A\| = \|A\|^2$ for every $A \in \mathscr{B}$. An example of a C*-algebra is the set $\mathscr{B}(H)$ of all bounded linear operators on a complex Hilbert space H. In this case (*) is the adjoint map and $\|\cdot\|$ is the operator norm.

An element A of a C*-algebra is *self-adjoint* if $A^* = A$. It is straightforward to show that the set of all self-adjoint elements of a C*-algebra form a distributive Segal algebra. In this case the symmetrized product $A \circ B = \frac{1}{2}[(A+B)^2 - A^2 - B^2]$ takes the simple form $A \circ B = \frac{1}{2}(AB + BA)$. A distributive Segal algebra is said to be *special* if it is isomorphic to the set of all self-adjoint elements of a C*-algebra. Important unsolved problems are whether every distributive Segal algebra is special and if not to characterize special Segal algebras. These problems appear to be very difficult and it seems unlikely that an arbitrary distributive Segal algebra is special. However, all Segal algebras that have been encountered in physical situations have been special. For this reason and also because the theory of C*-algebras has a rich mathematical development it is postulated that the Segal algebra corresponding to a physical system is special. It can be shown that any state ω of a special

Segal algebra can be extended to a positive (i.e. $\omega(AA^*) \geq 0$), normalized (i.e. $\omega(I) = 1$), linear functional on the C*-algebra. Having made this postulate, we can proceed to the study of C*-algebras interpreting the self-adjoint elements as observables and the positive, normalized, linear functionals as states.

One of the important consequences of this last postulate is that it provides the mechanisms for representing the elements of a Segal algebra as self-adjoint operators on a Hilbert space. This follows from the GNS construction (after Gelfrand and Naimark (1943) and Segal (1947)). We now develop the necessary material to understand this construction.

A map π of a C*-algebra $\mathcal{B}$ into the set $\mathcal{B}(H)$ of all bounded linear operators on a Hilbert space H is said to be a *representation* of $\mathcal{B}$ if

(1). $\pi(\alpha A + \beta B) = \alpha \pi(A) + \beta \pi(B)$;
(2). $\pi(AB) = \pi(A)\pi(B)$;
(3). $\pi(A^*) = \pi(A)^*$;

for all $A, B \in \mathcal{B}$ and complex numbers α, β. It can be shown that if π is a representation of $\mathcal{B}$, then $\|\pi(A)\| \leq \|A\|$ for every $A \in \mathcal{B}$. Thus a representation is automatically continuous. A representation $\pi: \mathcal{B} \to \mathcal{B}(H)$ is *cyclic* if there is a vector $\psi \in H$ such that the subspace $\pi(\mathcal{B})\psi = \{\pi(A)\psi : A \in \mathcal{B}\}$ is dense in H. In this case, ψ is said to be a *cyclic vector* for π. Cyclic representations play an important role in the theory; in particular, it is easy to show that any representation is the direct sum of cyclic representations. In many physical applications the vacuum state plays the part of a cyclic vector.

A positive, normalized, linear functional on a C*-algebra $\mathcal{B}$ is called a *state*. Let $\pi: \mathcal{B} \to \mathcal{B}(H)$ be a representation of $\mathcal{B}$ and let $\psi \in H$, $\|\psi\| = 1$. Then the functional $\omega(A) = \langle \psi, \pi(A)\psi \rangle$ is a state called a *vector state* associated with the representation π. A state is *pure* if it cannot be written as a convex combination of other states. Let $\pi: \mathcal{B} \to \mathcal{B}(H)$ and $\pi': \mathcal{B} \to \mathcal{B}(H')$ be two representations of $\mathcal{B}$. If there is an isomorphism $U: H \to H'$ such that $\pi'(A) = U\pi(A)U^{-1}$ then π and π' are *spatially* (or *unitarily*) *equivalent*. A closed subspace M of H is *invariant* with respect to $\pi: \mathcal{B} \to \mathcal{B}(H)$ if $\pi(A)M \subseteq M$ for every $A \in \mathcal{B}$. A representation $\pi: \mathcal{B} \to \mathcal{B}(H)$ is *irreducible* if the only invariant subspaces of H with respect to π are $\{0\}$ and H. Irreducible representations are the most economical in the sense that they cannot be written as a direct sum of representations.

THEOREM 2. (The GNS Construction) Let $\mathcal{B}$ be a C*-algebra and let ω be a state on $\mathcal{B}$. Then there exists a Hilbert space H and a cyclic representation $\pi_\omega: \mathcal{B} \to \mathcal{B}(H)$ with cyclic vector $\psi \in H$ such that $\omega(A) = \langle \psi, \pi(A)\psi \rangle$ for every $A \in \mathcal{B}$. If $\pi': \mathcal{B} \to \mathcal{B}(H')$ is another cyclic representation with cyclic vector $\psi' \in H'$ such that $\omega(A) = \langle \psi', \pi'(A)\psi' \rangle$ for every $A \in \mathcal{B}$, then π and π' are spatially equivalent. Furthermore, π_ω is irreducible if and only if ω is pure.

The GNS construction has important physical consequences, one of which is the following. It is not the Hilbert space and the self-adjoint operators on it as postulated in the classical approach of Section 2 that contains the essence of the physical system, but it is the C*-algebra generated by the Segal algebra. The Hilbert space and the operators on it corresponding to the observables depend upon the state of the system and can be obtained via the GNS construction. There may be many inequivalent representations of a C*-algebra and the 'right' one is determined by the state of the system.

The above observation overcomes, to a certain extent, one of the weaknesses of the classical approach mentioned in Section 2, namely where the Hilbert space and self-adjoint operators come from. Another weakness mentioned in Section 2 that is overcome is the continuity condition placed on the states of the classical approach. In the algebraic approach, no such condition is imposed. The states are defined algebraically in terms of the physically natural conditions of linearity, positivity and normalization. In fact, there are more states in the algebraic approach than those given by the density operators of the classical approach. Furthermore, these extra states actually occur in physical situations. An example of such a state can be given as follows. Let $\mathcal{A}$ be the Segal algebra of all bounded self-adjoint operators on an infinite dimensional Hilbert space. Then $\mathcal{A}$ contains an operator A with non-empty continuous spectrum $\sigma_c(A)$. Let $\lambda \in \sigma_c(A)$. It can be shown (Segal, 1947) that there exists a pure state ω such that $\omega(A) = \lambda$. Now this pure state cannot have the form $\omega(A) = \langle \psi, A\psi \rangle$ since then ψ would be an eigenvector of A with eigenvalue λ, which contradicts the fact that $\lambda \in \sigma_c(A)$. Such states as the above are delta function-like or Schwartz distribution-like elements that lie outside the Hilbert space. It can be shown (Emch, 1972), however, that all states of $\mathcal{A}$ can be approximated in a certain topology by density operator states.

Although we have now formulated the basic concepts of the algebraic approach, we have only scratched the surface of its later developments. We have not mentioned such important areas as physical equivalence, symmetry groups, representations of canonical commutation and anticommutation relations, quasilocal field theories and applications to concrete problems. For further study the reader is referred to (Emch, 1972) and the modern literature.

(C) Strengths and Weaknesses

One of the strengths of the algebraic approach is that it is based on axioms that have more physical relevance than in the classical approach. This is especially true of the axioms for a Segal algebra. It has also clarified the existence of inequivalent representations of the CCR and CAR. It has enjoyed some notable successes that we have not had space to explore and is responsible for important applications to such areas as statistical mechanics and solid-state physics.

Weaknesses of this approach include some of the later axioms, especially the jump from a Segal algebra to a C*-algebra. What, for example, is the physical significance of the product in a C*-algebra? Furthermore, even with all the mathematical power that has been brought to bear, a satisfactory field theory has still not be developed.

4. THE QUANTUM LOGIC APPROACH

In the quantum logic approach the propositions of a physical system are taken as primitive axiomatic elements. The propositions correspond to yes–no (or true–false) experiments on the physical system. For example, suppose the physical system consists of a single particle and let E be a region of space. If a_E is a counter which is activated if and only if the particle enters the region E, then a_E corresponds to a proposition. This proposition is true if and only if a_E is activated and the particle is in the region E. The propositions correspond to two-valued observables and it can be argued (we shall substantiate this later) that an observable can be decomposed into these simpler two-valued observables. Thus a treatment of propositions is general enough to describe all observables. The standard references on this approach are Jauch (1968), Mackey (1963) and Varadarajan (1968).

(A) Quantum Logics

Let $\mathcal{P}_0$ be a set of elements called *experimental propositions*. If $a \in \mathcal{P}_0$ then a is true or false depending upon the state of the system. But in quantum mechanics one cannot, in general, predict whether a proposition will be true even if the state is precisely known. All one can predict is the probability that a proposition will be true. Thus a state m can be thought of as a function from $\mathcal{P}_0$ to the unit interval $[0, 1]$ and $m(a)$ for $a \in \mathcal{P}_0$ is interpreted as the probability a is true when the system is in the state m. If $m(a) = 1$ then a is true with certainty in the state m. Now suppose $a, b \in \mathcal{P}_0$ and $m(a) + m(b) \leq 1$ for every state m. Since $m(a) = 1$ implies $m(b) = 0$, whenever a is true with certainty, b is false with certainty. In this case a and b can be interpreted as corresponding to non-interfering experiments and their truth or falsity can be verified simultaneously. In this case one can consider the experimental proposition c which is true with certainty precisely when a and b are both false with certainty. Then we should have $m(c) + m(a) + m(b) = 1$ for every state m. For example, in our counter experiment suppose E_1 and E_2 are disjoint regions of space. Then for any state m the probability the particle is in E_1 plus the probability the particle is in E_2 does not exceed unity, $m(a_{E_1}) + m(a_{E_2}) \leq 1$. Now if $E_3 = (E_1 \cup E_2)'$ (E' is the complement of the set E) we should have $m(a_{E_3}) + m(a_{E_2}) + m(a_{E_1}) = 1$. Such considerations also carry over to sequences of propositions.

A *proposition system* is a pair $(\mathcal{P}_0, \mathcal{M})$ where $\mathcal{P}_0$ is a non-empty set and $\mathcal{M}$ is a non-empty set of functions from $\mathcal{P}_0$ into $[0, 1]$ satisfying:

Axiom A. For any sequence $a_1, a_2, \ldots \varepsilon \mathcal{P}_0$ such that $m(a_i) + m(a_j) \leq 1$, $i \neq j$, for every $m \in \mathcal{M}$, there exists $b \in \mathcal{P}_0$ such that $m(b) + m(a_1) + m(a_2) + \ldots = 1$ for every $m \in \mathcal{M}$.

Axiom A is the only axiom that we shall impose on the system. We call the elements of $\mathcal{M}$ states and the element of $\mathcal{P}_0$ *experimental propositions*. Now suppose $a, b \in \mathcal{P}_0$ and $m(a) = m(b)$ for every $m \in \mathcal{M}$. Then a and b are physically indistinguishable and we write $a \sim b$. It is clear that $\sim$ is an equivalence relation and we denote the equivalence class containing a by $\bar{a}$. Furthermore, if $m \in \mathcal{M}$ we define $m(\bar{a}) = m(a)$. We denote the set of equivalence classes by $\mathcal{P}$ and call the elements of $\mathcal{P}$ *propositions*. We call the pair $(\mathcal{P}, \mathcal{M})$ a *quantum logic*. Thus a quantum logic is a pair $(\mathcal{P}, \mathcal{M})$ satisfying Axiom A (with $\mathcal{P}_0$ replaced by $\mathcal{P}$) together with the condition $m(\bar{a}) = m(\bar{b})$ for every $m \in \mathcal{M}$ implies $\bar{a} = \bar{b}$. The quantum logic will be the main framework of our study and for simplicity we shall drop the bars in the following.

Our next order of business is to prove the main structure theorem for quantum logics. But first we need some definitions. Let $(\mathcal{P}, \leq)$ be a partially ordered set with first and last elements 0 and 1, respectively. An *orthocomplementation* on $\mathcal{P}$ is a map $a \mapsto a'$ from $\mathcal{P}$ to $\mathcal{P}$ with the following properties:

(1). $a'' = a$ for every $a \in \mathcal{P}$;
(2). $a \leq b$ implies $b' \leq a'$;
(3). $a \vee a' = 1$ for every $a \in \mathcal{P}$.

In (3) $a \vee a'$ denotes the least upper bound of a and a'.

If $(\mathcal{P}, \leq)$ has an orthocomplementation ($'$), then $(\mathcal{P}, \leq, ')$ is called an *orthocomplemented poset*. It is easily verified that in an orthocomplemented poset if $a \vee b$ exists then so does $a' \wedge b'$ and $a' \wedge b' = (a \vee b)'$. If $a \leq b'$, we say that a and b are *orthogonal* and write $a \perp b$. An orthocomplemented poset $(\mathcal{P}, \leq, ')$ is σ-*orthocomplete* if the following holds:

(4). If $a_1, a_2, \ldots$ is a sequence of mutually orthogonal elements in $\mathcal{P}$ then $a_1 \vee a_2 \vee \ldots$ exists.

An orthocomplemented, σ-orthocomplete poset $(\mathcal{P}, \leq, ')$ is *orthomodular* if

(5). $a \leq b$ implies $b = a \vee (b \wedge a')$.

A *probability measure* on a σ-orthocomplete poset $(\mathcal{P}, \leq, ')$ is a map $m : \mathcal{P} \to [0, 1]$ which satisfies:

(a) $m(1) = 1$;
(b) if $a_1, a_2, \ldots$ is a sequence of mutually orthogonal elements of $\mathcal{P}$ then $m(\vee a_i) = \sum m(a_i)$.

A set $\mathcal{M}$ of probability measures on $(\mathcal{P}, \leq, ')$ is *order determining* if $m(a) \leq m(b)$ for every $m \in \mathcal{M}$ implies $a \leq b$.

Let us now return to the quantum logic $(\mathcal{P}, \mathcal{M})$. If $a, b \in \mathcal{P}$ define $a \leq b$ if $m(a) \leq m(b)$ for every $m \in \mathcal{M}$. The relation $a \leq b$ can be interpreted as a implies b. That is, b has a greater probability of being true than a. It follows that whenever a is true with certainty so is b. If $a \in \mathcal{P}$, since $m(a) \leq 1$ for every $m \in \mathcal{M}$, by Axiom A there exists $b \in \mathcal{P}$ such that $m(b) = 1 - m(a)$ for every $m \in \mathcal{M}$. We then write $b = a'$ and call b the *orthogonal complement* of a. We can interpret a' as the proposition which is true if and only if a is false. We have thus defined a relation $\leq$ on $\mathcal{P}$ and a map $('): \mathcal{P} \to \mathcal{P}$. We now prove our main structure theorem. This theorem is due to Maczynski (1974) and the proof is a simplification of his.

THEOREM 3. $(\mathscr{P}, \mathscr{M})$ is a quantum logic if and only if $(\mathscr{P}, \leq, ')$ is a σ-orthocomplete orthomodular poset and $\mathscr{M}$ is an order-determining set of probability measures on $\mathscr{P}$.

Proof. First suppose $(\mathscr{P}, \mathscr{M})$ is a quantum logic. It is clear from the definition that $(\mathscr{P}, \leq)$ is a poset. It is also clear that $(')$ satisfies conditions (1) and (2) of an orthocomplementation. For $a \in \mathscr{P}$, we have $m(a) + m(a') = 1$ for every $m \in \mathscr{M}$ so by Axiom A there is an element $0 \in \mathscr{P}$ such that $m(0) + m(a) + m(a') = 1$. It follows that $m(0) = 0$ for every $m \in \mathscr{M}$. Define $1 = 0'$. Notice that $0 \leq a \leq 1$ for every $a \in \mathscr{P}$ so 0 and 1 are the first and last elements of $\mathscr{P}$, respectively. If $b \geq a, a'$ then $m(a), m(a') \leq m(b)$. Hence $m(a) + m(b') \leq 1$ and $m(a') + m(b') \leq 1$. Then a, a', b' satisfy the condition of Axiom A so there exists $c \in \mathscr{P}$ such that $m(c) + m(a) + m(a') + m(b') = 1$ for every $m \in \mathscr{M}$. But then $m(c) = m(b') = 0$ for every $m \in \mathscr{M}$. Hence $m(b) = 1$ for every $m \in \mathscr{M}$ and $b = 1$. Hence (3) is satisfied and $(')$ is an orthocomplementation. Now $a \perp b$ if and only if $m(a) + m(b) \leq 1$. Thus if $a_1, a_2, \ldots$ is a sequence of mutually orthogonal elements, then by Axiom A there exists $b \in \mathscr{P}$ such that $m(b) + m(a_1) + m(a_2) + \ldots = 1$. Hence $m(b') = \sum m(a_i)$ for every $m \in \mathscr{M}$. It follows that $b' \geq a_1, a_2, \ldots$. Now suppose $c \geq a_1, a_2, \ldots$. Then $m(a_i) + m(c') \leq 1$ for every $m \in \mathscr{M}$. Hence the sequence $c', a_1, a_2, \ldots$ satisfies the condition of Axiom A so there exists $d \in \mathscr{P}$ such that $m(d) + m(c') + \sum m(a_i) = 1$. It follows that $b' = \vee a_i$ and $\mathscr{P}$ is σ-orthocomplete. Furthermore, since $m(\vee a_i) = m(b') = \sum m(a_i)$ and $m(1) = 1$, it follows that every $m \in \mathscr{M}$ is a probability measure on $\mathscr{P}$. It is obvious that $\mathscr{M}$ is order determining. To show that $\mathscr{P}$ is orthomodular, suppose $a \leq b$. Then $a \perp b'$ and since $a \leq a \vee b' = (a' \wedge b)', a \perp a' \wedge b$. Hence for every $m \in \mathscr{M}$

$$m[a \vee (a' \wedge b)] = m(a) + m(a' \wedge b) = m(a) + 1 - m(a \vee b')$$
$$= m(a) + 1 - m(a) - m(b') = m(b)$$

Therefore, $b = a \vee (a' \wedge b)$. Conversely, suppose $(\mathscr{P}, \leq, ')$ is a σ-orthocomplete orthomodular poset and $\mathscr{M}$ is an order-determining set of probability measures on $\mathscr{P}$. Then $m(a) = m(b)$ for every $m \in \mathscr{M}$ implies $a = b$. Suppose $a_1, a_2, \ldots$ is a sequence in $\mathscr{P}$ such that $m(a_i) + m(a_j) \leq 1$, $i \neq j$, for every $m \in \mathscr{M}$. Then $m(a_i) \leq 1 - m(a_j) = m(a_j')$ so $a_i \perp a_j$, $i \neq j$. Hence $b = \vee a_i$ exists and $m(b') + \sum m(a_i) = 1$. Thus Axiom A holds and $(\mathscr{P}, \mathscr{M})$ is a quantum logic. ∎

We thus see that there is no difference between a quantum logic and a σ-orthocomplete orthomodular poset with an order-determining set of probability measures. Notice that a quantum logic need not be a lattice (that is, $a \vee b$ and $a \wedge b$ need not exist). For example, let $\Omega = \{1, 2, 3, 4, 5, 6\}$ and let $\mathscr{P}$ be the collection of subsets of Ω with an even number of elements. Order $\mathscr{P}$ by inclusion and let $(')$ be the usual set complementation. For $i = 1, \ldots, 6$ define for $a \in \mathscr{P}$ $m_i(a) = 1$ if $i \in a$ and $m_i(a) = 0$ if $i \notin a$. Then if we let $\mathscr{M} =$

$\{m_i : i = 1, \ldots 6\}$ it is easily verified that $(\mathcal{P}, \mathcal{M})$ is a quantum logic. However, $\mathcal{P}$ is not a lattice since, for example, $\{1, 2, 3, 4\} \wedge \{2, 3, 4, 5\}$ does not exist.

We say that two propositions a, b are *compatible* (written $a \leftrightarrow b$) if there exist mutually disjoint propositions a_1, b_1 and c such that $a = a_1 \vee c$, $b = b_1 \vee c$. We shall see that compatible propositions are ones that can be verified simultaneously; that is, propositions whose experiments do not interfere. Notice, if $a \perp b$ then $a \leftrightarrow b$ and $0 \leftrightarrow a$, $1 \leftrightarrow a$ for every $a \in \mathcal{P}$. Physically, our interpretation of $a \leq b$ demands that $a \leftrightarrow b$ if $a \leq b$. This is indeed the case since if $a \leq b$ then by (5) $b = a \vee (b \wedge a')$ and $a = a \vee 0$ where $a \perp (b \wedge a')$.

We now show how observables can be defined. If x is an arbitrary observable and $E \in B(R)$ is a Borel set on R, then the pair (x, E) corresponds to the proposition: 'the observable x has a value in the set E'. Thus if $(\mathcal{P}, \mathcal{M})$ is a quantum logic, an *observable* can be thought of as a map $x : B(R) \to \mathcal{P}$. Furthermore, an observable should satisfy:

(1). $x(R) = 1$;
(2). If $E \cap F = \emptyset$, then $x(E) \perp x(F)$;
(3). If $E_i \in B(R)$ is a sequence of mutually disjoint sets, then $x(U E_i) = \vee x(E_i)$.

The reader can easily justify these three conditions.

Two observables x, y are *compatible* (written $x \leftrightarrow y$) if $x(E) \leftrightarrow y(F)$ for all $E, F \in B(R)$. We shall show later that observables which are compatible may be thought of physically as being simultaneously measurable.

The reader should note that we have constructed a generalized probability theory. Instead of being a Boolean σ-algebra of subsets of a set, our events (propositions) are more general, belonging to a less restrictive structure. The usual probability measures are replaced by states and the random variables by observables. Notice that if x is an observable and m a state, then the probability that x has a value in $E \in B(R)$ when the system is in state m is $m[x(E)]$. Before proceeding further, let us consider two examples of quantum logics.

EXAMPLE 1. Let Ω be a phase space and let $B(\Omega)$ be the Borel subsets of Ω. $B(\Omega)$ may be thought of as the set of mechanical events. It is easily checked that $B(\Omega)$, under set inclusion and complementation, is a σ-orthocomplete, orthomodular poset (in fact, it is a Boolean σ-algebra). The set of states $\mathcal{M}$ are the conventional probability measures on $\mathcal{B}(\Omega)$ and these are order determining so $(B(\Omega), \mathcal{M})$ is a quantum logic. If x is an observable, it follows from a theorem of Sikorski (1949)–Varadarajan (1962) that there is a measurable function $f : \Omega \to R$ such that $x(E) = f^{-1}(E)$ for every $E \in B(R)$. Thus observables are just inverses of dynamical variables. We thus see that the quantum logic generalizes classical mechanics and also the conventional Kolmogorov (1956) formulation of probability theory. It is easily checked that all events (propositions) and observables are compatible in this example.

EXAMPLE 2. Let H be a complex Hilbert space and let $\mathcal{P}$ be the collection of all closed subspaces of H. Ordering $\mathcal{P}$ by inclusion and defining the complement of a subspace as its orthocomplement it is easily checked that $\mathcal{P}$ is a σ-orthocomplete, orthomodular poset (in fact, a lattice). Furthermore, the set of states $\mathcal{M}$, by Gleason's theorem (see Section 2), are given by density operators and are order determining. Hence $(\mathcal{P}, \mathcal{M})$ is a quantum logic. Identifying closed subspaces with their orthogonal projections, an observable may be thought of as a projection-valued measure. Since, using the spectral theorem, there is a one-to-one correspondence between projection-valued measures and self-adjoint operators, we may identify observables with self-adjoint operators. Thus the quantum-logic approach, in this case, reduces to the classical approach of Section 2. It is straightforward to show that $a, b \in \mathcal{P}$ are compatible if and only if their corresponding projections commute. It follows that two observables are compatible if and only if their corresponding self-adjoint operators commute.

Let us now return to quantum logics. If x is an observable we call $\{x(E): E \in B(R)\}$ the *range* of x. It is easily verified that the range of an observable is a Boolean σ-algebra.

LEMMA 4. (Varadarajan, 1962) Two propositions are compatible if and only if they are in the range of a single observable.

This last lemma justifies the fact that compatible propositions are simultaneously verifiable, since to verify two compatible propositions one need measure only a single observable.

Now let x be an observable and let $u : R \to R$ be a Borel function. There is an operational significance for $u(x)$; namely, if x has the value $\lambda \in R$, then $u(x)$ has the value $u(\lambda)$. This is equivalent to saying that the proposition '$u(x)$ has a value in $E \in B(R)$' is the same as the proposition 'x has a value in $u^{-1}(E)$'. Motivated by this, we define $u(x)$ as $u(x)(E) = x[u^{-1}(E)]$ for all $E \in B(R)$. It is easily checked that $u(x)$ is an observable and that $u(x) \leftrightarrow x$.

THEOREM 5. (Varadarajan, 1962) Two observables x, y are compatible if and only if there exists an observable z and Borel functions u, v such that $x = u(z)$ and $y = v(z)$.

This last theorem shows that, physically, compatible observables are simultaneously measurable (i.e. non-interfering) since to measure two compatible observables one need only measure a single observable.

Space does not permit a comparison of the quantum logic approach to the algebraic approach of Section 3. However, we mention that it can be shown that the approaches are not equivalent. It can also be shown that the Segal algebra of Section 3 can be embedded in a weaker type of quantum logic than that considered here (Gudder and Boyce, 1970; Plymen, 1968).

(B) Quantum Systems

Although some illuminating and physically valuable results have been obtained in the study of quantum logics, their structure is mathematically so general that they have not been particularly useful for concrete calculations. A quantum logic is so general that it is far from the concreteness of the Hilbert space of the classical approach. What is needed is something like the GNS construction of the algebraic approach. However, such a construction is impossible unless more axioms are imposed on the quantum logic. Such steps have been taken (Piron, 1964; Zierler, 1961; Varadarajan, 1968) and theorems have been found which represent the propositions of certain types of restricted quantum logics as closed subspaces of a Hilbert space. However, many of the additional axioms do not have convincing physical justification. This point is, of course, arguable. The usual additional axioms are that $\mathcal{P}$ is a complete, atomic, semi-modular lattice.

There is another approach which does bring the Hilbert space forward without imposing additional artificial axioms on the quantum logic $(\mathcal{P}, \mathcal{M})$. This is to adjoin physical structures to $(\mathcal{P}, \mathcal{M})$ such as physical space, position observables and symmetry. After all, in the known physical systems there is always more than just the quantum logic. There is a space in which the system lives, usually some sort of symmetry involved and some kind of distinguished observable such as position. We now briefly explore this approach.

First of all, any physical system concerns a phenomenon that takes place in some kind of physical arena which we call physical space. Mathematically, we shall assume that *physical space* $\mathcal{S}$ is a locally compact Hausdorff space with second countability. (We include these mathematical esoterics for preciseness. For the definitions of these terms see any book on topology or the reader can assume $\mathcal{S} = R^3$ which is general enough for many discussions and which is the prototype of such spaces.) In a concrete physical situation, $\mathcal{S}$ might be R^3, or R^n, or perhaps four-dimensional space–time, or some region in these spaces. Now many of the propositions in $\mathcal{P}$ are concerned with the location of the physical system in $\mathcal{S}$. If such propositions can be verified in the laboratory we call the system *localizable*. We shall now define this term mathematically.

Let $B(\mathcal{S})$ denote the Borel sets in $\mathcal{S}$, and if $E \in B(\mathcal{S})$ let the proposition that the physical system is located in E be denoted by $X(E)$. Thus X is a map from $B(\mathcal{S})$ into $\mathcal{P}$. It is clear that X is an observable based on $\mathcal{S}$ so

(1). $X(\mathcal{S}) = 1$;
(2). If $E \cap F = \varnothing$, then $X(E) \perp X(F)$;
(3). $X(\cup E_i) = \vee X(E_i)$ if $E_i \cap E_j = \varnothing, i \neq j$.

We require that X be maximal in a certain sense. Specifically, let $R(X) \subseteq \mathcal{P}$ denote the range of X and let w be a probability measure with domain $R(X)$. We say that X is *maximal* if every probability measure w on $R(X)$ has a unique extension $\tilde{w} \in \mathcal{M}$. We say that a physical system is *localizable* if there exists a maximal observable (called a *position observable*) $X : B(\mathcal{S}) \to \mathcal{P}$.

There are physical systems that are not localizable. However, as indicated by the work of Jauch and Piron (1967) many of these systems can be handled using a weaker notion of position observable. In this section we shall henceforth only consider localizable systems.

266 Uncertainty Principle and Foundations of Quantum Mechanics

We now consider symmetries. A symmetry may be thought of intuitively as being a transformation that maps the system into another system which is physically identical with the original one except for a relabelling. If $a \in \mathcal{P}$, then after a symmetry transformation we get a new proposition Wa. Thus a symmetry induces a map $W: \mathcal{P} \to \mathcal{P}$. Since W just relabels the propositions, W is a bijection on $\mathcal{P}$ that preserves all the operations on $\mathcal{P}$; that is, W is an automorphism on $\mathcal{P}$. We denote the automorphisms on $\mathcal{P}$ by aut$(\mathcal{P})$ and notice that aut$(\mathcal{P})$ is a group.

Usually symmetries come from transformations on the physical space $\mathcal{S}$. We say that a group G is a *transformation group* on $\mathcal{S}$ if G is a locally compact topological group with second countability for which there exists a continuous map from $G \times \mathcal{S}$ onto $\mathcal{S}$ denoted by $(g, s) \to gs$ such that

(1). $s \to gs$ is a homeomorphism of $\mathcal{S}$ with itself for every $g \in G$;
(2). $g_1(g_2(s)) = (g_1 \cdot g_2)(s)$ for every $g_1, g_2 \in G$;
(3). if $s_1, s_2 \in \mathcal{S}$, there exists $g \in G$ such that $s_1 = gs_2$ (*transitivity*);
(4). $gs = s$ for every $s \in \mathcal{S}$ if and only if $g = e$, the identity element of G (*effectiveness*).

Now if a transformation group is a symmetry for the system it must induce an automorphism group on $\mathcal{P}$. Let $\mathcal{L} = (\mathcal{P}, \mathcal{M})$ be a quantum logic, $\mathcal{S}$ a physical space and X a position observable. A *symmetry group* on $(\mathcal{L}, \mathcal{S}, X)$ is a pair $\mathcal{G} = (G, W)$ where G is a transformation group on $\mathcal{S}$ and W is a group homomorphism $W: G \to$ aut$(\mathcal{P})$ (i.e. $W_{g_1 g_2} = W_{g_1} W_{g_2}$) such that

(W1). $g \to m(W_g(a))$ is continuous for every $m \in \mathcal{M}, a \in \mathcal{P}$;
(W2). $X(gE) = W_g(X(E))$ for every $g \in G, E \in B(\mathcal{S})$ (*covariance*).

Condition (W1) is a natural continuity requirement while (W2) is a covariant condition which gives the natural interpretation that $W_g(X(E))$ is the proposition that the system is located in the set gE. We call W a *projective representation* of G in $\mathcal{P}$. We thus see that $g \to W_g$ gives a generalization of a continuous unitary representation of a group and (W2) generalizes Mackey's imprimitivity relation (Mackey, 1968).

This completes the background for our extended axiomatic structure. We shall call a four-tuple $(\mathcal{L}, \mathcal{S}, X, \mathcal{G})$ where $\mathcal{L} = (\mathcal{P}, \mathcal{M})$ is a quantum logic, $\mathcal{S}$ a physical space, X a position observable and $\mathcal{G} = (G, W)$ a symmetry group, a *quantum system*. We take the viewpoint that the important physical properties of a physical system are described by a quantum system.

Let us consider an example. This is the usual formulation of a spinless, non-relativistic particle moving in one-dimensional space. The set of propositions $\mathcal{P}$ is the lattice of closed subspaces (or equivalently, the lattice of orthogonal projections) of the complex Hilbert space $L^2(R, \mu)$ where μ is Lebesque measure on R. Let $\mathcal{M}$ be the set of pure states of the form $m_f(P) = \langle f, Pf \rangle$ where $f \in L^2(R, \mu), \|f\| = 1$ and $f \geq 0$. Then $\mathcal{M}$ is an order-determining set of states and $(\mathcal{P}, \mathcal{M})$ is a quantum logic. Let G be the group of translations on R; that is, for $\alpha \in R, \lambda \to \lambda + \alpha$ is a transformation group on R.

Let $U_\alpha : L^2(R, \mu) \to L^2(R, \mu)$ be the map $(U_\alpha f)(\lambda) = f(\lambda - \alpha)$. Then U_α is a unitary operator and if we define $W_\alpha P = U_\alpha P U_\alpha^{-1}$ for every $P \in \mathcal{P}$, then $W_\alpha \in \mathrm{aut}\,(\mathcal{P})$ and (G, W) is a symmetry group. The position observable is given by $(X(E)f)(\lambda) = \chi_E(\lambda) f(\lambda)$ where χ_E is the characteristic function of $E \in B(R)$. We now show that X is maximal. Let ν be a probability measure on $R(X)$ and define $\nu_0(E) = \nu(X(E))$, $E \in B(R)$. Then ν_0 is a measure on $B(R)$ that is absolutely continuous relative to μ (i.e. $\mu(E) = 0$ implies $\nu_0(E) = 0$). Hence by the Radon–Nikodym theorem there is a unique $f \in L^1(R, \mu)$, $f \geq 0$ such that $\nu_0(E) = \int_E f \, d\mu$ for every $E \in B(R)$. Let $g = f^{1/2}$ so that

$$\nu_0(E) = \int_E g^2 \, d\mu = \int_R \chi_E g^2 \, d\mu = \langle g, X(E) g \rangle$$

Then $m_g \in \mathcal{M}$ and since $m_g(X(E)) = \nu(X(E))$ for every $E \in B(R)$, we see that m_g is the unique extension of ν in M.

This last example is canonical in a certain respect. We shall show that corresponding to any quantum system $(\mathcal{L}, \mathcal{S}, X, \mathcal{G})$ there is an underlying Hilbert space and constructs similar to those in the above example that mirror much of the axiomatic structure of $(\mathcal{L}, \mathcal{S}, X, \mathcal{G})$.

A σ-finite measure μ on $B(S)$ is *quasi-invariant* relative to G if $\mu(E) = 0$ if and only if $\mu(gE) = 0$ for every $g \in G$.

LEMMA 6. (Gudder, 1973c) Let $(\mathcal{L}, \mathcal{S}, X, \mathcal{G})$ be a quantum system. Then there is a non-zero σ-finite quasi-invariant measure μ on $B(\mathcal{S})$ such that for every $m \in \mathcal{M}$ the measure $E \to m(X(E))$ is absolutely continuous relative to μ.

The space $L^2(F, \mu)$ will serve as the underlying Hilbert space where F is a certain subset of $\mathcal{S}$. Two states m_1 and m_2 are *orthogonal* if there is $a \in \mathcal{P}$ such that $m_1(a) = m_2(a') = 0$. A set of vectors H_0 is said to *generate* a Hilbert space H if the closed linear hull of H_0 is H.

THEOREM 7. (Gudder, 1973c) Let $(\mathcal{L}, \mathcal{S}, X, \mathcal{G})$ be a quantum system. Then there exists $F \in B(\mathcal{S})$ with the following properties: (a) $\mu(F) \neq 0$, (b) if $E \in B(F)$ and $\mu(E) \neq 0$, then $X(E) \neq 0$, (c) there is a one-to-one map $m \to \hat{m}$ from M onto a generating set H_0 in the complex Hilbert space $L^2(F, \mu)$ that preserves orthogonality.

We shall now see that the Hilbert space $L^2(F, \mu)$ derived in Theorem 7 mirrors many of the structural properties of the quantum system $(\mathcal{L}, \mathcal{S}, X, \mathcal{G})$. Let us first consider the symmetry group $\mathcal{G} = (G, W)$. Let V_0 be the complex vector space generated by $\hat{\mathcal{M}} = \{\hat{m} : m \in \mathcal{M}\}$, where $m \to \hat{m}$ is the map given in Theorem 7. Now $W_g \in \mathrm{aut}\,(\mathcal{P})$ can be thought of as a map from $\mathcal{M}$ into $\mathcal{M}$ defined by $(W_g m)(a) = m(W_g(a))$, $m \in \mathcal{M}$, $a \in \mathcal{P}$, $g \in G$.

Then W_g induces a natural transformation $\hat{W}$ on $\hat{\mathcal{M}}$ defined by $\hat{W}_g \hat{m} = (W_g m)^\wedge$. This map is well-defined since $m \to \hat{m}$ is injective. We next extend $\hat{W}_g$ to V_0 by linearity. If $g \in G$ define μ_g by $\mu_g(E) = \mu[g^{-1}(E)]$ for every $E \in B(F)$. Then μ_g is absolutely continuous with respect to μ. Let $d\mu_g/d\mu$ be the Radon–Nikodym derivative.

THEOREM 8. (Gudder, 1973c) The map $g \to \hat{W}_g$ is a continuous unitary representation on V_0 and $(\hat{W}_g f)(\lambda) = f(g^{-1}\lambda)[d\mu_g/d\mu(\lambda)]^{1/2}$ for every $f \in V_0$.

Now $L^2(F, \mu) = \bar{V}_0$ the closure of V_0 and $\hat{W}_g$ can be extended to a unitary transformation on $L^2(F, \mu)$ which we also denote by $\hat{W}_g$. We thus see that the states $\mathcal{M}$ are represented by certain unit vectors $\hat{\mathcal{M}}$ in the Hilbert space $L^2(F, \mu)$ and that the symmetry group $\mathcal{G}$ is represented by a unitary representation $\hat{W}_g$ of G on $L^2(F, \mu)$. We represent X on $L^2(F, \mu)$ by

$$\hat{X}(E) = \text{proj. on the closed span of } \{\hat{m} : m(X(E)) = 1\}$$

THEOREM 9. (Gudder, 1973c) If $(\mathcal{L}, \mathcal{S}, X, \mathcal{G})$ is a quantum system, then $[\hat{X}(E)f](\lambda) = (\chi_E f)(\lambda)$, $f \in L^2(F, \mu)$.

Denote the lattice of all orthogonal projections on $L^2(F, \mu)$ by $\hat{\mathcal{P}}$. We thus see that $\hat{X}$ is a position observable on $\hat{\mathcal{P}}$. Now $\hat{W}_g$ induces an automorphism on $\hat{\mathcal{P}}$ defined by $\hat{W}^g P = \hat{W}_g P \hat{W}_{g^{-1}}$ for every $P \in \hat{\mathcal{P}}$.

THEOREM 10. (Gudder, 1973c) If $(\mathcal{L}, \mathcal{S}, X, \mathcal{G})$ is a quantum system, then $\hat{X}[g(E)] = \hat{W}^g \hat{X}(E)$ for all $g \in G$, $E \in B(F)$ and $(G, \hat{W}^g)$ forms a symmetry group on $\hat{\mathcal{P}}$.

Letting $\hat{\mathcal{L}} = (\hat{\mathcal{P}}, \hat{\mathcal{M}})$ and $\hat{\mathcal{G}} = (G, \hat{W}^g)$ we see that the structure of a quantum system $(\mathcal{L}, \mathcal{S}, X, \mathcal{G})$ is mirrored by the Hilbert space quantum system $(\hat{\mathcal{L}}, \mathcal{S}, \hat{X}, \hat{\mathcal{G}})$.

(C) Strengths and Weaknesses

One of the strengths of the quantum-logic approach is that its axioms are simple and physically justified. This approach has contributed to the understanding of many quantum-mechanical concepts. However, one of its weaknesses is that it is too general for use in concrete problems. The quantum systems studied above are an improvement but the representation of a quantum system $(\mathcal{L}, \mathcal{S}, X, \mathcal{G})$ as a Hilbert space quantum system $(\hat{\mathcal{L}}, \mathcal{S}, \hat{X}, \hat{\mathcal{G}})$ given above is not completely satisfactory for the following two reasons. Except for the propositions in the range of X, there is no isomorphism between $\mathcal{P}$ and $\hat{\mathcal{P}}$ so the propositions in general are not represented by $\hat{\mathcal{P}}$. Second, there is no provision for distinguishing between pure and mixed states since all the states in $\mathcal{M}$ are mapped onto pure states in $\hat{\mathcal{M}}$.

5. THE CONVEXITY APPROACH

In this approach the states are taken as the undefined primitive axiomatic elements. The important property of states, as far as this approach is concerned, is that they are closed under the formation of convex mixtures. Now it is easy to define a convex combination of elements in a linear space. However, the linear space is artificial and devoid of physical meaning for states. One cannot add states or multiply them by scalars to get other states. Only the operation of forming convex combinations of states has meaning. For this reason an abstract definition of convex mixtures is defined that is independent of the concept of linearity. This approach to convexity originated with Stone (1949) and von Neumann and Morgenstern (1944) and later developed by Mielnik (1968, 1969), Ludwig (1968), Davies and Lewis (1970) and others.

(A) Convex Structures

We begin with a framework due to Noll and Cain (1974). Let S be the set of states for a physical system. We would like to define a notion of mixing a finite number of elements of S according to a given recipe. These recipes are described by listing the finite number of elements that are to be mixed together with the proportion of each element. Thus a *recipe* can be thought of as a function $f: S \to [0, 1]$ such that

(1). $f(s) = 0$ except for finitely many s's;
(2). $\sum_{s \in S} f(s) = 1$

If we define the *support* supp f of a function f to be the set on which f does not vanish, then condition (1) is the same as supp f is finite. We define the *simplex* ΔS of S to be the set of recipes on S.

There is a natural map $\delta: S \to \Delta S$ whose values δ_s are given by $\delta_s(t) = 1$ if $t = s$ and $\delta_s(t) = 0$ if $t \neq s$. Thus δ_s is the characteristic function of the singleton set $\{s\}$. Notice that every recipe has the form $f = \sum_{i=1}^{n} \lambda_i \delta_{s_i}$, where $\lambda_i \geq 0, \sum_{i=1}^{n} \lambda_i = 1$. Furthermore, there is a natural map $\Gamma: \Delta\Delta S \to \Delta S$ given by $\Gamma(F) = \sum_{f \in \Delta S} F(f) f$. Finally, with every map $M: \Delta S \to S$ we can associate, in a natural way, a corresponding map $\bar{M}: \Delta\Delta S \to \Delta S$ defined as

$$\bar{M}(F)(s) = \sum \{F(f): f \in \Delta S, M(f) = s\}$$

We say that $M: \Delta S \to S$ is a *mixing operation* for S if M is surjective and satisfies $M \circ \Gamma = M \circ \bar{M}$. A less concise but more illuminating way of writing this last equation is

$$M(\sum \lambda_i f_i) = M(\sum \lambda_i \delta_{M(f_i)}) \qquad (4)$$

when $f_i \in \Delta S, \lambda_i \geq 0, \sum \lambda_i = 1$.

We can interpret M as follows. If $f \in \Delta S$ is a recipe then $M(f)$ is the state resulting from mixing the $s \in S$ in the relative amounts $f(s)$ prescribed by the recipe. Intuitively $M(f)$ means mix the states according to the recipe f. Condition (4) means that if we mix a set of mixtures, we obtain the same result as when we apply each mixture individually and then mix.

It is instructive to see that the usual notion of a mixture in a linear space satisfies (4). Suppose then that S is a real vector space. Then, in this case, a mixing operation $M: \Delta S \to S$ should satisfy $M(\sum \lambda_i \delta_{s_i}) = \sum \lambda_i s_i (\lambda_i \geq 0, \sum \lambda_i = 1)$, or more generally $M(\sum \lambda_i f_i) = \sum \lambda_i M(f_i)$ and $M(\delta_{s_i}) = s_i$. But then

$$M(\sum \lambda_i f_i) = \sum \lambda_i M(f_i) = \sum \lambda_i M(\delta_{M(f_i)}) = M(\sum \lambda_i \delta_{M(f_i)})$$

270 Uncertainty Principle and Foundations of Quantum Mechanics

Let us now return to the general case. It follows from condition (4) that if M is a mixing operation, then $M(f) = M(\delta_{M(f)})$ for every $f \in \Delta S$. Since M is surjective, given $s \in S$ there is $f \in \Delta S$ such that $M(f) = s$. Hence $M(\delta_s) = s$ for every $s \in S$. This is interpreted as meaning that a mixture for a recipe containing one ingredient is identical to that ingredient.

Let M be a mixing operation on S. We define a map from $[0, 1] \times S \times S$ into S, $(\lambda, s, t) \mapsto \langle \lambda, s, t \rangle$ as follows: $\langle \lambda, s, t \rangle = M[\lambda \delta_s + (1 - \lambda) \delta_t]$. We can interpret $\langle \lambda, s, t \rangle$ as a mixture of the states s and t in the ratio $\lambda : (1 - \lambda)$. The following lemma lists the important properties of $\langle \lambda, s, t \rangle$. This lemma is proved by a straightforward application of the definition.

LEMMA 11. If M is a mixing operation, the map $(\lambda, s, t) \mapsto \langle \lambda, s, t \rangle$ satisfies the following conditions:

(M1). $\langle 1, s, t \rangle = s$;
(M2). $\langle \lambda, s, s \rangle = s$;
(M3). $\langle \lambda, s, t \rangle = \langle 1 - \lambda, t, s \rangle$;
(M4). $\langle \lambda, s, \langle \mu, t, v \rangle \rangle = \langle \lambda + (1 - \lambda) \mu, \langle \lambda [\lambda + (1 - \lambda) \mu]^{-1}, s, t \rangle, v \rangle$ whenever $\lambda \mu \neq 0$.

We call a map $(\lambda, s, t) \mapsto \langle \lambda, s, t \rangle$ satisfying (M1)–(M4) a *binary mixing operation*. The next theorem shows that any mixing operation can be obtained by successive applications of a binary mixing operation.

THEOREM 12. (Noll and Cain, 1974) If $(\lambda, s, t) \mapsto \langle \lambda, s, t \rangle$ is a binary mixing operation, then there exists a unique mixing operation M such that $\langle \lambda, s, t \rangle = M[\lambda \delta_s + (1 - \lambda) \delta_t]$. Furthermore M can be obtained by a successive application of $\langle \lambda, s, t \rangle$.

Because of Theorem 12 we can work exclusively with the binary mixing operation $(\lambda, s, t) \mapsto \langle \lambda, s, t \rangle$ and we shall do so in the following. A mixing operation is *distinguishing* if the corresponding binary mixing operation satisfies

(M5). If $\langle \lambda, s, t_1 \rangle = \langle \lambda, s, t_2 \rangle$ for some $s \in S$ and $\lambda \neq 1$, then $t_1 = t_2$.

Distinguishability is a reasonable physical condition which we shall later see is equivalent to having enough observables to distinguish between states. We call a set with a distinguishing mixing operation a *convex structure*. The axiom of this approach is the following.

Axiom. The set of states for a physical system form a convex structure.

The standard example of a convex structure is a real vector space V in which $\langle \lambda, s, t \rangle = \lambda s + (1 - \lambda) t$. The reader can easily check that this gives a convex structure. When we consider convex sets in vector spaces we always assume they are equipped with the above convex structure.

It is convenient to also consider a framework which is much more general than a convex structure. A *convex prestructure* is a set S together with a function $(\lambda, s, t) \mapsto \langle \lambda, s, t \rangle$ from $[0, 1] \times S \times S$ into S. This concept is so general that *any* non-empty set is a convex prestructure. This is because no conditions are placed upon the map $(\lambda, s, t) \mapsto \langle \lambda, s, t \rangle$.

If S_1 and S_2 are convex prestructures, a map $A: S_1 \to S_2$ is *affine* if $A\langle \lambda, s, t\rangle_1 = \langle \lambda, As, At\rangle_2$ for all $\lambda \in [0, 1]$, $s, t \in S_1$. We say that S_1 and S_2 are *isomorphic* if there is an affine bijection from S_1 to S_2. An *affine functional* f is an affine map from a convex prestructure S to the real line R; that is $f(\langle \lambda, s, t\rangle) = \lambda f(s) + (1-\lambda)f(t)$ for all $\lambda \in [0, 1]$, s, t in S. We denote the set of affine functionals on S by S^* and say that S^* is *total* if for any $s, t \in S$ with $s \neq t$ there exists $f \in S^*$ such that $f(s) \neq f(t)$.

Suppose S is a convex prestructure that corresponds to the set of states for some physical system. Since a bounded observable has an expectation in every state, the bounded observables can be thought of as functionals on S. It is also physically reasonable that these functionals are affine. Furthermore, since a state is determined by the expectation values it gives to bounded observables it is reasonable to assume that S^* is total.

THEOREM 13. *A convex prestructure S is a convex structure if and only if S^* is total.*

Proof. For sufficiency it is a simple matter to show that conditions (M1)–(M5) hold if S^* is total. For example, for (M1), since $f(\langle 1, s, t\rangle) = f(s)$ for all $f \in S^*$ we have $\langle 1, s, t\rangle = s$. Necessity will follow from the second representation theorem proved later.

A *convex substructure* of a convex structure S is a subset $S_1 \subseteq S$ which satisfies $\langle \lambda, s, t\rangle \in S_1$ whenever $s, t \in S_1$, $\lambda \in [0, 1]$. A subset $F \subseteq S$ is a *face* if F is a convex substructure and if $\langle \lambda, s, t\rangle \in F$ for some $\lambda \in (0, 1)$ implies $s, t \in F$. An element $s \in S$ is an *extreme point* if $\{s\}$ is a face. Thus s is an extreme point if and only if s is not a mixture of other elements.

(B) Representation Theorems

In this section we give two vector space representation theorems for convex structures. But first we need some definitions concerning convex sets. Let S_0 be a convex subset of a real vector space V (i.e. $x, y \in S_0$ implies $\lambda x + (1-\lambda)y \in S_0$ for every $\lambda \in [0, 1]$). The *hyperplane*, *cone* and *subspace*, respectively, *generated* by S_0 are defined as follows:

$$H(S_0) = \left\{ \sum_{i=1}^{n} \lambda_i x_i : \sum_{i=1}^{n} \lambda_i = 1, x_i \in S_0 \right\}$$

$$K(S_0) = \left\{ \sum_{i=1}^{n} \lambda_i x_i : \lambda_i > 0, x_i \in S_0 \right\}$$

$$V(S_0) = \left\{ \sum_{i=1}^{n} \lambda_i x_i : \lambda_i \in R, x_i \in S_0 \right\}$$

272 Uncertainty Principle and Foundations of Quantum Mechanics

Two vector spaces are *isomorphic* if there exists a linear bijection from one to the other.

We first consider the question of uniqueness of representations.

THEOREM 14. (Uniqueness) Let S be a convex prestructure and let T_1 and T_2 be affine bijections from X onto convex subsets $T_1(S)$, $T_2(S)$ of two real vector spaces V_1 and V_2, respectively. If $0 \notin H(T_1(S))$ and $0 \notin H(T_2(S))$ then $V(T_1(S))$ and $V(T_2(S))$ are isomorphic.

Proof. The function $g = T_2 \circ T_1^{-1}$ is an affine bijection from $T_1(S)$ to $T_2(S)$. We first extend g to $K(T_1(S))$. First, if $y \in K(T_1(S))$ then $y = \sum \lambda_i x_i$, $\lambda_i > 0$, $x_i \in T_1(S)$. Hence $y = \sum_j \lambda_j \sum_i (\lambda_i / \sum_j \lambda_j) x_i = \lambda x$ where $\lambda > 0$ and $x \in T_1(S)$. We now show that the representation $y = \lambda x$ is unique. Indeed, suppose $y = \lambda x = \mu z$ where $\lambda, \mu > 0$ and $x, z \in T_1(S)$. Then if $\lambda \neq \mu$ we have $0 = \lambda x - \mu z = (\lambda - \mu)[\lambda(\lambda - \mu)^{-1} x - \mu(\lambda - \mu)^{-1} z]$. Since $0 \notin H(T_1(S))$ the second factor on the right-hand side is not 0. Hence $\lambda = \mu$ which is a contradiction. Thus $\lambda = \mu$ and hence $x = z$. Define $g(y) = \lambda g(x)$. It is easy to see that the extended $g: K(T_1(S)) \to K(T_2(S))$ is a bijection. The following shows that g is additive on $K(T_1(S))$.

$$g(\lambda x + \mu y) = g\{(\lambda + \mu)[(\lambda + \mu)^{-1} x + \mu(\lambda + \mu)^{-1} y]\}$$
$$= \lambda g(x) + \mu g(y) = g(\lambda x) + g(\mu y)$$

Also, g is homogeneous on $K(T_1(S))$ since

$$g(\lambda(\mu x)) = g(\lambda \mu x) = \lambda \mu g(x) = \lambda g(\mu x)$$

for $\lambda, \mu > 0$, $x \in T_1(S)$. We now extend g to $V(T_1(S))$. Suppose $y \in V(T_1(S))$ and $y = \sum \lambda_i x_i$, $\lambda_i \in R$, $x_i \in T_1(S)$. Then the positive and negative coefficients can be grouped so that $y = \lambda x - \mu z$ where $\lambda, \mu \geq 0$ and $x, z \in T_1(S)$. Thus y has the form $y = u - v$ where $u, v \in K(T_1(S))$. Define g on $V(T_1(S))$ by $g(y) = g(u) - g(v)$. This extended g is well-defined since if $u - v = u_1 - v_1$, $u_1, v_1 \in K(T_1(S))$, then $u + v_1 = u_1 + v$ so by the additivity of g on $K(T_1(S))$ we have $g(u) + g(v_1) = g(u_1) + g(v)$ and $g(u) - g(v) = g(u_1) - g(v_1)$. That $g: V(T_1(S)) \to V(T_2(S))$ is linear and bijective is now easily verified.

We say that a convex prestructure S *is represented* as a convex set S_0 in a real vector space V if there exists an affine bijection $T: S \to S_0$, with $0 \notin H(S_0)$ and $V(S_0) = V$. It follows from the uniqueness theorem that if S is represented as convex sets S_1, S_2 in vector spaces V_1 and V_2, respectively, then V_1 and V_2 are isomorphic. Thus representative convex sets and their vector spaces are unique up to an isomorphism. Furthermore, if $f \in S^*$, then $f \circ T^{-1}$ is an affine functional on S_0 and by a method similar to that used in the proof of Theorem 14, $f \circ T^{-1}$ has a unique extension into a linear functional plus a constant on $V(S_0)$.

THEOREM 15. (First Representation Theorem) A convex prestructure S can be represented as a convex set S_0 if and only if S^* is total.

Proof. Let $F: S \to S_0$ be an affine bijection where S_0 is a convex subset of V. It is well-known that the set of linear functionals V^* on V are total over V.

Restricting the elements of V^* to S_0 we get a total set of affine functionals for S_0. Now if $f \in V^*$, then $f \circ F \in S^*$ so S^* is total. Conversely, suppose S^* is total. For $x \in S$ define $J(x): S^* \to R$ by $J(x)f = f(x)$. Clearly S^* is a vector space under pointwise operations and $J(x) \in S^{**}$ so that $J(S) \subseteq S^{**}$. Now $J(S)$ is a convex set since for $J(x), J(y) \in J(S)$ and $\lambda \in [0, 1]$ we have $f \in S^*$,

$$[\lambda J(x) + (1-\lambda)J(y)]f = \lambda f(x) + (1-\lambda)f(y) = f(\langle \lambda, x, y \rangle) = J(\langle \lambda, x, y \rangle)f$$

so $\lambda J(x) + (1-\lambda)J(y) \in J(S)$. To show J is injective suppose $x \neq y \in S$. Then since S^* is total there is $f \in S^*$ such that $f(x) \neq f(y)$ so $J(x) \neq J(y)$. We now show that $0 \notin H(J(S))$. If $0 \in H(J(S))$ then there exist $\lambda_i \in R$, $x_i \in S$, $i = 1, \ldots, n$, with $\sum \lambda_i = 1$ such that $\sum \lambda_i J(x_i) = 0$. Then $\sum \lambda_i f(x_i) = 0$ for every $f \in S^*$. Letting $f_1 \equiv 1$ we obtain the contradiction $\sum \lambda_i = 0$.

Now let S be a convex structure. If S^* is total the last theorem represented S as a convex set in S^{**}. We now give a different representation which although isomorphic to the last one by the uniqueness theorem, has a form that is useful in many applications. A *cone* is a set $K = \{X, Y, Z, \ldots\}$ on which there is a binary operation $(X, Y) \mapsto X + Y$ and a scalar multiplication $(\lambda, X) \mapsto \lambda X$, $\lambda \in R^+$ (i.e. $\lambda \geq 0$), $X, Y \in K$ satisfying:

(1). $X + Y = Y + X$;
(2). $X + (Y + Z) = (X + Y) + Z$;
(3). if $X + Y = X + Z$, then $Y = Z$;
(4). $\lambda(X + Y) = \lambda X + \lambda Y$;
(5). $(\lambda + \mu)X = \lambda X + \mu X$;
(6). $\lambda(\mu X) = (\lambda \mu)X$;
(7). $1 \cdot X = X$

THEOREM 16. (Second Representation Theorem) *A convex structure S can be represented as a convex set.*

Proof. We first show that S can be extended to a cone. Let $P = \{(\lambda, x): \lambda > 0, x \in S\}$. We define addition and scalar multiplication on P by $(\lambda, x) + (\mu, y) = (\lambda + \mu)\langle \lambda(\lambda + \mu)^{-1}, x, y \rangle$ and $\lambda(\mu, x) = (\lambda\mu, x)$. A straightforward verification using the properties of a convex structure shows that P is a cone. We next show that P can be extended to a vector space. Let $V_0 = \{(X, Y): X, Y \in P\}$. Define the relation $(X, Y) \sim (X', Y')$ if and only if $X + Y' = Y + X'$. This is easily seen to be an equivalence relation on $P \times P$. Denote the equivalence class containing (X, Y) by $[(X, Y)]$ and let $V = \{[(X, Y)]: X, Y \in P\}$. Define addition on V by $[(X, Y)] + [(X', Y')] = [(X + X', Y + Y')]$. To show this operation is well-defined, suppose $(X, Y) \sim (X_1, Y_1)$ and $(X', Y') \sim (X_1', Y_1')$. Then $X + Y_1 = Y + X_1$ and $X' + Y_1' = Y' + X_1'$. Hence $X + X' + Y_1 + Y_1' = Y + Y' + X_1 + X_1'$ and $(X + X', Y + Y') \sim (X_1 + X_1', Y_1 + Y_1')$. Under addition V is an abelian group with zero $[(X, X)]$. Define a scalar multiplication by real numbers as follows. If $\lambda > 0$, then $\lambda[(X, Y)] = [(\lambda X, \lambda Y)]$; if $\lambda = 0$, then $\lambda[(X, Y)] = [(X, X)]$; and if $\lambda < 0$, then $\lambda[(X, Y)] = [(-\lambda Y, -\lambda X)]$. As with addition, this operation is well-defined. It is straightforward to show that V is a vector space. Now define the maps $A : S \to P$ and $B : P \to V$ by $Ax = (1, x)$ and $BX = [(X+Y, Y)]$. The second map is well-defined since $(X+Y, Y) \sim (X+Z, Z)$ for every $Y, Z \in P$. Hence $B \circ A : S \to V$. Now B is additive since

$$B(X+Z) = [(X+Z+Y, Y)] = [(X+Z+2Y, 2Y)]$$
$$= [(X+Y, Y)] + [(Z+Y, Y)] = BX + BZ.$$

Also B is homogeneous since for $\lambda > 0$,

$$B(\lambda X) = [(\lambda X + Y, Y)] = [(\lambda X + \lambda Y, \lambda Y)] = \lambda[(X+Y, Y)] = \lambda BX$$

Furthermore, A is an affine map since

$$A(\langle \lambda, x, y \rangle) = (1, \langle \lambda, x, y \rangle) = (\lambda + (1-\lambda), \langle \lambda, x, y \rangle)$$
$$= (\lambda, x) + (1-\lambda, y) = \lambda(1, x) + (1-\lambda)(1, Y)$$
$$= \lambda Ax + (1-\lambda)Ay$$

It follows that $B \circ A : S \to V$ is affine. It is easily checked that A and B are injective so $B \circ A$ is injective. Also it is clear that $B \circ A(S)$ is convex and that $V[B \circ A(S)] = V$. Finally, suppose $0 \in H[B \circ A(S)]$. Then there exist $X_i, Z_i \in P$ and $\lambda_i \in R$ with $\sum \lambda_i = 1$ such that $\sum \lambda_i [(X_i + Z_i, Z_i)] = 0$. Combining the positive coefficients and negative coefficients, there exist $\lambda, \mu > 0$, $\lambda \neq \mu$, $x, y \in S$, $Z \in P$ such that $[((\lambda, x) + Z, Z)] = [((\mu, y) + Z, Z)]$. This implies that $(\lambda, x) = (\mu, y)$. But then $\lambda = \mu$ which is a contradiction.

A distance can be defined in a very natural way in a convex structure S. If $x, y \in S$, the closeness of x to y can be measured by comparing mixtures $\langle \lambda, x_1, x \rangle$, $\langle \lambda, y_1, y \rangle$ of S. If x and y are very close we would expect to find a mixture containing mostly x equal to a mixture containing mostly y; that is, $\langle \lambda, x_1, x \rangle = \langle \lambda, y_1, y \rangle$ in which λ is very small. Conversely, if $\langle \lambda, x_1, x \rangle = \langle \lambda, y_1, y \rangle$ and λ is small we expect that x and y are close. Thus the parameters λ such that $\langle \lambda, x_1, x \rangle = \langle \lambda, y_1, y \rangle$ give a measure of the closeness of x and y. We thus define a distance function σ as follows:

$$\sigma(x, y) = \inf \{0 \leq \lambda \leq 1 : \langle \lambda, x_1, x \rangle = \langle \lambda, y_1, y \rangle, x_1, y_1 \in S\}$$

Notice that since $\langle \frac{1}{2}, x, y \rangle = \langle \frac{1}{2}, y, x \rangle$ we have $0 \leq \sigma(x, y) \leq \frac{1}{2}$ for all $x, y \in S$. It is sometimes useful to make a change of scale and define the distance function $\rho(x, y) = \sigma(x, y)[1 - \sigma(x, y)]^{-1}$. Then $0 \leq \rho(x, y) \leq 1$ for every $x, y \in S$.

Using a representation of S it is straightforward to show that σ and ρ are metrics. One of the important properties of σ and ρ is that they are invariant under isomorphisms. That is, if $A : S_1 \to S_2$ is an isomorphism, then $\sigma_2(Ax, Ay) = \sigma_1(x, y)$ and $\rho_2(Ax, Ay) = \rho_1(x, y)$ for all $x, y \in S_1$. There is also a relationship between ρ and transition probabilities. One might expect, since $0 \leq \rho(x, y) \leq 1$, that ρ has something to do with probabilities. Specifically if $\rho(x, y)$ is small one might expect the transition probability from x to y to be large while for large $\rho(x, y)$ a transition from x to y would be unlikely. This is indeed the case. In fact, in the classical approach, if ϕ and ψ are unit vectors corresponding to pure states x and y, it can be show that the transition probability $|\langle\phi,\psi\rangle|^2 = 1 - \rho(x,y)^2$. For more details and other results the reader is referred to (Gudder, 1973a, b).

We next briefly show how this approach can be carried further. Let S be the convex structure of states. Then by one of the representation theorems, S can be represented by a convex set S_0 in a real vector space $V = V(S_0)$. The metric ρ on S can be transferred to S_0 giving a metric ρ_0 on S_0. It can be shown that there exists a unique norm $\|\cdot\|$ on V such that $\|x - y\| = \rho_0(x,y)$ for every $x, y \in S_0$. In one interpretation the states are thought of as 'unit beams' and the cone $P = \{(\lambda, x) : \lambda > 0, x \in S\}$ of the second representation theorem is the space of beams. The functional $\tau_0 : P \to R^+$ defined by $\tau_0[(\lambda, x)] = \lambda$ is interpreted as giving the beam intensity. It is easy to see that τ_0 has a unique extension to a linear functional τ on V and that $\tau(X) = \|X\|$ for every $X \in P$. The triple (V, P, τ) is called a *base normed space* and is the basic framework for the operational quantum mechanics of Davies and Lewis (1970).

(C) Strengths and Weaknesses

The strengths and weaknesses of this approach are similar to those of the quantum-logic approach. The axioms are simple and physically motivated. Although the approach has important theoretical uses, its practical utility has not been exploited. An important unsolved problem in this respect is to characterize convex structures that are isomorphic to the set of density operators on a Hilbert space.

REFERENCES

Davies, E. B. and Lewis, J. T. (1970) 'An operational approach to quantum probability', *Commun. Math. Phys.*, **17**, 239–260.

Dirac, P. A. M. (1930) *The Principles of Quantum Mechanics*, Clarendon Press, Oxford.

Emch, G. G. (1972) *Algebraic Methods in Statistical Mechanics and Quantum Field Theory*, Wiley-Interscience, New York.

Gelfand, I. and Naimark, M. A. (1943) 'On the imbedding of normed rings in the ring of operators in Hilbert space', *Mat. Sb.N.S.*, **12** [54], 197–217.

Gleason, A. M. (1957) 'Measures on the closed subspaces of a Hilbert space', *J. Math. Mech.*, **6**, 885–894.

Gudder, S. (1973a) 'State automorphisms in axiomatic quantum mechanics', *Intern. J. Theoret. Phys.*, **7**, 205–211.

Gudder, S. (1973b) 'Convex structures and operational quantum mechanics', *Commun. Math. Phys.*, **29**, 249–264.

Gudder, S. (1973c) 'Quantum logics, physical space, position observables and symmetry', *Rep. Math. Phys.*, **4**, 193–202.

Gudder, S. and Boyce, S. (1970) 'A comparison of the Mackey and Segal models for quantum mechanics', *Intern. J. Theoret. Phys.*, **3**, 7–21.

Haag, R. and Kastler, D. (1964) 'An algebraic approach to quantum field theory,' *J. Math. Phys.*, **5**, 848–861.

Jauch, J. (1968) *Foundations of Quantum Mechanics*, Addison Wesley, Reading, Mass.

Jauch, J. and Piron, C. (1967) 'Generalized localizability', *Helv. Phys. Acta*, **40**, 559–570.

Jordan, P., von Neumann, J. and Wigner, E. (1934) 'On an algebraic generalization of the quantum mechanical formalism', *Ann. Math.*, **35**, 29–64.

Kolmogorov, A. N. (1956) *Foundations of the Theory of Probability*, Chelsea, New York.

Ludwig, G. (1968) 'Attempt of an axiomatic foundation of quantum mechanics and more general theories III', *Commun. Math. Phys.*, **9**, 1–12.

Mackey, G. W. (1963) *The Mathematical Foundations of Quantum Mechanics*, W. A. Benjamin Inc., New York.

Mackey, G. W. (1968) *Induced Representations and Quantum Mechanics*, W. A. Benjamin Inc., New York.

Maczynski, M. J. (1974) 'When the topology of an infinite-dimensional Banach space coincides with a Hilbert space topology', *Studio Math.*, **49**, 149–152.

Mazur, S. and Ulam, S. (1932) 'Sur les transformations isometriques d'espace vectoriels normes', *C.R. Acad. Sci. Paris*, **194**, 946–948.

Mielnik, B. (1968) 'Geometry of quantum states', *Commun. Math. Phys.*, **9**, 55–80.

Mielnik, B. (1969) 'Theory of filters', *Commun. Math. Phys.*, **15**, 1–46.

Noll, W. and Cain, R. N. (1974) 'Convexity, mixing, colors, and quantum mechanics', Preprint: Department of Mathematics, Carnegie-Mellon University, Pittsburgh, Pa.

Piron, C. (1964) 'Axiomatique quantique', *Helv. Phys. Acta*, **37**, 439–468.

Plymen, R. J. (1968) 'C*-algebras and Mackey's axioms', *Commun. Math. Phys.*, **8**, 132–146.

Schatten, R. (1950) *A Theory of Cross-Spaces*, Ann. Math. Studies 26, Princeton University Press, Princeton, N.J.

Segal, I. E. (1947) 'Postulates for general quantum mechanics', *Ann. Math.*, **48**, 930–948.

Sherman, S. (1956) 'On Segal's postulates for general quantum mechanics', *Ann. Math.*, **64**, 593–601.

Sikorski, R. (1949) 'On the inducing homomorphisms by mappings', *Fund. Math.*, **36**, 7–22.

Stone, M. H. (1930) 'Linear transformations in Hilbert space III. Operational methods and group theory', *Proc. Nat. Acad. Sci. U.S.A.*, **16**, 172–175.

Stone, M. H. (1949) 'Postulates for the barycentric calculus', *Ann. Mat. Pura Appl.*, (4) **29**, 25–30.

Varadarajan, V. S. (1962) 'Probability in physics and a theorem on simultaneous observability', *Commun. Pure Appl. Math.*, **15**, 189–217.

Varadarajan, V. S. (1968) *Geometry of Quantum Theory I*, Van Nostrand, Princeton, N.J.

Von Neumann, J. (1931) 'Die Eindeutigkeit der Schrödingerschen Operatoren', *Math. Ann.*, **104**, 570–578.

Von Neumann, J. (1932) *Grundlagen der Quantenmechanik*, Springer, Berlin; english translation by R. T. Beyer, Princeton University Press, Princeton, N.J., 1955.

Von Neumann, J. and Morgenstern, O. (1944) *Theory of Games and Economic Behavior*, Princeton University press, Princeton, N.J.

Wigner, E. P. (1931) *Gruppentheorie und ihre Anwendugn*, Vieweg, Braunschweig; English translation by J. J. Griffin, Academic Press, New York, 1959.

Zierler, N. (1961) 'Axioms for non-relativistic quantum mechanics', *Pac. J. Math.*, **11**, 1151–1169.

14

Intermediate Problems for Eigenvalues in Quantum Theory

WILLIAM STENGER
Ambassador College, Pasadena, U.S.A.

INTRODUCTION

In any general study of quantum theory one sooner or later becomes involved in a discussion of eigenvalue problems. In fact, eigenvalue problems provide not only a link with classical mechanics, but are actually, in a sense, typical of quantum mechanics even in classical problems. For instance, if we consider such classical problems as those of the vibrations of strings, membranes and plates and of the buckling of beams and plates, we immediately see quantum-like phenomena.

The frequencies of vibration and buckling loads occur only at *discrete* numerical values. Modes of vibration and buckling 'jump' from one state to another. As a curiosity, one could say that these classical problems are more *purely* quantum-like than quantum problems in that the phenomenon of a *continuous spectrum* does not occur in classical cases.

In all these problems, whether we consider frequencies of vibration, buckling loads or energy levels, the common ground is, of course, the eigenvalue problem. Therefore, it is not at all surprising that methods, techniques and theoretical results dealing with eigenvalue problems of classical mechanics can be carried over and applied to problems of quantum mechanics.

Weinstein's methods of *intermediate problems* and their variants, to which the present chapter is devoted, are particularly exemplary of this kind of development.

1. DEFINITIONS AND NOTATIONS

Let $\mathfrak{H}$ be a real or complex Hilbert space having the scalar product (u, v) and let H be a self-adjoint linear operator defined on a subspace $\mathfrak{D}$ dense in $\mathfrak{H}$. In problems discussed here, H is bounded below and the lower part of its spectrum consists of a finite or infinite number of isolated eigenvalues $\lambda_1 \leq \lambda_2 \leq \ldots$ each having finite multiplicity. Let λ_∞ denote the lowest point (if any) in the essential spectrum of H. The point λ_∞ could be a non-isolated eigenvalue of finite or infinite multiplicity, an isolated eigenvalue of infinite multiplicity or a spectral point which is not an eigenvalue. There may be point eigenvalues, even isolated point eigenvalues, which are above λ_∞. However, when we enumerate the eigenvalues $\lambda_1, \lambda_2 \ldots$, we mean the isolated eigenvalues that are below λ_∞. We shall denote by $u_1, u_2 \ldots$ a corresponding orthonormal sequence of eigenfunctions.

The selection of operators having these properties is motivated by the fact that many problems in classical and quantum mechanics involve operators of this type. Since the Schrödinger operators for hydrogen, helium, etc., have such spectra, we call this type of operator 'type-$\mathscr{S}$'.

2. INTERMEDIATE PROBLEMS

It is possible to solve exactly for the eigenvalues of only a few operators of type-$\mathscr{S}$. In most cases one must devise methods of estimating the eigenvalues. Since an approximation is useless without also specifying its accuracy, the best approximations for eigenvalues have come by means of complementary methods, that is methods which approximate the eigenvalues from above and below.

The *Rayleigh–Ritz Method* has been used widely to obtain approximations from above (*upper bounds*) to the eigenvalues of operators of type-$\mathscr{S}$. This method is fairly straightforward to apply and with the advent of high-speed computers has given results of remarkable accuracy. For a detailed discussion of the *Rayleigh–Ritz Method*, see the books of Gould (1966) and Weinstein and Stenger (1972).

The problem of finding *lower bounds* to eigenvalues is intrinsically much more difficult. The first major breakthrough in this area was made by Alexander Weinstein (1935, 1937) who introduced intermediate problems to determine lower bounds to the buckling load and frequencies of vibration of a clamped square plate. In solving these problems Weinstein used classical techniques involving natural boundary conditions and Lagrange multipliers. Soon the problems were reformulated in the language of Hilbert space.

Without going into detail, the basic idea of intermediate problems is as follows. Given an eigenvalue problem

$$Hu = \lambda u$$

we first find another eigenvalue problem, called the *base problem*,

$$Au = \lambda u$$

whose eigenvalues are all lower than those of H. We then build a sequence of *intermediate problems* depending on a finite number of functions which link the base problem to the given problem and whose eigenvalues are intermediate between those of H and those of A. Finally, we must solve for the eigenvalues of the intermediate problems, thereby obtaining lower bounds to the eigenvalues of H.

In the problems solved by Weinstein the given problem turned out to be of the form

$$Hu = Au - PAu = \lambda u, \qquad Pu = 0$$

where P is the orthogonal projection operator onto a subspace $\mathfrak{P}$ of $\mathfrak{H}$. The base problem here is $Au = \lambda u$. If we select a finite number of functions $p_1, p_2, \ldots, p_n$ from $\mathfrak{P}$ and let P_n be the orthogonal projection operator onto the subspace spanned by the functions p_i, $i = 1, 2, \ldots, n$ we can formulate the nth *intermediate problem*

$$Au - P_n Au = \lambda u, \qquad P_n u = 0$$

which has eigenvalues intermediate between those of A and of H. This is called an intermediate problem of the *first*-type. The eigenvalues of the intermediate problems are obtained from the *Weinstein determinant*

$$W(\lambda) = \det\{(R_\lambda p_i, p_k)\}, \qquad i, k = 1, 2, \ldots, n$$

where R_λ is the resolvent operator of A, i.e. $R_\lambda = (A - \lambda I)^{-1}$.

If no special assumptions are made on the choice of functions p_i, it is possible that an eigenvalue of the base problem is also an eigenvalue of the intermediate problem. We call such eigenvalues *persistent*. Since the determinant $W(\lambda)$ may be singular at a persistent eigenvalue, in numerical applications a so-called 'big' Weinstein determinant is used which avoids possible singularities. An analogous situation occurs in problems of the second-type discussed later, see Weinstein and Stenger (1972) for details.

Intermediate problems of the first type have been applied numerically to problems of classical mechanics and have also had theoretical applications, some of which are related to quantum theory, see for instance Stenger (1968). A complete discussion of the numerical and theoretical applications of the first type of intermediate problems is given in the book of Weinstein and Stenger (1972).

A *second*-type of intermediate problems was introduced by Aronszajn (1951). The basic pattern here is the same as in intermediate problems of the first-type, although the form of the problems is somewhat different. In particular, the *given* problem admits the decomposition

$$Hu = Au + Bu = \lambda u$$

where A is of type-$\mathscr{S}$ and B is positive. The *base* problem is $Au = \lambda u$ and the nth intermediate problem is given by

$$Au + \sum_{i=1}^{n} \sum_{j=1}^{n} (u, Bp_i) \beta_{ij} Bp_j = \lambda u$$

where $\{\beta_{ij}\}$ is the inverse matrix of $\{(Bp_i, p_j)\}$. For a suitable choice of α_j and q_j

the intermediate problem can be written in the more general, yet simpler, form

$$Au + \sum_{j=1}^{n} \alpha_j(u, q_j)q_j = \lambda u$$

The eigenvalues of the intermediate problems are obtained from the Weinstein determinant

$$V(\lambda) = \det\{\delta_{ik} + \alpha_i(R_\lambda q_i, q_k)\}, \qquad i, k = 1, 2, \ldots, n$$

This determinant has also been called the *modified Weinstein determinant* and the *Weinstein–Aronszajn determinant*.

Up to now all numerical applications of intermediate problems to quantum theory have involved problems of the second-type or variants of problems of the second-type, as is illustrated in subsequent sections.

It should be mentioned here that while intermediate problems are not a part of perturbation theory, the solution of problems of the second-type has led to a number of contributions to perturbation theory, see for instance Kuroda (1961), Kato (1966), Stenger (1969), Weinstein and Stenger (1972) and Weinstein (1974). Several attempts have also been made to 'reduce' intermediate problems of the first-type to those of the second-type, e.g. Kuroda (1961), Fichera (1965) and Kato (1966). While such a reduction can be made in certain cases under severe limitations, the result in every such case actually leads to a more complicated problem than the original, see Stenger (1970a, b) and Weinstein and Stenger (1972).

3. BAZLEY'S APPLICATION TO THE HELIUM ATOM

The first application of intermediate problems to quantum theory was given by Bazley (1959, 1960, 1961) who was then joined by Fox. Their collaboration, as well as individual research, produced many significant numerical and theoretical results, e.g. Bazley and Fox (1961a, b; 1962a, b, c; 1963a, b; 1964; 1966a, b, c), Fox (1972). A more complete bibliography, a survey of their work and tables of numerical values are given in Weinstein and Stenger (1972).

Bazley first considered the problem of estimating the eigenvalues of the Hamiltonian operator for the helium atom. We now give an overview of the application to helium, omitting details which may be found in the original papers and book cited above.

If we neglect nuclear motion, relativistic effects and the influence of spin and denote by (x_1, y_1, z_1) and (x_2, y_2, z_2) the coordinates of the two electrons, the Schrödinger equation for helium is

$$Hu = -\tfrac{1}{2}\Delta_1 u - \tfrac{1}{2}\Delta_2 u - (2/r_1)u - (2/r_2)u + (1/r_{12})u = \lambda u$$

where Δ_i is the Laplacian in the coordinates (x_i, y_i, z_i).

$$r_i = (x_i^2 + y_i^2 + z_i^2)^{\tfrac{1}{2}}, \qquad i = 1, 2$$

$$r_{12} = [(x_2 - x_1)^2 + (y_2 - y_1)^2 + (z_2 - z_1)^2]^{\tfrac{1}{2}}$$

While the domain of definition of H from the point of view of the physicist was historically only vaguely defined, Kato (1951a, b) considered the operator H on the Hilbert space of square-integrable functions (i.e., $\mathscr{L}^2$ space) over six-dimensional coordinate space. He showed that H admits there a unique self-adjoint extension, that is, that H is essentially self-adjoint. In other words he proved that the closure of H is self-adjoint, see also Kato (1966). Of course, the closure of H is no longer a differential operator in the usual sense, but it reduces to the differential operator for sufficiently regular functions. If we wanted to be notationally strict, we should use different symbols for the formal differential operator and the self-adjoint extension in Hilbert space. However, for the sake of this exposition we avoid encumbering the notation and use the same symbol to denote the formal operator and its corresponding Hilbert space operator.

It is by no means a foregone conclusion that H is of type-$\mathscr{S}$. It is therefore significant that Kato (1951a, b) showed that the spectrum of H begins with isolated eigenvalues, each of finite multiplicity.

Following the general pattern of intermediate problems, we first find a suitable base problem. In the present case the base problem is

$$Au = -\tfrac{1}{2}\Delta_1 u - \tfrac{1}{2}\Delta_2 u - (2/r_1)u - (2/r_2)u = \lambda u$$

This operator A is the Hamiltonian of a system composed of two independent hydrogen-like atoms and admits a unique self-adjoint extension having the same domain as H. The eigenvalues and eigenfunctions of A are well known, see Kemble (1958). The eigenvalues are given by

$$-2[(1/n_1^2)+(1/n_2^2)], \qquad n_1, n_2 = 1, 2, \ldots$$

with multiplicities $n_1^2 n_2^2$, and the corresponding eigenfunctions are products of hydrogenic wave functions. Since the continuous spectrum of A consists of the interval $[-2, \infty)$, the lower part of the spectrum begins with isolated eigenvalues, that is to say, A is also of type-$\mathscr{S}$.

If we now decompose the given operator H as $H = A + B$, where B is the non-negative operator given by

$$Bu = (1/r_{12})u$$

we are in a position to form intermediate problems of the second-type.

At this point one could attempt to solve intermediate problems with arbitrary functions p_i. Such an approach, however, would be fruitless since the resulting intermediate problems would not lend themselves to numerical solutions. In order to overcome this difficulty, Bazley introduced a *special choice* of functions which led to an algebraic problem, readily solvable by using computers. The special choice of Bazley in certain respects parallels the distinguished choice used earlier by Weinstein in problems of the first-type.

In order to form the special choice, we let $u_i^{(0)}$ denote the (known) eigenfunctions of the base problem $Au = \lambda u$. We choose vectors p_i such that

$$Bp_i = u_i^{(0)}, \qquad i = 1, 2, \ldots, n$$

In this way, the nth intermediate problem becomes

$$Au + \sum_{i=1}^{n} \sum_{j=1}^{n} \beta_{ij}(u, u_i^{(0)})u_j^{(0)} = \lambda u$$

where $\{\beta_{ij}\}$ is matrix inverse to $\{(Bp_i, p_j)\}$.

In the case of helium the spectrum of A begins with isolated eigenvalues

$$\lambda_k^{(0)} = -2[1 + (1/k^2)], \qquad k = 1, 2, \ldots$$

having eigenfunctions

$$u_1^{(0)} = -2(1/4\pi)R_{10}(r_1)R_{10}(r_2)$$
$$u_k^{(0)} = (1/\sqrt{24\pi})[R_{10}(r_1)R_{k0}(r_2) + R_{10}(r_2)R_{k0}(r_1)], \qquad k = 2, 3, \ldots$$

where the elements R_{k0} are the normalized hydrogen radial wave functions.

By observing that B is easily invertible and yields the special choice $p_i = r_{12}u_i^{(0)}$ ($i = 1, 2, 3$), Bazley solved the third intermediate problem and obtained the lower bounds $-3.063_7 \leq E(1^1S)$ and $-2.165_5 \leq E(2^1S)$ for the S-states of parahelium.

Bazley obtained an improved lower bound for $E(1^1S)$ by using the lower bound for $E(2^1S)$ in *Temple's formula*. This lower bound combined with the Rayleigh–Ritz upper bound computed by Kinoshita (1957) gives the quite accurate estimate $-2.9037474 \leq E(1^1S) \leq -2.9037237$.

While we have concentrated our attention here on the first and second eigenvalues, it should be noted that intermediate problems may be used to obtain lower bounds to an arbitrary number of eigenvalues at the lower part of the spectrum.

4. TRUNCATION OF THE BASE OPERATOR

In many numerical applications, if we take the most obvious or natural base problem and attempt to solve intermediate problems relative to that base problem, we are confronted with transcendental equations which cannot be readily solved by numerical means. However, these computational difficulties may be circumvented by the following *truncation* of the base operator.

The idea of truncating the base operator was introduced by Weinberger (1959) in problems of the first-type and was later developed by Bazley and Fox (1961b) in problems of the second-type where it was successfully applied to problems of quantum theory.

We begin by considering the spectral representation of the original base operator A, namely

$$Au = \sum_i \lambda_i^{(0)}(u, u_i^{(0)})u_i^{(0)} + \int_{\lambda_\infty - 0}^{\infty} \lambda \, dE_\lambda u$$

where the sigma may denote a finite sum or an infinite series. For a fixed positive integer N, we define the *truncation operator of order N* by

$$T_N u = \sum_{i=1}^{N} \lambda_i^{(0)}(u, u_i^{(0)})u_i^{(0)} + \lambda_{N+1} \int_{\lambda_{N+1} - 0}^{\infty} dE_\lambda u$$

We assume without loss of generality that $\lambda_N^{(0)} < \lambda_{N+1}^{(0)}$. The truncation operator is generally simpler than the original base operator since it consists of a negative-semidefinite operator of finite rank plus a multiple of the identity. Such an operator has only a finite number of distinct eigenvalues and no continuous spectrum.

The advantage gained by using the truncation operator is that the resolvent of T_N has the form

$$R_\lambda(T_N)p = \sum_{i=1}^{N} \frac{(p, u_i^{(0)})u_i^{(0)}}{\lambda_i^{(0)} - \lambda} + \frac{1}{\lambda_{N+1}^{(0)} - \lambda} \left[p - \sum_{i=1}^{n} (p, u_i^{(0)})u_i^{(0)} \right]$$

Therefore, in the intermediate problems the resulting Weinstein determinant is a rational function instead of a (generally) transcendental function, thus reducing the difficulty of numerically determining the eigenvalues of the intermediate problems.

Bazley and Fox (1961a) applied the method of truncation to the helium atom. Even with a truncation of order two and only the second intermediate problem they obtained an improvement of the lower bounds Bazley (1961) obtained by the special choice.

In this case the truncated base problem is given by

$$T_2 u = \lambda_1^{(0)}(u, u_1^{(0)})u_1 + \lambda_2^{(0)}(u, u_2^{(0)})u_2^{(0)}$$
$$+ \lambda_3^{(0)}[u - (u, u_1^{(0)})u_1^{(0)} - (u, u_2^{(0)})u_2^{(0)}]$$
$$= \lambda u$$

where $\lambda_1^{(0)}$, $\lambda_2^{(0)}$, $\lambda_3^{(0)}$, $u_1^{(0)}$ and $u_2^{(0)}$ are as given in Section 3. Letting $Bu = (1/r_{12})u$ as before and choosing functions

$$p_1 = [(1.5)^3/\pi] e^{-1.5}(r_1 + r_2)$$
$$p_2 = [5\sqrt{5})/\pi] r_{12} \exp[-(5)^{\frac{1}{2}}(r_1 + r_2)]$$

Bazley and Fox were able to form an intermediate problem whose eigenvalues could be computed by hand.

Another problem to which Bazley and Fox applied truncation was the radial Schrödinger equation. In order to solve this problem the given eigenvalue

problem
$$-d^2\psi/dx^2 - z[(1-e^{-\alpha x})/x]\psi = E\psi$$
is transformed into another eigenvalue problem in the following way.

While ordinarily one would consider E as the eigenvalue, here we fix the energy E and take the charge z as the eigenvalue. The numerical results may then be inverted to give the energy eigenvalues E. The reason for taking z as the eigenvalue is that the resulting base problem has a pure point spectrum.

We put $E = -k^2$ and introduce the transformations
$$t = 2kx, \quad u(t) = t^{-\frac{1}{2}}\psi(t)$$
to obtain
$$-\frac{d}{dt}\left(t\frac{du}{dt}\right) + \frac{t^2+1}{4t}u = \lambda(1-e^{-\alpha t/2k})u$$
where $\lambda = z/2k$. We now have an eigenvalue problem of the form
$$Au = \lambda(1-B)u$$
where
$$Au = -\frac{d}{dt}\left(t\frac{du}{dt}\right) + \frac{t^2+1}{4t}u$$
and
$$Bu = e^{-\alpha t/2k}u$$

We note that the base operator A has known eigenvalues,
$$\lambda_i^{(0)} = i, \quad i = 1, 2, \ldots$$
and normalized eigenfunctions
$$u_i^{(0)} = (t^{\frac{1}{2}}/i!i^{\frac{1}{2}})L_i'(t)\,e^{-t/2}, \quad i = 1, 2, \ldots$$
where L_i' is the first derivative of the ith Laguerre polynomial. The given problem has a pure point spectrum $\lambda_1 \leq \lambda_2 \leq \ldots$ diverging to infinity and satisfying
$$\lambda_i^{(0)} \leq \lambda_i, \quad i = 1, 2, \ldots$$
Here the intermediate problems are of the form
$$T_N u = \lambda(I - BP^n)u$$
where T_N is the truncation operator previously defined and P^n denotes a projection on functions $p_1, p_2, \ldots, p_n$, orthogonal with respect to the inner product $[u, v] = (u, Bv)$. In solving this problem Bazley and Fox put $p_j = u_j^{(0)}$ ($j = 1, 2, \ldots, n$) and obtained the eigenvalues as solutions of an algebraic system.

For this example, the value of α was fixed and the substitution $k = \alpha/2$ was made so that $E = -\alpha^2/4$. Upper bounds obtained by solving a fourth-order Rayleigh–Ritz problem based on the trial functions $u_1^{(0)}$, $u_2^{(0)}$, $u_3^{(0)}$ and $u_4^{(0)}$ together with the lower bounds obtained from the intermediate problems provided the estimates:

$$1.2587\alpha \leq z_1 \leq 1.2590\alpha$$

$$2.3944\alpha \leq z_2 \leq 2.4164\alpha$$

$$3.4207\alpha \leq z_3 \leq 3.5576\alpha$$

5. AN ANHARMONIC OSCILLATOR

Another variant in intermediate problems of the second-type namely, the *generalized special choice*, was used by Bazley and Fox (1961a) in estimating the eigenvalues of an anharmonic oscillator.

Here the differential equation is given by

$$-u'' + x^2 u + \varepsilon x^4 u = \lambda u, \quad -\infty < x < \infty$$

where $\varepsilon > 0$. Restricting our discussion to the even symmetry class, the base problem

$$Au = -u'' + x^2 u = \lambda u$$

has well-known eigenvalues

$$\lambda_i^{(0)} = 4i - 3, \quad i = 1, 2, \ldots$$

The corresponding eigenfunctions are the linear oscillation eigenfunctions

$$u_i^{(0)} = C_i \exp(-x^2/2) H_{2i-2}(x), \quad i = 1, 2, \ldots$$

Here $C_i = 2^{1-i}[(2i-2)!]^{-\frac{1}{2}} \pi^{-\frac{1}{4}}$ and H_j is the jth Hermite polynomial.

Letting B be the operator defined by

$$Bu = \varepsilon x^4 u$$

we once again have the given problem in the form

$$Au + Bu = \lambda u$$

We recall from Section 3 that a special choice of functions p_i is given by $Bp_i = u_i^{(0)}$. In this problem, however, Bazley and Fox introduced a generalized special choice given by

$$Bp_i = \sum_{j=1}^{m} \beta_{ij} u_i^{(0)}, \quad i = 1, 2, \ldots, n$$

It turns out that by putting $p_i = u_i^{(0)}$ and using a recurrence relation for Hermite polynomials that the symmetric matrix $\{\beta_{ij}\}$ is readily obtained. The eigenvalues of the intermediate problems are then computable as the roots of a linear system.

For various values of ε the intermediate problems yielded lower bounds to the first five eigenvalues, which complemented by the Rayleigh–Ritz upper bounds (also given by Bazley and Fox), demonstrated the accuracy of the method. In fact, when compared graphically, the results for the first eigenvalue computed by intermediate problems and computed by perturbation theory show the overwhelming superiority of intermediate problems over perturbation theory. The Rayleigh–Ritz upper bounds and intermediate lower bounds were actually indistinguishable on the graph used, while even for quite small values of ε the perturbation theory values were not even close.

6. FOX'S APPLICATION TO THE LITHIUM ATOM

One of the more significant advances in the applicability of intermediate problems to quantum mechanics was recently given by Fox (1972) who introduced a method of constructing intermediate operators (called *comparison* operators in Fox's terminology) which make it possible to compute lower bounds for the eigenvalues of the Schrödinger operators for atoms and ions having three or more electrons.

The basic pattern of given problem, base problem and intermediate problems, is also in Fox's work. However, the actual form of the intermediate problems is new and fundamentally different from previously used intermediate problems of the first- and second-types.

The difficulty in dealing with the Schrödinger equation of atoms more complicated than helium is that the lowest point in the essential spectrum of the base operator lies below (or very close to) the first eigenvalue of the given operator. In intermediate problems of the second-type the intermediate problem is formed by adding an operator of finite rank to the base operator. Since the essential spectrum of the base operator is invariant under the addition of any compact operator, the eigenvalues of the intermediate problems would not provide meaningful numerical results.

In order to overcome this difficulty Fox constructs intermediate problems by adding operators to the base operator which provide, even though these operators are non-compact, intermediate problems whose eigenvalues can be numerically determined. In order to accomplish this Fox used a technique of separation of variables. Moreover, he had to introduce and develop important results in the spectral theory for the separation of variables in Hilbert space, see Fox (1968, 1975).

Let us now illustrate the general concepts of Fox's method. We again suppose that the given problem is of the form

$$Hu = Au + Bu = \lambda u$$

Moreover, we assume that A can be resolved by elementary separation of variables and that the separation of variables used for A also allows B to be written as a certain sum relative to this separation of variables. A complete discussion of what such a decomposition involves would require us to go into some detail regarding tensor products of Hilbert spaces, see Fox (1968, 1972, 1975).

Instead, for the purpose of the present chapter, we consider the specific problem of the Schrödinger equation for the non-relativistic fixed-nucleus model for the lithium atom without spin interaction. Here the *given problem* is

$$Hu = -\tfrac{1}{2}\Delta u - (3/r_1)u - (3/r_2)u$$
$$- (3/r_3)u + (1/r_{12})u + (1/r_{13})u + (1/r_{23})u$$
$$= \lambda u$$

where Δ is the nine-dimensional Laplacian, r_i is the (Euclidean) distance from the nucleus to the ith electron ($i = 1, 2, 3$) and r_{ij} is the distance between the ith electron and the jth electron.

The *base problem*

$$Au = -\tfrac{1}{2}\Delta u - (3/r_1)u - (3/r_2)u - (3/r_3)u = \lambda u$$

factors (separation of variables) into three resolvable hydrogen-like operators, see Kemble (1958). In order to decompose B we consider the nine-dimensional coordinate system $(x_1, y_1, z_1, x_2, y_2, z_2, x_3, y_3, z_3)$ where (x_i, y_i, z_i) gives the position of the ith electron ($i = 1, 2, 3$). We let $\mathfrak{H}$ denote the $\mathscr{L}^2$ space of functions defined on $(x_1, y_1, z_1, x_2, y_2, z_2, x_3, y_3, z_3)$, let $\mathfrak{H}_{ij}$ denote the $\mathscr{L}^2$ space of functions defined on $(x_i, y_i, z_i, x_j, y_j, z_j)$, and let $\mathfrak{H}_i$ denote the $\mathscr{L}^2$ space of functions defined on (x_i, y_i, z_i).

If we now define B_{12} to be multiplication by r_{12}^{-1} in $\mathfrak{H}_{12}$ and let I_3 be the identity operator on $\mathfrak{H}_3$, we can form a tensor product operator $\hat{B}_{12} = B_{12} \times I_3$ which gives multiplication by r_{12}^{-1} in $\mathfrak{H}$.

The operators $\hat{B}_{13}$ and $\hat{B}_{23}$ may be formed in an analogous manner. Now the operator B can be decomposed into

$$B = \hat{B}_{12} + \hat{B}_{13} + \hat{B}_{23}$$

which is the decomposition necessary for Fox to form intermediate problems and apply his method.

Instead of approximating B by operators of finite rank, as in intermediate problems of the second type, here one approximates the operators B_{ij} by operators of finite rank, say B_{ij}^n. While B_{12}^n is an approximation to B_{12} of finite rank, the tensor product operator $\hat{B}_{12}^n = B_{12}^n \times I_3$ is a *non-compact* approximation to $\hat{B}_{12}$. Similar non-compact approximations to $\hat{B}_{13}$ and $\hat{B}_{23}$ may be formed, say $\hat{B}_{13}^n$ and $\hat{B}_{23}^n$. The intermediate operator which is then given by

$$A + \hat{B}_{12}^n + \hat{B}_{13}^n + \hat{B}_{23}^n$$

consists of the base operator A plus non-compact operators. This means that the essential spectrum may be displaced and lower bounds obtained.

It should be noted that once the decomposition of B is achieved the intermediate operators are formed in ways similar to those in problems of the second type, that is, by using special choices, generalized special choices and truncation.

These methods have been applied by Fox and Sigillito (1972a, b, c) to obtain bounds for the energy levels of radial lithium. The radial model is a simplification of the usual fixed-nucleus non-relativistic model based on the assumption that the electron distributions depend on the distances of the electrons from the nucleus only and not on the angular variables. This simplified model was used to test the methods numerically while avoiding the complexities of angular momenta.

The Hamiltonian for radial lithium is given by

$$H = -\sum_{i=1}^{3} (\Delta_{r_i}^2/2 + 3/r_i) + \sum_{i<j}^{3} 1/\rho_{ij}$$

where $\Delta_{r_i}^2$ is the radial Laplacian

$$\frac{1}{r_i^2} \frac{\partial}{\partial r_i} \left(r_i^2 \frac{\partial}{\partial r_i} \right)$$

and $\rho_{ij} = \max[r_i, r_j]$. The operator acts on functions of the three radial distances that are square integrable with respect to the weight function $r_1^2 r_2^2 r_3^2$.

Here the base operator is the sum of three resolvable one-electron hydrogenic Hamiltonians and the intermediate operators are formed by treating each pairwise coupling $1/\rho_{ij}$ separately as a two-electron operator. Then, the resulting intermediate problems can be solved numerically by diagonalizing Hermitian matrices.

For total spin S equal to $\frac{1}{2}$ and $\frac{3}{2}$ upper bounds computed by the Rayleigh–Ritz method together with lower bounds obtained from the eigenvalues of the intermediate operators were given by Fox and Sigillito (1972b) and are reproduced in Table 1.

Table 1. Bounds for energies of radial lithium

$S = \frac{1}{2}$	$S = \frac{3}{2}$
$-7.620 \le \lambda_1 \le -7.488$	$-5.220 \le \lambda_1 \le -5.204$
$-7.493 \le \lambda_2 \le -7.324$	$-5.169 \le \lambda_2 \le -5.149$
$-7.457 \le \lambda_3 \le -7.275$	$-5.160 \le \lambda_3 \le -5.170$
.	.
.	.
.	.
$-7.418 \le \lambda_* \le -7.252$	$-5.123 \le \lambda_* \le -5.109$

The bounds were subsequently improved by Fox and Sigillito (1972c). In particular, the lower bound for the first point in the essential spectrum for $S = \frac{1}{2}$ was increased to -7.294. This result, together with the lower bound -5.123

given in the table, shows that the method is indeed successful in displacing the essential spectrum, since the lowest points in the essential spectrum of the corresponding base operators are -9 and -5.625, respectively.

Finally, we would like to mention that a similar approach to constructing intermediate operators for the lithium atom was published by Reid (1972). However, Reid's contribution appears to be purely formal and does not touch on the subtleties of separation of variables in Hilbert space and the properties of the spectra of operators acting on tensor products of Hilbert spaces. The contributions of Fox to the spectral theory of such operators, on the other hand, are a necessary and major part of the application of intermediate problems to lithium and other atoms.

7. CONCLUDING REMARKS

In the brief exposition of the present chapter we have given an overview of the applicability of the methods of intermediate problems to quantum theory. As a result of our attempt to emphasize what we feel to be the highlights in this regard, we have necessarily omitted many contributions to intermediate problems for eigenvalues, which are important and interesting in their own right. For instance, we were not able to go into detail here about the work of Löwdin and his collaborators which includes applications of intermediate problems and closely related methods to problems of quantum chemistry. A fairly complete bibliography of their work may be found by referring to Löwdin, (1965, 1968), Stenger (1974) and Weinstein and Stenger (1972). On the other hand, we did devote a little more space to recent developments by Fox in applying intermediate problems to lithium, since the latter results appeared after the publication of Weinstein and Stenger (1972) and could not be included there.

Anyone interested in more details about solving intermediate problems of the first- and second-types, the applications of intermediate problems to classical mechanics, the relationships and various inequalities for eigenvalues and results in functional analysis connected with intermediate problems, is referred to the books Gould (1966) and Weinstein and Stenger (1972), to the large number of primary references cited in these books, and the more recent papers given in the references here.

REFERENCES

Aronszajn, N. (1951) 'Approximation methods for eigenvalues of completely continuous symmetric operators,' *Proc. Symp. Spectral Theory and Differential Problems*, Stillwater, Oklahoma, pp. 179–202.

Bazley, N. W. (1959) 'Lower bounds for eigenvalues with application to the helium atom,' *Proc. Nat. Acad. Sci. U.S.A.*, **45**, 850–853.

Bazley, N. W. (1960) 'Lower bounds for eigenvalues with application to the helium atom', *Phys. Rev.*, **129**, 144–149.

Bazley, N. W. (1961) 'Lower bounds for eigenvalues', *J. Math. Mech.*, **10**, 289–308.

Bazley, N. W. and Fox, D. W. (1961a) 'Lower bounds for eigenvalues of Schrödinger's equation', *Phys. Rev.*, **124**, 483–492.

Bazley, N. W. and Fox, D. W. (1961b) 'Truncations in the method of intermediate problems for lower bounds to eigenvalues', *J. Res. Nat. Bur. Std. Sec. B*, **65**, 105–111.

Bazley, N. W. and Fox, D. W. (1962a) 'Error bounds for eigenvectors of self-adjoint operators', *J. Res. Nat. Bur. Std. Sec. B.*, **66**, 1–4.

Bazley, N. W. and Fox, D. W. (1962b) 'A procedure for estimating eigenvalues', *J. Math. Phys.*, **3**, 469–471.

Bazley, N. W. and Fox, D. W. (1962c) 'Lower bounds to eigenvalues using operator decompositions of the form B^*B', *Arch. Rational Mech. Anal.*, **10**, 352–360.

Bazley, N. W. and Fox, D. W. (1963a) 'Error bounds for expectation values', *Rev. Mod. Phys.*, **35**, 712–715.

Bazley, N. W. and Fox, D. W. (1963b) 'Lower bounds for energy levels of molecular systems', *J. Math. Phys.*, **4**, 1147–1153.

Bazley, N. W. and Fox, D. W. (1964) 'Improvement of bounds to eigenvalues of operators of the form T^*T', *J. Res. Nat. Bur. Std. Sec. B.*, **68**, 173–183.

Bazley, N. W. and Fox, D. W. (1966a) 'Methods for lower bounds to frequencies of continuous elastic systems', *Z. Angew. Math. Phys.*, **17**, 1–37.

Bazley, N. W. and Fox, D. W. (1966b) 'Error bounds for approximations to expectation values of unbounded operators', *J. Math. Phys.*, **7**, 413–416.

Bazley, N. W. and Fox, D. W. (1966c) 'Comparison operators for lower bounds to eigenvalues', *J. Reine Angew. Math.*, **223**, 142–149.

Fichera, G. (1965) *Linear Elliptic Differential Systems and Eigenvalue Problems* (Lecture Notes in Mathematics), Springer, New York.

Fox, D. W. (1968) *Separation of variables and spectral theory for self-adjoint operators in Hilbert space*, Informal Report, Applied Mathematics Group, Applied Physics Laboratory, The Johns Hopkins University, Silver Spring, Maryland.

Fox, D. W. (1972) 'Lower bounds for eigenvalues with displacement of essential spectra', *Siam J. Math. Anal.*, **3**, 617–624.

Fox, D. W. (1975) 'Spectral measures and separation of variables', *J. Res. Nat. Bur. Std.*, (to appear).

Fox, D. W. and Sigillito, V. G. (1972a) 'Lower and upper bounds to energies of radial lithium', *Chem. Phys. Letters*, **13**, 85–87.

Fox, D. W. and Sigillito, V. G (1972b) 'Bounds for energies of radial lithium', *J. Appl. Math. Phys.*, **23**, 392–411.

Fox, D. W. and Sigillito, V. G. (1972c) 'New lower bounds for energies of radial lithium', *Chem. Phys. Letters*, **14**, 583–585.

Gould, S. H. (1966) *Variational Methods for Eigenvalue Problems: An Introduction to the Weinstein Method of Intermediate Problems*, 2nd. ed., University of Toronto Press.

Kato, T. (1951a) 'Fundamental properties of Hamiltonian operators of Schrödinger type', *Trans. Amer. Math. Soc.*, **70**, 195–211.

Kato, T. (1951b) 'On the existence of solutions of the helium wave equation', *Trans. Amer. Math. Soc.*, **70**, 212–218.

Kato, T. (1966) *Perturbation Theory for Linear Operators*, Springer, New York.

Kemble, E. C. (1958) *The Fundamental Principles of Quantum Mechanics*, Dover, New York.

Kinoshita, T. (1957) 'Ground state of the helium atom', *Phys. Rev.*, **105**, 1490.

Kuroda, S. T. (1961) 'On a generalization of the Weinstein–Aronszajn formula and the infinite determinant', *Sci. Papers College Gen. Ed. Univ. Tokyo*, **11**, 1–12.

Löwdin, P. O. (1965) 'Studies in perturbation theory XI. Lower bounds to energy eigenvalues, ground state, and excited states', *J. Chem. Phys.*, **43**, S175–S185.

Löwdin, P. O. (1968) 'Studies in perturbation theory XIII. Treatment of constants of motion in resolvent method, partitioning technique, and perturbation theory', *Intern. J. Quantum Chem.*, **2**, 867–931.

Reid, C. E. (1972) 'Intermediate Hamiltonians for the lithium atom', *Intern. J. Quantum Chem.*, **6**, 793–797.

Stenger, W. (1968) 'On the variational principles for eigenvalues for a class of unbounded operators', *J. Math. Mech.*, **17**, 641–648.

Stenger, W. (1969) 'On perturbations of finite rank', *J. Math. Anal. Appl.*, **23**, 625–635.

Stenger, W. (1970a) 'Some extensions and applications of the new maximum-minimum theory of eigenvalues', *J. Math. Mech.*, **19**, 931–944.

Stenger, W. (1970b) 'On Fichera's transformation in the method of intermediate problems', *Rend. Accad. Naz. Lincei.*, **48**, 302–305.

Stenger, W. (1974) 'Intermediate problems for eigenvalues', *Intern. J. Quantum Chem.*, **8**, 623–625.

Weinberger, H. F. (1959) *A Theory of Lower Bounds for Eigenvalues*, Tech. Note BN-183, IFDAM, University of Maryland, College Park, Maryland.

Weinstein, A. (1935) 'On a minimal problem in the theory of elasticity', *J. London Math. Soc.*, **10**, 184–192.

Weinstein, A. (1937) 'Études des spectres des équations aux derivées partielles de la théorie des plaques élastiques, *Mémor. Sci. Math.*, **88**.

Weinstein, A. (1974) 'On non-self-adjoint perturbations of finite rank', *J. Math. Anal. Appl.*, **45**, 1–11.

Weinstein, A. and Stenger, W. (1972) *Methods of Intermediate Problems for Eigenvalues: Theory and Ramifications*, Academic Press, New York.

15

Position Observables of the Photon

K. KRAUS
Physikalisches Institut der Universität Würzburg, Germany

1. INTRODUCTORY REMARKS

Quantum theory was initiated by Planck's discovery of the discontinuous character of light emission and absorption and Einstein's subsequent hypothesis of light quanta. From the interference phenomena of light it was already apparent that these light quanta (or photons, as we call them now) could not be particles of the simple kind considered in classical mechanics. A more precise description of the 'non-classical' behaviour of particles, however, was first given much later by quantum mechanics. Perhaps the most impressive deviation from 'classical' behaviour shows up in Heisenberg's famous uncertainty relation (Heisenberg, 1927)

$$\Delta X_i \cdot \Delta P_j \geq \tfrac{1}{2}\delta_{ij} \tag{1}$$

for the components of position $\mathbf{X}$ and momentum $\mathbf{P}$ of, for example, an electron.*

The photon itself, however, has not yet found its way into textbooks of quantum mechanics as an example for typical quantum properties of particles. Of course some simple interference or polarization experiments with light (which at sufficiently low intensities may be interpreted tentatively as experiments with 'single photons') are sometimes discussed in introductory textbooks. A more detailed treatment of one-photon quantum mechanics, however, is usually reserved for advanced texts, for example, on quantum electrodynamics.

This neglect of the photon is perhaps partly due to the following circumstance. It has been proved (Newton and Wigner, 1949; Wightman, 1962) that there is no self-adjoint (vector) operator $\mathbf{X}$ in the state space of the photon which, according to the rules of ordinary quantum mechanics, could be interpreted as a position observable. Usually this is taken to indicate that, simply, photons are not localizable at all. Accordingly an uncertainty relation like (1), which is so typical for massive quantum particles, could not even be formulated for photons (or other massless particles, e.g. neutrinos).

---
*As usual, we set $\hbar = c = 1$ in our system of units.

Quite apart from possible theoretical objections, however, the experiment itself seems to reject such a radical interpretation of the mentioned results. In fact single photons may be localized experimentally, at least above some energy threshold, by suitable detectors (counters, photographic emulsions, etc.), which moreover are very similar to the detectors for massive particles. Then, obviously, one has to ask how such experiments can be described theoretically. It is clear from what has been said before that such a description can be obtained only if the usual requirements for a position observable are somewhat relaxed.

A first proposal in this direction was made by Jauch and Piron (1967) and Amrein (1969). For reasons which will be explained later, a different approach is preferred here, which has already been sketched elsewhere (Kraus, 1971), and which is based on Ludwig's reformulation of quantum theory (Ludwig, 1970). It will be shown that, starting from a suitable generalization of the notion of observables as suggested by Ludwig's theory (see also Neumann, 1971), position observables for the photon may indeed be constructed. For these position observables we will then prove, among other things, the validity of the uncertainty relation (1).

Before discussing the localization problem, however, we will first have to review the usual quantum-mechanical description of single photons. This could be done in a most satisfactory way by starting from the representation theory of the Poincaré (i.e. the inhomogeneous Lorentz) group (Wigner, 1939). Moreover, group theory would allow a unified treatment of other elementary particles along with the photon. Since, however, the main purpose of the present paper is neither mathematical elegance nor complete generality, we have chosen a more elementary treatment which is particularly adapted to the photon. The subsequent construction and discussion of a position observable for the photon is also very elementary in this formalism.

The intentions of the present paper may be sketched as follows. First of all, we want to present the photon as just another example — perhaps a somewhat surprising one — for the universal validity of the celebrated uncertainty relation (1). Secondly, a natural and useful generalization of the concept of quantum-mechanical observables will be illustrated by the example of photon position. We feel that, for both of these purposes, a fairly low level of mathematical sophistication and rigour is sufficient. Thus, more advanced mathematical techniques will be used only when (and to the extent that) they really help clarifying the matter and improving the presentation, and mathematical subtleties of a more technical type will often be omitted.

2. STATE SPACE AND ELEMENTARY OBSERVABLES OF A SINGLE PHOTON

The simplest quantum mechanical description of a single free photon is the following. Pure states* correspond to unit vectors in the Hilbert space $\mathcal{H}$ of

---

*Throughout this paper we will consider pure states only. The discussion of state mixtures (density matrices) is irrelevant for our present investigation.

complex square-integrable vector functions $\mathbf{A}(\mathbf{k})$ of a real vector $\mathbf{k}$ which satisfy the transversality condition

$$\mathbf{k} \cdot \mathbf{A}(\mathbf{k}) = 0 \tag{2}$$

The scalar product in $\mathcal{H}$ is given by

$$\langle \mathbf{A}, \mathbf{A}' \rangle = \int d^3k \, \overline{\mathbf{A}(\mathbf{k})} \cdot \mathbf{A}'(\mathbf{k}) \tag{3}$$

with a bar denoting complex conjugation. A complete system of commuting observables is given by the three momentum components

$$P_j = k_j \tag{4}$$

(multiplication operators) and the helicity (spin component in the direction of $\mathbf{P}$)

$$\sigma = \frac{\mathbf{P}}{|\mathbf{P}|} \cdot \mathbf{s} = \frac{\mathbf{k}}{\omega} \cdot \mathbf{s} \tag{5}$$

Here $\omega = |\mathbf{k}|$, and the j-component s_j of the spin operator $\mathbf{s}$ acts as the matrix

$$(s_j)_{kl} = -i\varepsilon_{jkl} \tag{6}$$

on vector functions $\mathbf{A}$, with ε_{jkl} denoting the Levi–Civita symbol. From (5) and (6) the action of σ on vector functions $\mathbf{A}$ is easily calculated:

$$(\sigma \mathbf{A})(\mathbf{k}) = i \frac{\mathbf{k}}{\omega} \times \mathbf{A}(\mathbf{k}) \tag{7}$$

With momentum $\mathbf{P}$, the energy (Hamiltonian) H is also a multiplication operator:

$$H = |\mathbf{P}| = |\mathbf{k}| = \omega \tag{8}$$

The helicity operator σ has eigenvalues ± 1 in $\mathcal{H}$, with corresponding eigenspaces $\mathcal{H}_\pm$. In order to show this explicitly, we choose for each $\mathbf{k}$ a right-handed orthonormal set of polarization vectors

$$\mathbf{e}_1(\mathbf{k}), \mathbf{e}_2(\mathbf{k}) \quad \text{and} \quad \mathbf{e}_3(\mathbf{k}) = \frac{\mathbf{k}}{\omega}$$

By (7), then, vector functions of the form

$$\mathbf{A}_\pm(\mathbf{k}) = a_\pm(\mathbf{k})(\mathbf{e}_1(\mathbf{k}) \pm i \mathbf{e}_2(\mathbf{k})) \tag{9}$$

with arbitrary coefficients $a_\pm(\mathbf{k})$ are eigenfunctions of σ with eigenvalues ± 1. Moreover, by (2), each $\mathbf{A} \in \mathcal{H}$ may be decomposed as

$$\mathbf{A}(\mathbf{k}) = \mathbf{A}_+(\mathbf{k}) + \mathbf{A}_-(\mathbf{k})$$

with suitable $\mathbf{A}_\pm \in \mathcal{H}_\pm$. Photon states of the form $\mathbf{A}_+$ and $\mathbf{A}_-$ correspond to right and left circular polarized light, respectively.

For later use a natural (but 'unphysical') extension of the state space $\mathcal{H}$ will now be constructed. This simply amounts to dropping the condition (2), while the inner product (3) is left unchanged. The enlarged Hilbert space is called $\hat{\mathcal{H}}$. The definitions (4), (7) and (8) of operators $\mathbf{P}$, σ and H in $\mathcal{H}$ make sense also for vector functions $\mathbf{A} \in \hat{\mathcal{H}}$, and thus may be taken to define natural extensions $\hat{\mathbf{P}}$, $\hat{\sigma}$ and $\hat{H}$ of these operators to the space $\hat{\mathcal{H}}$. In addition to (9), $\hat{\mathcal{H}}$ also contains vector functions of the form

$$\dot{\mathbf{A}}_0(\mathbf{k}) = a_0(\mathbf{k})\mathbf{e}_3(\mathbf{k}) \tag{10}$$

which belong to the eigenvalue zero of $\hat{\sigma}$ and constitute the subspace $\mathcal{H}_0$ of $\hat{\mathcal{H}}$. Such functions are 'unphysical' since they do not describe photon states, and consequently the assignment of 'momentum', 'energy' and 'helicity' to such 'states' by the operators $\hat{\mathbf{P}}$, $\hat{H}$ and $\hat{\sigma}$ is purely formal.

The concrete realization of the photon state space $\mathcal{H}$ given above has the advantage of providing a one-to-one correspondence between photon states and transverse vector functions $\mathbf{A}(\mathbf{k})$. The transformation law of such functions under Poincaré transformations, however, looks rather complicated if written down explicitly. It is therefore better to describe it implicitly in terms of another realization of the space $\mathcal{H}$.

We consider the space $\mathcal{F}$ of complex four-vector functions

$$B^\nu(\mathbf{k}) = \{\mathbf{B}(\mathbf{k}), B^4(\mathbf{k})\} \tag{11}$$

with the additional requirements

$$k_\nu B^\nu = \mathbf{k} \cdot \mathbf{B} - \omega B^4 = 0 \tag{12}$$

(Lorentz condition; $k^4 \equiv \omega$) and

$$\int d\mu(\mathbf{k}) \overline{B_\nu(\mathbf{k})} B^\nu(\mathbf{k}) < \infty, \quad d\mu(\mathbf{k}) \equiv \frac{d^3 k}{\omega} \tag{13}$$

(square integrability).* With the inner product

$$\langle B, B' \rangle = \int d\mu(\mathbf{k}) \overline{B_\nu(\mathbf{k})} B'^\nu(\mathbf{k}) \tag{14}$$

$\mathcal{F}$ becomes something like a Hilbert space. In virtue of (12) the space part $\mathbf{B}$ of any $B^\nu \in \mathcal{F}$ may be decomposed according to

$$\mathbf{B} = \omega^{\frac{1}{2}} \mathbf{A} + \left(\frac{\mathbf{k}}{\omega} \cdot \mathbf{B}\right) \frac{\mathbf{k}}{\omega} = \omega^{\frac{1}{2}} \mathbf{A} + B^4 \frac{\mathbf{k}}{\omega} \tag{15}$$

into a transverse part $\omega^{\frac{1}{2}} \mathbf{A}$ (i.e. $\mathbf{k} \cdot \mathbf{A} = 0$) and a longitudinal component $(\mathbf{k}/\omega) \cdot \mathbf{B} \equiv B^4$. Then (14) immediately yields

$$\langle B, B' \rangle = \int d^3 k \overline{\mathbf{A}(\mathbf{k})} \cdot \mathbf{A}'(\mathbf{k}) \tag{16}$$

---

*Notation for four-vectors $a^\nu = \{\mathbf{a}, a^4\}$, $\nu = 1 \ldots 4$: $a_\nu = a^\nu$ for $\nu = 1, 2, 3$; $a_4 = -a^4$; $a_\nu b^\nu = \mathbf{a} \cdot \mathbf{b} - a^4 b^4$ (sum convention).

(The factor $\omega^{\frac{1}{2}}$ has been introduced in (15) since we want to have $d^3\mathbf{k}$, instead of $d\mu(\mathbf{k})$, in (16).) From (16), the positivity of the inner product (14) follows. Moreover, since only the transverse part of **B** enters (16), we see that all four-vector functions B^ν which differ only with respect to $(\mathbf{k}/\omega)\cdot\mathbf{B} = B^4$ represent the same vector in the Hilbert space defined by the inner product (14). In other words: Functions $B^\nu \in \mathcal{F}$ with vanishing transverse part $\omega^{\frac{1}{2}}\mathbf{A}$ of **B** are zero vectors with respect to the inner product (14). Such functions constitute a subspace $\mathcal{F}_0$ of $\mathcal{F}$. Then, not $\mathcal{F}$ itself, but the space $\mathcal{F}/\mathcal{F}_0$ of equivalence classes in $\mathcal{F}$ with respect to $\mathcal{F}_0$, is a Hilbert space. Such an equivalence class consists of all B^ν of the form

$$B^\nu = \left\{\omega^{\frac{1}{2}}\mathbf{A} + B^4 \frac{\mathbf{k}}{\omega}, B^4\right\} \tag{17}$$

with a given transverse **A** and arbitrary B^4.

By (3) and (16), the Hilbert spaces $\mathcal{H}$ and $\mathcal{F}/\mathcal{F}_0$ may be identified in an obvious way, as already indicated by the use of the same symbol **A** in both cases. Thus $\mathcal{F}/\mathcal{F}_0$ is also a realization of the photon state space. (This realization is formally analogous to the Fermi gauge in quantum electrodynamics, whereas the former one corresponds to the Coulomb gauge. There also exists a description of one-photon states which corresponds to the Gupta–Bleuler gauge, but — contrary to what happens in quantum electrodynamics — this gauge is not very useful here.)

A certain disadvantage of the new formalism in the fact that, due to the arbitrariness of B^4 in (17), the correspondence between photon states and four-vector functions $B^\nu(\mathbf{k})$ is no longer one-to-one. (In fact, $B^\nu(\mathbf{k})$ and

$$B'^\nu(\mathbf{k}) = B^\nu(\mathbf{k}) + k^\nu \chi(\mathbf{k})$$

with arbitrary $\chi(\mathbf{k})$, represent the same photon state. This, obviously, corresponds to a certain class of gauge transformations in classical electrodynamics.) This disadvantage is more than compensated, however, by the simple (four-vector) transformation law of $B^\nu \in \mathcal{F}$ under Poincaré transformations. For a Poincaré transformation consisting of a homogeneous orthochronous* Lorentz transformation Λ (with matrix $\Lambda^\nu_{.\mu}$, $\nu, \mu = 1\ldots 4$) and a subsequent four-translation a (with components a^ν, $\nu = 1\ldots 4$), this transformation law is simply

$$B^\nu(\mathbf{k}) \to (U(a, \Lambda)B)^\nu(\mathbf{k}) = e^{-ik_\nu a^\nu} \Lambda^\nu_{.\mu} B^\mu(\Lambda^{-1}\mathbf{k}) \tag{18}$$

(Here $\Lambda^{-1}\mathbf{k}$ is the space part of the four-vector resulting from $k^\nu = \{\mathbf{k}, \omega\}$ by the Lorentz transformation Λ^{-1}). As easily shown, equation (18) defines a representation of the Poincaré group on $\mathcal{F}$. From (18) and the Lorentz invariance

$$d\mu(\mathbf{k}) = d\mu(\Lambda^{-1}\mathbf{k})$$

of the measure $d\mu(\mathbf{k})$, the invariance under (18) of the inner product (14) in $\mathcal{F}$

---

*The behaviour of B^ν under Lorentz transformations with time reversal looks somewhat more complicated.

is easily proved. In particular, $U(a, \Lambda)$ transforms the space $\mathscr{F}_0$ into itself. Therefore it may also be interpreted as a transformation of the equivalence classes (17), and is unitary in $\mathscr{F}/\mathscr{F}_0$. Since $\mathscr{F}/\mathscr{F}_0 \equiv \mathscr{H}$, finally, $U(a, \Lambda)$ also yields a unitary transformation law for the transverse vector functions $\mathbf{A}(\mathbf{k})$ in our previous formalism. The explicit calculation of this transformation law is straightforward but unnecessary since we will not need it here. [For a particular case see equation (39).]

We conclude this Section with an elementary investigation of how helicity behaves under Poincaré transformations. The helicity operator σ may be transferred to $\mathscr{F}$ by defining, for any

$$B^\nu = \{\mathbf{B}, B^4\} = \left\{\omega^{\frac{1}{2}}\mathbf{A} + \frac{\mathbf{k}}{\omega}B^4, B^4\right\} \in \mathscr{F},$$

$$(\sigma B)^\nu = \left\{i\frac{\mathbf{k}}{\omega} \times \omega^{\frac{1}{2}}\mathbf{A}, 0\right\} = \left\{i\frac{\mathbf{k}}{\omega} \times \mathbf{B}, 0\right\} \tag{19}$$

It is obvious that this operator σ in $\mathscr{F}$ induces a transformation of $\mathscr{F}/\mathscr{F}_0 = \mathscr{H}$ which coincides with the operator σ in $\mathscr{H}$ previously defined by (7). By (19) the equivalence class of a given $B^\nu \in \mathscr{F}$ describes a photon state of helicity ± 1 if and only if

$$i\frac{\mathbf{k}}{\omega} \times \mathbf{B} = \pm \omega^{\frac{1}{2}}\mathbf{A} = \pm\left(\mathbf{B} - \frac{\mathbf{k}}{\omega}B^4\right)$$

or, with $\omega = k^4$,

$$i(\mathbf{k} \times \mathbf{B}) = \pm(k^4\mathbf{B} - B^4\mathbf{k})$$

The last equation can be rewritten as

$$i\varepsilon_{\kappa\lambda\mu\nu}k^\mu B^\nu = \pm(k_\kappa B_\lambda - k_\lambda B_\kappa) \tag{20}$$

with the Levi–Civita symbol $\varepsilon_{\kappa\lambda\mu\nu}$. The invariance of (20) under pure space–time translations follows trivially from (18). According to (18), both k^ν and B^ν behave like four-vectors under pure (orthochronous) Lorentz transformations* whereas, as well-known, $\varepsilon_{\kappa\lambda\mu\nu}$ is a pseudotensor. Thus equation (20) is also invariant under proper orthochronous Lorentz transformations, but under space reflection the left-hand side changes sign. Therefore the helicity eigenspaces $\mathscr{H}_\pm$ of $\mathscr{H}$ are invariant under proper Lorentz transformations, whereas space reflection interchanges $\mathscr{H}_+$ and $\mathscr{H}_-$.

Many elementary particles (e.g. electron, proton, neutrinos) may be characterized by the fact that their state space carries an irreducible representation of the proper Poincaré group. The state of such a particle is uniquely determined by the expectation values of all 'kinematic' observables, i.e. all infinitesimal generators of Poincaré transformations (energy, momentum, angular momentum, etc.). This is not so for the photon, since the helicity eigenspaces $\mathscr{H}_+$ and

---

*I.e. for $B'^\nu = (U(0, \Lambda)B)^\nu$ we have $B'^\nu(\mathbf{k}') = \Lambda^\nu_{\cdot\mu}B^\mu(\mathbf{k})$ with $\mathbf{k}'^\nu = \Lambda^\nu_{\cdot\mu}k^\mu$.

$\mathcal{H}_-$ reduce the representation $U(a, \Lambda)$ (for Λ proper). With respect to 'kinematic' observables, therefore, a coherent superposition

$$\alpha \mathbf{A}_+ + \beta \mathbf{A}_- \tag{21}$$

of normalized states $\mathbf{A}_\pm \in \mathcal{H}_\pm$ (with $|\alpha|^2 + |\beta|^2 = 1$) cannot be distinguished from the incoherent mixture of these states with the weights $|\alpha|^2$ and $|\beta|^2$, respectively. However, there are observables which permit such a distinction. For instance, suitable states of the form (21) (with $|\alpha|^2 = |\beta|^2 = \frac{1}{2}$) correspond to linear polarization, whereas the corresponding mixtures describe totally unpolarized light. Measurements of linear polarization thus do not belong to the 'kinematic' observables of the photon. The position observable to be constructed in the subsequent section will be another example for observables of this 'non-kinematic' type.

3. CONSTRUCTION OF A PHOTON POSITION OBSERVABLE

In order to be acceptable as a photon position operator* in the sense of usual quantum mechanics, a (vector) operator $\mathbf{X}$ on the photon state space $\mathcal{H}$ has to satisfy two requirements. First, its components X_j have to be self-adjoint, and have to commute with each other in order to be measurable together. Secondly, the behaviour of $\mathbf{X}$ under spatial rotations and translations (i.e. Euclidean transformations) is prescribed to be

$$U^*(\mathbf{a}, R)\mathbf{X}U(\mathbf{a}, R) = R\mathbf{X} + \mathbf{a} \tag{22}$$

Here $U(\mathbf{a}, R)$ is the restriction of the Poincaré group representation (18) to the Euclidean group, the elements of which consist of space rotations (or reflections) R and subsequent translations $\mathbf{a}$. Equation (22) is equivalent to the self-evident requirement

$$\langle \mathbf{X} \rangle_{U(\mathbf{a},R)\mathbf{A}} = R \langle \mathbf{X} \rangle_\mathbf{A} + \mathbf{a} \tag{23}$$

for the expectation values

$$\langle \mathbf{X} \rangle_\mathbf{A} = \langle \mathbf{A}, \mathbf{X}\mathbf{A} \rangle \tag{24}$$

of $\mathbf{X}$ in arbitrary states $\mathbf{A}$.

Another way of describing position measurements is as follows. Spatial localizability of the photon implies the existence of observables $E(\Delta)$, corresponding to largely arbitrary space regions Δ,† which take the value one (respectively zero) if at time $t = 0$ the photon is found (respectively not found) inside the region Δ. Measurements of such $E(\Delta)$ should be actually feasible, at least for certain regions Δ, by means of suitable counters. According to the

---

*Throughout this paper we will use the Heisenberg picture. The position observables to be discussed thus refer to position measurements at a fixed time, $t = 0$ say. (For the conserved quantities considered before, such specification of time was unnecessary.)
†Precisely: to all Borel sets Δ.

rules of ordinary quantum mechanics, these 'yes–no' observables have to be represented by projection operators on $\mathcal{H}$ (which, for simplicity, are also called $E(\Delta)$), and the probability that at time $t = 0$ a photon in state **A** 'triggers the counter $E(\Delta)$ in the region Δ' is

$$w_\mathbf{A}(\Delta) = \langle \mathbf{A}, E(\Delta)\mathbf{A} \rangle \tag{25}$$

This physical interpretation of $E(\Delta)$ immediately implies, with $\varnothing$ denoting the empty set and $\mathbb{R}^3$ denoting all of space,

$$\left.\begin{aligned} E(\varnothing) &= 0, \quad E(\mathbb{R}^3) = 1 \\ E\left(\bigcup_i \Delta_i\right) &= \sum_i E(\Delta_i) \quad \text{if} \quad \Delta_i \cap \Delta_j = \varnothing \quad \text{for} \quad i \neq j \end{aligned}\right\} \tag{26}$$

(The last relation follows from the additivity

$$w_\mathbf{A}\left(\bigcup_i \Delta_i\right) = \sum_i w_\mathbf{A}(\Delta_i)$$

of the probabilities (25) for mutually disjoint regions Δ_i.) A correspondence $\Delta \to E(\Delta)$ of space regions Δ and projection operators $E(\Delta)$ with the properties (26) is called a spectral measure on $\mathbb{R}^3$. Equations (26) imply that any two $E(\Delta)$ commute with each other, and that

$$\left.\begin{aligned} E\left(\bigcap_i \Delta_i\right) &= \prod_i E(\Delta_i) \\ E(\Delta_1 \cup \Delta_2) &= E(\Delta_1) + E(\Delta_2) - E(\Delta_1 \cap \Delta_2) \\ E(\Delta') &= 1 - E(\Delta) \end{aligned}\right\} \tag{27}$$

with Δ' denoting the complement of Δ.

The requirement corresponding to (23) is now

$$w_\mathbf{A}(\Delta) = w_{U(\mathbf{a},R)\mathbf{A}}(\Delta_{\mathbf{a},R}) \tag{28}$$

with

$$\Delta_{\mathbf{a},R} = R\Delta + \mathbf{a} \equiv \{\mathbf{x} \mid \mathbf{x} = R\mathbf{y} + \mathbf{a}, \mathbf{y} \in \Delta\}$$

(i.e. the region obtained from Δ by the rotation R and translation $\mathbf{a}$), and is also self-explanatory. Since the state **A** in (28) is arbitrary, this condition is equivalent to

$$U(\mathbf{a}, R)E(\Delta)U^*(\mathbf{a}, R) = E(\Delta_{\mathbf{a},R}) \tag{29}$$

A spectral measure on $\mathcal{H}$ which satisfies (29) is called Euclidean covariant with respect to the given representation $U(\mathbf{a}, R)$, or simply: covariant.

The equivalence of the two descriptions of position measurements follows from the fact that one may construct **X** if the $E(\Delta)$ are given, and vice versa. Assume first the spectral measure $E(\Delta)$ to be given. From the physical interpretation (5) of $\langle \mathbf{A}, E(\Delta)\mathbf{A} \rangle$ we conclude that, with $dE(\mathbf{X}) \equiv E(d^3\mathbf{x})$,

$\int x_i \langle \mathbf{A}, \mathrm{d}E(\mathbf{x})\mathbf{A}\rangle$ is the expectation value* of the jth photon coordinate in state $\mathbf{A}$. In order to represent this as $\langle \mathbf{A}, X_j\mathbf{A}\rangle$ with the component X_j of a position operator $\mathbf{X}$, we have to take

$$X_j = \int x_j \, \mathrm{d}E(\mathbf{x}) \tag{30}$$

This formula, indeed, defines three self-adjoint operators X_j, and provides a common spectral representation of them. The more familiar spectral representations

$$X_j = \int \lambda \, \mathrm{d}E_j(\lambda) \tag{31}$$

with one-dimensional spectral families $E_j(\lambda)$ follow from (30) if we define

$$E_j(\lambda) = E(\Delta_{j\lambda}), \quad \Delta_{j\lambda} = \{\mathbf{x} \mid x_j \leq \lambda\}$$

Therefore the operators X_j commute with each other in the sense that

$$[E_j(\lambda), E_k(\mu)] = 0 \quad \text{for all } j, k, \lambda \text{ and } \mu \tag{32}$$

which is somewhat stronger than 'naive' commutativity, i.e.

$$[X_j, X_k] = 0 \tag{33}$$

(on the dense domain where the left-hand side exists). Vice versa, any three self-adjoint operators X_j which commute in the sense of (32) possess a common spectral representation of the form (30). The projection operators $E(\Delta)$ of the corresponding spectral measure may be calculated explicitly as

$$E(\Delta) = \int_\Delta \mathrm{d}E(\mathbf{x}) = \int \chi_\Delta(\mathbf{x}) \, \mathrm{d}E(\mathbf{x}) \tag{34}$$

with the characteristic function

$$\chi_\Delta(\mathbf{x}) = \begin{cases} 1 & \text{for } \mathbf{x} \in \Delta \\ 0 & \text{for } \mathbf{x} \notin \Delta \end{cases}$$

of the region Δ. The last expression in (34) is simply the operator function $\chi_\Delta(\mathbf{X})$ of $\mathbf{X}$, in accordance with the physical meaning of $E(\Delta)$. Finally the covariance requirements (22) for $\mathbf{X}$ and (29) for $E(\Delta)$ may also be shown to be equivalent.

The result of Newton and Wigner (1949) and Wightman (1962) is, simply, that the photon does not possess a position observable with the required properties. [The non-existence of an operator $\mathbf{X}$ was first proved by Newton and Wigner (1949) who, however, needed some additional assumptions for their proof. Later on Wightman (1962) was able to prove the non-existence of a covariant spectral measure $E(\Delta)$ without any additional requirements.] We do not want to reproduce these proofs here, but will start instead with a naive attempt to construct a photon position operator $\mathbf{X}$ explicitly. The failure of this

---

*This makes sense as a Stieltjes integral.

attempt will then illustrate the 'no-go' theorem of Newton, Wigner and Wightman. Besides this, however, a suitable refinement of this construction will lead us directly to a (generalized) position observable of the photon.

By (3) and (4), $\mathbf{A}(\mathbf{k})$ may be interpreted as the photon wave function in the momentum representation.* In analogy to ordinary quantum mechanics, we thus attempt to define a position operator $\hat{\mathbf{X}}$ by

$$(\hat{X}_j \mathbf{A})(\mathbf{k}) = i \frac{\partial}{\partial k_j} \mathbf{A}(\mathbf{k}) \tag{35}$$

However, if applied to $\mathbf{A} \in \mathcal{H}$ these operators $\hat{X}_j$ destroy the transversality (2) since, in general

$$\mathbf{k} \cdot i \frac{\partial}{\partial k_j} \mathbf{A} = i \frac{\partial}{\partial k_j}(\mathbf{k} \cdot \mathbf{A}) - i A_j \tag{36}$$

is not zero if $\mathbf{k} \cdot \mathbf{A} = 0$. This difficulty is absent if we read (35) as defining operators $\hat{X}_j$ on the larger Hilbert space $\hat{\mathcal{H}}$ introduced in Section 2. Equation (35) then defines three self-adjoint $\hat{X}_j$ on $\hat{\mathcal{H}}$, which commute with each other in the sense of (32). In fact, $\hat{X}_j$ acts as multiplication by x_j on the position space wave functions $\tilde{\mathbf{A}}(\mathbf{x})$ obtained from $\mathbf{A}(\mathbf{k})$ by Fourier transformation, and the spectral projections $\hat{E}(\Delta)$ of $\hat{\mathbf{X}}$ then correspond to multiplication by the characteristic functions $\chi_\Delta(\mathbf{x})$ of the regions Δ. The difficulty indicated by (36) may now be circumvented as follows. Denoting by Φ the projection operator which projects $\hat{\mathcal{H}}$ onto its physical subspace $\mathcal{H}$ and by $\hat{O}|_{\mathcal{H}}$ the restriction to $\mathcal{H}$ of an operator $\hat{O}$ acting on $\hat{\mathcal{H}}$, we define operators X_j on $\mathcal{H}$ by†

$$X_j = \Phi \hat{X}_j|_{\mathcal{H}} \tag{37}$$

This definition is chosen such that, for $\mathbf{A} \in \mathcal{H}$,

$$\langle \mathbf{A}, X_j \mathbf{A} \rangle = \langle \mathbf{A}, \hat{X}_j \mathbf{A} \rangle \tag{38}$$

In this sense the operators X_j on $\mathcal{H}$ are substitutes for the $\hat{X}_j$ which lead out of $\mathcal{H}$. Since $\hat{X}_j$ is self-adjoint, (38) implies that $\langle \mathbf{A}, X_j \mathbf{A} \rangle$ is real; therefore the X_j are at least symmetric. (We claim that they are even self-adjoint, but since this property is unessential here we did not try to prove it.)

A little detour is appropriate if we want to discuss Euclidean transformations of these operators X_j. The transformation law of state functions $\mathbf{A}(\mathbf{k}) \in \mathcal{H}$ under Euclidean transformations follows from (18) as

$$(U(\mathbf{a}, R)\mathbf{A})(\mathbf{k}) = e^{-i\mathbf{k}\cdot\mathbf{a}} R\, \mathbf{A}(R^{-1}\mathbf{k}) \tag{39}$$

As expected, $\mathbf{A}$ behaves as a vector under space rotations. This transformation law may be extended quite naturally to the enlarged Hilbert space $\hat{\mathcal{H}}$ by taking

$$(\hat{U}(\mathbf{a}, R)\mathbf{A})(\mathbf{k}) = e^{-i\mathbf{k}\cdot\mathbf{a}} R\, \mathbf{A}(R^{-1}\mathbf{k}) \tag{40}$$

---

*Namely, $\overline{\mathbf{A}(\mathbf{k})} \cdot \mathbf{A}(\mathbf{k})$ is the probability density in momentum space corresponding to a (normalized) state $\mathbf{A} \in \mathcal{H}$. For this it is essential that the inner product (3) is defined with $d^3\mathbf{k}$ instead of $d\mu(\mathbf{k})$.

†It is easily proved that (37) yields operators whose domain of definition is dense in $\mathcal{H}$.

for $\mathbf{A} \in \hat{\mathcal{H}}$ as well. [For clarity of notation we have used the symbol $\hat{U}(\mathbf{a}, R)$ for the extension of $U(\mathbf{a}, R)$ to $\hat{\mathcal{H}}$. This extension is related to — and consistent with — our previous extensions of momentum and helicity operators to $\hat{\mathcal{H}}$; in fact, the latter may be expressed in terms of infinitesimal generators of $\hat{U}(\mathbf{a}, R)$.] Since under $\hat{U}(\mathbf{a}, R)$ both $\mathbf{k}$ and $\mathbf{A}$ transform as vectors, the longitudinal part $\mathbf{A}_0 = (\mathbf{k}/\omega)[(\mathbf{k}/\omega) \cdot \mathbf{A}]$ and the transverse part $\mathbf{A}_{tr} = \mathbf{A} - \mathbf{A}_0$ of an arbitrary $\mathbf{A} \in \hat{\mathcal{H}}$ transform separately under $\hat{U}(\mathbf{a}, R)$. This implies

$$[\hat{U}(\mathbf{a}, R), \Phi] = 0, \quad \hat{U}(\mathbf{a}, R)|_{\mathcal{H}} = U(\mathbf{a}, R) \tag{41}$$

i.e. Φ reduces $\hat{U}(\mathbf{a}, R)$, and the subrepresentation of $\hat{U}(\mathbf{a}, R)$ in $\mathcal{H}$ is $U(\mathbf{a}, R)$. A straightforward calculation with (35) and (40) yields

$$\hat{U}^*(\mathbf{a}, R)\hat{\mathbf{X}}\hat{U}(\mathbf{a}, R) = R\hat{\mathbf{X}} + \mathbf{a} \tag{42}$$

as to be expected from the vector character of $\hat{\mathbf{X}} = i\nabla_{\mathbf{k}}$. Together with (41) this immediately leads to a transformation law of the desired form (22) for the operator $\mathbf{X}$ defined by (37).

However, we know from Newton and Wigner (1949) and Wightman (1962) that our construction of a photon position operator has to fail somewhere. This failure is indeed easily seen. With (35), (36) and (37) we obtain explicitly, for $\mathbf{A} \in \mathcal{H}$,

$$X_j \mathbf{A} = \hat{X}_j \mathbf{A} - \frac{\mathbf{k}}{\omega}\left(\frac{\mathbf{k}}{\omega} \cdot \hat{X}_j \mathbf{A}\right) = i\left(\frac{\partial}{\partial k_j}\mathbf{A} + \frac{\mathbf{k}}{\omega^2}A_j\right) \tag{43}$$

From this we find by a trivial calculation that the commutativity condition (33) is violated by our $\mathbf{X}$. In the rest of this Section we will try to show that $\mathbf{X}$, in spite of not being an ordinary photon position operator, nevertheless may have something to do with photon position.

As already mentioned, the operator $\hat{\mathbf{X}}$ on $\hat{\mathcal{H}}$ has self-adjoint and mutually commuting components. By (42) it also satisfies the transformation law required for a position operator. Therefore the spectral measure $\hat{E}(\Delta)$ associated with $\hat{\mathbf{X}}$ is covariant with respect to $\hat{U}(\mathbf{a}, R)$. Any difficulties associated with the photon position operator would thus be absent if, instead of $\mathcal{H}$, the enlarged Hilbert space $\hat{\mathcal{H}}$ were the physical state space of the photon. This suggests the following tentative description of position measurements for photons. We consider $\hat{X}$ on $\hat{\mathcal{H}}$, or the corresponding spectral measure $\hat{E}(\Delta)$, as operators representing the photon position, which allows us to satisfy the usual requirements at least formally. We deviate from ordinary quantum mechanics, however, to the extent that not the whole Hilbert space $\hat{\mathcal{H}}$ but only the subspace $\mathcal{H}$ of it is interpreted as the state space of the photon. Accordingly we interpret, for physical states $\mathbf{A} \in \mathcal{H}$ (and only for them),

$$\langle \mathbf{X} \rangle_{\mathbf{A}} = \langle \mathbf{A}, \hat{\mathbf{X}}\mathbf{A} \rangle \tag{44}$$

as expectation value of the photon position, and

$$w_{\mathbf{A}}(\Delta) = \langle \mathbf{A}, \hat{E}(\Delta)\mathbf{A} \rangle \tag{45}$$

as probability for finding the photon in the space region Δ. These definitions satisfy the covariance requirements (23) and (28), as easily checked:

$$\langle \mathbf{X} \rangle_{U(\mathbf{a},R)\mathbf{A}} = \langle U(\mathbf{a}, R)\mathbf{A}, \hat{X} U(\mathbf{a}, R)\mathbf{A} \rangle$$
$$= \langle \hat{U}(\mathbf{a}, R)\mathbf{A}, \hat{X} \hat{U}(\mathbf{a}, R)\mathbf{A} \rangle$$
$$= \langle \mathbf{A}, (R\hat{X} + \mathbf{a})\mathbf{A} \rangle = R\langle \mathbf{X} \rangle_{\mathbf{A}} + \mathbf{a}$$

by (41) and (42); similarly, (28) follows from (41) and the covariance of $\hat{E}(\Delta)$.

The unphysical Hilbert space $\mathcal{H}$ can be eliminated altogether from this description. By (38), equation (44) may also be written as

$$\langle \mathbf{X} \rangle_{\mathbf{A}} = \langle \mathbf{A}, \mathbf{X}\mathbf{A} \rangle \tag{46}$$

with **X** defined by (37), and (45) may be reformulated as

$$w_{\mathbf{A}}(\Delta) = \langle \mathbf{A}, F(\Delta)\mathbf{A} \rangle \tag{47}$$

with operators $F(\Delta)$ on $\mathcal{H}$ defined by

$$F(\Delta) = \Phi \hat{E}(\Delta)|_{\mathcal{H}} \tag{48}$$

From $0 \le w_{\mathbf{A}}(\Delta) \le 1$ for all normalized states **A** we get

$$F(\Delta)^* = F(\Delta), \quad 0 \le F(\Delta) \le 1 \tag{49}$$

or in words: all $F(\Delta)$ are self-adjoint, non-negative and bounded in norm by one. They are in general not projection operators, except for particular regions Δ like $\varnothing$ and $\mathbb{R}^3$ (see below). This follows from a simple mathematical result:

> For two projection operators E_1 and E_2, $E_1 E_2 E_1$ is a projection operator if and only if E_1 and E_2 commute* (50)

Assume all $F(\Delta)$ to be projection operators, which by (48) means that all $\Phi \hat{E}(\Delta) \Phi$ are projection operators on $\mathcal{H}$. Thus, by (50), Φ commutes with all $\hat{E}(\Delta)$, and therefore also with $\hat{X}_j = \int x_j \, d\hat{E}(\mathbf{x})$. This, however, is a contradiction since, by (36), there are $\mathbf{A} \in \mathcal{H}$ with $\hat{X}_j \mathbf{A} \notin \mathcal{H}$.

The spectral measure $\hat{E}(\Delta)$ satisfies relations of the form (26) and (27). Together with (48) this leads to similar relations for the operators $F(\Delta)$. We obtain from (26)

$$F(\varnothing) = 0, \quad F(\mathbb{R}^3) = 1$$
$$F\left(\bigcup_i \Delta_i\right) = \sum_i F(\Delta_i) \quad \text{if} \quad \Delta_i \cap \Delta_j = \varnothing \quad \text{for} \quad i \ne j \tag{51}$$

Proof: Let $F = E_1 E_2 E_1$ be a projection operator. Then $F^2 = E_1 E_2 E_1 E_2 E_1 = F$. This implies, for $A = E_2 E_1 - E_1 E_2 E_1$, that $A^* A = 0$, and thus $0 = A = A^* = A^* - A = [E_1, E_2]$. The converse is well known.

and from (27) — or directly from (51) —

$$\left.\begin{array}{r}F(\Delta_1 \cup \Delta_2) = F(\Delta_1) + F(\Delta_2) - F(\Delta_1 \cap \Delta_2) \\ F(\Delta') = 1 - F(\Delta)\end{array}\right\} \quad (52)$$

The first relation of (27), however, has no simple analogue for the operators $F(\Delta)$. We also cannot conclude from (51) that the operators $F(\Delta)$ commute with each other.* The physical interpretation of (51) and (52) in terms of the probabilities (47) is obvious. Any correspondence $\Delta \to F(\Delta)$ of space regions Δ and operators $F(\Delta)$ with the properties (49) and (51) is called here, as usual, a POV (positive operator valued) measure on $\mathbb{R}^3$. Our POV measure $F(\Delta)$ satisfies the additional condition

$$U(\mathbf{a}, R) F(\Delta) U^*(\mathbf{a}, R) = F(\Delta_{\mathbf{a}, R}) \quad (53)$$

and is therefore called Euclidean covariant with respect to $U(\mathbf{a}, R)$. Equation (53) follows immediately, since the condition (28) is satisfied for $w_\mathbf{A}(\Delta)$ as given by (47), with $\mathbf{A}$ arbitrary. A (covariant) spectral measure, clearly, is a particular case of a (covariant) POV measure, distinguished by the additional property that $(F(\Delta))^2 = F(\Delta)$ for all Δ.

With this terminology, the formalism proposed here may be characterized by the fact that it describes the localization probabilities $w_\mathbf{A}(\Delta)$, via (47), in terms of a covariant POV measure $F(\Delta)$ instead of, as usual, a covariant spectral measure. The 'position operator' $\mathbf{X}$ introduced in addition is already determined by $F(\Delta)$. From $\hat{\mathbf{X}} = \int \mathbf{x} \, d\hat{E}(\mathbf{x})$ we obtain, for $\mathbf{A} \in \mathcal{H}$ and $dF(\mathbf{x}) \equiv F(d^3\mathbf{x}) \equiv \Phi \, d\hat{E}(\mathbf{x})|_{\mathcal{H}}$,

$$\mathbf{XA} = \Phi \hat{\mathbf{X}} \mathbf{A} = \Phi \int \mathbf{x} \, d\hat{E}(\mathbf{x}) \mathbf{A}$$

$$= \int \mathbf{x} \Phi \, d\hat{E}(\mathbf{x}) \mathbf{A} = \int \mathbf{x} \, dF(\mathbf{x}) \mathbf{A}$$

or, shortly,

$$\mathbf{X} = \int \mathbf{x} \, dF(\mathbf{x}) \quad (54)$$

and thus

$$\langle \mathbf{A}, \mathbf{XA} \rangle = \int \mathbf{x} \langle \mathbf{A}, dF(\mathbf{x}) \mathbf{A} \rangle \quad (55)$$

[For a physical interpretation of (55) compare the discussion of (30). As shown by (54), the POV measure $F(\Delta)$ provides a substitute for the non-existent common spectral representation of the three components X_j of $\mathbf{X}$.] The converse, however, is not true: There is no general procedure for reconstructing the POV measure $F(\Delta)$ from the operator $\mathbf{X}$ related to it by (54) (unless $F(\Delta)$ is known to be a spectral measure, which case was discussed above). This is due to the fact that a given operator $\mathbf{X}$ may have several representations of

---

*In fact they do not commute, for otherwise equation (54) below would imply commutativity of the components X_j of $\mathbf{X}$.

the form (54) with different POV measures $F(\Delta)$, as will be shown by means of an example in Section 6.

Moreover, as compared to the case of an 'ordinary' position operator (i.e. one belonging to a spectral measure), the knowledge of $\mathbf{X}$ is also less useful here from a physical point of view. Whereas in both cases the expectation values of position in arbitrary states $\mathbf{A}$ may be calculated in terms of $\mathbf{X}$ as $\langle \mathbf{A}, \mathbf{XA} \rangle$, the mean square deviations $\Delta_\mathbf{A} X_j$ are given by the familiar formula

$$(\Delta_\mathbf{A} X_j)^2 = \|(X_j - \langle X_j \rangle_\mathbf{A})\mathbf{A}\|^2 = \|X_j \mathbf{A}\|^2 - (\langle X_j \rangle_\mathbf{A})^2 \tag{56}$$

for an 'ordinary' position operator $\mathbf{X}$ only. For our position observable given by the POV measure $F(\Delta)$, the physical meaning of $\langle \mathbf{A}, dF(\mathbf{x})\mathbf{A} \rangle$ implies

$$(\Delta_\mathbf{A} X_j)^2 = \int (x_j - \bar{x}_j)^2 \langle \mathbf{A}, dF(\mathbf{x})\mathbf{A} \rangle, \quad \bar{x}_j \equiv \langle X_j \rangle_\mathbf{A} \tag{57}$$

With (48) we get from this

$$(\Delta_\mathbf{A} X_j)^2 = \int (x_j - \bar{x}_j)^2 \langle \mathbf{A}, d\hat{E}(\mathbf{x})\mathbf{A} \rangle$$
$$= \|(\hat{X}_j - \langle X_j \rangle_\mathbf{A})\mathbf{A}\|^2 = \|\hat{X}_j \mathbf{A}\|^2 - (\langle X_j \rangle_\mathbf{A})^2 \tag{58}$$

which does not reduce to (56) since, in general, $\hat{X}_j \mathbf{A} \neq X_j \mathbf{A}$. A formal description of position measurements for a photon in terms of the 'position operator' $\mathbf{X}$ is thus incomplete, in contrast to the description in terms of the POV measure $F(\Delta)$. On the other hand, $F(\Delta)$ is fixed uniquely by the operator $\hat{\mathbf{X}}$ on the extended Hilbert space, since $\hat{\mathbf{X}}$ uniquely determines $\hat{E}(\Delta)$. As exemplified by (44) and (58), important physical quantities may also be calculated directly in terms of $\hat{\mathbf{X}}$.

From the point of view of quantum mechanics in its usual form, our description of position measurements in terms of a POV measure $F(\Delta)$ looks at least rather unconventional. Moreover, the explicit construction of $F(\Delta)$ described above is quite heuristic. Before looking for a better theoretical justification of the formalism, however, we will first derive some physical consequences from it. If these consequences look reasonable, this may perhaps help to strengthen the subsequent, more theoretical arguments in favour of our approach.

4. POSITION–MOMENTUM UNCERTAINTY RELATION AND TIME DEPENDENCE OF POSITION MEASUREMENTS

We start from Schwarz's inequality

$$\|\mathbf{A}'\| \cdot \|\mathbf{A}''\| \geq |\langle \mathbf{A}', \mathbf{A}'' \rangle| \geq |\mathrm{Im}\langle \mathbf{A}', \mathbf{A}'' \rangle| = \tfrac{1}{2}|\langle \mathbf{A}', \mathbf{A}'' \rangle - \langle \mathbf{A}'', \mathbf{A}' \rangle| \tag{59}$$

for arbitrary vectors $\mathbf{A}'$ and $\mathbf{A}''$ in $\mathcal{H}$. From this we obtain, for a normalized state vector $\mathbf{A} \in \mathcal{H}$ in the domain of definition of both X_i and P_j (or, equivalently, of both $\hat{X}_i$ and $\hat{P}_j$), the estimate

$$\|(\hat{X}_i - \bar{x}_i)\mathbf{A}\| \cdot \|(P_j - \bar{p}_j)\mathbf{A}\| \geq \tfrac{1}{2} |\langle \hat{X}_i \mathbf{A}, \hat{P}_j \mathbf{A} \rangle - \langle \hat{P}_j \mathbf{A}, \hat{X}_i \mathbf{A} \rangle| \quad (60)$$

with $\bar{x}_i = \langle X_i \rangle_\mathbf{A}$, $\bar{p}_j = \langle P_j \rangle_\mathbf{A}$. (Take $\mathbf{A}' = (\hat{X}_i - \bar{x}_i)\mathbf{A}$ and $\mathbf{A}'' = (\hat{P}_j - \bar{p}_j)\mathbf{A} = (P_j - \bar{p}_j)\mathbf{A}$, and note that the terms with $\bar{x}_i$ and $\bar{p}_j$ cancel in $\langle \mathbf{A}', \mathbf{A}'' \rangle - \langle \mathbf{A}'', \mathbf{A}' \rangle$.) According to (58) and the analogue of (56) for the 'ordinary' observables P_j, the left-hand side of (60) is equal to $\Delta_\mathbf{A} X_i \cdot \Delta_\mathbf{A} P_j$. A simple calculation, using the explicit definitions (35) of $\hat{X}_i$ and (4) of $\hat{P}_j$, yields

$$\langle \hat{X}_i \mathbf{B}, \hat{P}_j \mathbf{B}' \rangle - \langle \hat{P}_j \mathbf{B}, \hat{X}_i \mathbf{B}' \rangle = i\delta_{ij} \langle \mathbf{B}, \mathbf{B}' \rangle \quad (61)$$

for arbitrary vectors $\mathbf{B}$ and $\mathbf{B}'$ in $\mathcal{H}$ belonging to the domain of definition of both $\hat{X}_i$ and $\hat{P}_j$. An alternative, more abstract proof of (61) uses the transformation property (42) of $\hat{\mathbf{X}}$ and the fact that $\hat{P}_j$ is the self-adjoint generator of translation along the jth axis, i.e.

$$e^{i\lambda \hat{P}_j} \hat{X}_i \, e^{-i\lambda \hat{P}_j} = \hat{X}_i + \lambda \delta_{ij} \quad (62)$$

as follows:

$$\langle \hat{X}_i \mathbf{B}, e^{i\lambda \hat{P}_j} \mathbf{B}' \rangle = \langle \mathbf{B}, \hat{X}_i \, e^{i\lambda \hat{P}_j} \mathbf{B}' \rangle = \langle \mathbf{B}, e^{i\lambda \hat{P}_j} (\hat{X}_i - \lambda \delta_{ij}) \mathbf{B}' \rangle$$
$$= \langle e^{-i\lambda \hat{P}_j} \mathbf{B}, \hat{X}_i \mathbf{B}' \rangle - \lambda \delta_{ij} \langle \mathbf{B}, e^{i\lambda \hat{P}_j} \mathbf{B}' \rangle$$

by (62), and thus

$$\langle \hat{X}_i \mathbf{B}, \hat{P}_j \mathbf{B}' \rangle - \langle \hat{P}_j \mathbf{B}, \hat{X}_i \mathbf{B}' \rangle$$
$$= \frac{1}{i} \frac{d}{d\lambda} (\langle \hat{X}_i \mathbf{B}, e^{i\lambda \hat{P}_j} \mathbf{B}' \rangle - \langle e^{-i\lambda \hat{P}_j} \mathbf{B}, \hat{X}_i \mathbf{B}' \rangle)|_{\lambda = 0}$$
$$= i \frac{d}{d\lambda} (\lambda \delta_{ij} \langle \mathbf{B}, e^{i\lambda \hat{P}_j} \mathbf{B}' \rangle)|_{\lambda = 0} = i\delta_{ij} \langle \mathbf{B}, \mathbf{B}' \rangle$$

From (60) and (61) we obtain Heisenberg's position–momentum uncertainty relation

$$\Delta_\mathbf{A} X_i \cdot \Delta_\mathbf{A} P_j \geq \tfrac{1}{2} \delta_{ij} \quad (63)$$

for all normalized photon states $\mathbf{A}$ of the type specified above. As apparent from (58) for $\Delta_\mathbf{A} X_i$, and from the analog of (56) for $\Delta_\mathbf{A} P_j$, such states are the only ones for which both $\Delta_\mathbf{A} X_i$ and $\Delta_\mathbf{A} P_j$ are finite. For all other states, therefore, (63) is satisfied in a trivial way.

Some readers might find our derivation of (63) rather pedantic. They might feel it would be easier to use the commutation relation

$$[\hat{X}_i, \hat{P}_j] = i\delta_{ij}$$

for an evaluation of the right-hand side of (60) in the form

$$\langle \hat{X}_i \mathbf{A}, \hat{P}_j \mathbf{A} \rangle - \langle \hat{P}_j \mathbf{A}, \hat{X}_i \mathbf{A} \rangle = \langle \mathbf{A}, [\hat{X}_i, \hat{P}_j] \mathbf{A} \rangle = i\delta_{ij}$$

308 Uncertainty Principle and Foundations of Quantum Mechanics

This short-cut calculation, clearly, is not perfectly rigorous. That it may even lead to wrong physical conclusions is explained elsewhere (Kraus, 1970).

As mentioned before (see footnote at beginning of Section 3), the POV measure $F(\Delta)$ describes position measurements at time $t = 0$ in a given inertial frame. Therefore the operators $U(a, \Lambda)F(\Delta)U^*(a, \Lambda)$ describe position measurements at time $t' = 0$ in a 'primed' inertial frame, generated from the original 'unprimed' one by the Poincaré transformation (a, Λ). Of particular interest is the case where (a, Λ) is a pure time translation by the amount t, which leads to the POV measure

$$\left. \begin{array}{l} F_t(\Delta) = U(t)F(\Delta)U^*(t) \\ U(t) = U(\{0, t\}, 1) = e^{iHt} \end{array} \right\} \quad (64)$$

As the corresponding measurements, obviously, may also be interpreted as position measurements at time t in the original inertial frame, equation (64) is nothing but the familiar time dependence of the observables $F_t(\Delta)$ in the Heisenberg picture.

The same transformation $F(\Delta) \to F_t(\Delta)$ is also obtained from

$$\left. \begin{array}{l} \hat{E}_t(\Delta) = \hat{U}(t)\hat{E}(\Delta)\hat{U}^*(t) \\ F_t(\Delta) = \Phi\hat{E}_t(\Delta)|_{\mathcal{H}} \end{array} \right\} \quad (65)$$

with a unitary operator $\hat{U}(t)$ on $\hat{\mathcal{H}}$ satisfying

$$[\hat{U}(t), \Phi] = 0, \quad \hat{U}(t)|_{\mathcal{H}} = U(t) \quad (66)$$

In fact, (65) and (66) imply

$$\begin{array}{rl} F_t(\Delta) = \Phi\hat{E}_t(\Delta)|_{\mathcal{H}} &= \Phi\hat{U}(t)\hat{E}(\Delta)\hat{U}^*(t)|_{\mathcal{H}} \\ &= U(t)\Phi\hat{E}(\Delta)|_{\mathcal{H}}U^*(t) \\ &= U(t)F(\Delta)U^*(t) \end{array}$$

in accordance with (64). The condition (66) is satisfied, for example, by

$$\hat{U}(t) = e^{i\hat{H}t} \quad (67)$$

with the 'natural' extension $\hat{H}$ of the Hamiltonian H described in Section 2. Of course, (66) has many other solutions besides (67), but this particular one is very convenient since it permits the explicit calculation of the self-adjoint operator

$$\hat{\mathbf{X}}_t = \hat{U}(t)\hat{\mathbf{X}}\hat{U}^*(t) \quad (68)$$

which corresponds to the spectral measure $\hat{E}_t(\Delta)$. Indeed, a simple calculation using the explicit expressions

$$\hat{\mathbf{X}} = i\nabla_\mathbf{k}, \quad \hat{U}(t) = e^{i\omega t}$$

yields

$$\hat{\mathbf{X}}_t = \hat{\mathbf{X}} + \hat{\mathbf{V}}t \quad (69)$$

with the (vector) multiplication operator

$$\hat{\mathbf{V}} = \frac{\hat{\mathbf{P}}}{\hat{H}} = \frac{\mathbf{k}}{\omega} \qquad (70)$$

on $\mathcal{H}$. Implicitly equation (69) contains the complete solution of the Heisenberg equation of motion (64) for $F_t(\Delta)$, since $\hat{X}_t$ uniquely determines its spectral measure $\hat{E}_t(\Delta)$ which in turn, via (65), yields $F_t(\Delta)$. This does not mean that (69) is really helpful for the explicit calculation of $F_t(\Delta)$'s. However, in most cases one is satisfied with a much less detailed description of position measurements at different times, and in such cases (69) may be applied directly.

Consider, for instance, the time-dependent expectation value $\langle \mathbf{X}_t \rangle_\mathbf{A}$ of position in a given state $\mathbf{A}$. We obtain

$$\langle \mathbf{X}_t \rangle_\mathbf{A} = \langle \mathbf{A}, \hat{\mathbf{X}}_t \mathbf{A} \rangle = \langle \mathbf{A}, \hat{\mathbf{X}} \mathbf{A} \rangle + \langle \mathbf{A}, \hat{\mathbf{V}} \mathbf{A} \rangle t$$
$$= \langle \mathbf{A}, \mathbf{X}_t \mathbf{A} \rangle = \langle \mathbf{A}, \mathbf{X} \mathbf{A} \rangle + \langle \mathbf{A}, \mathbf{V} \mathbf{A} \rangle t \qquad (71)$$

with

$$\mathbf{X}_t = \Phi \hat{\mathbf{X}}_t |_\mathcal{H} = \mathbf{X} + \mathbf{V} t \qquad (72)$$

and

$$\mathbf{V} = \Phi \hat{\mathbf{V}} |_\mathcal{H} = \frac{\mathbf{P}}{H} = \frac{\mathbf{k}}{\omega} \qquad (73)$$

a multiplication operator on $\mathcal{H}$ which, obviously, has to be interpreted as the photon velocity operator. Its components V_j satisfy the relations

$$-1 \le V_j \le 1, \quad |\mathbf{V}| \equiv (V_1^2 + V_2^2 + V_3^2)^{1/2} = 1 \qquad (74)$$

as to be expected from this interpretation.* As a $\mathbf{k}$ space average with weight function $\overline{\mathbf{A}(\mathbf{k})} \cdot \mathbf{A}(\mathbf{k})$ of the unit vectors $\mathbf{k}/\omega$, the vector $\langle \mathbf{V} \rangle_\mathbf{A} = \langle \mathbf{A}, \mathbf{V} \mathbf{A} \rangle$ (with components $\langle V_j \rangle_\mathbf{A} = \langle \mathbf{A}, V_j \mathbf{A} \rangle$) has a length $|\langle \mathbf{V} \rangle_\mathbf{A}|$ smaller than one.† Thus (71) implies

$$\frac{d}{dt} \langle \mathbf{X}_t \rangle_\mathbf{A} = \langle \mathbf{V} \rangle_\mathbf{A}, \quad |\langle \mathbf{V} \rangle_\mathbf{A}| < 1 \qquad (75)$$

i.e. the time-dependent average photon positions $\langle \mathbf{X}_t \rangle_\mathbf{A}$ lie on a straight time-like worldline.

For the time dependence of the mean square deviation of the jth photon coordinate, an obvious generalization of (58) together with (69) yields

$$\Delta_\mathbf{A} X_{jt} = \|(\hat{X}_{jt} - \bar{x}_{jt}) \mathbf{A}\| = \|(\hat{X}_j - \bar{x}_j) \mathbf{A} + (\hat{V}_j - \bar{v}_j) t \mathbf{A}\|$$

with

$$\bar{x}_j = \langle X_j \rangle_\mathbf{A}, \quad \bar{v}_j = \langle V_j \rangle_\mathbf{A}, \quad \bar{x}_{jt} = \langle X_{jt} \rangle_\mathbf{A} = \bar{x}_j + \bar{v}_j t$$

---

*Note that in our units the light velocity is equal to one.
†The value one is excluded since, for normalized $\mathbf{A}$, the weight function $\overline{\mathbf{A}(\mathbf{k})} \cdot \mathbf{A}(\mathbf{k})$ cannot be a delta function.

and thus
$$\Delta_\mathbf{A} X_{jt} \leq \|(\hat{X}_j - \bar{x}_j)\mathbf{A}\| + |t|\|(\hat{V}_j - \bar{v}_j)\mathbf{A}\|$$

But
$$\|(\hat{X}_j - \bar{x}_j)\mathbf{A}\| = \Delta_\mathbf{A} X_j$$

whereas
$$\|(\hat{V}_j - \bar{v}_j)\mathbf{A}\|^2 = \|(V_j - \bar{v}_j)\mathbf{A}\|^2 = (\Delta_\mathbf{A} V_j)^2$$
$$= \|V_j\mathbf{A}\|^2 - \bar{v}_j^2 \leq \|V_j\mathbf{A}\|^2 < 1$$

in virtue of (74).* Therefore we obtain the estimates
$$\Delta_\mathbf{A} X_{jt} \leq \Delta_\mathbf{A} X_j + \Delta_\mathbf{A} V_j |t|, \quad \Delta_\mathbf{A} V_j < 1 \tag{76}$$

which show that the growth in time of the mean square deviations of photon coordinates (or, if, translated into the Schrödinger picture, of the widths of a wave packet in the directions of the three coordinate axes) is also restricted by the velocity of light. By using rotational invariance [or a simple generalization of (58), compare equation (79)], a similar estimate may be derived for the width $\Delta_\mathbf{A}(\mathbf{e} \cdot \mathbf{X}_t)$ of the wave packet $\mathbf{A}$ in the direction of an arbitrary unit vector $\mathbf{e}$.

The relations (75) and (76) express a certain kind of causal behaviour of position measurements. If, for instance, there were wave packets $\mathbf{A}$ with average positions $\langle \mathbf{X}_t \rangle_\mathbf{A}$ moving faster than light, this would hardly be consistent with relativistic causality, since one could easily imagine the use of such wave packets as faster-than-light signals. Likewise, the existence of wave packets with average widths $\Delta_\mathbf{A} X_{jt}$ growing with superluminar velocity would look suspect from the point of view of causality, although it is not at all obvious how such wave packets could be used to exchange signals between space-like separated observers.

Another causality requirement for successive position measurements would be the following: If, in some state $\mathbf{A}$, the photon position at time $t = 0$ is certainly inside a region Δ, then at any time $t \neq 0$ the photon has to be with certainty in the region
$$\Delta_t = \{\mathbf{x} \mid |\mathbf{x} - \mathbf{y}| \leq t \text{ for all } \mathbf{y} \in \Delta\} \tag{77}$$

i.e.
$$\langle \mathbf{F}, F(\Delta)\mathbf{A} \rangle = 1 \text{ implies } \langle \mathbf{A}, F_t(\Delta_t)\mathbf{A} \rangle = 1 \text{ for all } t \tag{78}$$

As the corresponding property for classical particles is obvious, the postulate (78) seems to be well-founded, too. However, (78) is simply wrong, and is moreover wrong not only for photons and the POV measure $F(\Delta)$ considered here but, quite generally, for all relativistic elementary particles and for all conceivable position observables. More precisely, it turns out that $\langle \mathbf{A}, F_t(\Delta_t)\mathbf{A} \rangle$

---

*Namely, (74) implies $\|V_j\| = 1$, and thus $\|V_j \mathbf{A}\| \leq \|\mathbf{A}\| = 1$. $\|V_j\mathbf{A}\|^2 = \langle \mathbf{A}, V_j^2 \mathbf{A} \rangle = 1$ would mean that $\mathbf{A}$ is an eigenstate of V_j^2 of eigenvalue one, but the spectrum of $V_j^2 = k_j^2/\omega^2$ is purely continuous.

is strictly less than one for *all* states $\mathbf{A}$ with $\langle \mathbf{A}, F(\Delta)\mathbf{A}\rangle = 1$ and *all* times $t \neq 0$. In this generality, the result is due to Hegerfeldt (1974); we refer to his paper for the (surprisingly simple) proof. Of course, the requirement (78) is non-trivial only if there are at least some regions Δ and corresponding states $\mathbf{A}$ with $\langle \mathbf{A}, F(\Delta)\mathbf{A}\rangle = 1$. In our present case, however, it is easily seen that such states $\mathbf{A}$ indeed exist for an arbitrarily given region Δ.*

On the other hand, it is clear from (75) that only the 'tail' of the wave packet can be outside of Δ_t at time t. Indeed, its centre $\langle \mathbf{X}_t\rangle_\mathbf{A}$ is certainly in Δ_t for all t since, at least for convex regions Δ, it is in Δ for $t = 0$. A similar intuitive picture of the spreading of wave packets is suggested by the following estimate. With $\bar{\mathbf{x}} = \langle \mathbf{X}\rangle_\mathbf{A}$ and an arbitrary unit vector $\mathbf{e}$, we define by

$$(\delta(\mathbf{e}, t))^2 = \int (\mathbf{e} \cdot (\mathbf{x} - \bar{\mathbf{x}}))^2 \langle \mathbf{A}, dF_t(\mathbf{x})\mathbf{A}\rangle$$

a measure $\delta(\mathbf{e}, t)$ for the average spatial distance, in the direction of $\mathbf{e}$, of the wave packet at time t from its centre $\bar{\mathbf{x}}$ at time $t = 0$. Since

$$\delta(\mathbf{e}, t) = \|(\mathbf{e} \cdot (\hat{\mathbf{X}}_t - \bar{\mathbf{x}}))\mathbf{A}\|$$

(compare the derivation of (58)), we find from (69):

$$\delta(e, t) \leq \|(\mathbf{e} \cdot (\hat{\mathbf{X}} - \bar{\mathbf{x}}))\mathbf{A}\| + |t| \|(\mathbf{e} \cdot \hat{\mathbf{V}})\mathbf{A}\|$$

Now

$$\Delta_\mathbf{A}(\mathbf{e} \cdot \mathbf{X}) = \|(\mathbf{e} \cdot (\hat{\mathbf{X}} - \bar{\mathbf{x}}))\mathbf{A}\| \equiv \delta(\mathbf{e}, 0) \tag{79}$$

clearly, is the width of the wave packet, measured along the direction of $\mathbf{e}$, at time $t = 0$, whereas

$$\|(\mathbf{e} \cdot \hat{\mathbf{V}})\mathbf{A}\| = \|(\mathbf{e} \cdot \mathbf{V})\mathbf{A}\| \leq \|\mathbf{A}\| = 1$$

by (74). Thus, finally,

$$\delta(\mathbf{e}, t) \leq \Delta_\mathbf{A}(\mathbf{e} \cdot \mathbf{X}) + |t| \tag{80}$$

The physical meaning of this estimate is obvious. Estimates analogous to (75), (76) and (80) may also be derived, e.g., for the Newton–Wigner position operator of a particle with mass $m > 0$. [In this case the velocity operator is $\mathbf{P}/H = \mathbf{P} (\mathbf{P}^2 + m^2)^{-\frac{1}{2}}$.]

We feel that the violation of (78) should not be taken to indicate an 'acausality' until it is proved that its violation indeed makes possible, at least in principle, the exchange of faster-than-light signals. It seems not implausible at least that the impossibility of such signals already follows from estimates like (75) or (80). This problem, clearly, should be investigated further. In any case, however, the violation of (78) is a nice example of how misleading the classical particle picture may be in quantum mechanics.

*Since $\hat{E}(\Delta)$ acts as multiplication by $\chi_\Delta(\mathbf{x})$ on the Fourier transform $\tilde{\mathbf{A}}(\mathbf{x})$ of $\mathbf{A}(\mathbf{k})$, we have $\hat{E}(\Delta)\mathbf{A} = \mathbf{A}$, and thus $\langle \mathbf{A}, F(\Delta)\mathbf{A}\rangle = 1$, if $\tilde{\mathbf{A}}(\mathbf{x}) \equiv 0$ outside of Δ.

5. EFFECTS AND GENERALIZED OBSERVABLES

The simplest measurements are those for which only two different outcomes 'yes' and 'no', or one and zero, are possible. In usual quantum mechanics these 'yes–no' observables are represented by projection operators E on the state space $\mathcal{H}$, such that $\langle \mathbf{A}, E\mathbf{A}\rangle$ is the probability for the outcome one (i.e., 'yes') if E is measured in the normalized state $\mathbf{A} \in \mathcal{H}$. If applied to particle detectors, this formalism immediately leads to the description of particle position by a covariant spectral measure $E(\Delta)$. However, if quantum mechanical yes–no measurements are investigated more closely [e.g. by considering suitable models (Kraus, 1971 and 1974)], one realizes that they do not in general correspond to projection operators. Instead, a general yes–no measurement has to be described by an operator F on $\mathcal{H}$ with

$$F^* = F, \quad 0 \leq F \leq 1 \qquad (81)$$

the probability for the outcome 'yes' in state $\mathbf{A}$ being given again by $\langle \mathbf{A}, F\mathbf{A}\rangle$. The projection operators E are a very particular class of such operators F, and in fact any practically performable yes–no measurement most likely does not correspond to a projection operator. This observation may serve as the starting point for Ludwig's axiomatic reformulation of quantum theory (Ludwig, 1970). General yes–no experiments are called 'effects' in this theory, whereas the particular ones corresponding to projection operators are denoted as 'decision effects'.

The corresponding generalization of the notion of an observable is almost obvious. Usually a quantum-mechanical observable is taken to correspond to a self-adjoint operator X on $\mathcal{H}$, and the spectral measure $E(\Delta)$ on the real line obtained from the spectral representation

$$X = \int x \, dE(x) \qquad (82)$$

is interpreted as follows: For a given interval Δ, $E(\Delta)$ corresponds to the yes–no observable which takes the value one (respectively zero) if for the original observable a value x in (respectively outside of) Δ is measured. The generalization consists of admitting observables for which a general POV measure $F(\Delta)$ on the real line takes the rôle of $E(\Delta)$, with the same physical interpretation.*
From this point of view, then, the use of a covariant POV measure $F(\Delta)$ on $\mathbb{R}^3$ for the description of photon detectors looks quite natural.† The properties of such generalized observables have been illustrated in Section 4 by the example of the photon coordinates X_j. In particular, we have seen that the associated

---

*In particular, we do not interpret POV measures as describing inaccurate ('fuzzy') measurements, as done by Ali and Emch (1974).
†The operators $F(\Delta)$ are interpreted here as describing 'exact' photon positions, i.e. a 'click' in the 'counter' corresponding to $F(\Delta)$ is taken to indicate that the photon is really inside Δ. We feel that one could speak meaningfully of inaccurate position measurements only if there were another, 'more exact' position observable to compare with.

operator

$$X = \int x \, dF(x) \tag{83}$$

does not completely describe a generalized observable.*

One further point, however, is worth mentioning here. By their very definition, any two effects $F(\Delta_1)$ and $F(\Delta_2)$ belonging to the POV measure of a generalized observable can be measured together (e.g. simply by determining the value of the observable with sufficient precision). In Ludwig's terminology, such effects are called 'coexistent'. If one is dealing with a spectral measure $E(\Delta)$, the well-known necessary and sufficient condition for the 'coexistence' (usually called 'commensurability' in this case) of $E(\Delta_1)$ and $E(\Delta_2)$ is commutativity, which is indeed satisfied for any spectral measure (cf. Section 3). More generally, two effects F_1 and F_2 are coexistent if and only if there exist three effects F_1', F_2' and F_3 such that

$$F_1 = F_1' + F_3, \quad F_2 = F_2' + F_3, \quad F_1' + F_2' + F_3 \leq 1 \tag{84}$$

(See Ludwig (1970), or Kraus (1974) for a more elementary discussion.) Commutativity of F_1 and F_2 is sufficient to guarantee the validity of (84), but is necessary only if F_1 or F_2 or both are projection operators. For $F_1 = F(\Delta_1)$ and $F_2 = F(\Delta_2)$ belonging to a POV measure, (84) is satisfied in virtue of the measure property (51), which implies

$$F(\Delta_1) = F(\Delta_1 \cap \Delta_2') + F(\Delta_1 \cap \Delta_2)$$
$$F(\Delta_2) = F(\Delta_2 \cap \Delta_1') + F(\Delta_1 \cap \Delta_2)$$

and

$$F(\Delta_1 \cap \Delta_2') + F(\Delta_2 \cap \Delta_1') + F(\Delta_1 \cap \Delta_2)$$
$$= F((\Delta_1 \cap \Delta_2') \cup (\Delta_2 \cap \Delta_1') \cup (\Delta_1 \cap \Delta_2)) = F(\Delta_1 \cup \Delta_2) \leq 1$$

The attempt of Jauch and Piron (1967) and Amrein (1969) of constructing a photon position observable is closely related to the one discussed here. These authors, however, insist on the description of yes–no observables by projection operators, and therefore do not accept $F(\Delta)$ as describing a photon counter in the space region Δ. They take instead, for this purpose, the projection operator $E'(\Delta)$ onto the subspace of eigenvectors of $F(\Delta)$ belonging to the eigenvalue one. An equivalent definition is $E'(\Delta) = \Phi \cap \hat{E}(\Delta)|_{\mathcal{H}}$, with the projection operator $\Phi \cap \hat{E}(\Delta)$ onto the intersection of the subspaces $\mathcal{H} = \Phi \hat{\mathcal{H}}$ and $\hat{E}(\Delta) \hat{\mathcal{H}}$ of $\hat{\mathcal{H}}$. The covariance condition (29) and the first two relations of (26) are easily seen to be satisfied for the operators $E'(\Delta)$. The third (additivity) condition of (26) has thus to be violated, since otherwise $E'(\Delta)$ would be a covariant spectral measure. Because this additivity condition has a direct physical interpretation,

---

*There are even POV measures $F(\Delta)$ for which (83) makes sense only if applied to the zero vector, so that there is no operator X at all. However, such 'observables' are pathological also from the physical point of view.

we consider its violation as a serious disadvantage. Moreover, there are certain pairs of regions for which the corresponding operators $E'(\Delta)$ do not commute, and thus do not describe commensurable measurements.

6. THE UNIQUENESS PROBLEM

At first sight the method used in Section 3 for constructing the Euclidean covariant POV measure $F(\Delta)$ for the photon might look somewhat fortuitous. This is not the case, however, as the following discussion shows. We start with

THEOREM 1:

(1). Consider a Hilbert space $\mathcal{H}$ with a POV measure $F(\Delta)$ on $\mathbb{R}^3$. Then there exists an extended Hilbert space $\hat{\mathcal{H}} \supset \mathcal{H}$ with a spectral measure $\hat{E}(\Delta)$ on $\mathbb{R}^3$, such that

$$F(\Delta) = \Phi \hat{E}(\Delta)|_{\mathcal{H}}$$

Φ being the projection operator on $\hat{\mathcal{H}}$ with range $\mathcal{H}$.

(2). Let $F(\Delta)$ be covariant with respect to a continuous unitary representation $U(\mathbf{a}, R)$ of the Euclidean group on $\mathcal{H}$. Then there exists a continuous unitary representation $\hat{U}(\mathbf{a}, R)$ on $\hat{\mathcal{H}}$ which extends $U(\mathbf{a}, R)$, i.e.

$$[\hat{U}(\mathbf{a}, R), \Phi] = 0, \quad \hat{U}(\mathbf{a}, R)|_{\mathcal{H}} = U(\mathbf{a}, R)$$

such that $\hat{E}(\Delta)$ is covariant with respect to $\hat{U}(\mathbf{a}, R)$.*

(3). An extension $\hat{\mathcal{H}}$ of $\mathcal{H}$ as described under (1) is called minimal if $\hat{\mathcal{H}}$ is spanned by vectors of the form $\hat{E}(\Delta)\mathbf{A}$, with arbitrary regions Δ and arbitrary vectors $\mathbf{A} \in \mathcal{H}$. The space $\hat{\mathcal{H}}$ and the spectral measure $\hat{E}(\Delta)$ (and, for covariant $F(\Delta)$, also the representation $\hat{U}(\mathbf{a}, R)$) of a minimal extension are unique up to unitary equivalence. A non-minimal extension $\hat{\mathcal{H}}$ contains a subspace which reduces the spectral measure $\hat{E}(\Delta)$ (and the representation $\hat{U}(\mathbf{a}, R)$, if $F(\Delta)$ is covariant), and which is a minimal extension of $\mathcal{H}$.

This Theorem is also true for POV measures on arbitrary spaces (instead of $\mathbb{R}^3$) and for more general covariance groups. In this general form, parts (1) and (3) of the Theorem are due to Neumark (1943) [see also Riesz and Nagy (1956)] whereas part (2) has been proved recently by Neumann (1972).

If one wants to construct a covariant POV measure $F(\Delta)$ on a Hilbert space $\mathcal{H}$ with a given representation $U(\mathbf{a}, R)$ of the Euclidean group, Theorem 1 suggests the following procedure: First, look for a suitable extension $\hat{U}(\mathbf{a}, R)$ of $U(\mathbf{a}, R)$ to a larger Hilbert space $\hat{\mathcal{H}}$, such that on $\hat{\mathcal{H}}$ there exists a spectral measure $\hat{E}(\Delta)$ which is covariant with respect to $\hat{U}(\mathbf{a}, R)$; then, take $F(\Delta) = \Phi \hat{E}(\Delta)|_{\mathcal{H}}$. It is easily shown that this $F(\Delta)$ is indeed a covariant POV measure,

---

*If one is dealing with particles of half-integer spin, then U and $\hat{U}$ are representations not of the Euclidean group itself but of its covering group [cf., for example, Wightmann (1962)].

whereas Theorem 1 guarantees that every covariant POV measure may be constructed in this way. A further advantage of this construction is the fact that all representations $\hat{U}(\mathbf{a}, R)$ of the Euclidean group which admit a covariant spectral measure $\hat{E}(\Delta)$ are explicitly known up to unitary equivalence:

THEOREM 2: Consider a Hilbert space $\hat{\mathcal{H}}$ with a continuous unitary representation $\hat{U}(\mathbf{a}, R)$ of the Euclidean group and a covariant spectral measure $\hat{E}(\Delta)$ on $\mathbb{R}^3$. Then there is a unitary transformation which brings $\hat{\mathcal{H}}$, $\hat{U}(\mathbf{a}, R)$ and $\hat{E}(\Delta)$ to the following standard form:

(1). $\hat{\mathcal{H}}$ consists of all complex 'vector' functions $\mathbf{f}(\mathbf{k})$ of a real three-vector $\mathbf{k}$, with 'vector' components $f_\alpha(\mathbf{k})$, $\alpha \in I$ (some finite or infinite index set), which are square-integrable in the sense that

$$\int d^3k \sum_{\alpha \in I} |f_\alpha(\mathbf{k})|^2 < \infty$$

The inner product in $\hat{\mathcal{H}}$ is

$$\langle \mathbf{f}, \mathbf{f}' \rangle = \int d^3k \sum_{\alpha \in I} \overline{f_\alpha(\mathbf{k})} f'_\alpha(\mathbf{k})$$

(2). $\hat{U}(\mathbf{a}, R)$ is given by

$$(\hat{U}(\mathbf{a}, R)f)_\alpha(\mathbf{k}) = e^{-i\mathbf{k} \cdot \mathbf{a}} \sum_{\beta \in I} D_{\alpha\beta}(R) f_\beta(R^{-1}\mathbf{k}) \qquad (85)$$

with a continuous unitary representation of the rotation group by matrices $D(R)$ with matrix elements $D_{\alpha\beta}(R)$, $\alpha, \beta \in I$.

(3). $\hat{E}(\Delta)$ is the spectral measure of the self-adjoint 'position' operator

$$\hat{\mathbf{X}} = \int \mathbf{x} \, d\hat{E}(\mathbf{x})$$

whose components $\hat{X}_j$ are defined by*

$$(\hat{X}_j f)_\alpha(\mathbf{k}) = i \frac{\partial}{\partial k_j} f_\alpha(\mathbf{k})$$

This Theorem plays the crucial rôle in the paper of Wightman (1962), where a detailed proof is given.

Representations $\hat{U}(\mathbf{a}, R)$ of the Euclidean group of the form (85) are highly reducible. First of all, the unitary representation $D(R)$ of the rotation group may be decomposed in the usual way into irreducible representations $D_S(R)$ with fixed angular momentum quantum number ('spin') S,† of dimension

---

*In the 'position' representation, i.e. in terms of the Fourier transforms $\tilde{\mathbf{f}}(\mathbf{x})$ of $\mathbf{f}(\mathbf{k})$, the operators $\hat{X}_j$ and $\hat{E}(\Delta)$ act as multiplication by x_j and $\chi_\Delta(\mathbf{x})$, respectively.
†In our case each S is integer, whereas half-integer S occur in the case mentioned in the footnote to Theorem 1.

$2S+1$, which decomposes $\hat{U}(\mathbf{a}, R)$ into subrepresentations $\hat{U}_S(\mathbf{a}, R)$. For each subrepresentation $\hat{U}_S(\mathbf{a}, R)$, the representation space may be further decomposed into $2S+1$ subspaces with definite helicities $-S, -S+1 \ldots S-1, S$, which further reduce $\hat{U}_S(\mathbf{a}, R)$ since helicity is Euclidean invariant.* (For details see, for example, Amrein (1969).) On the other hand, the representation $U(\mathbf{a}, R)$ on the photon state space $\mathcal{H}$ may be decomposed into subrepresentations with helicities $+1$ and -1 (cf. Section 2). Thus Theorem 2 forbids the existence of a covariant spectral measure on $\mathcal{H}$ which would require, at least, the presence also of helicity zero states. The simplest way of extending $U(\mathbf{a}, R)$ to a representation $\hat{U}(\mathbf{a}, R)$ of the type (85) is, therefore, to add just this 'missing' subrepresentation of helicity zero. This leads to a representation $\hat{U}(\mathbf{a}, R)$ of the form (85), with $D(R) = R$ irreducible and belonging to $S = 1$, and exactly this has been done in Section 3. In the light of Theorems 1 and 2, therefore, the construction of $F(\Delta)$ in Section 3 appears quite natural.

However, it is obvious now that this construction is only the simplest but not a unique one. There are very many different possibilities of embedding $U(\mathbf{a}, R)$ into representations of the form (or unitary equivalent to) (85), which in general lead to different POV measures. We will illustrate this non-uniqueness by two simple examples.

For instance, we can embed $U(\mathbf{a}, R)$ by adding, besides the missing helicity zero states, two other subrepresentations of helicities $+2$ and -2, so that we obtain a representation $\hat{U}'(\mathbf{a}, R)$ with $D'(R)$ belonging to $S = 2$. (The primes serve to distinguish the present construction from the one considered in Section 3.) This representation may be realized concretely in the Hilbert space $\mathcal{H}'$ of complex symmetric traceless second-rank tensors $g_{ij}(\mathbf{k})$ ($i, j = 1, 2, 3$) with the inner product

$$\langle g, g' \rangle = \int d^3 \mathbf{k} \, \overline{g_{ij}(\mathbf{k})} g'_{ij}(\mathbf{k}) \tag{86}$$

(sum convention) and the tensor transformation law

$$(\hat{U}'(\mathbf{a}, R)g)_{ij}(\mathbf{k}) = e^{-i\mathbf{k}\cdot\mathbf{a}} R_{ii'} R_{jj'} g_{i'j'}(R^{-1}\mathbf{k}) \tag{87}$$

The photon state space $\mathcal{H}$ of transverse vector functions $\mathbf{A}(\mathbf{k})$ may be embedded isometrically in $\mathcal{H}'$ by identifying a given $\mathbf{A} \in \mathcal{H}$ with the tensor

$$g_{ij}(\mathbf{k}) = \frac{1}{\sqrt{2}} \left(\frac{k_i}{\omega} A_j(\mathbf{k}) + \frac{k_j}{\omega} A_i(\mathbf{k}) \right) \tag{88}$$

in $\mathcal{H}'$. As easily checked, the transformation law (87) for tensors g_{ij} of the particular form (88) is equivalent to the vector transformation law (39) for $\mathbf{A}$, so that, by the embedding (88), $U(\mathbf{a}, R)$ becomes a subrepresentation of $\hat{U}'(\mathbf{a}, R)$. The projection operator Φ' on $\mathcal{H}'$ with range $\mathcal{H}$ transforms a given

---

*The subrepresentations with definite helicities are still highly reducible, since the absolute value of momentum $\hat{\mathbf{P}} = \mathbf{k}$ is also Euclidean invariant.

tensor $g_{ij} \in \hat{\mathcal{H}}'$ into a vector $\mathbf{A} \in \mathcal{H}$ with components

$$A_i(k) = \sqrt{2}\left(\frac{k_j}{\omega}g_{ij}(\mathbf{k}) - \frac{k_ik_jk_l}{\omega^3}g_{jl}(\mathbf{k})\right) \tag{89}$$

Since this extension $\hat{\mathcal{H}}'$ of $\mathcal{H}$ is of the form required in Theorem 2, the self-adjoint operator $\hat{\mathbf{X}}' = i\nabla_{\mathbf{k}}$ with covariant spectral measure $\hat{E}'(\Delta)$ exists on $\hat{\mathcal{H}}'$, and leads to a covariant POV measure

$$F'(\Delta) = \Phi'\hat{E}'(\Delta)|_{\mathcal{H}} \tag{90}$$

and a position operator

$$\mathbf{X}' = \Phi'\hat{\mathbf{X}}'|_{\mathcal{H}} \tag{91}$$

on the photon states space $\mathcal{H}$. A straightforward calculation with (88), (89) and (91) leads to

$$X_j'\mathbf{A} = i\left(\frac{\partial}{\partial k_j}\mathbf{A} + \frac{\mathbf{k}}{\omega^2}A_j\right)$$

for an arbitrary $\mathbf{A} \in \mathcal{H}$. This coincides with $X_j\mathbf{A}$ as given by (43), and thus $\mathbf{X}' = \mathbf{X}$. On the other hand, the POV measure $F'(\Delta)$ is different from $F(\Delta)$ as constructed in Section 3 (see below). Since, therefore,

$$\mathbf{X} = \int \mathbf{x}\, dF(\mathbf{x}) = \int \mathbf{x}\, dF'(\mathbf{x})$$

but

$$F(\Delta) \neq F'(\Delta)$$

we have here an example for the non-uniqueness of the POV measure corresponding to a given operator $\mathbf{X}$.

As an example for a covariant POV measure $F''(\Delta)$ for which the corresponding position operator $\mathbf{X}''$ is different from $\mathbf{X}$, consider

$$F''(\Delta) = E_+F(\Delta)E_+ + E_-F(\Delta)E_- \tag{92}$$

with $F(\Delta)$ as in Section 3 and the projection operators $E_\pm$ onto the subspaces $\mathcal{H}_\pm$ of $\mathcal{H}$ belonging to the helicities ± 1. Obviously (92) defines a POV measure, whose covariance follows from $[E_\pm, U(\mathbf{a}, R)] = 0$. The corresponding position operator is

$$\mathbf{X}'' = \int \mathbf{x}\, dF''(\mathbf{x}) = E_+\mathbf{X}E_+ + E_-\mathbf{X}E_- \tag{93}$$

and is different from $\mathbf{X}$ since $[\mathbf{X}, E_\pm] \neq 0$ (as easily checked by direct calculation) whereas, clearly, $[\mathbf{X}'', E_\pm] = 0$.

Both examples may be easily generalized. Embedding of $U(\mathbf{a}, R)$ into $\hat{U}_S(\mathbf{a}, R)$ with $S = 1, 2, 3 \ldots$, as described above for $S = 1$ and 2, leads to an infinite sequence of covariant POV measures $F_S(\Delta)$, $S = 1, 2, 3 \ldots$, which are all different. We can simply show this as follows. It is known (and follows easily from Theorem 2) that, for any given S, all unitary operators $\hat{U}_S(\mathbf{a}, R)$ together with all spectral projections $\hat{E}_S(\Delta)$ form an irreducible set of operators on the representation space $\hat{\mathcal{H}}_S$. Statement (3) of Theorem 1 then implies that the extensions $\hat{\mathcal{H}}_S$ of $\mathcal{H}$ are minimal and, consequently, that $F_S(\Delta) \neq F_{S'}(\Delta)$ if $S \neq S'$.* Instead of (92) we may consider, more generally,

$$F''(\Delta) = \sum_{i,S} A_{iS}^* F_S(\Delta) A_{iS} \tag{94}$$

with $F_S(\Delta)$ as above and (finitely or infinitely many) operators A_{iS} satisfying

$$[A_{iS}, U(\mathbf{a}, R)] = 0, \quad \sum_{i,S} A_{iS}^* A_{iS} = 1 \tag{95}$$

It may be proved from (95) that (94) indeed defines a covariant POV measure.† Since the representation $U(\mathbf{a}, R)$ is highly reducible, there are very many different sets of operators A_{iS} which satisfy (95) and which, in general, will also lead to different POV measures $F''(\Delta)$. By Theorem 1, each $F''(\Delta)$ may also be obtained from a suitable extension of $\mathcal{H}$ and $U(\mathbf{a}, R)$. However, except for particularly simple cases like (92), such an extension is expected to look rather complicated.

We have seen that the covariant POV measure $F(\Delta) \equiv F_1(\Delta)$ constructed in Section 3 is very far from being unique. On the contrary, the diversity of possible candidates for a photon position observable might appear really bewildering. Moreover, any covariant POV measure leads to Heisenberg's position–momentum uncertainty relation (63) and to the photon velocity operator $\mathbf{V}$ given by (73), and is thus acceptable as a photon position observable also from this point of view. This follows from Theorem 1 by a straightforward generalization of the reasoning applied in Section 4. One could try to reduce this non-uniqueness, as done by Wightman (1962) for the particular case of spectral measures, by exploiting suitable additional postulates like time reversal invariance and 'smoothness' in momentum space. With such additional assumptions Wightman was able to prove uniqueness of the Newton–Wigner position observables. However, one cannot hope to obtain uniqueness by this method in the case of general POV measures unless Wightman's additional postulates are sharpened considerably because, for example, all $F_S(\Delta)$ and many $F''(\Delta)$ of the form (94) are both time reversal invariant and 'smooth' in momentum space.

Therefore we do not believe that the 'true' photon position observable, i.e. the one which describes real position measurements, can be determined by purely kinematic considerations. In this respect we fully agree with Wightman (1962), who wrote: 'All investigations of localizability for relativistic particles up to now ... construct position observables consistent with a given transformation law. It remains to construct complete dynamical theories ... and then

---

*Presumably, however, the corresponding position operators $\mathbf{X}_S$ are all equal. (At least $\mathbf{X}_1 = \mathbf{X}_2$, see above.)
†This is trivial if one is dealing with finitely many operators A_{iS}.

to investigate whether the position observables are indeed observable with the apparatus that the dynamical theories themselves predict.' At present the only candidate for a 'complete dynamical theory' of elementary particles is quantum field theory, and photon localization experiments should thus be investigated in the framework of quantum electrodynamics if one wants to go beyond pure kinematics. Such an investigation is also expected to allow a more profound treatment of the causality problem mentioned at the end of Section 4. Since quantum electrodynamics describes photons as 'quanta of a vector field', we are tempted to speculate that the 'vector' POV measure $F_1(\Delta)$ of Section 3 might be distinguished from this point of view. An additional argument in favour of $F_1(\Delta)$ is simplicity. It is therefore not unreasonable to consider $F_1(\Delta)$, in spite of its non-uniqueness, as describing actually realizable photon detectors.

The generalization of the present discussion to other elementary particles is almost obvious. For a massive particle, for instance, one finds that there exist infinitely many generalized position observables besides the usual Newton–Wigner position operator. The latter, however, is distinguished by the fact that it is the only 'ordinary' position observable.* For this reason, the non-uniqueness problem appears not to be so serious in this case. Like the photon, also the neutrino does not possess an 'ordinary' position observable (Wightman, 1962) whereas it is very easy to construct, via Theorems 1 and 2, generalized position observables. Again one of them is distinguished by simplicity. It is an additional advantage of the present approach as compared to the one of Amrein, Jauch and Piron that the latter does not provide a theoretical description of neutrino position measurements (Amrein, 1969).

Acknowledgment

I would like to thank Georg Reents and Michael Everitt for critical readings of the manuscript.

REFERENCES

Ali, S. T. and Emch, G. G. (1974) 'Fuzzy observables in quantum mechanics', *J. Math. Phys.*, **15**, 176–182.
Amrein, W. O. (1969) 'Localizability for particles of mass zero', *Helv. Phys. Acta*, **42**, 149–190.
Hegerfeldt, G. C. (1974) 'Remark on causality and particle localization', *Phys. Rev.*, **D10**, 3320–3321.
Heisenberg, W. (1927) 'Über den anschaulichen Inhalt der quantentheoretischen Kinematik und Mechanik', *Z. Physik*, **43**, 172–198.
Jauch, J. M. and Piron, C. (1967) 'Generalized localizability', *Helv. Phys. Acta*, **40**, 559–570.
Kraus, K. (1970) 'Note on azimuthal angle and angular momentum in quantum mechanics', *Amer. J. Phys.*, **38**, 1489–1490.

*Note that the 'decision effects' form a distinct class of 'effects' in Ludwig's theory (Ludwig, 1970). This implies a corresponding distinction of 'ordinary' (so-called 'decision') observables.

Kraus, K. (1971) 'General state changes in quantum theory', *Ann. Phys. (N.Y.)*, **64**, 311–335.

Kraus, K. (1974) 'Operations and effects in the Hilbert space formulation of quantum theory', *Lecture Notes in Physics*, (Springer-Verlag), **29**, 206–229

Ludwig, G. (1970) 'Deutung des Begriffs "physikalische Theorie" und axiomatische Grundlegung der Hilbertraumstruktur der Quantenmechanik durch Hauptsätze des Messens', *Lecture Notes in Physics*, (Springer-Verlag), **4**.

Neumann, H. (1971) 'Classical systems and observables in quantum mechanics', *Commun. Math. Phys.*, **23**, 100–116.

Neumann, H. (1972) 'Transformation properties of observables', *Helv. Phys. Acta*, **45**, 811–819.

Neumark, M. A. (1943) 'On a representation of additive operator set functions', *Doklady Acad. Sci. URSS*, **41**, 359–361.

Newton, T. D. and Wigner, E. P. (1949) 'Localized states for elementary systems', *Revs. Mod. Phys.*, **21**, 400–406.

Riesz, F. and Sz.-Nagy, B. (1956), *Vorlesungen über Funktionalanalysis*, Anhang. Berlin: VEB Deutscher Verlag der Wissenschaften.

Wightman, A. S. (1962) 'On the localizability of quantum mechanical systems', *Revs. Mod. Phys.*, **34**, 845–872.

Wigner, E. (1939) 'Unitary representations of the inhomogeneous Lorentz group', *Ann. Math.*, **40**, 149–204.

A New Theoretical and Experimental Outlook on Magnetic Monopoles

ERASMO RECAMI
Università di Catania, Italy
and
ROBERTO MIGNANI
Università dell' Aquila, Italy

Since experiments looking for magnetic monopoles have failed until now, and new experiments are going on, it should be interesting to know—and to take into account—the predictions of the mere special relativity on the subject.

We are going to show that the mere special relativity:

(1). Does *not* explicitly predict the existence of (slower-than-light) magnetic monopoles;

(2). It *does* explicitly predict, on the contrary, the existence of tachyonic (i.e. faster-than-light) 'monopoles';

(3). Their unit magnetic-charge appears predicted to be about one hundred times less than that usually assumed (Dirac, 1931, 1948; Schwinger, 1966) ($g = \pm e$, in Gaussian units);

(4). Many good features of the old hypothesis about magnetic monopoles (Dirac, 1931, 1948; Schwinger, 1966) are reproduced by simply taking account of Superluminal ($v^2 > c^2$) speeds. In particular, the existence of both subluminal ($v^2 < c^2$) and Superluminal 'electric' charges leads to *fully symmetrical* Maxwell equations (Mignani and Recami, 1974b), cf. equation (1) in the following, and possibly to the *Schwinger-type* relation: $eg = n\alpha\hbar$.

In fact, let us build anew the theory of special relativity *without* assuming *a priori* $|v| < c$ (Recami and Mignani, 1974a). In other words, let us start from the postulates:

(1). Principle of relativity: the laws of mechanics and of electromagnetism are covariant under a transition between two inertial frames, whose relative speed u is *a priori* $-\infty < u < +\infty$.

(2). Space is isotropic and space–time homogeneous. Moreover, negative-energy particles do *not* exist and for every observer, physical signals are transported only by positive-energy objects. The usefulness of the last sentence—even in *standard* Relativity!—has been shown by us, e.g. in Recami and Mignani (1974a).

There follows an 'extended relativity' (Recami and Mignani, 1974a and references therein), in which light speed is invariant with respect to *all* inertial frames, both subluminal (s) and Superluminal (S), and in which tachyons do *not* imply (see, e.g., Recami, 1973; Pavšič, Recami and Ziino, 1976) any causality violation. What is more, the 'extended relativity' proved to be useful even for standard particle physics, since for example it allowed the derivation of the 'crossing relations' for the relativistic reactions (Mignani and Recami, 1974e), and the CPT theorem (Mignani and Recami, 1974d). It leads, incidentally, to suitable redefinitions of the discrete symmetries.

The point we want to stress here is the following. If we consider the existence of *electric* charges both subluminal [with four-current $j_\mu(s) \equiv (\rho(s), \mathbf{j}(s))$] and Superluminal [whose four-current is $j_\mu(S) \equiv (\rho(S), \mathbf{j}(S))$], then the (generalized) Maxwell equations read (Mignani and Recami, 1974b), for $v^2 \gtreqless c^2$,

$$\left. \begin{array}{l} \text{div } \mathbf{D} = +\rho(s) \\ \text{div } \mathbf{B} = -\rho(S) \\ \text{rot } \mathbf{E} = -\partial \mathbf{B}/\partial t + \mathbf{j}(S) \\ \text{rot } \mathbf{H} = +\partial \mathbf{D}/\partial t + \mathbf{j}(s) \end{array} \right\} \begin{array}{l} [v^2 \gtreqless c^2; \\ s \leftrightarrow |v| < c; \\ S \leftrightarrow |v| > c] \end{array} \quad (1)$$

That is to say, faster-than-light electric charges are predicted by (extended) relativity to behave in a similar way *as* magnetic monopoles were supposed to do (apart from the different speed!)—cf. also Figure 1 in Mignani and Recami (1974b). In better words, a Superluminal electric, *positive* charge (e.g. with speed $V > c$, along the x axis) will bring into the field equations a contribution similar to that which was supposed to come from a magnetic *south* pole (with $v = c^2/V$, along x), (Recami and Mignani, 1974a). Thus, 'tachyonic electrons' will appear with north magnetic charge ($+g$), and 'tachyonic protons' with south magnetic charge ($-g$); and so on. Therefore, a Superluminal unit electric charge e will appear to us as a (tachyonic) 'monopole' with possibly the unit magnetic charge:

$$g = -e \quad \text{(in Gaussian units)} \quad (2)$$

so that in general we expect to have (when quantizing):

$$eg = \pm \alpha \hbar \quad (3)$$

where α is the fine-structure constant. It follows that *relativity seems to predict a magnetic strength unit about 100 times less than that usually assumed*.

In other words, extended relativity predicts only one charge (let us call it 'electromagnetic charge'), which behaves—if you like—as 'electric' when subluminal ($v^2 < c^2$) and as 'magnetic' when Superluminal ($v^2 > c^2$). Cf. again Figure 1 in Mignani and Recami (1974b). Also, *Maxwell's equations may be* written in a *fully symmetrical form* [cf. equation (1)], *without assuming (subluminal) monopole existence*.

What is more, the *universality* of electromagnetic interactions is recovered in extended relativity, since $|g|=|e|$, i.e. only one coupling constant essentially exists in our framework even *before* quantizing the theory.

When passing to quantum mechanics, on the contrary, we can say the following. If one assumes the existence of *subluminal* magnetic monopoles, then the simultaneous quantization of both electric and magnetic charges follows. This *might* suggest that even subluminal magnetic monopoles could exist, with their large unit charge. Notice, however, that the previous argument would be the *only* one in favour of subluminal magnetic charges, since, for example, in the present theory Maxwell equations *already* have a fully symmetrical form (moreover, that argument would become even weaker if we actually succeed—when quantizing our theory—in deriving a relation like equation (3), which would also yield *too* a charge quantization).

Here, let us mention only the following, in order to support our equation (2): (i) The Dirac relation $eg = n\hbar/2$ (or the analogous one by Schwinger) does come in the theory only when magnetic monopoles are supposed to be *subluminal*; (ii) On the contrary, if magnetic monopoles are considered to be Superluminal, then 'extended relativity' seems to yield the *alternative* relation $g = ne$.

In fact, let us eventually quantize our theory by using Mandelstam's method, i.e. by following Cabibbo and Ferrari (1962). In that approach, the field quantities describing the charges (in interaction with the electromagnetic field) are defined so that:

$$\phi(x, P') = \phi(x, P) \cdot \exp\left[-\frac{ie}{2}\int_S F_{\mu\nu}\, d\sigma_{\mu\nu}\right]$$

where S is a surface delimited by the two (space-like) paths P, P' considered, ending at point x. In other words, the field quantities ϕ are independent of the gauge chosen for the fourpotential A_μ but *are* path-dependent. When only (subluminal) electric charges are present, then $F_{\mu\nu} = A_{\nu/\mu} - A_{\mu/\nu}$ and equation (4) does *not* depend on the selected surface S (but depends merely on its boundary $P - P'$). However, if subluminal magnetic monopoles are present too, then $F_{\mu\nu} = A_{\nu/\mu} - A_{\mu/\nu} - i\varepsilon_{\mu\nu\rho\sigma}B_{\sigma/\rho}$ (where B_μ is a *second* fourpotential), and the following *condition* must be explicitly imposed:

$$\exp\left[-\frac{ie}{2}\oint_{S-S'} F_{\mu\nu}\, d\sigma_{\mu\nu}\right] = 1$$

wherefrom Dirac's relation $eg = n\hbar/2$ follows. At this point, it is immediate to realize that, if 'magnetic monopoles' cannot be put at rest, as in the case of *tachyon monopoles*, then equation (4) is *again* automatically satisfied, *without any recourse to Dirac's condition*.

CONCLUSIONS

According to (extended) relativity, all the experimental searches for magnetic monopoles should be done, or redone, by actually looking for 'tachyon monopoles'; i.e. taking into account the newly proposed kinematics (faster-than-light speeds) and the possibly much lower value of the apparent magnetic strength. In particular, the 'tachyon monopoles' will probably suffer in an electromagnetic field the 'Lorentz force' $\mathbf{F} = g\mathbf{H} - g\mathbf{V} \wedge \mathbf{E}$, where however, $V^2 > c^2$. Actually, Bartlett and Lahana (1972) have tried already to look for 'tachyon monopoles', but in vain because the basis of their theoretical assumptions—Cherenkov radiation *supposedly* emitted by tachyons in vacuum (!)—is incorrect as has been shown by us (Mignani and Recami, 1974a). More details can be found in Recami and Mignani (1976) and in the proceedings (to appear) of the interdisciplinary seminars on 'Tachyons and Related Topics' delivered at ERICE (September, 1976).

ACKNOWLEDGEMENT

The authors are grateful to Dr. S. Chissick and to Dr. E. Papp for their kind interest.

REFERENCES

Bartlett, D. F. and Lahana, M. D. (1972) *Phys. Rev.*, **D6,** 1817.
Cabibbo, N. and Ferrari, E. (1962) *Nuovo Cimento,* **23** 1147.
Dirac, P. A. M. (1931) *Proc. Roy. Soc.*, **A133,** 60.
Dirac, P. A. M. (1948) *Phys. Rev.*, **74,** 817.
Mignani, R. and Recami, E. (1974a) *Lett. Nuovo Cimento,* **9,** 362.
Mignani, R. and Recami, E. (1974b) *Lett. Nuovo Cimento,* **9,** 367.
Mignani, R. and Recami, E. (1974c) *Lett. Nuovo Cimento,* **11,** 417.
Mignani, R. and Recami, E. (1974d) *Lett. Nuovo Cimento,* **11,** 421.
Mignani, R. and Recami, E. (1974e) *Nuovo Cimento,* **A24,** 438.
Recami, E. and Mignani, R. (1976) *Physics Letters,* **62B,** 41.
Recami, E. (1973) *Annuario 73, Enciclopedia EST-Mondadori* (Milano), p. 85.
Recami, E. and Mignani, R. (1974a) *Rivista Nuovo Cimento,* **4,** 209–290; [Erratum] **4,** 398.
Recami, E. and Mignani, R. (1974b) *Lett. Nuovo Cimento,* **9,** 479.
Pavšič, M., Recami, E. and Ziino, G. (1976) *Lett. Nuovo Cimento,* in press.
Schwinger, J. (1966) *Phys. Rev.*, **144,** 1087.

17

Problems in Conformally Covariant Quantum Field Theory

W. RÜHL and B. C. YUNN
Universität Kaiserslautern, Germany

1. INTRODUCTION

The conformal group appeared in physics as early as 1909 when Cunningham (1909) and Bateman (1910) first noticed that Maxwell's equations are not only Lorentz covariant but also covariant under the larger conformal group. This consists of the usual Lorentz transformations, translations, dilations and special conformal transformations. Since then many attempts have been made to somehow utilize this group in physics (Kastrup, 1962, 1964; Wess, 1960; Fulton and coworkers, 1962; Mack and Salam, 1969).

We are particularly interested in the possibility of constructing a local quantum field theory which is also conformally covariant. One particular feature of such a field theory is that it possesses global operator product expansions of the Wilson type (Wilson, 1969). This may have far-reaching consequences in more realistic field theories involving non-zero masses as well.

The requirement of the conformal symmetry is so strong that the most general two- and three-point functions are determined completely up to arbitrary normalization constants. Therefore, for example, their analytic structures can be studied unambiguously. So far there are two non-perturbative approaches of analysing a conformally covariant quantum field theory. One is the so-called bootstrap approach which tries to construct general n-point functions from the skeleton graph expansion using the conformally covariant two- and three-point functions. This was initiated by Migdal (1971), Mack and Todorov (1973) and Polyakov (1969). In this approach one can indeed prove that every term in the expansions is ultraviolet convergent if one restricts the anomalous dimensions of the fields to a certain range. Thus the construction is term by term conformally covariant. The dimensions and coupling constants, however, are not free parameters, instead they are determined from self-consistency conditions which arise from integral equations for the two- and three-point functions. The main drawback of this bootstrap approach lies in the inability to handle the infinite series appearing in the expansions just as in conventional perturbation theory.

Another approach adopted by Mack (1974) to avoid this difficulty starts by writing down an infinite number of coupled integral equations for Euclidean Green's functions and solving them by making use of conformal partial wave expansions. A remarkable observation is that in this way one can diagonalize the whole set of integral equations and thereby reduce them to a set of algebraic equations for the partial wave amplitudes. A careful analysis shows that these waves must possess some poles, and the factorization property of their residues also follows when one considers them as analytic functions of the representation parameters. This in turn enables one to derive asymptotic operator product expansions with a certain additional assumption. This program, intensively pursued in recent years by Dobrev and coworkers (1975a, 1975b), has also some intrinsic difficulties. In particular imposing crossing symmetry on the partial wave amplitudes is difficult to carry out. In any case it remains to be seen whether the latter difficulty is easier to handle than the infinite summation problem in the bootstrap approach.

Formulating the conformally covariant quantum field theory directly in Minkowski space provides additional difficulties connected with the fact that the conformal group is larger than the group of causal automorphisms of Minkowski space. The difficulties already manifest themselves in free field theories. Until recently it was considered that one was either forced to move on Euclidean space or to restrict oneself only to infinitesimal transformations thus inventing a terminology like 'weak conformal invariance' (Hortaçsu and coworkers, 1972). Detailed studies (Schroer and Swieca, 1974; Kupsch and coworkers, 1975), on the free fields and some explicitly soluble interacting field theories in two-dimensional space–time has made it possible now to understand the structure involved in it. It is generally accepted that the necessity of the universal covering group of the conformal group is essential. The fields are subjected to a non-local Fourier decomposition on the centre of this group which generalizes the concept of decomposition in a creation and an annihilation part of the free field. Operator product expansions in Minkowski space have to be studied not in terms of fields but in terms of these non-local projections. This makes the whole scheme very complicated. An interesting question is whether one can somehow recombine all these components into a local expression, and also the question of the convergence of the operator product expansion is well worth pursuing further.

Our plan is as follows. In Section 2 some difficulties associated with the fact that the conformal group does not preserve the causal structure of the Minkowski space M_D (D = dimension) are discussed. The universal covering of a compactified Minkowski space (denoted M_D^{uc}) with its causal structure that is locally isomorphic to that of M_D and is invariant under the universal covering group of the conformal group is introduced and the possibility of defining a field theory on M_D^{uc} is also discussed. The transformation properties of quantized fields are investigated in Section 3. A significant role is played by the generating element Z of the centre in formulating the non-local decomposition of the field operators. The Thirring model is introduced as an explicit example of a conformally covariant field theory. In the subsequent Section 4 this model is used in our study of operator product expansions in Minkowski space. In two-dimensional space–time the conformal group and its covering are small enough to carry out necessary computations explicitly and this makes it easy to show the local structure of operators appearing in the expansions. In Section 5 we try to develop some general model independent ideas on the operator product expansions.

2. CAUSAL STRUCTURES

Causality as a geometric concept is a partial ordering in Minkowski space, or more general of a manifold. We are interested mainly in the Minkowski space M_4, but for the sake of constructing models other Minkowski spaces M_D with the characteristic form

$$x^2 = x_0^2 - x_1^2 - x_2^2 - \ldots - x_{D-1}^2 \tag{1}$$

are also of importance. Automorphisms of the manifold, in particular of M_D, that together with their inverses preserve the causal ordering are called causal automorphisms. Zeeman's theorem (Zeeman, 1964) asserts that the group of causal automorphisms of M_4 (relative time-like pairs of vectors are ordered into an 'earlier' and a 'later' vector, relative space-like pairs are not ordered) consists of orthochronous Lorentz transformations, translations and dilations. This group we call the 'Weyl group'. The conformal group possesses the Weyl group as a proper subgroup and thus violates the causal ordering. Zeeman's theorem can be generalized to Minkowski spaces with $D \geq 2$ easily. It has for a long time been interpreted as forbidding any extension of the space-time symmetry beyond the Weyl group, in particular excluding any internal symmetry combined with space-time symmetry.

The conformal group consists of products of inhomogeneous Lorentz transformations, of dilations

$$x_g^\mu = \lambda x^\mu, \quad \lambda > 0 \tag{2}$$

and of special conformal transformations

$$x_g^\mu = \sigma(b, x)^{-1}(x^\mu + b^\mu x^2) \tag{3}$$

$$\sigma(b, x) = 1 + 2b_\mu x^\mu + b^2 x^2 \tag{4}$$

It is obvious that M_D is not a homogeneous space for these transformations, since whatever we choose for b in (3), there are vectors $x \in M_D$ for which $\sigma(b, x) = 0$. Compactifying M_D evades this problem but leads to a manifold that trivially does not possess a causal ordering which extends the causal ordering of M_D, since the time axis is now closed at infinity.

A useful parametrization of M_D and $\overline{M}_D$, the compactification of M_D, is defined as follows (Rühl 1975). Introduce the Hermitian 2×2 matrix

$$X = x^0 \sigma_0 + \mathbf{x}\boldsymbol{\sigma}, \quad \sigma_0 = \mathbb{1} \tag{5}$$

and

$$U = (\sigma_0 + iX)(\sigma_0 - iX)^{-1} \quad (6)$$

$$U = e^{i\varphi/2} u, \quad \det u = 1, \quad 0 \leq \varphi < 2\pi \quad (7)$$

(for M_3 (M_2) define x_1 (x_1 and x_2) to be zero). Then the compactification $\overline{M}_D$ is obtained from M_D by adjoining all U with $\det(\sigma_0 + U) = 0$. As can be seen from (7) and the parametrization

$$u = u^0 \sigma_0 + \mathbf{u}\boldsymbol{\sigma}, \quad (u^0)^2 + \mathbf{u} \cdot \mathbf{u} = 1 \quad (8)$$

$\overline{M}_D$ has the topological structure

$$S_1 \times S_{D-1} \quad (9)$$

This is also true for $D > 4$.

Despite these difficulties with causality physical models are known that are quantum field theories in the proper sense and exhibit conformal covariance with an invariant vacuum, namely the free massless operator fields. Beyond these free theories a few models with interactions in M_2 are known to be conformally symmetric.

In the framework of quantum field theory and free field theories are limiting cases as comes out in the following fashion (Rühl, 1973). We assume that a conformally covariant quantum field theory in the sense of Wightman is given, the vacuum is invariant. We consider the state

$$\Phi_A^{\dot{B}}(x)|O\rangle \quad (10)$$

where $\Phi_A^{\dot{B}}(x)$ is any spinor field operator. Due to the spectrum condition this state can be analytically continued in x into the tube domain which is a homogeneous space for the conformal group and its universal covering group. If the conformal group acts on it, it transforms as an analytic representation, i.e. a representation of the discrete series (Rühl, 1973; Mack, 1975). Such representations are labelled by three parameters: j_1, j_2, d, where j_1 (j_2) is the undotted (dotted) spin as for a spinor representation of $SL(2, C)$, and d is the 'dimension' of the field. Since one wants d to assume not only integral or half-integral values, one has to study the universal covering group of the conformal group. This group is denoted G_D in the following.

The invariant two-point function

$$\langle O|\Phi_A^{\dot{B}}(x)[\Phi_C^{\dot{D}}(y)]^\dagger|O\rangle = |\gamma|^2 S_{AC}^{\dot{B}\dot{D}}(x-y) \quad (11)$$

is fixed by group theory alone up to a positive normalization constant to be an intertwining operator for the discrete series representations of G_D. It is a homogeneous distribution in $x-y$ of degree $-2d$. The requirement that this distribution is positive, is equivalent with the requirement that the discrete series representation involved admits an invariant norm in a Hilbert space, i.e.

it is unitary. This entails that the dimension d is bounded from below, namely

$$d > j_1 + j_2 + 2 \quad \text{if } j_1 j_2 \neq 0 \tag{12a}$$

$$d > j_1 + j_2 + 1 \quad \text{if } j_1 j_2 = 0 \tag{12b}$$

At the lower bound (12b) of d degenerate representations appear that, as we shall see, belong to free fields.

In fact, from (11) we deduce the vacuum expectation value of the causal commutator (or anticommutator). The usual connection between spin and statistics can be verified. If d assumes the lower bound value (12b), the commutator (anticommutator) function assumes the canonical form for a free massless field. Due to the theorem of Jost, Schroer and Pohlmeyer (Jost, 1961; Pohlmeyer, 1969), the field itself is a free massless field in this case.

Another conclusion can be drawn from (12b) and the homogeneity of the two-point function. A conformally covariant quantum field involves asymptotic states carrying particles if and only if it is a free massless field. This reduces the value of such field theories considerably. We can either regard them as models of academic interest only, some of whose properties can hopefully be carried over to more general quantum field theories with particles, or they appear at best as limiting theories in the Gell-Mann–Low sense of realistic quantum field theories (Gell-Mann and Low, 1954).

For free fields causality does not cause any problem. Their commutator (anticommutator) is a number-valued distribution and this is conformally covariant. All n-point functions can be expressed by these two-point functions and are automatically covariant. For a deeper inspection we make the ansatz

$$U_g \Phi_A^{\dot{B}}(x) U_g^{-1} = \sum_{A'B'} \mu(g, x)_{AB'}^{A'\dot{B}} \Phi_{A'}^{\dot{B}'}(x_g) \tag{13}$$

i.e. a 'local transformation' under the conformal group. It involves a singular multiplier $\mu(g, x)$ as soon as special conformal transformations (3), (4) participate in the group element g. It involves a negative power of $\sigma(b, x)$ (for the exact expression see Section 3) and thus is singular whenever x_g (3) is singular. In the case that we project both sides of (13) on the vacuum from the right, the singular multiplier is a boundary value of an antiholomorphic function in the forward tube domain, and if we project it on the vacuum from the left, we obtain a boundary value of a holomorphic function on the forward tube domain. It follows that the singular multiplier in (13) cannot be given a unique meaning, since the multipliers are different in either case. For free fields the projections on the vacuum can equivalently be performed by decomposing the field into its positive and negative frequency parts and to write an equation of the type (13) for each part separately. As multipliers we use the appropriate boundary values. As we shall show in the subsequent section, in the general case the field operator has to be harmonically analysed on the centre of G_D instead, and each Fourier component transforms as (13) with its specific multiplier composed of both boundary values in general.

Though it is not necessary it is quite useful both for technical and illustrative purposes, to formulate a conformally covariant quantum field theory by maintaining the form of the transformation law (13) without Fourier decomposition of the field and with a unique multiplier, by introduction of fields on the universal covering space $\mathbb{M}_D^{uc}$. This amounts essentially to letting φ in (7) assume all real values from $-\infty$ to $+\infty$. Instead of (9) we get the structure

$$\mathbb{R}_1 \times \mathbb{S}_{D-1} \tag{14}$$

In the case $D=2$ we have to take the universal covering with respect to both factors $\mathbb{S}_1$ and thus obtain

$$\mathbb{R}_1 \times \mathbb{R}_1 \tag{15}$$

In the latter case U (7) is a diagonal 2×2 matrix

$$U = \begin{pmatrix} e^{i\varphi_+} & 0 \\ 0 & e^{i\varphi_-} \end{pmatrix} \tag{16}$$

with

$$tg\frac{\varphi_\pm}{2} = x^0 \pm x^3 \tag{17}$$

In this case we let both $\varphi_\pm$ assume all real values.

The manifold $\mathbb{M}_D^{uc}$ has a remarkable property first discovered by Segal (Segal, 1971; Mayer, 1974). It possesses a conformally invariant (under G_D) causal ordering in the sense described above. For $D>2$ $\mathbb{M}_D^{uc}$ possesses an infinite number of sheets labelled $n = 0, \pm 1, \pm 2, \ldots$. A space-like vector of the zeroth sheet can be mapped by continuous variation of the group element on other space-like vectors on the same sheet, onto a point at infinity, and further on points of the $\pm$-first sheet that lie over time-like vectors of the zeroth sheet. If we identify these points on the $\pm$-first sheet with the points of the zeroth sheet, we have transformed space-like into time-like vectors and thus violated the causal ordering of $\mathbb{M}_D$ (the second vector of the pair can be taken to be the null-vector). However, on $\mathbb{M}_D^{uc}$ we may call these points on the $\pm$-first sheet obtained from space-like points on the zeroth sheet by means of special conformal transformations also space-like. It is then easy to see that the remainder of the manifold $\mathbb{M}_D^{uc}$ can be cast into a future and a past submanifold plus a light-cone. Thus we have succeeded in extending the causal structure from $\mathbb{M}_D$ onto $\mathbb{M}_D^{uc}$.

This way it is possible to define a quantum field $\Phi(\tilde{x})$ on $\mathbb{M}_D^{uc}$ with $\tilde{x}$ over x on the nth sheet, such that locality can be formulated by

$$[\Phi(\tilde{x}), \Phi(\tilde{y})] = 0 \tag{18}$$

whenever $\tilde{x}, \tilde{y}$ are relatively space-like say for a scalar field. This locality condition (18) can be postulated to be invariant under G_D.

Any Wightman m-point function

$$\langle O | \Phi_1(\tilde{x}_1) \Phi_2(\tilde{x}_2) \ldots \Phi_m(\tilde{x}_m) | O \rangle \tag{19}$$

is independent of the sheet number n if all fields are defined on the same sheet. Field theories on fixed sheets are isomorphic. In fact there is a unitary operator Z so that

$$\Phi(\tilde{x}) = Z^{n_1-n_2}\Phi(\tilde{\tilde{x}})Z^{-n_1+n_2} \tag{20}$$

if $\tilde{x}$ ($\tilde{\tilde{x}}$) lies on the n_1-st (n_2-nd) sheet over x. Any conformal transformation that does not lead any of the arguments of the m-point function (19) out of its sheet, leaves (20) unaltered.

From this one can deduce that Z commutes with all conformal transformations and thus represents an element of the centre of G_D. In fact, it represents the generating element of the centre. Of course, any local observable should be identical on all sheets and thus commute with Z.

Finally all Wightman m-point functions with arguments on arbitrary sheets can be obtained from the same function with all arguments on the zeroth sheet by analytic continuation. This and the previous assertions can be proved (Lüscher and Mack, 1975) by first requiring conformal covariance only under infinitesimal transformations, then continuing the Wightman (or time-ordered) functions into the Euclidean domain, where the generators of the conformal group can be implemented easily to the Euclidean conformal transformations, finally continuing back to the Minkowskian boundary, which then turns out to have the sheet structure just described.

In the case $D=2$ a few alterations are necessary. There is a doubly infinite sequence of sheets $n_+ = 0, \pm 1, \pm 2, \ldots$ and $n_- = 0, \pm 1, \pm 2, \ldots$ and corresponding operators Z_+ and Z_- as the intersheet isomorphisms. In this case G_D is the direct product of two groups

$$G_D = SU(1,1)_+^{uc} \times SU(1,1)_-^{uc} \tag{21}$$

The first (second) group acts on $\varphi_+(\varphi_-)$, according to

$$e^{i\varphi_g} = \frac{\alpha\, e^{i\varphi} + \beta}{\bar{\beta}\, e^{i\varphi} + \bar{\alpha}} \tag{22}$$

$$\varphi_g = \varphi + 2\,\arg(\alpha + \beta\, e^{-i\varphi}) \tag{23}$$

where $\arg \alpha$ is allowed to range over

$$-\infty < \arg \alpha < +\infty \tag{24}$$

in order to obtain the universal covering group of $SU(1,1)$.

3. LOCAL FIELDS AND THEIR TRANSFORMATIONS

Physically relevant unitary irreducible representations of G_4, the universal covering group of the conformal group, can be constructed in the usual fashion by inducing them from an appropriate subgroup. Requiring that these representations be realizable on spaces of functions of vectors x in Minkowski space, i.e. by classical fields, we are led to consider the stability subgroup of these vectors. For $x=0$ this subgroup consists of homogeneous Lorentz transformations, of dilations, and of special conformal transformations. The representations obtained in this fashion have been classified as follows (Mack and Salam, 1969).

Let κ_μ be the generators of the special conformal transformations (3) represented by a matrix acting on the classical field at $x=0$. Then κ_μ may be identically zero, this type of field representation is called Ia in Mack and Salam (1969). Secondly κ_μ may be non-zero but a finite dimensional matrix. Then it has to be nilpotent due to the abelian structure of the subgroup of special conformal transformations. Such representations are denoted Ib. Finally there are infinite dimensional matrices κ_μ, these are denoted type II. Representations of type Ia are the representations almost exclusively encountered in field theory. If we require in addition that the energy–momentum spectrum be restricted to the forward light-cone, we obtain the discrete series representations mentioned in Section 2 that were used there for the one-particle states (Rühl, 1973; Mack, 1975). We mention finally that the group G_2 and its representations that are used for the Thirring model in $\mathbb{M}_2$ have been studied by many authors. We shall only explain a few notations in this article but otherwise refer to an exhaustive presentation in the literature (Rühl and Yunn, 1975a).

The transformation property of a conformally covariant spinor field under special conformal transformations of G_4 is by representation theory

$$U_g \Phi_{A\dot{B}}(x) U_g^{-1} = \sigma(b, x)^{-d-j_1-j_2} \sum_{A'B'} D^{j_1}(\sigma_0 + X\hat{B})_{AA'} \Phi_{A'\dot{B}'}(x_g) D^{j_2}(\sigma_0 + \hat{B}X)_{\dot{B}'\dot{B}} \tag{25}$$

$$\hat{B} = b^0 \sigma_0 - \mathbf{b}\boldsymbol{\sigma} \tag{26}$$

for group elements sufficiently close to the unit element. D^j denote representation matrices of covariant spinor representations of $SL(2, C)$. As pointed out in the preceding section, the difficulty consists in interpreting the singular factor $\sigma^{-d-j_1-j_2}$, which for fixed x and a sufficiently small neighbourhood of the group unit of G_4 is regular and well defined by $\arg \sigma(b, x) = 0$. The solution to the general problem has been found by studying the Schroer model (Schroer and Swieca, 1974) and the Thirring model (Kupsch, Rühl and Yunn, 1975) in $\mathbb{M}_2$. We shall describe it now in general terms.

We assume that we have a Wightman-type field theory that is conformally covariant, i.e. there exists a unitary representation U_g of $\tilde{G}_D$ with

$$U_g |O\rangle = |O\rangle \tag{27}$$

The generating element of the centre of $\tilde{G}_D (D > 2)$ be represented by Z. We introduce the Fourier component Φ^τ, $0 \le \tau < 1$ on the centre of $\tilde{G}_D$

$$\Phi^\tau(x) = \sum_{n=-\infty}^{+\infty} Z^n \Phi(x) Z^{-n} e^{-2\pi i n \tau} \tag{28}$$

Obviously

$$Z\Phi^\tau(x)Z^{-1} = e^{2\pi i\tau}\Phi^\tau(x) \tag{29}$$

We continue the equation (25) away from the group unit so that $\tilde{x}$ on the zeroth sheet over $x \in \mathbb{M}_D$ moves over to the first sheet (minus first sheet). We consider the two boundary values

$$\sigma_\pm(b, x) = \lim_{y \to 0} \sigma(b, x \pm iy) \tag{30}$$

where y tends to zero in the forward light-cone. We obtain then on the first (minus first) sheet

$$\arg \sigma_\pm(b, x) = \mp\pi(\pm\pi) \tag{31}$$

In accordance with (29) we make therefore the ansatz

$$U_g \Phi^\tau_{A\dot{B}} U_g^{-1} = \sigma_+(b, x)^{-\frac{1}{2}d - j_1 + \tau} \sigma_-(b, x)^{-\frac{1}{2}d - j_2 - \tau}$$
$$\times \sum_{A'B'} D^{j_1}(\sigma_0 + X\hat{B})_{AA'} \Phi^\tau_{A'\dot{B}'}(x_g) D^{j_2}(\sigma_0 + \hat{B}X)_{\dot{B}'\dot{B}} \tag{32}$$

We can introduce field operators on $\mathbb{M}_D^{uc}$ by setting on the zeroth sheet

$$\Phi_{A\dot{B}}(x) = 2^{2(d-1)} |\det(\sigma_0 - iX)|^{-d - j_1 - j_2}$$
$$\times \sum_{A'B'} D^{j_1}_{AA'}(\sigma_0 - iX) f_{A'\dot{B}'}(U(X)) D^{j_2}_{\dot{B}'\dot{B}}(\sigma_0 - iX) \tag{33}$$

and requiring a 'local' transformation law

$$U_g f_{A\dot{B}}(U) U_g^{-1}$$
$$= |\det(A^\dagger + UB^\dagger)|^{-\frac{1}{2}d - j_1} |\det(CU + D)|^{-\frac{1}{2}d - j_2}$$
$$\times \sum_{A'B'} D^{j_1}(A^\dagger + UB^\dagger)_{AA'} f_{A'\dot{B}'}(U_g) D^{j_2}(CU + D)_{\dot{B}'\dot{B}} \tag{34}$$

Here A, B, C, D denote 2×2 matrices making up a 4×4 matrix

$$m = \begin{pmatrix} A & B \\ C & D \end{pmatrix} \tag{35}$$

$$H = \begin{pmatrix} -\sigma_0 & 0 \\ 0 & +\sigma_0 \end{pmatrix}, \quad m^\dagger H = Hm^{-1} \tag{36}$$

that belongs to $SU(2, 2)$. The group $SU(2, 2)^{uc}$ is isomorphic to G_4. The matrix m with

$$B = C = 0, \quad A = D = i\sigma_0 \tag{37}$$

generates the centre of $SU(2, 2)^{uc}$. We find from (34)

$$Zf_{A\dot{B}}(u, \varphi)Z^{-1} = e^{i\pi(j_2 - j_1)} f_{A\dot{B}}(-u, \varphi - 2\pi) \tag{38}$$

by means of

$$U_g = (AU+B)(CU+D)^{-1} \tag{39}$$

$$\varphi_g = \varphi - \arg \det (A^\dagger + UB^\dagger)$$
$$- \arg \det (CU+D) \tag{40}$$

Of course the field operator $f_{A\dot{B}}(U)$ can also be decomposed on the centre of G_4 by a formula such as (28). It follows then from (38)

$$Zf^\tau(U)Z^{-1} = e^{2\pi i \tau} f^\tau(U)$$
$$= e^{i\pi(j_2-j_1)} f^\tau(-u, \varphi - 2\pi) \tag{41}$$

Therefore $f^\tau(U)$ can be expanded in the canonical basis

$$f^\tau(u, \varphi) = \sum_{j=0,\frac{1}{2},1\ldots}^{\infty} \sum_{r,s=-j}^{+j} \sum_{q=-\infty}^{+\infty} a_{rs,q}^{j,\tau} e^{iq\varphi} D_{rs}^j(u) \tag{42}$$

with

$$\tfrac{1}{2}(j_2-j_1)+j-q \triangleq \tau \bmod 1 \tag{43}$$

This expansion is in many cases more advantageous than a Fourier decomposition into plane waves on $\mathbb{M}_4$.

The central element Z is an element of a one-parameter subgroup of G_D and as such can be written in the form

$$Z = e^{2\pi i T} \tag{44}$$

The self-adjoint operator T is not uniquely determined but only up to a self-adjoint operator with entire eigenvalues, such as a number operator in a free field model. T has always the character of an 'anomalous part' of such a number operator in the known models. The eigenvalues of T and its corresponding eigenspaces fix the irreducible components of a local operator. In the known models its spectrum is discrete and it has been suggested (Lüscher and Mack, 1975) that this be so in general. It has also been proved (Lüscher and Mack, 1975) that the spectrum of T can be assumed to be positive. We shall make use now of the hypothesis that the spectrum is discrete.

If λ_i denote the eigenvalues of T and $\Pi(\lambda_i)$ their respective projection operators on the eigenspaces, we define

$$\Phi_{\lambda_1\lambda_2} = \Pi(\lambda_1)\Phi\Pi(\lambda_2) \tag{45}$$

It follows then that

$$Z\Phi_{\lambda_1\lambda_2}Z^{-1} = e^{2\pi i(\lambda_1-\lambda_2)}\Phi_{\lambda_1\lambda_2} \tag{46}$$

and inserting this into (28)

$$\tau \triangleq \lambda_1 - \lambda_2 \bmod 1 \tag{47}$$

and finally

$$\Phi^\tau = \sum_{\substack{\lambda_1 \lambda_2 \\ \lambda_1 - \lambda_2 \hat{=} \tau}} \Phi_{\lambda_1 \lambda_2} \tag{48}$$

As an example we want to present now the Thirring model (Kupsch, Rühl and Yunn, 1975; Rühl, 1975). We shall use the formalism by means of fields on M_2^{uc}. The main tool in the construction of the Thirring field is the current operator (Dell'Antonio, Frishman and Zwanziger, 1972) that transforms as a free mass zero vector field. Its components J_+ and J_- depend on φ_+ and φ_- respectively only and, expanded in the canonical basis, are

$$J_\pm(\varphi_\pm) = \frac{1}{\pi} \sum_{m=0}^{\infty} (m+1)^{\frac{1}{2}} \{c_{\pm,m}^\dagger e^{i(m+1)\varphi_\pm} + c_{\pm,m} e^{-i(m+1)\varphi_\pm}\} + \frac{1}{\pi} Q_\pm \tag{49}$$

or

$$J_\pm(\varphi_\pm) = J_\pm^{(-)}(\varphi_\pm) + J_\pm^{(+)}(\varphi_\pm) + \frac{1}{\pi} Q_\pm \tag{50}$$

The operators $c_{\pm,m}$, $c_{\pm,m}^\dagger$ satisfy canonical commutation relations, e.g.

$$[c_{\sigma,m}, c_{\sigma',m'}^\dagger] = \delta_{\sigma\sigma'} \delta_{mm'} \tag{51}$$

and commute with the charge operators $Q_\pm$. We introduce the 'sources' of the current operators by

$$\Gamma_\pm^{(-)}(\varphi_\pm) = \pi \int_{+i\infty}^{\varphi_\pm} d\psi J_\pm^{(-)}(\psi) = \Gamma_\pm^{(+)}(\varphi_\pm)^\dagger \tag{52}$$

Then we define the Thirring field ($\gamma = 1, 2$) by Rühl and Yunn (1975b)

$$f_\gamma(\varphi) = 2^d \exp i \sum_\pm C_{\pm,\gamma} [\Gamma_\pm^{(-)}(\varphi_\pm) + \tfrac{1}{2} Q_\pm \varphi_\pm]$$
$$\times \sigma_\gamma \exp i \sum_\pm C_{\pm,\gamma} [\Gamma_\pm^{(+)}(\varphi_\pm) + \tfrac{1}{2} Q_\pm \varphi_\pm] \tag{53}$$

where the coefficients $C_{\pm,\gamma}$ are defined by

$$Q_\pm \sigma_\gamma = \sigma_\gamma [Q_\pm + C_{\pm,\gamma}] \tag{54}$$

σ_γ commutes with the source terms (52).

Under a space reflection we require that the two components of σ_γ interchange and that $Q_\pm$ goes into $Q_\mp$. This necessitates

$$C_{\pm,1} = C_{\mp,2} := C_\pm \tag{55}$$

which leaves two free parameters to the model. The spin and dimension of the field operator (53) are

$$d = \tfrac{1}{2}[C_+^2 + C_-^2] \tag{56}$$

$$s = \tfrac{1}{2}|C_+^2 - C_-^2| \tag{57}$$

By differentiation of (53) we find four equations

$$-i\frac{\partial}{\partial \varphi_\pm} f_\gamma(\varphi) = \pi C_{\pm,\eta}\left[\left(J_\pm^{(-)}(\varphi_\pm) + \frac{1}{2\pi}Q_\pm\right)f_\gamma(\varphi)\right.$$
$$\left. + f_\gamma(\varphi)\left(J_\pm^{(+)}(\varphi_\pm) + \frac{1}{2\pi}Q_\pm\right)\right] \quad (58)$$

two of which can be shown to be identical with the field equations (Thirring, 1958)

$$-i\partial_\mu \gamma^\mu \psi(x) = g\gamma^\mu [J_\mu^{(-)}(x)\psi(x) + \psi(x)J_\mu^{(+)}(x)] \quad (59)$$

whereas two equations are additional. The coupling constant g is

$$g = 2\pi C_- \quad (60)$$

The operator σ_γ is a constant field belonging to $d = s = 0$ as can be seen from (56), (57). It satisfies abnormal commutation relations (Klaiber, 1968; Lowenstein and Swieca, 1971)

$$\{\sigma_1, \sigma_2\} = \{\sigma_1, \sigma_2^\dagger\} = 0$$
$$[\sigma_1, \sigma_1^\dagger] = [\sigma_2, \sigma_2^\dagger] = \frac{1}{2\pi} \quad (61)$$

and has the Wightman functions

$$\langle O|(\sigma_1)^n(\sigma_1^\dagger)^{n'}|O\rangle = \langle O|(\sigma_2)^n(\sigma_2^\dagger)^{n'}|O\rangle$$
$$= (2\pi)^{-n}\delta_{nn'} \quad (62)$$

The transformation behaviour of the Thirring field (53) under the conformal group follows from the canonical transformation behaviour of the current, from the invariance of the vacuum state and the charge operators $Q_\pm$, and from an appropriate definition of the transformation behaviour of the σ-field. This takes account of the non-invariance of the subtraction point $\pm i\infty$ in (52), and is consistent with the invariance of the Wightman functions (62) and the commutators (61) (Rühl and Yunn, 1975b). We obtain

$$U_g f_\gamma(\varphi) U_g^{-1} = \prod_\pm \{\alpha_\pm e^{i\varphi_\pm} + \beta_\pm|^{-C_\pm^2\gamma}\} f_\gamma(\varphi_{+g}, \varphi_{-g}) \quad (63)$$

Moreover we have from (53), (54)

$$f_\gamma(\varphi_\pm + 2\pi, \varphi_\mp) = e^{i\pi Q_\pm^2} f_\gamma(\varphi) e^{-i\pi Q_\pm^2} \quad (64)$$

In fact the finite conformal transformations (63) can be obtained from the energy momentum tensor (Dell'Antonio and coworkers, 1972) in 'Sugawara form'

$$\Theta_{\mu\nu} = \pi : J_\mu J_\nu : -\tfrac{1}{2}\pi : J_\lambda J^\lambda : g_{\mu\nu} \quad (65)$$

by the canonical (that is: free massless field) formulae for the generators by exponentiation. Among these generators we find the combinations

$$T_\pm = \sum_{m=0}^{\infty} (m+1)c^+_{\pm,m}c_{\pm,m} + \tfrac{1}{2}Q^2_\pm \tag{66}$$

that create one-parameter subgroups containing the central elements Z_+ and Z_- of G_2 in agreement with (64).

From (54) we see that the eigenspaces of $Q_\pm$ belong to the eigenvalues

$$\lambda_\pm(n_1, n_2) = n_1 C_\pm + n_2 C_\mp \tag{67}$$

($n_{1,2} = 0, \pm 1, \pm 2, \ldots$). They are obtained by applying σ_1 n_1 times (respectively σ_1^+ $(-n_1)$ times) and σ_2 n_2 times (respectively σ_2^+ $(-n_2)$ times) to the vacuum state and operating with arbitrary polynomials of the currents on these states. It follows that the operator $f_\gamma(\varphi)\Pi(\lambda_\pm(n_1, n_2))$ has one covariant component only with

$$\tau_{\pm,\gamma} \triangleq C_{\pm,\gamma}(n_1 C_\pm + n_2 C_\mp) + \tfrac{1}{2}C^2_{\pm,\gamma} \mod 1 \tag{68}$$

A crucial property of the Thirring model is that any product of operators f_γ or their adjoints $f^\dagger_\gamma$ can be regularized by splitting off a regular factor R

$$\prod_i f_{\gamma i}(\varphi_i) \prod_j f^\dagger_{\gamma j}(\psi_j) = s_{\{\gamma_i\};\{\gamma_j\}}[\{\varphi_i\}; \{\psi_j\}] R_{\{\gamma_i\};\{\gamma_j\}}[\{\varphi_i\}; \{\psi_j\}] \tag{69}$$

that is C^∞, multilocal, and conformally covariant in all variables. The singular factor s is a covariant distribution. Identifying arguments in R leads to other local conformally covariant operators. Applying derivative operators to R before identifying arguments leads to local, but to conformally covariant differential operators only if the differential operator is itself covariant in the sense specified in the subsequent section. For the Thirring model the conformal analysis of operator products reduces therefore to the analysis of the regularized products.

4. COVARIANT OPERATOR PRODUCT DECOMPOSITIONS

Products of local operators

$$A(x)B(y)$$

and their singular behaviour if x approaches y have been studied for two different purposes. The first approach was motivated by phenomenology, namely the investigation of high energy asymptotic behaviour of certain matrix elements of such operator products (e.g. deep inelastic electron proton scattering). This approach aimed at asymptotic expansions of the type

$$A(x)B(y) \simeq \sum_{n=0}^{\infty} s_n(u)C_n(v), u = x - y, v = \tfrac{1}{2}(x+y) \tag{70}$$

either for $u \to 0$ ('short distance expansion' or Wilson expansion) or for $u^2 \to 0$

('light-cone expansion'), see Wilson (1969) and Wilson and Zimmermann (1972) respectively, Brandt and Preparata (1971) and Frishman (1971). Both kinds of expansions have been studied in the framework of perturbative quantum field theory (Zimmermann, 1970, 1973). The singularity of the function $s_n(u)$ decreases with increasing n, whereas $C_n(v)$ are local operators.

The second approach is fundamental (Polyakov, 1973; Efremov, 1968; Ferrara and co-workers, 1973; Bonora and co-workers, 1973; Swieca, 1974; Schroer and co-workers, 1975). If expansions of the type (70) for all local operators in a Wightman formalism together with all the two-point functions for these fields are given, then the structure of the quantum-field theory is fixed provided the validity of the expansion (70) is not only asymptotic but in the sense of weak convergence in some real or complex domain. In fact, all n-point functions can be reduced into two-point functions this way.

Within a conformally covariant quantum field theory this programme seems to have a chance to be set up successfully. In fact, fixing a normalization of the local fields by any *ad hoc* prescription, all two-point functions are uniquely determined. Moreover, requiring each term in the expansion (70) to be conformally covariant reduces the number of local fields and restricts the form of the singular functions. We investigate such a programme in this section, considering the Thirring model as a guide. We restrict the investigation to $\mathbb{M}_2$.

In order to derive an expansion of the Wilson-type whose terms are each covariant (or semicovariant, as we shall see), we intend to apply the tensor product decomposition theorem for the conformal group. First we project out covariant components $A^{\tau_A}(x)$, $B^{\tau_B}(y)$ of $A(x)$ and $B(y)$ as explained in the preceding section. For these components we make an ansatz

$$A^{\tau_A}(x) B^{\tau_B}(y) \simeq \sum_{n=0}^{\infty} \int dz Q(\chi_A, x; \chi_B, y | \chi(n), z) C_n^{\tau_C}(z) \tag{71}$$

where the kernel Q satisfies the covariance constraints. $C_n^{\tau_C}(z)$ is a covariant component of a local operator with

$$\tau_C \triangleq \tau_A + \tau_B \bmod 1 \tag{72}$$

$\chi_{A,B}$ and $\chi(n)$ denote the representations of G_D involved. Finally there remains the problem of recombining the components $C_n^{\tau_C}(z)$ to a local operator. The tensor product decomposition theorem is used both to derive the kernels Q and the operators $C_n^{\tau_C}(z)$, where for the latter part of the problem explicit knowledge of the quantum field model under investigation is presumed.

We outline the derivation of the expansion (71) for models in $\mathbb{M}_2$ by group theoretic arguments (Rühl and Yunn, 1975a, b, c). The first tool we need is an asymptotic completeness relation for covariant kernels of the second kind. We consider a space of C^∞ functions $f(\varphi)$, $-\infty < \varphi < +\infty$ with

$$f(\varphi + 2\pi) = e^{2\pi i \tau} f(\varphi), \qquad 0 \leq \tau < 1 \tag{73}$$

with an appropriate set of norms. We denote it $\mathscr{D}_\tau$. It carries a representation $\chi = (j, \tau)$ of $SU(1, 1)^{uc}$ if we define

$$T_g^\chi f(\varphi) = |\alpha\, e^{i\varphi} + \beta|^{2j-1} f(\varphi_g) \tag{74}$$

with φ_g as in (22), (23). If j is purely imaginary $\mathscr{D}_\tau$ can be completed to a Hilbert space with invariant scalar product

$$(f_1, f_2) = \int_0^{2\pi} \overline{f_1(\varphi)} f_2(\varphi)\, d\varphi \tag{75}$$

These representations form the principal series of $SU(1, 1)^{uc}$. If

$$\tfrac{1}{2} - j \mp \tau \triangleq 0 \bmod 1 \tag{76}$$

$\mathscr{D}_\tau$ possesses invariant subspaces $\mathscr{F}_\chi^{(\pm)}$ respectively, spanned by the canonical basis elements $e^{iq\varphi}$, $q - \tau \triangleq 0 \bmod 1$, with

$$\pm q = \tfrac{1}{2} - j + m, \quad m = 0, 1, 2, \ldots \tag{77}$$

They carry the discrete series representations.

Tensor products of spaces $\mathscr{D}_\tau$ can be mapped into spaces $\mathscr{D}_\tau$ by means of operators K

$$\mathscr{D}_{\tau_1} \times \mathscr{D}_{\tau_2} \xrightarrow{K} \mathscr{D}_{\tau_3}$$

which we call covariant if

$$K(T_g^{\chi_1} \times T_g^{\chi_2}) = T_g^{\chi_3} K \tag{78}$$

Such covariant operators can be given in the form of convolution kernels. They span themselves a two-dimensional linear space. As a basis we take the following two kernels ($\arg(2i \sin(\varphi - iO)) = -\arg(-2i \sin(\varphi + iO)) = \pi/2$ for $0 < \varphi < \pi$)

$$(2\pi)^3 K_1(\chi_3, \varphi_3 | \chi_1, \varphi_1; \chi_2, \varphi_2)$$
$$= [2i \sin \tfrac{1}{2}(\varphi_1 - \varphi_2 - iO)]^{-\frac{1}{2} - j_2 + \tau_2}$$
$$\times [-2i \sin \tfrac{1}{2}(\varphi_1 - \varphi_2 + iO)]^{-j_1 - j_3 - \tau_2}$$
$$\times [2i \sin \tfrac{1}{2}(\varphi_2 - \varphi_3 - iO)]^{-\frac{1}{2} + j_1 - j_2 + j_3}$$
$$\times [2i \sin \tfrac{1}{2}(\varphi_3 - \varphi_1 - iO)]^{-\frac{1}{2} + j_3 + \tau_1 + \tau_2}$$
$$\times [-2i \sin \tfrac{1}{2}(\varphi_3 - \varphi_1 + iO)]^{-j_1 + j_2 - \tau_1 - \tau_2} \tag{79}$$

$$(2\pi)^3 K_2(\chi_3, \varphi_3 | \chi_1, \varphi_1; \chi_2, \varphi_2)$$
$$= [2i \sin \tfrac{1}{2}(\varphi_1 - \varphi_2 - iO)]^{-\frac{1}{2} - j_1 - \tau_1}$$
$$\times [-2i \sin \tfrac{1}{2}(\varphi_1 - \varphi_2 + iO)]^{-j_2 - j_3 + \tau_1}$$
$$\times [2i \sin \tfrac{1}{2}(\varphi_2 - \varphi_3 - iO)]^{-\frac{1}{2} + j_3 - \tau_1 - \tau_2}$$
$$\times [-2i \sin \tfrac{1}{2}(\varphi_2 - \varphi_3 + iO)]^{+j_1 - j_2 + \tau_1 + \tau_2}$$
$$\times [2i \sin \tfrac{1}{2}(\varphi_3 - \varphi_1 - iO)]^{-\frac{1}{2} - j_1 + j_2 + j_3} \tag{80}$$

[In Rühl and Yunn (1975a, b, c) the kernel K_2 was denoted K_3!] Note that τ_3 is necessarily $\tau_1 + \tau_2$ mod 1.

If all representations are in the principal series, the operators are well defined. For other representations we obtain the operators K by analytic continuation in j. This method is used throughout, the spaces $\mathcal{D}_\tau$ that are independent of j are particularly useful for this purpose. A kernel K may develop poles and zeros during this continuation. Poles may either be related with the appearance of invariant subspaces $\mathcal{F}_\chi^{(\pm)}$ (τ-type poles) or at positions depending only on j_1, j_2, j_3 (j-type poles). Residues of j-type poles may turn out to be differential operators that are also denoted covariant. They can be used to construct new local covariant operators from multilocal covariant operators.

Dual covariant operators for the mapping

$$\mathcal{D}_{\tau_3} \xrightarrow{K^d} \mathcal{D}_{\tau_1} \times \mathcal{D}_{\tau_2}$$

$$K^d T_g^{\chi_3} = (T_g^{\chi_1} \times T_g^{\chi_2}) K^d \tag{81}$$

can be obtained from the covariant kernels by replacing $j_i \to -j_i$, $\tau_i \to -\tau_i$ ($i = 1, 2, 3$). Both types of kernels together satisfy a completeness relation

$$(2\pi)^3 \frac{\pi}{\sin \pi(j_1 - j_2 + \tau_1 + \tau_2)} \int d\mu(\chi_3)_{\text{PS}} \int_0^{2\pi} d\varphi_3$$

$$\times \{K_2^d(\chi_1, \varphi_1; \chi_2, \varphi_2 | \chi_3, \varphi_3) K_1(\chi_3, \varphi_3 | \chi_1, \varphi_1'; \chi_2, \varphi_2')$$

$$- K_1^d(\chi_1, \varphi_1; \chi_2, \varphi_2 | \chi_3, \varphi_3) K_2(\chi_3, \varphi_3 | \chi_1, \varphi_1'; \chi_2, \varphi_2')\}$$

+ discrete series terms

$$= \sum_{k_1 = -\infty}^{+\infty} e^{2\pi i \tau_1 k_1} \delta(\varphi_1 - \varphi_1' - 2\pi k_1) \sum_{k_2 = -\infty}^{+\infty} e^{2\pi i \tau_2 k_2} \delta(\varphi_2 - \varphi_2' - 2\pi k_2) \tag{82}$$

where $d\mu(\chi_3)_{\text{PS}}$ is the principal series part of the Plancherel measure ($j_3 = i\rho$)

$$d\mu(\chi_3)_{\text{PS}} = \frac{\rho \operatorname{sh} 2\pi\rho}{\operatorname{ch} 2\pi\rho + \cos 2\pi\tau_3} d\rho \tag{83}$$

Applying K_1 and K_2 to a regularized bilocal covariant operator, we obtain covariant and in general non-local operators (due to the integration, there is in addition always the non-locality from projecting on eigenspaces of T_+ and T_-). Inserting them into (82) allows us to reconstruct the original operators. Shifting the contour of (82) to any direction, yields contributions from the poles (four sequences of j-type and two of τ-type poles for either variables $\varphi_\pm$) plus a residual integral. Both with increasing and decreasing Re j_3 the degree of singularity of these contributions grows. Therefore we do not obtain a covariant Wilson expansion.

To overcome this difficulty we proceed as in the theory of Regge poles (Rühl, 1969). We exploit the symmetry of the integrand in (82) under $j_3 \to -j_3$ and replace $K^d_{1,2}$ by kernels of the second kind

$$K^d_1 = Q_{1a} + Q_{1b}$$
$$K^d_2 = Q_{2a} + Q_{2b} \qquad (84)$$

so that

$$\int_0^{2\pi} d\varphi_3 (Q_{2a}K_1 - Q_{1a}K_2)$$

goes into

$$\int_0^{2\pi} d\varphi_3 (Q_{2b}K_1 - Q_{1b}K_2)$$

under $j_3 \to -j_3$. In turn we have explicitly

$$Q_{1a} = aK^d_1 + b \exp\{-i\pi(j_1 - j_2 + \tau_1 + \tau_2) \operatorname{sign} \sin(\varphi_1 - \varphi_2)\} K^d_2$$
$$Q_{2a} = \exp\{i\pi(j_1 - j_2 + \tau_1 + \tau_2) \operatorname{sign} \sin(\varphi_1 - \varphi_2)\} Q_{1a} \qquad (85)$$

with

$$a = \frac{\sin \pi(\tfrac{1}{2} - j_1 + j_2 + j_3) \sin \pi(\tfrac{1}{2} + j_3 - \tau_1 - \tau_2)}{\sin 2\pi j_3 \sin \pi(j_1 - j_2 + \tau_1 + \tau_2)}$$

$$b = -\frac{\sin \pi(\tfrac{1}{2} + j_1 - j_2 + j_3) \sin \pi(\tfrac{1}{2} + j_3 + \tau_1 + \tau_2)}{\sin 2\pi j_3 \sin \pi(j_1 - j_2 + \tau_1 + \tau_2)} \qquad (86)$$

From (85) we see that the kernels Q_a are not globally covariant but only infinitesimally, namely whenever $\sin(\varphi_1 - \varphi_2) \neq 0$. We call them therefore 'semicovariant'. For $\varphi_1 \to \varphi_2$ we have asymptotically (as a distribution in φ_3)

$$|Q_{1a}| \simeq \operatorname{const} |2 \sin \tfrac{1}{2}(\varphi_1 - \varphi_2)|^{-\tfrac{1}{2} - \operatorname{Re} j_3 + j_1 + j_2} \qquad (87)$$

whence a decreasing singularity in the left half plane.

Shifting then the contour in (82) we obtain the asymptotic completeness relation

$$\sum_{k_1=-\infty}^{+\infty} e^{2\pi i \tau_1 k_1} \delta(\varphi_1 - \varphi'_1 - 2\pi k_1) \sum_{k_2=-\infty}^{+\infty} e^{2\pi i \tau_2 k_2} \delta(\varphi_2 - \varphi'_2 - 2\pi k_2)$$

$$= (2\pi)^2 \exp\{i\pi(\tfrac{1}{2} - j_2 + \tau_2) \operatorname{sign} \sin(\varphi_1 - \varphi_2)\}$$

$$\times \sum_{k=0}^{\infty} \frac{(-1)^{k+1}}{k!} \Gamma(2j_3 - k) j_3 [tg\pi(j_3 + \tau_1 + \tau_2) + tg\pi(j_3 - \tau_1 - \tau_2)]$$

$$\times \int_0^{2\pi} d\varphi_3 \int_0^{2\pi} d\varphi'_3 Q_{1a}(\chi_1, \varphi_1; \chi_2, \varphi_2 | \chi_3, \varphi_3) S(\chi_3, \varphi_3 | \chi^c_3, \varphi'_3)$$

$$\times \Delta(\chi^c_3, \varphi'_3 | \chi_1, \varphi'_1; \chi_2, \varphi'_2)|_{j_3 = j_3(k)} \qquad (88)$$

where

$$\chi^c_3 = (-j_3, \tau_3) \qquad (89)$$

and
$$j_3(k) = \tfrac{1}{2} - j_1 - j_2 + k \tag{90}$$

S is an intertwining operator, i.e. a continuous operator from $\mathcal{D}_{\tau_3}$ into $\mathcal{D}_{\tau_3}$ that intertwines χ_3^c and χ_3. Δ is a covariant differential operator

$$\int_0^{2\pi} d\varphi_1 \int_0^{2\pi} d\varphi_2 \Delta(\chi_3^c, \varphi_3 | \chi_1, \varphi_1; \chi_2, \varphi_2) g(\varphi_1, \varphi_2)$$
$$= Q_k\left(-i\frac{\partial}{\partial \varphi_1}, -i\frac{\partial}{\partial \varphi_2}\right) g(\varphi_1, \varphi_2)|_{\varphi_1 = \varphi_2 = \varphi_3} \tag{91}$$

where Q_k is a hypergeometric polynomial

$$Q_k(q_1, q_2) = \sum_{m=0}^{k} (-1)^m \binom{k}{m} (2j_1 - k)_m (\tfrac{1}{2} - j_2 - q_2)_m (2j_2 - k)_{k-m} (\tfrac{1}{2} - j_1 - q_1)_{k-m} \tag{92}$$

The asymptotic completeness relation (88) is applied twice, once to the variables labelled $(+)$, once to those labelled $(-)$ in a regularized covariant operator $R(\varphi_{1+}, \varphi_{1-}, \varphi_{2+}, \varphi_{2-})$. Of course we have first to project on a simultaneous eigenspace of T_+ and T_-. For a light-cone expansion it suffices to apply (88) to one variable only. In the completeness relation we have then the operators

$$O(\chi_3^c(k_+), \chi_3^c(k_-), \varphi_{3+}, \varphi_{3-}) \Pi(\lambda_+, \lambda_-)$$
$$= \prod_{\pm} \left\{ Q_{k\pm}\left(-i\frac{\partial}{\partial \varphi_{1\pm}}, -i\frac{\partial}{\partial \varphi_{2\pm}}\right) \right\} R(\varphi_{1+}, \varphi_{1-}, \varphi_{2+}, \varphi_{2-}) \Pi(\lambda_+, \lambda_-)|_{\varphi_{1\pm} = \varphi_{2\pm} = \varphi_{3\pm}} \tag{93}$$

Obviously the differential operators (91) do not depend on λ_+ and λ_-. One can show then that this way we obtain a semicovariant Wilson expansion in so far as the degree of singularity of the semicovariant kernels decreases with k_+ and k_-. But locality of the operators (93) is not yet guaranteed. Finally we multiply both sides of the expansion for R with the singular factor and get a semicovariant Wilson expansion for the operator product itself.

It ought to be mentioned that in certain degenerate cases, namely if discrete series representations occur, the series may be truly covariant termwise (Rühl and Yunn, 1975b; Swieca, 1974; Schroer and Swieca, 1975). This is due to the fact that one of the components in (85) drops out and covariance is restored this way. A famous degenerate case of this kind is obtained if the expansion operates on the vacuum state. The discrete series representations appear then as a consequence of the spectrum condition (Section 2).

5. PROBLEMS WITH LOCALITY

The Thirring model is particularly simple in several respects. Firstly we know that the field operators (53) and all those local operators derived from regularized multilocal products of it, have a certain simple behaviour under commutation with $Q_\pm$ (see (54)) and therefore with $\frac{1}{2}Q_\pm^2$. Thus projecting a product $A(x)B(y)$ on simultaneous eigenspaces of T_+ and T_-

$$A(x)B(y)\Pi(\lambda_+, \lambda_-) \tag{94}$$

fixes both pairs $\tau_{A\pm}$ and $\tau_{B\pm}$. Secondly we can regularize any product of such operators by splitting off a singular covariant factor, leaving a covariant bilocal C^∞ operator. Any such operator in $\mathbb{M}_2$ leads to a series (a 'family') of operators with increasing dimension

$$d = -j_+ - j_- + 1 \tag{95}$$

which for $\chi_3^C(k_\pm)$, gives

$$j_\pm = -j_3(k_\pm), \, d(k_+, k_-) = d_1 + d_2 + k_+ + k_- \tag{96}$$

i.e. the dimensions increase in integral steps. In fact, our asymptotic completeness relation is a reordered Taylor expansion and each derivation enhances the dimension by one (within the context of a Weyl symmetry). It is known from perturbation theory with respect to ε in a model in a space-time of $6-\varepsilon$ dimensions, that this set of operators is too small for a Wilson expansion in general (Mack, 1973).

Both properties of the Thirring model are not quite independent. If an operator product can be covariantly regularized as in (69), then this can make sense only if projections on eigenspaces from the right and the left fixes the transformation property of the regularized operator in both variables. In turn the transformation properties of the factors in the unregularized product must then also be fixed, namely in

$$\Pi(\lambda'_+, \lambda'_-)A(x)B(y)\Pi(\lambda_+, \lambda_-)$$
$$= \sum_{\lambda''_\pm} \Pi(\lambda'_+, \lambda'_-)A(x)\Pi(\lambda''_+, \lambda''_-)B(y)\Pi(\lambda_+, \lambda_-) \tag{97}$$

all $\lambda''_\pm$ must be equal modulo one.

We study the problem of recombining the different projections to a local operator first and then try to gain an idea of how the general case might look. In the Thirring model

$$\Pi(\lambda'_+, \lambda'_-)A(x)B(y)\Pi(\lambda_+, \lambda_-) = s(x, y)\Pi(\lambda'_+, \lambda')R(x, y)\Pi(\lambda_+, \lambda_-) \tag{98}$$

$\lambda'_\pm$ is either related to $\lambda_\pm$ by a function depending on the operators A, B

$$\lambda'_\pm = \mu_{\pm,A,B}(\lambda_\pm) \tag{99}$$

or both sides of (98) vanish identically. Moreover $\tau_{A\pm}$ and $\tau_{B\pm}$ are fixed and satisfy

$$\tau_{A\pm} + \tau_{B\pm} \triangleq \lambda'_\pm - \lambda_\pm \bmod 1 \tag{100}$$

Extracting the singular factor $s(x, y)$ changes the transformation property of A and B $j_{A\pm}$, $\tau_{A\pm}$, $j_{B\pm}$, $\tau_{B\pm}$ into that of R: $j'_{A\pm}$, $\tau'_{A\pm}$, $j'_{B\pm}$, $\tau'_{B\pm}$, explicitly

$$\tau_{A\pm} \triangleq \tau'_{A\pm} + t_\pm$$
$$\tau_{B\pm} \triangleq \tau'_{B\pm} - t_\pm \tag{101}$$

and

$$j_{A\pm} = j'_{A\pm} + \kappa_\pm$$
$$j_{B\pm} = j'_{B\pm} + \kappa_\pm \tag{102}$$

with some real parameters $t_\pm$ and $\kappa_\pm$. d_1 and d_2 in (96) refer to the transformation properties of the regularized operator and ought to be primed, too.

We write the decomposition of the regularized operator in the form ($\chi_A = \chi_{A+} \times \chi_{A-}$, etc.)

$$\Pi(\lambda'_+, \lambda'_-) R(x, y) \Pi(\lambda_+, \lambda_-)$$
$$\simeq \sum_{k_\pm=0}^{\infty} \int dz\, dz'\, Q(\chi'_A, x; \chi'_{B,y} | \chi(k_\pm), z) S(\chi(k_\pm), z | \chi^c(k_\pm), z')$$
$$\times \Pi(\lambda'_+, \lambda'_-) O(\chi^c(k_\pm), z') \Pi(\lambda_+, \lambda_-) \tag{103}$$

As emphasized in the preceding section, the operators $O(\chi^c(k_\pm), z)$ have a meaning without the projection operators applied to them, since the differential operator, by which it is obtained from $R(x, y)$, is independent of the eigenvalues $\lambda'_\pm$ and $\lambda_\pm$. It can be shown by explicit calculation that the semicovariant kernel Q in (103) (but not Q_{1a}!) depends at least in a neighbourhood of $x = y$ only on

$$\tau'_{A\pm} + \tau'_{B\pm} \triangleq \tau_{A\pm} + \tau_{B\pm} \triangleq \lambda'_\pm - \lambda_\pm \bmod 1 \tag{104}$$

Of course the same holds true for the intertwining operator S. In the Thirring model $\lambda'_\pm - \lambda_\pm$ is in general a function of $\lambda_\pm$. In some cases, however, namely if the operator product commutes with $Q_\pm$, this difference vanishes. Whenever the difference is a unique value modulo one, the summation over $\lambda''_\pm$ can be performed in (103) and it results in an expansion for the operator R in terms of local operators $O(\chi^c(k_\pm), z)$. In general the summation cannot be performed this way over the spectrum of T_+ and T_-. Then the non-local components of the operators $O(\chi^c(k_\pm), z)$ cannot be recombined to a local operator. An alternative formulation makes the kernel Q an operator such that the components of $O(\chi^c(k_\pm), z)$ combine to a local operator. The non-locality is then carried over to the kernel.

Finally we want to consider what happens to the kernel Q if we multiply it with the singular function $s(x, y)$ (98). If we insert

$$j(k_\pm) = \tfrac{1}{2} - j'_{1,\pm} - j'_{2,\pm} + k_\pm \tag{105}$$

Q depends (besides the coordinates) solely on the parameters $j'_{1\pm}$, $j'_{2\pm}$, $k_\pm$, and $\tau_{A\pm} + \tau_{B\pm}$. After multiplication with s we obtain an analogous function depending in the same fashion on the parameters $j_{1\pm}$, $j_{2\pm}$, $k_\pm + 2\kappa_\pm$, and $\tau_{A\pm} + \tau_{B\pm}$, except possibly an unessential k-dependent change in the normalization.

The asymptotic completeness relation (88) has been derived for C^∞ functions. It can be generalized to other classes of functions exhibiting singularities. This can be achieved by appropriate regularization techniques. Splitting off a singular covariant factor as in the Thirring model is just one method for just one class of functions. For this class of functions we have learnt that the whole family of local operators gets shifted in the dimension by a fixed amount and that the semicovariant kernels depend solely on the combinations $\lambda'_\pm - \lambda_\pm$.

Crucial to any model in $\mathbb{M}_2$ is therefore the regularization of the individual terms in the sum

$$\sum_{\lambda''_\pm} \Pi(\lambda'_+, \lambda'_-) A(x) \Pi(\lambda''_+, \lambda''_-) B(y) \Pi(\lambda_+, \lambda_-) \tag{106}$$

If all these terms can be regularized by extraction of the same covariant singular factor, we are in the same position as in the case of the Thirring model. The dependence on $\lambda''_\pm$ drops out completely and we can sum over it in a trivial fashion. A slightly more general situation arises if the individual terms in (106) can be expanded in a series each term of which can be covariantly regularized by factorization

$$\sum_{n=0}^{N_s(\infty)} s_n(x, y) \Pi(\lambda'_+, \lambda'_-) R_{n,\lambda''_+\lambda''_-}(x, y) \Pi(\lambda_+, \lambda_-) \tag{107}$$

One could call such expansion a 'covariant pre-Wilson-expansion'. It is assumed that the singular factors s_n are independent of $\lambda''_\pm$, and that the degree of the singularity of s_n decreases with increasing n. Each term in the pre-Wilson-expansion yields a family of operators

$$O_{n,\lambda''_+\lambda''_-}(\chi_n^c(k_\pm), z)$$

The intermediary projection operators can be eliminated by summation over $\lambda''_\pm$. Concerning the recombination to local operators we are still in the same position as in the case of the Thirring model.

If, however, in the pre-Wilson-expansion, the singular factors depend on $\lambda''_\pm$ in a non-trivial fashion, we have a new source for non-local operators appearing in the Wilson expansion. In any case, a pre-Wilson expansion (107) always leads to families of operators whose dimensions are non-integrally separated. Of course, it is also conceivable that no families of operators occur at all though this seems rather natural to us.

REFERENCES

Bateman, H. (1910) *Proc. London Math. Soc.*, **8**, 223.
Bonora, L., Cicciariello, S., Sartori, G. and Tonin, M. (1973) *Scale and Conformal Symmetry in Hadron Physics, Proc. Advanced School of Physics Frascati 1972*, R. Gatto (Ed.), John Wiley, New York.

Brandt, R. A. and Preparata, G. (1971) *Nucl. Phys.*, **B27**, 541.
Cunningham, E. (1909) *Proc. London Math. Soc.*, **8**, 77.
Dell'Antonio, G. F., Frishman, Y. and Zwanziger, D. (1972) *Phys. Rev.*, **D6**, 988.
Dobrev, V., Mack, G., Petkova, V. and Todorov, I. T. (1975a) *JINR Report E2-7977*; *Elementary Representations and Intertwining Operators for the Generalized Lorentz Group*, Institute for Advanced Study Preprint, Princeton.
Dobrev, V., Petkova, V., Petrova, S. and Todorov, I. T. (1975b) *Dynamical Derivation of Vacuum Operator Product Expansion in Euclidean Conformal Quantum Field Theory*, Institute for Advanced Study Preprint, Princeton.
Efremov, A. V. (1968) A model of Lie fields, Preprint P2-3731, JINR Dubna.
Ferrara, S., Gatto, R. and Grillo, A. F. (1973) *Springer Tracts in Modern Physics*, Vol. 67, Springer-Verlag, Berlin, p. 1.
Frishman, Y. (1971) *Ann. Phys. (N.Y.)*, **66**, 373.
Fulton, T., Rohrlich, F. and Witten, L. (1962) *Rev. Mod. Phys.*, **34**, 442.
Gell-Mann, M. and Low, F. (1954) *Phys. Rev.*, **95**, 1300.
Hortaçsu, M., Seiler, R. and Schroer, B. (1972) *Phys. Rev.*, **D5**, 2519.
Jost, R. (1961) 'Properties of Wightman functions', in *Lectures on Field Theory and the Many-Body Problem*, E. R. Caianello (Ed.), Academic Press, New York.
Kastrup, H. A. (1962) *Ann. Phys. (Leipzig)*, **7**, 388.
Kastrup, H. A. (1964) *Nucl. Phys.*, **58**, 561.
Klaiber, B. (1968) in *Quantum Theory and Statistical Physics, Lectures in Theoretical Physics*, Vol. X-A, A. O. Barut and W. E. Brittin (Eds.), Gordon and Breach, New York, p. 141.
Kupsch, J., Rühl, W. and Yunn, B. C. (1975) *Ann. Phys. (N.Y.)*, **89**, 115.
Lowenstein, J. H. and Swieca, J. A. (1971) *Ann. Phys. (N.Y.)* **68**, 172.
Lüscher, M. and Mack, G. (1975) *Comm. Math. Phys.*, **41**, 203.
Mack, G. (1973) in *Strong Interaction Physics, Lecture Notes in Physics*, Vol. 17, W. Rühl and A. Vancura (Eds.), Springer-Verlag, Berlin, p. 300.
Mack, G. (1974) in *Renormalization and Invariance in Quantum Field Theory*, E. R. Caianello (Ed.), Plenum Press, New York, p. 123.
Mack, G. (1975) All Unitary Ray Representations of the Conformal Group $SU(2,2)$ with Positive Energy, Universität Hamburg Preprint 1975, see this paper for the latest list of references on representations of the conformal group.
Mack, G. and Salam, A. (1969) *Ann. Phys. (N.Y.)*, **53**, 174, and references cited there.
Mack, G. and Todorov, I. T. (1973) *Phys. Rev.*, **D8**, 1764.
Mayer, D. H. (1974) Conformal Invariant Causal Structures on Pseudo-Riemannian Manifolds, Preprint Technische Hochschule Aachen, April 1974.
Migdal, A. A. (1971) *Phys. Letters*, **37B**, 386.
Pohlmeyer, K. (1969) *Comm. Math. Phys.*, **12**, 204.
Polyakov, A. M. (1969) *Sov. Phys. JETP*, **28**, 533.
Polyakov, A. M., (1973) Non-Hamiltonian Approach to the Quantum Field Theory at Small Distances, Preprint Landau Institute for Theoretical Physics, Chernogolovka 1973.
Rühl, W. (1969) *The Lorentz Group and Harmonic Analysis*, Benjamin, New York; see the references in this book for references to Toller's work.
Rühl, W. (1973) *Comm. Math. Phys.*, **30**, 287, **34**, 149.
Rühl, W. (1975) *Acta Physica Austriaca*, Suppl. XIV, 643.
Rühl, W. and Yunn, B. C. (1975a) Representations of the Universal Covering Group of $SU(1,1)$ and Their Bilinear and Trilinear Invariant Forms, Preprint Universität Kaiserslautern, June 1975, to appear in *J. Math. Phys.*
Rühl, W. and Yunn, B. C. (1975b) Operator Product Expansions in Conformally Covariant Quantum Field Theory, Part I: Strictly Covariant Expansions, Preprint Universität Kaiserslautern, October 1975.
Rühl, W. and Yunn, B. C. (1975c) Operator Product Expansions in Conformally Covariant Quantum Field Theory, Part II: Semicovariant Expansions, Preprint Universität Kaiserslautern, November 1975.
Schroer, B. and Swieca, A. (1974) *Phys. Rev.*, **D10**, 480.
Schroer, B., Swieca, J. A. and Völkel, A. H. (1975) *Phys. Rev.*, **D11**, 1509.
Segal, I. (1971) *Bull. Am. Math. Soc.*, **77**, 958.
Swieca, J. A. (1974) Conformal Operator Product Expansions in the Minkowski Region, Pontificia Universidade Católica do Rio de Janeiro, Preprint, May 1974.

Thirring, W. (1958) *Ann. Phys. (N.Y.)*, **3**, 91.
Wess, J. (1960) *Nuovo Cimento*, **18**, 1086.
Wilson, K. G. (1969) *Phys. Rev.*, **179**, 1499.
Wilson, K. and Zimmermann, W. *Comm. Math. Phys.* (1972) **24**, 871.
Zeeman, E. C. (1964) *J. Math. Phys.*, **5**, 490.
Zimmermann, W. (1970) *Lectures on Elementary Particles and Quantum Field Theory*, Brandeis University Summer Institute in Theoretical Physics 1970, Vol. 1, S. Deser, M. Grisaru and H. Pendleton (Eds.), MIT Press, Cambridge Mass., p. 395.
Zimmermann, W. (1973) 'Operator product expansions', *in Strong Interaction Physics, Lecture Notes in Physics*, Vol. 17, W. Rühl and A. Vancura (Eds.), Springer-Verlag, Berlin, p. 343.

18

The Construction of Quantum Field Theories

LUDWIG STREIT
Universität Bielefeld, Bielefeld, West Germany

1. THE PROBLEM

Heisenberg's uncertainty relations are the most compact formulation of the two-fold challenge presented by modern physics. As efforts failed to dispute their fundamental nature (Einstein's 'God does not throw dice' [Heisenberg (1969) gives a vivid eye-witness account of this struggle]) natural philosophy was called upon to cope with a radically new way of thinking (Heisenberg, 1960).

Mathematical physics on the other hand was faced with the task of enlarging the structures of classical theory in such a way that the uncertainty relations would find a place in them, or more precisely to base a consistent theory of quantum mechanics on the uncertainty relations, with classical mechanics as a macroscopic limit.

The philosophical revolution has not come to a close in the past 50 years, we shall not deal with it here. The physicist is reassured by the fact that the other, theoretical, challenge has been dealt with quite successfully. In the past 50 years quantum mechanics has become well established as a physically relevant and mathematically consistent theory [cf. for example Mackey's book (Mackey, 1963) for an axiomatic development of quantum theory on the basis of the uncertainty principle]. But not all is well. Einstein's theory of relativity, some 20 years before the advent of quantum mechanics even, amounted to yet another transgression beyond the domain of classical physics. At first glance these new territories appear to be quite disjoint: quantum theory takes the place of classical mechanics in the submicroscopic domain of atoms and nuclei, while special relativity does so in the realm of high velocities, comparable to that of light.

However, in any attempt to formulate a theory of elementary particles it is the uncertainty principle itself which points to the necessity of an amalgam between quantum mechanics and special relativity. If we insert subnuclear masses and dimensions in

$$\Delta x \Delta p \geq h$$

we find a velocity range $\Delta p/m$ which is by no means small compared to the speed of light.

And yet in these past 50 years and in spite of the hard work of what are now generations of physicists the construction of a relativistic quantum theory of interacting particles has not come to a close. Here we are faced with a problem that has turned out to be much more tenacious than its non-relativistic counterpart.

Two questions come to mind:

'Why not give up?'

and if this can be answered to satisfaction, then

'Why is it so hard?'

To answer the first one need only observe that we are dealing with two theories—quantum mechanics and special relativity—which are undoubtedly appropriate and powerful where only one and not both extensions of classical physics are called for, i.e. for submicroscopic phenomena as long as one may neglect the relativistic aspect and, respectively, for relativistic phenomena as long as quantum effects are unimportant. A fundamental theory—of elementary particles, if this concept should indeed survive—must deal with phenomena which are at the same time submicroscopic *and* relativistic, hence the quest for such a unified relativistic quantum theory is tantamount to the search for a fundamental theory of matter.

Now why is this so hard that it has defied the efforts of so many?—To clarify this a few generalities concerning the physical 'ansatz' are in order. Evidently different attacks on the problem have been based on different sets of assumptions, and one has frequently criticized the following list for being overly conservative, until the recent successes of constructive quantum field theory gave indications that they are reasonable.

As in the non-relativistic theory one assumes a Hilbert space description of the physical system with the (pure) states represented by unit vectors and the observables by a non-commutative algebra of operators.

A relativistic space–time structure is introduced if one considers *local* observables, i.e. operators with a space–time label, which

(1). transform covariantly under a suitable unitary representation of the Poincaré group with a unique invariant state (the vacuum).

(2). commute if they are affiliated with space-like regions of space time.

More specifically one can consider the algebras of (e.g. bounded) observables that are associated with given *space–time regions* [see Araki (1969) for a review of this 'algebraic approach']. For all its mathematical advantages this framework has not permitted the formulation of a dynamical ansatz. If on the other hand one tries to use classical relativistic dynamics as, for example, given by the equations of motion of electrodynamics as a guideline for the construction of a quantized relativistic theory one is immediately confronted with the concept of operators labelled by *space–time points*, i.e. to local relativistic quantum fields.

Apart from deviations often dictated by frustration the construction of such fields has been the goal of many elementary particle theorists for some 40 years. How would one go about this? As early as 1936 Heisenberg discussed the relevance of classical non-linear field theory for elementary particle physics (Heisenberg, 1939). But almost all of the pertinent research since then has followed a different course. A systematic understanding of non-linear wave equations is only now beginning to emerge (cf., for example, Reed, 1976), and as recent are most efforts to base quantization on their solutions (Dashen, 1974).

Instead, for lack of more powerful methods, non-interacting 'free' fields—the construction of which poses no insurmountable problems—were taken as a starting point. Interaction terms modelled after those of the classical theories were then added to the free Hamiltonian in the hope that it might be possible to treat the resulting dynamical changes perturbatively.

This program quickly ran into problems of convergence, ill-defined divergent expressions, etc. 'Subtraction physics' evolved first as the art of dropping infinite terms to obtain a finite remainder, later in a systematic way as renormalization theory. While this allowed precise predictions at least for quantized electrodynamics the questions of existence remained open. This was—and is—particularly serious for nuclear forces since they are so strong that the perturbative approximations must also fail.

Before embarking on a more detailed discussion of Hamiltonian quantum field theory it is worth while to pause and—with a good portion of hindsight—to isolate and exhibit the sources of these difficulties.

(A) The Infinite Volume Divergence

Addition of the space integral of an interaction energy density to the free Hamiltonian

$$H = H_0 + \int h_I(\mathbf{x}) \, d\mathbf{x}$$

gives rise to an operator which cannot be finite when applied to the vacuum Ω, since

$$H_0 \Omega = 0$$

so that

$$\|H\Omega\|^2 = \int (\Omega, h_I(\mathbf{x}) h_I(\mathbf{y}) \Omega) \, d\mathbf{x} \, d\mathbf{y}$$

with the integrand depending only on $\mathbf{x} - \mathbf{y}$ because of translation invariance.

A more refined argument leads to 'Haag's theorem' [Haag (1955), for a very general proof cf., for example, Emch (1972)] which says that the canonical variables of the problem with interaction cannot be equivalent to those appropriate for the free Hamiltonian. Note that 'all' representations of the canonical commutation relations

$$[q_i, q_k] = [p_i, p_k] = 0, \quad [q_i, p_k] = i\hbar\delta_{ik} \quad i, k = 1, \ldots, n$$

are unitarily equivalent (up to multiplicity and under reasonable technical assumptions). This important theorem of von Neumann (see, for example, Putnam, 1967) assures us that, for example, Heisenberg's matrix representation of these operators will never produce results that are different from those calculated in, say, Schrödinger's framework, where

$$q_i = x_i \quad p_k = -i\hbar\frac{\partial}{\partial x_k} \quad \text{in } L^2(d^n\mathbf{x})$$

In particular, any dynamical problem of quantum mechanics can be stated in these terms and solved by applying to an 'initial data' function from L^2 the unitary group generated by the Hamiltonian.

Not so in quantum field theory: von Neumann's uniqueness theorem breaks down as the number of degrees of freedom becomes infinite. There is then a vast and largely unexplored set of inequivalent representations for the canonical algebra, and Haag's theorem tells us that we have to find a non-standard one appropriate for the given Hamiltonian—not even the canonical algebras for free fields of two different masses are equivalent. We mention in passing that the situation is somewhat different if we formulate the initial value problem not on a space-like hyperplane of space–time like, for example,

$$\{(\mathbf{x}, t): t = 0\}$$

but on a light-like one such as

$$\{(\mathbf{x}, t): x_i + t = 0\}$$

But at this point little is known about the adequacy of such data for non-trivial theories cf., for example, Leutwyler and coworkers (1970) or for some recent results and further references Driessler (in press). That is we have to deal with the paradoxical situation that to state the initial values has become a non-trivial, dynamical question, and we have to solve it before we can even formulate the dynamical problem correctly. This discouraging paradoxon was bound to influence the directions of research. The decade from 1955 to 1965 was characterized by the strategy to learn about field theories not by construction but by postulating their existence and fundamental properties (locality, relativistic covariance, energy–momentan spectrum) as in the 'axiomatic' formulations of Lehmann, Symanzik and Zimmermann, of Wightman and— for local rings of bounded observables—of Araki, Haag and Kastler (cf., for example, Jost, 1965; Streater and Wightman, 1964; Emch, 1972). The constructive problem was generally relegated until after the advent of some 'totally new creative idea, a further essential change in our conceptions of the structural laws of matter' as one author put it in 1956. It was a fascinating episode of recent science history to observe how, ten years later, this turned out not to be the case. But let us first turn to the other obstacles that field quantization had to cope with.

(B) The Ultraviolet Divergences

There were expectations in the early days of quantum field theory that singularities such as the infinite self-energy of classical point charges would go away through quantization. But quite to the contrary virtually every second calculation of quantum electrodynamics included the process of throwing away an infinite term and interpreting the remainder as the 'correct result'. These procedures were formalized in the renormalization theory of Feynman, Dyson and Schwinger. Used as a recipe for calculations they allowed for the astounding numerical predictions of quantum electrodynamics while at the same time the meaning of the formal dynamical ansatz or the formulation of a meaningful one was further obscured.

(C) Series Divergences

The successes of renormalized perturbation theory as applied to quantum electrodynamics are even more impressive in the face of yet another type of divergence—series divergences. For non-linear interactions the convergence question of the perturbation series for, say, the Green's functions looks hopeless. Combinatorial considerations show a veritable explosion of the number of terms as the order of the perturbation increases. Also a glance at, for example, the quartic oscillator potential makes plausible that inverting the sign of the coupling constant changes the nature of the interaction so drastically that we should not expect analyticity in the neighbourhood of zero.

(D) Infrared Divergences

With these we come to the end of our list of difficulties. They arise from the long range of forces mediated by the exchange of massless particles. In momentum space these long distance problems become problems of small momenta (hence the name). Certain aspects can be studied in non-relativistic models: note the discussion of scattering from potentials with Coulomb tails. Also, in contradistinction to the other complications, this one is not intrinsic to *all* non-trivial local quantum field theories. It does not arise as long as we focus on theories without massless excitations, and we shall not consider it further.

As we turn to an account of recent progress in constructive quantum field theory we shall aim neither for mathematical rigour nor for any kind of completeness—this would be quite meaningless anyway in a situation of such rapid progress—but instead we shall try to communicate to the non-expert how the main structural problems are being tackled and what one can say about the evolving theory. (The references were selected correspondingly.)

II. HAMILTONIAN QUANTUM FIELD THEORY

A systematic exposition of the subject may be found in various texts (e.g., Schweber, 1961; Bjorken and Drell, 1965) we shall content ourselves here to present the most important concepts as generalizations of ones that are well-known from non-relativistic quantum mechanics to the case of infinitely many degrees of freedom.

We shall begin our short dictionary of quantum field theory language with the canonical variables which in field theory are indexed by points in s-dimensional space the expressions on the left-hand side refer to quantum mechanics and those on the right-hand side to quantum field theory in this and the following examples.

$[q_i, q_K] = [p_i, p_K] = 0$ $\qquad$ $[\varphi(\mathbf{x}), \varphi(\mathbf{y})] = [\pi(\mathbf{x}), \pi(\mathbf{y})] = 0$

$[q_i, p_K] = i\delta_{Ki}$ $\qquad$ $[\varphi(\mathbf{x}), \pi(\mathbf{y})] = i\delta^{(S)}(\mathbf{x} - \mathbf{y})$

Generic variables are obtained as follows

$(q, \lambda) = \sum_{i=1}^{n} \lambda_i q_i \quad \lambda \in \mathbb{R}^n$ $\qquad$ $\varphi(f) = \int f(\mathbf{x})\varphi(\mathbf{x}) \, d^S\mathbf{x} \quad \mathscr{S}_\varepsilon$

This 'smearing out' with smooth, rapidly decreasing functions has the extra advantage of making $\varphi(f)$ a less singular operator than $\varphi(\mathbf{x})$ is.

The equations of motion are

$\dot{q}_K = i[H, q_K] = p_K$ $\qquad$ $\dot{\varphi}(\mathbf{x}) = i[H, \varphi(\mathbf{x})] = \pi(\mathbf{x})$

$\ddot{q}_K = i[H, p_K]$ $\qquad$ $\ddot{\varphi}(\mathbf{x}) = i[H, \pi(\mathbf{x})]$

For the vacuum state we borrow a typical property of quantum mechanical ground states:

$\psi_0(x) \neq 0$ $\qquad$ ψ_0 cyclic for the field algebra $\mathscr{A}_\varphi$ i.e. $\Psi_\varphi \psi_0$ dense in the representation space

for almost all x

Such cyclic representations allow for a very compact and useful description via

$E(\lambda) = (\psi_0, e^{i(q,\lambda)} \psi_0)$ $\qquad$ $E[f] = (\Psi_0, e^{i\varphi(f)} \Psi_0)$

$= \int_{\mathbb{R}^n} e^{i(x,\lambda)} |\psi_0(x)|^2 \, d^n x$ $\qquad$ $= \int_{\mathscr{S}'} e^{i(\chi,f)} \, d\mu(\chi)$

E is the 'characteristic function' (= Fourier transform) of a probability measure $|\psi_0(x)|^2 \, d^n x$ on the vector space $\{x\} = \mathbf{R}^n$ dual to the λ.

E is the characteristic functional of a probability measure μ on the vector space $\mathscr{S}'$ of distributions dual to the space $\mathscr{S}$ (cf., e.g., Gelfand and Vilenkin, 1964; Hida, 1970).

A prominent example is furnished by the harmonic oscillator ground state and its field theoretical counterpart:

$$H_{osc} = \tfrac{1}{2} \sum_K (p_K^2 + \omega_K^2 q_K^2) - E_0$$

$$= \tfrac{1}{2}(p, p) + \tfrac{1}{2}(q, \omega^2 q) - E_0$$

$$H_{osc}\psi_0 = 0$$

$$H_0 = \tfrac{1}{2} \int d^s x : \pi^2(x) + (\nabla \varphi(x))^2 + m^2 \varphi^2(x):$$

$$= \tfrac{1}{2} \int d^s x : \pi^2 + \varphi \omega^2 \varphi :$$

with $\omega^2 = -\Delta_x + m^2$

$$H_0 \Psi_0 = 0$$

so that in this case

$$E(\lambda) = e^{-1/4(\lambda, \omega^{-1}\lambda)}$$

$$= e^{-1/2\|(q,\lambda)\psi_0\|^2}$$

$$E[f] = e^{-1/4(f, \omega^{-1}f)_{L^2}}$$

$$= e^{-1/2\|\varphi(f)\Psi_0\|^2}$$

In both cases we are dealing with multivariate Gaussian distributions of mean zero: their characteristic function(al)s are obtained by exponentiating their second moments. With

$$\omega = (-\Delta + m^2)^{1/2} \text{ so that } (f, \omega^{-1}f) = \int \tilde{f}^*(k) \frac{1}{\sqrt{K^2 + m^2}} \tilde{f}(k) \, d^s k$$

$E[f]$ is the generating functional of the Fock representation of a scalar relativistic free field of mass m.

$$\ddot{\varphi}(x) + \omega^2 \varphi(x) = 0$$

Creation and annihilation operators are introduced through

$$q_k = \frac{1}{\sqrt{2\omega_k}}(a_k^+ + a_k)$$

$$p_k = i\sqrt{\frac{\omega_k}{2}}(a_k^+ - a_k)$$

$$a_k \psi_0 \equiv 0$$

$$[a_k, a_{k'}^+] = \delta_{kk'}$$

$$\varphi(x) = (2\pi)^{-S/2} \int \frac{d^s k}{\sqrt{2\omega(K)}}(a^+(k) e^{-ikx} + a(k) e^{iKx})$$

$$\pi(x) = i(2\pi)^{-S/2} \int d^s k \sqrt{\frac{\omega(k)}{2}}(a^+(k) e^{-ikx} - a(k) e^{-iKx})$$

$$a(k)\Psi_0 \equiv 0 \qquad \omega(k) \doteq \sqrt{k^2 + m^2}$$

$$[a(k), a^+(k')] = \delta^{(S)}(k - k')$$

In both cases the double dots : : of Wick ordering signify ordering field operator products such that all annihilation operators a stand to the right of the creation operators a^+.

In particular this procedure makes $:\varphi^n(x):$ a well-defined local operator in the sense that

$$:\varphi^h:(f) = \int f(\mathbf{x}):\varphi^n(\mathbf{x}): d^s\mathbf{x}$$

is densely defined or even self-adjoint for suitable n and s.

As a consequence

$$gh_I(\mathbf{x}) \equiv g:\varphi^n(\mathbf{x}): \quad n > 2$$

has become the classical ansatz for the interaction energy density of a self-coupled scalar field. In the following we shall concentrate on models of this type—while self-interacting scalar fields may not be appropriate as a fundamental concept for particle physics, they provide the simplest model for the discussion of the basic mathematical problems inherent in any non-trivial field theoretical ansatz.

III. CUTOFFS: THE GUENIN-SEGAL STRATEGY

Cutoffs come to mind as a remedy of the basic problems: 'Putting the theory into a finite box' to avoid the infinite volume divergence, setting finite upper limits for momentum space integrals to eliminate ultraviolet divergences—these techniques have been employed from the early days of relativistic quantum field theory. From the theoretical point of view any such surgery amounts to a violation of basic symmetries and principles such as translation invariance or locality, from the practical point of view it was often a matter of luck or intuition to extract just those quantities which were not violently cutoff-dependent, or otherwise to find a (more or less) physical interpretation of the cutoff.

It is a major and characteristic achievement of constructive quantum field theory that one has learned to make the cutoffs reversible, by first introducing sufficiently many of them to be able to construct a well-defined model and then controlling the limits as the cutoffs are removed in such a way that a non-trivial relativistic quantum field theory emerges. Evidently on the practical side much was learned about which quantities do not depend catastrophically on cutoffs and hence are amenable to approximate computation.

The Guenin–Segal strategy [reviewed by Jaffe (1969)] presents the most transparent example of such reversible surgery. Its goal is to circumvent Haag's theorem, i.e. to deal with the infinite volume divergence of interaction Hamiltonians such as

$$H = H_0 + \int g:\varphi^n(\mathbf{x}): d^s\mathbf{x}$$

and it is based on one cutoff and two observations.

The cutoff is rather obvious. Since the infinite integral over the interaction energy density causes the problem we reduce the latter to zero at large distances through a space-dependent coupling $g(\mathbf{x})$

$$g(\mathbf{x}) = \begin{cases} g > 0 & |\mathbf{x}| < l \\ 0 & |\mathbf{x}| > l + \varepsilon \end{cases}$$

with a smooth transition between the regions of constant coupling strength $|\mathbf{x}| < l$ and of no interaction $|\mathbf{x}| > l + \varepsilon$. We denote the modified Hamiltonian by H_l.

The two observations that bring this cutoff under control exploit the locality of the interaction term and the continuity of vacuum expectation values.

(A) Locality

The equation of motion

$$\ddot{\varphi}(\mathbf{x}) = i[H_l, \pi(\mathbf{x})]$$

is insensitive to the values of $g(\mathbf{y})$ for $\mathbf{y} \neq \mathbf{x}$ since $\pi(x)$ commutes with the energy density at such points. Causal propagation of the field [a feature to be verified! (Jaffe, 1969, p. 126ff.)] then allows one to conclude that the time evolution of the field is insensitive to the cutoff in the causal dependence region of the constant coupling domain $|\mathbf{x}| < l$, i.e. we have a cutoff independent solution in the diamond

$$|\mathbf{x}| + |t| < l$$

in which we can imbed any bounded space–time region by choosing a sufficiently large, yet finite, cutoff parameter one. But this solution of the equations of motion is not all that is required. For the construction of physical states we next invoke the following.

(B) Continuity of the Vacuum Expectation Values

Looking for a physical vacuum which, formally, should be given to us as the lowest lying eigenstate of H, we run into the following problem. Consider the ground states [it is by no means trivial to verify their existence (Glimm and Jaffe, 1970)] Ω_l of the approximate Hamiltonians H_l: on the basis of Haag's theorem we cannot hope for Ω_l to have a non-trivial limit as the cutoff l is taken to infinity. However there is a subtle distinction between convergence of the vectors Ω_l in Fock space and that of the expectation functionals

$$\omega_l(A) \equiv (\Omega_l, A\Omega_l)$$

on the field algebra generated by the approximate vacua. The following heuristic argument supports this distinction: as the cutoff parameter l is increased, the ground state differs from the Fock vacuum (and all other Fock space vectors) over larger and larger regions until in the limit it becomes orthogonal to all of them ['van Hove's phenomenon', Guerra (1972)]

$$w - \lim_{l \to \infty} \Omega_l = 0$$

On the other hand it is plausible that the expectation value of local observables A changes only little if the state in question is altered at distances of the order of a large l, and less and less as l approaches infinity:

$$\lim_{l \to \infty} \omega_l(A) \equiv \omega(A)$$

is expected to exist.

There is then a well-known procedure ('GNS construction'*) to cast $\omega(A)$ in a Hilbert space form:

$$\omega(A) = (\Omega, A\Omega)$$

At first sight this may be confusing. Have we found a vector Ω where there was none before? This is not the case. Recall that we have found it impossible to construct the limiting vector Ω in *Fock space*. Here it occurs as a cyclic vector of a field which is inequivalent to that of the Fock representation.

One might say that by controlling the limiting state we have succeeded in constructing the theory. What then remains to be done is to verify the required properties, such as Poincaré invariance [for Lorentz covariance in Fock space cf. Cannon and Jaffe (1970) and for a 'Euclidean' proof Simon (1974)] and the desirable ones, like the existence of particles (Glimm and coworkers, 1974) and of non-trivial scattering processes (Osterwalder and Sénéor, 1975); Eckmann and Dimock (in press) between them.

We should emphasize that this construction actually predicts the particles of the theory—the resulting representation of the Poincaré group will not be equivalent to the original one in Fock space. In this sense, too, a *relativistic* quantum theory provides a more fundamental description of matter. The program that we have sketched for the construction of such theories includes many steps which we have barely mentioned here although they are technically very involved. It is a tremendously important step forward in the construction of a relativistic quantum theory of matter though that this program has been proven viable—if only for a sufficiently simplified class of models.

It will turn out to be very instructive for us to track down the cause of such restrictions. Recall that we had cutoff the interaction Hamiltonian in an effort to obtain a finite vector when applying it to the Fock vacuum:

$$\|H_I \Psi_0\| < \infty$$

If we express, for example, an interaction energy density

$$h_I(\mathbf{x}) = g(\mathbf{x}){:}\varphi^n(\mathbf{x}){:}$$

---

*For this 'reconstruction' of fields resp. bounded observables cf., for example, Jost (1965), Streater and Wightman (1964) and Emch (1972).

in terms of the creation and annihilation operators given in our 'dictionary' it is straightforward to calculate

$$\|H_I\Psi_0\|^2 = \text{const.} \int \prod_{\nu=1}^{n} \frac{d\mathbf{k}_\nu}{\omega(\mathbf{k}_\nu)} \left| \tilde{g}\left(\sum_\nu \mathbf{k}_\nu\right) \right|^2$$

Here $\tilde{g}$ denotes the Fourier transform of the cutoff function g. Whatever its exact form may be the integral is finite only in a model world where the $\mathbf{k}$-integration—and hence space—is one-dimensional. With increasing space–time dimensionality (and increasing interaction power n) the integral exhibits a higher and higher degree of divergence for large k—i.e. an 'ultraviolet' divergence that calls for renormalizations.

We have found that in such cases the space cutoff Hamiltonian may not see the Fock vacuum, technically the latter is not in the domain of H_I. Nor, as one can check, are any other simple Fock space vectors that one might think of (Glimm (1969)).

For such singular perturbations the domains of the Hamiltonians (the vectors of finite energy) are sensitive to the detailed features of the interaction, its specific form would have to be taken into account in their construction. This particular problem can be attacked with the help of approximate Møller operators. In non-relativistic quantum mechanics these serve to intertwine between the free and the interacting Schrödinger Hamiltonians, and consequently, between their domains. A viable adaptation of these ideas to the case at hand proceeds along the following steps: introduce a high momentum cutoff in the interaction Hamiltonian to make it well-defined—use Friedrichs' perturbative construction [for a review and references cf. Streit (1970)] to obtain approximate wave operators ('dressing transformations')—apply these 'dressing transformations' to suitable Fock space vectors to obtain state vectors that the interacting Hamiltonian can see—remove cutoffs to obtain states appropriate for the full, no cutoff interaction.

Technically the construction of such approximate dressing transformation and controlling the limits is extremely complicated, but two structurally interesting observations should be made before we embark on a more recent alternate approach. One can only hope to find intertwining transformations for operators with matching spectra. Friedrichs' construction actually generates these adjustments of the ground-state energy, mass gap, etc. These are the so-called renormalization counter-terms. In the limit as the cutoff is removed they would become infinite but as they cancel corresponding infinite ground-state energies, masses, etc., in the original Hamiltonian the overall renormalized energy operator has a finite limit. A particularly accessible subclass of models is formed by those where such asymptotically infinite renormalization terms occur only up to a finite perturbation theoretical order. These are the so-called *superrenormalizable* models, among them the 'Yukawa interaction' of fermions and bosons in two space–time dimensions (Y_2), and the quartic self-interactions of scalar mesons in a three-dimensional space–time (φ_3^4). At present the problem of going beyond this class and of tackling models like φ_4^4 in the physical four-dimensional space–time is still unsolved.

Secondly two cases must be distinguished regarding the limiting 'dressed states' as the momentum cutoff is removed. In the less singular case the limits can be performed within Fock space. Otherwise one must proceed as with the infinite volume limit that we have discussed above and construct a new field representation from limits of expectation values. In this latter case then, the ultraviolet divergences alone already call for a non-Fock representation of the field. Prominent examples are the Y_2 and φ_3^4 models, respectively [for a review cf. Hepp (1969)].

IV. EUCLIDEAN SPACE–TIME—AND BACK!

It has been observed frequently in various contexts of non-relativistic as well as relativistic quantum dynamics that the transition to imaginary time results in remarkable structural simplifications: one obtains the heat equation from the Schrödinger equation, correspondingly Wiener integrals instead of Feynman's path integral, better behaved kernels in the Bethe–Salpeter equation for relativistic two-particle amplitudes, and most importantly for us here, the transition from relativistic to 'Euclidean' quantum field theory brought about by switching from relativistic Minkowski space–time to a space–(imaginary) time with positive definite Euclidean metric gives us models of equilibrium statistical mechanics (Symanzik, 1969) which we are comparatively much better equipped to handle.

The central role that this latter transformation has recently begun to play in the development of relativistic quantum dynamics stems from the fact that Nelson (Simon, 1974, Chap. IV) and K. Osterwalder and R. Schrader (Simon, 1974, Chap. II) have given conditions under which it becomes reversible.

In the light of this discovery it has become an advantageous, and very effective, approach to the construction of quantum field theories to first establish the corresponding Euclidean theories and as many of their properties as possible by means of methods borrowed from statistical mechanics, and finally to check that they survive the transition back to the relativistic Minkowski space–time. As a recent example—among many others—we mention the work of (cf. the papers of Eckmann and Dimock, in press) on the existence of a non-trivial scattering matrix and its asymptotic series expansion. For in-depth reading on the 'Euclidean strategy' a monograph written by one of the leading experts in this field is available (Simon, 1974). In the present review we want to give an introductory sketch of the method and of its scope. To this end we need to introduce one more concept from quantum field theory, the time-ordered Green's functions. They are symmetric functions of n space–time arguments defined to equal the vacuum expectation value of the n-field time-ordered product of the field at space–time points

$$x_i = (\mathbf{x}_i, x_{0i})$$

$$\tau_n(x_1, \ldots, x_n) = (\Omega, \varphi(x_n), \ldots \varphi(x_n)\Omega) \quad \text{if } x_{01} > x_{02} > x_{03} \ldots > x_{0n}$$

and they are described most handily by their generating functional

$$\tau[f] = \sum_n \frac{i^n}{n!} \prod_{\nu=1}^n \int d^{S+1}x_\nu f(x_\nu) \tau_n(x_1, \ldots, x_n)$$

$$= (\Omega, T\, e^{i\int \varphi(x)f(x)\,d^{S+1}x_\nu}\Omega)$$

It is straightforward but very instructive to calculate this functional for the trivial case where φ is a free field of mass $m > 0$ in Fock space so that it obeys the Klein–Gordon equation of motion

$$(\partial_\mu \partial^\mu + m^2)\varphi(x) = 0$$

Defining its Green's function by

$$\Delta_F(x) = \frac{i}{(2\pi)^{S+1}} \int d^{S+1}p \frac{e^{-ip^\mu x_\mu}}{p_\mu p^\mu - m^2 + i\varepsilon}$$

one finds for the free-field functional τ_0

$$\tau[f] = \tau_0[f] = \exp -\tfrac{1}{2} \int dx\, dy f(x) \Delta_F(x-y) f(y)$$

Continuation to imaginary times yields the functional $\sigma[f]$. Minkowski inner products become Euclidean ones

$$p^\mu p_\mu \to -p^2$$

so that, in terms of the Fourier transform $\tilde f$

$$\sigma_0[f] = \exp -\tfrac{1}{2}(\tilde f, (p^2 + m^2)^{-1}\tilde f)$$

σ is the generating functional of the τ-functions continued to imaginary time, the so-called Schwinger functions S_n. Their interest lies in the fact that fairly explicit and very useful expressions for σ and the Schwinger functions can also be derived for an interacting field. It will be our main task in this section to do so in a heuristic fashion. The necessary mathematical arguments are presented in Simon (1974), Chapter V; as examples of recent extensions to more singular models such as the Yukawa model Y_2 or the quartic meson self-interaction φ_3^4 in three-dimensional space–time we quote McBryan (in press).

The crucial observation is that—in contradistinction to $\tau_0[f]$—$\sigma_0[f]$ is the exponential of a negative definite quadratic form, i.e. just like the generating functional of a free field at fixed time we may write it as the Fourier transform of a (Gaussian) probability measure on the space of generalized functions:

$$\sigma_0[f] = \int_{\mathscr{S}'(\mathbb{R}^{S+1})} e^{i(\chi, f)}\, d\mu_0(\chi)$$

Recall that—for finite dimensional vector spaces!—the Fourier transform of a Gaussian is again a Gaussian, with the inverse quadratic form in the exponent, i.e. formally

$$d\mu_0(\chi) = \text{const}\, e^{-1/2(\chi^2,\, (p^2+m^2)\chi^2)}\, d^\infty \chi$$

$$= \text{const}\, e^{-1/2\int d^{S+1}x(\chi^2 + (\nabla\chi)^2 + m^2\chi^2)}\, d^\infty \chi$$

Observe the space–time integral of the Hamiltonian density in the exponent, an extra factor

$$\mathcal{N}_g\, e^{-\int dx h g_1(x)}$$

ought to generate the measure $d\mu_g$ appropriate to the interaction mediated by the Hamiltonian

$$H = H_0 + \int d^s x\, g h_1(\mathbf{x})$$

Now evidently the infinite volume element $d^\infty \chi$ above is purely formal but if we add the interaction factor to the left-hand side there is at least a fighting chance for

$$d\mu_g = \mathcal{N}_g\, e^{-g\int dx h_1(x)}\, d\mu_0$$

to be well-defined since $d\mu_0$ is. $\mathcal{N}_g$ is just a normalization factor for the new measure:

$$\mathcal{N}_g^{-1} = \int e^{-\int dx g h_1(x)}\, d\mu_0$$

Not unexpectedly there is a Euclidean variant of Haag's theorem in our way but by now we know how to deal with this: we cutoff the interaction strength by making g space–time dependent and let $g \to$ const in the final expression. For our favourite model φ_2^4 where

$$h_1(x) : \varphi^4(x):,\quad s = 1$$

this leads to the probability measure

$$d\mu_g[\chi] = \lim_{g \to \text{const}} \frac{e^{-\int d^2 x g(x):\varphi^4(x):}\, d\mu_0[\chi]}{\int e^{-\int d^2 x g(x):\varphi^4(x):}\, d\mu_0[\chi]}$$

with the Schwinger function as its moments

$$S_n(x_n, \ldots, x_n) = \lim_{g \to \text{const}} \frac{\int \chi(x_n) \ldots \chi(x_n)\, e^{-\int d^2 x g(x):\varphi^4(x):}\, d\mu_0[\chi]}{\int e^{-\int d^2 x g(x):\varphi^4(x):}\, d\mu_0[\chi]}$$

For these quantities then one must verify the 'Osterwalder–Schrader axioms' (Simon, 1970, Chap. II) which are Euclidean analogues to those of Wightman and which guarantee that a corresponding relativistic quantum field theory obeying the latter exists.

The shorthand notation

$$\int \cdot\, d\mu_0 = \langle \cdot \rangle_0,\quad \int_v d^2 x h_1(x) = U_v$$

makes the similarity with infinite volume Gibbs states of equilibrium statistical mechanics even more transparent:

$$\sigma[f] = \lim_{v \to \infty} \langle e^{i(\chi, f)}\, e^{-g U_v} \rangle_0 / \langle e^{-g U_v} \rangle_0.$$

The following random collection of observations is meant to serve as an illustration—by no means an exhaustive one—of the wealth of information which this analogy opens up.

(1). The coupling constant g plays the role of an inverse temperature. High temperature expansions as in statistical mechanics have turned out to be useful to deal with the weak coupling regime of model quantum field theories (Simon, 1974).

(2). Physical masses, i.e. the lowest excitations of the system, can be discussed effectively in terms of inverse correlation lengths.

(3). The direct coupling of the random field at different space–time points is brought about by the gradient term of the *free* Hamiltonian. In a lattice approximation to where the random field is replaced by a discrete set of 'spin' variables X_i this coupling amounts to that of a nearest neighbour Ising ferromagnet. As a result various useful correlation inequalities can be proven for the Schwinger functions (Simon, 1974).

(4). As the coupling strength is increased φ_2^4-models exhibit phase transitions, long-range order, and the breaking of the $\varphi \to -\varphi$ symmetry (Glimm and coworkers, 1975; cf. also Glimm and Jaffe, in press). The proof uses mean field techniques and the Peierls argument from statistical mechanics. Here it becomes patent to what extent the Euclidean formulation has come into its own.

(5). The existence of a non-trivial φ_4^4 model has turned out to be closely related to the non-triviality of the four-dimensional Ising model at the critical point (Glimm and Jaffe, 1974; Schrader, 1975).

With this glimpse of the final goal—namely to establish non-trivial relativistic quantum theories for interacting particles in four-dimensional space–time—we close this 'introductory review'. We have tried to display a representative subset of the techniques and the trends of a field that has recently seen rapid development. At this point there is good reason to be optimistic about the emergence of relevant models for the subnuclear structure and interaction of matter. With this goal in mind the impressive amount not just of abstract existence proofs, but beyond these of structural insight and of sound computational techniques inherent in the recent development of constructive quantum field theory acquires a particular relevance.

REFERENCES

Araki, H. (1969) *in Local Quantum Theory*, R. Jost (Ed.), Academic Press, New York.
Bjorken, J. D. and Drell, S. D. (1965) *Relativistic Quantum Fields*, McGraw-Hill, New York.
Cannion, J. T. and Jaffe, A. (1970) *Comm. Math. Phys.*, **17**, 261.
Dashen, R., Hasslacher, B. and Neveu, A. (1974) *Phys. Rev.*, **D10**, 4138.
Eckmann, J. P. (in press) *in Quantum Dynamics: Models and Mathematics*, L. Streit (Ed.), Springer, Vienna.
Dimock, J. (in press) *in Quantum Dynamics: Models and Mathematics*, L. Streit (Ed.), Springer, Vienna.

Driessler, W. (in press) 'On the structure of fields and algebras on null-planes I, II; *Acta Phys. Austriaca.*
Emch, G. (1972) *Algebraic Methods in Statistical Mechanics and Quantum Field Theory*, John Wiley, New York.
Gelfand, T. M. and Vilenkin, N. Ya. (1964) *Generalized Functions*, Vol. 4, Chap. IV, Academic Press, New York.
Glimm, J. and Jaffe, A. (1970) *Ann. Math.*, **91,** 362.
Glimm, J. and Jaffe, A. (1974) *Phys. Rev. Lett.*, **33,** 440.
Glimm, J. and Jaffe, A. (in press) in *Quantum Dynamics: Models and Mathematics*, L. Streit (Ed.), Springer, Vienna.
Glimm, J., Jaffe, A. and Spencer, T. (1974) *Ann. Math.*, **100,** 583.
Glimm, J., Jaffe, A. and Spencer, T. (1975) *Comm. Math. Phys.*, **45,** 203.
Guerra, F. (1972) *Phys. Rev. Lett.*, **28,** 1213.
Haag, R. (1955) *Dan. Mat. Fys. Medd.*, **29,** no. 12.
Heisenberg, W. (1939) *Z. Physik*, **113,** 61.
Heisenberg, W. (1960) *Sprache und Wirklichkeit in der modernen Physik* in Gestalt und Gedanke, Folge 6.
Heisenberg, W. (1969) *Der Teil und das Ganze*, Chaps. 5–11, 17, Piper, Munich.
Hepp, K. (1969) *Théorie de la Renormalisation*, Springer, Berlin.
Hida, T. (1970) *Stationary Stochastic Processes*, Section 4, Princeton University Press, Princeton.
Jaffe, A. (1969) *Local Quantum Theory*, R. Jost (Ed.), Academic Press, New York.
Jost, R. (1965) *The General Theory of Quantized Fields*, Amer. Math. Soc., Providence.
Leutwyler, H., Klauder, J. R. and Streit, L. (1970) *Nuovo Cimento*, **66A,** 536.
Mackey, G. W. (1963) *Mathematical Foundations of Quantum Mechanics*, Benjamin, New York.
McBryan, D. A. (in press) *Quantum Dynamics: Models and Mathematics*, L. Streit (Ed.), Springer, Vienna.
Osterwalder, K. and Sénéor, R. (1975) 'The scattering matrix is non-trivial for weakly coupled $P(\varphi)_2$ models'. Preprint.
Putnam, C. R. (1967) *Commutation Properties of Hilbert Space Operators and Related Topics*, Springer, Berlin.
Reed, M. (1976) *Abstract Non-linear Wave Equations*, Springer, Berlin.
Schrader, R. (1975) 'A possible constructive approach to φ_4^4 I, II'. Berlin preprints.
Schweber, S. S. (1961) *An Introduction to Relativistic Quantum Field Theory*, Harper and Row, Evanston.
Simon, B. (1974) *The $P(\varphi)_2$ Euclidean (Quantum) Field Theory*, Princeton University Press, Princeton.
Streater, R. F. and Wightman, A. S. (1964) *PCT, Spin and Statistics, and All That*, Benjamin, New York.
Streit, L. (1970) *Acta Phys. Austriaca Suppl.* VII, 355.
Symanzik, K. (1969) in *Local Quantum Theory*, R. Jost (Ed.), Academic Press, New York.

19

Classical Electromagnetic and Gravitational Field Theories as Limits of Massive Quantum Theories

GORDON FELDMAN
The Johns Hopkins University, Baltimore, Maryland, U.S.A.

1. INTRODUCTION

The correspondence principle in quantum mechanics states, *inter alia*, that as Planck's constant h approaches zero the theory must approach the corresponding classical theory. This principle is meaningful if there exists a classical theory which corresponds to the particular quantum theory. If we examine quantum field theories we can apply the principle only to 'massless' theories, i.e. to field theories which on quantization describe particles of zero mass. One can see this in many ways. The simplest is to notice that a field equation involves derivatives of the field to which one must add a 'mass term'. Dimensional arguments require that this term be proportional to powers of mc/h (the inverse Compton wavelength, m being the mass). We see that taking the limit $h \to 0$ with m kept fixed is completely different from the limit $m \to 0$ and then $h \to 0$. Accordingly, a classical field theory of particles requires taking the $m \to 0$ limit before the $h \to 0$ limit, i.e. a classical field theory of particles, of necessity, describes massless particles. In fact as $m \to 0$ the parameter h disappears from the field equations. Two familiar examples of classical field theories are the electromagnetic (Maxwell theory) and gravitational (Einstein theory) field theories. We can interpret the Maxwell field (or photon field) as a relativistic field describing particles of mass, zero, and spin, one. The Einstein (or gravitational field) is a relativistic field describing particles of zero mass and spin two. One might expect that these classical field theories may be the limit of quantum field theories which describe massive particles of spin one and two. This problem has attracted some attention recently (Boulware and Deser, 1972; van Dam and Veltman, 1970). One examines quantum field theories describing massive particles of spin one and two coupled to sources and then performs the $m \to 0$ limit. The question is to discover whether this limit is smooth, i.e. does this limiting theory give rise to the same experimental consequences as the corresponding field theories describing massless particles of spin one and two coupled to the same sources, respectively. That there may be some problems connected with the $m \to 0$ limit we can see from the properties of the representations of the Poincaré group. Those irreducible representations which span the space of massive particle states also have a spin parameter, s, with degeneracy $2s+1$, i.e. a particle of mass m and spin s has $2s+1$ degrees of freedom. However, the irreducible representations corresponding to a massless particle also has a spin parameter s but only two degrees of freedom*. The implication of the above remarks is that the Hilbert space of physical states describing particles of mass $m \neq 0$ and $s \geq 1$ is somehow larger than the Hilbert space for the corresponding massless particles. Since degrees of freedom cannot disappear, the resolution to the problem must be in the fact that either the $m \to 0$ limits are not smooth (i.e. the two theories are different) or that the 'disappearing' degrees of freedom decouple or both. In this article we examine carefully the $m \to 0$ limits for spin one and spin two field theories to see what happens to the structure of the theories.

Most of the results have been obtained previously by Boulware and Deser (1972) and by van Dam and Veltman (1970). What we do in this article is to approach the problem by using different techniques. In Section 2 we examine the equations of motion for spin one and two fields in order to see how the $m \neq 0$ and $m = 0$ equations each describe the correct number of degrees of freedom. In Section 3 we solve the equations in the presence of sources by finding the propagators. We again compare and contrast the solutions for the massive and massless cases in order to see if and why the massive solution approaches the massless case. In Section 4, we find those Lagrangians which lead to the required equations of motion. We also make use of the Lagrangian to find the commutation relations for the independent degrees of freedom, in order to see again if the 'disappearing' degrees of freedom do or do not have smooth limits. In the Appendices we outline some of the projection operator techniques used in the paper.

2. THE EQUATIONS OF MOTION

In this section we discuss the equations of motion of massive and massless fields of spins one and two in the presence of external sources. We will demand ultimately that these equations be derivable from a Lagrangian. Therefore if the field for spin one is a vector field A_μ, its source j_μ, is also a vector and if the field for spin two is a symmetric tensor field $h_{\mu\nu}$, its source $T_{\mu\nu}$, is also a symmetric tensor. We discuss two problems in this section; (a) how the equations of motion for a vector field with four components describe only three dynamical variables for mass $m \neq 0$ and two dynamical variables for $m = 0$ and (b) how the equations of motion for a symmetric tensor field with ten components describe only five dynamical variables for $m \neq 0$ and two dynamical variables for $m = 0$.

---

*In addition to the operations of the Poincaré group we include the spatial inversion (or parity) operation. Of course for spin $s = 0$ there is only one degree of freedom.

We can write down the equations for both spins uniformly by making use of the Levi–Civita tensor density $\varepsilon_{\mu\nu\lambda\rho}$ with the usual antisymmetry properties*. Define the second order differential operator

$$D^{\beta\sigma\tau}_{\alpha\rho\lambda} \equiv \varepsilon_{\mu\alpha\rho\lambda}\varepsilon^{\nu\beta\sigma\tau}\partial_\nu\partial^\mu \tag{1}$$

where

$$\partial_\mu \equiv \frac{\partial}{\partial x^\mu} \tag{2}$$

We can write the equations for spin one and two by operating with D on either A_μ or $h_{\mu\nu}$ and saturating enough indices so that the resulting tensor transforms like A_μ or $h_{\mu\nu}$ respectively. Thus for spin one we form

$$\frac{1}{2!}D^{\beta\rho\lambda}_{\alpha\rho\lambda}A_\beta(x) \tag{3}$$

and for spin two

$$\frac{1}{1!}D^{\beta\sigma\lambda}_{\alpha\rho\lambda}h^\rho_\sigma(x) \tag{4}$$

The mass term will be proportional to the A_μ and $h_{\mu\nu}$ respectively. Accordingly for spin one we have

$$\frac{1}{2!}D^{\beta\rho\lambda}_{\alpha\rho\lambda}A_\beta(x) - m^2 A_\alpha(x) = j_\alpha(x) \tag{5}$$

and for spin two†

$$\frac{1}{1!}D^{\beta\sigma\lambda}_{\alpha\rho\lambda}h^\rho_\sigma(x) + m^2(h^\beta_\alpha(x) - a\delta^\beta_\alpha h(x)) = T^\beta_\alpha(x), \tag{6a}$$

or

$$\frac{1}{1!}D^{\beta\sigma\lambda}_{\alpha\rho\lambda}h^\rho_\sigma(x) + m^2(h^\beta_\alpha(x) - \delta^\beta_\alpha h(x)) = T^\beta_\alpha(x), \tag{6}$$

where

$$h = h^\lambda_\lambda, \tag{7}$$

and a, at present is arbitrary.

*We shall use units such that $\hbar = c = 1$. The metric $\eta_{\mu\nu}$ has only the diagonal elements $(1, -1, -1, -1)$, $\mu, \nu = 0, 1, 2, 3$, $i, j = 1, 2, 3$ and $\varepsilon_{0123} = +1$.

†Note that we can write the equation for a spin zero field $\phi(x)$ as

$$\frac{1}{3!}D^{\alpha\rho\lambda}_{\alpha\rho\lambda}\phi - m^2\phi = j$$

since

$$\frac{1}{3!}\varepsilon_{\mu\alpha\rho\lambda}\varepsilon^{\nu\alpha\rho\lambda} = -\delta^\nu_\mu$$

If we make use of the following identities

$$\frac{1}{1!}\varepsilon_{\mu\alpha\rho\lambda}\varepsilon^{\nu\beta\sigma\lambda} \equiv -\delta^{\nu\beta\sigma}_{\mu\alpha\rho} = -(\delta^{\nu}_{\mu}\delta^{\beta}_{\alpha}\delta^{\sigma}_{\rho} + \delta^{\beta}_{\mu}\delta^{\sigma}_{\alpha}\delta^{\nu}_{\rho}$$

$$+ \delta^{\sigma}_{\mu}\delta^{\nu}_{\alpha}\delta^{\beta}_{\rho} - \delta^{\beta}_{\mu}\delta^{\nu}_{\alpha}\delta^{\sigma}_{\rho} - \delta^{\nu}_{\mu}\delta^{\sigma}_{\alpha}\delta^{\beta}_{\rho} - \delta^{\sigma}_{\mu}\delta^{\beta}_{\alpha}\delta^{\nu}_{\rho}) \tag{8}$$

and

$$\frac{1}{2!}\varepsilon_{\mu\alpha\rho\lambda}\varepsilon^{\nu\beta\rho\lambda} = -\delta^{\nu\beta}_{\mu\alpha} = -(\delta^{\nu}_{\mu}\delta^{\beta}_{\alpha} - \delta^{\beta}_{\mu}\delta^{\nu}_{\alpha}) \tag{9}$$

we see that equation (5) for $m = 0$ is just Maxwell's equation and equation (6) is the Pauli–Fierz equation (Fierz and Pauli, 1939) for massive spin two.

Now we must see how many dynamical variables appear in equations (5), (6) and (6a). Any component of A_μ or $h_{\mu\nu}$, say ψ, and its time derivative $\partial_0 \psi$ can be assigned arbitrarily on some constant time surface, say $t = 0$. A dynamical variable will be those components of A_μ and $h_{\mu\nu}$ which appear in the equations involving second time derivatives. If in equation (5) we set the index $\beta = 0$, using (1) we find $\nu \ne 0$ and thus A_0 appears in the equations as a zeroth or first time derivative. We have then that only the A_i are dynamical variables and that the equations of motion determine A_0 in terms of the A_i. We apparently have not used the fact that $m \ne 0$. However if $m = 0$, equation (5) is invariant under a set of transformations—gauge transformations*

$$A_\beta \to A_\beta + \partial_\beta \Lambda \tag{10}$$

One may see this trivially by using (1) and the antisymmetry property of $\varepsilon^{\nu\beta\sigma\tau}$. We can choose Λ (i.e. find a gauge) so that one of the apparent dynamical variables A_i is identically zero. We are left with two dynamical variables for $m = 0$.

Another property that follows immediately from (5) for $m = 0$ is current conservation. That is, if we take ∂^α of equation (5) we must have, for $m = 0$,

$$\partial^\alpha j_\alpha = 0 \tag{11}$$

This is a consequence of the equations of motion—it is not a separate equation of motion. Again, it is proved trivially using the properties of $\varepsilon_{\mu\alpha\rho\lambda}$. For the case $m \ne 0$, we may be able to choose sources such that (11) follows, but in this case (11) will be an additional equation of motion.

We can now carry out the same procedure for the field $h_{\mu\nu}$. In equation (6a) if we set the index $\sigma = 0$, and using (1), $\nu \ne 0$ and therefore the four h_0^ρ cannot be dynamical variables. We are left with the six h_{ij} as possible dynamical variables. Consider, now, equation (6a) with $\alpha = \beta = 0$. It reads

$$F(\partial_i \partial_j h_k^l) + m^2[(1-a)h_0^0 - ah_i^i] = T_0^0 \tag{12}$$

The derivative terms F are only spatial derivatives of the h_{ij}. Accordingly, if we choose

$$a = 1 \tag{13}$$

*We assume of course that j_μ is gauge invariant.

equation (12) is a constraint equation on the six h_{ij} and we are left with five dynamical variables, the number required to describe a massive spin two field. The resulting equation, (6) is indeed the Pauli–Fierz equation (Fierz and Pauli, 1939).

Let us now turn to the case $m = 0$ for (6). Of course (12) with $m = 0$ again reduces the six h_{ij} to five independent variables. Again, as for spin one, equation (6) for $m = 0$ is invariant under a set of transformations—gauge transformations

$$h_{\mu\nu} \to h_{\mu\nu} + \partial_\mu \Lambda_\nu + \partial_\nu \Lambda_\mu \tag{14}$$

which one deduces trivially from the properties of D. We can choose Λ_i such that three of the remaining five h_{ij} are identically zero. This leaves us with two dynamical variables for $m = 0$, as required. Again it follows from the equations of motion that

$$\partial_\beta T_\alpha^\beta = 0 \tag{15}$$

if $m = 0$. This leads to the well-known problem of the consistency of these equations, if we identify T_α^β with the energy momentum tensor of matter and radiation and h_α^β with the gravitational field. This T_α^β cannot be conserved and we must add to it the energy momentum of the gravitational field itself which then leads to equations non-linear in the h_α^β. We can use this technique to lead us to the full Einstein equations for the gravitational field. See Deser (1970) for references. In this work we shall restrict ourselves to the linearized version. In doing so we are in effect assuming that T_α^β is proportional to some small coupling constant f and that we work only to $O(f^2)$, in which case the matter and radiation energy momentum tensor will be conserved.

3. THE PROPAGATORS

In this section we shall obtain the propagators for the classical fields by using the projection operator techniques outlined in Appendix A. It is convenient to take the Fourier transform of the equations of motion (5) and (6) and so work in momentum space. After this transformation and making use of the identities (8) and (9) we can write the equations of motion for $A_\mu(p)$ and $h_{\mu\nu}(p)$ the Fourier transformed fields as follows:

$$K_\mu^\nu A_\nu(p) = j_\mu(p) \tag{16}$$

and

$$K_{\mu\nu}^{\alpha\beta} h_{\alpha\beta}(p) = -T_{\mu\nu}(p) \tag{17}$$

where

$$K_\mu^\nu \equiv \frac{1}{2!} D_{\mu\rho\lambda}^{\nu\rho\lambda}(p) - m^2 \delta_\mu^\nu$$
$$= (p^2 - m^2)\delta_\mu^\nu - p_\mu p^\nu \tag{18}$$

and

$$K^{\alpha\beta}_{\mu\nu} \equiv -\frac{1}{2}\left\{\left[\frac{1}{1!}\eta_{\nu\sigma}\eta^{\beta\tau}D^{\sigma\alpha\lambda}_{\mu\tau\lambda}(p)+m^2(\delta^\alpha_\mu\delta^\beta_\nu-\eta_{\mu\nu}\eta^{\alpha\beta})\right]+(\alpha\leftrightarrow\beta)\right\}$$

$$=\frac{1}{2}\left\{(p^2-m^2)\delta^\alpha_\mu\delta^\beta_\nu-(\delta^\alpha_\mu p_\nu p^\beta+\delta^\beta_\mu p_\nu p^\alpha)+\eta_{\mu\nu}p^\alpha p^\beta+p_\mu p_\nu \eta^{\alpha\beta}\right.$$
$$\left.-(p^2-m^2)\eta_{\mu\nu}\eta^{\alpha\beta}+(\mu\leftrightarrow\nu)\right\} \tag{19}$$

where we have assumed $h_{\mu\nu}$ and $T_{\mu\nu}$ are symmetric tensors. If the tensors K have an inverse we can solve equations (16) and (17) for A_μ and $h_{\mu\nu}$ respectively, to read

$$A_\mu = G^\nu_\mu j_\nu \tag{20}$$

and

$$h_{\mu\nu} = -G^{\alpha\beta}_{\mu\nu} T_{\alpha\beta} \tag{21}$$

where in both cases

$$G = K^{-1} \tag{22}$$

i.e.

$$K^\lambda_\mu G^\nu_\lambda = \delta^\nu_\mu \tag{23}$$

and

$$K^{\lambda\rho}_{\mu\nu} G^{\alpha\beta}_{\lambda\rho} = \tfrac{1}{2}(\delta^\alpha_\mu \delta^\beta_\nu + \delta^\beta_\mu \delta^\alpha_\nu) \tag{24}$$

Since A_μ is a four-vector field we can find G^ν_μ by writing K^ν_μ in terms of its spin one and spin zero projection operators. These are easily found to be

$$P^{(1)\nu}_\mu \equiv \delta^\nu_\mu - \hat{p}_\mu \hat{p}^\nu \tag{25}$$

and

$$P^{(0)\nu}_\mu = \hat{p}_\mu \hat{p}^\nu \tag{26}$$

where

$$\hat{p}_\mu = p_\mu/(p^2)^{\frac{1}{2}} \tag{27}$$

and we have

$$(P^{(1)})^2 = P^{(1)} \tag{28}$$
$$(P^{(0)})^2 = P^{(0)} \tag{29}$$

and

$$P^{(1)} P^{(0)} = 0 \tag{30}$$

We can write

$$K^\nu_\mu = [(p^2 - m^2)P^{(1)} - m^2 P^{(0)}]^\nu_\mu \tag{31}$$

The inverse of K follows immediately giving

$$G^\nu_\mu = \left(\frac{1}{p^2-m^2}P^{(1)} - \frac{1}{m^2}P^{(0)}\right)^\nu_\mu = \frac{1}{p^2-m^2}\left(\delta^\nu_\mu - \frac{p_\mu p^\nu}{m^2}\right) \quad (32)$$

Again this shows that we are discussing a massive theory of spin one since only the spin one components have a pole in p^2 and therefore propagate in time—i.e. they are dynamical variables.

We see also that the limit $m \to 0$ is not straightforward. In fact for $m = 0$ the operator K^ν_μ does not have an inverse since it is proportional to a projection operator. This is precisely the manifestation of the gauge invariance of the massless theory. To solve our equations (16) for $m = 0$, one normally 'goes into' some particular gauge, i.e. we modify the equations of motion so as to introduce an operator $K^\nu_\mu(\lambda)$ which does have an inverse. The simplest set of gauges are the covariant gauges which depend on a parameter λ. We define

$$K^\nu_\mu(\lambda) \equiv \left(p^2 P^{(1)} + \frac{1}{\lambda}P^{(0)}\right)^\nu_\mu \quad (33)$$

which does have an inverse, which is

$$G^\nu_\mu(\lambda) = \left(\frac{1}{p^2}P^{(1)} + \lambda P^{(0)}\right)^\nu_\mu \quad (34)$$

Of course, gauge invariance implies that any physical result must be independent of the gauge, i.e. independent of λ.

We saw before that as a consequence of the equations of motion for the massless theory the current j_μ must be conserved which means

$$p^\mu j_\mu(p) = 0 \quad (35)$$

Substituting (34) into (20) and using (26) and (35) we can write for the case $m = 0$

$$A_\mu(p) = \frac{P^{(1)\nu}_\mu}{p^2}j_\nu(p) = \frac{j_\mu(p)}{p^2} \quad (36)$$

Any classical experiment which will involve the interaction of two sources—say $j^{(1)}_\mu$ and $j^{(2)}_\mu$ will depend on

$$j^{(1)\mu}A^{(2)}_\mu = \frac{j^{(1)\mu}j^{(2)}_\mu}{p^2} \quad (37)$$

and is indeed independent of the gauge.

We return now to the massive case and if in addition to the equations of motion (16) we postulate that the source is conserved—i.e. we assume equation (35) as a field equation we can write for the case $m \neq 0$

$$A_\mu(p) = \frac{1}{p^2-m^2}P^{(1)\nu}_\mu j_\nu(p) = \frac{j_\mu(p)}{p^2-m^2} \quad (38)$$

Assuming m is very small (specifically the Fourier components are such that $m^2 \ll p^2$) we find

$$A_\mu(p) \approx \frac{j_\mu(p)}{p^2} \tag{39}$$

This completes the proof, that as far as physical observations are concerned, the theory of a classical spin one field for small mass approaches the results for a massless spin one field.

Let us now turn to the spin two case. Since $h_{\mu\nu}$ is a symmetric tensor with ten components, we proceed by writing $K^{\alpha\beta}_{\mu\nu}$ in terms of its spin two, spin one and two spin zero projection operators. In order that we may find the inverse of K easily we must find those two spin zero projection operators which are orthogonal. This is carried out in Appendix A and we can write

$$K^{\alpha\beta}_{\mu\nu} = [(p^2 - m^2)Q^{(2)} - m^2 Q^{(1)} + X(p)Q^{(0)}_c + Y(p)\bar{Q}^{(0)}_c]^{\alpha\beta}_{\mu\nu} \tag{40}$$

where $Q^{(2)}$, $Q^{(1)}$, X, Y, $Q^{(0)}_c$ and $\bar{Q}^{(0)}_c$ are defined in the Appendix. Neither $X(p)$ nor $Y(p)$ vanish so that we can invert K to find

$$G^{\alpha\beta}_{\mu\nu} = \left(\frac{1}{p^2 - m^2}Q^{(2)} - \frac{1}{m^2}Q^{(1)} + \frac{1}{X(p)}Q^{(0)}_c + \frac{1}{Y(p)}\bar{Q}^{(0)}_c\right)^{\alpha\beta}_{\mu\nu} \tag{41}$$

Using the results of the Appendix we can also write

$$G^{\alpha\beta}_{\mu\nu} = \frac{1}{2}\frac{1}{(p^2 - m^2)}\left\{\left(\delta^\alpha_\mu \delta^\beta_\nu - \frac{1}{3}\eta_{\mu\nu}\eta^{\alpha\beta}\right) - \left(\frac{p_\mu p^\alpha \delta^\beta_\nu + p_\mu p^\beta \delta^\alpha_\nu}{2m^2}\right)\right.$$
$$\left. + \frac{p_\mu p_\nu \eta^{\alpha\beta} + \eta_{\mu\nu} p^\alpha p^\beta}{3m^2} + \frac{2}{3}\frac{p_\mu p_\nu p^\alpha p^\beta}{m^4} + (\mu \leftrightarrow \nu)\right\} \tag{42}$$

Since neither $X(p)$ nor $Y(p)$ vanish we see that only the spin two components have a pole in p^2 and thus only the five spin two fields are dynamical variables.

Again we see that we cannot take the limit $m \to 0$ in (41). As before, for $m = 0$, K is a combination of projection operators which do not span the space of symmetric second rank tensors. Therefore K does not have an inverse. In fact we find for $m = 0$

$$K^{\alpha\beta}_{\mu\nu} = (p^2 Q^{(2)} - 2p^2 Q^{(0)})^{\alpha\beta}_{\mu\nu} \tag{43}$$

where

$$Q^{(0)\alpha\beta}_{\mu\nu} = \tfrac{1}{3}(\eta_{\mu\nu} - \hat{p}_\mu \hat{p}_\nu)(\eta^{\alpha\beta} - \hat{p}^\alpha \hat{p}^\beta) \tag{44}$$

This is just a manifestation of the gauge invariance of the theory. We can solve for the field $h_{\mu\nu}$ by 'going into' a gauge. This means we modify the equations of motion so that the modified K will have an inverse. The simplest class of gauges are the covariant ones and we write

$$K^{\alpha\beta}_{\mu\nu}(\lambda_0, \lambda_1) = \left(p^2 Q^{(2)} - 2p^2 Q^{(0)} + \frac{1}{\lambda_1} Q^{(1)} + \frac{1}{\lambda_0}\bar{Q}^0\right)^{\alpha\beta}_{\mu\nu} \tag{45}$$

where $\bar{Q}^0$ must be orthogonal to $Q^{(0)}$ and is

$$\bar{Q}^{(0)\alpha\beta}_{\mu\nu} = \hat{p}_\mu \hat{p}_\nu \hat{p}^\alpha \hat{p}^\beta \tag{46}$$

The inverse of K is

$$G^{\nu\beta}_{\mu\nu}(\lambda_0, \lambda_1) = \left(\frac{1}{p^2} Q^{(2)} - \frac{1}{2p^2} Q^{(0)} + \lambda_1 Q^{(1)} + \lambda_0 \bar{Q}^0\right)^{\alpha\beta}_{\mu\nu} \tag{47}$$

Substituting for $Q^{(2)}$ and $Q^{(0)}$ from the Appendix we have

$$G^{\alpha\beta}_{\mu\nu}(\lambda_0, \lambda_1) = \frac{1}{p^2}\frac{1}{2}\left\{(\delta^\alpha_\mu - \hat{p}_\mu \hat{p}^\alpha)(\delta^\beta_\nu - \hat{p}_\nu \hat{p}^\beta) - \frac{1}{2}(\eta_{\mu\nu} - \hat{p}_\mu \hat{p}_\nu)(\eta^{\alpha\beta} - \hat{p}^\alpha \hat{p}^\beta) \right.$$
$$\left. + (\alpha \leftrightarrow \beta)\right\}$$
$$+ (\lambda_1 Q^{(1)} + \lambda_0 \bar{Q}^0)^{\alpha\beta}_{\mu\nu} \tag{48}$$

The result of any observation must be independent of the λ_i. If we have two sources $T^{(1)}_{\mu\nu}$ and $T^{(2)}_{\mu\nu}$, their interaction is proportional to

$$T^{(1)\mu\nu} h^{(2)}_{\mu\nu} = T^{(1)\mu\nu} G^{\alpha\beta}_{\mu\nu} T^{(2)}_{\alpha\beta} \tag{49}$$

We saw that for a massless theory the source must be conserved as a consequence of the field equations, i.e. we must have

$$p^\alpha T_{\alpha\beta} = 0 \tag{50}$$

Now from the properties of $Q^{(1)}$ and $\bar{Q}^0$ we have

$$T^{(1)\mu\nu} Q^{(1)\alpha\beta}_{\mu\nu} T^{(2)}_{\alpha\beta} = T^{(1)\mu\nu} \bar{Q}^{(0)\alpha\beta}_{\mu\nu} T^{(2)}_{\alpha\beta} = 0 \tag{51}$$

so that we may finally write for the interaction of two sources

$$T^{(1)\mu\nu} \frac{(\delta^\alpha_\mu \delta^\beta_\nu - \frac{1}{2}\eta_{\mu\nu}\eta^{\alpha\beta})}{p^2} T^{(2)}_{\alpha\beta} \tag{52}$$

Let us now return to the massive case. We find in the Appendix that for small m

$$Y \approx -2p^2 \tag{53}$$

$$X \approx \frac{3}{2}\frac{m^4}{p^2} \tag{54}$$

and

$$\frac{Q^{(0)}_c}{X} + \frac{\bar{Q}^{(0)}_c}{Y} \approx \frac{2p^2}{3m^4} \bar{Q}^0 \tag{55}$$

So that we have for m small ($\neq 0$)

$$G^{\alpha\beta}_{\mu\nu} \approx \left(\frac{1}{p^2} Q^{(2)} - \frac{1}{m^2} Q^{(1)} + \frac{2p^2}{3m^4} \bar{Q}^0\right)^{\alpha\beta}_{\mu\nu} \tag{56}$$

If we choose sources such that equation (50) is an equation of motion, then using equation (51) we have for the interactions between two sources when the spin two field has a small but non-zero mass

$$T^{(1)\mu\nu}\frac{Q^{(2)\alpha\beta}_{\mu\nu}}{p^2}T^{(2)}_{\alpha\beta} \tag{57}$$

which using (51) again gives

$$T^{(1)\mu\nu}\frac{(\delta^\alpha_\mu\delta^\beta_\nu-\frac{1}{3}\eta_{\mu\nu}\eta^{\alpha\beta})}{p^2}T^{(2)}_{\alpha\beta} \tag{58}$$

This result can only be the same as (52) if the sources are traceless. This is not normally the case. The energy momentum tensor for electromagnetic radiation is traceless while it is not for matter. This would give rise to a discrepancy in the bending of light experiment if gravitation were a spin two, small mass theory. [See van Dam and Veltman (1970) and Boulware and Deser (1972).]

By comparing equations (56) and (47) we see how we could modify the spin two massive theory to give the same results as the spin two massless theory. We need only add in a spin zero particle which in the limit of small m will contribute the term

$$-\frac{1}{2p^2}Q^{(0)\alpha\beta}_{\mu\nu}$$

to the propagator. This is most easily accomplished by choosing the a in equation (6a) not equal to one. We saw for $a \neq 1$, equation (6a) is an equation for six dynamical variables, one of which will be the extra spin zero particle. However, we see from the relative sign between the $Q^{(2)}$ and $Q^{(0)}$ term that the extra spin zero particle must be a ghost.

4. THE LAGRANGIAN AND CANONICAL VARIABLES

In this section we construct those Lagrangians which lead to the equations of motion (5) and (6). We do this in order to find the variables canonical to the independent dynamical variables. Having done so, we are able to pass to the $m \to 0$ limit in order to see what happens to the apparently vanishing degrees of freedom.

We shall discuss the spin one case first. Given the equations of motion (5) one can easily find a Lagrangian from which they are derived. We may write*

$$\mathcal{L}(x) = \tfrac{1}{2}A^\alpha \Delta^\beta_\alpha A_\beta + \tfrac{1}{2}m^2 A^\alpha \delta^\beta_\alpha A_\beta + j_\alpha A^\alpha \tag{59}$$

where

$$\Delta^\beta_\alpha = \frac{1}{2!}\overleftrightarrow{\partial}^\mu \varepsilon_{\mu\alpha\rho\lambda}\varepsilon^{\nu\beta\rho\lambda}\overrightarrow{\partial}_\nu \tag{60}$$

where

$$A^\alpha \overleftrightarrow{\partial}^\mu \text{ means } \partial^\mu A^\alpha$$

---

*This is the usual Lagrangian with kinetic energy term $-\tfrac{1}{4}F_{\mu\nu}F^{\mu\nu}$ where $F_{\mu\nu} = \partial_\mu A_\nu - \partial_\nu A_\mu$

This Lagrangian is not unique. By using the antisymmetry property of the Levi–Civita tensor we can add total derivatives to the Lagrangian by adding any multiple of*

$$\sum_{\alpha,\beta} A_\alpha \bar{\Delta}^{\alpha\beta} A_\beta \tag{61}$$

where

$$\bar{\Delta}^{\alpha\beta} = \frac{1}{2!} \bar{\partial}^\mu \varepsilon_{\mu\nu\rho\lambda} \varepsilon^{\alpha\beta\rho\lambda} \vec{\partial}^\nu \tag{62}$$

In fact there is no need that this extra piece be Lorentz invariant since it does not contribute to the equations of motion. More generally one can add to the Lagrangian $\mathscr{L}(x)$, $\mathscr{L}_\kappa(x)$ where

$$\mathscr{L}_\kappa(x) = \tfrac{1}{2} \sum_{\alpha\beta} \kappa_\alpha \kappa'_\beta A_\alpha \bar{\Delta}^{\alpha\beta} A_\beta \tag{63}$$

where κ_α, κ'_β are any set of parameters.

Since we have already discovered that the field A_0 is not a dynamical variable, we will choose that Lagrangian such that the variable canonical to A_0, namely Π^0 is identically zero. Now

$$\Pi_\alpha \equiv \frac{\delta \mathscr{L}_T}{\delta \partial^0 A^\alpha} \tag{64}$$

where

$$\mathscr{L}_T = \mathscr{L} + \mathscr{L}_\kappa \tag{65}$$

We have

$$\frac{\delta \mathscr{L}}{\delta \partial^\mu A^\alpha} = \frac{1}{2!} \varepsilon_{\mu\alpha\rho\lambda} \varepsilon^{\nu\beta\rho\lambda} \partial_\nu A_\beta \tag{66}$$

$$= -(\partial_\mu A_\alpha - \partial_\alpha A_\mu) \tag{67}$$

Thus

$$\frac{\delta \mathscr{L}}{\delta \partial^0 A^0} = 0 \tag{68}$$

If we choose

$$\mathscr{L}_T = \mathscr{L} \tag{69}$$

we find that

$$\Pi_0 = 0 \tag{70}$$

and

$$\Pi_i = -(\dot{A}_i - \partial_i A_0) \tag{71}$$

---

*We have of course been using the summation convention for summing over repeated indices. However, since in what follows we shall be writing down non-covariant additions to the Lagrangian we now specifically indicate summations where needed.

376 Uncertainty Principle and Foundations of Quantum Mechanics

Since A_0 is a dependent variable we use the equations of motion for A_0 to find Π_i entirely in terms of A_i and $\dot{A}_i$.

The equation of motion (5) for $\alpha = 0$ gives

$$A_0 = \frac{\partial_j \dot{A}_j}{m^2 - \nabla^2} - \frac{j_0}{m^2 - \nabla^2} \tag{72}$$

where

$$\nabla^2 = -\partial_i \partial^i \tag{73}$$

and finally

$$\Pi_i = -\Lambda_{ij} \dot{A}^j - \frac{\partial_i j_0}{m^2 - \nabla^2} \tag{74}$$

with

$$\Lambda_{ij} = \eta_{ij} - \frac{\partial_i \partial_j}{m^2 - \nabla^2} \tag{75}$$

The inverse of Λ is

$$\Lambda_{ij}^{-1} = \eta_{ij} + \frac{\partial_i \partial_j}{m^2} \tag{76}$$

For $m = 0$, we see that Λ does not have an inverse and is in fact a projection operator. It is precisely the helicity one projection operator. This indicates that for $m = 0$ there are only two canonical momenta which are the momenta canonical to the two helicity one dynamical variables.

We have been assuming that j_μ does not depend on the A_μ in which case j_0 will commute with the A_i. Accordingly, as far as the canonical commutation relations are concerned we can replace the Π_i by

$$\Pi_i^c = -\Lambda_{ij} \dot{A}_j \tag{77}$$

We will drop the superscript c whenever there is no confusion. The canonical commutation relations are

$$(A^i(\mathbf{x}), \Pi_j(\mathbf{y})) = i\delta_j^i \delta^3(\mathbf{x}-\mathbf{y}) \tag{78}$$

or using (76)

$$(A^i(\mathbf{x}), \dot{A}_j(\mathbf{y})) = -i\left(\delta_j^i - \frac{\partial^i \partial_j}{m^2}\right)\delta^3(\mathbf{x}-\mathbf{y}) \tag{79}$$

Of course we can write (79) only in the case when $m \neq 0$.

We can see what happens in the limit $m \to 0$ by writing (78) separately for the helicity one and helicity zero subspaces.

In Appendix B we show that we can write

$$A_i = A_i^{(1)} + A_i^{(0)} \tag{80}$$

where
$$A_i^{(1)} = R_{ij}^{(1)} A^j, \qquad A_i^{(0)} = R_{ij}^{(0)} A^j \tag{81}$$

and
$$R_{ij}^{(1)} = \left(\eta_{ij} + \frac{\partial_i \partial_j}{\nabla^2}\right) \tag{82}$$

$$R_{ij}^{(0)} = -\frac{\partial_i \partial_j}{\nabla^2} \tag{83}$$

where $R^{(1)}$ and $R^{(0)}$ are the helicity one and zero projection operators, respectively.

Now, from (77) and for m small
$$\Pi_i \approx -\left(\eta_{ij} + \frac{\partial_i \partial_j}{\nabla^2}\right) \dot{A}^j - \frac{m^2}{\nabla^2} \frac{\partial_i \partial_j}{\nabla^2} \dot{A}_j \tag{84}$$

$$= -\dot{A}_i^{(1)} + \frac{m^2}{\nabla^2} \dot{A}_i^{(0)}$$

The commutation relations (78) can be written
$$(A_i^{(1)}(\mathbf{x}), \dot{A}_j^{(1)}(\mathbf{y})) = -i R_{ij}^{(1)} \delta^3(\mathbf{x}-\mathbf{y}) \tag{85}$$

and
$$\left[A_i^{(0)}(\mathbf{x}), \frac{m^2}{\nabla^2} \dot{A}_j^{(0)}(\mathbf{y})\right] = i R_{ij}^{(0)} \delta^3(\mathbf{x}-\mathbf{y}) \tag{86}$$

Let us define
$$\phi \equiv \frac{m \partial_i A^i}{\nabla^2} \tag{87}$$

or
$$\partial_i \phi = -m A_i^{(0)} \tag{88}$$

Equation (86) becomes
$$(\phi(\mathbf{x}), \dot{\phi}(\mathbf{y})) = i \delta^3(\mathbf{x}-\mathbf{y}) \tag{89}$$

The commutation relations (85) are precisely those satisfied in the $m=0$ case by the independent helicity one fields. The canonical variables for the helicity zero are $\phi(\mathbf{x})$ and $\dot{\phi}(\mathbf{x})$ given by (87). This is verified by (89).

We now look at the equations of motion satisfied by $A_i^{(1)}$ and ϕ. From equation (5) for $\alpha = i$ we have
$$\Box A_i - \partial_i \partial_j A^j - \partial_i \dot{A}^0 + m^2 A_i = -j_i \tag{90}$$

where
$$\Box = \partial_\mu \partial^\mu \tag{91}$$

378 Uncertainty Principle and Foundations of Quantum Mechanics

We substitute for A_0 from (72) to obtain

$$\Box \Lambda_{ij} A^j = -\Lambda_{ij} j^j - \frac{\partial_i \partial_\mu j^\mu}{m^2 - \nabla^2} \tag{92}$$

We saw that for the case $m = 0$ current must be conserved and

$$\Lambda_{ij} \to R_{ij}^{(1)} \tag{93}$$

the helicity one projection operator, so that for $m = 0$ we have

$$\Box A_i^{(1)} = -j_i^{(1)} \tag{94}$$

and from (72)

$$A_0 = -\frac{\partial_j \dot{A}_j}{\nabla^2} + \frac{j_0}{\nabla^2} \tag{95}$$

For $m \neq 0$ but small we have

$$\Lambda_{ij} \approx R_{ij}^{(1)} - \frac{m^2}{\nabla^2} R_{ij}^{(0)} \tag{96}$$

We substitute into (92) and project out the helicity one and zero parts and use (18) to obtain

$$\Box A_i^{(1)} = -j_i^{(1)} \tag{97}$$

and

$$\Box \phi = \frac{-1}{m} \partial_\mu j^\mu + m \frac{\partial_i j^i}{\nabla^2} \tag{98}$$

If in the massive case we assume current conservation as a further equation of motion we find as $m \to 0$

$$\Box \phi = 0 \tag{99}$$

Accordingly, as $m \to 0$ the helicity one modes satisfy exactly the same equations of motion and commutation relations as the helicity one modes in the massless case. The helicity zero mode is, however, decoupled. This is the sense in which the third degree of freedom disappears.

We now turn to the case of spin two. From the equations of motion (6) we can deduce a Lagrangian, namely,

$$\mathcal{L}(x) = \tfrac{1}{2} h_\beta^\alpha \Delta_{\alpha\rho}^{\beta\sigma} h_\sigma^\rho - \frac{1}{2} m^2 h_\beta^\alpha (\delta_\alpha^\sigma \delta_\rho^\beta - \delta_\alpha^\beta \delta_\rho^\sigma) h_\sigma^\rho + T_\alpha^\beta h_\beta^\alpha \tag{100}$$

where

$$\Delta_{\alpha\rho}^{\beta\sigma} = \overleftrightarrow{\partial}^\mu \frac{\varepsilon_{\mu\alpha\rho\lambda} \varepsilon^{\nu\beta\sigma\lambda}}{1!} \overleftrightarrow{\partial}_\nu \tag{101}$$

Of course as in the spin one case this Lagrangian is not unique and we can add terms of the form

$$\tfrac{1}{2}\sum_{\alpha\beta,\rho\sigma} h_{\alpha\beta}\bar{\Delta}^{\alpha\beta\rho\sigma}h_{\rho\sigma}\kappa_\alpha^{(1)}\kappa_\beta^{(2)}\kappa_\rho^{(3)}\kappa_\sigma^{(4)} \tag{102}$$

where $\bar{\Delta}$ may be

$$\bar{\Delta}^{\alpha\beta\rho\sigma}=\overleftrightarrow{\partial}^\mu\varepsilon_{\mu\nu\gamma\lambda}\varepsilon^{\alpha\rho\gamma\lambda}\overleftrightarrow{\partial}^\nu\eta^{\beta\sigma} \tag{103}$$

or

$$\overleftrightarrow{\partial}^\mu\varepsilon_{\mu\nu\gamma\lambda}\varepsilon^{\alpha\beta\gamma\lambda}\overleftrightarrow{\partial}^\nu\eta^{\rho\sigma} \tag{104}$$

As in the spin one case we choose that Lagrangian such that the variables Π_0^α canonical to the redundant fields h_ρ^0 are identically zero. Consider

$$\frac{\delta\mathscr{L}}{\delta\partial^\mu h_\beta^\alpha}=\tfrac{1}{2}(\varepsilon_{\mu\alpha\rho\lambda}\varepsilon^{\nu\beta\sigma\lambda}\partial_\nu h_\sigma^\rho+\varepsilon_\mu^{\ \beta}{}_{\rho\lambda}\varepsilon_\alpha^{\ \nu\sigma\lambda}\partial_\nu h_\sigma^\rho) \tag{105}$$

where $\mathscr{L}(x)$ is given by (100). This gives

$$\frac{\delta\mathscr{L}}{\delta\partial^0 h_0^0}=0 \tag{106}$$

whereas,

$$\frac{\delta\mathscr{L}}{\delta\partial^0 h_0^\alpha}=\tfrac{1}{2}\varepsilon_{0\alpha\rho\lambda}\varepsilon^{\nu 0\sigma\lambda}\partial_\nu h_\sigma^\rho \tag{107}$$

$$=\tfrac{1}{2}(\delta_a^n\delta_r^s-\delta_a^s\delta_r^n)\partial_n h_s^r\delta_\alpha^a \tag{108}$$

or

$$\frac{\delta\mathscr{L}}{\delta\partial^0 h_0^a}=\tfrac{1}{2}(\partial_a h_r^r-\partial_r h_a^r) \tag{109}$$

Accordingly, we must add to $\mathscr{L}(x)$ some terms of the form (102) which will insure that

$$\Pi_0^\alpha=0 \tag{110}$$

We may do this if we add to $\mathscr{L}(x)$

$$\mathscr{L}_\kappa(x)=-h_{\alpha\beta}\overleftrightarrow{\partial}^\mu\frac{1}{2!}\varepsilon_{\mu\nu\gamma\lambda}\varepsilon^{\beta\sigma\gamma\lambda}\overleftrightarrow{\partial}^\nu h_\sigma^a$$

$$-h_{a0}\overleftrightarrow{\partial}^\mu\frac{1}{2!}\varepsilon_{\mu\nu\gamma\lambda}\varepsilon^{a0\gamma\lambda}\overleftrightarrow{\partial}^\nu h_r^r \tag{111}$$

$$=(\partial^\mu h_{a\mu}\partial^\nu h_\nu^a-\partial^\nu h_{a\mu}\partial^\mu h_\nu^a)$$

$$+(\partial^a h_{a0}\partial^0 h_r^r-\partial^0 h_{a0}\partial^a h_r^r) \tag{112}$$

Choosing

$$\mathscr{L}_T(x)=\mathscr{L}(x)+\mathscr{L}_\kappa(x)$$

we find,

$$\Pi_0^\alpha = \frac{\delta \mathcal{L}_T}{\delta \partial^0 h_\alpha^0} = 0 \tag{113}$$

The other six canonical momenta will be given by

$$\Pi_i^j = \frac{\delta \mathcal{L}_T}{\delta \partial^0 h_j^i} \tag{114}$$

From (105) we find

$$\frac{\delta \mathcal{L}}{\delta \partial^0 h_j^i} = \partial_0 h_i^j - \delta_i^j \partial_0 h_\rho^\rho + \delta_i^j \partial_\rho h_0^\rho - \tfrac{1}{2}(\partial_i h_0^j + \partial^j h_{i0}) \tag{115}$$

From (112) we obtain

$$\frac{\delta \mathcal{L}_\kappa}{\delta \partial^0 h_j^i} = -\tfrac{1}{2}(\partial_i h_0^j + \partial^j h_{i0}) + \delta_i^j \partial_r h_0^r \tag{116}$$

Finally

$$\Pi_{ij} = \dot{h}_{ij} + \eta_{ij}(2\partial_r h^{r0} - \dot{h}_r^r) - (\partial_i h_{j0} + \partial_j h_{i0}) \tag{117}$$

We eliminate the dependent variables, h_{i0} using the equations of motion. In equation (6) we set the indices $\alpha = i$ and $\beta = 0$ and using (8) we obtain

$$(-\nabla^2 + m^2)\Lambda_{ir} h^{r0} = T_{i0} + \partial_r \dot{h}_i^r - \partial_i \dot{h}_r^r \tag{118}$$

where Λ_{ir} is given by (75).

Equation (6) with $\alpha = \beta = 0$ gives the constraint equation on the six h_{ij} namely

$$\Lambda_{ij} h^{ij} = \frac{T_{00}}{\nabla^2 - m^2} \tag{119}$$

Now multiply (118) by Λ_{ji}^{-1} and use (119) to obtain

$$h_{0i} = \frac{1}{m^2 - \nabla^2}\left[\partial_r \dot{h}_i^r + T_{0i} + \frac{1}{m^2}\partial_i(\partial_\mu T^{\mu 0})\right] \tag{120}$$

We substitute h_{0i} into (117) to obtain

$$\Pi_{ij} = \Lambda_{ir}\Lambda_{js}\dot{h}^{rs} + \Lambda_{ij}\dot{h}_r^r + \frac{1}{\nabla^2 - m^2}\left[(\partial_i T_{0j} + \partial_j T_{0i}) + \frac{2}{m^2}\partial_i \partial_j (\partial_\mu T^{\mu 0})\right] \tag{121}$$

In writing our commutation relations we will assume that $T_{\mu\nu}$ is independent of h_{ij} and $\dot{h}_{ij}$. Since the h_{ij} are restricted by the constraints (119), we must find the Π_{ij} which are restricted to the same subspace. We make use of the helicity projection operators defined in Appendix B. A general six component tensor h_{ij} can be decomposed into its helicity components: two helicity two, two helicity one, one helicity zero and a second helicity zero.

$$h_{ij} = (R^{(2)} + R^{(1)} + R^{(0)} + \bar{R}^{(0)})_{ij}^{rs} h_{rs} \tag{122}$$

If we choose $\bar{R}^{(0)}$ to be the helicity zero projection operator,

$$\bar{R}_{ij}^{(0)rs} \equiv \frac{\Lambda_{ij}\Lambda^{rs}}{\Lambda_{mn}\Lambda^{mn}} \tag{123}$$

we may write using (119)

$$h_{ij} = (R^{(2)} + R^{(1)} + R^{(0)})_{ij}^{rs} h_{rs} + \frac{\Lambda_{ij}}{\Lambda_{mn}\Lambda^{mn}} \frac{T_{00}}{\nabla^2 - m^2} \tag{124}$$

Let us define the projection operator P_{ij}^{rs} such that

$$(R^{(2)} + R^{(1)} + R^{(0)})_{ij}^{rs} \equiv P_{ij}^{rs} \tag{125}$$

We may write the canonical commutation relations as

$$[\bar{h}_{ij}(\mathbf{x}), \bar{\Pi}^{rs}(\mathbf{y})] = iP_{ij}^{rs}\delta^3(\mathbf{x}-\mathbf{y}) \tag{126}$$

where

$$\left.\begin{array}{l}\bar{h}_{ij} = P_{ij}^{mn} h_{mn} \\ \bar{\Pi}_{rs} = P_{rs}^{kl} \Pi_{kl}\end{array}\right\} \tag{127}$$

Equivalently, we may write the commutation relations in the various helicity subspaces as

$$[h_{ij}^{(a)}, \Pi_{rs}^{(a)}] = iR_{ij,rs}^{(a)}\delta^3(\mathbf{x}-\mathbf{y}) \tag{128}$$

where

$$a = 0, 1, 2$$

and

$$h_{ij}^{(a)} = R_{ijrs}^{(a)} h^{rs}, \text{ etc.} \tag{129}$$

Using the results of Appendix B, we can find the commutation relations for small m.

Using (B.11) we may write (in coordinate space)

$$h_{ij}^{(1)} = \frac{1}{\nabla^2}(\partial_i h_j^{(1)} + \partial_j h_i^{(1)}) \tag{130}$$

where

$$h_i^{(1)} \equiv -\partial_r \left(\eta_{is} + \frac{\partial_i \partial_s}{\nabla^2}\right) h^{rs} \tag{131}$$

Note,

$$\partial^i h_i^{(1)} = 0 \tag{132}$$

so that the $h_i^{(1)}$ are the helicity one components of a spin one field. Also

$$\partial^j h_{ij}^{(1)} = -h_i^{(1)} \tag{133}$$

Substituting (B.31) into (128) we have

$$\left[h_{ij}^{(1)}(\mathbf{x}), \frac{m^2}{-\nabla^2}\dot{h}_{rs}^{(1)}(\mathbf{y})\right] = iR_{ijrs}^{(1)}\delta^3(\mathbf{x}-\mathbf{y}) \tag{134}$$

Now take $(\partial/\partial x_j)(\partial/\partial y_s)$ of (134) and we obtain

$$\left(h_i^{(1)}, \frac{m^2}{-\nabla^2}h_r^{(1)}\right) = \frac{i}{2}\nabla^2\left(\eta_{ir} + \frac{\partial_i\partial_r}{\nabla^2}\right)\delta^3(\mathbf{x}-\mathbf{y}) \tag{135}$$

Define

$$A_i \equiv \frac{\sqrt{2m}}{\nabla^2}h_i^{(1)} = -\frac{\sqrt{2m}}{\nabla^2}\partial_r\left(\eta_{is} + \frac{\partial_i\partial_s}{\nabla^2}\right)h^{rs} \tag{136}$$

then

$$(A_i(\mathbf{x}), \dot{A}_r(\mathbf{y})) = -i\left(\eta_{ir} + \frac{\partial_i\partial_r}{\nabla^2}\right)\delta^3(\mathbf{x}-\mathbf{y}) \tag{137}$$

$$= -iR_{ir}^{(1)}\delta^3(\mathbf{x}-\mathbf{y}) \tag{138}$$

where $R_{ir}^{(1)}$ is defined by (82). Using (B.34) and (B.25) we have for small m,

$$h_{ij}^{(0)} \approx \frac{\partial_i\partial_j}{\nabla^2}h^{(0)} \tag{139}$$

where

$$h^{(0)} = \frac{\partial_r\partial_s}{\nabla^2}h^{rs} \tag{140}$$

Similarly from (B.33)

$$\Pi_{ij}^{(0)} \approx \frac{3}{2}\left(\frac{m^2}{\nabla^2}\right)^2\frac{\partial_i\partial_j}{\nabla^2}\dot{h}^{(0)} \tag{141}$$

Substituting into (128) and projecting out with $S^{(0)}$ we have

$$\left[h_{ij}^{(0)}(\mathbf{x}), \frac{3}{2}\left(\frac{m^2}{\nabla^2}\right)^2\dot{h}_{rs}^{(0)}(\mathbf{y})\right] = i\frac{\partial_i\partial_j\partial_r\partial_s}{\nabla^4}\delta^3(\mathbf{x}-\mathbf{y}) \tag{142}$$

We now operate on (142) with

$$\frac{\partial}{\partial x_i}\frac{\partial}{\partial x_j}\frac{\partial}{\partial y_r}\frac{\partial}{\partial y_s}$$

to obtain

$$\left[\nabla^2 h^0, \frac{3}{2}\left(\frac{m^2}{\nabla^2}\right)^2\nabla^2\dot{h}^0\right] = i\nabla^4\delta^3(\mathbf{x}-\mathbf{y}) \tag{143}$$

Define
$$\phi \equiv \sqrt{\frac{3m^2}{2\nabla^2}} h^0 = \sqrt{\frac{3m^2}{2\nabla^2}} \frac{\partial_r \partial_s}{\nabla^2} h^{rs} \tag{144}$$

and (143) becomes
$$[\phi(\mathbf{x}), \dot{\phi}(\mathbf{y})] = i\delta^3(\mathbf{x}-\mathbf{y}) \tag{145}$$

Of course the commutation relation (128) for $a = 2$ is the same for both $m = 0$ and $m \neq 0$ i.e.
$$(h^{(2)}_{ij}, \Pi^{(2)}_{rs}) = iR^{(2)}_{ijrs} \delta^3(\mathbf{x}-\mathbf{y}) \tag{146}$$

As $m \to 0$ the dynamical variables associated with helicity one and zero are the fields A_i and ϕ defined by (136) and (144), respectively. Next we find their equations of motion. Operate on equation (6) with $\partial^\alpha \partial_\beta$ which gives
$$(\partial^\alpha \partial_\beta h^\beta_\alpha - \Box h) = \frac{1}{m^2} \partial^\alpha \partial_\beta T^\beta_\alpha \tag{147}$$

Next take the trace of (6) and use (147) to give
$$h^\mu_\mu = -\frac{1}{3m^2}\left(T^\mu_\mu - \frac{2}{m^2}\partial^\alpha\partial_\beta T^\beta_\alpha\right) \tag{148}$$

We use (147) and (148) in equation (6) when $\alpha = i, \beta = j$ to obtain
$$\Box h_{ij} - \partial_i \partial_\nu h^\nu_j - \partial_j \partial_\nu h^\nu_i + m^2 h_{ij}$$
$$= T_{ij} + \frac{\partial_i \partial_j}{3m^2}\left(T^\mu_\mu - \frac{2\partial_\mu \partial_\nu}{m^2} T^{\mu\nu}\right) - \frac{\eta_{ij}}{3}\left(T^\mu_\mu + \frac{\partial_\mu \partial_\nu}{m^2} T^{\mu\nu}\right) \tag{149}$$

We substitute for h_{oi} using (120) giving
$$(\Box + m^2)\left(\Lambda_{ir}\Lambda_{js} - \frac{\partial_i \partial_j \partial_r \partial_s}{(m^2 - \nabla^2)^2}\right) h^{rs}$$
$$= \left(\Lambda_{ir}\Lambda_{js} - \frac{\partial_i \partial_j \partial_r \partial_s}{(m^2 - \nabla^2)^2}\right) T^{rs} + \frac{\partial_i \partial_j}{3m^2}\left(T^\mu_\mu - 2\frac{\partial_\mu \partial_\nu}{m^2} T^{\mu\nu}\right)$$
$$- \frac{\eta_{ij}}{3}\left(T^\mu_\mu + \frac{\partial_\mu \partial_\nu}{m^2} T^{\mu\nu}\right)$$
$$+ \frac{1}{m^2 - \nabla^2}\left(\partial_i \partial_\mu T^\mu_j + \partial_j \partial_\mu T^\mu_i + \frac{1}{m^2}\partial_i \partial_j \dot{T}^{\mu 0}\right) \tag{150}$$

Using the formula for $\Lambda_{ir}\Lambda_{js}$ given by (B.24) the left side of equation (150) can be written for small m as
$$\Box\left(R^{(2)} - \frac{m^2}{\nabla^2} R^{(1)} - S^{(0)} + \bar{S}^0\right)_{ijrs} h^{rs} \tag{151}$$

Uncertainty Principle and Foundations of Quantum Mechanics

We may now take projections of (150) to obtain the equations for the various helicity fields. The $\bar{S}^0$ projection will just give the constraint equation (119) in the $m \to 0$ limit. Since

$$R_{ij}^{(2)rs}\partial_r\partial_s = R_{ij}^{(2)rs}\eta_{rs} = R_{ij}^{(2)rs}\partial_r = 0 \tag{152}$$

we have

$$\Box h_{ij}^{(2)} = T_{ij}^{(2)} \tag{153}$$

where

$$T_{ij}^{(2)} = R_{ij}^{(2)rs}T_{rs} \tag{154}$$

Using

$$R_{ij}^{(1)rs}\partial_r\partial_s = R_{ij}^{(1)rs}\eta_{rs} = 0 \tag{155}$$

and equations (133) and (136) we have for small m

$$\Box A_i \approx \frac{-1}{\sqrt{2}}\partial^j\frac{R_{ijrs}^{(1)}}{\nabla^2}\left[mT^{rs} + \frac{1}{m}\left(\partial^r\partial_\mu T^{\mu s} + \partial^s\partial_\mu T^{\mu r}\right)\right.$$
$$\left.+\frac{1}{m^2}\partial^r\partial^s\partial_\mu\dot{T}^{\mu 0}\right)\right] \tag{156}$$

If we choose sources which are conserved i.e.

$$\partial_\mu T^{\mu\nu} = 0 \tag{157}$$

then as $m \to 0$, equation (156) becomes

$$\Box A_i = 0 \tag{158}$$

This means the dynamical variables corresponding to the helicity one fields become decoupled in the $m \to 0$ limit if the source is conserved.

We now project out the helicity zero field from equation (150), by multiplying the equation by $S^{(0)}$, taking the trace and using the definition (144), to obtain for small m

$$\Box\phi = \sqrt{\frac{3}{2}\frac{m^2}{\nabla^2}}\partial_r\partial_s T^{rs} - \frac{1}{\sqrt{6}}\left(T_\mu^\mu - \frac{2\partial_\mu\partial_\nu}{m^2}T^{\mu\nu}\right) - \sqrt{\frac{3}{2}\frac{m^2}{\nabla^2}}\left(2\partial_r\partial_\mu T^{\mu r} - \frac{\nabla^2}{m^2}\partial_\mu\dot{T}^{\mu 0}\right) \tag{159}$$

Again, assuming (157) and letting $m \to 0$ we find

$$\Box\phi = -\frac{1}{\sqrt{6}}T_\mu^\mu \tag{160}$$

Here, we see that in the limit, the helicity zero field does not decouple if

$$T_\mu^\mu \neq 0 \tag{161}$$

Although the equation (153) together with the constraint equations (120) and (119) (with source conserved and $m = 0$) make up the content of the massless theory, the fact that the helicity zero field ϕ does not decouple shows that the $m \to 0$ theory is not the same as the massless theory.

5. CONCLUSIONS

The main conclusion of this work concerns the difference in behaviour between massive spin one and spin two theories as the masses approach zero. The limit as the mass approaches zero of a theory of a spin one field coupled to a conserved source gives the same observational results (in the classical limit or tree approximation limit) as the theory of massless spin one. However the same limit of a spin two theory coupled to a conserved source can give the same results only if the source is traceless in the limit. This is usually not the case if we expect the spin two source to be the energy–momentum tensor. These results have been obtained before. By making use of the properties of the Levi–Civita tensor and constructing projection operators in spin and helicity space we have followed through in detail the properties of the various dynamical variables as the mass m becomes small. From the equations of motion we have seen how the number of dynamical variables change from $2s+1$ to 2 (where $s = 1$ *or* 2) depending on whether $m \neq 0$ or $m = 0$. This is due to the gauge invariance which the $m = 0$ theories possess. The presence or absence of gauge invariance appears again in our construction of the propagators. For $m \neq 0$ the equations of motion can be inverted to give a unique propagator. For $m = 0$, due to the gauge invariance, the equations of motion are proportional to spin projection operators and can only be inverted by choosing some gauge. We see that for spin one, a conserved source allows the $m \to 0$ limit to be taken and the $m = 0$ results are reproduced for physical observables. For the spin two case a conserved source alone does not reproduce the same results as the $m = 0$ theory. One would have to assume the source to be traceless in addition. An examination of the dynamical variables and their canonical conjugates allows us to see what happens to the supposed disappearing degrees of freedom as $m \to 0$. For spin one we find that the extra helicity zero degree of freedom disappears only in the sense that it is decoupled if the source is conserved. For spin two the extra helicity one degrees of freedom are decoupled if the source is conserved but the helicity zero component does not disappear but is coupled to the trace of the source. This again shows why the $m \to 0$ limit for spin two does not reproduce the same observable results as the $m = 0$ case.

ACKNOWLEDGMENTS

The author would like to thank Dr. T. Fulton for many discussions which aroused his interest in the problem. He would also like to thank Dr. Abdus

APPENDIX A

In this Appendix we construct the spin projection operators in the space of second rank symmetric tensors in Lorentz space.

A symmetric tensor in space time has ten components and its spin decomposition is into spin two, spin one and two spin zero. Its spin two components $h^{(2)}_{\mu\nu}$, of which there are five, must satisfy

$$p^\mu h^{(2)}_{\mu\nu} = h^{(2)\mu}_\mu = 0 \tag{A.1}$$

Thus the projection operator $Q^{(2)\alpha\beta}_{\mu\nu}$ must also have the properties

$$p^\mu Q^{(2)\alpha\beta}_{\mu\nu} = Q^{(2)\mu\alpha\beta}_\mu = 0 \tag{A.2}$$

and it must be symmetric in (μ, ν) and (α, β). One easily finds

$$Q^{(2)\alpha\beta}_{\mu\nu} = \tfrac{1}{2}[(\delta^\alpha_\mu - \hat{p}_\mu \hat{p}^\alpha)(\delta^\beta_\nu - \hat{p}_\nu \hat{p}^\beta) + (\mu \leftrightarrow \nu)]$$
$$- \tfrac{1}{3}(\eta_{\mu\nu} - \hat{p}_\mu \hat{p}_\nu)(\eta^{\alpha\beta} - \hat{p}^\alpha \hat{p}^\beta) \tag{A.3}$$

$$(Q^{(2)})^2 = Q^{(2)} \tag{A.4}$$

i.e.

$$Q^{(2)\rho\sigma}_{\mu\nu} Q^{(2)\alpha\beta}_{\rho\sigma} = Q^{(2)\alpha\beta}_{\mu\nu} \tag{A.5}$$

The spin one component $h^{(1)}_{\mu\nu}$ must have the properties that

$$h^{(1)}_{\mu\nu} = \partial_\nu h^{(1)}_\mu + \partial_\mu h^{(1)}_\nu \tag{A.6}$$

and

$$p^\mu h^{(1)}_\mu = 0 \tag{A.7}$$

The projection operator $Q^{(1)\alpha\beta}_{\mu\nu}$ which must be orthogonal to $Q^{(2)\alpha\beta}_{\mu\nu}$ is found to be

$$Q^{(1)\alpha\beta}_{\mu\nu} = \tfrac{1}{2}[\hat{p}_\mu \hat{p}^\alpha(\delta^\beta_\nu - \hat{p}_\nu \hat{p}^\beta) + \hat{p}_\mu \hat{p}^\beta(\delta^\alpha_\nu - \hat{p}_\nu \hat{p}^\alpha) + (\mu \leftrightarrow \nu)]$$
$$= \tfrac{1}{2}[\hat{p}_\mu \hat{p}^\alpha \delta^\beta_\nu + \hat{p}_\mu \hat{p}^\beta \delta^\alpha_\nu + (\mu \leftrightarrow \nu)] - 2\hat{p}_\mu \hat{p}_\nu \hat{p}^\alpha \hat{p}^\beta \tag{A.8}$$

and again

$$(Q^{(1)})^2 = Q^{(1)} \tag{A.9}$$

and

$$Q^{(2)} Q^{(1)} = 0 \tag{A.10}$$

Since the symmetric tensor $h_{\mu\nu}$ is a reducible representation of the Lorentz group, there is no unique decomposition of the two spin zero parts. In fact there is generally a one parameter infinity of two orthogonal projection operators.

Let us define two projection operators,
$$A^{\alpha\beta}_{\mu\nu} = \tfrac{1}{4}\eta_{\mu\nu}\eta^{\alpha\beta} \tag{A.11}$$
and
$$B^{\alpha\beta}_{\mu\nu} = \hat{p}_\mu \hat{p}_\nu \hat{p}^\alpha \hat{p}^\beta \tag{A.12}$$
Now
$$A^2 = A \tag{A.13}$$
$$B^2 = B \tag{A.14}$$
$$AQ^{(2)} = AQ^{(1)} = BQ^{(2)} = BQ^{(1)} = 0 \tag{A.15}$$
However
$$AB \neq 0 \tag{A.16}$$
Let us define
$$C = AB + BA \tag{A.17}$$
That is
$$C^{\alpha\beta}_{\mu\nu} = \tfrac{1}{4}(\eta_{\mu\nu}\hat{p}^\alpha \hat{p}^\beta + \hat{p}_\mu \hat{p}_\nu \eta^{\alpha\beta}) \tag{A.18}$$
and we have the following identities
$$ABA = A/4 \tag{A.19}$$
$$BAB = B/4 \tag{A.20}$$
$$AC + CA = A/2 + C \tag{A.21}$$
$$BC + CB = B/2 + C \tag{A.22}$$
$$C^2 = (A + B + C)/4 \tag{A.23}$$
We find that the most general spin zero projection operator can be written
$$Q^{(0)}_c \equiv aA + bB - cC \tag{A.24}$$
where a and b are the two roots of
$$4x^2 - 2(c+2)x + c^2 = 0 \tag{A.25}$$
for any c. The roots will be real provided
$$2 \geq c \geq -\tfrac{2}{3} \tag{A.26}$$
For the case $c = 0$, either a or $b = 0$, i.e. A and B are projection operators. Another special case is
$$c = \tfrac{4}{3} \tag{A.27}$$

in which case
$$Q^{(0)}_{4/3} \equiv Q^{(0)} = \frac{4}{3}\left(A + \frac{B}{4} - C\right) \tag{A.28}$$

i.e.
$$Q^{(0)\alpha\beta}_{\mu\nu} = \tfrac{1}{3}(\eta_{\mu\nu} - \hat{p}_\mu \hat{p}_\nu)(\eta^{\alpha\beta} - \hat{p}^\alpha \hat{p}^\beta) \tag{A.29}$$

For a given $Q^{(0)}_c$ one can find a second projection operator $\bar{Q}^{(0)}_c$ orthogonal to $Q^{(0)}_c$
$$\bar{Q}^{(0)}_c = \bar{a}A + \bar{b}B - \bar{c}C \tag{A.30}$$

where we find
$$a + \bar{a} = b + \bar{b} = c + \bar{c} = \tfrac{4}{3} \tag{A.31}$$

Using (A.11) and (A.12) one can demonstrate that
$$(Q^{(2)} + Q^{(1)} + Q^{(0)}_c + \bar{Q}^{(0)}_c)^{\alpha\beta}_{\mu\nu} = \tfrac{1}{2}(\delta^\alpha_\mu \delta^\beta_\nu + \delta^\beta_\mu \delta^\alpha_\nu) \tag{A.32}$$

We now write the operator $K^{\alpha\beta}_{\mu\nu}$ defined by equation (19) in terms of these projection operators. First, we have
$$K^{\alpha\beta}_{\mu\nu} = [(p^2 - m^2)Q^{(2)} - m^2 Q^{(1)} - \tfrac{8}{3}(p^2 - m^2)A - \tfrac{2}{3}(p^2 + 2m^2)B$$
$$+ \tfrac{4}{3}(2p^2 + m^2)C]^{\alpha\beta}_{\mu\nu} \tag{A.33}$$

We must now find $X(p)$, $Y(p)$ and $c(p)$ such that the A, B, C part of (A.33) can be written
$$XQ^{(0)}_c + Y\bar{Q}^{(0)}_c \tag{A.34}$$

To do this we must solve the equations
$$aX + \bar{a}Y = -\tfrac{8}{3}(p^2 - m^2) \tag{A.35}$$
$$bX + \bar{b}Y = -\tfrac{2}{3}(p^2 + 2m^2) \tag{A.36}$$
$$cX + \bar{c}Y = -\tfrac{4}{3}(2p^2 + m^2) \tag{A.37}$$

where we also use (A.31) and (A.25). We find that X and Y are the two solutions for u of
$$u^2 + 2u(p^2 - m^2) - 3m^4 = 0 \tag{A.38}$$

Thus
$$XY = -3m^4 \tag{A.39}$$

This implies neither X nor Y vanishes for any value of p^2.
We take Y as

$$Y = -(p^2-m^2)+[(p^2-m^2)^2+3m^4]^{\frac{1}{2}} \qquad (A.40)$$

For small m,

$$Y \approx \frac{3m^4}{2p^2} \qquad (A.41)$$

and

$$X = -(p^2-m^2)-[(p^2-m^2)^2+3m^4]^{\frac{1}{2}} \qquad (A.42)$$
$$\approx -2p^2 \text{ for small } m \qquad (A.43)$$

In addition

$$c = -\frac{4}{3}\frac{p^2+2m^2}{X-Y}+\frac{2}{3} \qquad (A.44)$$

$$a = -\frac{4}{3}\frac{(p^2-m^2)}{X-Y}+\frac{2}{3} \qquad (A.45)$$

and

$$b = +\frac{2}{3}\frac{p^2-4m^2}{X-Y}+\frac{2}{3} \qquad (A.46)$$

Indeed as $m \to 0$,

$$c, a \to \tfrac{4}{3} \qquad (A.47)$$

and

$$b \to \tfrac{1}{3}$$
$$Q_c^{(0)} \to Q^0 \qquad (A.48)$$

where $Q^{(0)}$ is given by (A.29) and

$$\bar{Q}_c^{(0)} \to \bar{Q}^{(0)} = B \qquad (A.49)$$

Thus as $m \to 0$

$$XQ_c^{(0)} + Y\bar{Q}_c^{(0)} \to -2p^2 Q^{(0)} \qquad (A.50)$$

which produce equation (43).
For the inverse of K we have

$$G_{\mu\nu}^{\alpha\beta} = \left(\frac{1}{p^2-m^2}Q^{(2)} - \frac{1}{m^2}Q^{(1)} + \frac{Q_c^{(0)}}{X} + \frac{\bar{Q}_c^{(0)}}{Y}\right)_{\mu\nu}^{\alpha\beta} \qquad (A.51)$$

We look for the small m limit of

$$\frac{Q_c^{(0)}}{X} + \frac{\bar{Q}_c^{(0)}}{Y} \qquad (A.52)$$

We have
$$\frac{Q_c^{(0)}}{X} + \frac{\bar{Q}_c^{(0)}}{Y} = \frac{(aY+\bar{a}X)A + (bY+\bar{b}X)B - (cY+\bar{c}X)C}{XY} \tag{A.53}$$

Now
$$aY + \bar{a}X = +\tfrac{4}{3}(p^2 - m^2) + \tfrac{2}{3}(X+Y) = 0 \tag{A.54}$$
$$bY + \bar{b}X = -\tfrac{2}{3}(p^2 - 4m^2) + \tfrac{2}{3}(X+Y) = -2(p^2 - 2m^2) \tag{A.55}$$
$$cY + \bar{c}X = \tfrac{4}{3}(p^2 + 2m^2) + \tfrac{2}{3}(X+Y) = 4m^2$$

Using (A.39) we have
$$\frac{Q_c^{(0)}}{X} + \frac{\bar{Q}_c^{(0)}}{Y} = \frac{2}{3m^2}[(p^2 - 2m^2)B + 2m^2 C] \tag{A.56}$$

Thus for small m
$$\frac{Q_c^{(0)}}{X} + \frac{\bar{Q}_c^{(0)}}{Y} \approx \frac{2p^2}{3m^4} B = \frac{2p^2}{3m^4} \bar{Q}^{(o)} \tag{A.57}$$

This gives equation (55) for small m.

APPENDIX B

In this Appendix we construct the helicity projection operators in the space of vectors and second rank symmetric tensors in Euclidean three-space.

A vector A_i in three-space has three components: two helicity one components, $A_i^{(1)}$, and one helicity zero component $A_i^{(0)}$. The helicity one components are combinations of the transverse components of A_i and so must satisfy (in momentum space)
$$p^i A_i^{(1)} = 0 \tag{B.1}$$

The projection operator $R_{ij}^{(1)}$ into this space is
$$R_{ij}^{(1)} = (\eta_{ij} + \hat{p}_i \hat{p}_j) \tag{B.2}$$

where
$$\hat{p}_i \hat{p}^i = -1 \tag{B.3}$$

and
$$R_{ik}^{(1)} R^{(1)kj} = R_i^{(1)j} \tag{B.4}$$

The helicity zero projection operator $R_{ij}^{(0)}$ must be orthogonal to $R^{(1)}$ and is
$$R_{ij}^{(0)} = -\hat{p}_i \hat{p}_j \tag{B.5}$$

A symmetric tensor in three-space has six components and its helicity decomposition is into helicity two components (2), helicity one components (2) and two independent helicity zero components. The helicity two components $h_{ij}^{(2)}$ must satisfy the relations

$$p^i h_{ij}^{(2)} = h_i^{(2)i} = 0 \tag{B.6}$$

The projection operator $R_{ij}^{(2)rs}$ into this space is (in momentum space)

$$R_{ij}^{(2)rs} = \tfrac{1}{2}[(\delta_i^r + \hat{p}_i\hat{p}^r)(\delta_j^s + \hat{p}_j\hat{p}^s) + (i \leftrightarrow j)] - \tfrac{1}{2}(\eta_{ij} + \hat{p}_i\hat{p}_j)(\eta^{rs} + \hat{p}^r\hat{p}^s) \tag{B.7}$$

and, of course,

$$(R^{(2)})^2 = R^{(2)} \tag{B.8}$$

The helicity one components $h_{ij}^{(1)}$ must have the property

$$h_{ij}^{(1)} = \partial_i h_j^{(1)} + \partial_j h_i^{(1)} \tag{B.9}$$

and

$$p^i h_i^{(1)} = 0 \tag{B.10}$$

We find

$$R_{ij}^{(1)rs} = -\tfrac{1}{2}[\hat{p}_i\hat{p}^r(\delta_j^s + \hat{p}_j\hat{p}^s) + \hat{p}_i\hat{p}^s(\delta_j^r + \hat{p}_j\hat{p}^r) + (i \leftrightarrow j)] \tag{B.11}$$

$$(R^{(1)})^2 = R^{(1)} \tag{B.12}$$

and

$$R^{(1)}R^{(2)} = 0 \tag{B.13}$$

Just as in Lorentz space (in Appendix A) there will be an infinity of the two helicity zero projection operators which can be made up of linear combinations of the operators

$$\tfrac{1}{3}\eta_{ij}\eta^{rs},\ \hat{p}_i\hat{p}_j\hat{p}^r\hat{p}^s,\ -\tfrac{1}{3}(\eta_{ij}\hat{p}^r\hat{p}^s + \hat{p}_i\hat{p}_j\eta^{rs}) \tag{B.14}$$

However from equation (119) we know which helicity zero component of h_{ij} is not a dynamical variable, namely the one which is proportional to

$$\Lambda_{ij}h^{ij} \tag{B.15}$$

The helicity zero projection operator which projects this field out of h_{ij} is

$$\bar{R}_{ij}^{(0)rs} = \Lambda_{ij}\Lambda^{rs}/\Lambda_{mn}\Lambda^{mn} \tag{B.16}$$

Thus the helicity zero component which will be a dynamical variable will be obtained by finding that helicity zero projection operator $R^{(0)}$ which is orthogonal to $\bar{R}^{(0)}$. This operator is easily found to be

$$R_{ij}^{(0)rs} = \frac{m^4}{2Z}\left(\eta_{ij} + \frac{2p^2 + 3m^2}{m^2}\hat{p}_i\hat{p}_j\right)\left(\eta^{rs} + \frac{2p^2 + 3m^2}{m^2}\hat{p}^r\hat{p}^s\right) \tag{B.17}$$

where

$$Z \equiv (2p^4 + 4p^2m^2 + 3m^4) \tag{B.18}$$

Accordingly, the dynamical helicity variables are

$$h_{ij}^{(a)} \equiv R_{ij}^{(a)rs} h_{rs} \quad \text{with } a = 0, 1, 2 \tag{B.19}$$

and the variables canonical to these will be the

$$\Pi_{ij}^{(a)} \equiv R_{ij}^{(a)rs} \Pi_{rs} \tag{B.20}$$

with Π_{rs} given by (121)
We have

$$R_{ij}^{(a)rs} \Lambda_{rs} = 0 \tag{B.21}$$

since

$$R^{(0)} \bar{R}^0 = 0 \tag{B.22}$$

We may write

$$\Pi_{ij}^{(a)} = R_{ij}^{(a)rs} \Lambda_{rk} \Lambda_{sl} \dot{h}^{kl} \tag{B.23}$$

where we have neglected the terms depending on the source T_μ^ν since they will not contribute to the canonical commutation relations. We have

$$\Lambda_{rk} \Lambda_{sl} \dot{h}^{kl} = \left[R^{(2)} + \frac{m^2}{p^2 + m^2} R^{(1)} + \left(\frac{m^2}{p^2 + m^2}\right)^2 S^{(0)} + \bar{S}^{(0)} \right]_{rskl} \dot{h}^{kl} \tag{B.24}$$

where

$$S_{rs}^{(0)kl} = \hat{p}_r \hat{p}_s \hat{p}^k \hat{p}^l \tag{B.25}$$

and

$$\bar{S}_{rs}^{(0)kl} = \tfrac{1}{2}(\eta_{rs} + \hat{p}_r \hat{p}_s)(\eta^{kl} + \hat{p}^k \hat{p}^l) \tag{B.26}$$

Using (119) and neglecting source terms we may write

$$(\eta_{kl} + \hat{p}_k \hat{p}_l) \dot{h}^{kl} = \frac{m^2}{p^2 + m^2} \hat{p}_k \hat{p}_l \dot{h}^{kl} \tag{B.27}$$

Also from (B.21) we may write

$$R_{ij}^{(0)rs}(\eta_{rs} + \hat{p}_r \hat{p}_s) = \frac{m^2}{p^2 + m^2} R_{ij}^{(0)rs} \hat{p}_r \hat{p}_s \tag{B.28}$$

This gives

$$(R^{(0)} \bar{S}^0)_{ij}^{kl} \dot{h}_{kl} = \frac{1}{2} \left(\frac{m^2}{p^2 + m^2}\right)^2 (R^0 S^0)_{ij}^{kl} \dot{h}_{kl} \tag{B.29}$$

Finally we have

$$\Pi_{ij}^{(2)} = R_{ij}^{(2)rs} \dot{h}_{rs} \tag{B.30}$$

$$\Pi_{ij}^{(1)} = \frac{m^2}{p^2 + m^2} R_{ij}^{(1)rs} \dot{h}_{rs} \approx \frac{m^2}{p^2} R_{ij}^{(1)rs} \dot{h}_{rs} \tag{B.31}$$

for small m, and

$$\Pi_{ij}^{(0)} = \left[\left(\frac{m^2}{p^2+m^2}\right)^2 + \frac{1}{2}\left(\frac{m^2}{p^2+m^2}\right)^2\right](R^{(0)}S^{(0)})_{ij}^{rs}\dot{h}_{rs} \qquad (B.32)$$

Substituting (B.17) for $R^{(0)}$ and keeping only leading terms for small m we obtain

$$\Pi_{ij}^{(0)} \approx \frac{3}{2}\left(\frac{m^2}{p^2}\right)^2 S_{ij}^{(0)rs}\dot{h}_{rs} \qquad (B.33)$$

Similarly for small m we may write

$$h_{ij}^{(0)} \approx S_{ij}^{(0)rs}\dot{h}_{rs} \qquad (B.34)$$

REFERENCES

Boulware, D. G. and Deser, S. (1972) *Phys. Rev.*, **D6**, 3368.
Deser, S. (1970) *J. Gen. Rel. Grav.*, **1**, 9.
Fierz, M. and Pauli, W. (1939) *Proc. Roy. Soc.*, **173**, 211.
Van Dam, H. and Veltman, M. (1970) *Nucl. Phys.*, **B22**, 397.

Relativistic Electromagnetic Interaction Without Quantum Electrodynamics

JOHN H. DETRICH
University of Wisconsin, Madison, Wisconsin, U.S.A.

and

CLEMENS C. J. ROOTHAAN
University of Chicago, Chicago, Illinois, U.S.A., and
Ohio State University, Columbus, Ohio, U.S.A.

1. INTRODUCTION

In the extension of the Dirac theory of the electron to many-electron systems, the most obvious difficulty is a satisfactory treatment of relativistic electromagnetic interactions. The first such treatment was worked out many years ago by Breit (1929, 1930, 1932). A more elaborate derivation of Breit's results, based on quantum electrodynamics, was given by Bethe and Salpeter (1957). While quantum electrodynamics is thus capable of describing relativistic electromagnetic interactions, there are in principle and in practice still serious difficulties in this approach.

One reason for this situation is that relativistic effects are intrinsically far from simple. However, quantum electrodynamics complicates matters by yielding results which are not readily absorbed in the framework within which one naturally deals with many-electron problems. Thus quantum electrodynamics is organized around a perturbation treatment that regards quantum electrodynamical effects as *weak*. But in many-electron systems the quantum electrodynamical effects include the Coulomb interactions between electrons, and these cannot comfortably be regarded as weak. Again, quantum electrodynamics tends to sidestep the question of the influence of quantum electrodynamical effects on the wave function. Indeed, it may be inconsistent to take this up, since one of these effects is the interaction of an electron with its own field, and even a finite retarded self-action compromises the definition of a wave function (Feynman, 1948). Treatments of many-electron systems, on the other hand, customarily are formulated in terms of wave functions.

These considerations suggest that a vantage point somewhat different from that offered by quantum electrodynamics may be useful in treating relativistic electromagnetic interactions. We present in this paper a formalism that provides such an alternative approach.

Since one of our principal objectives is to deal with relativistic electromagnetic interactions in terms of a vocabulary that differs from the one customary in quantum electrodynamics, we begin in Section 1 and 2 with a brief outline of this vocabulary. Our extensive use of matrix elements is in the spirit of the original formulation of quantum mechanics by Heisenberg (1925), Born and Jordan (1925), Born, Heisenberg and Jordan (1926) and Dirac (1926). Section 3 reviews Dirac's treatment (1928) of a single relativistic electron.

In Section 4 we take up the actual treatment of relativistic electromagnetic interactions. Our considerations apply to an isolated system containing an arbitrary number of Dirac particles. Of course our assumption that the system is not subject to external disturbances is only an approximation, since real physical systems can always interact with external objects through emission or absorbtion of radiation or other means. Here we accept the limitation this approximation imposes on our treatment, since our interest is in interactions between particles. This is necessary in order to deal with a total energy for the system which is conserved and therefore capable of precise definition.

We deal here in terms of many-particle wave functions, and this implies that the number of particles remains constant. Thus no provision is made for the possible creation and/or annihilation of electron–positron pairs.

No additional approximation need be adopted in treating the relativistic electromagnetic interactions. We find that these interactions can be handled in closed form, using the version of classical electrodynamics advocated by Wheeler and Feynman (1945, 1949). Just as we desire, this formulation deals directly in terms of interparticle interactions, without reference to a mediating electromagnetic field. In such a theory there is nothing that compels one to include the action of a particle on itself, as Wheeler and Feynman demonstrate. Following them, we omit such self-actions, thereby avoiding all of the difficulties associated with those terms in conventional quantum electrodynamics.

Quantum mechanical treatments based on the electrodynamics of Wheeler and Feynman have previously been presented by Hoyle and Narlikar (1974), and also by Davies (1971, 1972). However there is little in this work that bears directly on our approach: while these treatments differ in some respects from conventional quantum electrodynamics, they adopt similar vantage points and vocabularies. Thus, from our point of view, they have essentially the same drawbacks as conventional quantum electrodynamics, including, rather unexpectedly, the occurrence of self-interactions.

In the remaining sections we take up the Pauli approximation. The results are significant in their own right, even though we simply reproduce long-accepted theoretical expressions. Careful examination of previous derivations reveals that they are somewhat inadequate. One problem is that these derivations start from an approximate treatment of the interactions. This is hardly to be avoided in a treatment based on quantum electrodynamics since this theory is organized around a perturbation treatment, but it places one in the awkward position of extracting one approximation from another. A more severe problem comes from the fact that quantum electrodynamics avoids evaluating electrodynamical effects on the wave function. To make up for this, one is obliged to adopt some more or less arbitrary assumptions for such effects. For example, Breit (1929, 1930, 1932) found that part of his approximate interaction term should not be included in the Hamiltonian used to determine the wave function, even though its expectation value contributes to the energy.

A derivation of the Pauli approximation which is free of such difficulties is presented in Section 6. In Section 7 we assess the significance of our results, and present a few very general suggestions for further work along these lines.

2. QUANTIZATION OF A NON-RELATIVISTIC HAMILTONIAN SYSTEM OF PARTICLES

For the classical description of a system of N particles, let the cartesian coordinates and conjugate momenta be denoted by $\mathbf{r}_\mu$ and $\mathbf{p}_\mu$, respectively, $\mu = 1, 2, \ldots, N$; we shall denote them collectively by

$$\left.\begin{array}{l} \mathbf{r}^N = (\mathbf{r}_1, \mathbf{r}_2, \ldots, \mathbf{r}_N) \\ \mathbf{p}^N = (\mathbf{p}_1, \mathbf{p}_2, \ldots, \mathbf{p}_N) \end{array}\right\} \quad (1)$$

There exists a Hamiltonian function $H(\mathbf{r}^N, \mathbf{p}^N)$; the time evolution of the system is governed by the canonical equations of motion

$$\left.\begin{array}{l} \dot{\mathbf{r}}_\mu = \dfrac{\partial H}{\partial \mathbf{p}_\mu} \\[2mm] \dot{\mathbf{p}}_\mu = -\dfrac{\partial H}{\partial \mathbf{r}_\mu} \end{array}\right\} \quad (2)$$

From these equations one can in principle calculate $\mathbf{r}_\mu(t)$ and $\mathbf{p}_\mu(t)$ for any time t, given a set of initial values $\mathbf{r}_\mu(0)$, $\mathbf{p}_\mu(0)$.

To describe the N-particle system quantum mechanically, we replace the coordinates and momenta by appropriate operators. It is customary to use for these operators the same symbols as for the classical quantities they replace. The coordinates and momenta must satisfy the well-known commutation relations, which can be written in the two equivalent forms

$$\left.\begin{array}{l} [\![\mathbf{p}_\mu, \mathbf{a} \cdot \mathbf{r}_\nu]\!] = -i\hbar \delta_{\mu\nu} \mathbf{a} \\ [\![\mathbf{a} \cdot \mathbf{p}_\mu, \mathbf{r}_\nu]\!] = -i\hbar \delta_{\mu\nu} \mathbf{a} \end{array}\right\} \quad (3)$$

where $[\![A, B]\!] = AB - BA$ is the *commutator* of the operators A and B, and $\mathbf{a}$ is any constant vector.

In general, algebraic equations between classical quantities are reinterpreted as operator equations. This process is usually not unique, since any product involving conjugate coordinates and momenta becomes dependent on the order of the factors, because of the commutation relations (3). Often the requirement that any observable must be represented by a Hermitian operator resolves the ambiguity. With that understanding the Hamiltonian operator is now defined in terms of the operators $\mathbf{r}_\mu, \mathbf{p}_\mu$ by the functional expression $H(\mathbf{r}^N, \mathbf{p}^N)$.

The replacement of the classical variables by operators leaves undefined the operators corresponding to time derivatives of those variables. The time derivative of an operator A which does not explicitly depend on the time is now defined by

$$\dot{A} = i\hbar^{-1}[\![H, A]\!] \tag{4}$$

If equation (4) is used to evaluate $\dot{\mathbf{r}}_\mu$ and $\dot{\mathbf{p}}_\mu$, one can prove, using the commutation relations (3), that the canonical equations (2) are now valid as operator equations.

In the Schrödinger representation, which we shall use throughout this paper, the coordinates are taken over without change as multiplicative operators. The momenta must then be defined by

$$\mathbf{p}_\mu = -i\hbar \nabla_\mu = -i\hbar \frac{\partial}{\partial \mathbf{r}_\mu} \tag{5}$$

in order that the commutation relations (3) are satisfied.

While classically the time evolution of the system is specified through the explicit functions of the time $\mathbf{r}_\mu(t)$ and $\mathbf{p}_\mu(t)$, quantum mechanically the time evolution is expressed in the *time-dependent wave function* $\tilde{\Psi}(\mathbf{r}^N, t)$ which must satisfy the *time-dependent Schrödinger equation*

$$i\hbar \frac{\partial \tilde{\Psi}}{\partial t} = H\tilde{\Psi} \tag{6}$$

This equation can be satisfied by wave functions representing *stationary states*, namely

$$\left.\begin{array}{c} \tilde{\Psi}(\mathbf{r}^N, t) = \Psi(\mathbf{r}^N) e^{-i\omega t} \\ \omega = E/\hbar \end{array}\right\} \tag{7}$$

where E is the energy of the stationary state represented by $\tilde{\Psi}$. The *time-independent wave function* Ψ satisfies the *time-independent Schrödinger equation*

$$H\Psi = E\Psi \tag{8}$$

which states that Ψ and E must be eigenfunction and eigenvalue, respectively, of the operator H.

The determination of the entire set of eigenvalues and corresponding eigenfunctions of a given Hamiltonian constitutes the central problem of non-relativistic quantum mechanics. Conceptually, the simplest systems are those for which the entire eigenvalue spectrum is discrete, each eigenvalue having only a finite number of linearly independent eigenfunctions; the harmonic oscillator is the best known example.

For the all-discrete case it is natural to denote the stationary state wave functions by

$$\tilde{\Psi}_{m\alpha}(\mathbf{r}^N, t) = \Psi_{m\alpha}(\mathbf{r}^N) e^{-i\omega_m t}$$
$$\omega_m = E_m/\hbar$$
(7a)

where m is an appropriate discrete index to label the distinct energies, while α labels degenerate wave functions if necessary. For the time-independent wave functions we have of course

$$H\Psi_{m\alpha} = E_m \Psi_{m\alpha} \quad (8a)$$

It is well-known that degeneracy is a necessary consequence of the symmetry of the problem. A symmetry operator is by definition a unitary operator which commutes with the Hamiltonian. The symmetry operators form a group; the wave functions $\Psi_{m\alpha}$ which belong to the energy E_m transform among themselves under symmetry operations, and the transformation matrices constitute a representation of the group. If we allow for the possibility $E_m = E_n$, $m \neq n$, then it is no loss of generality to assume that the representation associated with a level E_m is always *irreducible*; the particular representation to which the set $\Psi_{m\alpha}$ belongs is called its *symmetry species*. The normal situation is that the E_m are distinct; whenever $E_m = E_n$, $m \neq n$ occurs, it is called *accidental degeneracy*. Consideration of the symmetry properties of the wave function leads to the so-called *good quantum numbers*; a corresponding compound labelling of the energy and wave functions is usually adopted. For our present purposes, this is not necessary; we label the energies with a single index m, and recognize non-accidental degeneracy of each level by the wave function label $m\alpha$.

It is customary to postulate that the time-independent functions $\Psi_{m\alpha}$ constitute a *complete orthonormal* basis in the Hilbert space of functions used to describe the N-particle system at any given time t. The orthonormality is expressed by

$$\langle \Psi_{m\alpha} | \Psi_{n\beta} \rangle = \int d\mathbf{r}^N \Psi^*_{m\alpha}(\mathbf{r}^N) \Psi_{n\beta}(\mathbf{r}^N) = \delta_{m\alpha, n\beta} = \delta_{mn}\delta_{\alpha\beta} \quad (9)$$

where $\int d\mathbf{r}^N$ denotes $3N$-dimensional integration over all particle coordinates. Completeness of the base is conveniently expressed in Dirac notation by

$$\mathscr{I} = \sum_{m\alpha} |\Psi_{m\alpha}\rangle\langle\Psi_{m\alpha}| \quad (10)$$

where $\mathscr{I}$ is the identity operator; in ordinary functional notation completeness is expressed by

$$\Psi(\mathbf{r}^N) = \sum_{m\alpha} \Psi_{m\alpha}(\mathbf{r}^N)\langle\Psi_{m\alpha}|\Psi\rangle = \sum_{m\alpha} \Psi_{m\alpha}(\mathbf{r}^N) \int d\mathbf{r}^{N\prime} \Psi^*_{m\alpha}(\mathbf{r}^{N\prime})\Psi_{n\beta}(\mathbf{r}^{N\prime})$$
(10a)

where Ψ is an arbitrary function within certain reasonable constraints. This completeness postulate is fundamental to quantum mechanics; hence only

Hamiltonians which yield as their eigenfunctions complete sets can be used to describe physical systems.

Most systems of physical interest do not fall in the all-discrete category. The other extreme occurs for a free particle, where the entire spectrum consists of all $E \geq 0$, so that it is *all-continuous*. Most common is the *mixed discrete-continuous* case; for instance, the hydrogen atom has continuous eigenvalues $E \geq 0$, and discrete eigenvalues $E_m = -R/m^2$, $m = 1, 2, \ldots$, where R is the Rydberg constant. Clearly, equations (7a), (9), (10) and (10a) must now be reinterpreted. In general, the index m has a discrete and a continuous range, while α can be maintained as a discrete index. With this understanding, equation (7a) remains valid as it stands. In equation (9), if *either m or n* is in the discrete range, no change is required; but if *both m and n* are in the continuum, δ_{mn} must be replaced by a Dirac delta-function. In equations (10) and (10a) we must deal with the sum $\sum_{m\alpha} = \sum_m \sum_\alpha$. It is clear that $\sum_\alpha$ is to be retained as a discrete sum; however $\sum_m$ must be understood as a discrete sum and/or integration, depending on whether E_m is in the discrete or continuous range, respectively.

In our further deliberations we shall restrict ourselves to the all-discrete notation. This, however, should not limit the validity of our final results, since in the latter all references to stationary states will have disappeared.

We use the stationary states to define the *time-dependent* and *time-independent matrix elements* of an operator, namely

$$\left. \begin{aligned} \tilde{A}_{m\alpha,n\beta}(t) &= \langle \tilde{\Psi}_{m\alpha} | A | \tilde{\Psi}_{n\beta} \rangle = \int d\mathbf{r}^N \tilde{\Psi}^*_{m\alpha}(\mathbf{r}^N, t) A(\mathbf{r}^N, \mathbf{p}^N) \tilde{\Psi}_{n\beta}(\mathbf{r}^N, t) \\ A_{m\alpha,n\beta} &= \langle \Psi_{m\alpha} | A | \Psi_{n\beta} \rangle = \int d\mathbf{r}^N \Psi^*_{m\alpha}(\mathbf{r}^N) A(\mathbf{r}^N, \mathbf{p}^N) \Psi_{n\beta}(\mathbf{r}^N) \end{aligned} \right\} \quad (11)$$

Clearly, $\tilde{A}_{m\alpha,n\beta}$ and $A_{m\alpha,n\beta}$ are related by

$$\left. \begin{aligned} \tilde{A}_{m\alpha,n\beta}(t) &= A_{m\alpha,n\beta} e^{i\omega_{mn}t} \\ \omega_{mn} &= \omega_m - \omega_n \end{aligned} \right\} \quad (12)$$

For $m = n$ the time dependence drops out, and we write

$$\tilde{A}_{m\alpha,m\beta}(t) = A_{m\alpha,m\beta} = A_{m\alpha\beta} \quad (13)$$

We note that the matrix elements of the Hamiltonian are given by

$$H_{m\alpha,n\beta} = E_m \delta_{m\alpha,n\beta} \quad (14)$$

The matrix elements of the product of two operators are defined by

$$\left. \begin{aligned} (\widetilde{AB})_{m\alpha,n\beta} &= \sum_{p\gamma} \tilde{A}_{m\alpha,p\gamma} \tilde{B}_{p\gamma,n\beta} \\ (AB)_{m\alpha,n\beta} &= \sum_{p\gamma} A_{m\alpha,p\gamma} B_{p\gamma,n\beta} \end{aligned} \right\} \quad (15)$$

this is easily proved using equations (7a), (10) and (11).

For the time derivatives of the matrix elements we invoke the definition in terms of the commutator with the Hamiltonian, equation (4); we readily find

$$\left. \begin{array}{l} \dot{\tilde{A}}_{m\alpha,n\beta} = i\omega_{mn}\tilde{A}_{m\alpha,n\beta} \\ \dot{A}_{m\alpha,n\beta} = i\omega_{mn}A_{m\alpha,n\beta} \end{array} \right\} \qquad (16)$$

We note that the first equation (16) would also result if we defined $\dot{\tilde{A}}_{m\alpha,n\beta}(t)$ as $d\tilde{A}_{m\alpha,n\beta}(t)/dt$; this is of course the justification, in retrospect, for the definition of the time derivative of an operator, equation (4). The second equation (16) extends this definition by providing a time derivative for the matrix elements $A_{m\alpha,n\beta}$, even though these are time-independent quantities.

Quantization of classical relations can now be stated in terms of matrix elements, preferably the time-independent ones. One simply replaces classical quantities by the corresponding matrix elements, honouring the rules for products and time derivatives, equations (15) and (16); of course ambiguities due to the order of factors have to be resolved in the same manner as for the operators. Clearly, quantization in terms of matrix elements is completely equivalent to quantization in terms of operators, and vice versa.

From the matrix elements and wave functions one can recover the operators. In general

$$A = \sum_{m\alpha} \sum_{n\beta} |\Psi_{m\alpha}\rangle A_{m\alpha,n\beta} \langle \Psi_{n\beta}| \qquad (17)$$

For the Hamiltonian we have the special formula

$$H = \sum_{m\alpha} |\Psi_{m\alpha}\rangle E_m \langle \Psi_{m\alpha}| \qquad (18)$$

In addition to the operators for the positions and momenta of the particles, we shall need operators representing the charge and current densities associated with the particles. We define the charge density associated with the μth particle by means of

$$\rho_\mu(\mathbf{r}') = e_\mu \delta(\mathbf{r}' - \mathbf{r}_\mu) \qquad (19)$$

Here e_μ is the charge of the μth particle, and $\delta(\mathbf{r}' - \mathbf{r}_\mu)$ is the three-dimensional Dirac delta-function, so that the relation

$$f(\mathbf{r}) = \int d\mathbf{r}' \delta(\mathbf{r} - \mathbf{r}') f(\mathbf{r}') \qquad (20)$$

holds for any function $f(\mathbf{r})$ which is reasonably well-behaved. In equation (19), the space indicated by $\mathbf{r}'$ designates the position of an external electric probe, and thus may be regarded as the observer's space. In the following, we shall use $\mathbf{r}', \mathbf{r}''$ for electromagnetic probe positions, and $\mathbf{r}_\mu, \mathbf{r}_\nu$ for particle positions, or simply $\mathbf{r}$ for the position of a single particle.

In conjunction with the charge density we define the *current density* operator by means of

$$\mathbf{j}_\mu(\mathbf{r}') = (2c)g^{-1}\{\dot{\mathbf{r}}_\mu, \rho_\mu(\mathbf{r}')\} \qquad (21)$$

where in general $\{A, B\} = AB + BA$ is the *anticommutator* of the operators A and B, and the velocity $\dot{\mathbf{r}}_\mu$ is defined by equation (4), specifically

$$\dot{\mathbf{r}}_\mu = i\hbar^{-1}[H, \mathbf{r}_\mu] \tag{22}$$

Since $\dot{\mathbf{r}}_\mu$ and $\rho_\mu(\mathbf{r}')$ are both Hermitian, the symmetric product on the right-hand side of equation (21) guarantees that $\mathbf{j}_\mu(\mathbf{r}')$ is also Hermitian. Note that we have used the *electromagnetic* rather than the *electrostatic* definition of the current density; $\mathbf{j}_\mu(\mathbf{r}')$ has the same dimension as $\rho_\mu(\mathbf{r}')$.

Up to this point we have not made any use of the specific form of the Hamiltonian. For our system of point particles the *non-relativistic Hamiltonian* is

$$H(\mathbf{r}^N, \mathbf{p}^N) = T(\mathbf{p}^N) + V(\mathbf{r}^N) \tag{23}$$

where

$$T(\mathbf{p}^N) = \sum_\mu (2m_\mu)^{-1} \mathbf{p}_\mu \cdot \mathbf{p}_\mu \tag{24}$$

is the kinetic energy, and $V(\mathbf{r}^N)$ is the potential energy. We find easily for the velocity of the μth particle

$$\dot{\mathbf{r}}_\mu = m_\mu^{-1} \mathbf{p}_\mu \tag{25}$$

and therefore for the current density

$$\mathbf{j}_\mu(\mathbf{r}') = (2m_\mu c)^{-1} \{\mathbf{p}_\mu, \rho_\mu(\mathbf{r}')\} \tag{26}$$

We are now in a position to demonstrate that the charge and current densities obey the *equation of continuity*, which also may be called the law of charge conservation. Let $f_\mu = f(\mathbf{r}_\mu)$ be an arbitrary function of the position $\mathbf{r}_\mu$. Clearly f_μ commutes with $V(\mathbf{r}^N)$, hence

$$[H, f_\mu] - [T, f_\mu] = 0 \tag{27}$$

Using equations (4), (5) and (24) we can derive from this the operator equation

$$\dot{f}_\mu - (2m_\mu)^{-1}[\mathbf{p}_\mu \cdot (\nabla_\mu f_\mu) + (\nabla_\mu f_\mu) \cdot \mathbf{p}_\mu] = 0 \tag{28}$$

where $(\nabla_\mu f_\mu)$ is another function of position, that is, the operator ∇_μ does not act beyond the parentheses. On the other hand, taking matrix elements of equation (27) we find

$$(E_m - E_n)\langle \Psi_{m\alpha}|f_\mu|\Psi_{n\beta}\rangle - (2m_\mu)^{-1}\langle \Psi_{m\alpha}|[\mathbf{p}_\mu \cdot \mathbf{p}_\mu, f_\mu]|\Psi_{n\beta}\rangle = 0 \tag{29}$$

Obviously, equations (28) and (29) are completely equivalent.

If we take $f_\mu = \rho_\mu(\mathbf{r}')$ in equation (28), we can replace ∇_μ by $-\nabla'$, and obtain, using also equation (26),

$$\dot{\rho}_\mu(\mathbf{r}') + c\nabla' \cdot \mathbf{j}_\mu(\mathbf{r}') = 0 \tag{30}$$

We recognize equation (30) as the equation of continuity for the μth particle; note that charge conservation holds for each particle separately. The unusual notation $\dot{\rho}_\mu$ rather than $\partial \rho_\mu/\partial t$ is due to the fact that the observer's space $\mathbf{r}'$ was introduced as a parametric variable. Another useful form of the equation of continuity is obtained by taking the matrix elements of equation (30); the result is

$$ik_{mn}\rho_{\mu,m\alpha,n\beta}(\mathbf{r}') + \boldsymbol{\nabla}' \cdot \mathbf{j}_{\mu,m\alpha,n\beta}(\mathbf{r}') = 0 \tag{31}$$

where the *wavenumbers* k_{mn} are defined by

$$k_{mn} = \omega_{mn}/c = (E_m - E_n)/\hbar c \tag{32}$$

Finally, we make the observation that equations (27) to (31) are all equivalent. For instance, equation (29) is easily derived from equation (31) by multiplying the latter by $f(\mathbf{r}')$ and integrating over $\mathbf{r}'$. Note, however, that equations (27) to (29) are only meaningful with reference to the non-relativistic Hamiltonian; equations (30) and (31) contain the charge and current densities formally, and may therefore be valid for other Hamiltonians as well.

3. DIRAC THEORY FOR A SINGLE RELATIVISTIC PARTICLE

In the one-particle relativistic quantum mechanics of Dirac the wave function is generalized to a *four-component spinor*; in particular, for the stationary states we have

$$\tilde{\Psi}_{s,m\alpha}(\mathbf{r}, t) = \Psi_{s,m\alpha}(\mathbf{r})e^{-i\omega_m t}, \qquad s = 1, 2, 3, 4 \tag{33}$$

It is useful to consider the index s as another variable in the wave function; it is obviously a *discrete variable*.

For most formal considerations, the index s can be suppressed. When that is done, we understand that $\tilde{\Psi}$ and Ψ represent *column vectors*; the corresponding *Hermitian conjugate row vectors* are designated by $\tilde{\Psi}^*$ and Ψ^*, respectively. Product formation is to be treated according to the rules of matrix algebra; $\Psi^*\Phi$ implies summation over spinor components, and $\Psi\Phi^*$ is a 4×4 matrix. The latter is an example of an operator which acts on the discrete variable; in general, an operator A is a 4×4 matrix with elements A_{st}. If an operator does not act on the discrete variable, as, for instance, the position $\mathbf{r}$ or the momentum $\mathbf{p}$, the 4×4 identity matrix is implied as a factor to make it a genuine operator in the world of spinors; we say that the operator is *Dirac-diagonal*.

It is to be noted that the matrix structure of operators due to the spinor character of the wave functions is separate and distinct from the formation of matrix elements with respect to stationary state wave functions. The latter are defined, as before, by equations (11); however, with the present meaning of wave functions and operators, the integrals in equations (11) contain summations over spinor components as well as integrations.

Uncertainty Principle and Foundations of Quantum Mechanics

In the Dirac theory a central role is played by the set of *Dirac matrices* α_x, α_y, α_z, β satisfying

$$\left.\begin{array}{c} \alpha_x^* = \alpha_x, \quad \alpha_y^* = \alpha_y, \quad \alpha_z^* = \alpha_z, \quad \beta^* = \beta \\ \alpha_x^2 = \alpha_y^2 = \alpha_z^2 = \beta^2 = I \\ \{\alpha_x, \alpha_y\} = \{\alpha_x, \alpha_z\} = \{\alpha_y, \alpha_z\} = \{\alpha_x, \beta\} = \{\alpha_y, \beta\} = \{\alpha_z, \beta\} = 0 \end{array}\right\} \quad (34)$$

where I is the 4×4 identity matrix. The equations (34) express that the Dirac matrices are *unitary Hermitian* and anticommute with one another. While any choice of matrices satisfying equations (34) is acceptable, a commonly chosen representation is given by

$$\left.\begin{array}{c} \alpha_x = \begin{pmatrix} 0 & 0 & 0 & 1 \\ 0 & 0 & 1 & 0 \\ 0 & 1 & 0 & 0 \\ 1 & 0 & 0 & 0 \end{pmatrix}, \quad \alpha_y = \begin{pmatrix} 0 & 0 & 0 & i \\ 0 & 0 & -i & 0 \\ 0 & i & 0 & 0 \\ -i & 0 & 0 & 0 \end{pmatrix} \\ \\ \alpha_z = \begin{pmatrix} 0 & 0 & -1 & 0 \\ 0 & 0 & 0 & 1 \\ -1 & 0 & 0 & 0 \\ 0 & 1 & 0 & 0 \end{pmatrix}, \quad \beta = \begin{pmatrix} 1 & 0 & 0 & 0 \\ 0 & 1 & 0 & 0 \\ 0 & 0 & -1 & 0 \\ 0 & 0 & 0 & -1 \end{pmatrix} \end{array}\right\} \quad (35)$$

It can be shown that any other choice satisfying equations (34) can be transformed into the form (35) by a similarity transformation, which, with a corresponding transformation of the spinor wave functions, yields the same physical results.

With the help of the operators $\boldsymbol{\alpha}$, β we can now define the *Dirac Hamiltonian*

$$H_D = \beta m'c^2 + \boldsymbol{\alpha} \cdot \mathbf{p}c + V(\mathbf{r}) \quad (36)$$

where $V(\mathbf{r})$ is the potential energy of the particle. Note that both $\mathbf{p}$ and $V(\mathbf{r})$ do not act on the discrete variable. We wrote m' for the mass of the particle, rather than m, in order to avoid confusion with the wave function index m.

With the Dirac Hamiltonian in hand, and the interpretation of the wave functions as spinors, we can now translate most of the formalism of the preceding section into the proper one-particle relativistic equivalent. Up to and including equation (22) we only need to drop the individual particle labels, and replace H by H_D. The specific definition of the Hamiltonian, equation (23), is replaced by equation (36). *In lieu* of equations (25) and (26) we now get for the velocity and current density

$$\dot{\mathbf{r}} = \boldsymbol{\alpha} c \quad (37)$$

$$\mathbf{j}(\mathbf{r}') = \boldsymbol{\alpha} \rho(\mathbf{r}') \quad (38)$$

We can again demonstrate the validity of the equation of continuity. We note that a function of position $f(\mathbf{r})$ commutes with V and with $\beta m'c^2$, hence

$$[H_D, f] - c[\boldsymbol{\alpha} \cdot \mathbf{p}, f] = 0 \quad (39)$$

Using equations (4) and (5) we can derive from this the operator equation

$$\dot{f} - c\boldsymbol{\alpha} \cdot (\nabla f) = 0 \qquad (40)$$

On the other hand, taking matrix elements of equation (39) we obtain

$$(E_m - E_n)\langle \Psi_{m\alpha}|f|\Psi_{n\beta}\rangle - c\langle \Psi_{m\alpha}|[\boldsymbol{\alpha}\cdot\mathbf{p}, f]|\Psi_{n\beta}\rangle = 0 \qquad (41)$$

Equations (40) and (41) are completely equivalent.

The equation of continuity follows readily in this case if we specify $f = \rho(\mathbf{r}')$ in equation (40), make use of equation (38) for the current density, and observe that we can replace ∇ by $-\nabla'$:

$$\dot{\rho}(\mathbf{r}') + c\nabla' \cdot \mathbf{j}(\mathbf{r}') = 0 \qquad (42)$$

or, taking matrix elements,

$$ik_{mn}\rho_{m\alpha,n\beta}(\mathbf{r}') + \nabla' \cdot \mathbf{j}_{m\alpha,n\beta}(\mathbf{r}') = 0 \qquad (43)$$

Equations (42) and (43) are identical with the non-relativistic equations (30) and (31), if we omit the particle index μ in the latter. Note however that the resemblance is a formal one, since the current densities are given by different expressions, equations (26) and (38), in the two cases. This also accounts for the fact that equations (40) and (41) cannot be obtained from equations (28) and (29) by dropping the particle index μ.

4. QUANTIZATION OF A SYSTEM OF ELECTROMAGNETIC PARTICLES

The behaviour and mutual interaction of electrically charged particles can, in first approximation, be stated in terms of electrostatic forces only. If the external field is time-independent, we have a Hamiltonian system, and the quantization sketched in Section 2 applies. However if we want to do justice to the fact that we are dealing with currents as well as charges, and magnetic as well as electric fields, the system is no longer Hamiltonian: the energy cannot be expressed only in terms of the instantaneous positions and momenta of the particles, since mutual interaction of the particles involves retarded and/or advanced potentials.

Clearly it is a desirable goal to reformulate the problem in a Hamiltonian manner: this is the approach taken in the development of quantum electrodynamics. In that approach, new dynamical variables are introduced which describe the electromagnetic field, and the combined system of particles and field is considered, to which the rules of quantization are then applied. As is well known, this process builds in the self-energy of the particles, which presents considerable conceptual and calculational difficulties.

In this paper we present an alternative approach which avoids the difficulties just mentioned. It will be seen that our scheme yields an unambiguous and valid framework to deal with systems of electromagnetic particles beyond the electrostatic approximation, at least for moderate energies.

We wish to retain as much as possible the concepts and methods of the preceding sections. Although we do not have a Hamiltonian operator, we still assume that with respect to an external observer the system can be described in terms of stationary states, as expressed by equation (7a). Since we no longer have a Schrödinger or a Dirac equation, the wave functions and energies will have to be determined from some other principle; developing such a principle is one of the key objectives of this work.

Inasmuch as the energy is not a relativistic invariant, the assumption of stationary states is not a covariant one; rather we have chosen a specially simple form of representation for one Lorentz frame, namely the rest frame of the observer. Also, a consistent covariant formulation would require N time coordinates as companions to the N sets of space coordinates; our wave functions maintain N sets of space coordinates, but only one time.

While we thus adopt stationary state functions of the type (7a), we must of course demand that $\tilde{\Psi}_{m\alpha}$ and $\Psi_{m\alpha}$ are appropriate generalizations of proper relativistic one-particle wave functions. We shall restrict ourselves to Dirac particles: we permit different masses and charges, but each particle, if alone, would be represented by a four-component Dirac spinor.

In general, for N-particle wave functions the coordinate space is the direct product space of the N single particle spaces. Since the four components of a Dirac spinor may be considered to arise from a discrete variable capable of four values, the discrete aspect of the N-particle product space leads to a spinor with 4^N components. We label these components by the compound index

$$s^N = (s_1, s_2, \ldots, s_N) \tag{44}$$

where s_μ refers to the components with respect to the μth particle, so that $1 \leq s_\mu \leq 4$. The stationary state spinor wave functions are given by

$$\tilde{\Psi}_{s_1 s_2 \ldots s_N, m\alpha}(\mathbf{r}_1, \mathbf{r}_2 \ldots, \mathbf{r}_N; t) = \Psi_{s_1 s_2 \ldots s_N, m\alpha}(\mathbf{r}_1, \mathbf{r}_2, \ldots, \mathbf{r}_N) e^{-i\omega_m t} \tag{45}$$

or, in condensed notation

$$\tilde{\Psi}_{s^N, m\alpha}(\mathbf{r}^N, t) = \Psi_{s^N, m\alpha}(\mathbf{r}^N) e^{-i\omega_m t} \tag{45a}$$

As in the single-particle case, the spinor index s^N can usually be suppressed. When that is done, we understand $\tilde{\Psi}$ and Ψ to represent column vectors, the components being ordered by taking $s_1 s_2 \ldots s_N$ in dictionary order; $\tilde{\Psi}^*$ and Ψ^* are of course the complex conjugate row vectors. Again, matrix algebra applies: $\Psi^* \Phi$ is a scalar, and $\Psi \Phi^*$ is a $4^N \times 4^N$ matrix. In general, an operator A is a $4^N \times 4^N$ matrix with elements $A_{s_1 s_2 \ldots s_N, t_1 t_2 \ldots t_N}$, or, in condensed notation A_{s^N, t^N}. Taking the matrix elements of an operator with respect to the stationary state wave functions is again accomplished by equations (11), which implies complete summations over spinor components as well as integrations.

If a is any 4×4 matrix operator for a single particle with components a_{st}, we define the corresponding operator for the μth particle, a_μ, by

$$a_{\mu, s_1 s_2 \ldots s_N, t_1 t_2 \ldots t_N} = \delta_{s_1 t_1} \delta_{s_2 t_2} \cdots \delta_{s_{\mu-1} t_{\mu-1}} a_{s_\mu t_\mu} \delta_{s_{\mu+1} t_{\mu+1}} \cdots \delta_{s_N t_N} \quad (46)$$

In matrix notation, this is expressed by

$$a_\mu = I \times I \times \ldots I \times a \times I \times \ldots \times I \quad (46a)$$

where $\times$ indicates direct multiplication of matrices, and a occurs as the μth factor. It is easily seen that if a_μ and b_ν are any two operators so constructed, we have

$$[\![a_\mu, b_\nu]\!] = 0, \quad \mu \neq \nu \quad (47)$$

Among the one-particle operators which we can construct according to equation (46) are the generalized Dirac matrices α_μ, β_μ, the coordinates and momenta $\mathbf{r}_\mu$, $\mathbf{p}_\mu$, and the charge density $\rho_\mu(\mathbf{r}')$.

Since we do not have a Hamiltonian in hand, time derivatives of operators cannot be defined by equation (4), but we must in general use the matrix element form, equations (16), instead. However, for the velocity $\dot{\mathbf{r}}_\mu$ we desire a simpler definition in terms of fundamental operators associated with the μth particle; this requirement is dictated by the fundamental role played by the current density in electromagnetic phenomena. Taking our cue from the single particle Dirac formalism, we adopt the generalization of equation (37) as an *ad hoc postulate*, namely

$$\dot{\mathbf{r}}_\mu = \alpha_\mu c \quad (48)$$

Applying equation (21) we then find for the current density

$$\mathbf{j}_\mu(\mathbf{r}') = \alpha_\mu \rho_\mu(\mathbf{r}') \quad (49)$$

analogous to equation (38).

Another relation which cannot be derived in the absence of a Hamiltonian is the equation of continuity; because of the fundamental significance of charge conservation, we postulate its validity. Since $\dot{\rho}_\mu(\mathbf{r}')$ is not defined as an operator, we must take the matrix element form of the equation of continuity, namely

$$ik_{mn} \rho_{\mu, m\alpha, n\beta}(\mathbf{r}') + \nabla' \cdot \mathbf{j}_{\mu, m\alpha, n\beta}(\mathbf{r}') = 0 \quad (50)$$

This is formally identical with equation (31), but the current densities are defined differently in the two cases. If we multiply equation (50) by an arbitrary function $f(\mathbf{r}')$, integrate over $\mathbf{r}'$, and use equation (49), we obtain the equivalent form

$$(E_m - E_n)\langle \Psi_{m\alpha} | f_\mu | \Psi_{n\beta} \rangle - c \langle \Psi_{m\alpha} | [\![\alpha_\mu \cdot \mathbf{p}_\mu, f_\mu]\!] | \Psi_{n\beta} \rangle = 0 \quad (51)$$

If we take $f_\mu = 1$, equation (51) reconfirms part of the orthogonality relations, equation (9), namely for $E_m \neq E_n$. A more interesting result is obtained by taking $f_\mu \to \mathbf{r}_\mu$; we find, using the commutation relation (3)

$$i\omega_{mn} \mathbf{r}_{\mu, m\alpha, n\beta} = c \alpha_{\mu, m\alpha, n\beta} \quad (52)$$

408 Uncertainty Principle and Foundations of Quantum Mechanics

Hence $\dot{\mathbf{r}}_\mu$ also satisfies the general definition of time derivatives, equations (16), as of course it should.

We now proceed to the main task of this section, namely to write down a valid quantum mechanical expression for the energies E_m. Again, the corresponding operator, H, is not available; nevertheless it is possible to obtain a valid expression for E_m. We partition E_m according to

$$E_m = E_{D,m} + E_{I,m} \tag{53}$$

where $E_{D,m}$ is the many-particle generalization of the Dirac energy, and $E_{I,m}$ is the particle-particle interaction energy.

The Dirac energy is given by

$$E_{D,m} = \langle \Psi_{m\alpha} | H_D | \Psi_{m\alpha} \rangle \tag{54}$$

where

$$H_D = \sum_\mu (\beta_\mu m_\mu c^2 + \boldsymbol{\alpha}_\mu \cdot \mathbf{p}_\mu c) \tag{55}$$

is simply the many-particle sum of the individual particle Dirac Hamiltonians, without the potential energy. The latter has been omitted since we consider our system a *closed* system of electromagnetic particles; any additional energy is to be accounted for in the interaction energy. Obviously if $E_{D,m}$ were the only contribution to the energy (we could even add an external potential energy) we would have a Hamiltonian, namely H_D. We shall see shortly that the interaction energy $E_{I,m}$ *cannot* be written as the diagonal matrix element of an operator.

To formulate the electromagnetic interaction energy of the N-particle system we follow Wheeler and Feynman (1945, 1949). In this view direct particle-particle interaction is paramount, and the electromagnetic field plays a subordinate role. The interaction energy is then quantized, that is, reformulated in terms of matrix elements over charge and current densities; the electromagnetic field is never quantized independently. The deviations from conventional quantum electrodynamics are important; they may be summarized as follows.

(1). There is no such concept as 'the' electromagnetic field with degrees of freedom of its own. Instead, there is a collection of adjunct fields, each produced by an individual particle, and completely determined by the motion of that particle.

(2). The prevailing field acting on a given particle is determined by the sum of the fields produced by every particle *other* than the given particle. The interaction of a particle with its own field does not occur.

(3). The fields produced by the particles are always taken half-retarded and half-advanced. We note that this is the necessary and sufficient condition that energy and momentum remain conserved—and therefore defined—within a finite, but perhaps very large, volume. In classical electrodynamics, the half-retarded and half-advanced solution describes a system of particles that neither emits nor absorbs radiation.

In the light of these remarks, we now write down the classical electromagnetic interaction energy

$$E_I = \tfrac{1}{2} \sum_\mu \sum_{\nu \neq \mu} \int d\mathbf{r}'[\rho_\mu(\mathbf{r}', t)\phi_\nu(\mathbf{r}', t) - \mathbf{j}_\mu(\mathbf{r}', t) \cdot \mathbf{A}_\mu(\mathbf{r}', t)] \tag{56}$$

where in general $\rho_\mu(\mathbf{r}', t)$, $\mathbf{j}_\mu(\mathbf{r}', t)$, $\phi_\mu(\mathbf{r}', t)$, $\mathbf{A}_\mu(\mathbf{r}', t)$ are the classical time-dependent charge and current densities, and scalar and vector potentials associated with the μth particle. The latter are given by

$$\left. \begin{aligned} \phi_\mu(\mathbf{r}', t) &= \tfrac{1}{2} \int d\mathbf{r}'' R^{-1}[\rho_\mu(\mathbf{r}'', t - R/c) + \rho_\mu(\mathbf{r}'', t + R/c)] \\ \mathbf{A}_\mu(\mathbf{r}', t) &= \tfrac{1}{2} \int d\mathbf{r}'' R^{-1}[\mathbf{j}_\mu(\mathbf{r}'', t - R/c) + \mathbf{j}_\mu(\mathbf{r}'', t + R/c)] \end{aligned} \right\} \tag{57}$$

where we used the abbreviation

$$R = |\mathbf{r}' - \mathbf{r}''| \tag{58}$$

To properly quantize this scheme we must introduce the time-dependent matrix elements for the densities and potentials, as explained above, equations (11) and (12). When that is done, the quantized equations (57) can be reduced to relations between time-independent matrix elements by dividing by the exponential time factor; the result is

$$\left. \begin{aligned} \phi_{\mu,m\alpha,n\beta}(\mathbf{r}') &= \int d\mathbf{r}'' R^{-1} \cos(k_{mn}R) \rho_{\mu,m\alpha,n\beta}(\mathbf{r}'') \\ \mathbf{A}_{\mu,m\alpha,n\beta}(\mathbf{r}') &= \int d\mathbf{r}'' R^{-1} \cos(k_{mn}R) \mathbf{j}_{\mu,m\alpha,n\beta}(\mathbf{r}'') \end{aligned} \right\} \tag{59}$$

We now quantize equation (56) by introducing time-dependent matrix elements in the right-hand side, and observing the rule for product formation, equations (15). When that is done, the exponential time factors cancel, and we obtain for the quantum mechanical interaction energy

$$E_{I,m} = \tfrac{1}{2} \sum_\mu \sum_{\nu \neq \mu} \sum_{n\beta} \int d\mathbf{r}'[\rho_{\mu,m\alpha,n\beta}(\mathbf{r}')\phi_{\nu,n\beta,m\alpha}(\mathbf{r}') - \mathbf{j}_{\mu,m\alpha,n\beta}(\mathbf{r}') \cdot \mathbf{A}_{\nu,n\beta,m\alpha}(\mathbf{r}')] \tag{60}$$

We can eliminate the potentials using equations (59) to obtain

$$E_{I,m} = \tfrac{1}{2} \sum_\mu \sum_{\nu \neq \mu} \sum_{n\beta} \int d\mathbf{r}' \int d\mathbf{r}'' R^{-1} \cos(k_{mn}R)$$
$$\times [\rho_{\mu,m\alpha,n\beta}(\mathbf{r}')\rho_{\nu,n\beta,m\alpha}(\mathbf{r}'') - \mathbf{j}_{\mu,m\alpha,n\beta}(\mathbf{r}') \cdot \mathbf{j}_{\nu,n\beta,m\alpha}(\mathbf{r}'')] \tag{61}$$

In this expression, the symmetry of the interaction between particles is apparent.

We now combine the Dirac and interaction energies to obtain the total energy. Since we will soon need to consider the energies as functionals of trial wave functions, we display the wave functions and energies explicitly wherever they occur. We also need to generalize the formula in such a way that invariance with respect to unitary transformations of degenerate wave functions becomes guaranteed and transparent. The resulting formula is

$$E_m \delta_{\alpha\beta} = \langle \Psi_{m\alpha} | H_D | \Psi_{m\beta} \rangle$$

$$+ \tfrac{1}{2} \sum_\mu \sum_{\nu \neq \mu} \sum_{n\gamma} \int d\mathbf{r}' \int d\mathbf{r}'' R^{-1} \cos[(E_m - E_n)R/\hbar c]$$

$$\times [\langle \Psi_{m\alpha} | \rho_\mu(\mathbf{r}') | \Psi_{n\gamma} \rangle \langle \Psi_{n\gamma} | \rho_\nu(\mathbf{r}'') | \Psi_{m\beta} \rangle$$

$$- \langle \Psi_{m\alpha} | \mathbf{j}_\mu(\mathbf{r}') | \Psi_{n\gamma} \rangle \cdot \langle \Psi_{n\gamma} | \mathbf{j}_\nu(\mathbf{r}'') | \Psi_{m\beta} \rangle] \quad (62)$$

If the factor $\cos[(E_m - E_n)R/\hbar c]$ were absent in equation (62), we could carry out the summation over $n\gamma$ using equation (10); E_m would then reduce to the diagonal matrix element of an operator. Hence it is the retardation-advancement effect of the electromagnetic interaction, expressed in the factor $\cos[(E_m - E_n)R/\hbar c]$, which prevents the definition of a Hamiltonian for our relativistic many-particle system.

In a Hamiltonian case in general, if the energy expression is considered as a functional of a trial wave function, we can apply the variation principle to this expression: we demand that the energy be stationary to first order for any change in the wave function which preserves normalization. As is well known, this leads to the Schrödinger (or Dirac) equation. In the present case we do have an energy expression, but even formally there are important differences with the Hamiltonian case.

In the Hamiltonian case the energy E_m is expressed, *explicitly* and *bilinearly*, in terms of any one of the wave functions $\Psi_{m\alpha}$ belonging to the level m. Our equation (62) on the other hand is an *infinite* set of *implicit transcendental* equations, each equation containing all E_m and all $\Psi_{n\beta}$. We consider that equation (62) nevertheless *defines* the E_m as functionals; we must however expect that *each E_m is a functional of all $\Psi_{n\beta}$*.

We now adopt the variation principle in the following form. We demand that all E_m are stationary simultaneously to first order for any change in the *set* of trial wave functions $\Psi_{n\beta}$ which is constrained by orthonormality, equation (9), and completeness, equation (10). We furthermore expect that charge conservation, equation (51), holds for the solution of this variational problem.

A direct attempt to derive practical equations for the wave functions from this principle, as a replacement for the Schrödinger–Dirac equation, does not appear to be a simple matter. We shall see, however, that it leads to a straightforward and orderly procedure within the Pauli perturbation scheme.

5. PAULI APPROXIMATION FOR A SINGLE RELATIVISTIC PARTICLE

The Pauli approximation is based on the assumption that the wave functions and energies can be expanded using c^{-1} as the expansion parameter. Writing

$$\lambda = c^{-1} \quad (63)$$

we proceed from the assumption that the Hamiltonian, energies and wave functions are analytical functions of λ. For the sake of clarity, we shall in the following display the dependence on λ explicitly whenever appropriate; hence we shall write $H_D(\lambda)$, $E_m(\lambda)$, $\Psi_{m\alpha}(\lambda)$.

The Dirac Hamiltonian, and consequently the energies, are obviously of order λ^{-2}. It is convenient to introduce the *scaled* Hamiltonian and energies defined by

$$\eta(\lambda) = \lambda^2 H_D(\lambda) = M + P\lambda + V\lambda^2 \tag{64}$$

$$\varepsilon_m(\lambda) = \lambda^2 E_m(\lambda) \tag{65}$$

The scaled Dirac equation is

$$\eta(\lambda)\Psi_{m\alpha}(\lambda) = \varepsilon_m(\lambda)\Psi_{m\alpha}(\lambda) \tag{66}$$

In equation (64) we introduced the operators

$$M = \beta m' \tag{67}$$

$$P = \boldsymbol{\alpha} \cdot \mathbf{p} \tag{68}$$

M is the *rest mass* operator, and P may be considered the *momentum magnitude*, since the commutation properties of the Dirac matrices assure that

$$P^2 = \mathbf{p} \cdot \mathbf{p} \tag{69}$$

It is useful to establish first some properties of the solutions of the scaled Dirac equation (66) for $\lambda = 0$. In preparation for the many-particle case, we shall make a distinction between the operators M and β, although they are in this case proportional, see equation (67). We have

$$\left. \begin{array}{l} M\Psi_{m\alpha}(0) = \varepsilon_m(0)\Psi_{m\alpha}(0) \\ \beta\Psi_{m\alpha}(0) = \tau_m \Psi_{m\alpha}(0) \end{array} \right\} \tag{70}$$

Since β is unitary Hermitian, the eigenvalues are given by

$$\left. \begin{array}{l} \varepsilon_m(0) = \tau_m m' \\ \tau_m = \pm 1 \end{array} \right\} \tag{71}$$

We say that $\Psi_{m\alpha}(0)$ has *rest mass* $\pm m'$ for $\tau_m = \pm 1$. At this point the wave functions $\Psi_{m\alpha}(0)$ are still highly degenerate. In fact, all we can say so far is that for positive/negative rest mass the lower/upper pair of spinor components vanishes. This leaves completely undetermined the dependence on the space coordinate $\mathbf{r}$, and on the spin coordinate, the latter being the discrete index labelling the non-vanishing spinor components.

Because of the commutation properties of the Dirac matrices we have

$$\beta\eta(\lambda)\beta = \eta(-\lambda) \tag{72}$$

and therefore, from equation (66)

$$\eta(-\lambda)\beta\Psi_{m\alpha}(\lambda) = \varepsilon_m(\lambda)\beta\Psi_{m\alpha}(\lambda) \tag{73}$$

Clearly, the operator β maps the eigenfunctions of the level m of $\eta(\lambda)$ into those of some level, say n, of $\eta(-\lambda)$, such that $\varepsilon_m(\lambda) = \varepsilon_n(-\lambda)$. This mapping $m \to n$ remains valid for continuous changes in λ, and in particular holds for $\lambda = 0$. But for $\lambda = 0$ the mapping by β is stated in the second equation (70), so that $m = n$. Hence we have proved that

$$\varepsilon_m(\lambda) = \varepsilon_m(-\lambda) \tag{74}$$

and that $\beta \Psi_{m\alpha}(\lambda)$ is a linear transformation of $\Psi_{m\beta}(-\lambda)$. In Appendix A it is shown that the $\Psi_{m\alpha}(\lambda)$ may be chosen so that the transformation induced by β takes on the simple canonical form

$$\beta \Psi_{m\alpha}(\lambda) = \tau_m \Psi_{m\alpha}(-\lambda) \tag{75}$$

We assume in the following that equation (75) always holds.

We now put forward the *Pauli perturbation expansion* for the wave functions and energies, namely

$$\left. \begin{aligned} \Psi_{m\alpha}(\lambda) &= \sum_{p=0}^{\infty} \Psi_{m\alpha,p} \lambda^p \\ \varepsilon_m(\lambda) &= \sum_{p=0}^{\infty} \varepsilon_{m,p} \lambda^{2p} \end{aligned} \right\} \tag{76}$$

where we have limited the energy expansion to even powers of λ, because of equation (74). From equation (75) follows the important relation

$$\beta \Psi_{m\alpha,p} = \tau_m (-1)^p \Psi_{m\alpha,p} \tag{77}$$

The fact that the leading term in the scaled energy expansion is always finite points to a limitation of the Pauli perturbation expansion. Namely for a finite value of λ there are stationary states of *arbitrarily high* scaled energy; connecting these to a *finite* scaled energy for $\lambda = 0$ cannot be achieved by a uniformly convergent process. Hence the Pauli perturbation expansion is at best a semiconvergent process, and is practical perhaps only for states with energies close to the rest mass energies.

We now proceed to determine the wave functions and energies in more detail. Substitution of the expansions (76) into the scaled Dirac equation (66) should give us the necessary and sufficient equations to determine the wave functions and energies term by term for each power of λ. An immediate simplification is obtained from the relation

$$(M - \varepsilon_{m,0}) \Psi_{m\alpha,p} = \begin{cases} 0, & p \text{ even} \\ 2M \Psi_{m\alpha,p}, & p \text{ odd} \end{cases} \tag{78}$$

which is easily proved using equations (67), (70) and (77). The equation arising from λ^0 is identically satisfied; the next four are

$$\begin{aligned} 2M\Psi_{m\alpha,1} + P\Psi_{m\alpha,0} &= 0 \\ P\Psi_{m\alpha,1} + (V - \varepsilon_{m,1})\Psi_{m\alpha,0} &= 0 \\ 2M\Psi_{m\alpha,3} + P\Psi_{m\alpha,2} + (V - \varepsilon_{m,1})\Psi_{m\alpha,1} &= 0 \\ P\Psi_{m\alpha,3} + (V - \varepsilon_{m,1})\Psi_{m\alpha,2} - \varepsilon_{m,2}\Psi_{m\alpha,0} &= 0 \end{aligned} \qquad (79)$$

The first equation (79) is solved by

$$\Psi_{m\alpha,1} = K\Psi_{m\alpha,0} \qquad (80)$$

where K is the anti-Hermitian operator defined by

$$K = (2m')^{-1}\boldsymbol{\alpha}\cdot\mathbf{p}\beta \qquad (81)$$

It is interesting to note that equation (80) permits $\Psi_{m\alpha,1}$ to be calculated directly from $\Psi_{m\alpha,0}$, whereas the latter is still relatively undetermined as a solution of a highly degenerate eigenvalue problem.

Substituting $\Psi_{m\alpha,1}$ from equation (80) into the second equation (79) we obtain

$$(T + V)\Psi_{m\alpha,0} = \varepsilon_{m,1}\Psi_{m\alpha,0} \qquad (82)$$

where T is the Hermitian operator defined by

$$T = (2m')^{-1}\beta\mathbf{p}\cdot\mathbf{p} \qquad (83)$$

For wave functions of positive rest mass the operator β acts like the identity, and equation (82) becomes equivalent to the non-relativistic Schrödinger equation for a particle with spin one half, moving in a field represented by a spin-independent Hamiltonian.

We now proceed to calculate the lowest order relativistic correction to the energy, $\varepsilon_{m,2}$. Taking equation (80) and the last two equations (79) we can eliminate $\Psi_{m\alpha,1}$ and $\Psi_{m\alpha,3}$; the result is

$$(T + V - \varepsilon_{m,1})\Psi_{m\alpha,2} - [K(V - \varepsilon_{m,1})K + \varepsilon_{m,2}]\Psi_{m\alpha,0} = 0 \qquad (84)$$

Taking the scalar product with $\Psi_{m\alpha,0}$ and using equation (82) we find for the energy correction

$$\varepsilon_{m,2} = \langle \Psi_{m\alpha,0} | K(\varepsilon_{m,1} - V)K | \Psi_{m\alpha,0} \rangle \qquad (85)$$

The expression (85) can be transformed to yield terms which can be interpreted as representing specific physical effects. We eliminate $\varepsilon_{m,1}$ using equation (82), applying half of the operator $T + V$ to the right and the other half to the left; we obtain

$$\varepsilon_{m,2} = \langle \Psi_{m\alpha,0} | T' + V_{LS} | \Psi_{m\alpha,0} \rangle \qquad (86)$$

where

$$T' = \tfrac{1}{2}(K^2 T + TK^2) \tag{87}$$

$$V_{LS} = \tfrac{1}{2}[\![K,[\![K, V]\!]]\!] \tag{88}$$

The operator T' represents the *relativistic mass correction*. Using equations (81) and (83) one easily finds

$$T' = -(2m')^{-3}\beta(\mathbf{p}\cdot\mathbf{p})^2 \tag{89}$$

It is to be noted that T' as given by equation (89) is not properly Hermitian if the potential energy V is due in part to point charges. In that case the wave function $\Psi_{m\alpha,0}$ may have a mild discontinuity (cusp) at the site of a charge. Hermiticity of an operator involving $\mathbf{p}$ depends on partial integration, and the vanishing of the surface integrals occurring in that process. It turns out that for the wave function $\Psi_{m\alpha,0}$ the operator $\mathbf{p}\cdot\mathbf{p}$ causes no problem in this respect, but $(\mathbf{p}\cdot\mathbf{p})^2$ does. In a sense this difficulty was created artificially when we replaced $\varepsilon_{m,1}$ in equation (85) by $T+V$; T and V introduce compensating singular behaviour when operating on $\Psi_{m\alpha,0}$. In practice, a simple remedy for the non-Hermiticity of T' consists in applying one factor $\mathbf{p}\cdot\mathbf{p}$ to the right and the other one to the left.

For the further interpretation of V_{LS} we introduce the four-component generalization of the two-component Pauli spin vector

$$\boldsymbol{\sigma} = -\tfrac{1}{2}i\boldsymbol{\alpha}\times\boldsymbol{\alpha} \tag{90}$$

which satisfies

$$\boldsymbol{\sigma}\times\boldsymbol{\sigma} = 2i\boldsymbol{\sigma} \tag{91}$$

The angular momentum operator which represents the spin of the particle is

$$\mathbf{s} = \tfrac{1}{2}\hbar\boldsymbol{\sigma} \tag{92}$$

If $\mathbf{a}$ and $\mathbf{b}$ are any Dirac-diagonal vector operators, we have the useful relation

$$(\boldsymbol{\alpha}\cdot\mathbf{a})(\boldsymbol{\alpha}\cdot\mathbf{b}) = \mathbf{a}\cdot\mathbf{b} + i\boldsymbol{\sigma}\cdot(\mathbf{a}\times\mathbf{b}) \tag{93}$$

Equations (90), (92) and (93) are used to derive our final expression for V_{LS}:

$$V_{LS} = \tfrac{1}{2}m'^{-2}[(\boldsymbol{\nabla}V\times\mathbf{p})\cdot\mathbf{s} + \tfrac{1}{4}\hbar^2\Delta V] \tag{94}$$

The first term in equation (94) is the usual *spin–orbit coupling*; the second term, called the Darwin term, is often said to represent the *Zitterbewegung*. When V is due to point charges, ΔV vanishes except at the charge sites. Careful analysis shows that in this case ΔV yields a delta-function. Since the two terms of equation (94) really belong together, we shall in the following call V_{LS} the spin–orbit interaction operator.

Finally we note that the scheme presented so far does not make allowance for an external magnetic field, and the interaction of the spin with it: we introduced a potential V which can be specialized to Coulomb potentials due to external sources, but we did not introduce a corresponding vector potential. The reason for our omission is that it is not straightforward to introduce a vector potential from an external source without bringing about inconsistencies; in particular, the mapping $\Psi_{m\alpha}(\lambda) \leftrightarrow \Psi_{m\alpha}(-\lambda)$ with the operator β is no longer valid. Since the primary purpose of this paper is to treat interactions, we shall not dwell upon this any further. Actually, in the next section we will treat those magnetic and spin effects which arise from the interactions between electromagnetic particles, *all* of which are part of our quantum mechanical system.

6. PAULI APPROXIMATION FOR A RELATIVISTIC MANY-PARTICLE SYSTEM

We again consider the wave functions to depend parametrically on the variable $\lambda = c^{-1}$, and adopt the notation $E_m(\lambda)$, $\Psi_{m\alpha}(\lambda)$. We generalize the operators M, P, K, T, defined before for a single particle in equations (77), (68), (81) and (83), for the many-particle case:

$$\left.\begin{aligned} M &= \sum_\mu M_\mu = \sum_\mu \beta_\mu m_\mu \\ P &= \sum_\mu P_\mu = \sum_\mu \alpha_\mu \cdot \mathbf{p}_\mu \\ K &= \sum_\mu K_\mu = \sum_\mu (2m_\mu)^{-1} \alpha_\mu \cdot \mathbf{p}_\mu \beta_\mu \\ T &= \sum_\mu T_\mu = \sum_\mu (2m_\mu)^{-1} \beta_\mu \mathbf{p}_\mu \cdot \mathbf{p}_\mu \end{aligned}\right\} \quad (95)$$

Note that M, P, T are Hermitian, and K is anti-Hermitian. The following commutation and anti-commutation relations are easily verified:

$$\left.\begin{aligned} [\![M, T]\!] &= 0 \\ \{MP\} = \{M, K\} = \{P, K\} = \{P, T\} &= \{K, T\} = 0 \end{aligned}\right\} \quad (96)$$

The operators P and T can be written as commutators, namely

$$P = [\![K, M]\!] \quad (97)$$

$$T = -\tfrac{1}{2}[\![K, P]\!] = -\tfrac{1}{2}[\![K, [\![K, M]\!]]\!] \quad (98)$$

Equations (97) and (98) are specific examples of a particularly useful device: namely if we take the commutator of two operators, each of which is the sum of one-particle operators, the result is again a sum of one-particle operators. Note that there is no analogue of this for an anti-commutator.

The explicit formula for the scaled total energy, with λ displayed as a variable, is readily obtained from equation (62); we write

$$\varepsilon_m(\lambda)\delta_{\alpha\beta} = \langle \Psi_{m\alpha}(\lambda)|M+\lambda P|\Psi_{m\beta}(\lambda)\rangle$$
$$+ \tfrac{1}{2}\lambda^2 \sum_\mu \sum_{\nu\neq\mu} \sum_{n\gamma} \int d\mathbf{r}' \int d\mathbf{r}'' R^{-1} \cos\{[\varepsilon_m(\lambda)-\varepsilon_n(\lambda)]R/\hbar\lambda\}$$
$$\times [\langle \Psi_{m\alpha}(\lambda)|\rho_\mu(\mathbf{r}')|\Psi_{n\gamma}(\lambda)\rangle\langle \Psi_{n\gamma}(\lambda)|\rho_\nu(\mathbf{r}'')|\Psi_{m\beta}(\lambda)\rangle$$
$$- \langle \Psi_{m\alpha}(\lambda)|\mathbf{j}_\mu(\mathbf{r}')|\Psi_{n\gamma}(\lambda)\rangle \cdot \langle \Psi_{n\gamma}(\lambda)|\mathbf{j}_\nu(\mathbf{r}'')|\Psi_{m\beta}(\lambda)\rangle] \quad (99)$$

Similarly we rewrite the various conditions on the wave functions, with λ explicitly displayed. We obtain for the orthonormality and completeness

$$\langle \Psi_{m\alpha}(\lambda)|\Psi_{n\beta}(\lambda)\rangle = \delta_{m\alpha,n\beta} \quad (100)$$

$$\sum_{m\alpha} |\Psi_{m\alpha}(\lambda)\rangle\langle\Psi_{m\alpha}(\lambda)| = \mathscr{I} \quad (101)$$

Charge conservation is expressed by the two equivalent statements

$$[\varepsilon_m(\lambda)-\varepsilon_n(\lambda)]\rho_{\mu,m\alpha,n\beta}(\mathbf{r}',\lambda) - i\lambda\hbar\boldsymbol{\nabla}' \cdot \mathbf{j}_{\mu,m\alpha,n\beta}(\mathbf{r}',\lambda) = 0 \quad (102)$$

$$[\varepsilon_m(\lambda)-\varepsilon_n(\lambda)]\langle\Psi_{m\alpha}(\lambda)|f_\mu|\Psi_{n\beta}(\lambda)\rangle$$
$$-\lambda\langle\Psi_{m\alpha}(\lambda)|[\![P,f_\mu]\!]|\Psi_{n\beta}(\lambda)\rangle = 0 \quad (103)$$

where

$$\left.\begin{array}{l}\rho_{\mu,m\alpha,n\beta}(\mathbf{r}',\lambda) = \langle\Psi_{m\alpha}(\lambda)|\rho_\mu(\mathbf{r}')|\Psi_{n\beta}(\lambda)\rangle \\ \mathbf{j}_{\mu,m\alpha,n\beta}(\mathbf{r}',\lambda) = \langle\Psi_{m\alpha}(\lambda)|\mathbf{j}_\mu(\mathbf{r}')|\Psi_{n\beta}(\lambda)\rangle\end{array}\right\} \quad (104)$$

We emphasize again that f_μ is a function of the position $\mathbf{r}_\mu$ only.

Before proceeding further with equations (99) to (104) we introduce another useful operator, namely

$$\beta = \beta_1\beta_2\ldots\beta_N \quad (105)$$

obviously β is unitary Hermitian. We shall see shortly that this operator is the many-particle generalization of the one-particle operator β for the purpose of the mapping $\Psi(\lambda) \to \Psi(-\lambda)$. However, the many-particle β has no direct simple connection with the rest mass operator M; equation (67), or anything like it, does not hold for the many-particle case. In connection with other operators, the following commutation and anticommutation relations are useful:

$$\left.\begin{array}{l}[\![\beta,M]\!] = [\![\beta,T]\!] = 0 \\ \{\beta,P\} = \{\beta,K\} = 0 \\ [\![\beta,\rho_\mu(\mathbf{r}')]\!] = 0 \\ \{\beta,\mathbf{j}_\mu(\mathbf{r}')\} = 0\end{array}\right\} \quad (106)$$

As in the one-particle case, it is important to have in hand the wave functions and scaled energies for $\lambda = 0$. Clearly, for this limiting case our system becomes Hamiltonian, with the eigenvalue equation

$$M\Psi_{m\alpha}(0) = \varepsilon_m(0)\Psi_{m\alpha}(0) \tag{107}$$

Equation (107) is formally identical with equation (70) for the one-particle case. Actually, equation (107) is *separable* into N such one-particle problems. We take for the wave functions $\Psi_{m\alpha}(0)$ the *direct products* of the one-particle solutions; the eigenvalues are given by

$$\left.\begin{array}{l} \varepsilon_{m,0} = \sum_\mu \tau_{m,\mu} m_\mu \\ \\ \tau_{m,\mu} = \pm 1 \end{array}\right\} \tag{108}$$

To obtain all possible eigenvalues, one must in general take all possible combinations of $+$ and $-$ signs for the different values of μ; this corresponds to the individual particle spinors having positive and negative rest masses, respectively. We say that the spinors with $\tau_{m,\mu} = -1$ constitute *holes* in the wave function $\Psi_{m\alpha}(0)$. If any of the particles have equal masses, there is additional degeneracy for $\varepsilon_m(0)$, but that need not concern us at this moment. For the direct product wave functions we have

$$\left.\begin{array}{l} \beta\Psi_{m\alpha}(0) = \tau_m \Psi_{m\alpha}(0) \\ \\ \tau_m = (-1)^{\frac{1}{2}(N-\sum_\mu \tau_{m,\mu})} \end{array}\right\} \tag{109}$$

Note that according to equations (107) and (109) the $\Psi_{m\alpha}(0)$ are simultaneous eigenfunctions of M and β; this is of course possible because M and β commute, see equations (106). Evidently $\tau_m = 1$ or $\tau_m = -1$ holds for a wave function with an *even* or *odd* number of holes, respectively. Accordingly we call β the *hole parity operator*. The wave function $\Psi_{m\alpha}(\lambda)$ *does not* possess hole parity, but $\Psi_{m\alpha}(0)$ does possess it according to equations (109). If an operator commutes/anticommutes with β it will preserve/reverse hole parity.

We now return to equations (99) to (104) for $\lambda \neq 0$. With the help of equations (106) it is not difficult to see that equations (99) to (104) remain valid if we replace each wave function symbol Ψ by $\beta\Psi$, and change the sign of λ whenever λ occurs as an *argument* of ε, ρ, $\mathbf{j}$ or Ψ (not when λ appears *algebraically*). Suppressing indices for a moment, we can say therefore that the $\varepsilon(-\lambda)$ are the same functions of the $\beta\Psi(-\lambda)$ as the $\varepsilon(\lambda)$ are of the $\Psi(\lambda)$. Clearly the *physical* solutions for the two cases, which occur when all $\varepsilon(\lambda)$ and $\varepsilon(-\lambda)$ are stationary, must be mappings of each other. More precisely, there is a correspondence $m \leftrightarrow n$ so that $\varepsilon_m(-\lambda) = \varepsilon_n(\lambda)$, and the $\beta\Psi_{m\alpha}(\lambda)$ are linear transformations of the $\Psi_{n\beta}(-\lambda)$. Continuity for $\lambda = 0$ together with equations (109) establishes $m = n$; and Appendix A again justifies the canonical form of the transformation induced by β.

In summary, we have proved that for the many-particle case the dependence of the energies and wave functions on the expansion parameter λ has again the

418 Uncertainty Principle and Foundations of Quantum Mechanics

special properties expressed by equations (74) and (75). We conclude, then, that the Pauli perturbation expansion, as expressed by equations (76), also holds for the many-particle case, as does equation (77). The latter equation takes on new significance for the many-particle case; it expresses that the expansion functions $\Psi_{m\alpha,p}$ have definite hole parity, alternating for even and odd p.

For the charge and current densities the mapping by β yields the simple result

$$\left. \begin{array}{l} \rho_{\mu,m\alpha,n\beta}(\mathbf{r}', \lambda) = \tau_m \tau_n \rho_{\mu,m\alpha,n\beta}(\mathbf{r}', -\lambda) \\ \mathbf{j}_{\mu,m\alpha,n\beta}(\mathbf{r}', \lambda) = -\tau_m \tau_n \mathbf{j}_{\mu,m\alpha,n\beta}(\mathbf{r}', -\lambda) \end{array} \right\} \tag{110}$$

We now pursue in more detail the consequences of the perturbation expansion (76) with respect to orthonormality, completeness, charge and current densities, charge conservation and the energy, equations (99) to (104). For the orthonormality and completeness conditions the results are simple, namely

$$\sum_{q=0}^{p} \langle \Psi_{m\alpha,p-q} | \Psi_{n\beta,q} \rangle = \delta_{p0} \delta_{m\alpha,n\beta} \tag{111}$$

$$\sum_{m\alpha} \sum_{q=0}^{p} |\Psi_{m\alpha,p-q}\rangle\langle\Psi_{m\alpha,q}| = \delta_{p0} \mathcal{I} \tag{112}$$

For the charge and current densities we obtain the expansions

$$\left. \begin{array}{l} \rho_{\mu,m\alpha,n\beta}(\mathbf{r}', \lambda) = \sum\limits_{p=0}^{\infty} \rho_{\mu,m\alpha,n\beta,p}(\mathbf{r}')\lambda^p \\ \mathbf{j}_{\mu,m\alpha,n\beta}(\mathbf{r}', \lambda) = \sum\limits_{p=0}^{\infty} \mathbf{j}_{\mu,m\alpha,n\beta,p}(\mathbf{r}')\lambda^p \end{array} \right\} \tag{113}$$

where

$$\left. \begin{array}{l} \rho_{\mu,m\alpha,n\beta,p}(\mathbf{r}') = \sum\limits_{q=0}^{p} \langle \Psi_{m\alpha,p-q} | \rho_\mu(\mathbf{r}') | \Psi_{n\beta,q} \rangle \\ \mathbf{j}_{\mu,m\alpha,n\beta,p}(\mathbf{r}') = \sum\limits_{q=0}^{p} \langle \Psi_{m\alpha,p-q} | \mathbf{j}_\mu(\mathbf{r}') | \Psi_{n\beta,q} \rangle \end{array} \right\} \tag{114}$$

For the charge conservation condition we get different equations for the even and odd powers of λ, namely

$$\left. \begin{array}{l} \sum\limits_{q=0}^{p} (\varepsilon_{m,q} - \varepsilon_{n,q})\rho_{\mu,m\alpha,n\beta,2p-2q}(\mathbf{r}') - i\hbar \nabla' \cdot \mathbf{j}_{\mu,m\alpha,n\beta,2p-1}(\mathbf{r}') = 0 \\ \sum\limits_{q=0}^{p} (\varepsilon_{m,q} - \varepsilon_{n,q})\rho_{\mu,m\alpha,n\beta,2p-2q+1}(\mathbf{r}') - i\hbar \nabla' \cdot \mathbf{j}_{\mu,m\alpha,n\beta,2p}(\mathbf{r}') = 0 \end{array} \right\} \tag{115}$$

or the equivalent form

$$\left.\begin{array}{l}\sum_{q=0}^{p}(\varepsilon_{m,q}-\varepsilon_{n,q})\sum_{r=0}^{2p-2q}\langle\Psi_{m\alpha,2p-2q-r}|f_\mu|\Psi_{n\beta,r}\rangle\\[6pt]\quad-\sum_{r=0}^{2p-1}\langle\Psi_{m\alpha,2p-r-1}|[\![P,f_\mu]\!]|\Psi_{n\beta,r}\rangle=0\\[10pt]\sum_{q=0}^{p}(\varepsilon_{m,q}-\varepsilon_{n,q})\sum_{r=0}^{2p-2q+1}\langle\Psi_{m\alpha,2p-2q-r+1}|f_\mu|\Psi_{n\beta,r}\rangle\\[6pt]\quad-\sum_{r=0}^{2p}\langle\Psi_{m\alpha,2p-r}|[\![P,f_\mu]\!]|\Psi_{n\beta,r}\rangle=0\end{array}\right\} \quad (116)$$

In the first equations (115) and (116) the current term is meaningless for $p=0$; the correct interpretation is to omit the offending term.

In order to apply the perturbation expansion to the energy formula (99) it is convenient to introduce the functions

$$F_{nm\alpha\beta}(\mathbf{r}',\mathbf{r}'',\lambda)=$$

$$\tfrac{1}{2}\sum_\mu\sum_{\nu\neq\mu}\sum_\gamma[\rho_{\mu,m\alpha,n\gamma}(\mathbf{r}',\lambda)\rho_{\nu,n\gamma,m\beta}(\mathbf{r}'',\lambda)-\mathbf{j}_{\mu,m\alpha,n\gamma}(\mathbf{r}',\lambda)\cdot\mathbf{j}_{\nu,n\gamma,m\beta}(\mathbf{r}'',\lambda)] \quad (117)$$

Note that, because of equations (110)

$$F_{nm\alpha\beta}(\mathbf{r}',\mathbf{r}'',\lambda)=F_{nm\alpha\beta}(\mathbf{r}',\mathbf{r}'',-\lambda) \quad (118)$$

Using equation (117) we can rewrite the energy formula (99) in the simpler form

$$\varepsilon_m(\lambda)\delta_{\alpha\beta}=\langle\Psi_{m\alpha}(\lambda)|M+\lambda P|\Psi_{m\beta}(\lambda)\rangle$$
$$+\lambda^2\sum_n\int d\mathbf{r}'\int d\mathbf{r}''R^{-1}\cos\{[\varepsilon_m(\lambda)-\varepsilon_n(\lambda)]R/\hbar\lambda\}F_{nm\alpha\beta}(\mathbf{r}',\mathbf{r}'',\lambda)$$
$$(119)$$

We now apply the perturbation expansion (76) to equation (119). The Dirac term can be handled like the orthonormality condition, and poses no new problem. The interaction term is also straightforward with respect to the function $F_{nm\alpha\beta}(\mathbf{r}',\mathbf{r}'',\lambda)$, for which we have the expansion

$$F_{nm\alpha\beta}(\mathbf{r}',\mathbf{r}'',\lambda)=\sum_{p=0}^\infty F_{nm\alpha\beta,p}(\mathbf{r}',\mathbf{r}'')\lambda^{2p} \quad (120)$$

where

$$F_{nm\alpha\beta,p}(\mathbf{r}',\mathbf{r}'')=\tfrac{1}{2}\sum_{q=0}^{2p}\sum_\mu\sum_{\nu\neq\mu}\sum_\gamma[\rho_{\mu,m\alpha,n\gamma,2p-q}(\mathbf{r}')\rho_{\nu,n\gamma,m\beta,q}(\mathbf{r}'')$$
$$-\mathbf{j}_{\mu,m\alpha,n\gamma,2p-q}(\mathbf{r}')\cdot\mathbf{j}_{\nu,n\gamma,m\beta,q}(\mathbf{r}'')] \quad (121)$$

420 Uncertainty Principle and Foundations of Quantum Mechanics

The cosine factor in equation (119) poses a new problem, because of the manner in which λ appears in the argument. We find for this argument the expansion

$$[\varepsilon_m(\lambda) - \varepsilon_n(\lambda)]R/\hbar\lambda = R\hbar^{-1} \sum_{p=0}^{\infty} (\varepsilon_{m,p} - \varepsilon_{n,p})\lambda^{2p-1} \quad (122)$$

The first term in this expansion is of order λ if $\varepsilon_{m,0} = \varepsilon_{n,0}$, and of order λ^{-1} if $\varepsilon_{m,0} \neq \varepsilon_{n,0}$. The cosine will have to be treated quite differently for these two cases, and the summation over n in equation (119) has to be split up accordingly. For this purpose it is useful to define the sets $\mathscr{F}(m)$ and $\mathscr{G}(m)$ by means of

$$\left. \begin{array}{ll} n \subset \mathscr{F}(m), & \varepsilon_{n,0} = \varepsilon_{m,0} \\ n \subset \mathscr{G}(m), & \varepsilon_{n,0} \neq \varepsilon_{m,0} \end{array} \right\} \quad (123)$$

so that

$$\sum_n = \sum_{n \subset \mathscr{F}(m)} + \sum_{n \subset \mathscr{G}(m)} \quad (124)$$

If the argument of the cosine is of order λ, we can use the power series expansion of the cosine. Hence we have up to order λ^4

$$\cos\{[\varepsilon_m(\lambda) - \varepsilon_n(\lambda)]R/\hbar\} = 1 - \tfrac{1}{2}\hbar^{-2}(\varepsilon_{m,1} - \varepsilon_{n,1})^2 R^2 \lambda^2 + \mathcal{O}(\lambda^4), \quad n \subset \mathscr{F}(m) \quad (125)$$

On the other hand if the argument of the cosine is of order λ^{-1}, the power series expansion of the cosine is useless. We can however develop the *integral* in a power series of λ. We start from an asymptotic expansion which is proved in Appendix B, namely the expansion of the *operator* $R^{-1}\cos(kR)$ for $k \to \infty$:

$$R^{-1}\cos(kR) = -4\pi\delta(\mathbf{r}' - \mathbf{r}'') \sum_{p=0}^{\infty} k^{-2p-2} (\nabla' \cdot \nabla'')^p \quad (126)$$

We have to apply this formula, where k is itself an odd power series in λ starting with λ^{-1}. The result is, to order λ^4

$$R^{-1}\cos\{[\varepsilon_m(\lambda) - \varepsilon_n(\lambda)]R/\hbar\lambda\}$$
$$= -4\pi\hbar^2(\varepsilon_{m,0} - \varepsilon_{n,0})^{-2} \delta(\mathbf{r}' - \mathbf{r}'')\lambda^2 + \mathcal{O}(\lambda^4), \quad n \subset \mathscr{G}(m) \quad (127)$$

We can now put together the energy expansion up to order λ^4. Because of hole parity, we need to consider only even powers; we obtain

$$\begin{aligned}
\varepsilon_{m,0}\delta_{\alpha\beta} &= \langle \Psi_{m\alpha,0}|M|\Psi_{m\beta,0}\rangle \\
\varepsilon_{m,1}\delta_{\alpha\beta} &= \sum_{q=0}^{2} \langle \Psi_{m\alpha,2-q}|M|\Psi_{m\beta,q}\rangle + \sum_{q=0}^{1} \langle \Psi_{m\alpha,1-q}|P|\Psi_{m\beta,q}\rangle \\
&\quad + \sum_{n\subset\mathscr{F}(m)} \int d\mathbf{r}' \int d\mathbf{r}'' R^{-1} F_{nm\alpha\beta,0}(\mathbf{r}',\mathbf{r}'') \\
\varepsilon_{m,2}\delta_{\alpha\beta} &= \sum_{q=0}^{4} \langle \Psi_{m\alpha,4-q}|M|\Psi_{m\beta,q}\rangle + \sum_{q=0}^{3} \langle \Psi_{m\alpha,3-q}|P|\Psi_{m\beta,q}\rangle \\
&\quad + \sum_{n\subset\mathscr{F}(m)} \int d\mathbf{r}' \int d\mathbf{r}'' R^{-1} F_{nm\alpha\beta,1}(\mathbf{r}',\mathbf{r}'') \\
&\quad - \tfrac{1}{2}\hbar^{-2} \sum_{n\subset\mathscr{F}(m)} (\varepsilon_{m,1}-\varepsilon_{n,1})^2 \int d\mathbf{r}' \int d\mathbf{r}'' R F_{nm\alpha\beta,0}(\mathbf{r}',\mathbf{r}'') \\
&\quad - 4\pi\hbar^2 \sum_{n\subset\mathscr{G}(m)} (\varepsilon_{m,0}-\varepsilon_{n,0})^{-2} \int d\mathbf{r}' \int d\mathbf{r}'' \delta(\mathbf{r}'-\mathbf{r}'') F_{nm\alpha\beta,0}(\mathbf{r}',\mathbf{r}'')
\end{aligned} \quad (128)$$

In the last term of $\varepsilon_{m,2}$ we could of course have carried out one integration; however the expression as given will turn out to be more convenient in the development which follows.

We are now ready to apply the variation principle. In the case of a perturbation expansion, the proper procedure is to apply the variation process successively for each power of λ. We demand of course that the wave functions are constrained by orthonormality; we expect that our variational solutions can be chosen so that they are eigenfunctions of β, and that they satisfy the completeness and the charge conservation conditions.

We apply the variation principle to the first equation (128), which arises from λ^0; we also must honour the corresponding orthonormality constraint, equation (111) for $p = 0$. This is the usual Hamiltonian variational problem, and we obtain of course again equation (107). The solutions $\Psi_{m\alpha,0}$ are taken as direct products of one-electron spinors; $\varepsilon_{m,0}$ is the rest mass of $\Psi_{m\alpha,0}$. If there are identical particles, proper linear combinations of direct products can always be taken so that the $\Psi_{m\alpha,0}$ have appropriate symmetry properties with respect to permutations of particles. We feel confident that the eigenfunctions of M span the entire Hilbert space, so that the completeness condition (112) is satisfied for $p = 0$. In Appendix C it is shown that the charge conservation conditions which depend on $\Psi_{m\alpha,0}$ are also satisfied.

We now turn to the second equation (128). First we evaluate the interaction term. For the charge and current densities we have the obvious identities

$$\left.\begin{aligned}
[\![M, \rho_\mu(\mathbf{r}')]\!] &= 0 \\
\mathbf{j}_\mu(\mathbf{r}') &= (2m_\mu)^{-1}[\![M, \beta_\mu \mathbf{j}_\mu(\mathbf{r}')]\!]
\end{aligned}\right\} \quad (129)$$

Taking matrix elements with the zero-order wave functions we obtain

$$(\varepsilon_{m,0} - \varepsilon_{n,0})\rho_{\mu,m\alpha,n\beta,0}(\mathbf{r}') = 0$$

$$\mathbf{j}_{\mu,m\alpha,n\beta,0}(\mathbf{r}') = (2m_\mu)^{-1}(\varepsilon_{m,0} - \varepsilon_{n,0})\langle\Psi_{m\alpha,0}|\beta_\mu \mathbf{j}_\mu(\mathbf{r}')|\Psi_{n\beta,0}\rangle \quad (130)$$

We conclude from equations (130) that

$$\rho_{\mu,m\alpha,n\beta,0}(\mathbf{r}') = 0, \quad n \subset \mathcal{G}(m)$$
$$\mathbf{j}_{\mu,m\alpha,n\beta,0}(\mathbf{r}') = 0, \quad n \subset \mathcal{F}(m) \quad (131)$$

Using these relations we may write

$$\sum_{n\subset\mathcal{F}(m)} F_{nm\alpha\beta,0}(\mathbf{r}',\mathbf{r}'')$$

$$= \tfrac{1}{2}\sum_{n\subset\mathcal{F}(m)}\sum_\mu \sum_{\nu\neq\mu}\sum_\gamma \rho_{\mu,m\alpha,n\gamma,0}(\mathbf{r}')\rho_{\nu,n\gamma,m\beta,0}(\mathbf{r}'')$$

$$= \tfrac{1}{2}\sum_\mu \sum_{\nu\neq\mu}\sum_{n\gamma}\langle\Psi_{m\alpha,0}|\rho_\mu(\mathbf{r}')|\Psi_{n\gamma,0}\rangle\langle\Psi_{n\gamma,0}|\rho_\nu(\mathbf{r}'')|\Psi_{m\beta,0}\rangle$$

$$= \tfrac{1}{2}\sum_\mu \sum_{\nu\neq\mu}\langle\Psi_{m\alpha,0}|\rho_\mu(\mathbf{r}')\rho_\nu(\mathbf{r}'')|\Psi_{m\beta,0}\rangle$$

$$= \tfrac{1}{2}\sum_\mu \sum_{\nu\neq\mu} e_\mu e_\nu \langle\Psi_{m\alpha,0}|\delta(\mathbf{r}'-\mathbf{r}_\mu)\delta(\mathbf{r}''-\mathbf{r}_\nu)|\Psi_{m\beta,0}\rangle \quad (132)$$

In obtaining the final result (132), we first dropped the current density term on account of the second equation (131); next we extended the summation over states to *all* states, on account of the first equation (131); next we used the completeness relation, equation (112) for $p = 0$, to carry out the sum over states in closed form; and last, we used the definition of the charge density, equation (19). Using this result in the interaction term in the second equation (128), we can carry out the integrations, and obtain

$$\sum_{n\subset\mathcal{F}(m)}\int d\mathbf{r}' \int d\mathbf{r}'' R^{-1} F_{nm\alpha\beta,0}(\mathbf{r}',\mathbf{r}'') = \langle\Psi_{m\alpha,0}|V|\Psi_{m\beta,0}\rangle \quad (133)$$

where V is the usual *Coulomb interaction* operator, namely

$$V = \tfrac{1}{2}\sum_\mu \sum_{\nu\neq\mu} e_\mu e_\nu r_{\mu\nu}^{-1} \quad (134)$$

with the common abbreviation

$$r_{\mu\nu} = |\mathbf{r}_{\mu\nu}| = |\mathbf{r}_\mu - \mathbf{r}_\nu| \quad (135)$$

If we substitute the Coulomb energy expression (133) into the second equation (128), we note that the same result would have been obtained if the scaled Dirac Hamiltonian had had an additional term $\lambda^2 V$. Hence up to order λ^2 and the energy $\varepsilon_{m,1}$ we *do* have a Hamiltonian formulation as a valid description of the electromagnetic interaction. The second equation (128) now becomes explicitly

$$\varepsilon_{m,1}\delta_{\alpha\beta} = \langle\Psi_{m\alpha,2}|M|\Psi_{m\beta,0}\rangle + \langle\Psi_{m\alpha,1}|M|\Psi_{m\beta,1}\rangle + \langle\Psi_{m\alpha,0}|M|\Psi_{m\beta,2}\rangle$$
$$+ \langle\Psi_{m\alpha,1}|P|\Psi_{m\beta,0}\rangle + \langle\Psi_{m\alpha,0}|P|\Psi_{m\beta,1}\rangle + \langle\Psi_{m\alpha,0}|V|\Psi_{m\beta,0}\rangle \quad (136)$$

We eliminate $\Psi_{m\alpha,2}$ and $\Psi_{m\beta,2}$ by taking the orthonormality condition, equation (111), for $p = 2$, multiplying by $\varepsilon_{m,0}$, and subtracting the result from equation (136). Making use also of the commutator expressions for P and T, equations (97) and (98), and of equation (107), we obtain after some manipulation

$$\varepsilon_{m,1}\delta_{\alpha\beta} = \langle\Psi_{m\alpha,1} - K\Psi_{m\alpha,0}|M - \varepsilon_{m,0}|\Psi_{m\beta,1} - K\Psi_{m\beta,0}\rangle + \langle\Psi_{m\alpha,0}|T+V|\Psi_{m\beta,0}\rangle \quad (137)$$

We now apply the variational procedure to equation (137). We demand that $\varepsilon_{m,1}$ be stationary for variations in $\Psi_{m\alpha,0}$ and $\Psi_{m\alpha,1}$, maintaining the relevant orthonormality constraints, equation (111) for $p = 0$ and $p = 1$. As far as the variation is concerned, we specifically do *not* impose the other known conditions which the wave functions have to satisfy: completeness, charge conservation, $\Psi_{m\alpha,0}$ eigenfunctions of M and β, $\Psi_{m\alpha,1}$ eigenfunction of β. It will be seen, however, that the variational solutions obtained permit a choice so that all these other conditions are indeed satisfied. The variations with respect to $\Psi_{m\alpha,1}$ and $\Psi_{m\alpha,0}$ yield, after the usual determination of the Lagrange multipliers

$$(M - \varepsilon_{m,0})(\Psi_{m\alpha,1} - K\Psi_{m\alpha,0}) = 0 \quad (138)$$

$$(T+V)\Psi_{m\alpha,0} = \varepsilon_{m,1}\Psi_{m\alpha,0} \quad (139)$$

Since T and V commute with both M and β, equation (139) permits a solution so that $\Psi_{m\alpha,0}$ is a simultaneous eigenfunction of M, β, and $T+V$. We can also say that equation (139) removes a substantial part of the degeneracy inherent in the $\Psi_{m\alpha,0}$ up to this point. For the *no-hole* solutions, equation (139) is equivalent to the non-relativistic Schrödinger equation for N particles with spin one half.

The general solution of equation (138) is

$$\Psi_{m\alpha,1} = K\Psi_{m\alpha,0} + \Psi'_{m\alpha,0} \quad (140)$$

where $\Psi'_{m\alpha,0}$ is *any* function of rest mass $\varepsilon_{m,0}$. The requirement of hole parity however demands that we restrict the solutions (140) to those where $\Psi_{m\alpha,0}$ and $\Psi'_{m\alpha,0}$ have *opposite* hole parity. Hence we must have

$$\Psi'_{m\alpha,0} = 0 \quad (141)$$

unless there exists an *even* subset of the N particles with the same mass as some *odd* subset. Obviously this is a rather special case of accidental degeneracy. It cannot occur for a system of identical particles; since this is our primary concern, we assume from now on that equation (138) holds, so that

$$\Psi_{m\alpha,1} = K\Psi_{m\alpha,0} \quad (142)$$

Using equation (142), we easily verify the relevant orthonormality and completeness conditions, equations (111) and (112) for $p=1$. For charge conservation we again refer to Appendix C.

We now turn to the third equation (128), which arises from the terms of order λ^4. To evaluate the contribution due to $F_{nm\alpha\beta,1}$ we start from the identities

$$\left.\begin{aligned}[\![K,\rho_\mu(\mathbf{r}')]\!] &= (2m_\mu)^{-1}[\![M, \beta_\mu[\![K, \rho_\mu(\mathbf{r}')]\!]]\!] \\ [\![M, [\![K, \mathbf{j}_\mu(\mathbf{r}')]\!]]\!] &= 0 \end{aligned}\right\} \quad (143)$$

The proofs are elementary. We note further that from equations (114) and (142) follows

$$\left.\begin{aligned}\rho_{\mu,m\alpha,n\beta,1}(\mathbf{r}') &= -\langle\Psi_{m\alpha,0}|[\![K, \rho_\mu(\mathbf{r}')]\!]|\Psi_{n\beta,0}\rangle \\ \mathbf{j}_{\mu,m\alpha,n\beta,1}(\mathbf{r}') &= -\langle\Psi_{m\alpha,0}|[\![K, \mathbf{j}_\mu(\mathbf{r}')]\!]|\Psi_{n\beta,0}\rangle \end{aligned}\right\} \quad (144)$$

Hence taking matrix elements of equations (143) with the zero-order wave functions yields

$$\left.\begin{aligned}\rho_{\mu,m\alpha,n\beta,1}(\mathbf{r}') &= -(2m_\mu)^{-1}(\varepsilon_{m,0}-\varepsilon_{n,0})\langle\Psi_{m\alpha,0}|\beta_\mu[\![K, \rho_\mu(\mathbf{r}')]\!]|\Psi_{n\beta,0}\rangle \\ (\varepsilon_{m,0}-\varepsilon_{n,0})\mathbf{j}_{\mu,m\alpha,n\beta,1}(\mathbf{r}') &= 0 \end{aligned}\right\} \quad (145)$$

from which we conclude

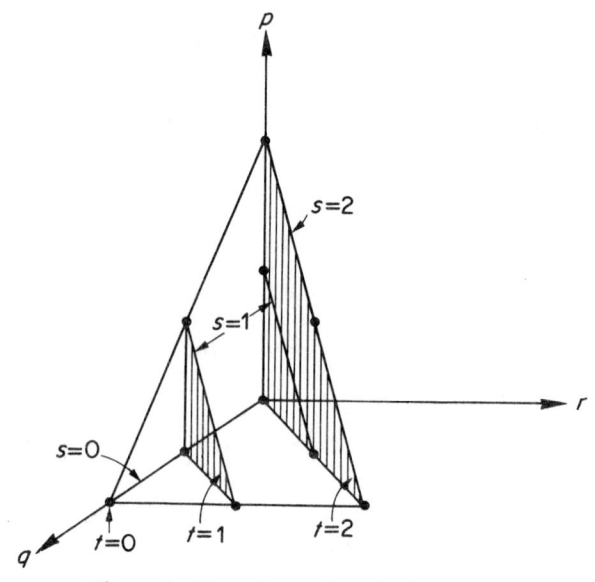

$$\left.\begin{aligned}\rho_{\mu,m\alpha,n\beta,1}(\mathbf{r}') &= 0, & n \subset \mathscr{F}(m) \\ \mathbf{j}_{\mu,m\alpha,n\beta,1}(\mathbf{r}') &= 0, & n \subset \mathscr{G}(m) \end{aligned}\right\} \quad (146)$$

Figure 1 Transformation of triple sum

With the help of equations (112), (114), (121), (131), (142) and (146) we proceed to simplify $F_{nm\alpha\beta,1}$, using techniques similar to those used to derive equation (132). We obtain

$$\sum_{n \subset \mathcal{F}(m)} F_{nm\alpha\beta,1}(\mathbf{r}', \mathbf{r}'')$$

$$= \tfrac{1}{2} \sum_{n \subset \mathcal{F}(m)} \sum_{\mu} \sum_{\nu \neq \mu} \sum_{\gamma} [\rho_{\mu,m\alpha,n\gamma,2}(\mathbf{r}')\rho_{\nu,n\gamma,m\beta,0}(\mathbf{r}'')$$

$$+ \rho_{\mu,m\alpha,n\gamma,0}(\mathbf{r}')\rho_{\nu,n\gamma,m\beta,2}(\mathbf{r}'')$$

$$- \mathbf{j}_{\mu,m\alpha,n\gamma,1}(\mathbf{r}') \cdot \mathbf{j}_{\nu,n\gamma,m\beta,1}(\mathbf{r}'')]$$

$$= \tfrac{1}{2} \sum_{\mu} \sum_{\nu \neq \mu} \sum_{n\gamma} \left[\sum_{p=0}^{2} \rho_{\mu,m\alpha,n\gamma,2-p}(\mathbf{r}')\rho_{\nu,n\gamma,m\beta,p}(\mathbf{r}'') \right.$$

$$- \rho_{\mu,m\alpha,n\gamma,1}(\mathbf{r}')\rho_{\nu,n\gamma,m\beta,1}(\mathbf{r}'')$$

$$\left. - \mathbf{j}_{\mu,m\alpha,n\gamma,1}(\mathbf{r}') \cdot \mathbf{j}_{\nu,n\gamma,m\beta,1}(\mathbf{r}'') \right]$$

$$= \tfrac{1}{2} \sum_{\mu} \sum_{\nu \neq \mu} \sum_{n\gamma} \left[\sum_{p=0}^{2} \sum_{q=0}^{2-p} \sum_{r=0}^{p} \langle \Psi_{m\alpha,2-p-q} | \rho_\mu(\mathbf{r}') | \Psi_{n\gamma,q} \rangle \langle \Psi_{n\gamma,r} | \rho_\nu(\mathbf{r}'') | \Psi_{m\beta,p-r} \rangle \right.$$

$$- \langle \Psi_{m\alpha,0} | [K, \rho_\mu(\mathbf{r}')] | \Psi_{n\gamma,0} \rangle \langle \Psi_{n\gamma,0} | [K, \rho_\nu(\mathbf{r}'')] | \Psi_{m\beta,0} \rangle$$

$$\left. - \langle \Psi_{m\alpha,0} | [K, \mathbf{j}_\mu(\mathbf{r}')] | \Psi_{n\gamma,0} \rangle \cdot \langle \Psi_{n\gamma,0} | [K, \mathbf{j}_\nu(\mathbf{r}'')] | \Psi_{m\beta,0} \rangle \right]$$

$$= \tfrac{1}{2} \sum_{\mu} \sum_{\nu \neq \mu} \sum_{n\gamma} \left[\sum_{t=0}^{2} \sum_{s=0}^{t} \sum_{r=0}^{s} \langle \Psi_{m\alpha,t-s} | \rho_\mu(\mathbf{r}') | \Psi_{n\gamma,s-r} \rangle \langle \Psi_{n\gamma,r} | \rho_\nu(\mathbf{r}'') | \Psi_{m\beta,2-t} \rangle \right.$$

$$- \langle \Psi_{m\alpha,0} | [K_\mu, \rho_\mu(\mathbf{r}')] | \Psi_{n\gamma,0} \rangle \langle \Psi_{n\gamma,0} | [K_\nu, \rho_\nu(\mathbf{r}'')] | \Psi_{m\beta,0} \rangle$$

$$\left. - \langle \Psi_{m\alpha,0} | [K_\mu, \mathbf{j}_\mu(\mathbf{r}')] | \Psi_{n\gamma,0} \rangle \cdot \langle \Psi_{n\gamma,0} | [K_\nu, \mathbf{j}_\nu(\mathbf{r}'')] | \Psi_{m\beta,0} \rangle \right]$$

$$= \tfrac{1}{2} \sum_{\mu} \sum_{\nu \neq \mu} \left[\sum_{t=0}^{2} \langle \Psi_{m\alpha,t} | \rho_\mu(\mathbf{r}')\rho_\nu(\mathbf{r}'') | \Psi_{m\beta,2-t} \rangle \right.$$

$$\left. - \langle \Psi_{m\alpha,0} | [K_\mu, [K_\nu, \rho_\mu(\mathbf{r}')\rho_\nu(\mathbf{r}'') + \mathbf{j}_\mu(\mathbf{r}') \cdot \mathbf{j}_\nu(\mathbf{r}'')]] | \Psi_{m\beta,0} \rangle \right] \quad (147)$$

In deriving this result we converted the triple sum over p, q, r by the substitution

$$\left. \begin{array}{l} p = 2 + r - t \\ q = s - r \end{array} \right\} \quad (148)$$

the corresponding transformation of the summation limits is

$$\sum_{p=0}^{2} \sum_{q=0}^{2-p} \sum_{r=0}^{p} = \sum_{t=0}^{2} \sum_{s=0}^{t} \sum_{r=0}^{s} \quad (149)$$

as illustrated in Figure 1.

426   Uncertainty Principle and Foundations of Quantum Mechanics

Integration of (147) over $\mathbf{r}'$ and $\mathbf{r}''$ with the factor R^{-1} readily yields for the first interaction term of $\varepsilon_{m,2}$

$$\sum_{n\subset\mathscr{F}(m)}\int d\mathbf{r}'\int d\mathbf{r}'' R^{-1}F_{nm\alpha\beta,1}(\mathbf{r}',\mathbf{r}'')=\sum_{t=0}^{2}\langle\Psi_{m\alpha,t}|V|\Psi_{m\beta,2-t}\rangle$$

$$-\tfrac{1}{2}\sum_{\mu}\sum_{\nu\neq\mu}e_\mu e_\nu\langle\Psi_{m\alpha,0}|[K_\mu,[K_\nu,r_{\mu\nu}^{-1}(1+\boldsymbol{\alpha}_\mu\cdot\boldsymbol{\alpha}_\nu)]]|\Psi_{m\beta,0}\rangle \qquad (150)$$

In order to evaluate the second interaction term of $\varepsilon_{m,2}$ we need to absorb the factor $(\varepsilon_{m,1}-\varepsilon_{n,1})^2$ into the matrix elements. Such a relation is conveniently provided by the equation of continuity, namely the first equation (115) for $p=1$, $n\subset\mathscr{F}(m)$, yielding

$$(\varepsilon_{m,1}-\varepsilon_{n,1})\rho_{\mu,m\alpha,n\beta,0}(\mathbf{r}')=i\hbar\boldsymbol{\nabla}'\cdot\mathbf{j}_{\mu,m\alpha,n\beta,1}(\mathbf{r}'),\quad n\subset\mathscr{F}(m) \qquad (151)$$

With the help of this we obtain

$$-\tfrac{1}{2}\hbar^{-2}\sum_{n\subset\mathscr{F}(m)}(\varepsilon_{m,1}-\varepsilon_{n,1})^2 F_{nm\alpha\beta,0}(\mathbf{r}',\mathbf{r}'')$$

$$=-\tfrac{1}{4}\hbar^{-2}\sum_{n\subset\mathscr{F}(m)}\sum_{\mu}\sum_{\nu\neq\mu}\sum_{\gamma}(\varepsilon_{m,1}-\varepsilon_{n,1})^2 \rho_{\mu,m\alpha,n\gamma,0}(\mathbf{r}')\rho_{\nu,n\gamma,m\beta,0}(\mathbf{r}'')$$

$$=-\tfrac{1}{4}\sum_{n\subset\mathscr{F}(m)}\sum_{\mu}\sum_{\nu\neq\mu}\sum_{\gamma}[\boldsymbol{\nabla}'\cdot\mathbf{j}_{\mu,m\alpha,n\gamma,1}(\mathbf{r}')][\boldsymbol{\nabla}''\cdot\mathbf{j}_{\nu,n\gamma,m\beta,1}(\mathbf{r}'')]$$

$$=-\tfrac{1}{4}\sum_{\mu}\sum_{\nu\neq\mu}\sum_{n\gamma}\langle\Psi_{m\alpha,0}|[K,\boldsymbol{\nabla}'\cdot\mathbf{j}_\mu(\mathbf{r}')]|\Psi_{n\gamma,0}\rangle\times\langle\Psi_{n\gamma,0}|[K,\boldsymbol{\nabla}''\cdot\mathbf{j}_\nu(\mathbf{r}'')]|\Psi_{m\beta,0}\rangle$$

$$=-\tfrac{1}{4}\sum_{\mu}\sum_{\nu\neq\mu}\langle\Psi_{m\alpha,0}|[K_\mu,[K_\nu,[\boldsymbol{\nabla}'\cdot\mathbf{j}_\mu(\mathbf{r}')][\boldsymbol{\nabla}''\cdot\mathbf{j}_\nu(\mathbf{r}'')]]]|\Psi_{m\beta,0}\rangle \qquad (152)$$

Integrating (152) over $\mathbf{r}'$ and $\mathbf{r}''$ with the factor R, we use Gauss's theorem twice, for $\boldsymbol{\nabla}'$ and $\boldsymbol{\nabla}''$. We get then for the second interaction term of $\varepsilon_{m,2}$

$$-\tfrac{1}{2}\hbar^{-2}\sum_{n\subset\mathscr{F}(m)}(\varepsilon_{m,1}-\varepsilon_{n,1})^2\int d\mathbf{r}'\int d\mathbf{r}''RF_{nm\alpha\beta,0}(\mathbf{r}',\mathbf{r}'')$$

$$=-\tfrac{1}{4}\sum_{\mu}\sum_{\nu\neq\mu}\int d\mathbf{r}'\int d\mathbf{r}''\langle\Psi_{m\alpha,0}|[K_\mu,[K_\nu,[\mathbf{j}_\mu(\mathbf{r}')\cdot\boldsymbol{\nabla}'][\mathbf{j}_\nu(\mathbf{r}'')\cdot\boldsymbol{\nabla}'']R]]|\Psi_{m\beta,0}\rangle$$

$$=-\tfrac{1}{4}\sum_{\mu}\sum_{\nu\neq\mu}e_\mu e_\nu\langle\Psi_{m\alpha,0}|[K_\mu,[K_\nu,(\boldsymbol{\alpha}_\mu\cdot\boldsymbol{\nabla}_\mu)(\boldsymbol{\alpha}_\nu\cdot\boldsymbol{\nabla}_\nu)r_{\mu\nu}]]|\Psi_{m\beta,0}\rangle$$

$$=\tfrac{1}{4}\sum_{\mu}\sum_{\nu\neq\mu}e_\mu e_\nu\langle\Psi_{m\alpha,0}|[K_\mu,[K_\nu,r_{\mu\nu}^{-1}\boldsymbol{\alpha}_\mu\cdot\boldsymbol{\alpha}_\nu$$
$$-r_{\mu\nu}^{-3}(\boldsymbol{\alpha}_\mu\cdot\mathbf{r}_{\mu\nu})\times(\boldsymbol{\alpha}_\nu\cdot\mathbf{r}_{\mu\nu})]]|\Psi_{m\beta,0}\rangle \qquad (153)$$

The third interaction term is evaluated in a similar manner: we need now to absorb $(\varepsilon_{m,0}-\varepsilon_{n,0})^{-2}$ into the matrix elements. The second equation (130) provides the required relation, namely

$$(\varepsilon_{m,0}-\varepsilon_{n,0})^{-1}\mathbf{j}_{\mu,m\alpha,n\beta,0}(\mathbf{r}') = (2m_\mu)^{-1}\langle\Psi_{m\alpha,0}|\beta_\mu\mathbf{j}_\mu(\mathbf{r}')|\Psi_{n\beta,0}\rangle \qquad (154)$$

To prepare for the completion of the summation over all states we need one more relation. Starting from

$$\beta_\mu\mathbf{j}_\mu(\mathbf{r}') = (2m_\mu)^{-1}[\![M, \mathbf{j}_\mu(\mathbf{r}')]\!] \qquad (155)$$

we obtain, taking matrix elements

$$\langle\Psi_{m\alpha,0}|\beta_\mu\mathbf{j}_\mu(\mathbf{r}')|\Psi_{n\beta,0}\rangle = (2m_\mu)^{-1}(\varepsilon_{m,0}-\varepsilon_{n,0})\mathbf{j}_{\mu,m\alpha,n\beta,0}(\mathbf{r}') \qquad (156)$$

from which we conclude

$$\langle\Psi_{m\alpha,0}|\beta_\mu\mathbf{j}_\mu(\mathbf{r}')|\Psi_{n\beta,0}\rangle = 0, \qquad n \subset \mathscr{F}(m) \qquad (157)$$

Using equations (112), (121), (131), (154) and (157) we obtain

$$-4\pi\hbar^2 \sum_{n\subset\mathscr{G}(m)} (\varepsilon_{m,0}-\varepsilon_{n,0})^{-2} F_{nm\alpha\beta,0}(\mathbf{r}',\mathbf{r}'') =$$

$$= 2\pi\hbar^2 \sum_{n\subset\mathscr{G}(m)} \sum_\mu \sum_{\nu\neq\mu} \sum_\gamma (\varepsilon_{m,0}-\varepsilon_{n,0})^{-2}\mathbf{j}_{\mu,m\alpha,n\gamma,0}(\mathbf{r}')\cdot\mathbf{j}_{\nu,n\gamma,m\beta,0}(\mathbf{r}'')$$

$$= -\tfrac{1}{2}\pi\hbar^2 \sum_{n\subset\mathscr{G}(m)} \sum_\mu \sum_{\nu\neq\mu} \sum_\gamma (m_\mu m_\nu)^{-1}$$
$$\times\langle\Psi_{m\alpha,0}|\beta_\mu\mathbf{j}_\mu(\mathbf{r}')|\Psi_{n\gamma,0}\rangle\cdot\langle\Psi_{n\gamma,0}|\beta_\nu\mathbf{j}_\nu(\mathbf{r}'')|\Psi_{m\beta,0}\rangle$$

$$= -\tfrac{1}{2}\pi\hbar^2 \sum_\mu \sum_{\nu\neq\mu} \sum_{n\gamma} (m_\mu m_\nu)^{-1}$$
$$\times\langle\Psi_{m\alpha,0}|\beta_\mu\mathbf{j}_\mu(\mathbf{r}')|\Psi_{n\gamma,0}\rangle\cdot\langle\Psi_{n\gamma,0}|\beta_\nu\mathbf{j}_\nu(\mathbf{r}'')|\Psi_{m\beta,0}\rangle$$

$$= -\tfrac{1}{2}\pi\hbar^2 \sum_\mu \sum_{\nu\neq\mu} (m_\mu m_\nu)^{-1}\langle\Psi_{m\alpha,0}|\beta_\mu\beta_\nu\mathbf{j}_\mu(\mathbf{r}')\cdot\mathbf{j}_\nu(\mathbf{r}'')|\Psi_{m\beta,0}\rangle \qquad (158)$$

Integration over $\mathbf{r}'$ and $\mathbf{r}''$ with the factor $\delta(\mathbf{r}'-\mathbf{r}'')$ yields for the third interaction term of $\varepsilon_{m,2}$

$$-4\pi\hbar^2 \sum_{n\subset\mathscr{G}(m)} (\varepsilon_{m,0}-\varepsilon_{n,0})^{-2} \int d\mathbf{r}' \int d\mathbf{r}''\delta(\mathbf{r}'-\mathbf{r}'')F_{nm\alpha\beta,0}(\mathbf{r}',\mathbf{r}'')$$

$$= -\tfrac{1}{2}\pi\hbar^2 \sum_\mu \sum_{\nu\neq\mu} (e_\mu e_\nu/m_\mu m_\nu)\langle\Psi_{m\alpha,0}|\delta(\mathbf{r}_\mu-\mathbf{r}_\nu)\beta_\mu\beta_\nu\boldsymbol{\alpha}_\mu\cdot\boldsymbol{\alpha}_\nu|\Psi_{m\beta,0}\rangle \qquad (159)$$

We now collect the interaction contributions to $\varepsilon_{m,2}$, equations (150), (153) and (159), and substitute them into the third equation (128). The result is

$$\varepsilon_{m,2}\delta_{\alpha\beta} = \sum_{q=0}^{4} \langle \Psi_{m\alpha,4-q}|M|\Psi_{m\beta,q}\rangle + \sum_{q=0}^{3} \langle \Psi_{m\alpha,3-q}|P|\Psi_{m\beta,q}\rangle$$

$$+ \sum_{q=0}^{2} \langle \Psi_{m\alpha,2-q}|V|\Psi_{m\beta,q}\rangle + \langle \Psi_{m\alpha,0}|V'+B+B'|\Psi_{m\beta,0}\rangle \quad (160)$$

where

$$V' = -\tfrac{1}{2}\sum_{\mu}\sum_{\nu\neq\mu} e_\mu e_\nu [K_\mu, [K_\nu, r_{\mu\nu}^{-1}]] \quad (161)$$

$$B = -\tfrac{1}{4}\sum_{\mu}\sum_{\nu\neq\mu} e_\mu e_\nu [K_\mu, [K_\nu, r_{\mu\nu}^{-1}\alpha_\mu \cdot \alpha_\nu + r_{\mu\nu}^{-3}(\alpha_\mu \cdot r_{\mu\nu})(\alpha_\nu \cdot r_{\mu\nu})]] \quad (162)$$

$$B' = -\tfrac{1}{2}\pi\hbar \sum_{\mu}\sum_{\nu\neq\mu} (e_\mu e_\nu/m_\mu m_\nu)\delta(\mathbf{r}_\mu - \mathbf{r}_\nu)\beta_\mu\beta_\nu\alpha_\mu \cdot \alpha_\nu \quad (163)$$

We note that the Coulomb operator V has again joined the Dirac Hamiltonian in a natural way. The remaining terms with the operators V', B and B' are evaluated with the zero-order wave functions, hence they do not participate in the further determination of the wave functions. If one retraces the origin of the term V', it is seen to represent a higher order residual Coulomb interaction. On the other hand B and B' are due to magnetic interaction and retardation; they represent *radiative particle–particle interactions*.

Before applying the variation principle to equation (160) we carry out a few more simplifications. Among other things, we eliminate the third and fourth order wave functions. We take the orthonormality condition (111) for $p=2$ and $p=4$, multiply by $\varepsilon_{m,1}$ and $\varepsilon_{m,2}$, respectively, and subtract these equations from equation (160). We observe further that

$$\left.\begin{array}{c}(M-\varepsilon_{m,0})\Psi_{m\alpha,1} + P\Psi_{m\alpha,0} = 0 \\ P\Psi_{m\alpha,1} + (V-\varepsilon_{m,1})\Psi_{m\alpha,0} = -\tfrac{1}{2}(M-\varepsilon_{m,0})K^2\Psi_{m\alpha,0}\end{array}\right\} \quad (164)$$

which are easily proved; the net result of all this is

$$\varepsilon_{m,2}\delta_{\alpha\beta} = \langle \Psi_{m\alpha,2} - \tfrac{1}{2}K^2\Psi_{m\alpha,0}|M-\varepsilon_{m,0}|\Psi_{m\beta,2} - \tfrac{1}{2}K^2\Psi_{m\beta,0}\rangle$$

$$+ \langle \Psi_{m\alpha,0}|K(\varepsilon_{m,1}-V)K + \tfrac{1}{4}K^2(\varepsilon_{m,0}-M)K^2 + V' + B + B'|\Psi_{m\beta,0}\rangle \quad (165)$$

We now apply the variation principle to equation (165). By the usual techniques we find

$$(M-\varepsilon_{m,0})(\Psi_{m\alpha,2} - \tfrac{1}{2}K^2\Psi_{m\alpha,0}) = 0 \quad (166)$$

The general solution of equation (166) is

$$\Psi_{m\alpha,2} = \tfrac{1}{2}K^2\Psi_{m\alpha,0} + \Psi''_{m\alpha,0} \quad (167)$$

where $\Psi''_{m\alpha,0}$ is *any* function of rest mass $\varepsilon_{m,0}$. The second-order energy is clearly not affected by $\Psi''_{m\alpha,0}$; the latter can only be determined fully by applying the variation principle to higher order.

It is of course necessary that $\Psi_{m\alpha,2}$ as given by equation (167) does not violate the orthonormality, completeness and charge conservation conditions. It is easy to show that orthonormality and completeness are satisfied provided that

$$\langle \Psi''_{m\alpha,0} | \Psi_{n\beta,0} \rangle + \langle \Psi_{m\alpha,0} | \Psi''_{n\beta,0} \rangle = 0 \tag{168}$$

That the applicable charge conservation conditions are satisfied is again demonstrated in Appendix C.

On account of equation (166) the second-order energy is given by the last term of equation (165) only. We simplify this expression further by eliminating $\varepsilon_{m,0}$ and $\varepsilon_{m,1}$ using equations (107) and (139), applying half of the operators M and $T+V$ to the right and half to the left. After some more manipulation, the result can be written in the simple form

$$\varepsilon_{m,2} \delta_{\alpha\beta} = \langle \Psi_{m\alpha,0} | T' + V_{LS} + B + B' | \Psi_{m\beta,0} \rangle \tag{169}$$

where

$$T' = \tfrac{1}{4} [\![K, [\![K, T]\!]]\!] \tag{170}$$

$$V_{LS} = \tfrac{1}{2} [\![K, [\![K, V]\!]]\!] + V' \tag{171}$$

The operators T and V_{LS} defined by equations (170) and (171) are the proper many-particle generalizations of the corresponding operators for the single-particle case; in fact if we specialize equations (170) and (171) for a single particle, equations (87) and (88) result. It is not unimportant that the operator T' is now in general defined as a double commutator where all the participating operators are sums of one-particle operators; this guarantees that T' is again a sum of one-particle operators, which of course it must be if it is to represent the relativistic mass correction. Straightforward evaluation of the double commutator yields

$$T' = -\sum_\mu (2m_\mu)^{-3} \beta_\mu (\mathbf{p}_\mu \cdot \mathbf{p}_\mu)^2 \tag{172}$$

For the evaluation of matrix elements with T' the same caution applies as for the single-particle case: one factor $\mathbf{p}_\mu \cdot \mathbf{p}_\mu$ should operate to the right, the other one to the left.

We now proceed with the evaluation of the spin–orbit interaction operator V_{LS}, equation (171). If we expand the operators K in the double commutator, we find that the cross terms due to K_μ and K_ν, $\mu \neq \nu$, exactly cancel against the operator V', so that

$$V_{LS} = \tfrac{1}{2} \sum_\mu [\![K_\mu, [\![K_\mu, V]\!]]\!] \tag{173}$$

430 Uncertainty Principle and Foundations of Quantum Mechanics

Next we use in equation (173) the explicit definition of K_μ and obtain, in analogy to the single-particle result, equation (94),

$$V_{LS} = \tfrac{1}{2} \sum_\mu m_\mu^{-2}[(\nabla_\mu V \times \mathbf{p}_\mu) \cdot \mathbf{s}_\mu + \tfrac{1}{4}\hbar^2 \Delta_\mu V] \tag{174}$$

where the spin of the μth particle is of course given by

$$\mathbf{s}_\mu = \tfrac{1}{2}\hbar \boldsymbol{\sigma}_\mu = -\tfrac{1}{4}i\hbar \boldsymbol{\alpha}_\mu \times \boldsymbol{\alpha}_\mu \tag{175}$$

The last step consists of the substitution of the explicit form of the Coulomb energy, equation (134), into equation (174). We perform the differentiations, obtaining

$$\Delta_\mu r_{\mu\nu}^{-1} = -4\pi \delta(\mathbf{r}_\mu - \mathbf{r}_\nu) \tag{176}$$

$$\nabla_\mu r_{\mu\nu}^{-1} \times \mathbf{p}_\mu = -r_{\mu\nu}^{-3} \mathbf{l}_{\mu\nu} \tag{177}$$

where

$$\mathbf{l}_{\mu\nu} = \mathbf{r}_{\mu\nu} \times \mathbf{p}_\mu \tag{178}$$

is the orbital angular momentum of the μth particle *around* the νth particle. Note that $\mathbf{l}_{\mu\nu} \neq \mathbf{l}_{\nu\mu}$. Inserting these results in equation (174) we obtain our final expression for the spin–orbit interaction operator, namely

$$V_{LS} = -\tfrac{1}{2} \sum_\mu \sum_{\nu \neq \mu} (e_\mu e_\nu / m_\mu^2)[r_{\mu\nu}^{-3} \mathbf{l}_{\mu\nu} \cdot \mathbf{s}_\mu + \pi \hbar^2 \delta(\mathbf{r}_\mu - \mathbf{r}_\nu)] \tag{179}$$

We turn now to the operator B' given by equation (163). Since B' contains a delta-function, it contributes in general a contact term. However for the case of primary interest to us, namely the no-hole functions, its contribution vanishes. This follows easily from the fact that β_μ anticommutes with α_μ, and that for a no-hole function we have $\beta_\mu \Psi_{m\alpha,0} = \Psi_{m\alpha,0}$ for *all* μ.

Finally, we evaluate the operator B, equation (162). We shall call it the *Breit operator*, although that term is often used for the operator which we would obtain if we omitted K_μ, K_ν and the commutator brackets in equation (162). For the evaluation of B the following auxiliary relations are useful:

$$\left.\begin{aligned}
(\boldsymbol{\alpha}_\mu \cdot \mathbf{a})(\boldsymbol{\alpha}_\mu \cdot \mathbf{b}) &= \mathbf{a} \cdot \mathbf{b} + i\boldsymbol{\sigma}_\mu \cdot (\mathbf{a} \times \mathbf{b}) \\
\mathbf{p}_\mu r_{\mu\nu}^k &= r_{\mu\nu}^k \mathbf{p}_\mu - i\hbar k r_{\mu\nu}^{k-2} \mathbf{r}_{\mu\nu} \\
\mathbf{p}_\mu \cdot \mathbf{r}_{\mu\nu} &= \mathbf{r}_{\mu\nu} \cdot \mathbf{p}_\mu - 3i\hbar \\
(\mathbf{a} \cdot \mathbf{p}_\mu)\mathbf{r}_{\mu\nu} &= \mathbf{r}_{\mu\nu}(\mathbf{a} \cdot \mathbf{p}_\mu) - i\hbar \mathbf{a} \\
\mathbf{p}_\mu(\mathbf{a} \cdot \mathbf{r}_{\mu\nu}) &= (\mathbf{a} \cdot \mathbf{r}_{\mu\nu})\mathbf{p}_\mu - i\hbar \mathbf{a} \\
\mathbf{p}_\mu \times \mathbf{r}_{\mu\nu} &= -\mathbf{r}_{\mu\nu} \times \mathbf{p}_\mu
\end{aligned}\right\} \tag{180}$$

where $\mathbf{a}$ and $\mathbf{b}$ are arbitrary vectors which commute with the Dirac matrices and momentum operators. The proofs of equations (180) are elementary.

With the help of equations (180) we find for the inner commutator in equation (162)

$$[\![K_\nu, r_{\mu\nu}^{-1}\boldsymbol{\alpha}_\mu \cdot \boldsymbol{\alpha}_\nu + r_{\mu\nu}^{-3}(\boldsymbol{\alpha}_\mu \cdot \mathbf{r}_{\mu\nu})(\boldsymbol{\alpha}_\nu \cdot \mathbf{r}_{\mu\nu})]\!]$$
$$= (2m_\nu)^{-1}[\![\boldsymbol{\alpha}_\nu \cdot \mathbf{p}_\nu \beta_\nu, r_{\mu\nu}^{-1}\boldsymbol{\alpha}_\mu \cdot \boldsymbol{\alpha}_\nu + r_{\mu\nu}^{-3}(\boldsymbol{\alpha}_\mu \cdot \mathbf{r}_{\mu\nu})(\boldsymbol{\alpha}_\nu \cdot \mathbf{r}_{\mu\nu})]\!]$$
$$= -m_\nu^{-1}\boldsymbol{\alpha}_\mu \cdot [r_{\mu\nu}^{-1}\mathbf{p}_\nu + r_{\mu\nu}^{-3}\mathbf{r}_{\mu\nu}(\mathbf{r}_{\mu\nu} \cdot \mathbf{p}_\nu) - \hbar r_{\mu\nu}^{-3}\mathbf{r}_{\mu\nu} \times \mathbf{p}_\nu]\beta_\nu \qquad (181)$$

Substituting this result into equation (162) we obtain

$$B = \tfrac{1}{8}\sum_\mu \sum_{\nu \neq \mu} (e_\mu e_\nu/m_\mu m_\nu)$$
$$\times [\![\boldsymbol{\alpha}_\mu \cdot \mathbf{p}_\mu \beta_\mu, \boldsymbol{\alpha}_\mu \cdot [r_{\mu\nu}^{-1}\mathbf{p}_\nu + r_{\mu\nu}^{-3}\mathbf{r}_{\mu\nu}(\mathbf{r}_{\mu\nu} \cdot \mathbf{p}_\nu) - \hbar r_{\mu\nu}^{-3}\mathbf{r}_{\mu\nu} \times \mathbf{p}_\nu]\beta_\nu]\!] \qquad (182)$$

Further application of equations (180) and some more algebra yields our final result:

$$B = B_{LL} + B_{LS} + B_{SS} \qquad (183)$$

where

$$B_{LL} = -\tfrac{1}{2}\sum_\mu \sum_{\nu \neq \mu} (e_\mu e_\nu/m_\mu m_\nu)\beta_\mu\beta_\nu[r_{\mu\nu}^{-1}\mathbf{p}_\mu \cdot \mathbf{p}_\nu + \tfrac{1}{2}r_{\mu\nu}^{-3}\mathbf{l}_{\mu\nu} \cdot \mathbf{l}_{\nu\mu}] \qquad (184)$$

$$B_{LS} = -\sum_\mu \sum_{\nu \neq \mu} (e_\mu e_\nu/m_\mu m_\nu)\beta_\mu\beta_\nu r_{\mu\nu}^{-3}\mathbf{l}_{\mu\nu} \cdot \mathbf{s}_\nu \qquad (185)$$

$$B_{SS} = -\tfrac{1}{2}\sum_\mu \sum_{\nu \neq \mu} (e_\mu e_\nu/m_\mu m_\nu)\beta_\mu\beta_\nu[3r_{\mu\nu}^{-5}(\mathbf{r}_{\mu\nu} \cdot \mathbf{s}_\mu)(\mathbf{r}_{\mu\nu} \cdot \mathbf{s}_\nu) - r_{\mu\nu}^{-3}\mathbf{s}_\mu \cdot \mathbf{s}_\nu] \qquad (186)$$

The operators B_{LL}, B_{LS} and B_{SS} are said to represent *orbit–orbit*, *spin–other orbit*, and *spin–spin* interactions, respectively.

VII. SUMMARY AND DISCUSSION

We have presented a quantum mechanical description of a closed system of electromagnetic particles, using a variation principle based on the classical formula for electromagnetic interaction. The interaction is expressed in terms of the dynamical variables of the particles only; it takes due account of magnetic as well as Coulombic interaction, and allows for retardation.

For a closed system a complete description can be given in terms of stationary states. The energy is conserved, the number of particles is conserved, etc. Unlike quantum electrodynamics, our complete description can be given in terms of wave functions which populate and define a conventional Hilbert space.

The stationary states have a prescribed harmonic time dependence; by their very nature, they are not subject to time evolution. Thus although the energy is well defined, a real Hamiltonian which governs the time evolution of a non-stationary state has not been defined. It turns out that one of the effects of retardation is to prevent the definition of a Hamiltonian in this regime.

While the variation principle we have given is fairly general, we have in this paper pursued only a rather limited application, namely the first few terms in the Pauli perturbation expansion. This is of course the scheme which is most suitable for atoms and molecules on a terrestrial scale.

In the Pauli perturbation expansion we have the benefit of a Hamiltonian in the few lowest orders, yielding also the Schrödinger equation in a natural way. This happy state of affairs is due to the fact that retardation, which compromises the definition of a Hamiltonian, does not affect the wave functions and energies in the few lowest orders. We can continue to calculate wave functions and energies to higher order, but these results *cannot* be calculated from a perturbation expansion of any Hamiltonian problem whatsoever. In particular the Breit operator, which is used to calculate the first genuinely relativistic effects, cannot be added to a Hamiltonian which would also determine the wave function.

Aside from the drawbacks due to the lack of a Hamiltonian, we have gained considerably in other ways over the conventional approach which uses quantum electrodynamics. The most significant one is the absence of particle self-energies. But also the maintenance of a wave function within a relativistic regime is a definite asset, even though it makes a covariant formulation difficult. After all, we should expect that for terrestrial atoms and molecules the observer's Lorentz rest frame provides a most suitable and therefore privileged vantage point. It is also noted that in our perturbation scheme the zero-order wave functions render a full account of the bulk of the electromagnetic interaction, namely the electrostatic Coulomb force. In contrast, in quantum electrodynamics the zero-order wave functions, if any, are usually plane waves, and therefore poor starting points in a perturbation scheme which deals with atoms and molecules.

There are many avenues along which the present treatment can be fruitfully extended. One thing which comes to mind is the calculation of transition probabilities. This seems at first sight a contradiction to the notion of stationary states. We want to point out, however, that our stationary states already describe radiative interactions, namely those which are internal to the system. If we define a system under study and an observer together as one larger closed system, the radiative interactions between the two subsystems can be reformulated as an interaction between the object and an electromagnetic field. In this way classical electromagnetic fields reappear in a meaningful way; they are never used however to mediate the internal behavior of a quantum mechanical system. Another matter which needs further clarification is charge conservation. We proved its validity for the Pauli perturbation expansion, up to the order to which we carried it through. Obviously a formal proof based on the variation principle as formulated in Section 4 is most desirable. We intend to pursue these and other related matters in the future.

APPENDIX A

For our present purpose it is most convenient to collect the degenerate wave functions $\Psi_{m\alpha}$ in a *row vector*, suppressing the index α:

$$\Psi_m = (\Psi_{m1}, \Psi_{m2}, \ldots, \Psi_{md_m}) \tag{A.1}$$

Considering that as spinors the $\Psi_{m\alpha}$'s are already column vectors, Ψ_m is in fact a rectangular matrix of 4^N rows and d_m columns, where d_m is the degeneracy of the level m. Naturally Ψ_m^* designates the Hermitian conjugate matrix. Orthonormality of the set Ψ_m is now conveniently expressed by

$$\langle \Psi_m | \Psi_m \rangle = I_m \tag{A.2}$$

where I_m is the identity matrix of the appropriate dimension.

The fact that β maps the functions $\Psi_{m\alpha}(\lambda)$ into linear combinations of $\Psi_{m\beta}(-\lambda)$ is now stated by

$$\beta \Psi_m(\lambda) = \tau_m \Psi_m(-\lambda) U_m(-\lambda) \tag{A.3}$$

where $\tau_m U_m(-\lambda)$ is the matrix of transformation. Making use of the orthonormality (A.2) and $\beta^2 = I$ we easily establish

$$\left. \begin{array}{r} U_m^*(\lambda) U_m(\lambda) = I_m \\ U_m(-\lambda) = U_m^*(\lambda) \end{array} \right\} \tag{A.4}$$

that is, $U_m(\lambda)$ is unitary and $U_m(-\lambda)$ is its inverse. The factor τ_m was introduced for convenience and guarantees that

$$U_m(0) = I_m \tag{A.5}$$

Now every unitary matrix U can be written in the form

$$U = e^{iH} \tag{A.6}$$

where H is Hermitian. In fact if the eigenvalues of U are $e^{i\varphi_1}, \ldots, e^{i\varphi_d}$, and W is the unitary matrix which diagonalizes U, we have

$$\left. \begin{array}{r} W^* U W = e^{i\varphi} \\ H = W \varphi W^* \end{array} \right\} \tag{A.7}$$

where φ is the real diagonal matrix

$$\varphi = \begin{pmatrix} \varphi_1 & \cdots & 0 \\ \vdots & & \vdots \\ 0 & \cdots & \varphi_d \end{pmatrix} \tag{A.8}$$

Accordingly we may put forward

$$U_m(\lambda) = e^{iH_m(\lambda)} \tag{A.9}$$

From the second equation (A.4) we conclude that

$$H_m(-\lambda) = -H_m(\lambda) + 2\pi k I_m \tag{A.10}$$

where k is any integer. Because of equation (A.5) we see that k must be even; obviously it is no loss of generality to take $k = 0$, so that

$$H_m(-\lambda) = -H_m(\lambda) \tag{A.11}$$

We now define
$$V_m(\lambda) = e^{\frac{1}{2}iH_m(\lambda)} \tag{A.12}$$
We note that $V_m(\lambda)$ has the same special properties as $U_m(\lambda)$ as expressed in equations (A.4) and (A.5). Furthermore we have
$$V_m(\lambda)V_m(\lambda) = U_m(\lambda) \tag{A.13}$$
that is, we can consider $V_m(\lambda)$ to be the square root of $U_m(\lambda)$.

Finally, constructing new wave functions by means of
$$\Psi'_m(\lambda) = \Psi_m(\lambda) V_m(\lambda) \tag{A.14}$$
we find
$$\beta \Psi'_m(\lambda) = \tau_m \Psi'_m(-\lambda) \tag{A.15}$$
that is, $\Psi'_m(\lambda)$ is a *canonical set*.

APPENDIX B

We observe that
$$\left. \begin{array}{l} \Delta r^{-1} \cos(kr) = -k^2 r^{-1} \cos(kr) \\ \dfrac{d}{dr} r^{-1} \cos(kr) = -r^{-2}[\cos(kr) + kr \sin(kr)] \end{array} \right\} \tag{B.1}$$

We apply Green's theorem under the assumption that $\mathbf{r} = 0$ is a singular point, and that the integrals considered vanish sufficiently strongly for $r \to \infty$. We have
$$\int d\mathbf{r}[u(\mathbf{r})\Delta v(\mathbf{r}) - v(\mathbf{r})\Delta u(\mathbf{r})]$$
$$= -\lim_{\rho \to 0} \int d\omega \rho^2 \left[u(\boldsymbol{\rho}) \frac{\partial v(\boldsymbol{\rho})}{\partial \rho} - v(\boldsymbol{\rho}) \frac{\partial u(\boldsymbol{\rho})}{\partial \rho} \right] \tag{B.2}$$
where ω is the solid angle around the singularity $\mathbf{r} = 0$. Letting
$$v(\mathbf{r}) = r^{-1} \cos(kr) \tag{B.3}$$
we obtain
$$\int d\mathbf{r} r^{-1} \cos(kr)(k^2 + \Delta)u(\mathbf{r})$$
$$= -\lim_{\rho \to 0} \int d\omega \left[\cos(k\rho) + k\rho \sin(k\rho) + \rho \cos(k\rho) \frac{\partial}{\partial \rho} \right] u(\boldsymbol{\rho})$$
$$= -4\pi u(0) = -4\pi \int d\mathbf{r} \delta(\mathbf{r}) u(\mathbf{r}) \tag{B.4}$$
Since $u(\mathbf{r})$ is quite arbitrary, we obtain the operator equation
$$r^{-1} \cos(kr) = -k^{-2}[4\pi\delta(\mathbf{r}) + r^{-1} \cos(kr)\Delta] \tag{B.5}$$
We iterate equation (B.5) to obtain an asymptotic series for $k \to \infty$, namely
$$r^{-1} \cos(kr) = -4\pi\delta(\mathbf{r}) \sum_{\nu=0}^{\infty} k^{-2p-2}(-\Delta)^p \tag{B.6}$$

carrying out the substitution $\mathbf{r} \to \mathbf{r}' - \mathbf{r}''$, $r \to R = |\mathbf{r}' - \mathbf{r}''|$, $\Delta \to -\nabla' \cdot \nabla''$ we obtain the desired result, equation (126).

APPENDIX C

In the variational determination of the wave function $\Psi_{m\alpha,1}$, $q = 0, 1, 2$ we did not impose the constraints arising from charge conservation. We expect that these conditions are satisfied automatically, as it should be in any good theory; we show below that this is indeed the case.

Before proceeding we note that at one point we made use of charge conservation, namely to evaluate the second interaction term of $\varepsilon_{m,2}$. This term only contributes to the Breit interaction, and does not participate in the variational chain of events. Hence the fact that charge conservation had at that point not yet been established does not affect the determination of the wave functions as presented; proof of the validity of charge conservation at this point fills in the gap in the derivation of the Breit operator.

There are two equivalent expressions of charge conservation, equations (115) or (116); for our proof we choose equations (116). We must demonstrate their validity for $p = 0, 1$ with the stipulation $n \subset \mathscr{F}(m)$ for the second equation (116) with $p = 1$, since this constitutes all the relations in which only $\Psi_{m\alpha,q}$, $q = 0, 1, 2$ occur.

To carry out the required algebra we need the operator relation

$$[\![A, [\![B, C]\!]]\!] + [\![B, [\![C, A]\!]]\!] + [\![C, [\![A, B]\!]]\!] = 0 \qquad (C.1)$$

We also define a new operator

$$K' = \sum_{\mu} (2m_{\mu})^{-2} \boldsymbol{\alpha}_{\mu} \cdot \mathbf{p}_{\mu} \qquad (C.2)$$

which satisfies

$$K = [\![K', M]\!] \qquad (C.3)$$

For the function f_{μ}, which is Dirac-diagonal and a function of $\mathbf{r}_{\mu}$ only, we have the useful relations

$$\left. \begin{aligned} [\![V, f_{\mu}]\!] &= 0 \\ [\![V, [\![K, f_{\mu}]\!]]\!] &= 0 \end{aligned} \right\} \qquad (C.4)$$

$$\left. \begin{aligned} [\![M, f_{\mu}]\!] &= 0 \\ [\![M, [\![T, f_{\mu}]\!]]\!] &= 0 \\ [\![M, [\![K, [\![K, f_{\mu}]\!]]\!]]\!] &= 0 \\ [\![M, [\![K, f_{\mu}]\!]]\!] &= -[\![P, f_{\mu}]\!] \\ [\![T, f_{\mu}]\!] &= -[\![K, [\![P, f_{\mu}]\!]]\!] = [\![P, [\![K, f_{\mu}]\!]]\!] \\ [\![T, [\![K, f_{\mu}]\!]]\!] &= -\tfrac{1}{2}[\![K, [\![T, f_{\mu}]\!]]\!] \\ &\quad -\tfrac{1}{2}[\![M, [\![K, [\![K, [\![K, f_{\mu}]\!]]\!]]\!]]\!] \\ [\![K, [\![T, f_{\mu}]\!]]\!] &= -[\![M, [\![K', [\![T, f_{\mu}]\!]]\!]]\!] \end{aligned} \right\} \qquad (C.5)$$

436 Uncertainty Principle and Foundations of Quantum Mechanics

Equations (C.4) follow from the fact that V is Dirac-diagonal, and that V, f_μ and $[\![K, f_\mu]\!]$ do not contain momentum operators. The first two equations (C.5) follow from the simple observation that M commutes with any expression which contains any β_μ's and an even number of α_μ's. The remaining equations (C.5) follow from the preceding ones, the operatr relations (C.1) and (C.3), and equations (95)–(98).

We now proceed with the proofs of the charge conservation conditions. First equation (116), $p = 0$:

$$(\varepsilon_{m,0} - \varepsilon_{n,0})\langle\Psi_{m\alpha,0}|f_\mu|\Psi_{n\beta,0}\rangle = \langle\Psi_{m\alpha,0}|[\![M, f_\mu]\!]|\Psi_{n\beta,0}\rangle = 0 \qquad (C.6)$$

Second equation (116), $p = 0$:

$$(\varepsilon_{m,0} - \varepsilon_{n,0})\sum_{q=0}^{1}\langle\Psi_{m\alpha,1-q}|f_\mu|\Psi_{n\beta,q}\rangle - \langle\Psi_{m\alpha,0}|[\![P, f_\mu]\!]|\Psi_{n\beta,0}\rangle$$

$$= \langle\Psi_{m\alpha,0}|-[\![M, [\![K, f_\mu]\!]]\!] - [\![P, f_\mu]\!]|\Psi_{n\beta,0}\rangle = 0 \qquad (C.7)$$

First equation (116), $p = 1$:

$$(\varepsilon_{m,0} - \varepsilon_{n,0})\sum_{q=0}^{2}\langle\Psi_{m\alpha,2-1}|f_\mu|\Psi_{n\beta,q}\rangle$$

$$+ (\varepsilon_{m,1} - \varepsilon_{n,1})\langle\Psi_{m\alpha,0}|f_\mu|\Psi_{n\beta,0}\rangle - \sum_{q=0}^{1}\langle\Psi_{m\alpha,1-q}|[\![P, f_\mu]\!]|\Psi_{n\beta,q}\rangle$$

$$= \langle\Psi_{m\alpha,0}|\tfrac{1}{2}[\![M, [\![K, [\![K, f_\mu]\!]]\!]]\!] + [\![T + V, f_\mu]\!] + [\![K, [\![P, f_\mu]\!]]\!]|\Psi_{n\beta,0}\rangle$$

$$+ \langle\Psi_{m\alpha,0}|[\![M, f_\mu]\!]|\Psi''_{n\beta,0}\rangle + \langle\Psi''_{m\alpha,0}|[\![M, f_\mu]\!]|\Psi_{n\beta,0}\rangle = 0 \qquad (C.8)$$

Second equation (116), $p = 1$, $n \subset \mathscr{F}(m)$:

$$(\varepsilon_{m,1} - \varepsilon_{n,1})\sum_{q=0}^{1}\langle\Psi_{m\alpha,1-q}|f_\mu|\Psi_{n\beta,q}\rangle - \sum_{q=0}^{2}\langle\Psi_{m\alpha,2-q}|[\![P, f_\mu]\!]|\Psi_{n\beta,q}\rangle$$

$$= \langle\Psi_{m\alpha,0}|-[\![T + V, [\![K, f_\mu]\!]]\!] - \tfrac{1}{2}[\![K, [\![K, [\![P, f_\mu]\!]]\!]]\!]|\Psi_{n\beta,0}\rangle$$

$$- \langle\Psi_{m\alpha,0}|[\![P, f_\mu]\!]|\Psi''_{n\beta,0}\rangle - \langle\Psi''_{m\alpha,0}|[\![P, f_\mu]\!]|\Psi_{n\beta,0}\rangle$$

$$= \langle\Psi_{m\alpha,0}|[\![M, -[\![K', [\![T, f_\mu]\!]]\!] + \tfrac{1}{2}[\![K, [\![K, [\![K, f_\mu]\!]]\!]]\!]]\!]|\Psi_{n\beta,0}\rangle$$

$$+ \langle\Psi_{m\alpha,0}|[\![M, [\![K, f_\mu]\!]]\!]|\Psi''_{n\beta,0}\rangle + \langle\Psi''_{m\alpha,0}|[\![M, [\![K, f_\mu]\!]]\!]|\Psi_{n\beta,0}\rangle$$

$$= (\varepsilon_{m,0} - \varepsilon_{n,0})[\langle\Psi_{m\alpha,0}|-[\![K', [\![T, f_\mu]\!]]\!] + \tfrac{1}{2}[\![K, [\![K, [\![K, f_\mu]\!]]\!]]\!]|\Psi_{n\beta,0}\rangle$$

$$+ \langle\Psi_{m\alpha,0}|[\![K, f_\mu]\!]|\Psi''_{n\beta,0}\rangle + \langle\Psi''_{m\alpha,0}|[\![K, f_\mu]\!]|\Psi_{n\beta,0}\rangle] = 0 \qquad (C.9)$$

ACKNOWLEDGEMENT

This work was assisted in part by the Louis Block Foundation at the University of Chicago, and by the National Science Foundation.

REFERENCES

Bethe, H. A. and Salpeter, E. E. (1957) *Quantum Mechanics of One- and Two-electron Atoms*, Springer-Verlag, Berlin.

Born, M. and Jordan, P. (1925) *Zeit. Phys.*, **34,** 858.
Born, M., Heisenberg, W. and Jordan, P. (1926) *Zeit. Phys.*, **35,** 557.
Breit, G. (1929) *Phys. Rev.*, **34,** 553.
Breit, G. (1930) *Phys. Rev.*, **36,** 383.
Breit, G. (1932) *Phys. Rev.*, **39,** 616.
Davies, C. W. (1970) *Proc. Cambridge Phil. Soc.*, **68,** 751.
Davies, C. W. (1971) *J. Phys.*, **A4,** 836.
Davies, C. W. (1972) *J. Phys.*, **A5,** 1025.
Dirac, P. A. M. (1926) *Proc. Roy. Soc.*, **A109,** 642.
Dirac, P. A. M. (1928) *Proc. Roy. Soc.*, **A117,** 610.
Feynman, R. P. (1948) *Rev. Mod. Phys.*, **20,** 367.
Heisenberg, W. (1925) *Zeit. Phys.*, **33,** 879.
Hoyle, F. and Narlikar, J. V. (1974) *Action at a Distance in Physics and Cosmology*, W. H. Freeman, San Francisco.
Wheeler, J. A. and Feynman, R. P. (1945) *Rev. Mod. Phys.*, **17,** 157.
Wheeler, J. A. and Feynman, R. P. (1949) *Rev. Mod. Phys.*, **21,** 425.

PART 4

Applied Quantum Mechanics

The Uncertainty Principle and the Structure of White Dwarfs

HUGH M. VAN HORN
University of Rochester, New York, U.S.A.

1. INTRODUCTION

A description of the state of matter in the interior of a white dwarf star requires the use of statistical mechanics. In the classical case, every point, $\{\mathbf{x}_i, \mathbf{p}_i : i = 1, N\}$, in the phase space of a system of N particles, corresponds to a distinct physical state of the system, and the goal of the theory is to calculate the probability that the system is in a given infinitesimal volume element $d^3x_1 d^3p_1 \ldots d x_N^3 d^3 p_N$ of the phase space. Heisenberg's (1927) uncertainty principle, however, implies quantization of the phase space: since it is impossible to localize the position and momentum of a particle simultaneously to an accuracy better than that given by the uncertainty relations, e.g. $\Delta x \Delta p_x \gtrsim h$ (where $h = 6.625 \times 10^{-27}$ erg s is Planck's constant), it is meaningless to consider an arbitrarily fine subdivision of the phase space. That is, the phase space can be thought of as being subdivided into 'cells' of volume h^{3N}, but it is not correct to pass to the limit of infinitesimal phase space volume elements. It is important to note that, historically, the significance of h for the statistical theory of black-body radiation had been recognized already by Planck (1900), while discussions based on the uncertainty principle did not appear until much later. Nevertheless, the uncertainty principle is important in demonstrating that quantization occurs on a still more fundamental level than the energy quantization first introduced by Planck.

This phase-space quantization has important consequences for the statistical mechanics of subatomic particles, and the development of quantum statistical mechanics has been essential to our understanding of the structure of the white dwarfs. It is the purpose of this paper to describe the development of this understanding and to show the role played by the uncertainty principle in this process. The historical background of the subject up to about 1920 is first reviewed briefly in Section 2. In Section 3 the development of our current state of knowledge of the mechanical structure of white dwarfs is described, and the thermal structure and evolution of these stars is discussed in Section 4. Some of the more important recent developments in the theory are reviewed in Section 5, and a brief summary in Section 6 concludes the paper.

A number of sources provide excellent reviews of various aspects of the physics and astrophysics of this problem. Among the most useful discussions of the state of physical theory during the early part of this century are those of Harnwell and Livingood (1933), Born (1935), ter Haar (1967) and Hermann (1971). Slater (1973) provides more detail for the important period 1924–26. The discovery of the unusual nature of the white dwarfs is described by Aitken (1935) and by Schatzmann (1958). The baffling puzzle presented by these stars in the early 1920's is well-expressed by Eddington (1926), and the resolution of this puzzle is described Chandrasekhar (1939). Mestel (1965), Marshak (1966) and Van Horn (1971) discuss the growth of our understanding of the thermal structure of these stars, and more recent developments in both observation and theory are reviewed by Weidmann (1968), Greenstein (1969a) and Ostriker (1971).

2. HISTORICAL BACKGROUND

The discovery that Sirius is a double star was made by Bessel in 1844, and the faint companion, Sirius B, was first detected by Clark in 1862 (cf. Burnham, 1906; Aitken, 1935). During this same period, Kirchhoff (1860) formalized the laws of cavity radiation, and Maxwell (1890), Boltzmann (1896) and Gibbs (1902) laid the foundations for statistical mechanics. The study of cavity (or black-body) radiation continued through the latter part of the nineteenth century, and the Stefan–Boltzmann radiation law, $F_{rad} = \sigma T^4$ (Stefan, 1879; Boltzmann, 1884), and Wien's displacement law, $\lambda_{max} T = $ constant (Wien, 1893), were discovered. The electron was found to be a fundamental particle of matter by Thomson (1897).

In the first year of the twentieth century, Planck (1900) published his epochal quantum hypothesis, which gave the first theoretical formula in agreement with the experimentally-determined spectrum of black-body radiation. Planck's idea did not gain wide acceptance, however, until Einstein (1905) showed that literal acceptance of the physical reality of light quanta (photons) provided the explanation of the photoelectric effect. This was the first serious disagreement with the wave theory of light propounded by Maxwell (1881), and provided the first evidence for the mysterious wave-particle duality that was to plague physicists for the next two decades.

By 1910, the year after Millikan (1909) first measured the charge of the electron, enough observational data had been amassed on the Sirius system (orbital period ≈ 50 years) to define the orbits of both stars and thus permit a determination of the masses (Boss, 1910; Aitken, 1935). The primary, Sirius A, was found to have a mass $M_A \approx 2.3 \, M_\odot$, while that of the faint secondary was found to be $M_B \approx 1.0 \, M_\odot$, where $M_\odot = 1.989 \times 10^{33}$ g is the solar mass.

The luminosities of both stars were also known by this time: $L_A \approx 40 L_\odot$, $L_B \approx L_\odot/400$, where $L_\odot = 3.90 \times 10^{33}$ erg s^{-1} is the solar luminosity. There was as yet no clear indication of anything particularly noteworthy about the system, except that the luminosities of the two stars differed by a factor $\sim 10^4$ while the masses differed only by about a factor of two. Still, from the Stefan–Boltzmann law, the low luminosity of Sirius B could be understood if the surface temperature were sufficiently low: ~ 1300 K, equal to the effective temperature of the sun (5800 K) divided by $(400)^{1/4}$.

In physics, the next few years saw the experimental verification of Einstein's theory of the photoeffect (Hughes, 1912; Richardson and Compton, 1912; Millikan, 1916), thus confirming the wave–particle duality for electromagnetic radiation. The same period saw major advances in the theoretical understanding of atomic structure. This had its origin in the systematic relationships discovered between the spectral lines of hydrogen by Balmer (1885), which led to the Rydberg–Ritz combination principle (Ritz, 1908). Subsequently, the Lyman (1906), Paschen (1908) and other series were discovered. With Rutherford's (1911) finding that the atom possessed a nucleus of a size much smaller than that of the atom as a whole, the puzzle began to fall into place. Two years later Bohr (1913) published his model of the atom, in which the orbits of electrons about atomic nuclei were also postulated to be quantized, just as were Planck's 'oscillators'. Experimental confirmation of the existence of the discrete atomic energy levels predicted by Bohr's model came the following year in the experiment of Franck and Hertz (1914).

Another success of the concept of quantization which was discovered at this time was the explanation of the rapid decrease of the specific heats of solids at very low temperatures. Already in 1907, Einstein (1907a) had shown that such behavior was a consequence of the quantization of atomic vibrations in solids, although Einstein's formula predicted an exponential decrease of C_V, in disagreement with the experimentally-determined relation $C_V \propto T^3$. Debye (1912; see also Born and von Karman, 1912) showed that the T^3-law was a consequence of the quantization of the elastic vibrations of the solid. This is relevant to the problem of the thermal structure and evolution of white dwarfs, to which we shall return below.

In the year 1915, the discovery of the baffling nature of the companion of Sirius came with the first measurement, by Adams (1915), of the surface temperature of the star. Adams found $T_e \sim 8000$ K; hotter than the sun, and *much* higher than the temperature expected. Together with the Stefan–Boltzmann law and the known luminosity, this led to an estimate of the stellar radius of $R_B \sim 0.026 R_\odot$, where $R_\odot = 6.96 \times 10^{10}$ cm is the solar radius. Here was a star, hotter than the sun, with a solar mass squeezed into a volume with a radius only about twice that of the earth. The mean density of this 'white dwarf' was thus $\sim 10^5$ g cm^{-3}, to which the scientists of the day appended the comment 'which is absurd' (Eddington, 1926, p. 171). Subsequent measurements have shown that the temperature of Sirius B is in fact much greater than Adams' estimate (the observations are quite difficult because of the close proximity of the enormously brighter star Sirius A). The most recent determinations, to which we shall subsequently return, yield $T_e \approx 32{,}000$ K.

Astronomers had thus discovered a major puzzle, the solution of which had to await the development of quantum statistical mechanics within the following decade.

3. THE MECHANICAL STRUCTURE OF WHITE DWARFS

After the hiatus caused by World War I, progress in physical science became even more rapid. Einstein (1916) published his general theory of relativity during the war, and shortly thereafter the predicted gravitational deflection of light was detected (Dyson and coworkers, 1920a, b). Einstein (1907b) had also shown, during his developmental work on general relativity, that the frequency of light would be redshifted in a strong gravitational field. If Adams were correct about the small radius of Sirius B, it would thus be possible to detect the gravitational redshift in the stellar spectrum: $\Delta\lambda/\lambda = GM/Rc^2 \approx 0.8 \times 10^{-4}$ for the white dwarf, implying a redshift of about 0.4 Å, equivalent to that produced by a recessional velocity of ~ 25 km s^{-1}. The greatest apparent separation of Sirius A and B in their orbit occurred around 1920, making measurements of the redshift possible. The observations were carefully carried out, and Adams (1925) was able to report the detection of a gravitational redshift equivalent to about a 19 km s^{-1} velocity, thus confirming the small radius and high mean density of this star. More recent measurements (Greenstein and coworkers, 1971) have revised the redshift upward to 89 ± 16 km s^{-1}, implying an even smaller stellar radius.

The years 1924–26 were a hectic period in physics, which saw the birth of modern quantum theory (cf. ter Haar, 1967; Hermann, 1971). The wave-particle duality manifested by the photons of electromagnetic radiation was extended to material particles by de Broglie (1923, 1924), and shortly thereafter Schrödinger (1926) published his equation describing the behavior of the wavefunction ψ of the matter waves. The interpretation of $|\psi|^2$ as a probability density came soon after (Born, 1926), and the experimental confirmation of the existence of matter waves followed with the electron diffraction studies of Davisson and Germer (1927) and the α-particle scattering work of Thomson (1928).

The fundamentals of quantum statistics were established during this same period. Bose (1924) and Einstein (1924) first provided a derivation of photon statistics by applying quantum concepts to the statistical mechanics of a 'photon gas'. The theory of electron statistics was worked out by Fermi, who was apparently little influenced by the work of Bose and Einstein, and also by Dirac. A brief account is given by Rasetti in Fermi's *Collected Papers* (Fermi, 1962). Fermi had considered this problem in connection with his earlier work on the constant of entropy, and Pauli's (1925) work on the exclusion principle provided the necessary clue to the answer, which was published the following year (Fermi, 1926 and 1962, p. 178 ff; Dirac 1926). A thorough discussion in the context of the theory of white-dwarf structure has been given by Chandrasekhar (1939; see also Landau and Lifshitz, 1958, pp. 152 ff, 330 ff). An elementary treatment, sufficient to illustrate the essential points, also given by Chandrasekhar, is as follows.

Consider a gas of N electrons confined, at zero temperature, within a box of volume V, and imagine that the Coulomb interactions have been switched off, so that we have a 'non-interacting electron gas'. The quantum states accessible to a single electron can thus be uniquely identified by its momentum eigenvalue, $\mathbf{p}$, and the z-component of its spin ($\pm\frac{1}{2}$). In accordance with the uncertainty principle, there are

$$2\frac{d^3x\,d^3p}{h^3} \tag{1}$$

such quantum states in an element $d^3x\,d^3p$ of phase space, where the factor of two comes from the two possible spin states. However, according to the exclusion principle (Pauli, 1925) no more than one electron may occupy a given quantum state. Thus even at zero temperature almost all of the electrons are excluded from the state with zero kinetic energy and momentum, and since at zero temperature only the lowest energy levels can be filled, the maximum kinetic energy, ε_0, or momentum, p_0, per electron is determined by the condition

$$N = \int_V 2\frac{d^3x\,d^3p}{h^3} = \frac{V}{(2\pi\hbar)^3} \cdot \frac{8\pi}{3} p_0^3 \tag{2}$$

where $\hbar = 1.054 \times 10^{-27}$ erg sec is Planck's constant divided by 2π. Thus the Fermi momentum p_0 is proportional to $n_e^{1/3}$, where $n_e = N/V$ is the average electron density.

The total energy and pressure of the system can also be calculated, given a relation between the particle kinetic energy and momentum. For non-(special) relativistic electrons, for which the energy is $\varepsilon = p^2/2m_e$, the total internal energy is simply

$$E = \int_V \varepsilon \cdot 2\frac{d^3x\,d^3p}{h^3} = \frac{V}{(2\pi\hbar)^3} \cdot \frac{4\pi}{m_e} \frac{1}{5} p_0^5 \tag{3}$$

where $m_e = 9.108 \times 10^{-28}$ g is the electron rest mass. As with any thermodynamic system, the pressure can be computed* from the relation $P = -(\partial F/\partial V)_{T,N}$, where the Helmholtz free energy $F \equiv E - TS$ reduces to E at zero temperature. Thus, since $V \cdot p_0^5 \propto V^{-2/3}$, we have

$$P = -\frac{\partial E}{\partial V}\bigg|_{N,T=0} = \frac{1}{(2\pi\hbar)^3} \cdot \frac{8\pi}{3} \cdot \frac{1}{5}\frac{p_0^5}{m_e} \tag{4}$$

The zero-temperature electron pressure is therefore $P \propto p_0^5 \propto n_e^{5/3}$.

*Alternatively one may employ the definition of pressure as the average rate of momentum transfer across unit area:

$$P = \frac{1}{3}\int pv(p) \cdot 2\frac{d^3x\,d^3p}{h^3} = \frac{V}{(2\pi\hbar)^3} \frac{8\pi}{3} \int_0^{p_0} \frac{\partial\varepsilon}{\partial p} p^3\,dp$$

Corresponding to a density $\rho \sim 10^5$ g cm^{-3}, such as that estimated for the white dwarf Sirius B, equation (4) predicts a pressure $P \sim 10^{22}$ dyne cm^{-2}. This is much greater than the central pressure of the sun, and the existence of such high pressures—even at zero temperature—prompted Fowler (1926) to suggest that electron degeneracy pressure was the principal mechanism of support for white dwarfs. To test the validity of this hypothesis it is necessary to construct a stellar model. Since a star is simply a fluid held in equilibrium under the combined action of its self-gravitation and its own internal pressure, the mechanical structure of a stellar model is described by the equations of mass conservation and of hydrostatic equilibrium:

$$\frac{dM_r}{dr} = 4\pi r^2 \rho \qquad (5a)$$

$$\frac{dP}{dr} = -\frac{\rho G M_r}{r^2} \qquad (5b)$$

subject to the boundary conditions

$$M_r = 0 \quad \text{at } r = 0, \qquad P = 0 \quad \text{at } r = R \qquad (5c)$$

Here M_r is the mass interior to radius r, and R is the radius of the stellar surface (containing the total mass M).

If the equation of state (EOS) given by equation (4) is used, equations (5) form a closed system, which can be solved by numerical integration (cf. Chandrasekhar, 1939, p. 149 ff). Dimensional analysis is sufficient, however, to demonstrate the order of magnitude of the stellar radius corresponding to a stellar mass $M \sim M_\odot$, appropriate for Sirius B. One obtains $GM^2/R^4 \sim [(M/R^3)/H]^{\frac{5}{3}} \cdot \hbar^2/m_e$, or

$$R \sim \frac{\hbar^2}{m_e} \frac{(M/H)^{-\frac{1}{3}}}{GH^2} \sim 10^{-2} R_\odot (M/M_\odot)^{-\frac{1}{3}} \qquad (6)$$

where $H = 1.66044 \times 10^{-24}$ g is the unit of atomic mass. This is precisely the correct order of magnitude for the white dwarf, thus confirming the importance of degeneracy pressure for these objects.

At a density $\sim 10^6$ g cm^{-3}, however, the Fermi energy ε_0 of the electrons becomes comparable with $m_e c^2$, and equations (3), (4) and (6) no longer apply, as first pointed out by Anderson (1929) and Stoner (1930). The correct (special) relativistic generalization of the theory is obtained simply by replacing ε in equation (3) with the relativistic expression $\varepsilon = (p^2 c^2 + m_e^2 c^4)^{\frac{1}{2}}$ and recalculating the EOS and stellar models. The first complete discussion was given by Chandrasekhar (1935). Relativistic effects cause a gradual 'softening' of the EOS at high densities: $\varepsilon \to pc$ implies $P \propto p_0^4 \propto n_e^{\frac{4}{3}}$ rather than the proportionality to $n_e^{\frac{5}{3}}$ valid at non-relativistic densities. This has drastic consequences for the stellar structure (Chandrasekhar, 1931): for $P \propto \rho^{\frac{4}{3}}$, the pressure force and gravitational force vary in the same way with stellar radius; hence a fully relativistic white dwarf cannot achieve equilibrium by adjusting its radius, as can other stars. Thus there exists a critical mass, the 'Chandrasekhar limit', above which no stable white dwarf can exist. The numerical value is

$$M_{\text{lim}} = 1.44 \, M_\odot (\mu_e/2)^{-2} \tag{7}$$

where μ_e, the 'mean molecular weight per electron', equals A/Z for a pure element.

Another correction to the EOS of a fully degenerate electron 'gas' was pointed out by Salpeter (1961). He noted that when Coulomb interactions are taken into account, the energy and pressure of a zero-temperature plasma are slightly altered from the values appropriate to a non-interacting gas. The main effect is due to the interaction between the electrons and the distribution of ions responsible for overall charge neutrality of the system. Because the electron energies are so large, the interactions are only a small perturbation, and the electron density thus remains nearly uniform (as it is in the non-interacting approximation). The point ions, however, arrange themselves in such a way as to stay as far apart from each other as possible (i.e. on a lattice), thus minimizing the energy of the system. Each ion is located at the centre of a (nearly spherical) polyhedron in which the electrons are nearly uniformly distributed. The interaction energy of this configuration can be closely approximated by the energy of the corresponding Wigner–Seitz (1933) sphere:

$$\varepsilon_{\text{coul}} = -\frac{9}{10} \frac{Ze^2}{a} \tag{8}$$

where a is the radius of a sphere containing Z electrons: $n_e \equiv Z/\frac{4}{3}\pi a^3$. In addition, Salpeter calculated corrections due to the exchange energy of the electrons and to departures from uniformity of the electron density near the ions, and he estimated the importance of quantum effects due to the ions.

Because the Coulomb attraction between the electrons and ions produces a negative correction to the energy, the total pressure is *de*creased from that of a non-interacting electron gas. The stellar radius is thus smaller, for a given stellar mass, the order of magnitude of the (composition-dependent) correction being given by

$$\frac{\varepsilon_{\text{coul}}}{\varepsilon_0} \sim \begin{matrix} -Z^{\frac{2}{3}}(n_e a_0^3)^{-\frac{1}{3}}, & \text{non-relativistic} \\ -Z^{\frac{2}{3}}(e^2/\hbar c), & \text{relativistic} \end{matrix} \tag{9}$$

The correction also decreases the limiting mass, with the greatest effect being found for the equilibrium composition of 'cold catalysed matter', where the limiting mass is decreased to $1.02 \, M_\odot$. These effects are shown in Figure 1, following Hamada and Salpeter (1961).

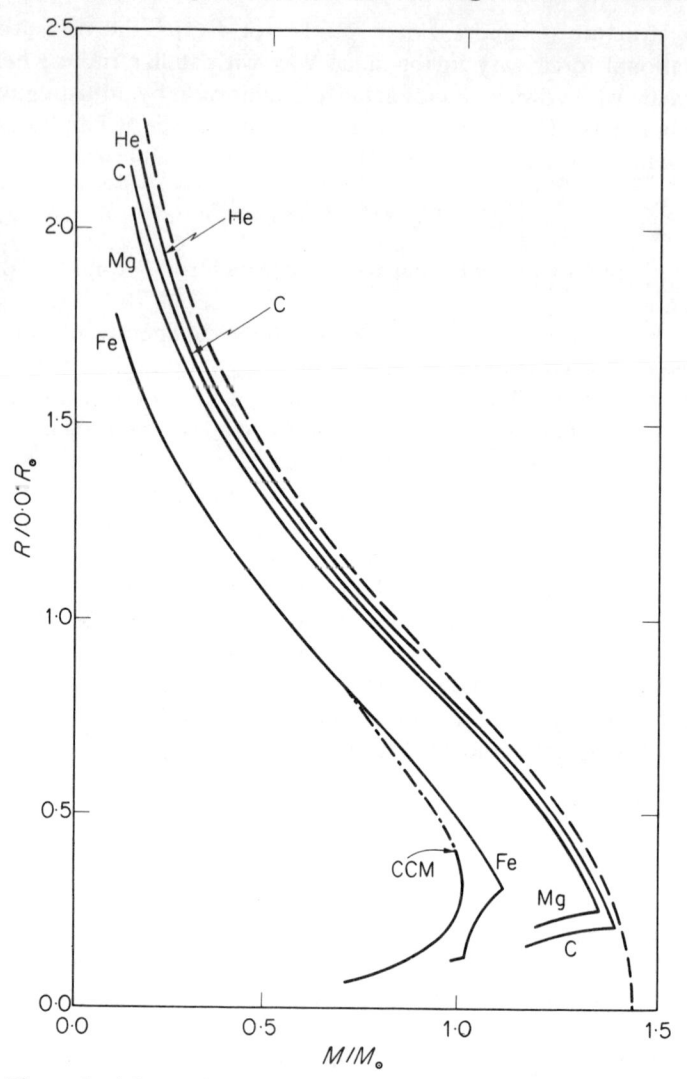

Figure 1 Mass-radius relations for zero-temperature degenerate dwarfs. The dashed curve, which terminates at the Chandrasekhar limit ($1.44M_\odot$), gives the Chandrasekhar (1939) relation for $\mu_e = 2$. The solid curves give the relations obtained by Hamada and Salpeter (1961) for the compositions indicated. The curve labelled CCM is for "cold catalyzed matter"; i.e. material in which all the electron captures have been taken into account that are necessitated by the high electron Fermi energies even at zero temperature. Models with radii less than the radius of the maximum mass for a given composition are dynamically unstable to gravitational collapse, and do not represent physically realizable objects. [Adapted, with permission, from T. Hamada and E. E. Salpeter, Ap. J., 134 (1961): 683, Copyright by the University of Chicago Press.]

4. THE THERMAL STRUCTURE AND EVOLUTION OF WHITE DWARFS

From the fact that a white dwarf is optically detectable (i.e. that the stellar luminosity $L \neq 0$), it follows that there exist temperature gradients in these objects and hence that the internal temperatures are different from zero. The uncertainty principle, as reflected in the quantization of phase space, has important consequences not just for the mechanical structure of white dwarfs but for their thermal structure as well.

It is evident that equations (5) are not sufficient for the discussion of hot white dwarfs but must be supplemented by equations describing the generation and flow of heat within the star. These equations are (cf. Cox and Giuli, 1968, p. 675):

$$\frac{dL_r}{dr} = 4\pi r^2 \rho \left(\varepsilon - T \frac{\partial s}{\partial t} \right) \tag{10a}$$

$$\frac{dT}{dr} = -\frac{3}{4ac} \frac{\kappa \rho}{T^3} \cdot \frac{L_r}{4\pi r^2} \tag{10b}$$

subject to the boundary conditions

$$L_r = 0 \quad \text{at } r = 0, \qquad T = 0 \quad \text{at } r = R \tag{10c}$$

Here L_r is the luminosity produced interior to radius r; ε is the net rate of energy generation per unit mass; s is the entropy per gram of the stellar matter; $a = 7.564 \times 10^{-15}$ erg cm^{-3} deg^{-4} is the radiation density constant; and κ is the effective opacity of the stellar material. Equations (5) and (10) must be solved together in order to obtain a self-consistent stellar model, and to do this it is necessary to specify as functions of two independent thermodynamic state variables (e.g. ρ and T) the pressure $P(\rho, T)$, opacity $\kappa(\rho, T)$, energy generation rate $\varepsilon(\rho, T)$ and entropy $s(\rho, T)$.

Even in the approximation of the non-interacting electron gas, a more rigorous treatment of the statistical physics is necessary in order to calculate the EOS when $T \neq 0$. When this is done (cf. Fowler, 1936, p. 42 ff; Chandrasekhar, 1939, p. 386; Landau and Lifshitz, 1958, p. 152 ff), the number density and pressure of the partially degenerate electrons are given by [compare equations (2)-(4)]

$$N = \int_V [e^{(\varepsilon-\mu)/kT} + 1]^{-1} 2 \frac{d^3p \, d^3x}{h^3} = \frac{V}{(2\pi\hbar)^3} \cdot \frac{8\pi}{3} \int_0^\infty \frac{3p^2 \, dp}{e^{(\varepsilon-\mu)/kT} + 1} \tag{11a}$$

and

$$\begin{aligned} PV \equiv -\Omega &= kT \int_V \ln[1 + e^{-(\varepsilon-\mu)/kT}] 2 \frac{d^3p \, d^3x}{h^3} \\ &= \frac{V}{(2\pi\hbar)^3} \cdot \frac{8\pi}{3} \int_0^\infty \frac{(d\varepsilon/dp) p^3 \, dp}{e^{(\varepsilon-\mu)\mu T} + 1} \end{aligned} \tag{11b}$$

The entropy can be obtained from the thermodynamic relation $s = -(\partial\Omega/\partial T)_{V,\mu}$ (cf. Landau and Lifshitz, 1958, p. 158).

For given T, N and V, equation (11a) serves to define μ, the chemical potential ($\equiv$Gibbs free energy per electron), once the relation $\varepsilon = \varepsilon(p)$ is specified. For $\mu/kT \gg 1$ (high degeneracy) equations (11a, b) reduce to equations (2) and (4), respectively, and $\mu \approx \varepsilon_0$, the Fermi energy. The principal effect of the more detailed treatment of the statistics necessary to include the temperature-dependence thus is to incorporate the Fermi–Dirac distribution function,

$$f(\varepsilon) = [e^{(\varepsilon-\mu)/kT} + 1]^{-1} \tag{12}$$

into the thermodynamics.

Other corrections to the EOS of course must also be taken into account, just as in the zero-temperature case. At high temperatures the main correction is due to the kinetic energy of the ions, which can there be adequately approximated by a non-interacting Maxwell–Boltzmann gas. (The ions are so much more massive than the electrons that they can be regarded as non-degenerate for white dwarfs.) At low temperatures, the Coulomb interactions dominate, as we have seen above. We shall return below to a discussion of the regime in which $kT \sim \varepsilon_{\text{coul}}$, which has interesting consequences for white-dwarf evolution.

We next consider the effective opacity of stellar matter. In normal stars, energy is transported either by convection or by radiative diffusion. However, under conditions of strong degeneracy, conduction by the degenerate electrons becomes highly efficient, as in metals. The quantum theory of metallic conduction was worked out by Sommerfeld (1928) and Sommerfeld and Bethe (1933), who extended the classical free-electron model of metals to Fermi–Dirac statistics. With the realization of the importance of electron degeneracy in white dwarfs, it was natural to extend these calculations further to include these objects as well.

An excellent review of electron conduction in white dwarfs is given by Marshak (1966). Assuming the electron mean free path to be determined by elastic collisions between electrons and ions, Marshak (1940) had applied the theory of Sommerfeld and Bethe to calculate the thermal conductivity of white-dwarf matter. For convenience of application to equation (10b), he expressed his results in the form of a so-called 'conductive opacity', κ_c. This is possible because equation (10b) can be cast in the form of a heat conduction equation: heat flux $(L_r/4\pi r^2)$ proportional to $-\nabla T$, with the coefficient of proportionality being the thermal conductivity, K.

Classically, $K \sim k \cdot \lambda \cdot n_e \cdot (kT/m_e)^{\frac{1}{2}}$, while when (non-relativistic) degeneracy is taken into account one finds $K \sim k \cdot \lambda \cdot n_e (kT/\varepsilon_0) \cdot (p_0/m_e)$ (cf. Fowler, 1936, p. 404 ff), where λ is the electron mean free path. For electron–ion Coulomb collisions one may roughly estimate $\lambda \sim (n_i \sigma)^{-1}$, where n_i is the number density of ions and $\sigma \sim (Ze^2/\varepsilon_0)^2$ is a rough measure of the electron–ion Coulomb scattering cross-section. Thus degeneracy affects the thermal conductivity in several ways. Because the Fermi energy is $\varepsilon_0 \gg kT$, the fraction of electrons able to participate in the scattering is reduced [$n_e(kT/\varepsilon_0)$ versus n_e], but the speed of propagation is increased [(p_0/m) versus $(kT/m)^{\frac{1}{2}}$], and the mean free path is greatly increased [$\lambda \propto \varepsilon_0^2$ versus $\lambda \propto (kT)^2$]. The increase in electron mean free path can be thought of as being due directly to the inhibition of scattering by the Pauli principle. For non-relativistic electrons it is a straightforward matter to show that $K \sim k(n_e/Z^2 n_i)n_e a_0^2 kT/\hbar$, where a_0 is the Bohr radius. Thus $K \propto n_e T$, and from equation (10b), $\kappa_c \equiv (4ac/3)T^3/\rho K \propto T^2/\rho^2$. Detailed calculations (Marshak, 1940; Mestel, 1950; Lee, 1950) give, approximately

$$\kappa_c \approx 2.8 \times 10^{41} X_R I T^2 / n_e^2 \text{ cm}^2 \text{ g}^{-1} \tag{13}$$

where X_R is the fractional abundance of heavy elements (≈ 1 for the interior of a white dwarf) and I is the dimensionless Coulomb logarithm. More recent calculations by Hubbard (1966) and by Hubbard and Lampe (1969) have shown how the (divergent) Coulomb logarithm is naturally cut off by plasma screening and correlation effects, and calculations by Kovetz and Shaviv (1973) [see also Canuto (1970) for an earlier treatment] have extended the conductivities to cases where the electrons are relativistic.

A comparison of radiative and conductive opacities shows (cf. Marshak, 1940; 1966) that radiative energy transport dominates in the thin non-degenerate envelope of a white dwarf ($\mu/kT < 0$), while electron conduction is far more important in the degenerate core ($\mu/kT > 0$). Electron conduction is so highly efficient that the degenerate core is approximately isothermal, at a temperature $\sim 10^7$ K for $L \sim 10^{-2} L_\odot$. Under these circumstances, most of the temperature drop occurs in the thin, non-degenerate outer layers of the star. If energy production can be neglected in these layers (as turns out to be the case), L is constant there. Also, M_r may be well-approximated by M, the total stellar mass, and equations (5b) and (10b) can be integrated directly with the assumption of a perfect gas law EOS ($P = nkT$) and Kramers' law of radiative opacity ($\kappa = \kappa_0 \rho/T^{3.5}$). Terminating the calculation at the degeneracy boundary then gives a relation between L and the core temperature T_c (cf. Marshak, 1940; Mestel, 1952; Schwarzschild, 1958; Van Horn, 1971):

$$L \approx 1.7 \times 10^{-3} \frac{M}{M_\odot} \left(\frac{T_c}{10^7 \text{ K}}\right)^{3.5} \text{ erg s}^{-1} \tag{14}$$

With the discovery that the cores of white dwarfs were isothermal, with $T_c \sim 10^7$ K, the question of the energy sources in these stars became acute. Bethe and Critchfield (1939) had already worked out the proton chain and Bethe (1939) had discovered the carbon cycle. Both of these sequences of nuclear reactions would have produced energy at far greater rates than observed for the white dwarfs if hydrogen were present in significant quantities in the cores of these stars. Lee (1950) and Mestel (1952) showed that white-dwarf models which derived their luminosity from nuclear burning would be secularly unstable for the carbon cycle, contrary to observation, and

Ledoux and Sauvenier-Goffin (1950) found such models to be pulsationally unstable as well, again contrary to observation for all white dwarfs then known. These results clearly showed that nuclear energy sources could not be the answer, but left a puzzle concerning the source of the observed luminosity.

The solution was provided by Mestel (1952), who showed that a consistent interpretation of these stars followed from the simple assumption that the heat stored in the stellar core during the prior evolution was now slowly leaking out through the non-degenerate surface layers. With this assumption, ε in equation (10a) could be set to zero, and the time derivative of the entropy could be approximately written as

$$T\frac{\partial s}{\partial t} \approx T\left(\frac{\partial s}{\partial T}\right)_\rho \frac{\partial T}{\partial t} = C_v \frac{\partial T}{\partial t} \approx \frac{3}{2}\frac{k}{AH}\frac{\partial T_c}{\partial t} \qquad (15)$$

Here the specific heat C_v has been approximated by that of an ideal gas of ions of atomic mass A, and T has been approximated by T_c, which is taken to be constant throughout the star. Note that the electronic heat capacity and the term $T(\partial s/\partial \rho)_T(\partial \rho/\partial t)$ have been neglected because of the high degeneracy of the star. With the aid of equation (15), equation (10a) can be integrated directly over the star to give

$$L \approx -\frac{3}{2}\frac{k}{AH}M\frac{\partial T_c}{\partial t} \qquad (16)$$

and equations (14) and (16) can be combined to solve for both L and T_c as functions of time. Mestel's assumptions thus led to a self-consistent picture of the *evolution* of a white dwarf, which has since remained unchanged, except in numerical details.

5. RECENT DEVELOPMENTS

Up to this point, we have discussed equations (5) and (10) as if they could be solved separately. Although this does provide insight into the physics of the problem it does not constitute a self-consistent solution. In 1964, however, Henyey and coworkers (1964) introduced an automatic method for the numerical solution of these equations, which was to revolutionize the study of stellar evolution. These techniques were quickly applied to the study of pre- and post-main sequence evolution, and Vila (1965, 1966, 1967) became the first to extend them to white dwarfs. Also during the early sixties, calculations of neutrino energy loss rates, due to various processes expected in late stellar evolution, were carried out: photo-neutrinos (Pontecorvo, 1959), pair-annihilation neutrinos (Chiu and Morrison, 1960), and plasma neutrinos (Adams and coworkers, 1963). These were included in Vila's calculations and were found to be important during the pre-white-dwarf phases, but to have little effect in the white-dwarf stage because of the lower temperatures there. Additional computational studies of white-dwarf evolution have since been carried out for a variety of other stellar masses and compositions (cf. Savedoff and coworkers, 1969; Beaudet and Salpeter, 1969; Chin and Stothers, 1971), which have further corroborated Mestel's (1952) basic concept of white-dwarf cooling, while extending it in various ways.

The past decade also has seen significant advances in our detailed knowledge of the physical properties of white-dwarf matter. Neutrino rates have been revised (Beaudet and coworkers, 1967; Festa and Ruderman, 1969), but as noted above this has little consequence for white dwarfs. Also as noted previously, the conductive opacity calculations have been greatly improved by Hubbard (1966), Hubbard and Lampe (1969) and Kovetz and Shaviv (1973). An important feature of these calculations was the utilization of an accurate ion pair-correlation function, first calculated for a classical, one-component plasma by Brush and coworkers (1966, hereafter termed BST). This work also formed the basis for the most recent investigation of electron-screening effects on thermonuclear reaction rates [DeWitt, Graboske and Cooper 1973; Graboske and coworkers, 1973; see Salpeter and Van Horn (1969) for an earlier treatment]. These rates may be relevant for studies of some white dwarfs, especially those undergoing mass-accretion (novae, dwarf novae, etc.), although not for the majority.

The result of BST which has probably stimulated the greatest interest and investigation, however, was their discovery that an ideal, one-component plasma crystallizes at a temperature given by $\Gamma \sim 100$, where

$$\Gamma \equiv \varepsilon_{\text{coul}}/kT \tag{17}$$

Although it had been appreciated earlier that a zero-temperature star should be in a crystalline state (Salpeter, 1961; Abrikosov, 1960; Kirzhnits, 1960), the temperature of the phase transition had not been accurately estimated. The temperature given by equation (17), however, falls precisely in the range inferred for the cores of typical white dwarfs and indicates that phenomena associated with crystallization may play a role in white-dwarf evolution.

Mestel and Ruderman (1967) first pointed out that one consequence of crystallization is the validity of the Debye approximation for the specific heat at low temperatures. This is yet another instance of the application of the uncertainty principle to white dwarfs through the quantization of phase space. Debye (1912) had shown that one could regard the thermal motions of ions about their lattice sites as being comprised of quantized vibrations of the entire lattice ('phonons'). Since phonons, like photons, obey Bose–Einstein statistics, application of quantum statistical mechanics leads to the following expression for the total energy content of the phonons:

$$E = \sum_{\alpha=1}^{3} \frac{V}{(2\pi\hbar)^3} 4\pi \int_0^{\varepsilon_\alpha^{\max}} \frac{\varepsilon_\alpha p^2 \, dp}{e^{\varepsilon_\alpha/kT} - 1} \tag{18a}$$

Here $\varepsilon^{\max}$ is determined by the condition that the total number of available phonon modes equal the number of degrees of freedom of the lattice:

$$\sum_{\alpha=1}^{3} \frac{V}{(2\pi\hbar)^3} \cdot 4\pi \int_0^{\varepsilon_\alpha^{\max}} p^2 \, dp = 3N_i \tag{18b}$$

where N_i is the total number of ions in volume V, and the sum extends over the three independent modes of lattice vibration (one longitudinal and two transverse). If the three modes are lumped together, using an average sound speed a_s for the lattice, the usual acoustic dispersion relation $\omega = a_s k$ (or $\varepsilon = a_s \cdot p$) leads to (cf. Landau and Lifshitz, 1958, p. 187 ff)

$$\varepsilon^{\max} = \left(\frac{3}{4\pi}\frac{N_i}{V}\right)^{\frac{1}{3}} \cdot 2\pi\hbar a_s \equiv k\Theta \tag{19a}$$

$$E = 3N_i kT \cdot 3\left(\frac{T}{\Theta}\right)^3 \int_0^{\Theta/T} \frac{z^3\,dz}{e^z - 1} \equiv 3N_i kT \cdot D(\Theta/T) \tag{19b}$$

where Θ is the Debye temperature. For high temperatures ($T \gg \Theta$), equation (19b) gives $E \approx 3N_i kT$ and $C_v = (\partial E/\partial T)_V \approx 3N_i k$, while at low temperatures ($T \ll \Theta$), we find $E \approx 3N_i kT(T/\Theta)^3 \cdot \pi^4/5$ and $C_V \approx 12N_i k(T/\Theta)^3 \cdot \pi^4/5$.

Since the luminosity of a white dwarf is just equal to the time rate of decrease of the thermal energy (cf. equation (16)), a white dwarf in which $T_c \ll \Theta$ can be expected to cool increasingly rapidly. Calculations of this late stage of evolution (Mestel and Ruderman, 1967; Van Horn, 1968; Ostriker and Axel, 1969; Vila, 1969) have confirmed this and have shown that a white dwarf more massive than $\sim 0.7\,M_\odot$ may cool off completely in a time $<10^{10}$ years. Although there is some indication of this effect in the observational data [cf. the possible deficiency of red degenerate stars: Greenstein (1969b); however see also Hintzen and Strittmatter (1974)], it is premature to regard it as established.

Another effect is the release of the latent heat of crystallization accompanying the first-order transition to the solid phase. This provides a new energy source for a cooling white dwarf, the consequences of which were first explored by Van Horn (1968). Subsequent investigations of the equation of state of a crystallizing plasma (Kovetz and Shaviv, 1970; Shaviv and Kovetz, 1972; Hansen, 1973; Pollock and Hansen, 1973; Hansen, Pollock and McDonald, 1974; Lamb, 1974; Lamb and Van Horn, 1975; Kovetz and Shaviv, 1976) have confirmed the basic physics of the phase transition and have greatly improved the numerical accuracy of the thermodynamic quantities (especially Shaviv and Kovetz, 1972; Hansen and coworkers, 1973 and Lamb, 1974). On the basis of these improved data, the direct effect of the latent heat release upon the cooling rates of white dwarfs appears to be too small to be discernible in the current observational data (cf. Figure 2).

Our discussion of white-dwarf structure cannot be concluded without some discussion of the recent advances in understanding and calculation of envelopes and atmospheres for these stars. Böhm (1968, 1969, 1970) first emphasized the importance of convection in white dwarf envelopes and pointed out the significance for the process of Debye cooling of the lower core temperatures resulting from the high efficiency of convection. The envelope convection zone is also an ionization zone, however, and the effects of electron degeneracy and strong (Coulomb) interactions upon the ionization equilibrium

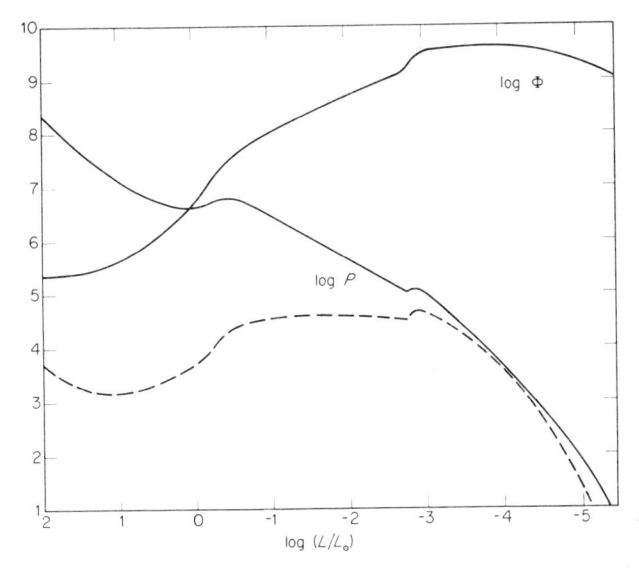

Figure 2 The luminosity function Φ and the "discovery function" $P \propto \Phi \cdot (L/L_\odot)^{3/2}$ for $1 M_\odot$, ^{12}C white dwarfs, as derived by Lamb and Van Horn (1975). The dashed curve illustrates the effect upon P of the bolometric correction (proportional to the logarithmic difference between visual and total luminosities), which is large both at high and at low luminosities. The dashed curve is proportional to the theoretical probability of observing a nearby white dwarf, while Φ is proportional to the theoretical space density of such stars. The relative depression of the curves at high luminosities ($\log L/L_\odot \gtrsim -0.5$) is due to neutrino emission. The increase near $\log L/L_\odot \sim -3$ is caused by the onset of crystallization at the center, and the decline, below $\log L/L_\odot \sim -4.5$, is due to Debye cooling, enhanced by envelope convection. [Adapted, with permission, from D. Q. Lamb and H. M. Van Horn, Ap. J., 200 (1975): 306, Copyright by the University of Chicago Press.]

constitute a very complex problem. Numerical calculations of the equation of state and opacity in this regime are beginning to become available (Carson and coworkers, 1968; Graboske and coworkers, 1969; Magni, 1972; Carson 1975), but there are still some unresolved differences among the results. Nevertheless, calculations using these data to construct white dwarf convective envelope models have begun to elucidate some of the effects of convection upon these stars: e.g. the effect of convection extending into the partially transparent atmosphere (Böhm and Grenfell, 1973), the effects of differences in equations of state, convection theories and composition (Fontaine and coworkers, 1974; Fontaine and Van Horn, 1976), and the effects of convection upon the distribution of spectral types of white dwarfs (Strittmatter and

Wickramasinghe, 1971; Baglin and Vauclair, 1973; d'Antona and Mazzitelli, 1975). The evolutionary calculations of Lamb (1974, see also Lamb and Van Horn, 1975) include the effects of envelope convection in detail.

The construction of accurate model atmospheres for the white dwarfs has now been started also. Strittmatter and Wickramasinghe (1971) have computed continua and hydrogen-line profiles and have applied their results to a preliminary interpretation of the evolution of DA (H-line) white dwarfs into DB (He-line) stars. Shipman (1972) computed further models and used them both to extend the suggestion of Strittmatter and Wickramasinghe to include evolution from DB white dwarfs into type DC (featureless continua) as well as to provide more accurate determinations of the effective temperatures and surface gravities of these stars. Bues (1973) has computed spectra for a variety

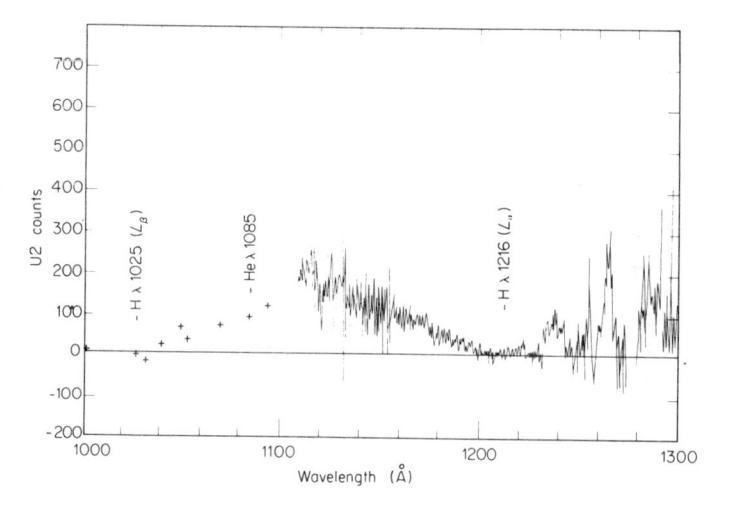

Figure 3 The far ultraviolet spectrum of the white dwarf Sirius B, obtained by Savedoff *et al.* (1976) using the *Copernicus* satellite by subtracting the spectrum of Sirius A from that of A plus B. The data longward of $\lambda 1110$ Å were obtained with the U2 spectrophotometer (0.16 Å resolution), while the crosses are from the U1 spectrometer (0.02 Å resolution) scaled to the level of the U2 data. The ordinate scale gives the U2 counts per 0.16 Å per 14-second integration time. The fluctuations in the U2 data are noise (and not real spectral features), which becomes large longward of $\sim \lambda 1240$ Å due to the rapid increase of Sirius A at these wavelengths. The positions of the centers of Lyman α, Lyman β and the helium $\lambda 1085$ Å line are indicated. From this data the effective temperature and radius of Sirius B were found to be $T_{\text{eff}} = 27{,}000°\text{K} \pm 6000°\text{K}$ and $R/R_\odot = 0.009 \pm 0.002$, in good agreement with the results of Greenstein, Oke, and Shipman (1971) and with Hamada-Salpeter (1961) models for compositions of carbon or magnesium (cf. Fig. 1). [Reprinted, with permission, from M. P. Savedoff, *et al.*, Ap. J. (Letters), 207, L 45, (1976). Copyright by the University of Chicago Press.]

of non-DA types and quite recently Wehrse (1975) has calculated spectra of high quality for the DA stars as well. It will be necessary for further detailed model atmosphere calculations to be carried out to extend the interpretation of the data, especially for very hot, high gravity, degenerate stars. Savedoff and coworkers (1976) have recently obtained the first far ultraviolet spectrum of a white dwarf (cf. Figure 3), and have confirmed the conclusion of Greenstein, Oke and Shipman (1971) from visual spectra that the effective temperature of this star is $T_{\text{eff}} \approx 32{,}000$ K. Greenstein and Sargent (1974) have deduced even higher temperatures for the hot subdwarfs, which are thought to be the immediate precursors of the white dwarfs, and Lampton and coworkers (1976) have discovered an extreme ultraviolet source which appears to be coincident with the white dwarf HZ43 and has an estimated temperature in excess of 100,000 K.

6. SUMMARY AND CONCLUSIONS

I have attempted to trace the development of our current understanding of white-dwarf structure, and to show the impact of advances in physical theory upon this process. The uncertainty principle, through the quantization of phase space, affects many facets of this understanding: The basic mechanical structure of a white dwarf depends in an essential way upon the properties of a degenerate electron gas. Electron degeneracy also affects the thermal conductivity, which in turn controls the temperature distribution in a white dwarf. The heat content, and thus the evolution, of a white dwarf after core crystallization is affected by the phonon energy distribution, which is another point of contact with the uncertainty principle. In addition, the calculations of most properties of matter are affected insofar as it is necessary to employ quantum mechanics, of which the uncertainty principle is a cornerstone.

Although I have tried to provide a relatively detailed account, it is impossible in these few pages to be complete. In particular, I have not even mentioned the exciting new discoveries of magnetic fields and rapid (100–1000 sec) pulsations in some of these stars, and I have ignored many other recent observational and theoretical advances as well. Nevertheless it is my hope that this review may provide some sense of the course of development of our current ideas about the structure of white dwarfs.

ACKNOWLEDGMENTS

I am grateful to E. E. Salpeter, M. P. Savedoff, R. E. Marshak and J. L. Greenstein for their various roles in introducing me to the delights of the astrophysical study of white dwarfs. I have also benefited immensely from stimulating interactions with K.-H. Böhm, G. Fontaine, C. J. Hansen, D. Q. Lamb and L. Mestel, and I thank M. P. Savedoff for his helpful comments on an earlier version of this paper. This work has been supported by the National Science Foundation (U.S.A.) under grant AST 74-13257 A02.

REFERENCES

Abrikosov, A. A. (1960) *Sov. Phys. JETP*, **12**, 1254.
Adams, W. S. (1915) *Publ. Astron. Soc. Pacific*, **27**, 236.
Adams, W. S. (1925) *Proc. Nat. Acad. Sci.*, **11**, 382.
Adams, J. B., Ruderman, M. A. and Woo, H. C. (1963) *Phys. Rev.*, **129**, 1383.
Aitken, R. G. (1935) *The Binary Stars*, McGraw-Hill, New York; reprinted 1964 by Dover Publications, New York.
Anderson, W. (1929) *Z. Phys.*, **54**, 433.
Baglin, A. and Vauclair, G. (1973) *Astron. Ap.*, **27**, 307.
Balmer, J. J. (1885) *Ann. Physik*, **25**, 80.
Beaudet, G., Petrosian, V. and Salpter, E. E. (1967) *Astrophys. J.*, **150**, 979.
Beaudet, G. and Salpeter, E. E. (1969) *Astrophys. J.*, **155**, 203.
Bethe, H. A. (1939) *Phys. Rev.*, **55**, 434.
Bethe, H. A. and Critchfield, C. L. (1939) *Phys. Rev.*, **54**, 248.
Bohr, N. (1913) *Phil. Mag.*, **26**, 1, 476, 857.
Böhm, K.-H. (1968) *Astrophys. Space Sci.*, **2**, 375.
Böhm, K.-H. (1969) in *Low Luminosity Stars*, S. Kumar, Ed., Gordon and Breach, New York, p. 393.
Böhm, K.-H. (1970) *Astrophys. J.*, **162**, 919.
Böhm, K.-H. and Grenfell, T. C. (1973) *Astron. Ap.*, **28**, 79.
Boltzmann, L. (1884) *Wied. Ann.*, **22**, 31, 291.
Boltzmann, L. (1896) *Vorlesungen über Gastheorie* translated 1964 as *Lectures on Gas Theory* by S. G. Brush, University of California Press, Berkeley.
Born, M. (1926) *Z. Phys.*, **38**, 803.
Born, M. (1935) *Atomic Physics*, Blackie and Son. Ltd., Glasgow.
Born, M. and von Karman, T. (1912) *Physik. Z.*, **13**, 297.
Bose, S. N. (1924) *Z. Phys.*, **26**, 178.
Boss, L. (1910) *Preliminary General Catalogue*, Carnegie Institution, Washington, p. 265.
Brush, S. G., Sahlin, H. L. and Teller, E. (1966) *J. Chem. Phys.*, **45**, 2102.
Bues, I. (1973) *Astron. Ap.*, **28**, 181.
Burnham, S. W. (1906) *A General Catalogue of Double Stars*, Carnegie Institution, Washington, pt. 2, p. 467.
Canuoto, V. (1970) *Astrophys. J.*, **159**, 641.
Carson, T. R. (1975) personal communication.
Carson, T. R., Mayers, D. F. and Stibbs, D. W. N. (1968) *M.N.R.A.S.*, **140**, 483.
Chandrasekhar, S. (1931) *M.N.R.A.S.*, **91**, 456.
Chandrasekhar, S. (1935) *M.N.R.A.S.*, **95**, 207.
Chandrasekhar, S. (1939) *An Introduction to the Study of Stellar Structure*, University of Chicago Press, Chicago; reprinted 1957 by Dover Publications, New York.
Chin, C.-W. and Stothers, R. (1971) *Astrophys. J.*, **163**, 555.
Chiu, H.-Y. and Morrison, P. (1960) *Phys. Rev. Letters*, **5**, 573.
Cox, J. P. and Giuli, R. T. (1968) *Principles of Stellar Structure*, Gordon and Breach, New York.
d'Antona, F. and Mazzitelli, I. (1975) *Astron. Ap.*, **42**, 165.
Davisson, C. and Germer, L. H. (1927) *Phys. Rev.*, **30**, 707.
de Broglie, L. (1923) *Nature*, **112**, 50.
de Broglie, L. (1924) 'Recherches sur la Théorie des Quanta', thesis, Masson et Cie, Paris.
Debye, P. (1912) *Ann. Physik*, **39**, 789.
DeWitt, H. E., Graboske, H. C. and Cooper, M. S. (1973) *Astrophys. J.*, **181**, 439.
Dirac, P. A. M. (1926) *Proc. Roy. Soc.*, **A112**, 661.
Dyson, F. W., Eddington, A. S. and Davidson, C. (1920a) *Phil. Trans. Roy. Soc.*, **220A**, 291.
Dyson, F. W., Eddington, A. S. and Davidson, C. (1920b) *Mem. Roy. Ast. Soc.*, **62**, 291.
Eddington, A. S. (1926) *Internal Constitution of the Stars*, Cambridge University Press, Cambridge.
Einstein, A. (1905) *Ann. Physik*, **17**, 132.
Einstein, A. (1907a) *Ann. Physik*, **22**, 180.
Einstein, A. (1907b) *Jahrb. Radioakt.*, **4**, 411.
Einstein, A. (1916) *Ann. Phys.*, **49**, 769.
Einstein, A. (1924) *Berliner Berichte*, p. 261.
Fermi, E. (1926) *Z. Phys.*, **36**, 902.

Fermi, E. (1962) *Collected Papers*, E. Segre, E. Amaldi, H. L. Anderson, E. Persico, F. Rasetti, C. S. Smith and A. Wattenberg, Eds., University of Chicago Press, Chicago, Vol. 1.
Festa, G. G. and Ruderman, M. A. (1969) *Phys. Rev.*, **180**, 1227.
Fontaine, G. and Van Horn, H. M. (1976) *Astrophys. J. Suppl.*, **31**, 476.
Fontaine, G., Van Horn, H. M., Böhm, H.-C. and Grenfell, T. C. (1974) *Astrophys. J.*, **193**, 205.
Fowler, R. H. (1926) *M.N.R.A.S.*, **87**, 114.
Fowler, R. H. (1936) *Statistical Mechanics*, Macmillan, New York.
Franck, J. and Hertz, G. (1914) *Verhandl Deut. Physik. Ges.*, **16**, 512.
Gibbs, J. W. (1902) *Elementary Principles in Statistical Mechanics*, Charles Scribner's Sons, New York.
Graboske, H. C., DeWitt, H. E., Grossman, A. S. and Cooper, M. S. (1973) *Astrophys. J.*, **181**, 457.
Graboske, H. C., Harwood, D. J. and Rogers, F. J. (1969) *Phys. Rev.*, **186**, 210.
Greenstein, J. L. (1969a) *Comments Ap. Space Sci.*, **1**, 62.
Greenstein, J. L. (1969b) in *Low Luminosity Stars*, S. Kumar, Ed., Gordon and Breach, New York.
Greenstein, J. L., Oke, J. B. and Shipman, H. L. (1971) *Astrophys. J.*, **169**, 563.
Greenstein, J. L. and Sargent, A. (1974) *Astrophys. J. Suppl.*, **28**, 157.
Hamada, T. and Salpeter, E. E. (1961) *Astrophys. J.*, **134**, 683.
Hansen, J. P. (1973) *Phys. Rev.*, **8A**, 3096.
Hansen, J. P., Pollock, E. L. and McDonald, I. R. (1974) *Phys. Rev. Letters*, **32**, 277.
Harnwell, G. P. and Livingood, J. J. (1933) *Experimental Atomic Physics*, McGraw-Hill, New York.
Heisenberg, W. (1927) *Z. Phys.*, **43**, 172.
Henyey, L. G., Forbes, J. E. and Gould, N. L. (1964) *Astrophys. J.*, **139**, 306.
Hermann, A. (1971) *The Genesis of Quantum Theory (1899–1913)*, MIT Press, Cambridge, Mass.
Hintzen, P. and Strittmatter, P. A. (1974) *Astrophys. J. (Letters)*, **193**, L111.
Hubbard, W. B. (1966) *Astrophys. J.*, **146**, 858.
Hubbard, W. B. and Lampe, M. (1969) *Astrophys. J. Suppl.*, **18**, 297.
Hughes, A. L. (1912) *Phil. Trans. Roy. Soc.*, **212**, 205.
Kirchhoff, G. (1860) *Ann. Physik*, **109**, 292.
Kirzhnits, D. A. (1960) *Soviet Phys. JETP*, **11**, 365.
Kovetz, A. and Shaviv, G. (1970), *Astron. Ap.*, **8**, 398.
Kovetz, A. and Shaviv, G. (1973) *Astron. Ap.*, **28**, 315.
Kovetz, A. and Shaviv, G. (1976) preprint.
Lamb, D. Q. (1974) Ph.D. Thesis, University of Rochester.
Lamb, D. Q. and Van Horn, H. M. (1975) *Astrophys. J.*, **200**, 306.
Lampton, M., Margon, B., Paresce, F., Stern, R. and Bowyer, S. (1976), *Astrophys. J. (Letters)*, **203**, L71.
Landau, L. D. and Lifshitz, E. M. (1958) *Statistical Physics*, translated by E. and R. F. Peierls, Pergamon Press, London.
Ledoux, P. and Sauvenier-Goffin, E. (1950) *Astrophys. J.*, **111**, 611.
Lee, T. D. (1950) *Astrophys. J.*, **111**, 625.
Lyman, T. (1906) *Astrophys. J.*, **23**, 181.
Magni, G. (1972) Internal Report #35, Laboratorio di Astrofisica Spaziale (CNR) Frascati.
Marshak, R. E. (1940) *Astrophys. J.*, **92**, 321.
Marshak, R. E. (1966) in *Perspectives in Physics*, R. E. Marshak, Ed., Wiley-Interscience, New York.
Maxwell, J. C. (1881) *A Treatise on Electricity and Magnetism*, Clarendon Press, Oxford.
Maxwell, J. C. (1890) *The Scientific Papers of James Clark Maxwell*, Cambridge University Press, Cambridge; reprinted 1952 by Dover Publications, New York.
Mestel, L. (1950) *Proc. Cambridge Phil. Soc.*, **46**, 331.
Mestel, L. (1952) *M.N.R.A.S.*, **112**, 583.
Mestel, L. (1965) in *Stars and Stellar Systems VIII, Stellar Structure*, L. H. Aller and D. B. McLaughlin, Eds., University of Chicago Press, Chicago, p. 297.
Mestel, L. and Ruderman, M. A. (1967) *M.N.R.A.S.*, **136**, 27.
Millikan, R. A. (1909) *Phys. Rev.*, **29**, 560.
Millikan, R. A. (1916) *Phys. Rev.*, **7**, 355.
Ostriker, J. P. (1971) *Ann. Revs. Ast. Ap.*, **9**, 353.
Ostriker, J. P. and Axel, L. (1969) in *Low Luminosity Stars*, S. Kumar, Ed., Gordon and Breach, New York, p. 357.

Paschen, F. (1908) *Ann. Phys.*, **27**, 537.
Pauli, W. (1925) *Z. Phys.*, **31**, 765.
Planck, M. (1900) *Verhandl Deut. Physik. Ges.*, **2**, 202, 237.
Pollock, E. L. and Hansen, J. P. (1973) *Phys. Rev.*, **8A**, 3110.
Pontecorvo, B. M. (1959) *Soviet Phys. JETP*, **9**, 1148.
Richardson, O. W. and Compton, K. T. (1912) *Phil. Mag.*, **24**, 575.
Ritz, W. (1908) *Ann. Phys.*, **25**, 660; *Ap. J.*, **28**, 237.
Rutherford, E. (1911) *Phil. Mag.*, **21**, 669.
Salpeter, E. E. (1961) *Astrophys. J.*, **134**, 669.
Salpeter, E. E. and Van Horn, H. M. (1969) *Astrophys. J.*, **155**, 183.
Savedoff, M. P., Van Horn, H. M. and Vila, S. C. (1969) *Astrophys. J.*, **155**, 221.
Savedoff, M. P., Van Horn, H. M., Wesemael, F., Auer, L. H., Snow, T. P. and York, D. G. (1976) *Astrophys. J. (Letters)*, **207**, L45.
Schatzman, E. (1958) *White Dwarfs*, North-Holland Publishing Co., Amsterdam.
Schrödinger, E. (1926) *Ann. Physik (4)*, **79**, 361, 489; **80**, 437; **81**, 109.
Schwarzschild, M. (1958) *Structure and Evolution of the Stars*, Princeton University Press, Princeton.
Shaviv, G. and Kovetz, A. (1972) *Astron. Ap.*, **16**, 72.
Shipman, H. L. (1972) *Astrophys. J.*, **177**, 723.
Slater, J. C. (1973) in *Wave Mechanics*, W. C. Price, S. S. Chissick, and T. Ravensdale, Eds., Wiley, New York.
Sommerfeld, A. (1928) *Z. Physik*, **47**, 1.
Sommerfeld, A. and Bethe, H. A. (1933) *Hdb. d. Physik*, **24**, (2), 333.
Stefan, J. (1879) *Wien. Ber. (2)*, **79**, 391.
Stoner, E. C. (1930) *Phil. Mag.*, **9**, 944.
Strittmatter, P. A. and Wickramasinghe, D. T. (1971) *M.N.R.A.S.*, **152**, 47.
ter Haar, D. (1967) *The Old Quantum Theory*, Pergamon Press, Oxford.
Thomson, G. P. (1928) *Proc. Roy. Soc.*, **A117**, 600.
Thomson, J. J. (1897) *Phil. Mag.*, **44**, 293.
Van Horn, H. M. (1968) *Astrophys. J.*, **151**, 227.
Van Horn, H. M. (1971) in *White Dwarfs*, I.A.U. Symp. No. 42, W. J. Luyten, Ed., Reidel, Dordrecht, p. 97.
Vila, S. C. (1965) Ph.D. Thesis, University of Rochester.
Vila, S. C. (1966) *Astrophys. J.*, **146**, 437.
Vila, S. C. (1967) *Astrophys. J.*, **149**, 613.
Vila, S. C. (1969) in *Low Luminosity Stars*, S. Kumar, Ed., Gordon and Breach, New York, p. 351.
Wehrse, R. (1975) *Astron. Ap.*, **39**, 169.
Weidemann, V. (1968) *Ann. Revs. Ast. Ap.*, **6**, 351.
Wien, W. (1893) *Preuss. Akad. Wissens. Sitz. Ber.*, p. 55.
Wigner, E. and Seitz, F. (1933) *Phys. Rev.*, **43**, 804.

Applications of Model Hamiltonians to the Electron Dynamics of Organic Charge Transfer Salts

MARK A. RATNER
Northwestern University, Illinois, U.S.A.
and
JOHN R. SABIN and SAMUEL B. TRICKEY
University of Florida, Florida, U.S.A.

1. INTRODUCTION

Many of the charge-transfer* salts based on planar aromatic organic electron-acceptor molecules, such as tetracyanoquinodimethan or chloranil, crystallize in segregated stacks of donor and acceptor molecules. These stacks provide large overlap of valence orbitals, and as a result the salts can become rather good conductors, though the conductivity is usually activated due to disorder. To describe magnetic excitations in these systems, the Heisenberg (1926, 1928) model has been applied quite successfully (Soos, 1974). In particular, salts have been observed which exhibit both delocalized and localized spin excitations, depending on the magnitude of the Heisenberg exchange integral J. The electrical properties of these systems can be discussed in terms of modified Hubbard (1963) or Pariser–Parr–Pople (Parr, 1963; Linderberg and Öhrn, 1968, 1973) models, which have been shown (Van Vleck, 1966; Anderson, 1963; Linderberg and Öhrn, 1968) to be equivalent (to second order) to the Heisenberg (1926, 1928; Dirac, 1929) model.

We shall discuss the general nature of the excitations in these systems, certain experimental results and the numerical estimation of tunnelling, exchange and repulsion parameters. Comments about vibronic effects and geometry dependence will also be included. Finally, questions of dimensionality are considered very briefly, in connection with the low-temperature phase transitions which these systems undergo. Similarities to other molecular systems are indicated.

---
*We will use the term charge-transfer salts generically; it also will include the ion–radical salts, which are simply charge-transfer salts with essentially complete charge transfer (Soos, 1974).

We wish to discuss the application of various model Hamiltonians, related to that derived by Heisenberg in 1926–28, to the description of transport properties of organic charge transfer salts. This has been an area of active research for the last score of years, and has received recent impetus from the experimental discoveries of the electronic phase transitions and extremely high conductivities in some salts containing the electron-acceptor tetracyanoquinodimethan (TCNQ, I). The entire subject has been recently reviewed (Soos, 1974; Keller, 1975; Garito and Heeger, 1974), and it is not our purpose to catalogue either the experimental findings or their explicit theoretical interpretation. Rather, we shall point out the extreme utility and signal importance of these models in understanding the phenomenology of transport, in semiquantitative estimation of the relevant parameters, and even in the design of new experimental systems which it is hoped will exhibit particular transport properties.

TCNQ

(I)

The systems which we will consider are all salts consisting of organic electron donors and acceptors, usually in a rather simple stoichiometric ratio (e.g. 1:1 donor:acceptor in tetrabutylammonium TCNQ, 2:3 in acridinium TCNQ). The acceptor species, and usually the donor species as well, is a planar or nearly planar aromatic hydrocarbon containing a delocalized, polarizable π-electron system. Frequently the strong unidirectional ('stacking') interactions perpendicular to the molecular planes result in a crystal habit characterized by one-dimensional stacking of donors and acceptors.* This structural unidimensionality leads to interesting and highly specific pseudo one-dimensional transport properties. The Heisenberg model, however, when the exchange parameters are chosen to be anisotropic, quite well characterizes the spin excitations in this system and has even been used successfully to interpret the effects of the actual three-dimensionality on the pseudo-one-dimensionality of the transport properties (e.g. Plumlee and coworkers, 1975; Plumlee, 1975). We shall discuss specific instances where these three-dimensional effects have been observed and successfully described.

The Heisenberg (1926, 1928) model deals with interaction between itinerant or localized pseudospins. In understanding these organic systems, a crucial

---

*This is not necessary; if the conductivity is of p-type, the donors should stack but the acceptors do not have to. Possible cases whose conductivity is high ($\sigma \cong 10^3 \, \Omega^{-1} \, cm^{-1}$ at room temperature) include TTF bromide (La Placa and coworkers, 1975) and TTF hexacyanobutadiene (Wudl and Southwick, 1974) (for TTF see structure (II) below).

distinction must be borne in mind. In application of the Heisenberg or related model to a system such as alkali metals, the localized electronic state can be characterized in terms of a Wannier function which is essentially a single hydrogen-like atomic orbital whose characteristic energy is therefore of order 13 eV. In these organic systems, however, the local basis functions are molecular orbitals which are in turn delocalized over the π-electron system of the molecule. Therefore, the energy spacing between these local functions can become extremely small; even more important, electron delocalization within the molecule can lead to highly reduced coulombic repulsions between electrons of opposite spin in the same orbital on the same site—the electrons have enough space to avoid one another (Garito and Heeger, 1974; Ratner and coworkers, 1973). Therefore the localized spin should be thought of as occupying not an atomic Wannier state but instead a molecular Wannier state. This one-electron function is often called the lowest unoccupied molecular orbital (LUMO). Thus when describing the electronic or spin transport in these systems we will always use these molecular states as a basis. Our considerations will therefore be restricted to intermediate energy phenomena (susceptibility, EPR spectrum, conductivity, electronic spectra) which are largely determined by the behaviour of the outermost valence electrons.

In the next section we will review briefly the relevant molecular electronic structure concepts which will facilitate discussion of the characteristics of the localized electrons. We then outline the types of experimental results which have been obtained on these systems, and whose description in terms of model Hamiltonians is the topic of this review. Specifically, Section 3 is devoted to the Heisenberg Hamiltonian itself and its application to the magnetic properties of these organic salts. Section 4 delineates very briefly some of the structural aspects of these materials, and then considers the conductivity process in them, as well as the various types of electronic excitation which can be observed. A unified description of these phenomena is afforded by the use of the Hubbard (1963) model, which is, to second order, equivalent to the Heisenberg model. In the last section we discuss some recent experimental results and some current theoretical efforts.

2. MOLECULAR ELECTRONIC STRUCTURE AND TRANSPORT PROPERTIES

To avoid difficulties in nomenclature, we will define the mobile electrons of the system, i.e. those which can migrate from the donor to the acceptor site, as *transfer electrons*.* In consonance with this usage we refer to the donor and acceptor one-electron molecular levels as donor transfer levels and acceptor transfer levels.

*Demarking electrons in this way is of course contrary to the spirit of the Pauli principle. Our characterization here is in terms of a one-electron picture, and the antisymmetrization process is always understood.

The organic molecules with which we shall be concerned are planar or nearly planar. The symmetry element corresponding to reflection in this plane separates the atomic basis functions into those which are symmetric under this operation and those which are antisymmetric; the former are conventionally called σ and the latter π.* For many experimental properties the response characteristics of the σ- and π-electrons can be separated. This can be understood on the basis of separate response time scales, π-electron energy level differences being of the order of 1 eV, while σ-electron differences tend to be at least ten times as large. For discussing transport properties of the charge-transfer salts, the σ-electrons may be assumed able to adjust to external perturbations essentially independently of the π-electrons. Since transport properties are dominated by π-electron behaviour, one can, to an excellent approximation, neglect any role of the σ-electrons. Actually, one can see that the σ- core acts like the nuclear motion: interaction effects with the π-electrons are small because of differences in the characteristic time scale, but can be reckoned using perturbation theory in the rare instances when this becomes necessary (Lykos and Parr, 1956; Linderberg and Öhrn, 1968; Fischer, 1970). The simplification resulting from this σ-π separation is quite considerable: of the 104 electrons in the TCNQ molecule, for instance, only 24 are π-electrons. Formally, this means that in defining a model Hamiltonian to describe transport processes, one can limit the basis functions which appear in the expansion of the field operator to those of π-type. Experimentally, the approximate validity of the σ-π separation is confirmed by dielectric function measurements on graphite (Taft and Philipp, 1965) as well as by the great success of π-electron models in describing isolated organic molecules (Parr, 1963).

One can then limit this expansion basis to those π-basis orbitals located on each atom which are occupied in the isolated atom. Then, after symmetrical orthogonalization of the basis functions or the neglect of overlap, one can write the molecular Hamiltonian in the usual way as:

$$\mathcal{H} = \sum_\mu \sum_{i,j} \beta_{ij} a^\dagger_{i\mu} a^\dagger_{j\mu} \sum_{\mu,\mu'} \sum_{ijkl} V_{ijkl} a^\dagger_{i\mu} a^\dagger_{j\mu'} a_{l\mu'} a_{k\mu} \tag{1}$$

Here $a^\dagger_{i\mu}$ creates an electron in atomic basis function u_i with spin component μ. The matrix elements β_{ij} and V_{ijkl} are nearly always chosen semi-empirically; this is necessitated by the neglect of the σ-electron system. The most successful of these semi-empirical model Hamiltonians, the so-called Pariser–Parr–Pople (PPP) model, has been widely and effectively applied to the treatment of both chemical and physical properties of π-electron molecules.

The usual calculations with the above Hamiltonian (1) are done using SCF or CI methods. For our purposes, we will employ an SCF or molecular-orbital scheme. We will thus define a molecular orbital φ_p by

$$\varphi_p = \sum_i C_{pi} u_i \tag{2}$$

---

*Strictly speaking, σ and π refer to angular momenta around the local bond axis. Our usage is that conventional in describing organic molecules. The two definitions are equivalent except in certain pathological cases such as the —C≡N moiety, which has four π-electrons.

where the molecular-orbital coefficients C_{pi} are chosen using the variational principle. The molecular orbital ground state wave function can then be written as a single determinant:

$$|g\rangle = \prod_{p=\text{occ}} a^{\dagger}_{p\mu} |\text{vacuum}\rangle \qquad (3)$$

where the product is limited to occupied π-molecular orbitals. The highest energy occupied molecular orbital (HOMO) will be the donor level and the lowest unoccupied molecular orbital (LUMO) will be the acceptor level. We will denote the energy of the LUMO, φ_L by ε_L, and similarly ε_H for the HOMO, φ_H. In fact, even within an SCF description, the energy levels ε_p and wavefunctions φ_p will of course change upon ionization or electron capture. In our model Hamiltonians we will nevertheless assume implicitly that the energy level ε_p is independent of occupation number. This assumption, while correct for the $1s$ level of the hydrogen atom for which the Heisenberg model was originally derived, is quite incorrect for the molecular-orbital levels of π-electron molecules. This is a shortcoming in the theory, which must be remembered when attempting to make quantitative correlations of model Hamiltonian parameters with experimental data.*

The π-electron energy level scheme for a semiempirical (CNDO) (Pople and Beveridge, 1970) model of TCNQ is indicated in Figure 1a. Similarly, a CNDO energy level scheme for the typical donor molecule tetrathiofulvalene (II, TTF) is indicated in Figure 1b. Two conclusions are readily apparent from these levels. Firstly, the optical transition is, as indicated above, of the order of volts. The relevant orbitals are delocalized over the π-system and the particle/hole excited state is stable; narrow absorption lines are characteristic of these molecules. Secondly, the Koopmans' theorem ionization potential of TTF is rather small (8.4 eV), whereas the low-lying unoccupied (LUMO) orbital of TCNQ is indicative of a relatively high electron affinity. These are precisely the characteristics required for efficient formation of charge-transfer salts with interesting electron-transport properties: stable, fairly low excited states, delocalized transfer levels, high electron affinity of the acceptor and low ionization potential of the donor (Soos, 1974). We will see that nearly all of the important charge transfer systems fulfill all of these criteria.

For certain processes, in particular radiative and non-radiative decay of excited states, electron–vibration interaction effects can be quite important in π-electron systems. A number of discussions of this problem have been given, both for the isolated molecule and for electron–phonon interaction in organic solids. Bari (1974) and others have suggested that small polaron behaviour may

---

*The alternative, however, is a full SCF calculation of a crystal, which is nearly too horrible to imagine. (Pottsork and Twitchell, 1975) Soos and coworkers (Klein and Soos, 1971; Strebel and Soos, 1970) have in fact shown that within a minimum basis set for the crystal, consisting of one function per site, one can define many-electron creation operators which obey Fermi statistics and create or destroy actual eigenstates of the neutral or ionic molecular species at each lattice site. This interpretation, which is valid only in a minimum basis, renders the frozen core approximation which we have sketched here more palatable.

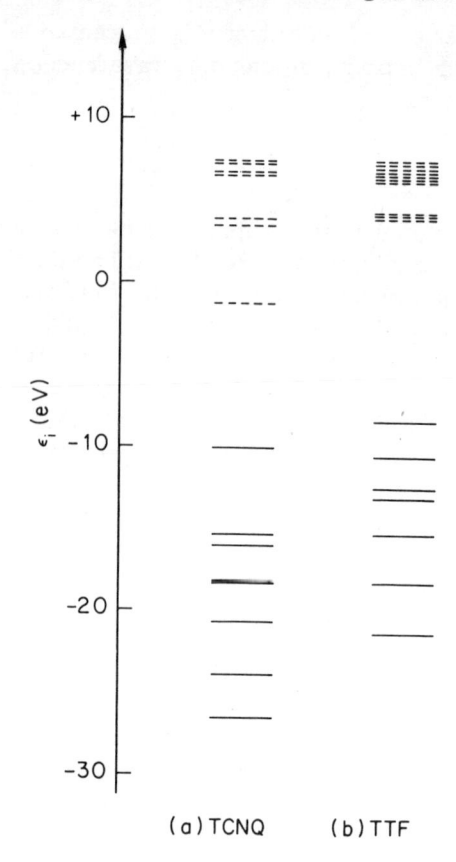

Figure 1 π-Type orbital energies of:
(a) TCNQ
(b) TTF
calculated by CNDO/2 at the experimental geometries

be observed in some transport processes in the charge-transfer solids, and estimates of the coupling constants (Bright and coworkers, 1974; Krogh-Jespersen and Ratner, 1975) have been made. For most systems, however, polaron effects do not seem important, apart from the usual Franck–Condon sort of interaction; a typical example is afforded by the discussion of the temperature dependence of the magnetic resonance spectrum of maleonitriledithiolato (mnt, III) complexes of divalent copper given by Hoffman and coworkers (Plumlee and coworkers, 1975; Plumlee, 1975). We will therefore concentrate on purely electronic transport and will neglect the role of vibrational interactions.

Several severe materials problems are encountered in preparation of the charge-transfer salts. The isolated donor and acceptor molecules can usually be prepared rather straightforwardly, and their purification is not unduly difficult. The preparation of the charge-transfer species, however, is extremely difficult

TTF
(II)

mnt
(III)

in many cases, and single crystals are not readily available for some of the most interesting systems. A large variety of preparative procedures has been employed, the most successful being crystallization using slow diffusional mixing in a high dielectric solvent such as acetonitrile. Due to the quasi-one-dimensional structure of many of the most interesting compounds, the role of crystalline defects, stacking faults, etc., is very important in these systems. Indeed, the difficulty in reproducing certain transport measurements* is partly due to problems in preparing reasonable crystals. Consequently, a disordered systems model has been applied to the transport properties of the charge-transfer systems (Brenig and coworkers, 1972; Bloch and coworkers, 1972). Since our major emphasis is on the understanding of transport processes in terms of simple model Hamiltonians, we shall not discuss the disordered system aspects of these materials.

The most important transport measurements on the charge-transfer salts have been paramagnetic resonance studies of the spin properties, optical studies and electrical conductivity studies. Of these, the spin measurements are most fully interpreted; they can be characterized straightforwardly in terms of the Heisenberg model (Hoffman and Hughes, 1970; Soos, 1974). The optical measurements are often more difficult to interpret, since the interesting charge transfer excitations can be in the infrared where overlap of the vibrational spectra makes identification impossible. Nevertheless, some important optical studies have been completed and are generally interpreted in terms of the Hubbard model (Torrance and coworkers, 1975). Most of the flurry of recent

---

*Considerations of chemical stability may also be important. Indeed, van Duyne and coworkers (Suchanski and coworkers, 1975) have shown that many TCNQ salts when prepared in the usual manner, are oxidized by atmospheric oxygen to the yellow, stable, ionic decay product (DCTC). The presence of this ion has confused several optical measurements, distorted several semiempirical calculations, and may possibly seriously effect conductivity measurements.

DCTC

activity in this area has been an outgrowth of suggestions by various workers (Little, 1964; Bardeen, 1973; Epstein and coworkers, 1973) that Fröhlich (1954) type superconductivity might be observed in these pseudo-one-dimensional systems. Consequently a large number of electrical conductivity measurements have been made using various procedures (two probe, four probe, microwave, etc.); the exact interpretation of these measurements is still not absolutely clear, but a Hubbard model involving partial charge transfer, complicated by phonon interaction effects and incipient three dimensionality at low temperature seems (Keller, 1975) capable, at least qualitatively, of describing the observations thus far.

3. SPIN PROPERTIES AND THE HEISENBERG HAMILTONIAN

Many derivations of the Heisenberg Hamiltonian have been given. The most helpful in general is based on a consideration using perturbation theory of weak electrostatic interactions between non-overlapping basis functions. We will, instead, consider the two site, two particle case, first treated by Heitler and London (1927) in an attempt to see how this interaction arises in the simplest case.

Consider then (Anderson, 1963; Haug, 1972), a two particle, two site system, such as the hydrogen molecule. Denoting the two $1s$ basis functions on either side as u_r and u_l, one can construct four states as:

$$\psi_S = \tfrac{1}{2}[u_l(1)u_r(2) + u_r(1)u_l(2)][\alpha(1)\beta(2) - \beta(1)\alpha(2)] \tag{4}$$

$$\psi_{T,m} = \frac{1}{\sqrt{2}}[u_l(1)u_r(2) - u_r(1)u_l(2)] \begin{cases} \alpha(1)\alpha(2) & m = 1 \\ \beta(1)\beta(2) & m = -1 \\ \dfrac{1}{\sqrt{2}}[\alpha(1)\beta(2) + \beta(1)\alpha(2)] & m = 0 \end{cases}$$

These are the Heitler–London, or valence-bond states. The singlet ψ_s is a bonding state, while the degenerate triplets are antibonding. The energies are given by

$$E_S = E_0 + \frac{C + A}{1 + Q^2} + \frac{1}{R} \tag{5}$$

$$E_T = E_0 + \frac{C - A}{1 - Q^2} + \frac{1}{R} \tag{6}$$

where the relevant interaction integrals are defined (Hartree atomic units) by:

$$E_S = \langle \psi_S | \mathcal{H} | \psi_S \rangle / \langle \psi_S | \psi_S \rangle \tag{7}$$

$$E_T = \langle \psi_{T,m} | \mathcal{H} | \psi_{T,m} \rangle / \langle \psi_{T,m} | \psi_{T,m} \rangle \tag{8}$$

$$Q = \langle u_l | u_r \rangle \tag{9}$$

$$C = \left\langle u_l(1) u_r(2) \left| \frac{1}{r_{12}} - \frac{1}{r_{l2}} - \frac{1}{r_{r1}} \right| u_l(1) u_r(2) \right\rangle \tag{10}$$

$$A = \left\langle u_l(1) u_r(2) \left| \frac{1}{r_{12}} - \frac{1}{r_{l2}} - \frac{1}{r_{r1}} \right| u_r(1) u_l(2) \right\rangle \tag{11}$$

while R is the internuclear distance and E_0 is the energy of the separated atoms. The two electron terms in C and A are respectively coulomb and exchange repulsions. At the risk of triviality, we point out again that the coulombic repulsion is responsible for the exchange term, which arises from antisymmetry. The singlet/triplet energy difference is then:

$$\Delta E = E_S - E_T = 2 \frac{A - CQ^2}{1 - Q^4} \tag{12}$$

which will vary in sign according to the sign of A: when $A < 0$, as it is for H_2, the singlet state lies lower, which is equivalent to antiferromagnetism, so that a negative A corresponds to antiferromagnetic behaviour.

It is now convenient to write the Hamiltonian in its state representation as:

$$\mathcal{H} = H_0 + H' + H_{ex} \tag{13}$$

Here H_0 is the energy operator for the isolated atoms, H' is the coulombic repulsion operator and H_{ex} is the operator describing electron exchange. These last two operators arise as stated from the coulombic repulsion: if an antisymmetrized product wave function is employed, H_{ex} should be set equal to zero, whereas if a Hartree product wave function is employed, the form of (13) is correct. The form of H_{ex} is obtained by noting that

$$\left. \begin{array}{c} E_S \\ E_T \end{array} \right\} = E_0 + C \pm A \tag{14}$$

if the overlap integrals Q are neglected. We then rewrite

$$\pm A = A[1 - S(S+1)] \tag{15}$$

where S is the total spin, 0 or 1, for singlet and triplet, respectively. Or, since $S(S+1)$ is the eigenvalue of $\mathcal{S}^2$, H_{ex} can be written:

$$H_{ex} = A(1 - \mathcal{S}^2) \tag{16}$$

We also note that the two parts of the Hamiltonian commute, with H_0 and H' depending only on spatial variables and H_{ex} depending only on spin variables.

The Heisenberg form results if the spin operator $\mathscr{S}$ is replaced by the individual spin operators:

$$\mathscr{S} = \sigma_1 + \sigma_2$$
$$\mathscr{S}^2 = \sigma_1^2 + \sigma_2^2 + 2(\sigma_1 \cdot \sigma_2) \tag{17}$$

The first two terms on the right in (17) are constant, and therefore:

$$H_{ex} = -2A(\sigma_1 \cdot \sigma_2) \tag{18}$$

This form makes it transparent that negative A is equivalent to opposite spins, or antiferromagnetic coupling. If we take truly orthogonal orbitals and therefore have vanishing overlap,* the last two terms of A, equation (11), drop out so that we can write, using the usual notation, for the two site problem:

$$H_{ex} = -2J(\sigma_1 \cdot \sigma_2) \tag{19}$$

where

$$J = \left\langle u_l(1) u_r(2) \left| \frac{1}{r_{12}} \right| u_r(1) u_l(2) \right\rangle \tag{20}$$

This form of J is clearly positive so that it would predict ferromagnetic coupling in the hydrogen molecule. Thus we see that the presence of non-orthogonal basis functions is crucial to the correct Heitler–London interpretation of chemical binding as due to an effectively attractive interaction between opposite spin electrons. In solids, however, the simplification of orthogonal orbitals is often made, in which case the correct J must be very carefully considered; of course, density-of-states effects can lead to antiferromagnetic behaviour, even in the case of much more general Hamiltonians (Lieb and Mattis, 1962; Van Vleck, 1966). In general, the notation J actually refers to our parameter A, and to follow accepted usage we will hereafter call the exchange parameter J. Considerations of the role of J in determining ferromagnetic behaviour, pioneered by Heisenberg (1926, 1928) have been given by Anderson (1963) and by Herring (1966).

The two site problem makes clear the source of the exchange terms and thus the utility of the Heisenberg models. If we consider a crystal, include tight-binding exchange (only between nearest neighbours), assume that the exchange is isotropic and assume exactly one spin per site, we obtain the Heisenberg Hamiltonian

$$H_{ex} = -J \sum_{m,n} (\sigma_m \cdot \sigma_n) \tag{21}$$

where σ_m is the spin operator at the mth site. Since the remaining terms, $H_0 + H'$ are spin independent, the Heisenberg exchange term should describe the spin properties of the system completely when the assumptions underlying

---

*This means that the overlap integral, Q, is vanishing, not that the wavefunctions have no spatial overlap.

its derivation are correct. Of these assumptions, those which are crucial in the analysis of charge-transfer systems are isotropy of J and unit spin population per site. The anisotropy of J can be easily taken into account: one simply writes, for pseudo-one-dimensional crystals,

$$H_{ex} = -J' \sum_{m,n} (\sigma_{mx}\sigma_{nx} + \sigma_{my}\sigma_{ny}) - J \sum_{m,n} \sigma_{mz}\sigma_{nz} \qquad (22)$$

where the easy axis exchange, J, is considerably larger in absolute value than the perpendicular component, J'.

The assumption of unit spin population per site obtains less often in charge-transfer crystals. When the Madelung and ionization potential terms allow unit electron transfer, such as in the alkali metal TCNQ salts, analysis of the magnetic properties using a simple exchange model can be carried out with little difficulty. For systems in which non-negligible donor *and* acceptor band widths occur, or in which charge transfer is incomplete for some other reason, the Heisenberg terms will become less important than terms corresponding to charge-transfer excitations, and analysis of the spin properties becomes more difficult. These systems include many of the most interesting charge-transfer salts, such as tetrathiofulvalene (TTF, II)–TCNQ. This material was originally thought (Coleman and coworkers, 1973) to have 1.00 electrons transferred per site, corresponding to 'complete' charge transfer. Recent experimental (Wozniak and coworkers, 1975; Grobman and coworkers, 1974; Torrance and coworkers, 1975; Coppens, 1975) and theoretical (Klymenko and coworkers, 1975) work suggests strongly that the charge transfer is closer to 0.5 than to 1.0 in TTF–TCNQ. The systems exhibiting incomplete charge transfer are among the most exciting presently being studied, but the experimental work and its interpretation are often less clear cut than in the 1:1 systems. These systems are discussed in Section 4.

To our knowledge, there are no charge-transfer systems which exhibit ferromagnetic exchange. Thus, neglecting impurities, surface effects, etc., these crystals are expected to be diamagnetic at very low temperatures and to exhibit ordinary Curie–Weiss behaviour at higher temperatures. This has indeed been found to be the case experimentally. As a result, bulk susceptibility studies are not particularly difficult to understand in terms of the simple exchange model. As the temperature is increased, electrons can be promoted above the Fermi surface into paramagnetic states so that an activated paramagnetic susceptibility can be observed. For example, the TMPD(IV)–ClO_4 crystal is a weak semiconductor which displays activated paramagnetism with an activation energy of less than 0.01 eV. In the case of incomplete charge transfer, the activation energies can be larger. In the case of alternating stacks or of dimerization within stacks, the activation energy can be much larger; for instance it is 0.39 eV in the dimerized RbTCNQ. Figure 2 shows an experimental susceptibility plot for triethylammonium $(TCNQ)_2$; it is taken from the pioneering work of the DuPont Group (Kepler, 1963). The background is due

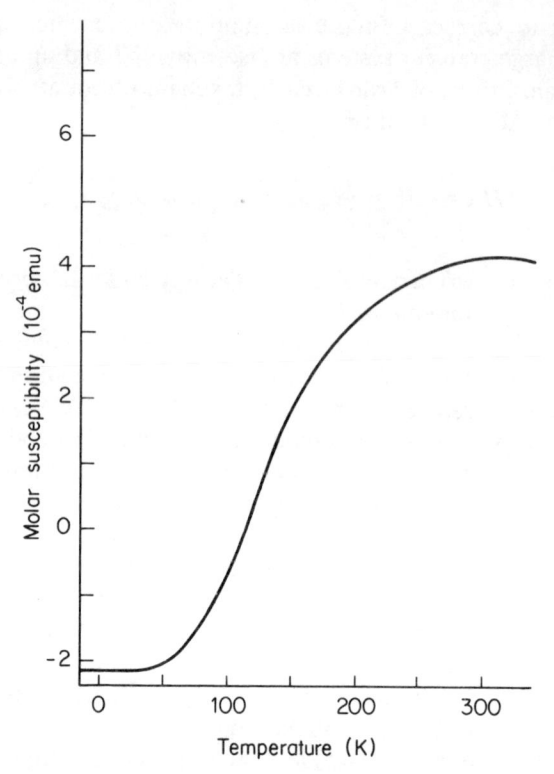

Figure 2 Temperature dependence of magnetic susceptability of $(Et_3NH)(TCNQ)_2$ after correction for paramagnetic impurities. After Kepler (1963)

to free-radical impurities. The temperature increase can be fitted nicely by assuming a singlet–triplet Boltzmann distribution, with an exchange splitting $J = 0.034$ eV. The susceptibility of low and intermediate conductivity charge-transfer and ion radical salts nearly always displays this behaviour. More elaborate theoretical models (Adler and Brooks, 1967) can be used to fit even more accurately (Keller, 1975).

An example of a case in which an essentially complete understanding of the spin dynamics has been achieved is found in the bis(tetrabutylammonium) and bis(tetraethylammonium) salts of $Cu(mnt)_2^{2-}$, studied by Hoffman and coworkers (Plumlee and coworkers, 1975).* The large cationic species in these materials are positively charged and therefore exactly 1.00 transfer electrons per site are found on the $Cu(mnt)_2^{2-}$ species. The copper is nominally in a d^9 configuration, with a spin of 1/2. The structure consists of isolated stacks of

---

*Although this is, strictly speaking, an organometallic rather than an organic species, the theoretical discussion and indeed the dynamics are the same as for many of the organic charge transfer salts. In addition a large number of related organometallic species is known (Interrante, 1975; Keller, 1975) and these represent another large class of pseudo-one-dimensional systems whose behaviour can be characterized in terms of Heisenberg and related model Hamiltonians.

nearly planar $Cu(mnt)_2^{2-}$ species (III), with space group $C_i^1-P\bar{1}$ for the tetrabutylammonium salt. The $Cu(mnt)_2^{2-}$ ions stacked along the C-axis of the crystal form a regular linear chain of crystallographically equivalent $S = 1/2$ species. The Cu–Cu distance along this axis is found to be 9.403 Å. The planes are offset slightly from one another; the plane-to-plane perpendicular distance being only 6.91 Å. The cations do not interleave the anionic chain, but do effectively separate the chains. The susceptibility, in the range above 13 K, exhibits Curie–Weiss behaviour with extremely small (less than 1 K) Weiss constant, which indicates very weak Heisenberg exchange.

In discussing their EPR observations on these systems Hoffman and coworkers (Plumlee and coworkers, 1975) considered the Heisenberg model which contains, in accordance with the observed structure, two different exchange parameters, as in equation (22). The C-axis J was taken as 0.0107 cm^{-1} at 4.2 K in the tetrabutylammonium species, while for the tetraethylammonium, where the Cu–Cu distance is much smaller (7.595 Å), a larger exchange parameter of 2.0 cm^{-1} was used. Weak temperature dependence is observed for these exchange parameters, probably because thermal motion, which is rapid on the EPR time scale, averages over some geometrical configurations in which the exchange can be comparatively larger [for discussion of this effect in transition metal complexes, cf. Hay and coworkers, (1975)]. For the tetrabutylammonium salt, the interstack exchange parameter, J', has no experimental consequences and can therefore be neglected, whereas the tetraethylammonium salt exhibits line broadening due to J', to which Hoffman and coworkers have assigned the value 0.06 cm^{-1}. The other important Hamiltonian term affecting the EPR lineshape is the nuclear hyperfine interaction. For the tetrabutylammonium salt, the principal nuclear hyperfine components are 165 G and 41 G, and the relative size of the hyperfine and exchange terms depends on the angular orientation of the crystal in the applied magnetic field. For the tetraethylammonium case, the hyperfine term still contains the same strong angular dependence, but in this case the larger exchange is always the dominant process.

The observed EPR lineshapes in these crystals mirror the relative sizes of the hyperfine and exchange terms. When J is greater than the magnetic interactions, which is the most common case, a single exchange-narrowed EPR line of width,

$$\Gamma \approx \frac{\langle \Delta H^2 \rangle}{J} \qquad (23)$$

where ΔH^2 is the second moment of the dipolar interactions, will be observed. This is the case for the tetraethylammonium salt. More interesting behaviour is observed in the tetrabutylammonium salt. Here the hyperfine interactions are large enough that the magnetic inequivalence of the $Cu(mnt)_2^{2-}$ sites, due to differing projections of the nuclear spin, $I(Cu) = 3/2$, is not averaged by large exchange. Therefore the observed line exhibits hyperfine splittings. These are

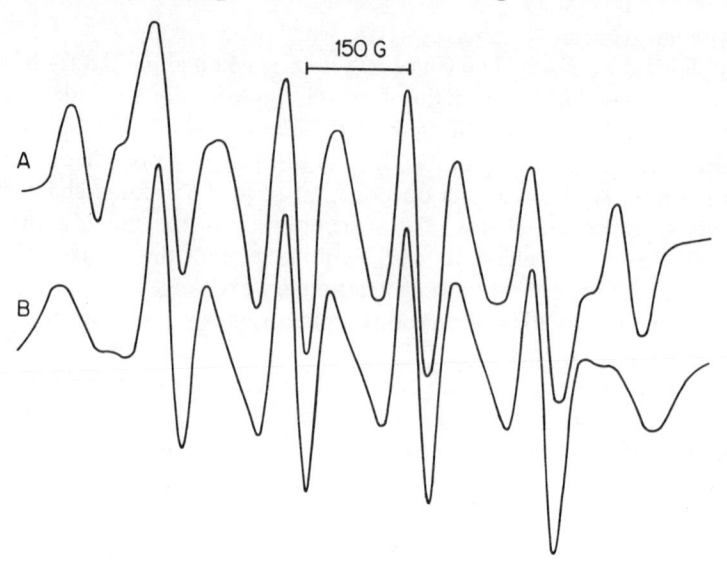

Figure 3 EPR spectrum (A) and computer simulation (B) for $[N(n-Bu)_4]_2[Cu(mnt)_2]$ at 4.2 K with $H_0 \| z$ ($\Theta = 0$). After Plumlee (1975)

statistically distributed according to the probabilities of observing short segments with a given nuclear spin complexion. Hoffman and coworkers have carried out a very detailed analysis of the EPR lineshape, and have been successful in explaining all of the observed vagaries. As an example, we present in Figure 3, the observed EPR spectrum, at 4.2 K, of this salt, along with a computer simulation calculated using the hyperfine and exchange values cited above with no adjustable parameters.

The work [on the $Cu(mnt)_2^{-2}$ species] represents a very unusual case in that the hyperfine and exchange components are of the same size and the dipolar splitting component is extremely small due to crystal morphology. In the general case, however, it is still true that a descrition using only the Heisenberg model plus dipolar and intramolecular terms (and charge-transfer terms when necessary) (Soos, 1974) accounts successfully for the observed EPR lineshapes.

4. ELECTRON DYNAMICS

Due to the interplay of Madelung forces, strongly directional van der Waals forces and kinetic effects during crystallization, organic charge transfer materials exhibit a very wide range of crystal structures (Herbstein, 1971; Soos, 1974). Although all of the materials are characterized by face-to-face stacks of planar molecules, the geometry of the stacks, the interstack distance and even the stack composition, vary widely. In Section 3 we discussed the specific case of $Cu(mnt)_2^{2-}$ salts, which crystallize as segregated, homogeneous uniform stacks of single Cu(mnt)$_2^{2-}$ anions. A large number of other stacking arrangements is also observed. For example, in RbTCNQ, the TCNQ stacks contain anion dimers. In tetramethylparaphenylenediamine [TMPD(IV)]–TCNQ, the stacks contain alternating TMPD$^+$ and TCNQ$^-$ ions. In morpholinium–TCNQ, the overall stoichiometry is 2:3. Systems of these structures exhibit interesting magnetic and electronic behaviour. We wish to concentrate here, however, on

TMPD

(IV)

those systems which crystallize in uniform, homogeneous, segregated stacks. Examples include the best electrical conductors among the charge-transfer species, such as TTF–TCNQ, tetraselenofulvalene–TCNQ, and N-methylphenazinium (NMP) (V)–TCNQ. For simplicity, we will consider only

NMP

(V)

electrical conductivity and electronic excitation along the anionic (TCNQ) stack. These are known from thermoelectric power measurements to be responsible for the conductivity of many of these species (not the selenium case). The analysis can be carried out in terms of a generalized site representation theory (Strebel and Soos, 1970; Soos, 1974) which is essentially identical to the Hubbard model for electronic motion and ferromagnetism in narrow band systems. The basic idea is that coulombic repulsion between different molecular units is extremely small due to delocalization of the electronic charge over these large planar π-electron molecules, and that a reasonable description of the electronic behaviour at low energy is therefore afforded by including only electronic repulsions on a single site and band motion between sites. This is essentially a tight-binding model. It is clearly very closely related to the Heisenberg interaction, in that only one- and two-site electronic terms are included.

The Hubbard (1963) model describing the dynamics of n electrons distributed over N sites in one dimension can be written in the case of a non-degenerate (s orbital, or single molecular basis function) band as:

$$\mathcal{H} = \sum_{\mu} \sum_{i,j}' t_{ij} a_{i\mu}^\dagger a_{j\mu} + U \sum_i^N a_{i\uparrow}^\dagger a_{i\uparrow} a_{i\downarrow}^\dagger a_{i\downarrow} \qquad (24)$$

Here $a_{i\mu}^\dagger$ creates an electron with spin μ in the basis function localized at site i; again these basis functions are really molecular eigenstates or, within an SCF picture, one-electron molecular orbitals. The Coulomb parameter, U, describes repulsion between two transfer electrons located at the same site i, while the tunnelling, or band-width parameter t_{ij}, is generally restricted to include only nearest-neighbour interactions. The set of basis functions is nearly always assumed to be orthogonal.* For most systems, the U parameter is considerably larger than the t parameter: for a chain of hydrogen atoms, for instance, this one site repulsion, U, is of the order 13 eV, while the band-width parameter, t, for reasonable separations, is considerably smaller. In the case of a narrow band metal, there will be a band, of width $4t$ with energy centred at U, and another with energy centred at zero. The upper band, which arises from motion of an electron between already occupied sites, will be thermally inaccessible for reasonable atomic or metallic U's. For the organic salts, however, U, as discussed in Section 2, is considerably reduced due to electronic delocalization around the large π-system. Thus, at room temperature, we might expect the presence of a non-zero Boltzmann population of dianionic sites.*

To glean some insight into the relationship between the Heisenberg and Hubbard models, let us, following Van Vleck (1966), consider this case of exactly one electron per site. The lowest states will then be those for which only one electron occupies each site and no doubly occupied sites occur. This configuration has exactly zero energy, since there are no polar terms and the transfer term operating on the non-polar state introduces a polar component. Thus this configuration has zero energy independent of the spin distribution. We now consider the effects of excited states. In the case of two sites, the excited state of polar character can exist only if the electrons are antiparallel; thus the Heitler–London singlet and triplet states, which are degenerate in the Hubbard model, are no longer degenerate when second order perturbation theory is used to correct the energies. If we examine the energies of the hydrogen molecule Heitler–London singlet and triplet states [equation (4)] in

---

*This assumption is not true for the highly conductive charge-transfer salts such as TTF–TCNQ. There is in fact fairly large overlap of the p–π type atomic basis functions and therefore of the molecular site basis functions which appear in the Hubbard Hamiltonian. Formally, it may well be that correlated solutions to the Hubbard model Hamiltonian utilizing non-orthogonal basis functions will be simpler than those which make the orthogonality assumption (Arai, 1968). Indeed, for describing intramolecular interactions, schemes in which the presence of finite overlap is crucial can be very successful (Linderberg and Öhrn, 1973).
*Which as indicated in Section 2, can lead to atmospheric oxidation.

the Hubbard model, they are, as stated both equal to zero. In second order, however, we obtain:

$$E_S^{(2)} = 0 - t^2/u \qquad (25)$$

$$E_T^{(2)} = 0 - 0 \qquad (26)$$

Thus, to second order, the singlet is stabilized by interaction with polar excited states. Van Vleck goes on to argue that this state of affairs (stabilization of antiferromagnetic ground states by interaction with excited levels) will obtain whenever only exchange between equivalent functions occurs, and that Hund's rule or Slater interatomic exchange (Slater, 1936) is necessary for ferromagnetism to occur. For our purposes, a more interesting, though less important, observation is that the matrix element which gives the second-order interaction of (25) can be replaced by the effective Hamiltonian

$$\mathcal{H}_{\text{eff}} = \frac{4t^2}{u} \sum_{j=i\pm 1} (\sigma_i \cdot \sigma_j - \tfrac{1}{4}) \qquad (27)$$

Notice that, if this expression is compared with the Heisenberg model [equation (21)], the effective second order exchange parameter of the Hubbard model is seen to be:

$$J_{\text{eff}} = \frac{-4t^2}{u} \qquad (28)$$

and that the interaction is therefore antiferromagnetic. Thus, the Hubbard model to second order* is seen to be equivalent to the Heisenberg model; a more general proof is given by Linderberg and Öhrn (1968).

Many of the analyses of the Hubbard model start by considering the half-filled band case (equivalently the case of complete charge transfer, or the case of one electron per site). In many organic charge transfer systems, as indicated in Section 3, this is, in fact, expected to be the case, and for a long time it was assumed that all of these systems exhibit complete charge transfer. On chemical grounds this assumption is not correct; if the effective Hubbard repulsion parameter, U, is small enough, the equilibrium constant expression will indicate a finite number of partial charge transfer states as well as of dianions. Banding effects modify this conclusion, but the supposition of exactly half-filled bands is not correct in the general case. For example, Torrance and coworkers (1975) have examined the optical spectrum of a large number of the TCNQ salts, and have argued that the three bands often observed (cf. the spectrum of TTF-TCNQ; Figure 4) correspond to intramolecular excitation, charge-transfer excitation to a dianion, and excitation of a transfer electron to a previously empty site. The last of these cannot occur in the half-filled band case. In addition, the Mott-Hubbard gap (Hubbard, 1963; Mott, 1974), which

---

*Esterling and Dubin (1972) have argued that the Hubbard model has, in fact, never been solved to higher than second order for any realistic case.

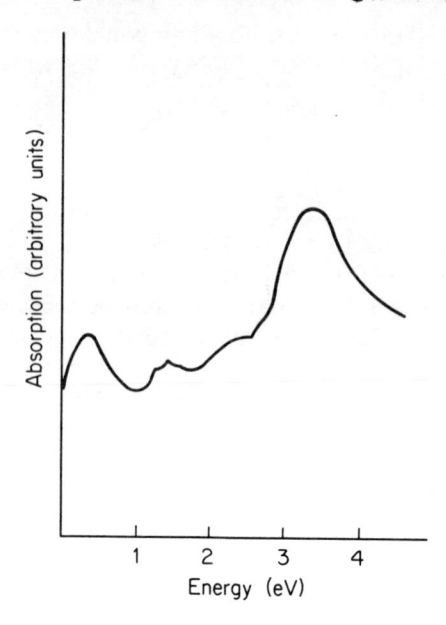

Figure 4 Optical Spectrum of TTF–TCNQ. After Torrance and coworkers (1975)

is caused by the interatomic repulsion, U, will result in insulating behaviour for the exactly half-filled band at zero temperature, in the absence of excitation or collective excitation effects; essentially, for an electron to pass from one site to another, it must overcome the repulsion of the resident electron. On the basis of the optical work of Torrance and coworkers (1975), of photoelectron spectroscopy work by Grobman and coworkers (1974), of vibrational spectroscopy by Wozniak and coworkers (1975), of density difference work (Coppens, 1975), and of analogous system studies (La Placa and coworkers, 1975), of theoretical arguments (Klymenko and coworkers, 1975), and of the simple chemical equilibrium argument above,* we must conclude that in several of the organic charge transfer salts, including the highly conductive ones based on TTF and its analogues, charge transfer is incomplete.

Conductivity studies on the charge-transfer systems are quite difficult to make for a number of experimental reasons including crystal growth and electrode attachment. Nevertheless, reproducible measurements have been made on a number of systems. The NMP–TCNQ system, for instance, has been studied extensively. Structurally, this material is characterized by uniform, isolated, well-separated, face-to-face stacks of TCNQ anions and NMP cations. What is found is that the conductivity in the range above 200 K is indeed metallic, corresponding to a one-dimensional metal. In the low-temperature range, as is expected from the Mott–Hubbard picture, the salt is

---

*We are grateful to R. P. VanDuyne for discussions of this point.

an antiferromagnetic insulator whose band gap is essentially U. The conductivity can be explained adequately using the Hubbard model (Epstein and coworkers, 1972), complicated somewhat by electron–phonon and disorder effects.

The data for the highly conductive ($\sigma \cong 10^3 \, \Omega^{-1} \, cm^{-1}$ at room temperature) (Keller, 1975; Ferraris and coworkers, 1973; Coleman and coworkers, 1973) TTF–TCNQ are more difficult to interpret. It seems clear that charge transfer is closer to one half than to one per site. The mechanism by which conductivity in this system is enhanced, as compared to NMP, is not entirely understood. Suggestions include enhanced polarizability leading to screening, electron–exciton coupling, disorder and most simply an increased Hubbard band-width parameter, t. What is clear, is that the phase transition occurs at 60 K, below which the material is a small band gap semiconductor, and above which it is a one-dimensional metal.* The phase transition is related to electron–phonon effects, and is probably simply a Peierls transition corresponding to sublattice formation within the TCNQ stack.

In view of the interest in preparing highly conductive, or even superconductive, organic charge transfer salts, one can gain from simple Hubbard model considerations some insight (Garito and Heeger, 1974) into the molecular properties upon which the design of such a system must be based. As we have stressed, the π-electron systems allow the Hubbard repulsion parameter U to become quite small; large U produces a large Mott–Hubbard gap in this half-filled band case. High electron affinity of the acceptor and low ionization potential of the donor, plus unsymmetrical ionic charge distribution to increase Madelung stabilization, are also desiderata. The Hubbard band-width parameter, t, should be as large as possible, which will be accomplished by face to face arrangement, and short interplanar distances; however, superexchange or Madelung effects may again become important. Thus, in the highly conductive TCNQ salts the planes are displaced with respect to one another, resulting in a parallelepiped-type stack; whereas in the insulating alkali metal TCNQ salts the TCNQ anions are arranged in a columnar fashion, completely eclipsing one another. Estimates of the Hubbard t and U parameters can be made using semiempirical calculations (Jonkman and Kommandeur, 1968; Bieber and Andre, 1975; Fulda and coworkers, 1975; Ratner and coworkers, 1973), although electron correlation effects may become very important. Reliable estimates of the Hubbard parameters can be derived from spectroscopic studies (Torrance and coworkers, 1975), or from polarographic reduction (although solvent effects will be extremely important). The extent of charge transfer, easily measured in solution (Tomkiewicz and coworkers, 1974) is difficult to obtain in the crystals, although photoemission studies (Grobman and coworkers, 1974) and vibrational spectroscopy (Wozniak and coworkers,

---

*Heeger and Garito (Keller, 1975) have recently argued that above 60 K, TTF–TCNQ is not a metal, but a collective-mode semiconductor, that is, that a Peierls–Fröhlich 'super conductor' has indeed been prepared. In addition, very recent diffuse scattering results (Denoyer and coworkers, 1975) have observed a Peierls transition in TTF–TCNQ.

1975) help elucidate this question. The actual design of organic metals is under active study in a number of laboratories, especially those at the University of Pennsylvania, IBM Laboratories and John Hopkins University. Historically the work on organic charge transfer salts was pioneered by the DuPont group (Melby and coworkers, 1962) and later by McConnell and his students (McConnell and coworkers, 1965; Nordio and coworkers, 1966).

5. REMARKS AND DISCUSSION

The field of organic solids is par excellence a borderline between chemistry and physics. The earliest workers based their arguments on chemical expectations of intermolecular interaction. These were amplified and enriched by the introduction of model Hamiltonian concepts by a number of workers (Nordio and coworkers, 1966; Hoffman and Hughes, 1970; Klein and Soos, 1971; Strebel and Soos, 1970; Epstein and coworkers, 1972). The estimation and prediction of transport properties in terms of the Heisenberg–Hubbard models has led to deepened understanding of the fundamental interactions involved, and of the role of molecular engineering in designing organic metals with desired properties. In fact, it now appears that the major impediment to the attainment of extremely high low-temperature conductivities arises from electron–phonon effects which result in displacive phase transitions to a sublattice, insulating or semiconducting low-temperature phase. Peierls (1955) showed that one-dimensional systems are unstable with respect to a lattice distortion with wave vector $2k_F$, where k_F is the Fermi momentum. These Peierls transitions have actually been observed in certain systems (Glaser, 1974). A great deal of recent work has been devoted to the synthesis of materials which will be stabilized with respect to these displacive phase transitions. The stabilization can be brought about by multiple chemical bonds, by disorder, by increased compressibilities or by incipient three dimensionality. These electron–phonon effects are usually described by another simple model, the Fröhlich (1950) interaction, but this is not a purely electronic problem, and as such lies outside the scope of this discussion.

The observation that columnar stack materials are insulators whereas parallelepiped stacks may be metallic, mentioned in Section 4, has led to new synthetic work by the IBM group. Theoretically, Berlinsky (1975) has shown using extended Hückel theory, that the non-eclipsed, parallelepiped stacking is most favourable energetically. These calculations, however, ignore electron–electron repulsion entirely, and also do not consider Madelung terms, so that their result, while suggestive, will certainly require elaboration. Kaufman (1975) has investigated dimer interactions using paramagnetic resonance methods, and has utilized simple molecular orbital arguments to discuss the stabilization (Kaufman and coworkers, 1975). A possible role of superexchange in these systems has been considered with respect to magnetic properties (Keller, 1975), but not in connection with conductivity. The actual evaluation of the Hubbard t parameter (or of the Heisenberg J parameter) from *ab initio* calculation has not yet been successful (cf. however the references in Section 4).

The evaluation of the intramolecular repulsion U is in principle quite simply given by configuration interaction molecular orbital calculations on the isolated species. However, Madelung forces and polarization effects from nearby molecules can reduce considerably these calculated parameters, which are of the order 4 eV. Thus, the effective Hubbard parameters are rather more difficult to obtain than the isolated parameters; full correlated self-consistent band structure calculations with inclusion of polarization effects constitute a rather forbidding undertaking. Experimentally, the U parameter should be quite straightforward to find polarographically or using optical spectroscopy.

Although the actual understanding of transport processes in organic charge transfer systems is still far from complete, the employment of model Hamiltonian concepts has to some extent transmuted our knowledge from the black art stage to the point at which reasonable classification and estimation of experimental data and of different crystals are possible. The signal importance of the Heisenberg and Hubbard models for this understanding is a tribute to the versatility and generality of the models which were pioneered by the work of Heisenberg exactly a half-century ago, and to the resourcefulness of the several laboratories working in this field.

ACKNOWLEDGEMENTS

Mark Ratner is indebted to E. Engler, F. Kaufman, B. A. Scott, B. D. Silverman, J. Torrance and R. P. VanDuyne for helpful discussion and for preprints of their work prior to publication, and to K. Plumlee and B. Hoffman for detailed discussions on the $Cu(mnt)_2^{2-}$ systems.

Acknowledgement is made for partial support of this work to NSF (SBT and JRS) and to the Materials Research Center of Northwestern University (MAR).

REFERENCES:

Adler, D. and Brooks, H. (1967) 'Theory of semiconductor-to-metal transitions', *Phys. Rev.*, **155**, 826–840.
Anderson, P. W. (1963) 'Theory of magnetic exchange interactions: magnetic exchange in insulators and semiconductors, *Solid State Physics*, **14**, 99–214.
Arai, T. (1968) 'Cumulant expansion of localized-electron model for antiferromagnetic insulators', *Phys. Rev.*, **165**, 706-724.
Bardeen, J. (1973) 'Superconducting fluctuations in one-dimensional organic solids', *Solid State Commun*, **13**, 357–361.
Bari, R. A. (1974) 'Small-polaron effects on the DC conductivity and thermoelectric power of the one-dimensional mott semiconductor', *Phys. Rev.*, **B9**, 4329–4339.
Berlinsky, A. J. and Carolan, J. F. (1975) 'A molecular orbital approach to the theory of crystalline TTF–TCNQ', Preprint. Compare also *Solid State Commun*, **15**, 795–801 (1974).
Bieber, A. and Andre, J. J. (1975) 'Electronic Spectra of Dimers of TMPD, CA and TCNQ' *Chemical Physics*, **7**, 137–142.

Bloch, A. N., Weisman, R. B. and Varma, C. M. (1972) 'Identification of a class of disordered one-dimensional conductors', *Phys. Rev. Letters*, **28**, 753–756.

Brenig, W., Dohler, G. H. and Heyszeneau, H. (1972) '$T^{-1/3}$ Hopping in organic charge-transfer crystals', *Physics Letters*, **A39**, 175–176.

Bright, A. A., Chaikin, P. M. and McGhie, A. R. (1974) 'Photoconductivity and small-polaron effects in tetracyanoquinodimethan', *Phys. Rev.*, **B6**, 3560–3568.

Coleman, L. B. and coworkers (1973) 'Superconducting fluctuations and the Peierls instability in an organic solid', *Solid State Commun.*, **12**, 1125–1131.

Coppens, P. (1975) 'Direct Evaluation of charge transfer in the TTF–TCNQ complexes at 100 K by numerical integration of X-ray diffraction amplitudes', *Phys. Rev. Letters*, **35**, 98–100.

Denoyer, F., Comes, F., Garito, A. I. and Heeger, A. J. (1975) 'X-ray-diffuse scattering evidence for a phase transition in tetrathiafulvalene–tetracyanoquinodimethan (TTF–TCNQ)', *Phys. Rev. Letters*, **35**, 445–449.

Dirac, P. A. M. (1929) 'Quantum mechanics of many-electron systems', *Proc. Roy. Soc.*, **123A**, 714–732.

Epstein, A. J., Etemad, S., Garito, A. F. and Heeger, A. J. (1972) 'Metal-Insulator transition and antiferromagnetism in a one-dimensional organic solid', *Phys. Rev.*, **B5**, 952–977.

Esterling, D. M. and Dubin, H. C. (1972) 'Moment-Generated Solution to the Hubbard narrow-energy-band model', *Phys. Rev.*, **B6**, 4276–4283.

Ferraris, J. and coworkers (1973) 'Electron transfer in a new highly-conducting donor acceptor complex', *J. Am. Chem. Soc.*, **95**, 948–949.

Fischer, S. F. (1970) 'On the sigma-pi separability problem', *Int. J. Quantum Chem.*, **3S**, 651–657.

Fröhlich, H. (1954) 'On the theory of super-conductivity: the one-dimensional case', *Proc. Roy. Soc.*, **A223**, 296–305.

Fröhlich, H., Pelzer, H. and Zienau, S. (1950) 'Properties of Slow Electrons in polar materials', *Phil. Mag.*, **41**, 221–242.

Fulda, P. and coworkers (1975) 'Investigation of the electron structure of the TTF-TCNQ system. I. TCNQ and TTF monomers and dimers in the all-valence electron approximation', *Chem. Phys.*, **7**, 267–278.

Garito, A. F. and Heeger, A. J. (1974) 'The design and synthesis of organic metals', *Accts. Chem. Res.*, **7**, 232–240.

Glaser, W. (1974) 'Experimental studies of the electron–phonon interaction in one-dimensional conducting systems', *Festkörper-probleme*, **14**, 205–228.

Grobman, W. and coworkers (1974) 'Valence electronic structure and charge-transfer in tetrathiofulvalinium tetracyanoquinodimethan (TTF–TCNQ) from photoemission spectroscopy', *Phys. Rev. Letters*, **32**, 534–537.

Haug, A. (1972) *Theoretical Solid-State Physics*, Pergamon, London.

Hay, P. J., Thibeault, J. C. and Hoffmann, R. (1975) 'Orbital interactions in metal dimer complexes', *J. Am. Chem. Soc.*, **97**, 4884–4899.

Heisenberg, W. (1926) 'Mehrkörperproblem und Resonanz in der Quantenmechanik', *Z. Phys.*, **38**, 411–426.

Heisenberg, W. (1928) 'Zur Theorie der Ferromagnetismus', *Z. Phys.*, **49**, 619–636.

Heitler, W. and London, F. (1927) 'Wechselwirkung neutraler Atome und homöopolare Bindung nach Quantenmechanik', *Z. Phys.*, **44**, 455–472.

Herbstein, F. H. (1971) 'Crystalline π-molecular compounds: chemistry, spectroscopy and crystallography', *Perspectives in Structural Chemistry*, **3**, 166–395.

Herring, C. (1966) 'Exchange interaction among itinerant electrons', in *Magnetism, A Treatise on Modern Theory and Materials*, (G. T. Rado and H. Suhl, ed.), Academic Press, New York, Vol. 4.

Hoffman, B. M. and Hughes, R. C. (1970) 'ESR of TMPD–TCNQ: spin excitations of the Heisenberg regular linear chain', *J. Chem. Phys.*, **52**, 4011–4023.

Hubbard, J. (1963) 'Electron correlations in narrow energy bands', *Proc. Roy. Soc.*, **A276**, 238–251.

Interrante, L. V. (1975) *Extended Interactions between Metal Ions*, American Chemical Society, Washington.

Jonkman, H. T. and Kommandeur, J. (1972) 'The UV spectra and their calculation in TCNQ and its mono and divalent anion', *Chem. Phys. Letters*, **15**, 496–499.

Kaufman, F. (1975) 'Molecular criteria of solid state structure in organic charge-transfer salts. I. The effect of methyl substitution on stacking interactions of p-phenylenediamine cation radicals', *J. Am. Chem. Soc.*, in press.

Kaufman, F., Krogh-Jespersen, K. and Ratner, M. A. (1976) 'Electronic interactions in dimers of aromatic molecules', *Chem. Phys. Letters*, submitted.

Keller, H. J. (1975) *Low-Dimensional Cooperative Phenomena*, Plenum, New York.

Kepler, R. G. (1963) 'Magnetic properties of a new class of highly conductive organic solids', *J. Chem. Phys.*, **39**, 3528–3532.

Klein, D. J. and Soos, Z. G. (1971) 'Site representation for charge-transfer excitations in molecular crystals', *Mol. Phys.*, **20**, 1013–1024.

Klymenko, V. E. (1975) 'On the charge-transfer states of quasi-one-dimensional donor–acceptor crystals,. Preprint, L. Y. Karpov Institute, Kiev.

Krogh-Jespersen, K. and Ratner, M. A. (1975) 'Estimation of intramolecular polaron coupling constants for tetracyanoquinodimethan and its anion', *Chem. Phys. Letters*, submitted.

La Placa, S. J., Corfield, P. W. R., Thomas, R. and Scott, B. A. (1975) 'Non integral charge transfer in an organic metal: The structure and stability range of (TTF)Br 0.71–0.76, *Solid State Comm.*, **17**, 635–638.

Lieb, E. and Mattis, D. (1962) 'Theory of ferromagnetism and the ordering of electronic energy levels', *Phys. Rev.*, **125**, 164–172.

Linderberg, J. and Öhrn, Y. (1968) 'Derivation and analysis of the Pariser–Parr–Pople model', *J. Chem. Phys.*, **49**, 716–727.

Linderberg, J. and Öhrn, Y. (1973) *Propagators in Quantum Chemistry*, Academic Press, London.

Little, W. A. (1964) 'Possibility of synthesizing on organic superconductor, *Phys. Rev.*, **A134**, 1416–1424.

Lykos, P. G. and Parr, R. G. (1956) 'On the pi electron approximation and its possible refinement', *J. Chem. Phys.*, **24**, 1166–1173.

McConnell, H. M., Hoffman, B. M. and Metzger, R. M. (1965) 'Charge-transfer in molecular crystals', *Proc. Nat. Acad. Sci. (U.S.)*, **53**, 46–50.

Melby, L. R. and coworkers (1962) 'Substituted quinodimethans II: Anion-radical derivatives of 7,7,8,8-Tetracyanoquinodimethan', *J. Am. Chem. Soc.*, **84**, 3374–3387.

Mott, N. F. (1974) *Metal–Insulator Transition*, Taylor and Francis, London.

Nordio, P. L., Soos, Z. G. and McConnell, H. M. (1966) 'Spin excitations in ionic molecular crystals', *Annu. Revs. Phys. Chem.*, **17**, 237–260.

Parr, R. G. (1963) *Quantum Theory Of Molecular Electronic Structure*, Benjamin, New York.

Peierls, R. E. (1955) *Quantum Theory of Solids*, Oxford University Press, Oxford.

Plumlee, K. W. (1975) Thesis, Northwestern University.

Plumlee, K. W., Hoffman, B. M., Ibers, J. A. and Soos, Z. (1975) 'Weak exchange in the Heisenberg linear chain: structure and EPR of $[N(n-Bu)_4]_2 \cdot [Cu(mnt)_2]$', *J. Chem. Phys.*, **63**, 1926–1942.

Pople, J. A. and Beveridge, D. (1970) *Approximate Molecular Orbital Theory*, McGraw-Hill, New York.

Pottsork, E. N. and Twitchell, J. B. (1975), unpublished.

Ratner, M. A., Sabin, J. R. and Ball, E. E. (1973) 'SCF calculation of the effective parameters for the Hubbard model of TCNQ charge-transfer salts', *Mol. Phys.*, **26**, 1177–1184.

Slater, J. C. (1936) 'The ferromagnetism of nickel', *Phys. Rev.*, **49**, 537–545; 931–37.

Soos, Z. G. (1974) 'Theory of the π-molecular charge-transfer crystals,' *Annu. Rev. Phys. Chem.*, **25**, 121–153.

Strebel, P. J. and Soos, Z. G. (1970) 'Theory of charge transfer in aromatic donor-acceptor crystals', *J. Chem. Phys.*, **53**, 4077–4090.

Suchanski, M. R. and Van Duyne, R. P. (1975) 'Resonance Raman spectroelectro-chemistry. IV. The oxygen decay chemistry of tetracyanoquinodimethan dianion, *J. Am. Chem. Soc.*, in press.

Taft, E. A. and Philipp, H. R. (1965) 'Optical properties of graphite', *Phys. Rev.*, **138**, A197–202.

Tomkiewicz, Y. and coworkers (1974) 'Charge transfer equilibria in TTF–TCNQ solutions', *J. Chem. Phys.*, **60**, 5111–5112.

Torrance, J., Scott, B. A. and Kaufman, F. (1975) 'Optical properties of charge-transfer salts of tetracyanoquinodimethan', *Solid State Commun.*, in press.

Van Vleck, J. H. (1966) 'Some elementary thoughts on the Slater intra-atomic exchange model for ferromagnetism, *Quantum Theory of Atoms, Molecules and the Solid State*, (P.-O. Löwdin, ed.), Academic Press, New York, pp. 475–484.

Wozniak, W. T. and coworkers (1975) 'Vibrational Spectra of the TTF–TCNQ: Evidence for TTF° and TCNQ° in thin films', *Chem. Phys. Letters*. **33**, 33–36.

Wudl, F. and Southwick, E. W. (1974) *J. Chem. Soc. Chem. Comm.*, 254–255.

23

Alpha-clustering in Nuclei

PETER E. HODGSON
Nuclear Physics Laboratory, Oxford, England

1. INTRODUCTION

On the simplest view, atomic nuclei are composed of A nucleons, N of them neutrons and Z protons. The success of the shell model shows that each of these nucleons can be assigned a definite quantum state and its motion described by a single-particle wave function. This simple model enables many features of nuclei to be understood. Realistic shell model calculations require knowledge of the interactions between nucleons, and these are often obtained in a phenomenological way by postulating a force of a particular form and then choosing the values of the parameters to fit selected nucleon–nucleon data.

On a deeper level, the nucleon–nucleon forces can be understood in terms of exchange of pions and other strongly interacting mesons and this enables the forces to be calculated for nucleon–nucleon separations large enough for the interaction to be dominated by one-pion exchange. At smaller separations the interaction becomes too complicated for much quantitative progress to be possible, so one has to be satisfied with a phenomenological interaction. Phenomena specifically dependent on meson effects become manifest only at high energies (several hundred MeV) and for most practical purposes concerned with low energy nuclear phenomena of the type discussed here it is sufficient to think entirely in terms of neutrons and protons and the phenomenological interactions between them.

Although the shell model wave function contains potentially all the information concerning the low-energy behaviour of nuclei, there are nevertheless particular ways of looking at this description that can be useful in certain circumstances. These can first of all be described in a rather classical way, as a preliminary to the quantum-mechanical description that is essential for a full account of the phenomena we want to understand.

It is convenient to begin by imagining the nucleus as a collection of nucleons moving around in a well-defined region of space of radius approximately $R = 1.25\,A^{\frac{1}{3}}$ fm. If we could take a photograph of these nucleons with a long exposure we would see the averaged density distribution of the nucleons. The result would show an almost constant density of nucleons inside the nucleus, with a rather sharp exponential fall-off at the edges. The rate of fall-off is very similar for light and heavy nuclei, so that apart from the differences in curvature the surface region is rather similar for all nuclei.

If now we could take an instantaneous snapshot of the nucleus we would see the positions of the individual nucleons and hence their distribution within the nucleus. There are infinitely many such distributions that are consistent with the same time-averaged density distribution. In principle they could be regularly arrayed like the atoms in a crystal; they could be randomly distributed like the molecules in a gas; or they could have any one of an infinite number of possible correlations of two, three or more nucleons.

The idea that nucleons in the nucleus may cluster together was developed by Wheeler (1937), who showed that they may be regarded as forming all possible groups and that these are continually being broken up and reformed in new ways. The total wave function may thus be expressed in terms of a series of the cluster wave functions, with appropriate coefficients; this is the basis of the resonating group formalism.

This possibility that nucleons show some clustering tendencies is the one that interests us here. Is there some tendency for the nucleons in the nucleus to form clusters, and if so what do we know about them? The likelihood of the formation of a particular cluster depends on its binding energy: the higher the binding energy the more likely it is for a cluster to form. The binding energies of some possible clusters are: deuteron 2.22 MeV, triton 8.5 MeV, helion 7.7 MeV, alpha-particle 28.2 MeV. Thus of all possible clusters of a few nucleons the alpha-particle has by far the highest binding energy, so that this is the most likely cluster.

The binding energy of a cluster determines its size, and for it to form a recognizable entity inside the nucleus this must be less than the average spacing of the same number of unclustered nucleons. Nucleons have a radius of about 0.4 fm, and on the average are separated by about 1.5 fm; and this difference allows them to move sufficient distances between collisions to establish an orbit and hence justify the assignment of quantum numbers to each nucleon. More precisely, orbits are possible because many collisions are forbidden due to the corresponding final states being already occupied. A deuteron has a radius of about 2.2 fm, and so would not be recognizable as a cluster inside the nucleus. On the contrary an alpha-particle has a radius of 1.6 fm and so forms a recognizable cluster.

It is important to recognize that an alpha-particle cluster in the nucleus is not the same as a free alpha-particle. It is naturally distorted by the fields of the surrounding nucleons and may be violently changed by close interactions between them. This qualification is always to be understood when reference is made to an alpha-particle in the nucleus. At high excitations this distinction is particularly important, and in this case the term 'quartet' is used to refer to any two-proton two-neutron correlation.

This dynamic aspect of nuclear structure, as distinct from the static aspect considered so far, would become evident if we could take a series of instantaneous snapshots of the nucleus. We would then see that the clusters, if any are formed, are in a state of dynamic equilibrium with the rest of the nucleus. Groups of nucleons come together to form a cluster, move together for a while, and then are broken up by a close collision with another nucleon. The average lifetime of a cluster is related to its kinetic energy and to its binding energy: the faster it moves and the tighter it is bound the longer, on the average, it will retain its identity in the nucleus.

This picture of the possible alpha-clustering in nuclei enables us to imagine that each nucleon spends a certain fraction of its time as part of an alpha-particle, or in any of the other less energetically-favoured clusters. There is thus a certain probability of finding alpha-particles in the nucleus, and this may be expected to depend on the distance from the nuclear centre.

It is indeed possible to develop the simple shell model of individual nucleons moving in their independent quantum orbits in a way that includes the possibility of alpha-clustering. On this model the alpha-particles are assumed to move in quantum orbits just like the nucleons. Each alpha has its set of quantum numbers obtained by appropriate addition of the quantum numbers of its constituent nucleons. In almost every respect, as will appear in the following sections, the alpha-particles show similar behaviour to the corresponding nucleons: resonance phenomena in elastic and inelastic scattering; particle-transfer reactions, and the systematic nuclear dependence of the potentials corresponding to the orbits.

The first approach to the problem of seeing whether alpha-clusters occur in nuclei is through calculations of the properties of nuclear matter and nuclear structure, starting from phenomenological representations of the nucleon–nucleon interaction derived from studies of the free nucleon–nucleon interactions. The results of these calculations are described in Section 2.

One of the first phenomena studied in the early days of nuclear physics is that of the alpha-decay of nuclei, and this is likely to be influenced by the presence of alpha-clusters in the nucleus. The theories of alpha-decay and their relation to alpha-clustering are summarized in Section 3.

The elastic and inelastic scattering of alpha-particles by nuclei is in many ways similar to that of nucleon scattering and shows resonance phenomena connected with the presence of alpha-clusters. This is summarized in Section 4.

The possibility of transferring an alpha-particle to or from a nucleus is strongly correlated with the presence of alpha-clusters, and the analysis of these reactions is considered in Section 5. The same applies to alpha-particle knock-out reactions, which are discussed in Section 6. Finally Section 7 is devoted to what we can learn about alpha-clustering from the interaction of pions and kaons with nuclei.

In all this work a full, and in some cases a more complete description can always, at least in principle, be obtained solely in terms of the participating nucleons, taking no account of alpha-structure. The important point is whether it is possible to obtain a simpler description of the phenomena in terms of cluster structure that is still sufficiently accurate to be of some use. To state a very simple example, the elastic scattering of alpha-particles by nuclei can be very well described simply in terms of an alpha–nucleus optical potential, taking no account of the internal structure of the alpha-particle. The question we are concerned with is whether a similar description is possible and useful when we consider groups of nucleons inside the nucleus, or groups of nucleons being transferred from one nucleus to another.

Another way of looking at this distinction is in terms of the most appropriate co-ordinate system to describe a particular process. It is well-known that this may make all the difference between a physically transparent description showing clearly the main features of the process and suggesting further questions and experiments, and a description so complicated that it is impossible to see physically what is going on. The question we are interested in is whether the cluster description can play a simplifying role in a sufficient variety of physical processes for it to be considered a permanent feature of nuclear structure.

These remarks indicate that the question of establishing the presence of alpha-clusters in nuclei is not as simple as might appear, even in principle. Even in the simple shell model of uncorrelated nucleons there is a definite non-zero overlap between the wave functions of two neutrons and two protons, and this could be interpreted as an alpha-cluster. It is indeed possible to write the wave functions of light nuclei in completely equivalent ways using either the shell-model or the (fully antisymmetrized) alpha-cluster formalism (Perring and Skyrme, 1956). Thus if we are to speak physically of alpha-clustering it must simply be a degree of $2n-2p$ correlation that exceeds the amount inherent in the nucleon shell model.

There are thus several possible ways of defining the degree of alpha-clustering in the nucleus:

(1). The probability per unit volume of finding two neutrons and two protons in the appropriate quantum states. This is precise, but is always non-zero and thus may not correspond to anything that could physically be recognized as a cluster.

(2). The amount by which this probability exceeds the overlap given by the shell model. Physically this is better, but is imprecise as the shell model may be formulated in many different ways. In particular, the wider the basis, the higher the alpha-probability as defined in (1).

On the whole, the former definition is preferable because it is unambiguous, and it will be used here.

Arima (1975) has distinguished between a *shell model cluster*, which can be well described by the spatial and $SU3$ symmetry within a truncated shell model space, and a *localized cluster*, which is separated from the remainder of the nucleus by a distance greater than the average distance given by the shell model and has a different size parameter from that of the shell model cluster. These definitions are useful in detailed considerations of surface clustering. There is much evidence for shell model clusters, especially in light nuclei, but not much for localized clusters. As will be shown later, clustering occurs in different ways in light and heavy nuclei.

2. NUCLEAR STRUCTURE CALCULATIONS

One of the most direct ways of studying alpha-clustering in nuclei is to consider infinite nuclear matter bound by the phenomenological nucleon–nucleon forces, and excluding the Coulomb forces. For a particular array of nucleons it is then possible to calculate the relative stability of ordinary nuclear matter and nuclear matter in which the nucleons are condensed into alpha-clusters.

Such a calculation has been made by Brink and Castro (1973), who considered two types of infinite nuclear matter: one consisting of nucleons represented by plane waves and the other consisting of groups of four nucleons (alpha-clusters) in a regular crystalline array. If a particular form of the nucleon–nucleon interaction is assumed, the total binding energy of the system may be calculated by standard quantum-mechanical procedures. It is then possible to see whether the array of alpha-particles is more stable than the nucleons, and this in turn has implications for the structure of finite nuclei.

In such a calculation it is essential to ensure that the results do not depend too critically on the assumption made, so Brink and Castro used two extreme types of nuclear force (the Brink–Boeker and the Skyrme) and three types of crystalline array (the simple, body-centred and face-centred cubic lattices). In each case it was found that for the nucleon densities corresponding to those in the centre of heavy nuclei the nucleon matter was more stable than the alpha-cluster matter. This suggests that the alpha-clustering inside heavy nuclei is rather small.

The calculations were then repeated for a series of different nucleon densities, and it was found that when the density was reduced to about one-third of that inside heavy nuclei the alpha-cluster matter becomes the more stable. Thus as the density of nuclear matter is reduced there comes a point where it condenses into alpha-particles. This is to be expected from the high binding energy of the alpha-particle, and the condensation takes place as soon as the density is low enough to prevent the interactions with all the nearby nucleons from breaking up any incipient clusters that may be formed.

The actual density at which the condensation occurs naturally depends on the nuclear force and on the form of the crystal lattice assumed in the calculation, but since the ones used were deliberately chosen to be rather extreme, any other realistic force or lattice will give results similar to those already obtained. Thus the general conclusion is not affected by the details of the assumptions made in the calculation.

It is natural to apply these results to the nuclear surface, where the density falls rapidly to zero, and to infer that in the outer region where the density is low some degree of alpha-clustering will occur. This argument is very plausible qualitatively, although the complications of a finite system with a curved surface and the presence of the Coulomb interaction alters the value of the critical condensation density. Nevertheless this work shows in a simple way that we may expect to find some degree of alpha-clustering in the nuclear surface, but not in the interior.

When we consider actual nuclei, even their gross properties show evidence of some form of alpha-clustering. Thus the variation with atomic number of the binding energy per nucleon shows marked minima for light nuclei consisting of $2n$ neutrons and $2n$ protons, where n is an integer, for example, ^{12}C, ^{16}O, ^{20}Ne and so on. This strongly suggests that two-neutron two-proton groups in nuclei tend to coalesce to form alpha-clusters and that in some respects these nuclei can be considered as formed of 3,4,5,... alpha-particles. In addition, many of the features of the low-energy spectra of light nuclei can be understood using the cluster model (Sheline and Wildermuth, 1960). Less pronounced evidence of alpha-structure is also found throughout the periodic table (Danos and Spicer, 1970; Danos and Gillet, 1971). Alpha-particle or quartet structure is found in the excited states of nuclei from ^{12}C to ^{52}Fe (Arima and coworkers, 1970).

For quantitative calculations of alpha-clustering it is necessary to define the quantity S_L that gives the probability of finding an alpha-cluster in the nucleus A in a state given by the relative wave function $\phi_{NL}(R_\alpha - R_{A'})$, where $A' = A - \alpha$. This is given by the square of the overlap between the internal wave function $\phi_i(A)$ of the nucleus A and the product of the wave functions of the internal and relative motions of the alpha-particle and the remainder of the nucleus A' (Arima, 1973):

$$S_L = |\langle \phi_i(A-\alpha)\phi_i(\alpha)\phi_{NL}(R_\alpha - R_{A'})|\phi_i(A)\rangle|^2 \qquad (1)$$

This quantity can be calculated from any nuclear model that gives explicit expressions for the wave functions, and several such calculations have been made (Rotter, 1968; Akiyama and coworkers, 1969; Kurath, 1973; Kurath and Towner, 1974; Draayer, 1975; Hecht and Braunschweig, 1975). For example, a calculation for the nucleus ^{20}Ne by Ichimura and coworkers (1973) assuming an $(sd)^4$ configuration and the harmonic oscillator shell model gives a maximum value of $S_L = 0.23$ for the $SU_3(8, 0)$ representation. On the other hand, very small values of S_L are found for jj-coupling wave functions of the $(d_{5/2})^4$ configuration. However as soon as configuration mixing is introduced into the shell model wave function (Akiyama and coworkers, 1969; Strottman and Arima, 1973: McGrory, 1971) values of S_L are obtained that are similar to the SU_3 limit.

These calculations show that configuration mixing concentrates the alpha-particle strength into a very few states, an effect that accounts for the selectivity of alpha-transfer reactions (see Section 5). Configuration mixing also enhances the alpha-spectroscopic factors in the alpha-decay of heavy nuclei (Mang, 1957, 1960: Harada, 1961, 1962). The results of Harada's calculation of the effect of configuration mixing in plutonium are shown in Figure 1.

It is also necessary to know the alpha-clustering probability as a function of radial distance. This may be obtained by defining the alpha-probability amplitude

$$X_L(R) = \binom{A}{4}^{\frac{1}{2}} \langle \phi_i(A')\phi_i(\alpha)|\phi_i(A)\rangle \qquad (2)$$

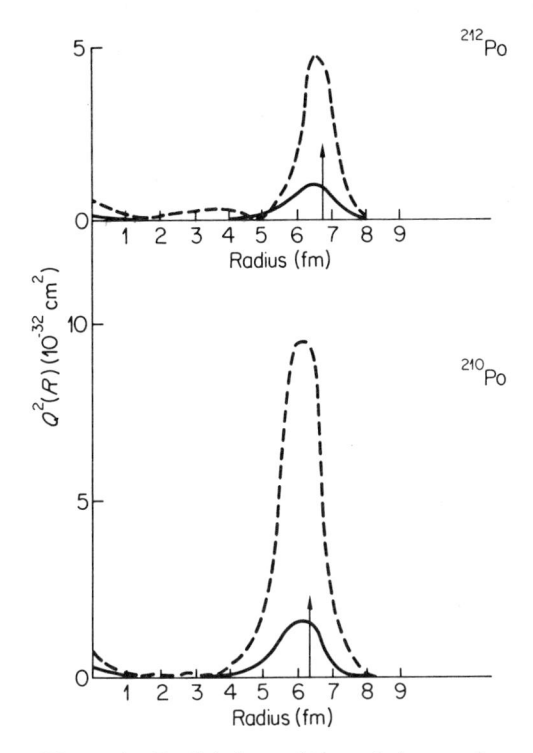

Figure 1 Radial dependence of the overlap integral of the wave functions of four nucleons in the top levels of ^{212}Po and ^{210}Po and that of the alpha-particle, using the harmonic oscillator shell model. The full curves are for pure configurations and the dashed curves are for mixed configurations. The arrows indicate the nuclear radius (Harada, 1962)

This quantity can also be expressed as a series of harmonic oscillator wave functions

$$X_L(R) = \sum_\lambda a_L(\lambda, 0)\sqrt{S_L(\lambda\, 0)}\phi_{NL}^{HO}(R) \qquad (3)$$

The square of $X_L(R)$ gives the probability of finding an alpha-particle cluster at a distance R from the core nucleus A'. The results of some calculations of $X_L(R)$ for ^{20}Ne are shown in Figure 2. The pure harmonic oscillator wave function $\sqrt{0.23}\phi_{08}^{HO}(R)$ is compared with the mixed wave function $X_L(R)$ and it is notable that the mixed wave function has a greater amplitude in the region of the nuclear surface.

The overlap of four single-particle wave functions for deformed nuclei gives information about the distribution of the alpha-clusters over the nuclear surface. Some calculations were made for even–even nuclei by Sharma (1967)

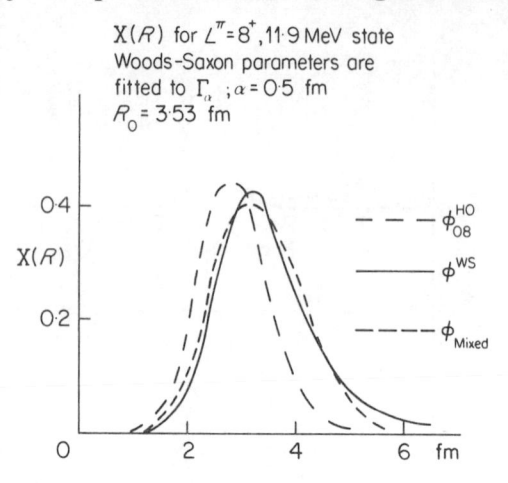

$X(R)$ for $L^\pi = 8^+$, 11.9 MeV state
Woods-Saxon parameters are fitted to Γ_α; $a = 0.5$ fm
$R_0 = 3.53$ fm

- - - - ϕ_{08}^{HO}
———— ϕ^{WS}
- - - - ϕ_{Mixed}

Figure 2 Comparison between wave functions (alpha-cluster probability amplitudes) $X(R)$ for the 8^+ (11.9 MeV) state of ^{20}Ne. $\phi_{08}^{HO}(R)$ is the pure harmonic oscillator wave function and ϕ_{mixed} the wave function obtained after including configuration mixing. The wave function ϕ^{WS} is obtained as the solution of a Schrödinger wave equation with a Saxon–Woods potential whose parameters are adjusted to give the measured alpha-decay widths (Arima, 1973)

and some of his results are shown in Figure 3. It is notable that most of these distributions are markedly non-uniform; sometimes the alpha-particles are more likely to be concentrated at the poles and sometimes at the equator. These distributions agree quite well with those found from the alpha-decay of aligned nuclei (Steenberg and Sharma, 1960). Nuclear structure calculations are thus able to provide estimates of the alpha-clustering probability in the various shells. For low-lying states the clustering is inhibited by the Pauli principle, but it increases with the excitation energy, becoming nearly unity in the region of the nuclear surface. For heavy nuclei the alpha-clustering is much less. The alpha-clustering in the ground states of nuclei may be measured using pick-up and knock-out reactions as described in Sections 5 and 6. Greater clustering is expected for some of the higher excited states, particularly those of a $4p4h$ character, and this may be measured by stripping reactions not only for light nuclei, but also for medium weight and possibly heavy nuclei as well.

Light nuclei consisting of even numbers of neutrons and protons have many features that can be quite well understood by assuming that they are formed of alpha-clusters bound by suitable potentials (Arima and coworkers, 1972; Ikeda and coworkers, 1972; Abe and Takigawa, 1972). Such models may be divided into two types. In the first the wave function of each alpha-particle is

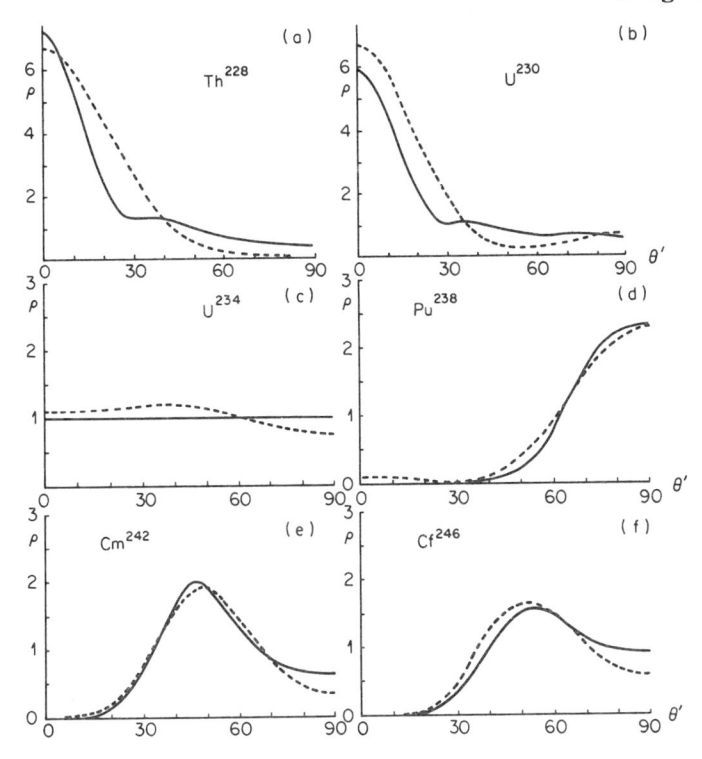

Figure 3 Surface density distributions of alpha-particles with respect to the polar angle. The full curves are the results of shell model wave function overlap calculations by Sharma (1967) and the dashed curves show semi-empirical distributions from the decay of aligned nuclei (Steenberg and Sharma, 1960)

constructed from that of its constituent nucleons, using a phenomenological nucleon–nucleon potential, and then the nucleus is built of the appropriate number of interacting alpha-particles. In the second-type of model the internal structure of the alpha-particles is completely ignored, and the nucleus is considered to consist simply of a number of alpha-particles interacting through a phenomenological alpha–alpha potential. Both types of alpha-particle model have been studied for several years and each has proved able to account for important aspects of nuclear structure. This model was originally introduced by Margenau and has been further developed by Brink (1966). In this form of the model the basic antisymmetrized many-body wave function $\phi(x, R_i, b_i)$ is a Slater determinant of Gaussian single-particle wave functions representing internally unexcited alpha-particles in harmonic oscillator potentials with separated centres R_i. The b_i are the corresponding oscillator lengths and the x are all particle co-ordinates. The $4N$ nuclei are natural candidates for the model, and Brink and coworkers (1970) have used it to calculate the ground- and excited-state properties and the mixing of alpha-configurations for several

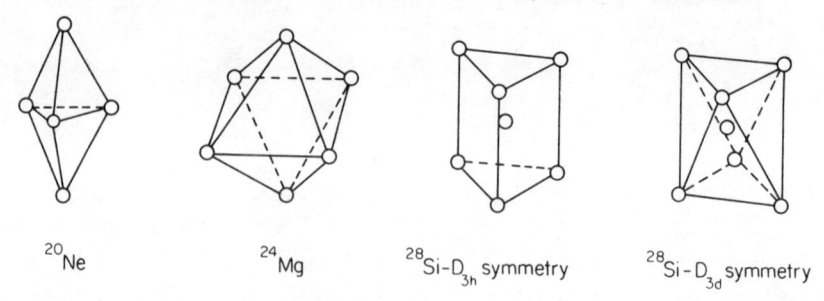

²⁰Ne ²⁴Mg ²⁸Si–D_{3h} symmetry ²⁸Si–D_{3d} symmetry

Figure 4 Alpha-particle configurations for some $4N$ nuclei (Brink and coworkers, 1970)

light $4N$ nuclei. Using the Volkov (1964, 1965) and Brink–Boeker (1967) effective two-body interactions they found fair agreement with the nuclear radii, quadrupole moments and binding energies of these nuclei, and with the energies of the $8p$–$8h$ band in ^{16}O. Some of their alpha-particle configurations are shown in Figure 4. Later calculations by Khadkikar (1971) using the density-dependent Skyrme interaction gave considerably less clustering, showing that the large alpha-clustering previously obtained was due to the strong p-state repulsion required for saturation with density-independent forces.

It has been shown by Dutta and Tomusiak (1971) that both the ground state form factor and the inelastic form factor for the $0^+ \to 0^+$ (6.06 MeV) transition in ^{16}O are well accounted for by the fully antisymmetrized alpha-cluster model.

The simpler form of alpha-particle model of the nucleus has also been extensively used to calculate many of the properties of light $4N$ even–even nuclei. A phenomenological alpha–alpha interaction is obtained from analyses of alpha–alpha elastic scattering, and the Schrödinger equation solved to give the nuclear energy levels. This model is quite successful in giving a useful insight into those nuclear properties that involve the constituent alpha-groups, but of course is unable to account for any phenomena involving nucleon excitations.

Grin and Leinson (1971) have developed a formalism for the calculation of many-particle excited states in nuclei and their results for the energies, quadrupole moments and shapes of many-particle excited states in ^{12}C and ^{16}O give clear evidence of cluster structure and confirm the results of calculations made with the alpha-cluster models.

The alpha-particle model has been used by Mendez-Moreno and coworkers (1974) to calculate the spectra and form factors of ^{12}C and ^{16}O, and they found that the empirical α–α interactions give good qualitative results.

In more detailed models of light nuclei, which take the individual nucleons into account, great simplification is introduced by explicitly including two-neutron two-proton quartet configurations. It is particularly necessary to do this to account for the low-lying even parity states of doubly even nuclei in the sd and pf shells. These $2n$–$2p$ configurations are characterized by a large overlap between the constituent single-particle wave functions and they are relatively easy to construct in light nuclei in the framework of the *LS* coupling scheme. In the *p* and *sd* shells alpha-clustering is a natural feature of almost any calculation, and it is most clearly seen using the *SU*3 classification of states. In heavy nuclei the situation is different because the spin–orbit interaction makes it necessary to use the *JJ* coupling scheme.

Particularly successful calculations of this type have been made in the *sd* and *fp* shells using the $2n$–$2p$ coupling in the aligned scheme of Bohr and Mottelson. It is found that there exists a simplified shell model space constructed on the basis of these configurations that contains the main features of four-particle correlations and gives results for the low-lying states that are comparable with those obtained with much greater computational effort using the full shell model diagonalization on the basis of single-particle wave functions (Jaffrin, 1970, 1972; Richert, 1972). The quartet model has been generalized by Harvey (1973) to include any number of particle–hole excitations, and has been applied in the framework of the Hartree–Fock theory to calculate the energies of states in the *sd* shell (Macdonald and coworkers, 1972).

More detailed calculations may be made using the self-consistent-field theories of nuclear structure, in particular those derived from the Hartree–Fock theory for atoms. These have been made for nuclei either by using an effective nucleon–nucleon interaction with parameters adjusted to give the best overall agreement with selected nuclear properties, or by a more fundamental calculation starting from the phenomenological nucleon–nucleon interaction itself.

Detailed Hartree–Fock calculations have now been made for a series of light nuclei, and some of the results are shown in Figure 5. It is clear that the density distributions suggest that the nucleons tend to bunch together in groups of four. It is not immediately evident that these groups are indeed alpha-particles, although this seems very likely.

3. ALPHA-DECAY

The phenomenon of alpha-particle radioactive decay was the first in nuclear physics to receive a quantum-mechanical explanation when Gamov accounted for the Geiger–Nuttall law in terms of barrier penetration. Since that time there have been many studies of alpha-decay, the aim being to account for the measured decay rates in terms of the structures of the initial and final nuclei. Most natural alpha-emitters are heavy nuclei, but a very similar problem is the alpha-decay of nuclei in excited states, which can occur throughout the periodic table.

Many detailed calculations by Mang and his colleagues (Mang, 1964; Poggenburg and coworkers, 1969) showed that the alpha-decay rate is a product of two factors, one depending on the barrier penetration and the other on the structure of the nuclei. The barrier penetration depends extremely

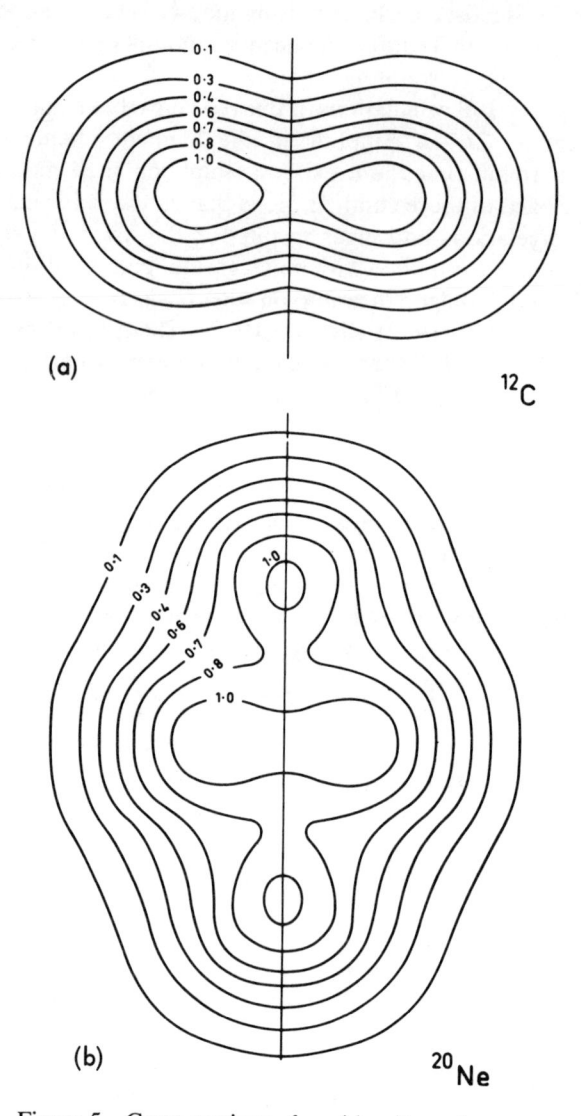

Figure 5 Cross-sections of equidensity surfaces cut by a plane containing the symmetry axis (vertical line) of axially symmetric Hartree–Fock nuclear densities. (a) The oblate solution of ^{12}C, (b) the prolate solution of ^{20}Ne (Ripka, 1968)

sensitively on the energy, and essentially accounts for the very wide range of half-lives of alpha-emitters; their relative values can be calculated quite accurately but the absolute values are more uncertain as they depend on details of the potential. The structure factors were evaluated using the Nilsson wave functions, and later configuration mixing was included by adding the pairing force using the BCS superfluidity formalism. This work showed the importance of taking nucleon–nucleon correlations into account (Mang, 1957, 1960; Harada, 1961, 1962). On the whole it was found that the theory can account very well for the relative alpha-decay rates, but the absolute values were uncertain by a substantial factor.

Alpha-decay calculations in heavy nuclei are complicated by the large number of nucleons involved, so a simpler way to test the theory is to apply it to the alpha-decay of excited states of light nuclei, whose structure is relatively well-understood. A series of such calculations have been made of the alpha-decay widths of excited states of ^{20}Ne.

The alpha-decay widths of the 6^+ (8.78 MeV) and 8^+ (11.95 MeV) states of ^{20}Ne were calculated by Yazaki (1973) starting from the Schrödinger equation for the $\phi_i(A)$. His values were about three to five times smaller than those found experimentally.

Insight into the reason for this discrepancy is provided by the calculations of Arima and Yoshida (1972) for the same nucleus. They used the R-matrix theory, in which the alpha-decay widths Γ_α are given by

$$\Gamma_\alpha = 2P\gamma^2 \tag{4}$$

where P is the penetration factor and γ^2 the reduced width

$$\gamma^2 = \frac{\hbar^2}{2\mu_\alpha R_c^2}|X_L(R = R_c)|^2 \tag{5}$$

where R_c is a matching radius. In the internal region they used the harmonic oscillator wave function multiplied by the appropriate spectroscopic amplitude

$$X_L(R) \approx \sqrt{0.23}\phi_{NL}^{HO}(R) \quad (2N+L=8) \tag{6}$$

In the external region they used a more realistic wave function obtained as a solution of the wave equation with a Saxon–Woods potential with parameters fixed to maximize the overlap with the harmonic oscillator wave function in the internal region. This calculation also gave values of the alpha-width that were about five times too small, consistent with Yazaki's results. They therefore increased the radius of the Saxon–Woods potential, which makes $X_L(R)$ peak at greater values of R, until the alpha-widths agreed with the experimental values. This Saxon–Woods wave function for the 8^+ (11.95 MeV) state is also shown in Figure 2, and it is notable that it is essentially the same as the wave function obtained by including configuration mixing. Thus a quantum-mechanical calculation of alpha-decay widths for ^{20}Ne using configuration-mixed wave functions gives results in accord with the experimental values. Similar results have been obtained for heavy nuclei by Mang (1975).

This work shows that alpha-decay widths can now be understood very well provided good wave functions are available for the initial and final nuclei.

Before it decays a radioactive nucleus is very similar to a compound nucleus in a nuclear reaction, so the decay rate to one final state is given by

$$W = \phi \frac{\langle D \rangle}{2\pi\hbar} T_L \tag{7}$$

where ϕ is the probability that four nucleons form an alpha-cluster, $\langle D \rangle$ is the average separation of the decaying states and T_L the barrier penetration factor. For alpha-particles the level density is one-quarter of the nucleon level density g so $\langle D \rangle = 4/g$.

Bonetti and Milazzo-Colli (1974) have used this expression to analyse the decay of about 100 even–even radioactive nuclei. These can only decay with $L = 0$, and in most cases to only one final state. The penetration factors T_0 were calculated from several potentials of exponential or Saxon–Woods form that are in accord with the data on the elastic scattering of alpha-particles by nuclei. The expression (7) then gives the alpha-particle preformation probability ϕ, and the values obtained are shown as a function of N in Figure 6. These are in good accord with those obtained from analyses of (n, α) and (p, α) reactions using the precompound emission model (see Section 6). It is notable that ϕ is near unity for nuclei far from closed shells and falls to around 1 per cent for nuclei with 126 neutrons. The expression (7) was used by Chang (1967) to calculate the alpha-reduced widths of even mass polonium nuclei, and good agreement was found for nuclei with $A > 210$.

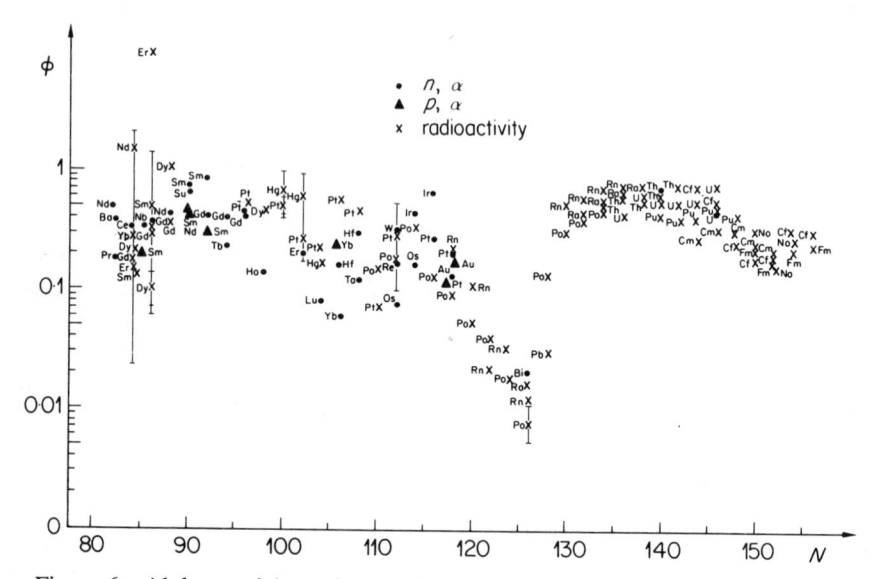

Figure 6 Alpha-particle preformation probability ϕ as a function of N determined by analyses of natural alpha-radioactivity and (n, α) and (p, α) reactions (Bonetti and Milazzo-Colli, 1974)

4. RESONANCES IN ELASTIC AND INELASTIC SCATTERING

If the elastic scattering cross-section of nucleons by nuclei is measured as a function of incident energy, resonance behaviour is found when the energy in the compound system corresponds to that of one of its excited states. These resonances are particularly prominent at low energies on light nuclei where there are few, if any, open reaction channels. Many examples of such phenomena have been studied, in particular for neutron and proton elastic scattering from ^{12}C and ^{16}O. Other resonances are observed in addition to those attributable to the excitation of single-particle states, and some of these may be interpreted as core-excited states, in which the single nucleon is coupled to a low-lying excited state of the target. Both types of resonances may be treated theoretically using a coupled-channels formalism, and assuming a suitable optical potential and model for the nuclear excitation a good quantitative understanding of the resonance behaviour in each partial wave can be obtained (Hodgson, 1971).

Exactly similar phenomena have been found in the elastic scattering of alpha-particles by nuclei, and they may be interpreted in the same way in terms of alpha-particle states. Thus the elastic scattering of alpha-particles by ^{12}C at low energies shows clear resonances due to the normal parity $(\pi = (-)^J)$ states of the compound nucleus ^{16}O (Figure 7) and these can be analysed by the coupled-channels formalism in just the same way as the corresponding nucleon interactions. One such analysis has been made by Terasawa and coworkers (1973) using the weak coupling model for the alpha plus ^{12}C system and the rotational model for the excited states of ^{12}C. With a phenomenological alpha–nucleus interaction with spin and angular momentum dependent components they are able to obtain a good overall understanding of the energy variation of the phase shifts, as shown in Figures 7 and 8. A detailed study of the excitation function yields much information on the alpha-particle structure of ^{16}O. Examination of the positive parity states of ^{16}O shows that they may be divided into two groups: those with large alpha-widths θ_α^2 having a configuration in which the alpha-particle is orbiting the target nucleus in its ground state (these are called 'single-particle like' states), and those with small amplitudes corresponding to more complicated excitations based on the lowest 2^+ state of ^{12}C. States of the first type are interpreted as simple potential resonances. In a similar way, if the interaction of the alpha-particle with the excited nucleus is similar to that with the nucleus in its ground state, the alpha-particles in the inelastic channel resonate at approximately the same energies. These resonances give rise, through the channel coupling, to sharp or intermediate resonances in the elastic channel, and these are the resonances of the second type. Such resonances can be excited even at incident energies below 4.4 MeV if the resonating state is a bound state of the (^{12}C+α) system. The coupled channels formalism gives a unified description of both types of states.

In these calculations, the optical potentials and nuclear deformation parameters are chosen to give the best overall fit to the energies of the

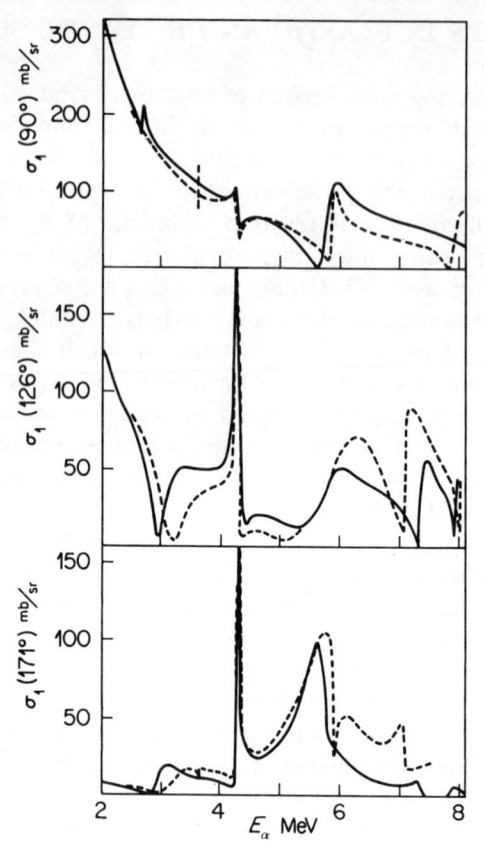

Figure 7 Comparison between the experimental excitation functions (dashed lines) and those calculated from the coupled-channels model (full lines) for the $^{12}C(\alpha, \alpha)^{12}C$ reaction at 90°, 126° and 171° (Terasawa and coworkers, 1973)

resonances and to their alpha-decay widths θ_α^2. The good qualitative agreement obtained for a considerable body of data with rather few parameters provides strong support for the overall correctness of the model. The data are so detailed and accurate that it is possible to study the dependence of the potential on the spin I of the target nucleus, the angular momentum L of the relative motion between the alpha-particle and the target nucleus, and the total angular momentum J. The potential is written in the form

$$V_\alpha(r) = -[V_0 + \mathbf{I} \cdot \mathbf{I} V_I + \mathbf{L} \cdot \mathbf{L} V_L + \mathbf{J} \cdot \mathbf{J} V_J] f(r) \tag{8}$$

and comparison with the data enables the potential depths V_0, V_I, V_L and V_J to be found for an assumed form factor $f(r)$.

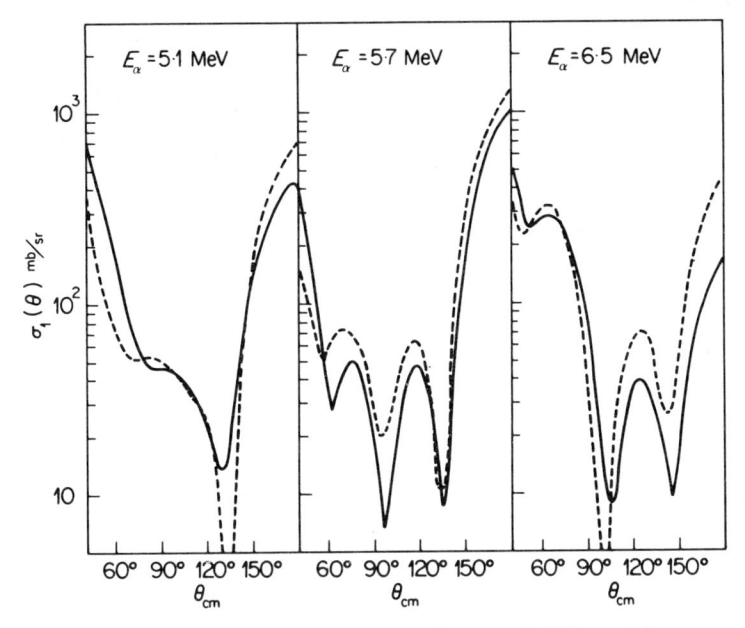

Figure 8 Comparison between the experimental differential cross-sections (dashed lines) and those calculated from the coupled channels model (full lines) for the $^{12}C(\alpha, \alpha)^{12}C$ reaction at 5.1, 5.7 and 6.5 MeV (Terasawa and coworkers, 1973)

For heavier nuclei and for higher energies these resonances become more difficult to find partly because the resonance phenomena in a particular partial wave form a small part of the total cross-section and also because the resonances are broadened by the presence of many reaction channels. It might, however, still be possible to find resonant behaviour in alpha-particle scattering of medium and heavy nuclei at energies much less than that of the Coulomb barrier, since in this case the above difficulties are not so serious. The difficulties with this work are that the energies are substantially less than the potential barrier so that the penetration factors are small, and also that in this region the states may be sharp and thus difficult to find. The angular momenta of these states may be found by seeing how the form of the resonance in the excitation function varies with energy, particularly how it behaves at the zeros of the Legendre polynomials. The results can also be interpreted as unbound alpha-particle states in an optical potential (Dudek and Hodgson, 1971).

For energies above the Coulomb barrier the alpha-structure of the nuclear surface gives rise to characteristic anomalies in the differential cross-section for elastic scattering, particularly in the backward direction (Gruhn and Wall, 1966; Bobrowska and coworkers, 1969; Stock and coworkers, 1972; Sewell and coworkers, 1973). It is found that for many nuclei, particularly those with $4n$ nucleons, the cross-section is as much as ten times that given by an optical potential fitted to the data in the forward direction. Closer examination shows that the extra cross-section has an angular distribution characteristic of a particular L-value, so that the overall features of the data can be fitted with a scattering amplitude obtained by summing the optical model amplitude and a Legendre polynomial of degree L, with appropriate amplitude. The characteristic L-value increases with energy and the excitation function in the backward direction shows resonances at energies proportional to $L(L+1)$ (Rinat, 1972).

The prominence of the enhancement of the backward cross-sections depends on the structure of the target nucleus, and Oeschler and coworkers (1972) find that they follow the rules:

(1). The $N = Z$ or $N = Z + 1$ nuclei all show back angle enhancements.

(2). Higher isotopes of these nuclei show a back-angle enhancement if the excess neutrons are all in the same major shell as the outer nucleus of the $N = Z, Z + 1$ isotopes.

(3). A strong reduction of the back-angle cross-section occurs if two excess neutrons enter the next higher shell not occupied in the $N = Z$ core.

Some insight into this anomaly may be obtained in terms of the 'glory' effect in optics (Bryant and Jarmie, 1968; Bobrowska and coworkers, 1969; Budzanowski and coworkers, 1969), or as backward diffraction scattering (Gubkin, 1970), but a detailed understanding can only come from a quantum-mechanical analysis. The explanations proposed so far fall into three distinct classes, viz. potential scattering, intermediate-structure resonances and exchange processes, and these will now be discussed, following the review of Fuchs (1973).

It is natural to see first of all if the phenomena can be obtained by some simple modification of the optical potential. The rapid fall of the cross-section with increasing angle in the backward direction in a normal optical model calculation requires an almost complete cancellation of the amplitudes due to the different partial waves, so any mechanism that upsets this cancellation will give an enhanced backward cross-section. This may be done, for example, by allowing the absorption to depend on L, and several analyses with an L-dependent absorption showed that this is able to account for the backward peaking (Bisson and coworkers, 1970; Chatwin and coworkers, 1970). Some physical basis to this approach is provided by considering the reaction channels that are open in the decay of the compound nucleus, and this enables us to understand to some extent why the anomaly varies from one nucleus to another (Eberhard, 1970).

The anomalies show characteristic resonance effects when they are examined as a function of incident energy and these can be interpreted within the same model as the unbound resonant states of the one-body potential. This is essentially the 'molecular rotation' theory of McVoy (1971, 1972) that attributes the backward peaking to the resonances in each partial wave with one oscillator quantum. The compound system is thought of as a target plus alpha-particle 'molecule' oscillating radially, the rest of the energy remaining in the angular motion and giving the resonances an approximately $L(L+1)$ energy dependence.

This approach is quite successful in particular cases, but it is found difficult to account for the behaviour of the cross-sections around 90° where the standard optical model and anomalous amplitudes interfere. It is also found impossible to reproduce the observed energy variation of the angular distributions with one set of optical parameters; in particular the L-value dominating the backward anomaly increases less rapidly with energy than that of the resonances in potential scattering (Oeschler and coworkers, 1973; Stock and coworkers, 1972).

This last observation suggests that the structure responsible for the anomalies has a larger moment of inertia than a circulating alpha-particle, and has suggested that it is attributable to intermediate structure resonances. According to this model, the incident alpha-particle excites the target nucleus and thus becomes more strongly bound to it, as in the 'quasi-bound molecule' of Fink and coworkers (1972). This system can then rotate as a whole with a moment of inertia greater than the alpha-particle alone (Rinat, 1972). It is possible that the nucleus is excited by the promotion of an alpha-particle into a higher orbit, and the rotating ^{8}Be model of Budzanowski (1972) is an example of such an excitation.

All these explanations have the common feature that the anomalous scattering is ascribed to a resonance phenomenon, and so the scattering amplitude may conveniently be described by the Regge pole formalism. (Högaasen, 1967; McVoy, 1971; Ceuleneer and coworkers, 1969; Rinat, 1972; Cowley and Heymann, 1970).

A more physical explanation can be obtained in terms of exchange processes. The exchange effects in the scattering of alpha-particles by light nuclei have also been studied by Honda and coworkers (1963), and they find that they are significant only when the mass of the projectile is comparable to that of the target, and when the separation energy of the alpha from the target is small.

This process has been used to analyze the elastic scattering of alpha-particles by ^{16}O made by Noble and Coelho (1971) using the alpha-cluster model for ^{12}C and ^{16}O. They attribute it to the heavy-particle transfer process shown diagrammatically in Figure 9, and evaluated the cross-section at several energies

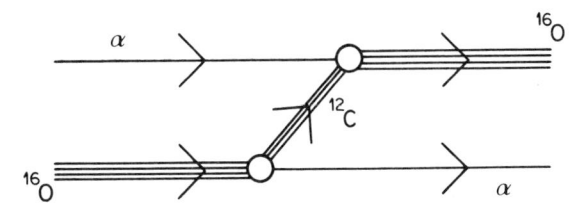

Figure 9 Diagrammatic representation of the heavy-particle transfer process that accounts for the backward scattering of alpha-particles by ^{16}O

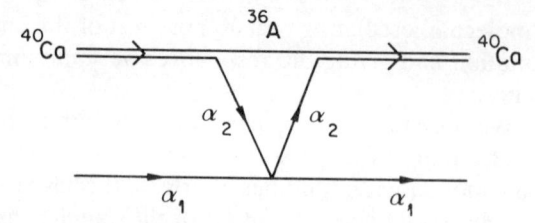

Figure 10 Diagrammatic representation of the alpha-particle knock-out process

from 30 to 50 MeV using the plane wave Born approximation. The main contribution comes from the transfer of ^{12}C in its lowest 2^+, 3^- and 4^+ states, with very little contribution from the ground and first excited 0^+ state. The Blair strong absorption model was used to take account of the distortion effects in the direct process. On the whole the calculation accounts qualitatively for the observed backward peaking but the period of angular oscillation is too small, probably due to the limit on the angular momentum of the available clusters. Further calculations with this model have been made by Coelho (1973a, b).

An associated process is the scattering by bound alpha-clusters, illustrated in Figure 10. It has been argued (Gaul and coworkers, 1969; Stock and coworkers, 1972) that large-angle scattering requires large momentum transfer which cannot result from scattering by a single target nucleon but can occur by scattering from a bound alpha-cluster (Schmeing and Santo, 1970) or by quartets (Arima and Gillet, 1971; Eichler and Yamamura, 1972). More detailed calculations by Thompson (1971) showed however that such processes have a very small effect on the elastic scattering of alpha-particles.

A very similar process is cluster knock-out, which has the same graph as scattering on bound clusters, but with the indices commuted as in Figure 11. According to this model the incoming alpha-particle collides with an alpha-particle on the surface of the target nucleus and knocks it out, while itself being captured by the target core. The amplitude of this exchange knock-on process has been calculated by Agassi and Wall (1973) in the distorted wave Born approximation. Since it is concentrated in some surface grazing partial waves it gives a good qualitative description of the backward peaking, but the value of the alpha-particle preformation probability was treated as an adjustable parameter. In this calculation, the wave function of the bound alpha-particle is an eigenfunction of the real part of the optical potential describing the elastic scattering, and its quantum numbers (N, L) are obtained assuming conservation of oscillator quanta in the Talmi transformation, i.e.

$$\mathcal{N} = 2(N-1) + L = \sum_i \{2(n_i - 1) + l_i\} \tag{9}$$

where (n_i, l_i) are the quantum numbers of the nucleons forming the alpha-particle, assumed to have no internal energy. A phenomenological alpha–alpha

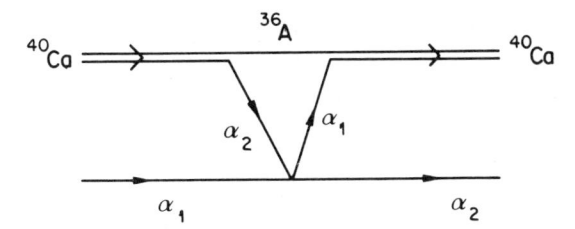

Figure 11 Diagrammatic representation of the alpha-particle knock-out exchange process

interaction is used and the model gives overall qualitative agreement with the experimental data.

The enhanced L-value corresponds closely to that of the classical grazing angle collision, $L \approx kR$, and this provides an explanation for the energy peaking of the enhancement in the backward directions. The interaction occurs preferentially for grazing collisions because for smaller impact parameters the incident particle interacts more strongly with the nuclear surface and so is absorbed to form a compound nucleus, while for larger impact parameters it misses the nucleus and so does not interact at all. The bound alpha-particles in the nuclear surface have a range of angular momenta from 0 to $\mathcal{N}$, where $\mathcal{N}$ is given by the above expression. The alpha-particles nearest to the surface are those with the highest $L = \mathcal{N}$, so these are the ones that interact to form the backward enhancement. Now there is a triangular relation between L, N and the multipolarity of the α–α interaction, but since the effects of the latter decrease rapidly with increasing multipolarity the greatest contribution comes from partial waves with $L = \mathcal{N} = kR$. This is the energy for which the L for grazing collisions corresponds to that of the alpha-particle in the nuclear surface. The experimental results for a number of reactions show that this relation is approximately obeyed (Hodgson, 1974).

These different exchange processes all have a common origin in the Pauli principle, and are automatically included in the calculation if we antisymmetrize the total wave function. Since this is usually very complicated it is customary to calculate one or two exchange amplitudes and to neglect the rest, which may lead to serious error. A method of including antisymmetrization has been developed by Sünkel and Wildermuth (1972), who represent its effects by a non-local, state-depending potential. This shifts the potential resonances in the odd and even partial waves in different directions, thus disturbing the delicate cancellation at large angles and giving enhanced backward cross-sections in agreement with experiment.

These calculations indicate that while the alpha-particle exchange processes probably make a large contribution to the enhanced backward cross-sections, the other exchange processes contribute significantly as well. This has the unfortunate result that it is not possible to extract from these data in a simple way information on alpha-clustering in the nucleus.

There are thus basically two physical explanations of the enhanced backward cross-sections, namely the intermediate-structure resonances with target excitation and potential scattering taking full account of antisymmetrization. A decision between these explanations requires more experimental data and detailed calculations.

The elastic scattering of helions by nuclei also shows anomalous enhancements in the backward direction, but there are differences from alpha-particle scattering that provide a further test of the model. The localization of the anomaly in angular-momentum space is not so sharp as for alpha-particles, so that an L-dependent absorption or a Regge pole description is not so satisfactory. The anomalies do not occur for all the same nuclei as for alpha-particles and show different dependence on incident energy; the knock-on exchange model is able to give a qualitative explanation of these differences (Morsch and Breuer, 1973).

The analysis of alpha-particle inelastic scattering can also be used to investigate the validity of this model. In the case of scattering by spin zero targets the peaks in the excitation functions are expected to occur at the same energies as for elastic scattering, and this has been confirmed by measurements at 29 MeV on ^{40}Ca (Schmeing and Santo, 1970). The relation between the elastic and inelastic scattering for other targets has been calculated by Ceuleneer and coworkers (1969) from a Regge pole model, but has not yet been tested experimentally. Marked enhancements in the backward direction have also been observed in the cross-sections for the inelastic scattering of alpha-particles by some nuclei (Schmeing and Santo, 1970). These results have been fitted by Cramer and coworkers (1973) using an L-dependent potential. However the nature of the L-dependent parameters is uncertain and it has been argued that the L-dependent potential should only be used to analyse elastic scattering data (Lega and Macq, 1974; Oeschler and coworkers, 1972).

A more detailed model of the inelastic scattering of alpha-particles that is able to account for the back-angle enhancements has been proposed by Ruzzene and Amos (1974). This model assumes that the reaction can proceed by a two-step process through a quasi-molecular intermediate state in which the alpha-particle is thought of as orbiting the target nucleus. At each incident energy the optimum J-value of the molecular state is obtained by a fitting procedure, and it is then found that the quasi-molecular intermediate states form a rotational band identical to that required to explain the anomalous structure in the angular distributions of the elastically scattered alpha-particles. They made calculations for alpha-particles scattered by ^{24}Mg, ^{28}Si and ^{40}Ca and used the distorted wave theory and the second Born approximation to evaluate the contributions due to the intermediate excitations. Some of their results are given in Figures 12 and 13 and show good accord with the main features of the cross-sections in the backward direction.

This reaction must proceed by a two-step process because the effect is found in transitions to unnatural parity states which cannot be reached by a one-step

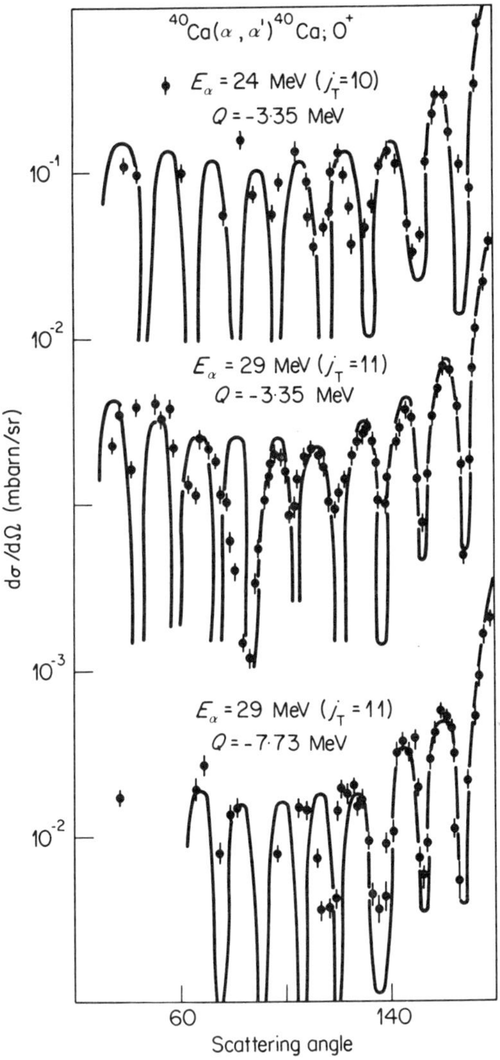

Figure 12 Two-step analyses of the differential cross-section for the inelastic scattering of alpha-particles to 0^+ states in ^{40}Ca. The angular momenta of the intermediate quasi-molecular states are shown in brackets after the projectile kinetic energy (Ruzzene and Amos, 1974)

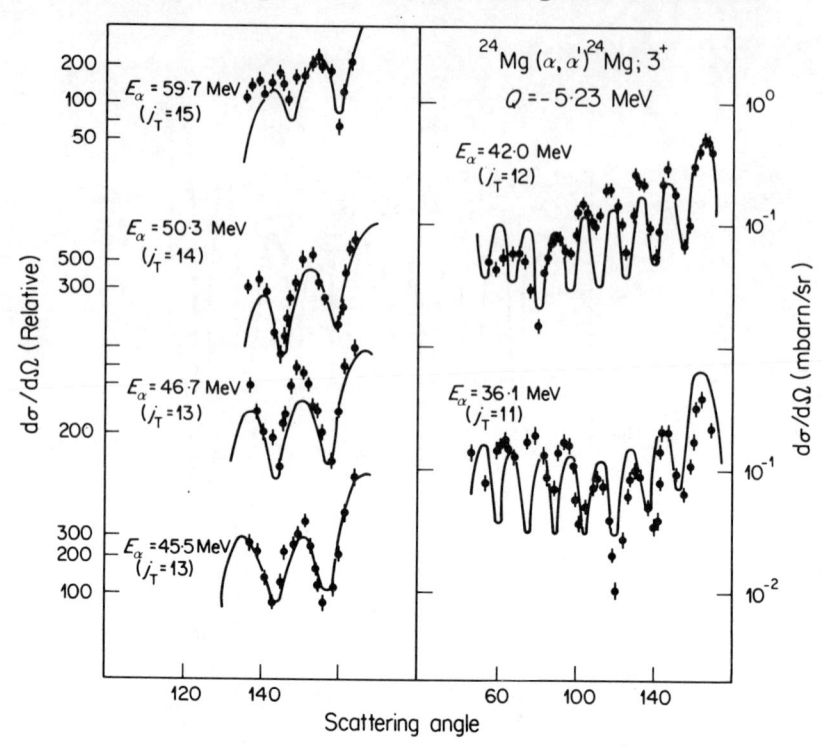

Figure 13 Differential cross-section for the inelastic scattering of alpha-particles by ^{24}Mg with excitation of the 3^+ state at 5.23 MeV compared with two-step calculations assuming intermediate quasi-molecular states of the angular momenta shown (Ruzzene and Amos, 1974)

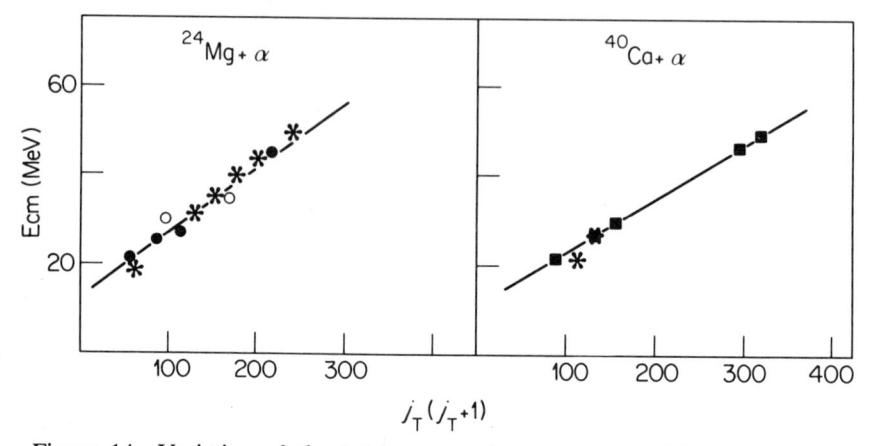

Figure 14 Variation of the total centre of mass energy with total angular momentum of the quasi-molecular states$^{(*)}$ required in the two step analysis of the inelastic scattering of alpha-particles by ^{24}Mg and ^{40}Ca. The squares for ^{40}Ca and filled circles for ^{24}Mg give the results of elastic scattering and the open circles show the results of analyses of heavy ion scattering (Ruzzene and Amos, 1974)

process. The energies of the quasi-molecular states agree very well, as shown in Figure 14, with those given by the rotational model expression

$$E = E_H + \frac{\hbar^2}{2\mathcal{J}} J_T(J_T + 1)$$

where $E_H = 13$ MeV is the band head energy and the moment of inertia factor is 0.1 MeV.

5. CLUSTER-TRANSFER REACTIONS

One of the most direct ways of finding out about alpha-clusters in the nucleus is by reactions that either add or take away a cluster from the nucleus. Reactions transferring two or more nucleons from one nucleus to another can take place in many different ways: by capture to form a compound nucleus followed by statistical emission, or by direct transfer either individually or as a cluster, or in some intermediate combinations. There is the additional possibility of multi-step processes by inelastic excitation of the target or the residual nucleus.

Surprisingly enough many reactions, especially those between heavy ions, often have simple cluster transfer as the dominant process (Lemaire, 1973). The energy spectra of the outgoing particles show that very few states are excited, even in regions of high level density. This is to be expected for cluster transfer which is restricted to few final states by stringent selection rules, whereas many more states can be excited by the more complicated transfer processes.

As a simple example of cluster transfer, Figure 15 shows the states of ^{14}C excited by the $^{12}C(^{12}C, ^{10}C)^{14}C$ and $^{13}C(^{11}B, ^{10}B)^{14}C$ reactions. The single-nucleon transfer adds a $d_{\frac{5}{2}}$ particle to the $p_{\frac{1}{2}}$ state of the target, giving the 3^- and 2^- states at 6.73 and 7.34 MeV, while for the two-nucleon transfer only the 3^- state is excited. This is readily understood if the two nucleons are transferred as a $T = 1$, $S = 0$ cluster, so that the excitation of the 2^- state is forbidden by conservation of angular momentum and parity (Anyas-Weiss, and coworkers, 1973). This preference for cluster transfer provides a powerful tool for nuclear spectroscopy, and many studies of the states of light nuclei with a particularly simple structure have been made in this way.

Cluster-transfer reactions have been identified for two and three nucleons but are expected to be particularly favoured for the transfer of two neutrons and two protons forming an alpha-particle.

Several investigations of $(^6Li, d)$ and similar reactions (Bethge and coworkers, 1967; Meier-Ewert and coworkers, 1968; Comfort and coworkers, 1970; Bassani and coworkers, 1971; Bingham and coworkers, 1971; Artemov and coworkers, 1971) showed that they are highly selective, only populating a few states of the residual nucleus. In particular the reaction on ^{12}C populates preferentially two rotational bands in ^{16}O: a positive parity band based on the O^+ state at 6.05 MeV, and a negative parity band based on the 1^-

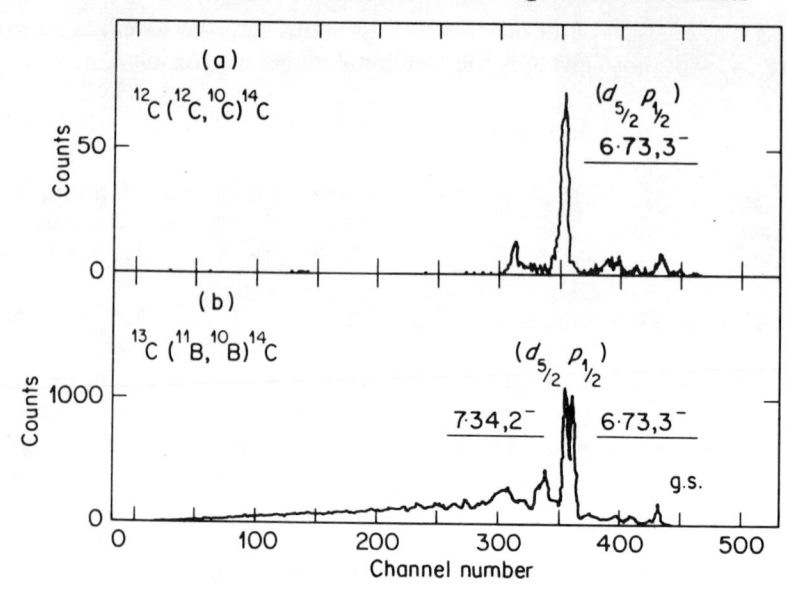

Figure 15 Energy spectrum showing states of ^{14}C excited in the ^{12}C(^{12}C, ^{10}C)^{14}C and ^{13}C(^{11}B, ^{10}B)^{14}C reactions at forward angles ($\approx 7°$) at 114 MeV. Comparison of these spectra indicates that the former reaction takes place by cluster transfer (Anyas-Weiss and coworkers, 1973)

state at 9.6 MeV. Such results are very suggestive of a direct alpha-transfer process, but since such selectivity could conceivably arise in other ways it is first of all necessary to establish what is the dominant reaction mechanism among the many possible competing processes.

The problem of identifying an alpha-transfer reaction can conveniently be considered in two stages, firstly the distinction between direct and compound nuclear processes and secondly the identification of alpha-transfer among all the possible direct processes.

Many studies of (^{6}Li, d), (^{7}Li, t) and similar reactions have shown that measurements of the excitation functions, angular distributions and de-excitation γ-rays enable the compound nucleus components to be detected and determined (Hodgson, 1971; Balamuth, 1971, 1973; Bethge and coworkers, 1967; Meier–Ewert and coworkers, 1968; Loebenstein and coworkers, 1967; Détraz and coworkers, 1971). When studying alpha-transfer it is always desirable to choose the incident energy high enough for the compound nucleus cross-sections to be negligible. This is usually quite practicable; for example studies of the (d, ^{6}Li) reaction on some light nuclei at 15 MeV by Denes and coworkers (1966) showed that at these energies the direct processes predominate.

When it has been established that a reaction takes place by a direct process, it remains to find out if it takes place by alpha-transfer. Such reactions would be particularly favoured if alpha-clusters are already existing in nuclei. They can in principle take place in many different ways so it is necessary to consider the relative probabilities of the different mechanisms to see which is the more likely.

The problem of identifying alpha-transfer reactions can be approached theoretically or phenomenologically. It is in principle possible to calculate the probabilities of alpha-transfer on the one hand and successive nucleon transfer or any other possible competing mechanism on the other, and to compare the result with the experimental data. Such calculations, besides being complicated, require a knowledge of the appropriate fractional parentage coefficients and nuclear structure wave functions that is not always available (Arima and coworkers, 1973).

A phenomenological study can be more easily carried out by seeing if four-particle transfer reactions selectively populate particular nuclear states. If the nucleons are transferred individually there is much more flexibility in the reaction, so it is possible to populate many of the states of the final nucleus. If on the other hand the nucleons are transferred in the form of an alpha-particle only a very few states of a particular structure that includes a complete alpha-particle can be populated. This gives rise to selection rules: for example, it is not possible to populate unnatural parity states in single-step alpha-transfer reactions. Thus if it is found that these states are not excited it is a good indication that the reaction proceeds by direct alpha-transfer.

These simple considerations can be modified if the reaction can take place in several distinct stages. For example, the target nucleus may be excited by an inelastic interaction before the transfer reaction takes place, and similar excitation can occur in the exit channel. Techniques have recently been developed for calculating the contributions of such processes, and in some cases they make a substantial contribution to the cross-section, particularly if the direct reaction is inhibited. Multistep contributions can also occur by transfer reactions, so that instead of the four particles being transferred in one stage, particles go backwards and forwards in several different ways. Rather few calculations of the contribution of such processes to transfer reactions have been made, but these also indicate that they may in some circumstances be appreciable.

The probability of a cluster-transfer reaction is proportional to the nuclear overlap factor, that gives the probability that the nucleus is found in the (core + cluster) configuration. This probability may be estimated for light nuclei by the $SU(3)$ model and some results of McGrory (1973) are given in Table 1.

These figures give the probabilities that states of ^{20}Ne are found in the (^{16}O+[4], (8, 0) cluster) configurations, and thus immediately assist the interpretation of the ^{16}O(^{6}Li, d)^{20}Ne reaction. The intensities for the five states of ^{20}Ne have been used by Ichimura and coworkers (1973) to calculate the strengths in the former reaction and are in fair accord with experiment. Several types of alpha-transfer reactions have been studied, and some of them are listed in Table 2. The cross-sections of such reactions are proportional to the product of the spectroscopic factors for break-up of the incident particle and of

Table 1. Probabilities that the nucleus is found in the (core + cluster) configuration estimated for light nuclei by the $SU3$ model (McGrory, 1973)

$[f], (\lambda, \mu)$	^{20}Ne [4], (8, 0)	
J	E(MeV)	P
0	0	0.78
2	1.45	0.82
4	3.86	0.76
6	8.18	0.80
8	12.35	0.72

Table 2. Probabilities of finding various nuclei in the (alpha + core) configuration

Reaction	Nucleus	Probability in LS coupling
(^{6}Li, d)	^{6}Li	1.13
(^{7}Li, t)	^{7}Li	1.19
(^{8}Be, α)	^{8}Be	1.50
(^{12}C, ^{8}Be)	^{12}C	0.68
(^{16}O, ^{12}C)	^{16}O	0.30
(^{20}Ne, ^{16}O)	^{20}Ne	0.23

References. Kurath, 1973; Ichimura and coworkers, 1973; Arima and Strottman, 1973.

the residual nucleus into an alpha-plus core, and certain dynamical factors. The probability of finding the incident particle in the alpha-plus core configuration is given in the table. The data in this table are also applicable to the corresponding pickup reactions.

The (^{6}Li, d) and (^{7}Li, t) reactions have been studied mainly for light nuclei (Denes and coworkers, 1966) since the cross-sections decrease rapidly with increasing A due to the dynamical factors and have only recently been detected for nuclei heavier than ^{40}Ca. The (^{12}C, ^{8}Be) reaction has the advantage of a large probability for ^{12}C → ^{8}Be + α and the (^{16}O, ^{12}C) reaction has been extensively used. This reaction suffers from the disadvantage that the ^{12}C can be excited, but even though the breakup ^{16}O → ^{12}C* + α is more probable than ^{16}O → ^{12}C + α it does not seem to occur experimentally at low energies due to dynamical factors (DeVries, 1973a, b), though it is found at higher energies.

The experimental data on these reactions frequently show a strong excitation of four-particle four-hole states while the unnatural parity states are only weakly excited. For example the ^{12}C(^{12}C, ^{8}Be)^{16}O reaction at 62.6 MeV strongly populates the 4p4h rotational band starting at 6.07 MeV (Wozniak and coworkers, 1972). Similar results are found for the (^{6}Li, d) and (^{7}Li, t) reactions. On the whole, it is found that the cluster transfer seems to increase in relative importance with the incident energy (Hildenbrand and coworkers, 1970).

The reactions on light nuclei usually excite final states that can be resolved experimentally, thus facilitating the analysis. In heavier nuclei the level density increases much more rapidly with excitation energy, and since alpha-particle states are likely to have excitation energies of at least a few MeV they are most frequently found in regions of high density of states. In these circumstances the alpha-states have to be detected in the midst of a dense background of other states, and this will only be possible if the excitation process is so selective that the corresponding peaks stand clear of the background due to all the other states. This seems to be the case for some alpha-transfer reaction between heavy ions.

An example of such reactions is provided by the work of Faraggi and coworkers (1971a, b) and Faívre and coworkers (1970) who have shown that alpha-particle or 'quartet' states may be excited in alpha-transfer reactions such as ^{54}Fe(^{16}O, ^{12}C)^{58}Ni; these states stand out as clear peaks in regions of excitation energy with high level densities (cf. Robson, 1972). If this interpretation of these states is correct it should be possible, by analogy with the corresponding situation for nucleon single-particle states, to excite these states by alpha-particle elastic scattering on the same nucleus, but so far no clear evidence for such processes has been obtained.

It is difficult to obtain reliable spectroscopic information from the (^{16}O, ^{12}C) reaction because of the strong Q-value dependence of the cross-sections and the uncertainty concerning the configuration of the four transferred nucleons (Bonche and coworkers, 1972). The (^{6}Li, d) reaction is preferable in these respects and has the additional advantage of a higher energy resolution.

Studies of the (^{6}Li, d) reaction on ^{58}Ni and ^{64}Ni by Gutbrod and Markham (1972) showed a strong selectivity and the zero-range distorted wave theory gave a good account of the differential cross-sections.

The (^{6}Li, d) reaction at 19.5 MeV on a range of nuclei from ^{10}B to ^{40}Ca was measured by Gutbrod and coworkers (1971), and for all except the lightest nucleus the finite-range distorted wave theory gave a fairly good account of the differential cross-sections. The cross-sections are very sensitive to the internal wave function of the ^{6}Li nucleus; as shown in Table 3 the shell model wave

Table 3. Comparison of theoretical and experimental cross-sections in μb/sr at the first maximum for the (d, ^{6}Li) reaction at 19.5 MeV on several nuclei. The transferred angular momentum is L and the spectroscopic factor is S (Gutbrod and coworkers, 1971)

Target	L	S	$\frac{d\sigma}{d\Omega}$(shell model)	$\frac{d\sigma}{d\Omega}$(cluster model)	$\frac{d\sigma}{d\Omega}$(experiment)
^{11}B	0	0.206	200	400	600
	2	0.615			
^{12}C	0	0.759	100	500	500
^{16}O	0	0.333	6	140	140
^{19}F	1	0.088	8	170	140
^{40}Ca	0	0.0087	0.04	50	40

function gives cross-sections that are far too low, whereas the cluster model wave functions give cross-sections similar to the experimental results.

Detailed finite-range calculations of the (^{6}Li, d), (^{7}Li, t), (h, ^{7}Be) transfer reactions have been made by Kubo and Hirata (1972) and by Kubo (1972) and they concluded that the applicability of the distorted wave theory is due to the strong cluster structure of the projectiles, the concentration of the reaction at the nuclear surface due to their strong absorption, and the simplicity of the alpha-transfer approximation.

A similar analysis of the ^{24}Mg(^{7}Li, t)^{28}Si reaction from 12–20 MeV by Rosner and coworkers (1974) showed that the finite-range distorted wave theory is able to give a fair account of the differential cross-sections, but it was not possible to extract spectroscopic factors due to the difficulty of calculating realistic form factors and the uncertainties in the distorting potentials. It was nevertheless possible to draw some qualitative conclusions from the observed selection rules. Thus if the four transferred nucleons constitute an alpha-cluster the reaction should populate all three low-lying 0^+ states in the final nucleus if the $SU(3)$ model is applicable. However only one 0^+ state is excited, and this may be connected with the change from prolate to the oblate deformation in the middle of the sd shell, which affects the description of the states in terms of the $SU(3)$ model.

The ^{40}Ca(^{6}Li, d) reaction at 32 MeV was studied by Strohbusch and coworkers (1972) and they found clearly structured angular distribution with cross-sections similar to those of the ^{40}Ca(^{16}O, ^{12}C) reaction that are well described by alpha-transfer distorted wave calculations.

The ^{18}O(^{7}Li, t)^{22}Ne reaction at 12 MeV has been studied by Scholz and coworkers (1972). Analysis of the angular distributions with the Coulomb-distorted plane wave model enabled them to determine the angular momentum transfers and to extract the alpha-cluster widths for the strong transitions. Significant alpha-clustering was found for ten states in the final nucleus. Many other studies of (^{7}Li, t) reactions on light nuclei also show that alpha-cluster transfer is the dominant reaction process (Middleton and coworkers, 1968a, b; Scholz and coworkers, 1965, 1972; Neogy and coworkers, 1970). These reactions are very useful in obtaining spectroscopic information concerning rotational bands, which are preferentially excited in alpha-transfer reactions.

Another test of the presence of the alpha-transfer process and the validity of the alpha-transfer approximation is to see whether the spectroscopic factors obtained from the analyses of different alpha-transfer reactions on the same nucleus give the same results. Such a test has been made by De Vries (1973a, b) who made a finite-range recoil DWBA analysis of the cross-sections of the (^{16}O, ^{12}C) and (^{6}Li, d) reactions on ^{40}Ca. The results are shown in Figure 16, and the overall agreement with the data and the consistency of the spectroscopic factors indicates that most of the reaction takes place by the alpha-transfer process. Similar comparisons have been made by Strohbusch and coworkers (1972, 1974), and by Morrison (1971).

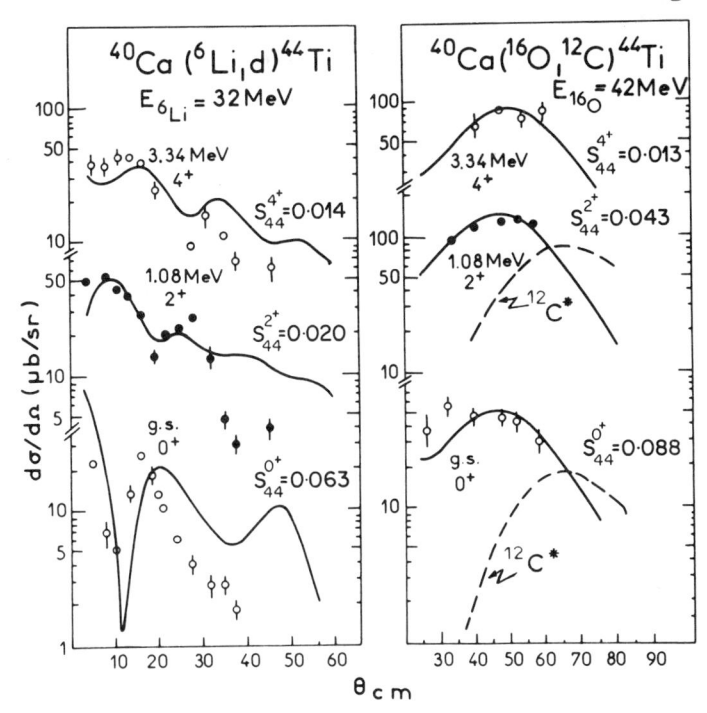

Figure 16 Cross-sections for the alpha-stripping reactions ^{40}Ca(^{6}Li, d)^{44}Ti and ^{40}Ca(^{16}O, ^{12}C)^{44}Ti compared with finite-range recoil DWBA calculations. The overall agreement and the similarity of the spectroscopic factors for the two reactions supports the assumption that they proceed by direct alpha-transfer (DeVries, 1973a)

A detailed study of the usefulness of the (^{6}Li, d) reaction as a spectroscopic tool in the f–p shell has been made by Fulbright and coworkers (1975). They measured the cross-sections for several nuclei at ^{6}Li energies around 30 MeV and compared their results with zero-range cluster-transfer distorted wave calculations. As shown in Figure 17 these agree well with the data provided a rather deep optical potential ($V = 250$ MeV) is used for the ^{6}Li particles. In most cases the angular distributions are sufficiently characteristic of the angular momentum transfer that they suffice to determine it. The angular distributions are relatively insensitive to changes in Q-value and target, increasing the utility of this reaction as a spectroscopic tool.

The (d, ^{6}Li) reaction has recently been used by Becchetti and coworkers (1975) to make a systematic study of alpha-pickup in a range of nuclei from ^{12}C to ^{238}U. They irradiated these nuclei with 35 MeV deuterons, and observed the ^{6}Li emitted at the angle corresponding to the maximum in the cross-section for $L = 0$ transfer.

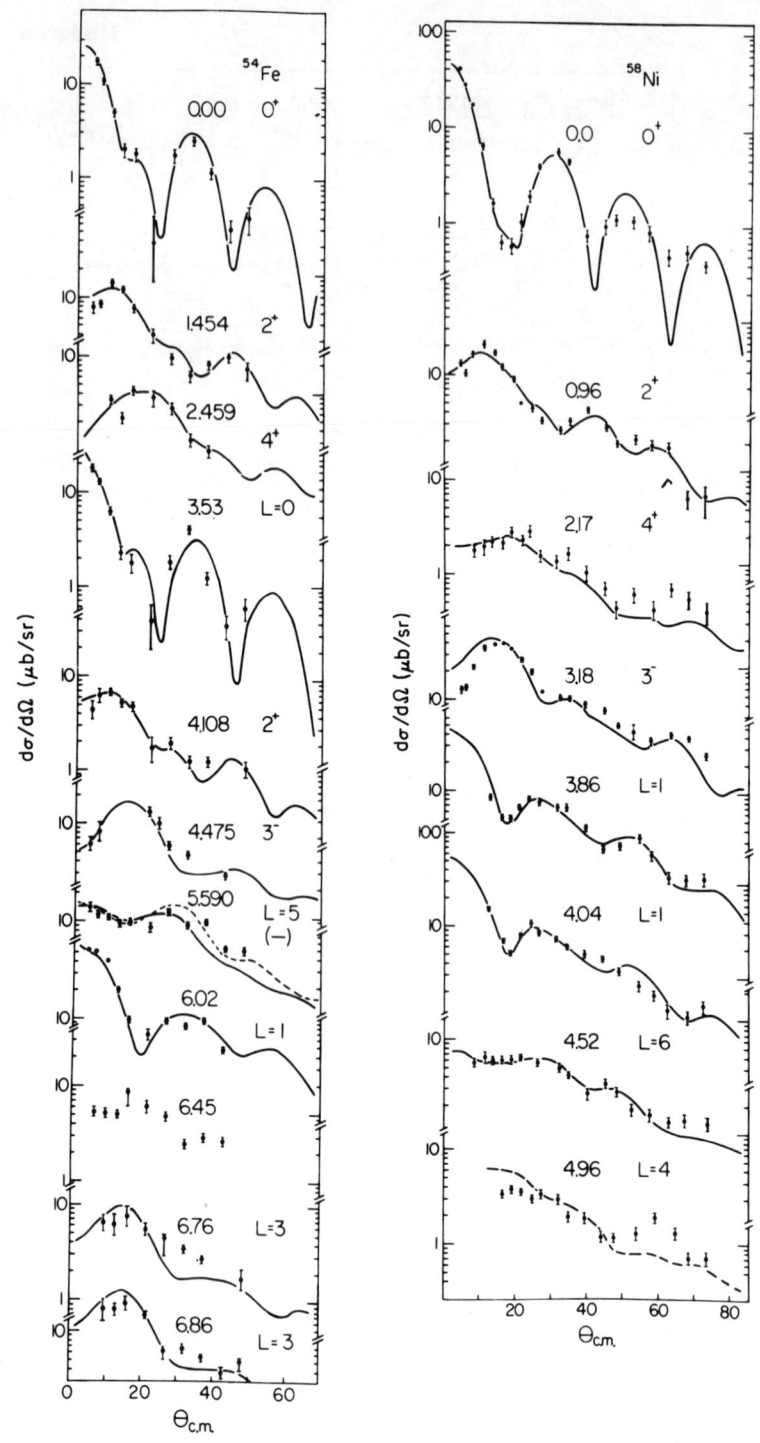

Figure 17 Differential cross-sections for the (^{6}Li, d) reaction on ^{54}Fe and ^{58}Ni at 30 MeV compared with zero-range cluster-transfer distorted wave calculations (Fulbright and coworkers, 1975)

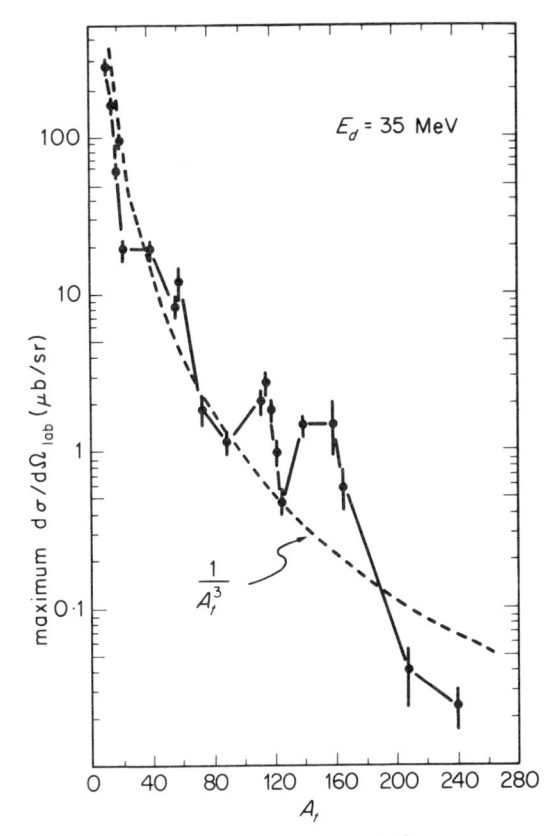

Figure 18 Cross-section of the $(d, {}^6Li)$ reaction at the maximum of the $L = 0$ transfer angular distribution (Becchetti and coworkers, 1975)

Some of their results are plotted in Figure 18 as a function of target mass, and it is clear that the cross-section falls rather steeply as the mass increases (roughly as A_t^{-3}), with superposed fluctuations that are connected with the nuclear shell structure.

To study these variations in more detail, they calculated the cross-section for alpha-particle pickup using the distorted wave theory. This contains a factor S_α that is a measure of the probability of finding an alpha-particle in the target nucleus. The resulting values of S_α, normalized to unity for ${}^{16}O$, are shown in Figure 19 as a function of A.

The fluctuations of S_α with A can be correlated with the nuclear shell structure in a very simple way. The minima correspond to nuclei with closed neutron or proton shells, while the maxima correspond to nuclei with open shells. Thus the maximum around $A = 120$ corresponds to open neutron shells and that around 160 to open neutron and proton shells.

The assumption that the reaction takes place by a one-body transfer of a group of two protons and two neutrons in its intrinsic ground state is a very

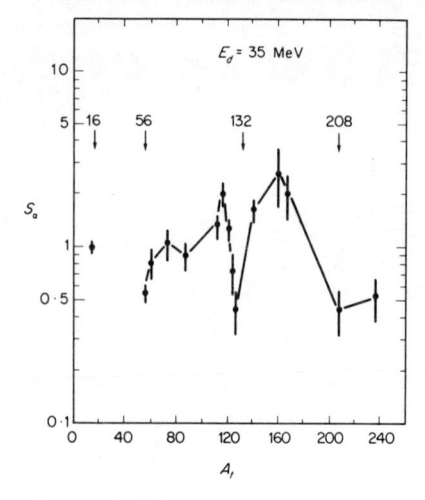

Figure 19 Alpha-particle spectroscopic factors S_α obtained from analyses of $(d, ^6\text{Li})$ reactions normalized to unity at ^{16}O and plotted as a function of target mass (Becchetti and coworkers, 1975)

special one. A full treatment of four-nucleon transfer (Rotter, 1969) represents the state of the nucleons in terms of centre-of-mass and intrinsic co-ordinates and the matrix element of the reaction is a coherent sum of terms representing different internal and centre-of-mass states. It is therefore not possible to express the cross-section as a product of reaction and structure factors, so that the structure amplitudes must be known before the cross-section can be calculated. The component corresponding to simple alpha-transfer is however favoured because it corresponds to the maximum surface amplitude for the centre-of-mass motion.

The excitation of unnatural parity states in ^{16}O by the $^{12}\text{C}(^{10}\text{B}, ^6\text{Li})^{16}\text{O}$ reaction could be due to the transferred alpha-particle being in an excited state (Hildenbrand and coworkers, 1970). Calculations of the cross-section for the reaction $^{12}\text{C}(^{16}\text{O}, ^{20}\text{Ne})^8\text{B}$ have been made by Yoshida (1973) taking the possibility of excited alpha-clusters into account, but as shown in Figure 20 he finds that their contribution to the total reaction amplitude is rather small.

Several distorted wave calculations have recently been made to try to account for the cross-sections for the $^{32}\text{S}(^{16}\text{O}, ^{12}\text{C})^{36}\text{Ar}$ reaction to the ground and first excited state of the residual nucleus that have been measured by Braun-Munzinger and coworkers (1973). These angular distributions have a strongly oscillatory structure and the question is whether they can be accounted for by a simple alpha-transfer process or whether it is necessary to include the possibility of excited alpha-transfer as well.

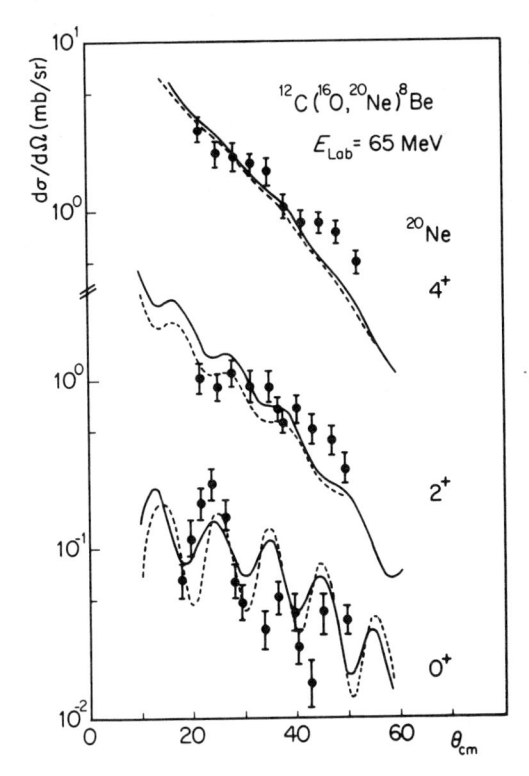

Figure 20 Differential cross-section for the alpha-transfer reaction $^{12}C(^{16}O, ^{20}Ne)^8B$ at 65 MeV to three states in ^{20}Ne compared with distorted wave calculations (a) including only the alpha-transfer amplitude (dashed lines) and (b) including the effect of excited alpha-transfer (full lines) (Yoshida, 1973).

Further studies of this problem were made by Charlton and Robson (1974), who found that it is possible to fit the data if the transferred alpha-particle is in its excited 2^+ state at 28.5 MeV. Then Werby and Tobocman (1974) showed that the oscillatory structure can be fitted with unexcited alpha-transfer provided the distorting optical potentials are weakly absorbing. More recently, Erskine and coworkers (1975) were able to obtain a fairly good fit to the data using the simple alpha-cluster transfer model without using weakly absorbing potentials or taking into account the possibility of excited alpha-transfer. A subsidiary calculation showed that the cross-section for this latter process is extremely small.

All this work shows how sensitive the alpha-transfer cross-sections are to the rather poorly understood optical potentials. It seems likely that further progress will only be made when these potentials are put on a proper physical

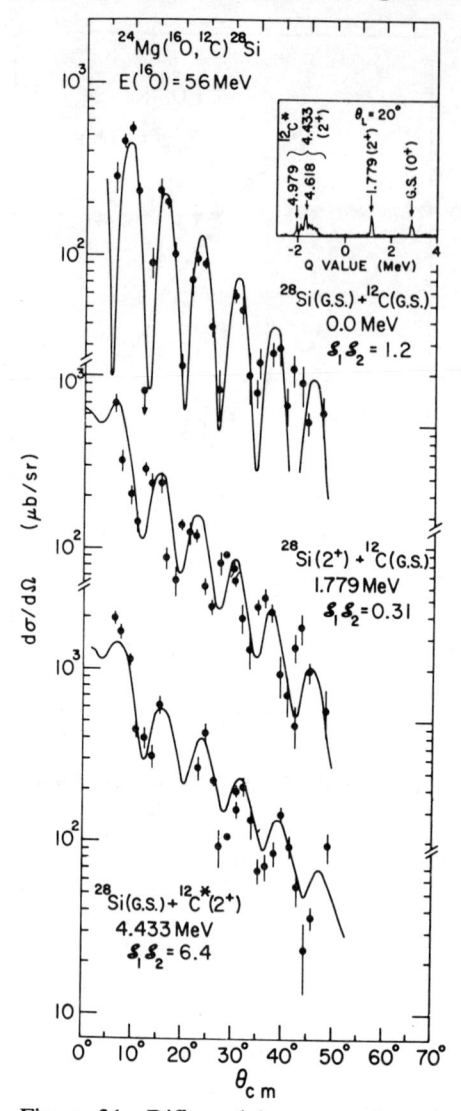

Figure 21 Differential cross-section of the reaction ^{24}Mg(^{16}O, ^{12}C)^{28}Si at 56 MeV to several final states compared with distorted wave alpha-transfer calculations. The insert shows a portion of the energy spectrum measured at $\theta_L = 20°$ (Erskine and coworkers, 1975)

basis, with their characteristics determined from the contributing nucleon–nucleon interactions, instead of relying on highly ambiguous phenomenological analyses. Even at the present time, however, it does seem unlikely that excited alpha-transfer contributes appreciably to these reactions.

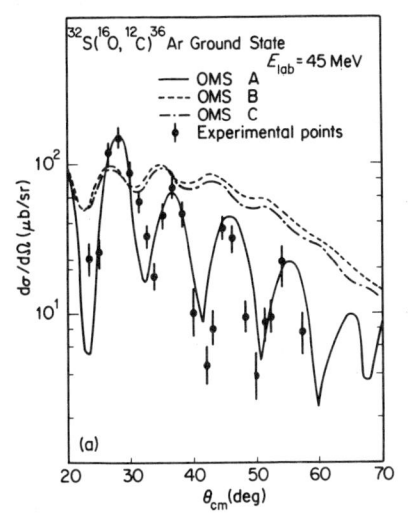

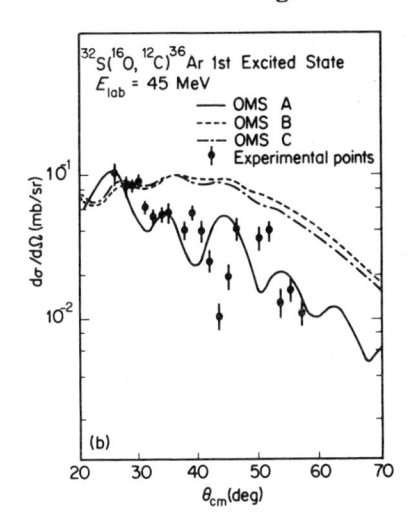

Figure 22 Differential cross-sections for the alpha-transfer reaction $^{32}S(^{16}O, ^{12}C)^{36}Ar$ to the ground state and to the first excited state compared with distorted wave calculations using three sets of potentials (Werby and Tobocman, 1974)

Erskine and coworkers (1975) also measured and analysed the differential cross-section of the reaction $^{24}Mg(^{16}O, ^{12}C)^{28}Si$ at 56 MeV, and as shown in Figure 21 they are able to account for their results very well using the simple alpha-transfer distorted wave theory. This work made it possible to compare the measured ratios of spectroscopic factors with those obtained theoretically. Thus they find $S(^{28}Si^*2^+)/S(^{28}Si_{gs}) = 0.26$ compared with 0.18 obtained from ($^6Li, d$) analysis (Draayer and coworkers, 1974) and 0.23 from the $SU(3)$ model (Draayer, 1975). For the ratio $S(^{12}C^*)/S(^{12}C_{gs})$ they found 5.3, in good agreement with the theoretical value of 5.5 (Rotter, 1968; Kurath, 1973). The value of 1.4 for this ratio obtained from analysis of the $^{12}C(^{16}O, ^{12}C)^{16}O$ data (De Vries, 1973b) is much lower, but this difference is probably attributable to inelastic-interference and resonance-interference effects (Badawy and coworkers, 1974).

Alpha-transfer reactions on radioactive nuclei are particularly interesting because they make it possible to study the inverse process to alpha-decay, and comparison of the corresponding spectroscopic amplitudes provides a useful way of evaluating the theories of the two processes. The reaction $^{208}Pb(^{16}O, ^{12}C)^{212}Po$ at 93 MeV has recently been studied for this purpose by De Vries and coworkers (1975). The differential cross-sections for the transitions to the ground and 0.727 MeV states of ^{212}Po had the characteristic bell-shape of a peripheral reaction, and were well fitted by finite range recoil distorted wave calculations. This analysis of the data gave a spectroscopic ratio $S(0.727 \text{ MeV } 2^+)/S(g.s.) = 0.64 \pm 0.13$. Analysis of the reduced alpha-decay widths using optical potential penetrabilities gave a value 0.61 ± 0.24.

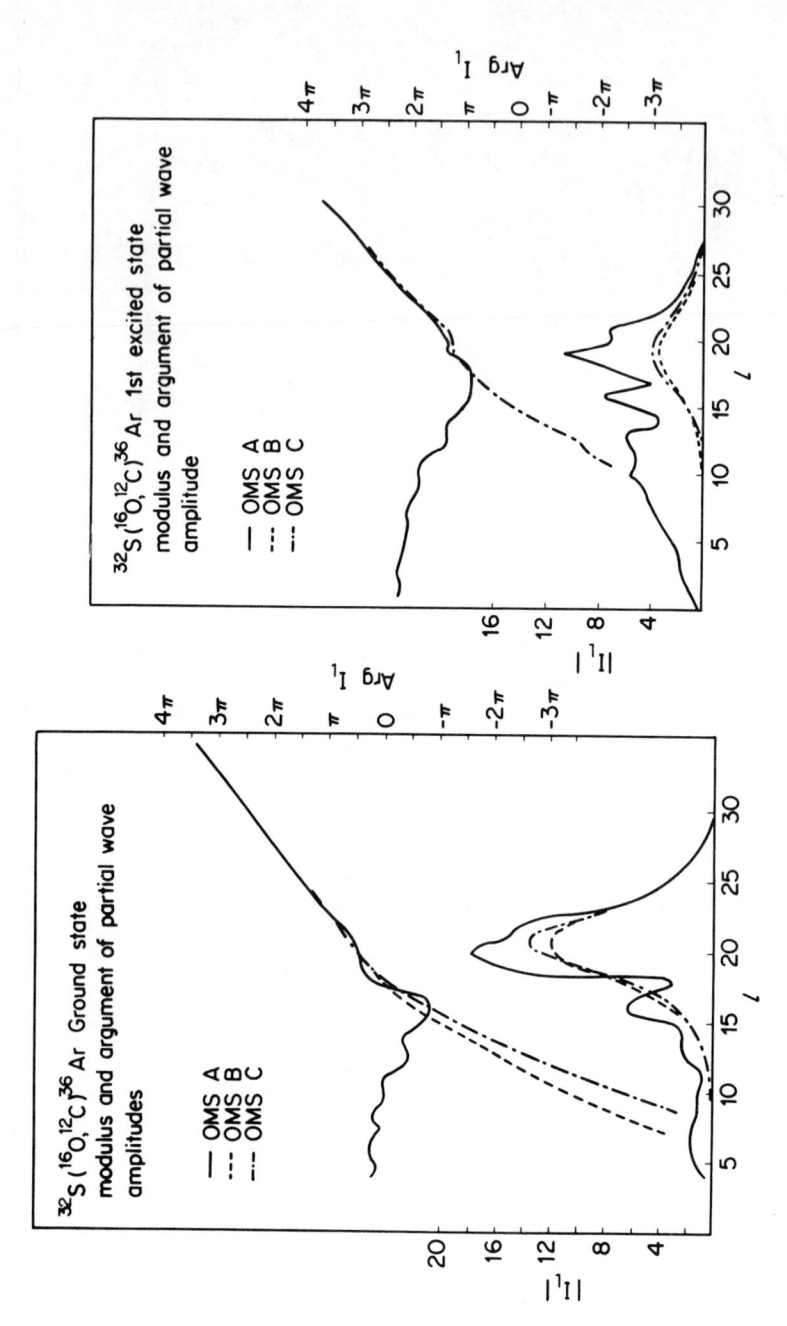

Figure 23 Modulus and phase of the partial wave amplitudes for the $^{32}S(^{16}O, {}^{12}C)^{36}Ar$ reaction to the ground and first excited states (Werby and Tobocman, 1974)

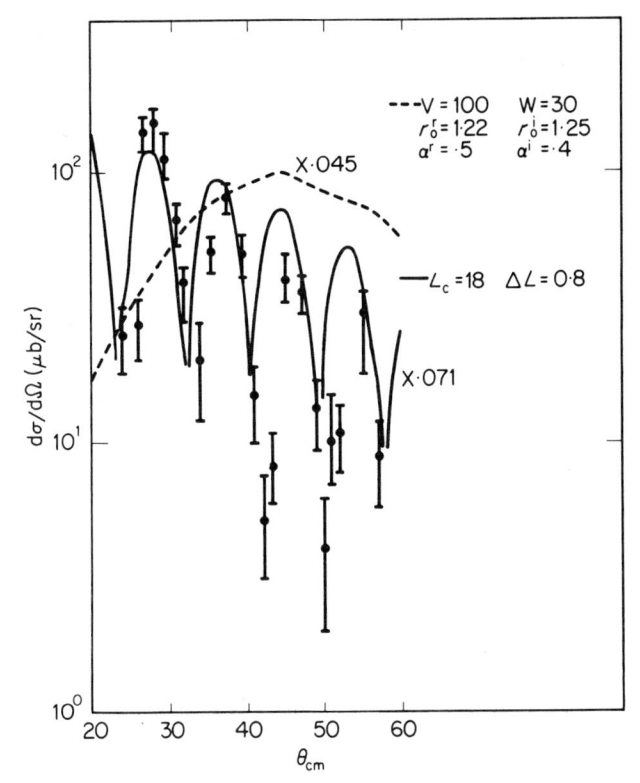

Figure 24 Differential cross-section for the reaction ^{32}S(^{16}O, ^{12}C)^{36}Ar at 30 MeV to the ground state of the residual nucleus compared with distorted wave calculations assuming the transferred alpha-particle to be in its ground (dashed line) and in its excited 2^+ state at 28.5 MeV (full line). The parameters of the optical potential are included in the figure and the numbers on the curves are the spectroscopic factors obtained by normalizing the calculated curves to the experimental data (Charlton and Robson, 1974)

Although these numbers are not very precise, this is a satisfactory confirmation of the overall correctness of the calculations. The comparison is made in terms of ratios for the two states because the extreme sensitivity of the penetrability to the optical potentials makes it difficult to obtain reliable absolute values of alpha-decay widths, and the cluster approximation introduces a similar uncertainty in the calculated values of the alpha-transfer cross-sections.

6. KNOCK-OUT REACTIONS

The presence of alpha-particles in the nuclear surface enhances the probability of reactions like $(p, p\alpha)$ and $(\alpha, 2\alpha)$ in which an incident particle apparently knocks an alpha-particle out of the target nucleus, and early experimental studies (Hodgson, 1958; DuBost and coworkers, 1964; Igo and coworkers, 1963) enabled the proportion of alpha-particles in the nuclear surface to be estimated. In a similar way the presence of these alpha-particles enhances the contribution of the knock-out mechanism over the successive pickup mechanism in (n, α) and (p, α) reactions.

The simple knock-out model has been used by Veselić and Tudorić-Ghemo (1968) to analyse the energy and angular distributions of the alpha-particles from the reaction ^{103}Rh(n, α) at 14.4 MeV. They found a value of 0.3 for the probability of a surface nucleon to be a constituent of an alpha-particle, a value similar to that obtained by Ostrumov and coworkers (1959) from high-energy protons on nuclei and by Kulišić and coworkers (1964) from (n, α) reactions on several nuclei.

The use of the $(p, p\alpha)$ and $(\alpha, 2\alpha)$ knock-out reactions to determine the degree of alpha-clustering is simple in principle but in practice encounters severe difficulties that greatly limit its usefulness. The basic idea is that the incident particle collides with the alpha-cluster and knocks it out of the nucleus. Such quasi-free scattering events should be easy to detect because the energies and angles of the emitted particles satisfy stringent kinematical conditions imposed by the conservation laws. From the cross-section for such reactions, and using that of the corresponding free p-α or α-α interaction, it is easy in principle to determine the alpha-clustering probability.

Unfortunately the trajectories of the incident and outgoing particles are severely distorted by the nuclear field so that to be able to recognize the quasi-free events by their characteristic kinematics requires incident energies of several hundred MeV, and then the total cross-section of the quasi-elastic events is low and the probability of breaking up the alpha-cluster becomes appreciable. There are also fundamental difficulties due to the presence of short-range correlations (Hodgson, 1975). Furthermore the available experimental energy resolution at these incident energies is insufficient to permit the separation of the reactions going to different states of the final nucleus, and this makes it impossible to determine the energies of the alpha-cluster states.

A simple classical model of alpha-emission at high energies may be made by assuming appropriate density distributions for the nucleon and alpha-clusters in the nucleus, and using experimental values of the nucleon–nucleon and nucleon–alpha cross-sections. A Monte Carlo calculation of the cascade through the nucleus then gives the frequencies, energies and angular distributions of the emitted nucleons and alpha-particles for comparison with the experimental data. The initial assumptions concerning the alpha-cluster distribution can then be altered to optimize the agreement with the experimental data.

A calculation of this type was made by Cohen (1966) who found that some data for the interaction of 155 MeV protons with ^{197}Au is quite well reproduced using a value of 5.7 for the number of alpha-clusters.

Calculations by Griffy and coworkers (1966) have shown that high energy electron scattering can be used to study the cluster structure of light nuclei, and apply this formalism to the reaction ^{6}Li$(e, e'd)^4$He.

The momentum distribution of the α-clusters inside the nucleus can be obtained by measuring the $(\alpha, 2\alpha)$ cross-section at high energies. Any departures from the kinematics of the two-body collision are due partly to the initial momentum distribution of the α-cluster and partly to the distributions of the incoming and outgoing waves by the nuclear field, and if the latter is evaluated the former can be determined. This method has been applied to the ^{6}Li$(\alpha, 2\alpha)$ and ^{7}Li$(\alpha, 2\alpha)$ reactions at 56 MeV by Pizzi and coworkers (1968), but they were unable to obtain the momentum distribution due to lack of knowledge of the $(\alpha, 2\alpha)$ cross-section.

A distorted wave analysis of the ^{6}Li$(\alpha, 2\alpha)^2$H reaction at 70.3 MeV by Jain and Sarma (1974) gave a good fit to the differential cross-section and an effective clustering probability of 0.45 ± 0.18, consistent with the value found from analyses of the ^{6}Li$(d, d\alpha)^2$H and ^{6}Li$(\alpha, 2\alpha)^2$H reactions at 27 and 55 MeV (Jain and coworkers, 1973).

The interpretation of (n, α) and (p, α) reactions depends on the incident energy. At low energies the compound nucleus process dominates, and this is well accounted for by statistical theories. At high energies direct reactions dominate, and their cross-sections can be calculated by the distorted wave theory. The two most likely mechanisms are alpha-knock-out, which can provide information on the probability of finding alpha-particles in the nucleus, and successive nucleon pickup which cannot. The complications of the formalism for the latter reaction makes it difficult to make reliable calculations, but such studies as have been made tend to indicate that it is favoured relative to alpha-knock-out. In other cases both processes contribute to some extent, thus further complicating the interpretation.

Data on (n, α) and (p, α) reactions in the intermediate energy region have, however, been successfully interpreted in terms of alpha-clusters in the nucleus by the Milan group (Milazzo-Colli and coworkers, 1972, 1973, 1974; Colli-Milazzo and coworkers, 1972; Braga-Marcazzan and coworkers, 1973). They measured the cross-section for a series of (n, α) and (p, α) reactions on heavy nuclei, and some typical results are shown in Figure 25. In these data the cross-sections of reactions to individual final states are not resolved, so it is necessary to seek a statistical explanation of the results.

The statistical evaporation of particles from an excited nucleus is now well-understood, together with the variation of nuclear temperature with excitation energy, so that the energy spectrum of the alpha-particles from the compound nucleus can be easily calculated. When this is done, it is found to give a spectrum that accounts very well for the alpha-particles of lower energies (the full curve in the figure), but falls far short of the intensity found at higher energies.

These alpha-particles of higher energy can be understood as coming from decays taking place before full statistical equilibrium has been established. The

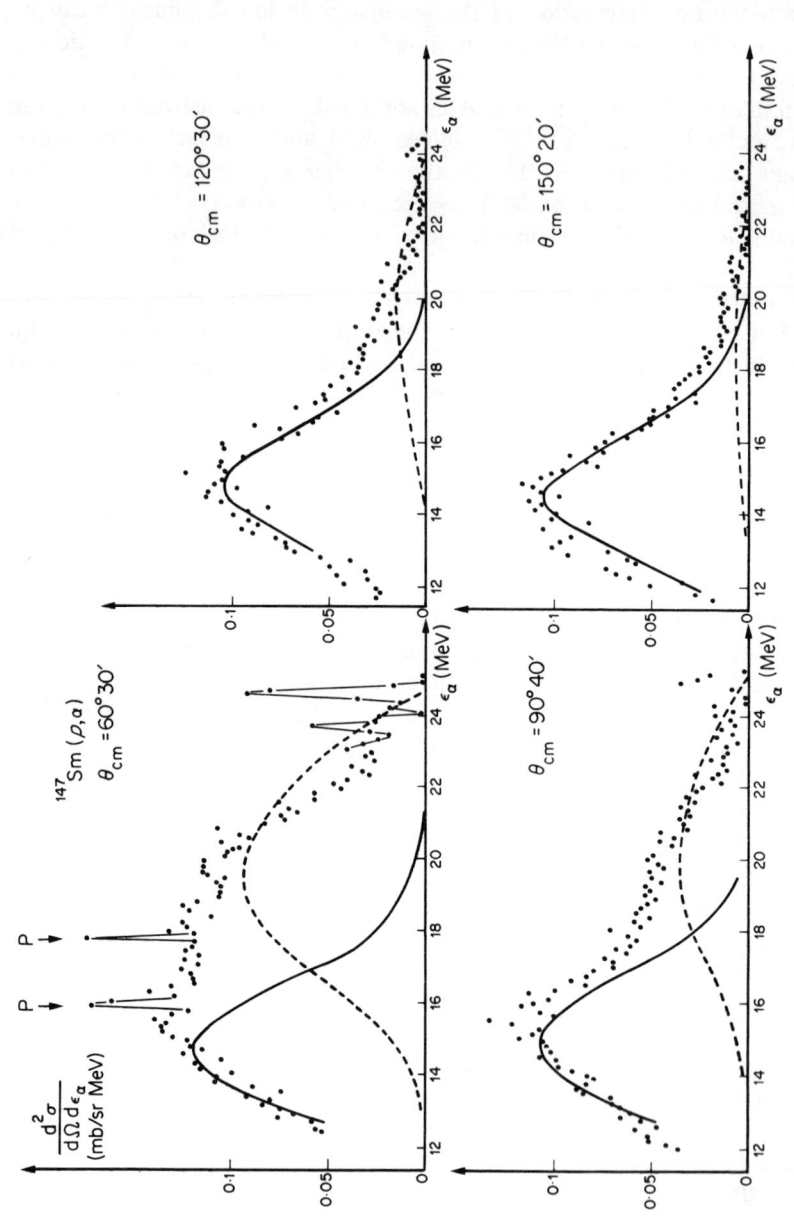

Figure 25 Energy distribution of alpha-particles from the reaction ^{147}Sm$(p, \alpha)^{144}$Pm at 18 MeV for four angles compared with calculations using the evaporation (full lines) and precompound emission of preformed alpha-particles (dashed lines) (Milazzo-Colli and coworkers, 1974)

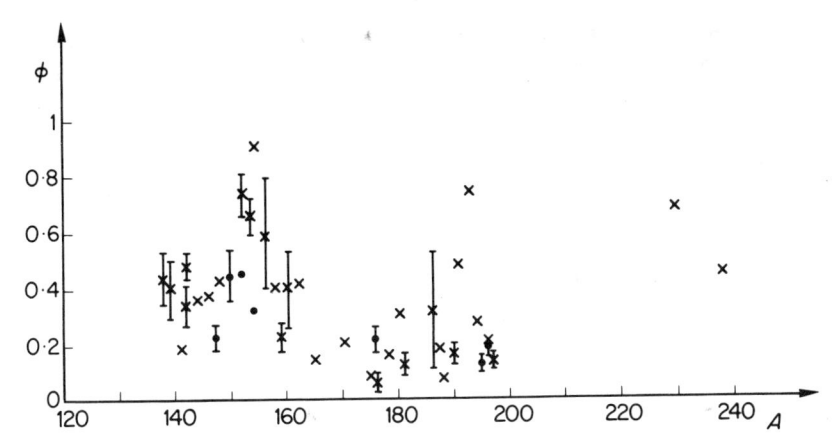

Figure 26 Values of the alpha-particle preformation probability ϕ obtained from analyses of precompound emission in (p, α) ($\cdot$) and (n, α) (×) reactions (Milazzo-Colli and coworkers, 1974)

model is that the incident proton interacts with an alpha-particle on the nuclear surface to form a local 'hot-spot', or pre-equilibrium excited state consisting of the incident proton, the alpha-particle and the alpha-hole. This state can decay either by alpha-emission or by the excitation of more nucleons leading eventually to the full statistical sharing of energy in the compound nucleus. In the case of immediate alpha-emission the alpha-particles naturally have, on the average, a higher energy than those from the later decay of the compound nucleus, as the temperature of the 'hot-spot' is much greater than that of the compound nucleus. This process of alpha-particle emission is called 'pre-equilibrium emission'.

The theory of pre-equilibrium emission has been developed by Griffin (1966) and others, and enables the energy spectrum of the alpha-particles to be calculated. The spectrum found in this way is shown by the dashed curve in the figure, and accounts very well for the more energetic alpha-particles not given by the statistical theory.

All the parameters of the theory, including the proton-reaction cross-section, the single particle level densities, the excitation energies of the residual and compound nuclei and the matrix element describing the effective two-body interaction are already known, and only the parameter describing the probability ϕ for the incoming proton to strike a preformed alpha-particle in the nucleus instead of a nucleon remains to be determined by normalizing the theory to the experimental data. This probability ranges from 0.1 to 0.8 for different nuclei, and it is an important test of the model to see if it can be calculated accurately from the structures of the nuclei concerned.

Some of the values of this preformation probability obtained from analyses of (n, α) and (p, α) reactions are given in Figure 26, and show a variation with shell structure similar to those found from analyses of alpha-decay (Figure 6) and of $(d, {}^6Li)$ reactions (Figure 19).

The experimental data for (n, α) reactions at 18 MeV on a series of rare earth nuclei have been analysed by Glowacka and coworkers (1975) in terms of the statistical, pre-equilibrium and knock-on models. They found that the statistical model is unable to account for the data, while the pre-equilibrium and knock-on models are qualitatively successful.

7. PION AND KAON REACTIONS

There are several ways in which pion and kaon capture could provide information on alpha-clustering in nuclei. If a slow kaon is captured, the reaction products depend on whether the primary interaction takes place with one or several nucleons, and this is dependent on the clustering probability. If alpha-particles are present already in the nucleus they are likely to be emitted as a result of the large amount of energy given to the nucleus in slow pion or kaon capture, or they could be knocked out of the nucleus by fast pions or kaons.

When a slow kaon approaches a nucleus it spirals slowly down the atomic Bohr orbits. Since the transitions go preferentially to S-states the kaon moves essentially in circular orbits and thus encounters first the extreme edge of the nucleus (Rook, 1962, 1963). The interaction of a kaon with a cluster of nucleons is qualitatively different from its interaction with a single nucleon, and examination of the outgoing particles enables these cases to be distinguished. The possible reactions are:

$$K^- + p \to \Sigma^{\pm,0} + \pi^{\mp,0} \quad K^- + n \to \Sigma^{-,0} + \pi^{0,-}$$
$$\to \Lambda^0 + \pi^0 \qquad \qquad \to \Lambda^0 + \pi^-$$

and

$$K^- + 2p \to \Sigma^+ + n \quad K^- + np \to \Sigma^- + p \quad K^- + 2n \to \Sigma^- + n$$
$$\to \Sigma^0 + p \qquad \qquad \to \Sigma^0 + n$$
$$\to \Lambda^0 + p \qquad \qquad \to \Lambda^0 + n$$

The two-nucleon interactions are thus characterized by the absence of a pion, and in addition the Σ's have substantially greater energy, about 200 MeV, compared with less than 60 MeV for the single-nucleon interactions. This allows the two types of interaction to be separated. A comparison between the one-nucleon and two-nucleon interaction rates shows that in deuterium less than one per cent of the K^- captures are attributable to the two-nucleon process, while the corresponding figures for helium and nuclear emulsion are around 14 and 20 per cent respectively. If two nucleons can be sufficiently closely correlated for the two-nucleon interaction to occur only when they are within an alpha-cluster, then these figures can be immediately interpreted in terms of the degree of alpha-clustering, and since the capture process occurs predominantly in the nuclear surface they suggest a high degree of alpha-clustering in that region.

However more detailed calculations by Wycech (1967) confirmed by Aslam and Rook (1970) showed that the observed interaction ratios can be obtained to within a factor of about two without assuming the presence of alpha-clusters. The simple argument is appreciably modified by considerations of relative phase space in deuterium and nuclei (Rook, 1968) and by the $Y_0^*(1405)$ resonance in the K^--p interaction, which enables the K^- to interact with a pair of well-separated nucleons. The experiments are still subject to considerable uncertainties and a number of approximations are made in the calculations, so all that can be said at present is that this work provides evidence neither for nor against the idea of alpha-clusters on the nuclear surface. These calculations indicate that kaon capture is not confined to the surface nucleons, a conclusion supported by the experimental work of Bugg and coworkers (1974).

Further evidence of cluster structure has been sought in the absorption of negative pions by nuclei. This has been studied theoretically for the light nuclei ^{12}C, ^{14}N and ^{16}O by Kopaleĭshvili and Machabeli (1971) and they find that it depends strongly on the cluster structure of the nucleus. They assumed that the pion is absorbed by an np pair in an alpha-cluster, and found that the angular-correlation between the deuteron momentum and the combined momentum of the two reactions depends strongly on the isolation of the cluster. The analysis of experimental data on the energy spectra and angular correlations of the nn and np pairs produced in the capture of negative pions by carbon nuclei indicate a considerable (more than 50 per cent) contribution from the absorption on alpha-clusters (Kolybasov and Tsepov, 1972).

The capture of negative pions by ^{16}O nuclei has recently been studied by Lewis and coworkers (1973). Since energy and momentum are conserved the pion cannot be captured by a single nucleon but must interact with several nucleons at once, so a detailed study of the particles emitted provides information on nucleon correlations in the nucleus. They looked at the γ-ray spectrum emitted by a ^{16}O target irradiated with slow pions and found a strong peak due to the 4.44 MeV γ-ray from the decay of the first 2^+ excited state of ^{12}C. It is thus natural to suppose that the pion capture has broken the ^{16}O into an alpha-particle and an excited ^{12}C nucleus, and indeed it is known that this mode of breaking up is about six times more likely than the corresponding process leaving ^{12}C in its ground state.

Since the alpha-particle has a momentum distribution relative to the ^{12}C in the original oxygen nucleus, the recoiling carbon nuclei have the same distributions superposed on the recoil momentum determined by the rest mass of the pion. This momentum distribution produces a Doppler broadening of the shape of the γ-ray line, and measurement of this broadening gives the momentum distribution of the alpha-particles.

This momentum distribution can also be calculated from a simple harmonic oscillator model, thus providing a check on the interpretation of the data. In such a model, the momentum distribution of a particle moving with angular momentum L with respect to the centre-of-mass of the harmonic oscillator potential has the form $P(K) \propto K^{2L} \exp(-K^2/2Q_L^2)$, and fits to the data give $Q_2 = 75 \pm 5$ MeV c^{-1}. This may be compared with analyses of the $^{16}O(p, p\alpha)$ knock-out reaction that give $Q_0 \approx 75$ MeV c^{-1}. The values of Q_0 and Q_2 are expected to be similar, so this is an indication of the consistency of the two analyses.

This experiment essentially gives the distribution of the sum of the momenta of the four nucleons emitted when pions are captured by ^{16}O. These four nucleons are most probably combined to form an alpha-particle because the alternative interpretation of absorption on a deuteron followed by evaporation of two nucleons would give the characteristic γ-rays from the decay of ^{12}B, and these are not observed. Other possible origins of the observed γ-rays were excluded by repeating the experiment with different targets.

It is thus very reasonable to interpret these observations by the alpha-particle absorption model and this gives further support to the presence of alpha-clusters in nuclei. Similar results have been obtained for other nuclei besides oxygen.

The total cross-sections for the scattering of pions by nuclei from ^6Li to ^{16}O in the region of the N* resonance around 170 MeV has been analysed by Yam (1975) using the Glauber formalism with a cluster model of the nucleus. He finds that the major features of the data are well reproduced using a simple parametrization of the cluster structure. In this work the use of the cluster model bypasses a difficult calculation involving infinitely many scatterings on strongly correlated nucleons by using the scattering data on the clusters which act as approximations to the correlated nucleons. The single scattering term already gives a good approximation to the data in the cluster model whereas it is quite inadequate in the nucleon model. The success of these calculations provides evidence from strong interactions in support of clustering in light nuclei.

There have been many other studies of the interaction of slow and fast kaons and pions with nuclei. It is unfortunately difficult to detect the secondary particles directly because the low cross-sections require the use of thick targets that absorb most of the emitted alpha-particles. Measurements of the energy spectra of the γ-rays emitted by the residual nuclei enable the residual nuclei to be identified, and this permits some conclusions concerning the interaction and hence on the probability of finding alpha-clusters in the nucleus. The main difficulty is in the interpretation of the results. Thus it is frequently found that the residual nuclei are predominantly those reached from the initial nuclei by the emission of one or more alpha-particles, but it is always possible that the nucleons are emitted separately and the predominance of $(A - n\alpha)$ nuclei is due essentially to the more favourable energetics. It is thus essential to make detailed statistical model analyses of the results to distinguish between these possibilities. Even if the four nucleons from a reaction to an $(A - \alpha)$ nucleus are emitted separately they could still have come from an alpha-cluster in the target nucleus that was broken up by the interaction. This process could be distinguished from independent statistical emission if the angular distributions of the emitted nucleons were known.

In one of the first of these experiments, Barnes and coworkers (1972) identified many of the nuclear γ-rays emitted after the capture of negative kaons by Ni, Cu, Si and Al. The distribution of residual nuclei suggested a direct mechanism which removes one, two or three alpha-particles from Ni, and a triton or a triton plus one or two alpha-particles from Ca. The results seemed to be inconsistent with the simple statistical spread that might be expected from an evaporation process, which tends to give successive neutrons, or several neutrons and one or two protons.

In another experiment, Jackson and coworkers (1973) found that when ^{60}Ni and ^{28}Si were bombarded with 500 MeV c^{-1} negative pions, γ-rays were emitted characteristic of the decay of the first excited states of ^{56}Fe, ^{52}Cr and ^{48}Ti, these nuclei being ^{60}Ni less one, two and three alpha-particles respectively. Similar results were found for ^{28}Si. It seems to be difficult to account for these results by an evaporation process.

The interaction of 220 MeV negative pions with a range of nuclei from carbon to vanadium has been studied by Lind and coworkers (1974). It is very difficult to detect directly the particles emitted from these interactions so, as in the previous work, the prompt γ-rays were studied instead. These are emitted immediately after the particles resulting from the initial interaction. Some of the γ-ray spectra are shown in Figure 27 and it is notable that there are several peaks common to the spectra from different nuclei.

The energies of these γ-rays identify the nucleus from which they were emitted, and it is then found that in the case of even–even targets the most likely residual nuclei correspond to removing one, two or several alpha-particles (or the equivalent numbers of nucleons) from the target. In the case of the odd-A nuclei the γ-rays correspond to the removal of a triton or an alpha-particle or a triton plus an alpha-particle from the target nucleus. The cross-sections for these processes are comparable to the inelastic scattering cross-sections. It is found that the proportion of multiple emissions of alpha-particles compared with single emissions increases with nuclear size.

The experimental technique does not distinguish between the emission of an alpha-particle and the emission of two neutrons and two protons, since only the residual nucleus is identified. The relative likelihood of the two processes, and indeed of all possible processes and hence residual nuclei, can be calculated from statistical theory. It is found that this gives results completely different from those found experimentally so that multiple alpha-emission cannot be explained by the statistical theory.

These results have now been confirmed by studies of the interactions of both positive and negative pions with ^{27}Al and ^{28}Si by Ashery and coworkers (1974). The negative pions had an energy of 70 MeV and the positive pions had energies of 25, 70 and 100 MeV. They also found that there is a large cross-section for removing an 'alpha' particle and leaving the nucleus in its first excited state. This occurs even below 25 MeV, which is the threshold for the

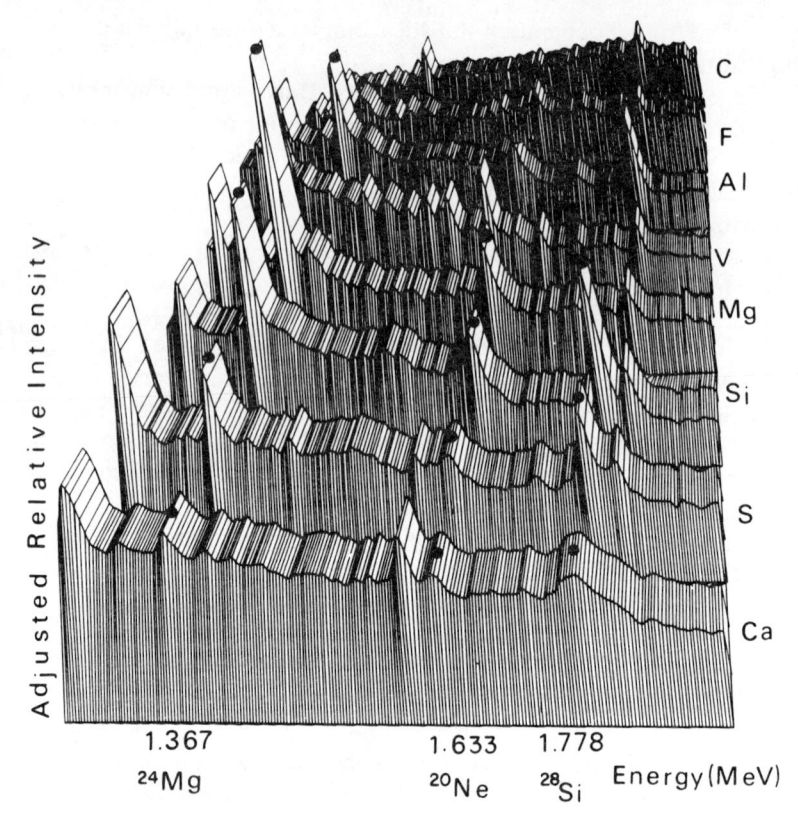

Figure 27 Spectra of prompt γ-rays emitted after the interaction of 220 MeV negative pions with a range of nuclei. The lines marked with a solid dot appear in many spectra and correspond to the multiple removal of 'alpha-particles' (Lind and coworkers, 1974)

emission of two protons and two neutrons, and this strongly supports the hypothesis of cluster emission.

It is found that at 70 MeV the cross-section for 'alpha' emission is very similar to that for the elastic scattering of pions by free alpha-particles, thus suggesting that the pions are simply knocking out pre-existing alpha-particles from the target nuclei. Substantial cross-sections are also found for reactions corresponding to the emission of deuteron clusters.

These experiments on the interaction of fast pions with nuclei thus give results similar to those found in studies of the interaction of slow pions or kaons. There seems to be some process that takes place in the initial stages of the interaction that leads to the preferential emission of alpha-particles. This in turn suggests that there are alpha-particles in the nuclear surface, a hypothesis for which there is already much evidence.

It is important to continue this work, and in particular to try to detect and identify directly the particles emitted in the early stages of the interaction. This is a difficult experimental problem but if it can be solved it will shed new light on the problem of alpha-particles on the nuclear surface.

The absorption of negative pions at rest and at 60 MeV on several nuclei from O to Nb has been studied by Ullrich and coworkers (1974), who found that alpha-cluster removal is comparable to other reaction channels.

A similar investigation was made by Chang and coworkers (1974) of the γ-rays emitted after the irradiation of ^{56}Fe and ^{58}Ni by 100 MeV protons, and they also found that some of the strongest γ-rays came from residual nuclei equivalent to the target minus one to three alpha-particles. However they also found that the cross-sections for these reactions were just what would be expected from a pre-equilibrium nucleon–nucleon cascade followed by evaporation of the remaining nucleus, without assuming any participation of alpha-particles in the pre-equilibrium process. The favouring of $(T-n\alpha)$ residual nuclei in this case is thus a result of the energetics of the particle emission and provides no evidence for alpha-clustering. It is possible that the results of Foley and coworkers (1962) on targets with $A \approx 25$ irradiated with 140 MeV protons, can be explained in a similar way.

A possible explanation of the emission of clusters from nuclei following slow negative kaon capture has been given by Barshay (1975). He suggests that when the negative kaon is captured by a surface proton there is a large probability of forming a virtual excited baryon with isotopic spin and hypercharge zero, the $Y_0^*(1405 \text{ MeV})$, which lies only about 27 MeV below the sum of the rest masses of proton and negative kaon. If this system experiences enhanced three-body interactions with neighbouring nucleons it might lead to the formation of a cluster that could be emitted subsequently when the excited baryon decays. This enhancement of the three-body forces for the excited baryon is relative to that experienced by three nucleons, and Barshay gives several reasons why this might occur. This explanation of cluster emission has yet to be quantitatively tested. A similar explanation may apply also to cluster emission following slow negative pion capture, where the initial interaction has a high probability of producing a low-lying excited baryon state, the $\Delta(1236 \text{ MeV})$.

8. UNIFIED MODEL OF ALPHA-CLUSTERING IN NUCLEI

The work described in the previous sections shows that alpha-clustering in nuclei is expected to occur and that evidence indicating alpha-clustering is provided by many different types of nuclear reaction. It is desirable to unify this diverse information to ensure that it is all consistent with the expected degree of alpha-clustering. This can of course be done, at least in principle, in the framework of the shell model, treating the nucleons individually. In the present context, however, we want to know to what extent it is useful and consistent to construct a theory of alpha-clustering that treats the alpha-particles as discrete entities and then gives a unified understanding of the reactions that are sensitive to alpha-clustering in nuclei.

Such a model can easily be constructed by analogy with the simple nucleon shell model. It is notable that all the phenomena that can be understood by the nucleon shell model, in particular the optical model analysis of elastic scattering and of the resonances due to coupling to inelastic channels as well as the nucleon transfer data, are paralleled by similar phenomena for alpha-particles (Hodgson 1974).

This suggests that alpha-clustering phenomena can be understood by grafting onto the nucleon shell model the possibility that nucleons in outer orbits can combine to form alpha-particles having a life long enough for their influence to be felt. These orbits have the quantum numbers obtained by the Talmi transformation (9) from those of the constituent nucleons and alpha-transfers can take place to and from these orbits just as nucleon transfers can take place to and from the nucleon orbits. This model has been mentioned already at several points, and in this section some of its predictions and results are collected together.

According to this model the alpha-particles are bound in an alpha-core potential, whose eigenvalues and eigenfunctions give the centroid energies and wave functions of the alpha-states, just as for the nucleons. It is simplest to assume that this potential has the Saxon–Woods form, and once its radius and diffuseness parameters are fixed the depth of the potential is determined by the alpha-transfer data. The only remaining parameter, the probability of alpha-formation, is given by the spectroscopic factor for alpha-transfer reactions. This parameter is the only one in the model that has nothing corresponding to it in the analogous nucleon case.

Many studies of nucleon-transfer reactions (Millener and Hodgson, 1973; Malaguti and Hodgson, 1973) showed that the nucleon potential determined from an overall analysis of many reactions varies in a systematic way from nucleus to nucleus, and preliminary analysis of alpha-transfer reactions showed that this is the case for the alpha-potentials as well (Hodgson, 1974).

This model has been applied to the alpha-particle resonant states in medium weight nuclei by Dudek and Hodgson (1971), and they find that the energies and ordering of the eigenstates is qualitatively in accord with experiment.

A detailed study of this model as applied to light nuclei has been made by Buck and coworkers (1975). They determined the alpha-potential by folding the densities of the nuclear core and the alpha-particle

$$V(r) = -\frac{2\pi\hbar^2}{M}\bar{f}\int \rho_A(\mathbf{r}-\mathbf{r}')\rho_B(\mathbf{r}')\,d\mathbf{r}' \qquad (10)$$

where $\bar{f}$ is a real depth parameter. The densities ρ_A and ρ_B were taken from electron-scattering analyses. It turns out that this potential has several advantages over the Saxon–Woods potential; in particular it can reproduce the energies of the states of low-lying rotational bands with essentially the same depth parameter.

Many studies of alpha-transfer reactions (see Section 5) have shown that they populate preferentially the rotational bands in light nuclei. This strongly

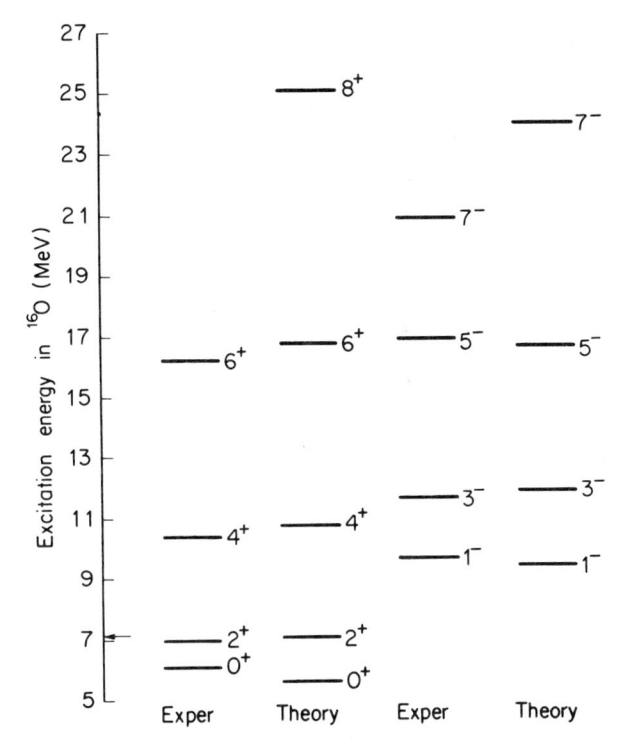

Figure 28 Comparison between the theoretical and experimental energies of the $K^\pi = 0^-$ and 0^- rotational bands in ^{16}O. The theoretical energies are calculated from an alpha-^{12}C folded potential with depth parameter $\bar{f} = 1.425$ fm for the 0^+ band and $\bar{f} = 1.55$ fm for the 0^- band. The arrow indicates the energy of the $\alpha + ^{12}$C threshold in ^{16}O (Buck and coworkers, 1975)

indicates that these states have core+alpha-structure, and it is therefore natural to interpret such states as eigenstates of the potential (10). Buck and coworkers (1975) therefore applied their model to calculate the energy spectra of the $K^\pi = 0^+$ and 0^- bands in ^{16}O and ^{20}Ne, and the results are shown in Figures 28 and 29. The parameter $\bar{f}$ was adjusted for each *band*, and then the model gives a good overall fit to the energies of all the states in that band. Since the widths are very strongly energy dependent, $\bar{f}$ was adjusted for each *state* to give the widths shown in Tables 4 and 5. The required changes in $\bar{f}$ from the mean (band) values are only a few per cent. The correspondence between the measured and calculated widths indicates the qualitative reliability of the model. Since the calculated widths are very sensitive to the addition of an imaginary part to the depth parameter $\bar{f}$ it is not correct to interpret as spectroscopic factors the ratios of experimental to calculated widths in this table.

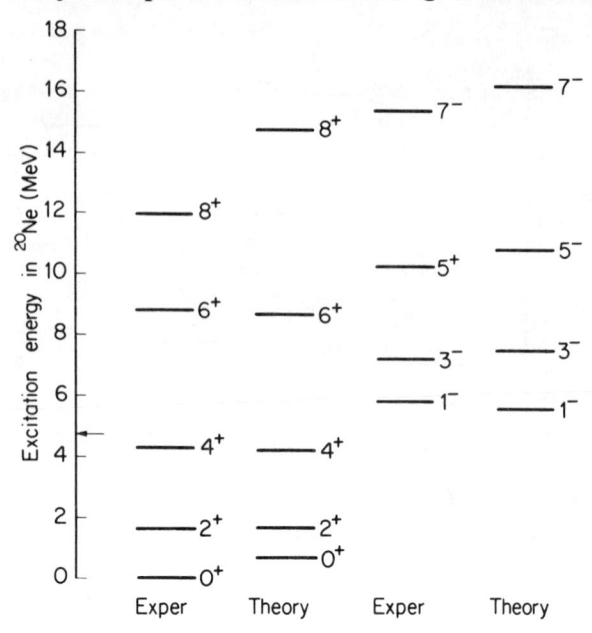

Figure 29 Comparison between the theoretical and experimental energies of the $K^\pi = 0^+$ and 0^- rotational bands in ^{20}Ne. The theoretical energies are calculated from an alpha-^{16}O folded potential, with depth parameter $\bar{f} = 1.237$ fm for the ground state band and $\bar{f} = 1.325$ fm for the negative parity band. The arrow indicates the energy of the $\alpha + {}^{16}$O threshold in ^{20}Ne (Buck and coworkers, 1975)

The potentials give the alpha-cluster wave functions, and hence the alpha-core separations and root-mean-square radii $\langle r^2 \rangle^{\frac{1}{2}}$ given for the ground-state band of ^{20}Ne in Table 6. The values labelled $\rho^{(1)}$ and $\rho^{(2)}$ are obtained by Mosley and Fortune (1974) and Arima and Yoshida (1974) respectively using Saxon–Woods potentials to generate the alpha-cluster wave functions, and $\rho^{(3)}$ is obtained from the Hartree–Fock calculations of Lee and Cusson (1972). The model also gives values of $\langle r^2 \rangle^{\frac{1}{2}}$ that agree well with the empirical value of 2.9 fm given by Elton (1961). All these models predict an anti-centrifugal stretching effect, i.e. the higher spin states correspond to a lower value of ρ and hence a smaller $\langle r^2 \rangle^{\frac{1}{2}}$. In the folding model this does not require a change in the radius of the potential itself.

As a final test of the model, Buck and coworkers (1975) calculate the electromagnetic transition rates between the states of the ground-state band of ^{20}Ne, and these are compared in Table 7 with the data and with shell model and rotational model calculations. In this calculation the effective charge was adjusted to $e_{\text{eff}} = 1.245 e$ and gives values of $B(E2)$ quite close to those of the complete shell model calculation which used $e_{\text{elf}} = 1.583$.

Table 4. Theoretical and experimental alpha-widths for ^{16}O (Buck and coworkers, 1975)

J^π	Γ_α^{exp}(KeV)	Γ_α^{th}(KeV)	$\bar{f}$(fm)
	K = 0$^+$ band		
4$^+$	27	17.5	1.436
6$^+$	125	238	1.432
8$^+$	—	385	1.425
	K = 0$^-$ band		
1$^-$	510	675	1.536
3$^-$	1200	1750	1.5575
5$^-$	700	≈ 2000	1.539
7$^-$	750	776	1.668

Table 5. Theoretical and experimental alpha-widths for ^{20}Ne (Buck and coworkers, 1975)

J^π	Γ_α^{exp}(KeV)	P_α^{th}(KeV)	$\bar{f}$(fm)
	K = 0$^+$ band		
6$^+$	0.11	0.21	1.232
8$^+$	0.035	0.108	1.291
	K = 0$^-$ band		
1$^-$	>0.013	0.021	1.313
3$^-$	8	6.7	1.3274
5$^-$	141	81	1.339
7$^-$	280	183	1.345

Table 6. Comparison of various calculated values of the alpha-core separation and RMS radii for the ground state band of ^{20}Ne (Buck and coworkers, 1975) (all in fm)

$J\pi$	ρ	$\rho^{(1)}$	$\rho^{(2)}$	$\rho^{(3)}$	$\langle r^2 \rangle^{\frac{1}{2}}$
0$^+$	3.67	3.82	3.75	3.86	2.96
2$^+$	3.68	3.85	3.78	3.84	2.97
4$^+$	3.58	3.81	3.76	3.76	2.95
6$^+$	3.44	3.78	3.76	3.67	2.92
8$^+$	3.06	3.48	3.56	3.54	2.85

Table 7. Theoretical and experimental electromagnetic transition rates for the ground state band of ^{20}Ne (Buck and coworkers, 1975). (Units of e^2 fm)

Transition	B(E2)$_{exp}$	B(E2)$_{th}$	B(E2)$_{SM}$	B(E2)$_{RM}$
2$^+ \to$ 0$^+$	57.3	57.3	52.6	57.6
4$^+ \to$ 2$^+$	71.0	73.8	64.5	82.4
6$^+ \to$ 4$^+$	66.0	62.7	53.4	90.8
8$^+ \to$ 6$^+$	24.0	28.9	32.8	95.6

These results show that the alpha-cluster model is able to give a good overall account of the energies and widths of the states of ^{16}O and ^{20}Ne, and the alpha-core separations, RMS radii and electromagnetic transition rates. It has considerable predictive power, since all this information can be calculated as soon as the depth parameter is fixed by comparison with the energy of a single state. It remains to be seen to what extent it will also be successful for other nuclei. It would be particularly interesting to use the alpha-cluster wave functions in analyses of alpha-transfer reactions, and hence obtain information on the alpha-spectroscopic factors and formation probabilities, which could be compared with values obtained in other ways.

9. CONCLUSIONS

Taken together, the many lines of evidence summarized in this review provide strong evidence for a significant degree of alpha-clustering in nuclei. Nuclear matter calculations indicate that such clustering becomes very likely at the low nucleon densities in the surface region, and this is supported by self-consistent-field calculations of the structure of light nuclei.

The probability of alpha-decay of nuclear states, the resonance in alpha-particle elastic and inelastic scattering, and the cross-sections for alpha-transfer and alpha-knock-out reactions, together with the mechanism of pion and kaon reactions are all sensitive to the presence of alpha-clustering.

There are indeed few aspects of nuclear physics that are not affected to some degree by alpha-clustering, and the concept has been used to solve many problems not discussed here. For example, Brown and coworkers (1973) have shown that the Nolen–Schiffer anomaly can be understood by taking into account the presence of alpha-particles in the nuclear surface, and Strobel and Geramb (1973) have found that in the ^{12}C$(d, h)^{11}$B* reaction the continuum can be explained by the interaction of the incident deuteron with a correlated pair of nucleons in the target, and these could well be constituents of alpha-particles. The alpha-structure of nuclei has also been used to explain anomalies in photoabsorption (Delsanto and coworkers, 1973).

In recent years there has been a marked increase in the precision of alpha-clustering calculations, so that it becomes possible to look for numerical convergence between the results of calculations based on measurements of quite different phenomena. At the same time the increasing sophistication of the calculations has shown the inadequacy of the simple classical ideas used in the introduction. Nevertheless, whatever mental picture we use, it remains true that many phenomena may be analysed very economically in terms of cluster coordinates with little loss of accuracy.

In some cases the arguments are still qualitative, or there are other possible interpretations of the phenomena that can only be excluded by further measurements and calculations, so that there still remains much work to be done to understand fully the phenomenon of alpha-clustering in nuclei.

ACKNOWLEDGEMENTS

I am grateful to Dr. A. A. Pilt for many detailed comments on the first draft of this review.

REFERENCES

Abe, Y. and Takigawa, N. (1972) *Suppl. Progr. Theoret. Phys.*, **52**, 228.
Agassi, D. and Wall, N. S. (1973) *Phys. Rev.*, **C7**, 1368; (see also Wall, N. S. (1972) Marburg Symposium).
Akiyama, Y., Arima, A. and Sebe, T. (1969) *Nucl. Phys.*, **A138**, 273.
Anyas-Weiss, N., Becker, J., Belote, T. A., Cardinal, C. U., Cornell, J. C., Fisher, P. S., Hudson, P. N., Menchaca-Rocha, A., Panagiotou, A. P. and Scott, D. K. (1973) Preprint.
Arima, A., Gillet, V. and Ginocchio, J. (1970) *Phys. Rev. Letters*, **25**, 1043.
Arima, A. (1973) *Proceedings of the International Conference on Nuclear Physics (Munich)*, J. de Boer and H. J. Mang, Eds., Vol. 2, 184 (North-Holland).
Arima, A. (1975) *Proceedings of the International Conference on Clustering in Nuclei (Maryland)*.
Arima, A. and Gillet, V. (1971) *Ann. Phys.*, **66**, 117.
Arima, A. and Strottman, D. (1973) *Nuclear Physics Theoretical Group Report No. 46* (Oxford).
Arima, A. and Yoshida, S. (1972) *Phys. Letters*, **40 B**, 15.
Arima, A. and Yoshida, S. (1974) *Nucl. Phys.*, **A219**, 475.
Arima, A., Broglia, R. A., Ichimura, M. and Schäfer, K. (1973) *Nucl. Phys.*, **A215**, 109.
Arima, A., Horiuchi, H., Kubodera, K. and Takigawa, N. (1972) *Advances in Nuclear Physics*, M. Baranger and E. Vogt, Eds., Vol. 5, 345 (Plenum Press).
Artemov, K. P., Goldberg, V. Z., Petrov, I. P., Rudakov, V. P., Serikov, I. N. and Timofeev, V. A. (1971) *Phys. Letters*, **37 B**, 61.
Ashery, D., Zaider, M., Shamai, Y., Cochavi, S. Moinester, M. A., Yavin, A. I. and Alster, J. (1974) *Phys. Rev. Letters*, **32**, 943.
Aslam, K. and Rook, J. R. (1970) *Nucl. Phys.*, **B20**, 159.
Badawy, I., Berthier, B., Charles, P., Fernandez, B. and Gastebois, J. (1974) *Proceedings of the International Conference on Reactions Between Complex Nuclei*, Nashville, Tennessee, Vol. I.
Balamuth, D. P. (1971) *Phys. Rev.*, **C3**, 1565.
Balamuth, D. P. (1973) *Phys. Rev.*, **C8**, 1185.
Barnes, P. D., Eisenstein, R. A., Lam, W. C., Miller, J., Sutton, R. B., Eckhause, M., Kane, J., Welsh, R. E., Jenkins, D. A., Powers, R. J., Kunselman, R., Redwine, R. P., Segal, R. E. and Schiffer, J. P. (1972) *Phys. Rev. Letters*, **29**, 230.
Barshay, S. (1975) *Phys. Letters*, **55 B**, 457.
Bassani, G., Pappalardo, G., Saunier, N. and Traore, B. M. (1971) *Phys. Letters*, **34 B**, 612.
Becchetti, F. D., Chua, L. T., Janecke, J. and Van der Molen, A. M. (1975) *Phys. Rev. Letters*, **34**, 225.
Bethge, K., Meier-Ewert, K., Pfeiffer, K. and Bock, R. (1967) *Phys. Letters*, **24B**, 663.
Bingham, H. G., Fortune, H. J. Garrett, J. D. and Middleton, R. (1971) *Phys. Rev. Letters*, **26**, 1448.
Bisson, A. E., Eberhard, K. A. and Davis, R. H. (1970) *Phys. Rev.*, **C1**, 539.
Bobrowska, A., Budzanowski, A., Grotowski, K., Jarczyk, L., Micek, S., Niewodniczański, H., Strzałkowski, A. and Wróbel, Z. (1969) *Nucl. Phys.*, **A126**, 361.
Bonche, P., Giraud, B., Cunsolo, A., Lemaire, M. C., Mermaz, M. C. and Quebert, J. L. (1972) *Phys. Rev.*, **C6**, 577.
Bonetti, R. and Milazzo-Colli, L. (1974) *Phys. Letters*, **49B**, 17.
Braga-Marcazzan, G. M., Milazzo-Colli, L. and Signorini, C. (1973) *Nuovo Cimento Letters*, **6**, 357.
Braun-Munsinger, P., Bohne, W., Gelbke, C. K., Grochulski, W., Harney, H. L. and Oeschler, H. (1973) *Phys. Rev. Letters*, **31**, 1423.
Brink, D. M. (1966) International School of Physics 'Enrico Fermi' XXXVI, 274 (Academic Press).
Brink, D. M. and Boeker, E. (1967) *Nucl. Phys.*, **A91**, 1.
Brink, D. M. and Castro, J. J. (1973) *Nucl. Phys.*, **A216**, 109.
Brink, D. M., Friedrich, H., Weiguny, A. and Wong, C. W. (1970) *Phys. Letters*, **33 B**, 143.

Brown, G. E., Horsfjord, V. and Liu, K. F. (1973) *Nucl. Phys.*, **A205**, 73.
Bryant, H. C. and Jarmie, N. (1968) *Ann. Phys.*, **47**, 127.
Buck, B., Dover, C. B. and Vary, J. P. (1975) *Phys. Rev.*, **C11**, 1803.
Budzanowski, A. (1972) Marburg Symposium 148.
Budzanowski, A., Dudek, A., Dymarz, R., Grotowski, K., Jarczyk, L., Niewodniczanski, H. and Strazalkowski, A. (1969) *Nucl. Phys.*, **A126**, 369.
Bugg, W. M., Condo, G. T., Hart, E. L. and Cohn, H. O. (1974) *Phys. Rev.*, **C9**, 1215.
Ceuleneer, R., Demeur, M. and Reignier, J. (1969) *Nucl. Phys.*, **89**, 177.
Chang, C. C., Wall, N. S. and Fraenkel, Z. (1974) *Phys. Rev. Letters*, **33**, 1493.
Chang, F. C., (1967) *Phys. Rev.*, **155**, 1299.
Charlton, L. A. and Robson, D. (1974) *Phys. Rev. Letters*, **32**, 946.
Chatwin, R. A., Eck, J. S., Robson, D. and Richter, A. (1970) *Phys. Rev.*, **C1**, 795.
Coelho, H. T. (1973a) *Phys. Rev.*, **C7**, 2340.
Coelho, H. T. (1973b) *Phys. Rev.*, **C8**, 93.
Cohen, J. P. (1966) *Nucl. Phys.*, **84**, 316.
Colli-Milazzo, L. and Marcazzan-Braga, G. M. (1972) *Phys. Letters* **38B**, 155.
Comfort, J. R., Morrison, G. C., Zeidman, B. and Fortune, H. T. (1970) *Phys. Letters* **32 B**, 685.
Cowley, A. A. and Heymann, G. (1970) *Nucl. Phys.*, **A146**, 465.
Cramer, J. G., Eberhard, K. A., Eck, J. S. and Trombik, W. (1973) *Phys. Rev.*, **C8**, 625.
Danos, M. and Gillet, V. (1971) *Phys. Letters*, **34 B**, 24.
Danos, M. and Spicer, B. M. (1970) *Z. Physik*, **237**, 320.
Delsanto, P. P., Barrett, R. F. and Wahsweiler, H. G. (1973) *Phys. Letters*, **44 B**, 433.
Denes, L. J., Daehnick, W. W. and Drisko, R. M. (1966) *Phys. Rev.*, **148**, 1097.
Détraz, C., Zafiratos, C. D., Moss, C. E. and Zaidins, C. S. (1971) *Nucl. Phys.*, **A177**, 258.
DeVries, R. M. (1973a) *Phys. Rev. Letters*, **30**, 666.
DeVries, R. M. (1973b) *Nucl. Phys.*, **A212**, 207.
DeVries, R. M., Shapira, D., Davies, W. G., Ball, G. C., Forster, J. S. and McLatchie, N. (1975) *Phys. Rev. Letters*, **35**, 835.
Draayer, J. P. (1975) *Nucl. Phys.*, **A237**, 157.
Draayer, J. P., Gove, H. E., Trentelman, J. P., Anantaraman, N. and Collins, H. (1974) Proceedings of the International Conference on Nuclear Structure and Spectroscopy (Amsterdam) 1974.
DuBost, H., Lefort, M., Peter, J. and Tarrago, X. (1964) *Phys. Rev.*, **B136**, 1618.
Dudek, A. and Hodgson, P. E. (1971) *J. Phys.*, **32**, C5.
Dutta, S. and Tomusiak, E. L. (1971) *Phys. Letters*, **35B**, 554.
Eberhard, K. A. (1970) *Phys. Letters*, **33 B**, 343.
Eberhard, K. A. (1973) Louvain-Cracow Seminar, 1973.
Eichler, J. and Yamamura, M. (1972) *Nucl. Phys.*, **A182**, 33.
Erskine, J. R., Henning, W., Kovar, D. G., Greenwood, L. R. and DeVries, R. M. (1975) *Phys. Rev. Letters*, **34**, 680.
Faivre, J. C., Faraggi, H., Gastebois, J., Harvey, B. G., Lemaire, M. C., Loiseaux, J. M., Mermaz, M. C. and Papineau, A. (1970) *Phys. Rev. Letters*, **24**, 1188.
Faraggi, H., Jaffrin, A., Lemaire, M. C., Mermaz, M. C., Faivre, J. C., Gastebois, J., Harvey, B. G., Loiseaux, J. M. and Papineau, A. (1971a) *Ann. Phys.*, **66**, 905.
Faraggi, H., Lemaire, M. C., Loiseaux, J. M., Mermaz, M. C. and Papineau, A. (1971b) *Phys. Rev.*, **C4**, 1375.
Fink, H. J., Scheid, W. and Greiner, W. (1972) *Nucl. Phys.*, **A188**, 259.
Foley, K. J., Clegg, A. B. and Salmon, G. L. (1962) *Nucl. Phys.*, **37**, 23.
Fuchs, H. (1973) Proceedings of the 1st Louvain-Cracow Seminar, 89.
Fulbright, H. W., Strohbusch, U., Markham, R. G., Lindgren, R. A., Morrison, G. C., McGuire, S. C. and Bennett, C. L. (1975) *Phys. Letters*, **53B**, 449.
Gaul, G., Lüdecke, H., Santo, R., Schmeing, H. and Stock, R. (1969) *Nucl. Phys.*, **A137**, 177.
Głowacka, L., Jaskóta, M., Turkiewicz, J., Zemto, L., Koztowski, M. and Osakiewicz, W. (1975) *Nucl. Phys.*, **A244**, 117.
Griffy, T. A., Oakes, R. J. and Schwartz, H. M. (1966) *Nucl. Phys.*, **86**, 313.
Griffin, J. J. (1966) *Phys. Rev. Letters*, **17**, 478.
Grin, Yu. T. and Leinson, L. B. (1971) *Phys. Letters*, **37 B**, 253.
Gruhn, C. R. and Wall, N. S. (1966) *Nucl. Phys.*, **81**, 161.
Gubkin, I. A. (1970) *Soviet J. Nucl. Phys.*, **11**, 336.
Gutbrod, H. H. and Markham, R. G. (1972) *Phys. Rev. Letters*, **29**, 808.

Gutbrod, H. H., Yoshida, H. and Bock, R. (1971) *Nucl. Phys.*, **A165**, 240.
Harada, K. (1961) *Progr. Theoret. Phys.*, **26**, 667.
Harada, K. (1962) *Progr. Theoret. Phys.*, **27**, 430.
Harvey, M. (1973) *Nucl. Phys.*, **A202**, 191.
Hecht, K. T. and Braunschweig, D. (1975) *Nucl. Phys.*, **A244**, 365.
Hildenbrand, K. D., Gutbrod, H. H., von Oertzen, W. and Bock, R. (1970) *Nucl. Phys.*, **A157**, 297.
Hodgson, P. E. (1958) *Nucl. Phys.*, **8**, 1.
Hodgson, P. E. (1971) *Nuclear Reactions and Nuclear Structure*, Clarendon Press, Oxford, Chap. 14. 5.
Hodgson, P. E. (1974) International Symposium on Correlations in Nuclei (Balatonfüred) 281.
Hodgson, P. E. (1975) *Rep. Progr. Phys.*, **38**, 847.
Högaasen, J. (1967) *Nucl. Phys.*, **A90**, 261.
Honda, T., Kudo, Y. and Ui, H. (1963) *Nucl. Phys.*, **44**, 472.
Ichimura, M., Arima, A., Halbert, E. C. and Terasawa, T. (1973) *Nucl. Phys.*, **A204**, 225.
Igo, G., Hansen, L. F. and Gooding, T. J. (1963) *Phys. Rev.*, **131**, 337.
Ikeda, K., Marumori, T., Tamagaki, R. and Tanaka, H. (1972) *Suppl. Progr. Theoret. Phys.*, **52**, 1.
Jackson, H. E., Meyer-Schützmeister, L., Wangler, T. P., Redwine, R. P., Segel, R. E., Tonn, J. and Schiffer, J. P. (1973) *Phys. Rev. Letters*, **31**, 1353.
Jaffrin, A. (1970) *Phys. Letters*, **32B**, 448.
Jaffrin, A. (1972) *Nucl. Phys.*, **A196**, 577.
Jain, A. K. and Sarma, N. (1974) *Nucl. Phys.*, **A233**, 145.
Jain, A. K., Grossiord, J. Y., Chevallier, M., Gaillard, P., Guichard, A., Gusakow, M. and Pizzi, J. R. (1973) *Nucl. Phys.*, **A216**, 519.
Khadkikar, S. B. (1971) *Phys. Letters*, **36 B**, 451.
Kolybasov, V. M. and Tsepov, V. A. (1972) *Soviet J. Nucl. Phys.*, **14**, 418.
Kopaleishvili, T. I. and Machabeli, I. Z. (1971) *Soviet J. Nucl. Phys.*, **12**, 286.
Kubo, K. I. (1972) *Nucl. Phys.*, **A187**, 205.
Kubo, K. I. and Hirata, M. (1972) *Nucl. Phys.*, **A187**, 186.
Kulisic, P., Ajdacic, V., Cindro, N., Lalovic, B. and Strohal, P. (1964) *Nucl. Phys.*, **54**, 17.
Kurath, D. (1973) *Phys. Rev.*, **C7**, 1390.
Kurath, D., and Towner, I. S. (1974) *Nucl. Phys.*, **A222**, 1.
Lee, H. D., and Cusson, R. Y. (1972) *Ann. Phys. (N.Y.)*, **72**, 353.
Lega, J. and Macq, P. C. (1974) *Nucl. Phys.*, **A218**, 429.
Lemaire, M. C. (1973) *Phys. Rep.*, **7**, 279.
Lewis, C. W., Ullrich, H., Engelhardt, H. D. and Boschitz, E. T. (1973) *Phys. Letters*, **47 B**, 339.
Lind, V. G., Plendl, H. S., Funsten, H. O., Kossler, W. J., Lieb, B. J., Lankford, W. F. and Buffa, A. J. (1974) *Phys. Rev. Letters*, **32**, 479.
Loebenstein, H. M., Mingay, D. W., Winkler, H. and Zaidins, C. S. (1967) *Nucl. Phys.*, **A91**, 481.
Macdonald, N., Morrison, J. and Watt, A. (1972) *Nucl. Phys.*, **A182**, 183.
Malaguti, F. and Hodgson, P. E. (1973) *Nucl. Phys.*, **A215**, 243.
Mang, H. (1957) *Z. Phys.*, **148**, 572.
Mang, H. (1960) *Phys. Rev.*, **119**, 1069.
Mang, H. J. (1964) *Ann. Rev. Nucl. Sci.*, **14**, 1.
Mang, H. J. (1975) Maryland Conference 1975.
McGrory, J. B. (1973) *Phys. Letters*, **47 B**, 481.
McVoy, K. W. (1971) *Phys. Rev.*, **C3**, 1104.
McVoy, K. W. (1972) Marburg Symposium 217.
Meier-Ewart, K., Bethge, K. and Pfeiffer, K. O. (1968) *Nucl. Phys.*, **A110**, 142.
Mendez-Moreno, R. M., Moreno, M. and Seligman, T. H. (1974) *Nucl. Phys.*, **A221**, 381.
Middleton, R., Rosner, B., Pullen, D. J. and Polsky, L. (1968a) *Phys. Rev. Letters*, **20**, 118.
Middleton, R., Polsky, L. M., Holbrow, C. H. and Bethge, K. (1968b) *Phys. Rev. Letters*, **21**, 1398.
Milazzo-Colli, L. and Braga-Marcazzan, G. M. (1972) *Phys. Letters*, **36 B**, 447.
Milazzo-Colli, L. and Braga-Marcazzan, G. M. (1973a) *Rev. Nuovo Cimento*, **3**, 535.
Milazzo-Colli, L. and Braga-Marcazzan, G. M. (1973b) *Nucl. Phys.*, **A210**, 297.
Milazzo-Colli, L., Braga-Marcazzan, G. M., Milazzo, M. and Signorini, C. (1974) *Nucl. Phys.*, **A218**, 274.
Millener, D. J. and Hodgson, P. E. (1973) *Nucl. Phys.*, **A209**, 59.
Morrison, G. C. (1971) *J. Phys. (Paris) Colloq.*, **32**, C6-69.
Morsch, H. P. and Breuer, H. (1973) *Nucl. Phys.*, **A208**, 255.
Mosley, C. A. and Fortune, H. T. (1974) *Phys. Rev.*, **C9**, 775.

Neogy, P., Scholz, W., Garrett, J. and Middleton, R. (1970) *Phys. Rev.*, **C2**, 2149.
Noble, J. V. and Coelho, H. T. (1971) *Phys. Rev.*, **C3**, 1840.
Oeschler, H., Fuchs, H. and Schröter, H. (1973) *Nucl. Phys.* **A202**, 513.
Oeschler, H., Schröter, H., Fuchs, H., Baum, L., Gaul, G., Lüdecke, H., Santo, R. and Stock, R. (1972) *Phys. Rev. Letters*, **28**, 694.
Ostrumov, V. D. and Filov, R. A. (1959) *JETP (Soviet Phys.)*, **57**, 642.
Perring, J. K. and Skyrme, T. H. R. (1956) *Proc. Phys. Soc. A (London)*, **69 A**, 600.
Pizzi, J. R., Bouche, R., Gaillard, M., Gaillard, P., Guichard, A., Gusakow, M., Leonhardt, J. L. and Ruhla, C. (1968) *Phys. Letters*, **28 B**, 32.
Poggenburg, J. K., Mang, H. J. and Rasmussen, J. O. (1969) *Phys. Rev.*, **181**, 1697.
Richert, J. (1972) *Phys. Letters*, **42 B**, 395.
Rinat, A. S. (1972) *Phys. Letters*, **38 B**, 281; Marburg Symposium, 245.
Ripka, G. (1968) *Advances in Nuclear Physics*, M. Basanger and E. Vogt, Eds., Vol. 1, 183.
Robson, D. (1972) *Comm. NPP.*, **5**, 16.
Rook, J. R. (1962) *Nucl. Phys.*, **39**, 479.
Rook, J. R. (1963) *Bucl. Phys.*, **43**, 363.
Rook, J. R. (1968) *Nucl. Phys.*, **B6**, 543.
Rosner, B., Wittner, K., Bethge, K. and Tserruya, I. (1974) *Nucl. Phys.*, **A218**, 1.
Rotter, I. (1968) *Forsch. der Phys.*, 16, 195.
Rotter, I. (1969) *Nucl. Phys.*, **A135**, 378.
Ruzzene, F. and Amos, K. (1974) *Z. Phys.*, **271**, 359.
Schmeing, H. and Santo, R. (1970) *Phys. Letters*, **33B**, 219.
Scholz, W., Neogy, P., Bethge, K. and Middleton, R. (1965) *Phys. Rev. Letters*, **22**, 949.
Scholz, W., Neogy, P., Bethge, K. and Middleton, R. (1972) *Phys. Rev.*, **C6**, 893.
Sewell, P. T., Hafele, J. C., Foster, C. C., O'Fallon, N. M. and Fulmer, C. B. (1973) *Phys. Rev,*, **C7**, 690.
Sharma, R. C. (1967) *Nucl. Phys.*, **A96**, 410.
Sheline, R. K. and Wildermuth, K. (1960) *Nucl. Phys.*, **21**, 196.
Steenberg, N. R. and Sharma, R. C. (1960) *Can. J. Phys.*, **38**, 290.
Stock, R., Gaul, G., Santo, R., Bernas, M., Harvey, B., Hendrie, D., Mahoney, J., Sherman, J., Steyaert, J. and Zisman, M. (1972) *Phys. Rev.*, **C6**, 1226.
Strobel, G. L. and Geramb, H. V. (1973) *Nucl. Phys.*, **A210**, 67.
Strohbusch, U., Fink, C. L., Zeidman, B., Horoshko, R. N., Fulbright, H. W. and Markham, R. (1972) *Phys. Rev. Letters*, **29**, 735.
Strohbusch, U., Fink, C. L., Zeidman, B., Horoshko, R. N., Fulbright, H. W. and Markham, R. (1974) *Phys. Rev.*, **C9**, 965.
Strottman, D. and Arima, A. (1973) Private communication to Arima.
Sunkel, W. and Wildermuth, K. (1972) *Phys. Letters*, **41 B**, 439.
Terasawa, T., Tanifuji, M. and Mikoshiba, O. (1973) *Nucl. Phys.*, **A203**, 225.
Thompson, W. G. (1971) *Part. and Nuc.*, **2**, 47.
Ullrich, H., Boschitz, E. T., Engelhardt, H. D. and Lewis, C. W. (1974) *Phys. Rev. Letters*, **33**, 433.
Veselic, D. and Tudoric-Ghemo, J. (1968) *Nucl. Phys.*, **A110**, 225.
Volkov, A. B. (1964) *Phys. Letters*, **12**, 118.
Volkov, A. B. (1965) *Nucl. Phys.*, **74**, 33.
Werby, and Tobocman, W. (1974) *Phys. Rev.*, **C10**, 1022.
Wheeler, J. A. (1937) *Phys. Rev.*, **52**, 1083.
Wilkinson, D. H. (1961) Proceedings of the Rutherford Jubilee International conference, J. B. Birks, Ed., p. 339.
Wozniak, G. J., Harney, H. L., Wilcox, K. H. and Cerny, J. (1972) *Phys. Rev. Letters*, **28**, 1278.
Wycech, S. (1967) *Acta. Phys. Pol.*, **32**, 161.
Yam, Y. Y. (1975) *Phys. Rev.*, **C11**, 73.
Yazaki, K. (1973) *Progr. Theoret. Phys.*, **49**, 1205.
Yoshida, S. (1973) *Phys. Letters*, **47B**, 411.

The references to the Marburg Symposium are to the Proceedings of the Symposium on Four-Nucleon Correlations and Alpha-Rotator Structure held in Marburg from 30th October to 2nd November 1972 and edited by R. Stock.

Commutation Relations, Hydrodynamics and Inelastic Scattering by Atomic Nuclei

LINDSAY J. TASSIE
The Australian National University, Canberra, Australia

1. INTRODUCTION

A simple hydrodynamical model of the nucleus (Tassie, 1956, 1958), and generalizations of it, have been widely used to analyse the inelastic scattering of electrons by nuclei (Überall, 1971) using the Born approximation and the distorted wave Born approximation (Onley and coworkers, 1963; Tuan and coworkers, 1968). In this model, the nucleus was considered as composed of an inhomogeneous fluid whose flow was assumed to be irrotational and incompressible. Boridy and Feshbach (1974) have used this model successfully to fit high energy proton scattering data using the first term of the multiple scattering series for the optical potential and the distorted wave impulse approximation.

Various sum rules have been obtained for inelastic scattering form factors (Kao and Fallieros, 1970; Noble, 1971; Deal and Fallieros, 1973a, b; Ui and Tsukamoto, 1974: Reinhard and Dreschsel, 1975: Tassie, 1975) by considering the matrix elements of the double commutator of operators for form factors and the nuclear Hamiltonian using the basic Heisenberg commutation relations of quantum mechanics. Deal and Fallieros (1973a) showed for isoscalar transitions that the result of the hydrodynamical model can be obtained from a sum rule together with the assumption that a single doorway state dominates the inelastic scattering. It has been shown (Tassie, 1975) that the form factor for isoscalar excitation of a single state can be obtained by inverting the sum rules, and that at small momentum transfer the inelastic form factor is the same as that given by the hydrodynamical model, so that the intuitive assumption of irrotational and incompressible flow of the nuclear fluid now has a formal quantum mechanical justification based on the use of the Heisenberg commutation relations.

The sum rules and the hydrodynamical model can be applied to any many-body system composed of identical particles, but the discussion will be confined to applications to the nucleus. Because the nucleus is composed of both neutrons and protons, the treatment is confined to isoscalar transitions and it is assumed that the distribution of neutrons in the ground state is the same as that of protons, so that the difference between protons and neutrons can be ignored.

Only excitations which can be described in terms of the transition density are considered. Thus, for electron scattering, only longitudinal electric excitation is treated. Magnetic and transverse electric excitation are not dealt with as they depend on the current distribution within the nucleus.

2. THE HYDRODYNAMICAL MODEL (Tassie, 1956, 1973)

The nuclear fluid is described by a density $\rho(\mathbf{r}, t)$ and a velocity $\mathbf{v}(\mathbf{r}, t)$, and it is assumed that the motion of the nuclear fluid is incompressible and irrotational. From the equation of continuity

$$\nabla \cdot (\rho \mathbf{v}) + \partial \rho / \partial t = 0 \qquad (1)$$

together with the assumption of incompressibility

$$d\rho/dt = \mathbf{v} \cdot \nabla \rho + \partial \rho / \partial t = 0 \qquad (2)$$

and the assumption of irrotational flow

$$\mathbf{v} = \nabla \Phi \qquad (3)$$

it follows that

$$\nabla^2 \Phi = 0 \qquad (4)$$

with solutions

$$\Phi(\mathbf{r}, t) = \sum_{\substack{l=2 \\ m}} \gamma_{lm}(t) r^l Y_{lm}(\theta, \phi) \qquad (5)$$

(A term with $l=1$ would correspond to motion of the centre of mass of the nucleus.)

It is assumed that the nucleus is vibrating about a spherical equilibrium shape for which the density distribution is $\rho^0(r)$. Because the motion is incompressible, the density can be written as

$$\rho(\mathbf{r}, t) = \rho^0(r_0) \qquad (6)$$

where r_0 is the equilibrium position of the element of nuclear fluid with position $\mathbf{r}$ at time t. In general

$$r - r_0 = \sum_{lm} a_{lm}(r, t) Y_{lm}(\theta, \phi) \qquad (7)$$

Pal (1972) has shown that writing the equation of a surface of constant density as

$$F(\mathbf{r}, t) = 0 \qquad (8)$$

the condition of matching the normal component of the velocity at the surface is

$$\nabla F \cdot \nabla \Phi = -\partial F/\partial t \tag{9}$$

The function in equation (8) is given by

$$F(\mathbf{r}, t) = r - \sum_{lm} a_{lm}(r, t) Y_{lm} - r_0 \tag{10}$$

To first order in the amplitude of the oscillation about spherical equilibrium, equation (9) yields

$$\partial \varphi/\partial r = \sum_{lm} \dot{a}_{lm} Y_{lm} \tag{11}$$

with solution

$$a_{lm}(r, t) = l\alpha_{lm}(t) r^{l-1} \tag{12}$$

where

$$\dot{\alpha}_{lm} = \gamma_{lm} \tag{13}$$

Then

$$\rho(\mathbf{r}, t) = \rho^0 \left(r - \sum_{lm} a_{lm}(\mathbf{r}, t) Y_{lm}(\theta, \phi) \right)$$

$$= \rho^0(r) - \sum_{lm} l\alpha_{lm}(t) r^{l-1} (d\rho^0/dr) Y_{lm}(\theta, \phi) \tag{14}$$

to first order in α_{lm}.

For a transition from initial state $|i\rangle$ to a final state $|f\rangle$ ($i \neq f$), the transition density is

$$\langle f|\rho(\mathbf{r})|i\rangle = -\sum_{lm} l\langle f|\alpha_{lm}|i\rangle r^{l-1} (d\rho^0/dr) Y_{lm}(\theta, \phi) \tag{15}$$

Making the approximation of small oscillations, the Hamiltonian is

$$H = T + V$$

with

$$T = \tfrac{1}{2} \sum_{lm} B_l |\gamma_{lm}|^2 \tag{16}$$

$$V = \tfrac{1}{2} \sum_{lm} C_l |\alpha_{lm}|^2 \tag{17}$$

which is of the form of a Hamiltonian for a set of uncoupled harmonic oscillators. Using

$$T = \tfrac{1}{2} m \int \rho(\mathbf{r}) v^2 \, d^3 r \tag{18}$$

yields

$$B_l = (4\pi)^{-1} Aml(2l+1)\langle r^{2(l-1)}\rangle \tag{19}$$

where

$$\langle r^n \rangle = A^{-1} \int \rho^0(r) r^n \, d^3r \tag{20}$$

In this approximation of small oscillations, the harmonic oscillator approximation, all the transition strength for a particular multipole l from the gound state goes into one state—a giant multipole state—which is the one-phonon state $|1\rangle$ for that l with energy

$$\hbar\omega = \hbar(C_l/B_l)^{\frac{1}{2}} \tag{21}$$

and

$$\langle 1|\alpha_{lm}|0\rangle = (\hbar/2\omega B_l)^{\frac{1}{2}} \tag{22}$$

so that

$$\langle 1|\rho(\mathbf{r})|0\rangle = -\{2\pi\hbar/[Am\omega l(2l+1)\langle r^{2(l-1)}\rangle]\}^{\frac{1}{2}} lr^{l-1}(d\rho^0/dr) Y_{lm}(\theta,\phi) \tag{23}$$

3. SUM RULES

We consider a system of A particles with the Hamiltonian

$$H = T + V \tag{24}$$

with

$$T = \sum_{i=1}^{A} p_i^2/2m \tag{25}$$

The eigenstates of the Hamiltonian are

$$H|n\rangle = E_n|n\rangle \tag{26}$$

Let

$$A = \sum_i A_i(\mathbf{r}_i), \quad B = \sum_i B_i(\mathbf{r}_i) \tag{27}$$

Then, from the Heisenberg commutation relations

$$[p_{i\mu}, x_{j\nu}] = -i\hbar\delta_{ij}\delta_{\mu\nu}, \quad \mu, \nu = 1, 2, 3$$

it follows that

$$[A, [H, B]] = [A, [T, B]] = m^{-1}\hbar^2 \sum_i \nabla_i A_i \cdot \nabla_i B_i \tag{28}$$

where it is assumed that V is velocity independent, or at most linear in momentum as for instance for a single-particle spin–orbit coupling, so that

$$[A, [V, B]] = 0 \tag{29}$$

We introduce the operator

$$F(\mathbf{q}) = \sum_{i=1}^{A} \exp(i\mathbf{q} \cdot \mathbf{r}_i) \tag{30}$$

Then

$$F_{nm}(\mathbf{q}) = \langle n|F(q)|m\rangle \tag{31}$$

is the isoscalar scattering form factor for a transition from state $|m\rangle$ to state $|n\rangle$. From equation (28),

$$[F(\mathbf{q}'), [H, F(q)]] = -(\hbar^2/m)\mathbf{q}' \cdot \mathbf{q} F(\mathbf{q}'+\mathbf{q}) \tag{32}$$

Taking the matrix element of equation (32) between two eigenstates of the Hamiltonian and inserting the complete sets of states (26) within the left-hand side

$$\sum_n [(E_n - E_i)F_{fn}(\mathbf{q}')F_{ni}(\mathbf{q}) + (E_n - E_f)F_{fn}(\mathbf{q})F_{ni}(\mathbf{q}')] = -(\hbar^2/m)\mathbf{q}' \cdot \mathbf{q} F_{fi}(\mathbf{q}'+\mathbf{q}) \tag{33}$$

Equation (33) is a slight generalization of a progenitor sum rule given by Noble (1971), and the various particular sum rules (Sachs and Austern, 1951; Kao and Fallieros, 1970; Deal and Fallieros, 1973a, 1973b; Ui and Tsukamoto, 1974; Reinhard and Drechsel, 1975; Tassie, 1975) can be obtained from it, although in practice it is usually easier to derive the various sum rules separately by considering the appropriate double commutator for each case.

The operator $F(q)$ is not Hermitian, but

$$F^+(\mathbf{q}) = F(-\mathbf{q}) \tag{34}$$

so that

$$F_{nm}(\mathbf{q}) = F^*_{mn}(-\mathbf{q}) \tag{35}$$

The treatment of invariance under time reversal can be simplified for systems consisting of an even number of fermions by choosing basis states which are eigenstates of the time reversal operator with eigenvalue +1 (see Section 1B-2 of Bohr and Mottelson, 1969). Then from invariance under time reversal,

$$F_{nm}(\mathbf{q}) = F_{mn}(\mathbf{q}) \tag{36}$$

Equation (36) does not hold for a system of an odd number of fermions, as the stationary states are not eigenstates of the time reversal operator, and for such systems a slightly more complicated formalism must be used (see for instance Bohr and Mottelson, 1969) which leads to the same sum rules.

Using equation (36), for $f = i$ equation (33) yields

$$\sum_n (E_n - E_i) F_{in}(\mathbf{q}') F_{ni}(\mathbf{q}) = -(\hbar^2/2m) \mathbf{q}' \cdot \mathbf{q} F_{ii}(\mathbf{q}' + \mathbf{q}) \tag{37}$$

which is the progenitor sum rule of Noble (1971).

Note that under the parity operator P, defined by

$$P \mathbf{r}_i P = -\mathbf{r}_i \tag{38}$$

we have

$$P F(\mathbf{q}) P = F(-\mathbf{q}) \tag{39}$$

so that from invariance under reflections,

$$\langle n|F(\mathbf{q})|m\rangle = \pi_n \pi_m \langle m|F(\mathbf{q})|n\rangle^* \tag{40}$$

using equation (35). π_n is the parity of the state $|n\rangle$. Then from the result of invariance under time reversal, equation (36),

$$F_{nm}(\mathbf{q}) = \pi_n \pi_m (F_{nm}(\mathbf{q}))^* \tag{41}$$

so that $F_{nm}(\mathbf{q})$ is pure imaginary for a transition involving a parity change and is real for a transition without parity change.

The operators for form factors of definite multipolarity are

$$F^J(q) = [4\pi(2J+1)]^{\frac{1}{2}} \sum_{i=1}^{A} j_J(qr_i) Y_{J0}(\theta_i, \phi_i) \tag{42}$$

and

$$F(\mathbf{q}) = \sum_J i^J F^J(q) \tag{43}$$

where the Z-axis is taken along the $\mathbf{q}$-direction. We define

$$Q_{J\alpha} = \sum_{i=1}^{A} Q_{J\alpha,i} \tag{44}$$

$$Q_{J\alpha,i} = r_i^{J+2\alpha} Y_{J0}(\theta_i, \phi_i) \tag{45}$$

For $J > 1$, the Q_{J0} are the isoscalar electric multipole moment operators. The isoscalar electric monopole operator is Q_{01}.

The further treatment will be confined to nuclei with zero spin in the ground state, and we consider the excitation of states with definite spin J. The treatment can be extended to nuclei with non-zero spin in the ground state. We take $E = 0$ as the energy of the ground state $|0\rangle$.

The following sum rules can be obtained by using equation (43) in the progenitor sum rule (37) and equating powers of q and q', or more easily by using equation (28) with the operators (42) and (44) (Ui and Tsukamoto, 1974: Tassie, 1975).

$$\langle 0|Q_{J\beta}HQ_{J\alpha}|0\rangle = \tfrac{1}{2}\langle 0|[Q_{J\beta},[H,Q_{J\alpha}]]|0\rangle$$
$$= (\hbar^2/2m)(4\pi)^{-1}[J(2J+2\alpha+2\beta+1)+4\alpha\beta]A\langle r^{2J+2\alpha+2\beta-2}\rangle \quad (46)$$

$$\langle 0|Q_{J\alpha}HF^J(q)|0\rangle = \tfrac{1}{2}\langle 0|[Q_{J\alpha},[H,F^J(q)]]|0\rangle$$
$$= -(\hbar^2/2m)[(2J+1)/4\pi]^{\tfrac{1}{2}}\int d^3r\, r^{J+2\alpha-2}j_J(qr)$$
$$\times\left[2\alpha(2\alpha+2J+1)\rho_{00}+(J+2\alpha)r\frac{d\rho_{00}}{dr}\right] \quad (47)$$

where

$$\langle r^n\rangle = A^{-1}\left\langle 0\left|\sum_{i=1}^{A}r_i^n\right|0\right\rangle \quad (48)$$

and

$$\rho_{00} = \langle 0|\rho(\mathbf{r})|0\rangle \quad (49)$$

where

$$\rho(\mathbf{r}) = \sum_{i=1}^{A}\delta(\mathbf{r}-\mathbf{r}_i) \quad (50)$$

For $J\neq 0$, $\alpha = 0$ equation (47) can be written (Tassie, 1958)

$$\langle 0|Q_{J0}HF^J(q)|0\rangle = (\hbar^2/2m)[(2J+1)/4\pi]^{\tfrac{1}{2}}Jq^J\left(-q^{-1}\frac{d}{dq}\right)^{J-1}F_{00}(q) \quad (51)$$

where

$$F_{00}(q) = \int j_0(qr)\rho_{00}(r)\,d^3\mathbf{r} \quad (52)$$

is the form factor for elastic scattering. Similarly for $\alpha \neq 0$,

$$\langle 0|Q_{J\alpha}HF^J(q)|0\rangle = (\hbar^2/2m)[(2J+1)/4\pi]^{\tfrac{1}{2}}[(J+2\alpha)q(d/dq)+J(J+1)]$$
$$\times q^J\left(-q^{-1}\frac{d}{dq}\right)^J\left(-\frac{d^2}{dq^2}-2q^{-1}\frac{d}{dq}\right)^{\alpha-1}F_{00}(q) \quad (53)$$

For instance, for monopole transitions $J = 0$ with $\alpha = 1$, equation (53) reduces to

$$\langle 0|Q_{01}HF^0(q)|0\rangle = (\hbar^2/2m)(4\pi)^{-\tfrac{1}{2}}2q(d/dq)F_{00}(q) \quad (54)$$

which is equivalent to the result given by Kao and Fallieros (1970).

The form factor (42) and density (50) are related by

$$F^J(q) = [4\pi(2J+1)]^{\frac{1}{2}} \int d^3r\, \rho(\mathbf{r}) j_J(qr) Y_{J0}(\theta, \phi) \qquad (55)$$

and analogously to (47) we have

$$\langle 0|Q_{J\alpha} H\rho(r)|0\rangle = -(\hbar^2/2m) r^{J+2\alpha-2}[2\alpha(2\alpha+2J+1)\rho_{00}$$
$$+ (J+2\alpha) r(d\rho_{00}/dr)] Y_{J0}(\theta, \phi) \qquad (56)$$

For $J \neq 0$, $\alpha = 0$, equation (56) reduces to the sum rule of Deal and Fallieros (1973a) and has the spatial dependence of the transition density (23) of the hydrodynamical model. Assuming that

$$\langle 0|Q_{J0}|n\rangle \neq 0$$

for only a single state $n = 1$, then equations (46) and (56) yield the hydrodynamical result (23) for $\langle 1|\rho(\mathbf{r})|0\rangle$, in both spatial dependence and magnitude.

4. INVERSION OF THE SUM RULES (Tassie, 1975)

The density operator (50) is a sum of single-particle operators, and defining the no-particle-no-hole state as the ground state, the only off-diagonal matrix elements of the density operator from the ground state are to one-particle-one-hole states. To the extent that inelastic scattering by the nucleus can be described in terms of the transition density, as for instance in Born approximation or distorted wave Born approximation or the approximation used by Boridy and Feshbach (1974), then inelastic scattering excites only one-particle-one-hole states.

From equation (42) and

$$j_J(z) = z^J[(2J+1)!!]^{-1}\{1 - [1!(2J+3)]^{-1}\tfrac{1}{2}z^2 + \ldots\} \qquad (57)$$

the states $Q_{J\alpha}|0\rangle$ give all the states excited by electron scattering. Then any final eigenstate of the nucleus can be written

$$|f\rangle = \sum_\alpha b_\alpha Q_{J\alpha}|0\rangle + |\text{remainder}\rangle \qquad (58)$$

where $|\text{remainder}\rangle$ does not contribute to inelastic electron scattering. For instance, $|\text{remainder}\rangle$ will include two-particle-two-hole states and so on.

Equations (47) for all α constitute an infinite set of sum rules for the scattering form factor. In order to be able to invert these sum rules to obtain the form factor for the excitation of a single final state in terms of a sum over the sum rules, we introduce another set of operators

$$M_{J\alpha} = \sum_\beta A_{\alpha\beta} Q_{J\beta} \qquad (59)$$

chosen so that
$$\langle 0|M_{J\alpha}HM_{J\beta}|0\rangle = 0, \quad \text{for } \alpha \neq \beta \tag{60}$$
From equations (59), (57) and (42),
$$F^J(q) = \sum_\beta f_\beta(q) M_{J\beta} \tag{61}$$
Then, using equations (61) and (60),
$$\sum_\gamma \langle f|M_{J\gamma}|0\rangle\langle 0|M_{J\gamma}HF^J(q)|0\rangle/\langle 0|M_{J\gamma}HM_{J\gamma}|0\rangle$$
$$= \sum_{\gamma\beta} \langle f|M_{J\gamma}|0\rangle\langle 0|M_{J\gamma}HM_{J\beta}|0\rangle f_\beta(q)/\langle 0|M_{J\gamma}HM_{J\gamma}|0\rangle$$
$$= \left\langle f\left|\sum_\gamma f_\gamma(q) M_{J\gamma}\right|0\right\rangle$$
$$= F^J_{f0}(q) \tag{62}$$

which is the form factor for the excitation of the state $|f\rangle$.

Equation (62) can be written as
$$F^J_{f0}(q) = E_f^{-\frac{1}{2}} \sum_\gamma c_\gamma \langle 0|M_{J\gamma}HF^J(q)|0\rangle/\langle 0|M_{j\gamma}HM_{J\gamma}|0\rangle^{\frac{1}{2}} \tag{63}$$
where
$$c_\gamma = E_f^{\frac{1}{2}}\langle f|M_{J\gamma}|0\rangle/\langle 0|M_{J\gamma}HM_{J\gamma}|0\rangle^{\frac{1}{2}} \tag{64}$$
Note that
$$|c_\gamma|^2 = E_f|\langle f|M_{J\gamma}|0\rangle|^2/\left\{\sum_n E_n|\langle n|M_{J\gamma}|0\rangle|^2\right\} \tag{65}$$
is the fraction of the sum rule for $M_{J\gamma}$ contributed by the state $|f\rangle$.

A set of $M_{J\alpha}$ is constructed by taking
$$M_{J\alpha} = \sum_{\beta=0}^{\alpha} A_{\alpha\beta} Q_{J\beta} \tag{66}$$
with $A_{\alpha\alpha} = 1$. The condition (60) is satisfied by
$$\langle 0|Q_{J\beta}HM_{J\alpha}|0\rangle = 0, \quad \text{for } \beta < \alpha \tag{67}$$
which for given α provides α simultaneous linear equations to determine the α coefficients $A_{\alpha\beta}$, $\beta < \alpha$. Then
$$\langle 0|M_{J\gamma}HF^J(q)|0\rangle \propto q^{J+2\gamma} \quad \text{as } q \to 0 \tag{68}$$
so that the term with $\gamma = 0$ predominates at small q. Since
$$M_{J0} = Q_{J0} \tag{69}$$

the first term in the form factor is

$$\langle 0|Q_{J0}HF^J(q)|0\rangle$$

which has the same q-dependence as the form factor given by the hydrodynamical model (Tassie, 1956, 1958). Thus at small q, the inelastic form factor is that of the hydrodynamical model. The remaining terms will give corrections which become more important as q increases.

Solving the equations (67) the final expression for the transition density is given by (writing out only the first two terms)

$$\langle f|\rho|0\rangle = (2\pi\hbar^2/mE_f A)^{\frac{1}{2}}(2J+1)^{-\frac{1}{2}}Y_{J0}(\theta,\phi)\{-c_0 J^{\frac{1}{2}}\langle r^{2J-2}\rangle^{-\frac{1}{2}}r^{J-1}(d\rho_{00}/dr)$$
$$-c_1[(2J+1)(J(2J+5)+4)\langle r^{2J+2}\rangle - J(2J+3)^2\langle r^{2J}\rangle^2\langle r^{2J-2}\rangle^{-1}]^{-\frac{1}{2}}$$
$$\times\{(2J+1)r^J[2(2J+3)\rho_{00}+(J+2)r(d\rho_{00}/dr)]$$
$$-(2J+3)\langle r^{2J}\rangle\langle r^{2J-2}\rangle^{-1}Jr^{J-1}(d\rho_{00}/dr)\}+\ldots\} \quad (70)$$

The first term in the transition density has the same spatial dependence as the transition density of the hydrodynamical model, equation (23).

If the density distribution of the ground state is taken to be a uniform distribution of radius R, the inelastic form factor is

$$F^J_{f0}(q) = [3E_f^{-1}(\hbar^2/2m)(2J+1)A]^{\frac{1}{2}}R^{-1}\Big\{c_0 J^{\frac{1}{2}}j_J(qR)$$
$$+\sum_{\alpha=1}(-1)^\alpha c_\alpha(2J+4\alpha+1)^{\frac{1}{2}}j_{J+2\alpha}(qR)\Big\} \quad (71)$$

which we write as

$$F^J_{f0}(q) = [3E_f^{-1}(\hbar^2/2m)(2J+1)A]^{\frac{1}{2}}R^{-1}\sum_\alpha c_\alpha F_\alpha \quad (72)$$

For electric quadrupole transitions, $J=2$, the F_α for $\alpha=0$ to 3 are shown in Figure 1. Assuming that all the c_α are of the same order of magnitude, it is seen that taking the summation to $\alpha=2$ is adequate up to $qR=4$. The predominance of the first term up to $qR=3$ is an indication of why the hydrodynamical model has been so useful in analysing electron-scattering data.

Because the recoil of the nucleus has been neglected, the motion of the centre of mass of the nucleus has not been treated correctly here. Corrections to the sum rules due to the motion of the centre of mass have been given by Deal (1973a, b) and for heavy nuclei are small except for isoscalar electric dipole transitions.

5. THE EFFECT OF THE FINITE SIZE OF THE PROTON

The results of the hydrodynamical model and the sum rules apply to the density of nucleons within the nucleus. For the calculation of electron scattering, the

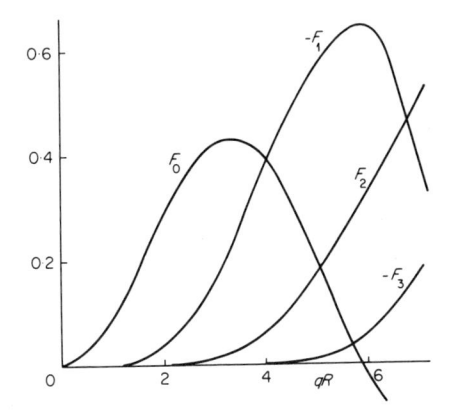

Figure 1 F_α, the terms in equation (72) for the inelastic form factor for electric quadrupole transitions for a uniform density distribution

effect of the finite size of the proton should be included (Tassie, 1960). The charge density of the nucleus is given by

$$\rho^c(\mathbf{r}) = \int \rho_p(\mathbf{r}-\mathbf{r}')\rho(\mathbf{r}')\,d^3r' \tag{73}$$

where ρ_p is the charge density of the proton. (More properly we should replace ρ_p by the isoscalar charge density of the nucleon, but except for very large momentum transfer the contribution from the charge density of the neutron can be neglected.) $\rho_{00}(r)$ must be determined by unfolding the proton charge density from the charge density of the nuclear ground state, which can be determined from elastic electron scattering, and inserted in equation (23) of the hydrodynamical model or the more general equation (70), and then the proton charge density must be folded in again to yield the transition charge density with which to calculate inelastic electron scattering. Applying this procedure to equation (51), which is the hydrodynamical result, yields for the inelastic form factor for electron scattering

$$F_{f0}^c(q) \propto F_p(q)q^J\left(-q^{-1}\frac{d}{dq}\right)^{J-1}[F_{00}^c(q)/F_p(q)] \tag{74}$$

where $F_p(q)$ is the charge form factor of the proton and $F_{00}^c(q)$ is the form factor for elastic scattering of electrons by the nucleus.

6. CORRECTIONS TO THE HYDRODYNAMICAL MODEL

Deviations of observed inelastic electron scattering from the prediction of the hydrodynamical model can in principle show that at least some of the c_α for $\alpha \neq 0$ are not zero. Some experiments (for instance Eisenstein and coworkers,

1969; Nagao and Torizuka, 1971; Heisenberg and coworkers, 1971; Friedrich, 1972; Johnston and Drake, 1974) have indicated deviations from the hydrodynamical model, but there have been two sources of error in the analysis of such experiments. One is that the finite size of the proton has been neglected. The other is that a fairly simple functional form has been used for the charge distribution of the ground state, which may not be sufficiently accurate. The calculation of Boridy and Feshbach (1974) of the excitation of the 3^- level at 2.62 MeV in ^{208}Pb by 1.04 GeV protons used a density distribution (Negele, 1970) for the ground state which gave a good fit to the elastic electron scattering, and they found a very good agreement between the experimental results and the scattering calculated using the hydrodynamical model.

To conclusively establish deviations from the hydrodynamical model for a single level, the charge density of the ground state should be varied to fit both the inelastic and elastic scattering. However, a simple prediction of the hydrodynamical model is that all levels of the same spin and parity in the same nucleus should have the same inelastic form factors. So that the observation of differences between the inelastic form factors for two levels of the same spin and parity in the same nucleus (Neuhausen and coworkers, 1971; Nagao and Torizuka, 1971; Friedrich, 1972; Lightbody, 1972) do show that corrections are needed to the hydrodynamical model, and thus at least some $c_\alpha \neq 0$ for $\alpha > 0$.

Deal (1973a) has calculated inelastic scattering using what he calls the single doorway and double doorway approximation. The single doorway is just the first term in the sum in equation (70), which is the same as the hydrodynamical model, and the double doorway approximation is equivalent to taking the two terms shown in equation (70). For the electron excitation of the 2^+ level in ^{12}C at 4.43 MeV, Deal (1973a) obtains a good fit to the experimental data up to $q = 0.7$ fm^{-1} with the single doorway approximation, and up to $q = 1.8$ fm^{-1} with the double doorway approximation. The data up to $q = 2.2$ fm^{-1} deviates from the double doorway approximation, indicating the need to include a third term in equation (70).

The existence of deviations from the hydrodynamical model can have unfortunate consequences for the determination of spin and parities of nuclear states by inelastic electron scattering. Equation (71) for the uniform charge distribution shows that higher order terms in the form factor for excitation of a state of spin J have the same q-dependence as the first term for excitation of states of higher spin, $J+2, J+4$, etc. as also noted by Ui and Tsukamoto (1974) for $\alpha = 1$ and 2. Although this result does not hold for an arbitrary density distribution, the actual density distribution for heavy nuclei is roughly like the uniform distribution, and so qualitatively one expects a similarity between the higher order terms for spin J and lower order terms for states with higher spin. A well-known special case is the similarity for electric monopole and quadrupole transitions,

$$\langle 0|M_{01}HF^J(q)|0\rangle = -5^{-\frac{1}{2}}\langle 0|M_{20}HF^J(q)|0\rangle \tag{75}$$

which is valid for an arbitrary density distribution.

If the form factors for two transitions are the same, it is only in first Born approximation that the scattering must be the same. In distorted wave Born approximation, the effects of distortion depend on the multipolarity of the transition, and this gives some help in determining spins. However, some caution is needed in allocating spins to nuclear states according to the shape of the angular distribution of inelastic electron scattering.

The treatment given here strictly only holds for isoscalar transitions, but if we ignore corrections for exchange forces (Inopin and Roshchupkin, 1973), the results can be used for isovector transitions. Then for an isovector electric dipole transition the form factor for a uniform charge distribution is

$$F_{f0}^1(q) = K[c_0 j_1(qR) - c_1 7^{\frac{1}{2}} j_3(qR) + \ldots] \tag{76}$$

In Born approximation the scattering is proportional to

$$|F_{f0}^1(q)|^2 = K^2 \{c_0^2 [j_1(qR)]^2 - 2c_0 c_1 7^{\frac{1}{2}} j_1(qR) j_3(qR) + \ldots\} \tag{77}$$

and the first term is the same as the result of the model of Goldhaber and Teller (1948) or the hydrodynamical model. However the second term of equation (77) has a similar q-dependence (for qR not too large) to the first term for an electric quadrupole transition

$$[j_2(qR)]^2$$

as is shown in Figure 2. In analysing electron excitation of the giant isoscalar quadrupole resonance, the contribution of the giant isovector dipole resonance has to be subtracted, and subtracting the dipole contribution according to the hydrodynamical model can leave a contribution with a q dependence similar to

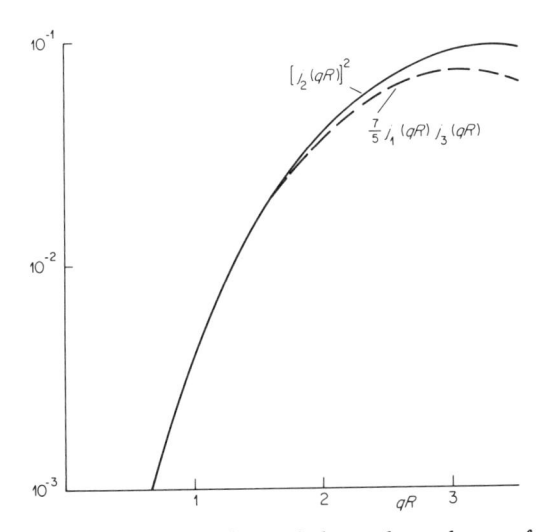

Figure 2 Comparison of the q dependence of $[j_2(qR)]^2$ and $j_1(qR)j_3(qR)$, showing the difficulty of distinguishing between electric quadrupole and additional contribution to electric dipole electron scattering

a quadrupole contribution. This shows a need for caution in making multipolarity assignments on the basis of the hydrodynamical model.

7. GIANT RESONANCES

Let us call a state which saturates a sum rule, and which is an eigenstate of the nuclear Hamiltonian, a perfect giant state. It has been shown (Tassie, 1975) that if a state saturates one of the sum rules, i.e. $|c_\alpha| = 1$ for some α, then $c_\alpha = 0$ for all $\gamma \neq \alpha$ and there is only one non-zero term in the sum in equation (63). The inelastic form factor and the transition density are uniquely determined for a perfect giant state and for $\alpha = 0$ are given by the hydrodynamical model.

Koo and Tassie (1976) have derived a selection rule for perfect giant multipole states from the sum rule of Reinhard and Drechsel (1975). For spin of the initial state $J_0 = 0$ and J_f odd, the sum rule of Reinhard and Drechsel yields

$$(E_f - E_0) \sum_n F^l_{fn}(q) F^l_{n0}(q) = 0 \tag{78}$$

and taking the limit $q \to 0$

$$(E_f - E_0) \sum_n \langle f|Q_{l0}|n\rangle\langle n|Q_{l0}|i\rangle = 0 \tag{79}$$

If there is a perfect giant 2^l-pole state

$$|n\rangle = |g\rangle$$

which saturates the sum rule for Q_{l0} so that there is only a single term in the sum in (79), then

$$\langle f|Q_{l0}|g\rangle = 0 \tag{80}$$

and the perfect giant 2^l-pole state cannot decay to a state of odd spin by emission of a 2^l-pole photon.

Of course, the observed giant resonances neither completely saturate a sum rule nor are eigenstates of the nuclear Hamiltonian (Walcher, 1973), so that at best one expects the hydrodynamical model to give a good approximation for their transition density and for the selection rule of equation (80) to be only approximate.

8. CONCLUDING REMARKS

There are basically two reasons for the success of the hydrodynamical model based on irrotational and incompressible flow in describing electron scattering. The first is that inelastic electron scattering excites only the one-particle-one-hole parts of the excited nuclear states. The second is that the form factor for inelastic scattering at low momentum transfer q is that of the hydrodynamical model as has been shown by the algebraic method using the Heisenberg commutation relations. This algebraic method is then a quantum-mechanical proof that no matter how compressible or how viscous a fluid is, it flows incompressibly and irrotationally if it is pushed gently enough (i.e. with small enough momentum transfer).

Although the application to nuclei has been considered, the same methods can be used for any system of identical particles. The results of the hydrodynamical model should apply to the inelastic scattering at small momentum transfer from any system of identical particles. In particular, the form factor (63) should be valid for inelastic scattering by atoms.

REFERENCES

Bohr, A. and Mottelson, B. R. (1969) *Nuclear Structure, Vol. I Single-Particle Motion*, Benjamin, New York.
Boridy, E. and Feshbach, H. (1974) 'High energy proton scattering by medium and heavy nuclei', *Phys. Letters*, **B50**, 433–437.
Deal, T. J. (1973a) 'Sum rules, nuclear structure, and form factors for isoscalar electroexcitations', Ph.D. thesis, Brown University.
Deal, T. J. (1973b) 'Isoscalar dipole resonances', *Nucl. Phys.*, **A217**, 210–220.
Deal, T. J. and Fallieros, S. (1973a) 'Models and sum rules for nuclear transition densities', *Phys. Rev.*, **C7**, 1709–1710.
Deal, T. J. and Fallieros, S. (1973b) 'Electroexcitation form factors and fixed-multipolarity sum rules, *Phys. Letters*, **B44**, 224–226.
Eisenstein, R. A., Madsen, D. W., Theissen, H., Cardman, L. S. and Bockelman, C. K. (1969) 'Electron-scattering studies on Ca^{40} and Ca^{48}', *Phys. Rev.*, **188**, 1815–1830.
Friedrich, J. (1972) 'Investigation of some low-lying states in ^{208}Pb by inelastic electron scattering', *Nucl. Phys.*, **A191**, 118–136.
Goldhaber, M. and Teller, E. (1948) 'On nuclear dipole vibrations', *Phys. Rev.*, **74**, 1046–1049.
Heisenberg, J., McCarthy, J. S. and Sick, I. (1971) 'Inelastic electron scattering from several Ca, Ti and Fe isotopes', *Nucl. Phys.*, **A164**, 353–366.
Inopin, E. V. and Roshchupkin, S. N. (1973a) 'Effect of exchange forces on the sum rule for electron scattering by nuclei', *Sov. J. Nuc. Phys.*, **17**, 526–527.
Inopin, E. V. and Roshchupkin, S. W. (1973b) *Yad. Fiz.*, **17**, 1008–1011.
Johnston, A. and Drake, T. E. (1974) 'A study of ^{24}Mg by inelastic electron scattering', *J. Phys.*, **A7**, 898–935.
Kao, E. L. and Fallieros, S. (1970) 'Monopole sum rule and the electroexcitation of the first excited state in ^{16}O', *Phys. Rev. Letters*, **25**, 827–828.
Koo, W. K. and Tassie, L. J. (1976) 'A selection rule for giant resonances', to be published.
Lightbody, J. W. (1972) 'Electron scattering from one- and two-phonon vibrational states', *Phys. Letters*, **B38**, 475–479.
Nagao, M. and Torizuka, Y. (1971) 'Electron excitation of low-lying states in ^{208}Pb', *Phys. Letters*, **B37**, 383–385.
Negele, J. W. (1970) 'Structure of finite nuclei in the local density approximation', *Phys. Rev.*, **C1**, 1260–1321.
Neuhausen, R., Lightbody, J. W., Fivozinsky, S. P. and Penner, S. (1971) 'Electron scattering from the isotopes ^{64}Zn, ^{66}Zn, ^{68}Zn and ^{70}Zn', *Bull. Amer. Phys. Soc.*, **16**, 561–562.
Noble, J. V. (1971) 'Progenitor sum-rules in nuclear physics', *Ann. Phys.*, **67**, 98–113.
Onley, D. S., Griffy, T. A. and Reynolds, J. T. (1963) 'Partial wave analysis of the inelastic scattering of electrons by nuclei. II. Application to the liquid drop model', *Phys. Rev.*, **129**, 1689–1690.
Pal, M. K. (1972) 'A semiclassical treatment of the dynamic ellipsoidal shapes in nuclei', *Nucl. Phys.*, **A183**, 545–581.

Reinhard, P. G. and Drechsel, D. (1975) 'Application of sum rules to the electroexcitation of anharmonic vibrator nuclei', *Phys. Letters*, **B56,** 17–20.

Sachs, R. G. and Austern, N. (1951) 'Consequences of gauge invariance for radiative transitions', *Phys. Rev.*, **81,** 705–709.

Tassie, L. J. (1956) 'A model of nuclear shape oscillations for γ-transitions and electron excitation', *Australian J. Phys.*, **9,** 407-418.

Tassie, L. J. (1958) 'Electron excitation of collective nuclear transitions', *Australian J. Phys.*, **11,** 481–489.

Tassie, L. J. (1960) 'Inelastic electron scattering and nuclear compressibility', *Nuovo Cimento*, **18,** 523–531.

Tassie, L. J. (1973) 'Finite deformations of the nucleus', *Australian J. Phys.*, **26,** 433–440.

Tassie, L. J. (1975) 'Electron excitation of isoscalar nuclear transitions', *Nucl. Phys.*, **A248,** 465–476.

Tuan, S. T., Wright, L. E. and Onley, D. S. (1968) 'A computer program for analysis of inelastic electron scattering from nuclei', *Nucl. Instrum. Methods*, **60,** 70–76.

Überall, H. (1971) *Electron Scattering from Complex Nuclei, Part B*, Academic Press, New York.

Ui, H. and Tsukamoto, T. (1974) 'Form factor sum rule and giant multipole states', *Progr. Theoret. Phys.*, **51,** 1377–1386.

Walcher, T. (1973) 'Giant resonances', *Proceedings of the International conference on Nuclear Physics, Munich 1973*, J. de Boer and H. J. Mang, Eds., Vol. 2, pp. 510–527, North-Holland, Amsterdam.

25

Heisenberg's Contribution to Physics

DAVID BOHM
Birkbeck College, London

The modern quantum theory has brought about a revolutionary transformation, not only in the content of our knowledge of physics, but also in our overall philosophical views concerning the nature of matter and man's relationship to the ultimate object of his investigations. The death of Professor Werner Heisenberg last week virtually marks an end to this remarkable era in the development of physics. It is therefore appropriate at this point to try to give a brief evaluation of the contributions of Professor Heisenberg, who played such a key part in giving shape both to the mathematical-physical principles of the quantum theory and to its philosophical interpretation.

To appreciate the magnitude of the changes brought about by the quantum theory, one has to recall that at the end of the 19th century it seemed that the general outlines of physics were pretty well settled. It was thought that the whole of nature could be adequately accounted for and explained in terms of the concept of material bodies interacting through fields. All motions of such bodies and all changes in the fields were assumed to be continuous and calculable in terms of initial conditions by means of suitable differential and partial differential equations. Thus, the whole of the universe was thought to be reducible to some sort of extended mechanical system, which was in principle completely describable in unlimited detail in terms of classical concepts, such as the positions and momenta of particles, field strengths and so on, and whose behaviour was totally determinate as well as in principle independent of its being observed or known.

Planck's study of the properties of radiation constituted the first evidence of the overall inadequacy of this structure of thought. He showed that radiation is emitted in the form of discrete quanta, with energies proportional to their frequencies. This completely denies the basic assumption of continuity of all physical processes, which is essential to classical concepts. Such discrete features of radiation were confirmed in the photoelectric effect (where light causes electrons to be ejected from a metal), as predicted by Einstein. Bohr then extended this notion of discreteness to the motions of material particles, notably electrons in hydrogen atoms. Later, with Somerfeld, he formulated a more general principle along the same lines: that of the quantization of the 'action' (a classical concept with the units of energy times time). With the aid of the *correspondence principle*, which asserted that discrete quantum properties must blend into the appropriate continuous classical properties when many quanta were involved, it then became possible to give a new and relatively orderly account, not only to spectroscopy as a whole, but also to ionization processes, and to many features of the properties of solids, liquids and gases (particularly specific heats).

Nevertheless, it was clear that the new quantum ideas were basically of an *ad hoc* character and depended heavily on empirical rules. They were moreover totally at variance with the very classical concepts in which they were expressed. Their successful application in a wide range of phenomena ultimately led to the view that a radically new approach would be needed.

It was in this context that Heisenberg began the work leading to his fundamental discoveries. What seems to have struck him was the perception that the attempt to discuss matter in terms of classical concepts tends to take our attention away from the essentially novel features of the actual fact of the quantum properties of matter as described above. It appears that, to some extent, he was guided by a positivist philosophy, which was a part of the prevailing intellectual background of the age. As a result, he expressed what he saw in the form of a requirement that one must avoid bringing into the theory quantities (for example the precise details of particle orbits) that are not actually observable or empirically determinable. However (as he himself later stated) there are important elements in Heisenberg's thinking that run contrary to the positivist philosophy, so that it may well be that the positivist form in which his ideas were put is incidental, and that the essential point was a perception of the need for a radically new mathematical and physical account of the fact as whole.

SPECTRAL LINES ARE THE FACT

By focusing attention on spectral line intensities (which were a key novel feature of the fact in the quantum domain) Heisenberg was led to formulate the dynamical laws of particles in terms of a new mathematical structure, that of matrices. These obeyed new kinds of laws, which, however, correspond with those of classical physics in the limit of high quantum numbers [where the matrix elements approached the fourier coefficients of the coordinates of the particles in their (periodic) orbits]. Meanwhile, de Broglie and Schrödinger, working on a very different line, brought out a new wave theory of matter, which was interpreted statistically by Born. It was soon shown that this approach is mathematically equivalent to that of Heisenberg, and from the transformation theory which made possible the demonstration of equivalence, the whole subject received a systematic and orderly formulation. From this point on, the quantum theory went from one achievement to another, being successfully applied in an extremely wide range of fields, to the removal of long standing puzzles, as well as in the prediction of many new kinds of phenomena.

In the same spirit in which he tried to free our thinking of its irrelevant attachment to classical concepts in the atomic domain, Heisenberg later proposed that the interactions of elementary particles be studied through 'S-matrices'. To understand what is behind such an approach, one may note that whereas quantum mechanics implies the need to cease to use a continuous description of the motions of material particles, it still uses a 'wave equation' which is expressed in terms of continuous coordinates. In his 'S-matrix' theory, Heisenberg outlined a schematic form in which the results of scattering particles from each other can in principle be expressed without the assumption of a continuous wave. Although this scheme has served as an inspiration for much work in physics, it cannot be said, as yet, to have led to a definitive new theory. Moreover Heisenberg himself appears to have, to some extent, turned away from this approach to his attempts to make a non-linear quantum field theory based on continuous functions and involving the introduction of a fundamental length to nature. Nevertheless, the idea behind the 'S-matrix' is still significant, in the sense that there is something that is not entirely fitting in the use of continuous differentiable functions in a theory in which the movement of matter has to be regarded as essentially not continuous.

As important as were these achievements (as well as many others, such as, for example, the introduction of 'isotopic spin' which expressed the idea that proton and neutron were different states of the same particle) it may safely be said that Heisenberg is best known for his philosophical contributions, which centre on the *uncertainty relations*. (For a discussion of such philosophical issues, see Heisenberg, 1971.)

From his new matrix mechanics, it is possible to show that two variables that would in classical physics be 'canonically conjugate' (an example of such a pair is position and momentum) cannot be simultaneously defined with complete precision. Rather, there is a reciprocal relationship between them, such that the product of their uncertainties must be greater than Planck's constant, h. Heisenberg accounted for the existence of such a limit by means of his famous hypothetical experiment, in which the position of an electron was measured very accurately by means of a microscope (see extract below from his 1930 book *Physical Principles of the Quantum Theory*). He showed in effect, that the quantum needed for such a measurement would disturb the observed electron in an unpredictable and uncontrollable way, and that as a result, it is impossible to assign to the electron simultaneous values of position and momentum with greater accuracy than that specified by the uncertainty relations.

Clearly this is a result of fundamental significance. For not only do classical concepts require complete continuity; as we have seen, they also imply complete determinism. Although the indeterminacy is negligible in the context of ordinary experience, it is evidently of the greatest importance for the basic principles of physics. Even more important, perhaps, is the implication that the act of observation appears to play a key part; in the sense that it not only discloses or reveals the attributes of the electron, but also in the sense that through the irreducible disturbance it helps actually to make or produce these

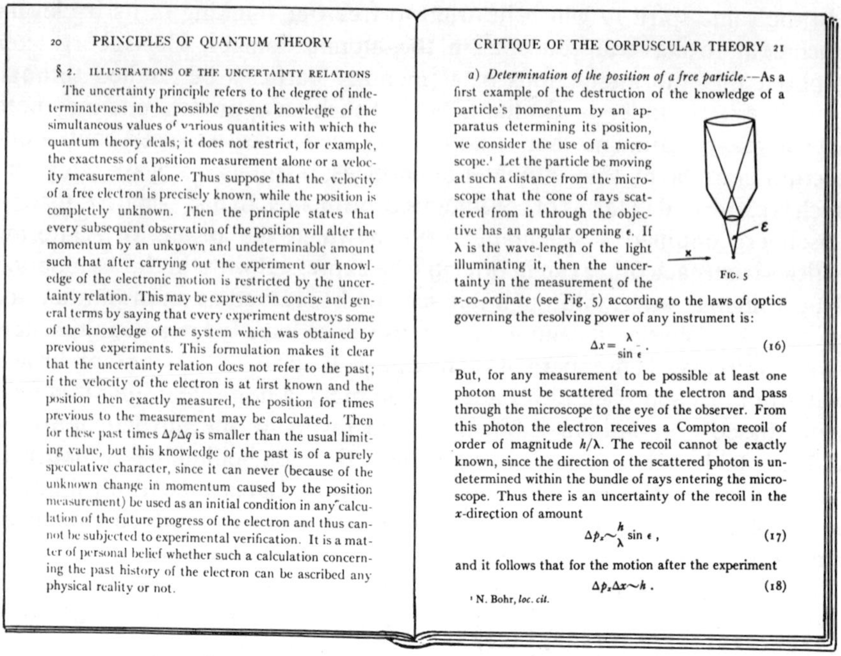

20 PRINCIPLES OF QUANTUM THEORY

§ 2. ILLUSTRATIONS OF THE UNCERTAINTY RELATIONS

The uncertainty principle refers to the degree of indeterminateness in the possible present knowledge of the simultaneous values of various quantities with which the quantum theory deals; it does not restrict, for example, the exactness of a position measurement alone or a velocity measurement alone. Thus suppose that the velocity of a free electron is precisely known, while the position is completely unknown. Then the principle states that every subsequent observation of the position will alter the momentum by an unknown and undeterminable amount such that after carrying out the experiment our knowledge of the electronic motion is restricted by the uncertainty relation. This may be expressed in concise and general terms by saying that every experiment destroys some of the knowledge of the system which was obtained by previous experiments. This formulation makes it clear that the uncertainty relation does not refer to the past; if the velocity of the electron is at first known and the position then exactly measured, the position for times previous to the measurement may be calculated. Then for these past times $\Delta p \Delta q$ is smaller than the usual limiting value, but this knowledge of the past is of a purely speculative character, since it can never (because of the unknown change in momentum caused by the position measurement) be used as an initial condition in any calculation of the future progress of the electron and thus cannot be subjected to experimental verification. It is a matter of personal belief whether such a calculation concerning the past history of the electron can be ascribed any physical reality or not.

CRITIQUE OF THE CORPUSCULAR THEORY 21

a) *Determination of the position of a free particle.*—As a first example of the destruction of the knowledge of a particle's momentum by an apparatus determining its position, we consider the use of a microscope.[1] Let the particle be moving at such a distance from the microscope that the cone of rays scattered from it through the objective has an angular opening ϵ. If λ is the wave-length of the light illuminating it, then the uncertainty in the measurement of the x-co-ordinate (see Fig. 5) according to the laws of optics governing the resolving power of any instrument is:

FIG. 5

$$\Delta x = \frac{\lambda}{\sin \epsilon} . \qquad (16)$$

But, for any measurement to be possible at least one photon must be scattered from the electron and pass through the microscope to the eye of the observer. From this photon the electron receives a Compton recoil of order of magnitude h/λ. The recoil cannot be exactly known, since the direction of the scattered photon is undetermined within the bundle of rays entering the microscope. Thus there is an uncertainty of the recoil in the x-direction of amount

$$\Delta p_x \sim \frac{h}{\lambda} \sin \epsilon , \qquad (17)$$

and it follows that for the motion after the experiment

$$\Delta p_x \Delta x \sim h . \qquad (18)$$

[1] N. Bohr, *loc. cit.*

Figure 1 Two key pages from the English edition of Heisenberg's *The Physical Principles of the Quantum Theory*, published by the University of Chicago Press. © University of Chicago Press and reproduced with permission.

very attributes. This means that the classical notion of a world that is essentially independent of the actions of the human observer is ultimately denied. Rather, the human being not only comes to know the world, but in doing this, he participates in an essential way.

A great many questions arise as to the deeper meaning of all this. Heisenberg proposed that the indeterminacy described above be regarded as a universal principle, thus implying that no new kinds of concepts are possible that could restore determism, in one way or another. In this he was, at least in some sense, in agreement with Bohr. Indeed, Heisenberg has generally subscribed explicitly to Bohr's philosophical principle of complementarity. Nevertheless, it is not at all clear that his thought does not contain aspects that are implicitly in contradiction with Bohr.

Bohr's view is actually a parallel to that of Kant, who says that all that we can know are the 'phenomena' (the appearances) and that the essence underlying the appearance (sometimes called the 'thing-in-itself') cannot be known. Thus, Bohr says that an experiment is characterized by a set of conditions and by a set of results. Both are described in terms of classical concepts. The form of the experimental conditions and the content (meaning) of the results are a single whole, not further analysable. Quantum theory provides nothing but an algorithm, connecting one set of (classically describable) phenomena to another, in a statistical way. One cannot properly say that there is a quantum object (for example an electron), but rather that this latter is only an abstract term arising in the description of the phenomena.

When Heisenberg accounts for the uncertainty principle as due to a disturbance, he is implicitly accepting that the electron is an object which exists independently of the observation, but adding that it is influenced by the observation in an unknowable way. Likewise, when he suggests that the electron's properties may be regarded as 'potentialities', which are realised differently according to the different experimental conditions, he is implying that through this concept, he is grasping the essence of the electron, and not merely its phenomenal appearance or manifestation. In this regard, he is close to the practice of most physicists today, who seem to feel that they are discovering (or who at least hope to discover) something about a reality that is beyond the phenomena. For example, in von Neumann's view, a measurement establishes the 'quantum state' of the electron, a state that is to some extent actually produced in the action of the measurement process, but that nevertheless continues to exist independently until the next interaction with a measuring apparatus. So Heisenberg, together with most physicists, appears tacitly to reject Bohr's view (and along with it many of the implications of the positivist philosophy which was very much in evidence in Heisenberg's early work).

Further unclarities with regard to such questions arise when we consider the well-known hypothetical experiment of Einstein, Rosen and Podolsky (See Bohm and Aharonov, 1070). From this experiment, one can show how it is possible, in general, to observe the state of a particle without interacting with it in any way. In such an observation, the uncertainty relationships still hold, but now they cannot be accounted for by means of a disturbance. Thus, Heisenberg's microscope experiment breaks down, when applied to this case. To Bohr, this presented no insuperable problem, since for him, there is never anything but the phenomena, and there is never any need to account for the phenomena (for example by attributing their uncertainty to a disturbance) as long as we can consistently describe them and predict their statistical distributions.

It would seem that in clearing up this whole question, there could well be room for fundamentally new developments in the conceptual structure of physics. Nevertheless, in spite of such unclarities, Heisenberg's matrix formulation of the quantum laws along with his insight that there is an irreducible act of participation in the process of observation will almost certainly still have a key significance, whatever new discoveries may be made in the future.

ACKNOWLEDGEMENT

This material first appeared in *New Scientist* London, the weekly review of Science and Technology.

REFERENCES

Bohm, D. and Aharonov, Y. *Phys. Rev.*, **108**, 1070. Heisenberg, W. (1971) *Physics and Beyond: Encounters and Conversations*, Allen and Linisin, London.

Author Index

Adams, W. S.   443, 444
Arnold, V. I.   165

Baglin, A.   456
Ballentine, I. L.   109
Bargmann, V.   154
Beadet, G.   453
Beck, G.   31
Belinfante, F. J.   133
Bergmann, P. G.   163, 164, 181
Bethe, H. A.   395, 450, 451
Blokhintsev, D. I.   31, 109, 130
Böhn, K.-H.   454, 455
Bohr, N.   116, 118, 562
Born, M.   147
Boulware, D. G.   365, 366, 374
Brauer, R.   236
Broyles, A. A.   33
Bues, I.   456

Caratheodory, C.   174
Chandrasekhar, S.   442, 445, 446, 447, 448, 449
Coish, H. R.   32

Dam, H. van   365, 366, 374
d'Antona, F.   456
Darling, B. T.   32
de Broglie, L.   147
Deser, S.   365, 366, 374
Dirac, P. A. M.   163, 164, 181, 239

Eddington, A. S.   9, 442, 443
Einstein, A.   109, 442, 443, 444
Engelmann, F.   30

Fick, E.   30
Flint, H. T.   30, 31, 32
Fock, V. A.   31
Fontaine, G.   455
Fürth, R.   31

Gantmacher, F. R.   237
Gerasimchuk, A. J.   30
Gien, T. T.   34
Glaser, W.   31
Greenman, M.   46
Greenstein, J. L.   442, 444, 454, 457
Griffith, R. W.   31

Heisenberg, W.   11, 31, 32, 109, 116, 117, 147, 227, 239
Hellund, E. J.   32
Hill, E. L.   32
Hubbard, W. B.   451, 453

Jordan, P.   147

Kadyshevsky, V. G.   32
Kálnay, A. J.   30
Kim, D. Y.   32
Kovetz, A.   451, 453, 454

Lamb, D. Q.   454, 455, 456
Lampton, M.   457
Landau, L.   31
Levy-Leblond, J. M.   240, 241
Lippmann, B. A.   35

Mackey, G. W.   231
Mandelstram, L.   109
March, A.   29
Markov, M.   31, 32
Marshak, R. E.   442, 450, 451
Mazzitelli, I.   456
Mestel, L.   442, 451, 452, 453, 454
Möglich, F.   31
Morinaga, K.   236
Morris, A. O.   236
Motz, L.   32
Moyal, J. E.   133, 149, 150, 157

Neumann, J. V. von   110, 112, 116, 119, 227, 230

Olkhovsky, V. S.   30
Ostriker, J. P.   442, 454

Papp, E. W. R.   30, 40
Pauli, W.   130
Pavlopoulos, T. G.   32
Peierls, R.   31
Penrose, R.   32
Planck, M.   46

Ramakrishnan, A.   228, 236
Recami, E.   30
Remak, B.   31
Richardson, O. W.   31

Robertson, H. P. 117
Rohrlich, F. 45
Rompe, R. 31
Ruark, A. E. 31
Ruderman, M. A. 453, 454

Salpeter, E. E. 395, 447, 448, 453
Santhanam, T. S. 227
Savedoff, M. P. 452, 456, 457
Schames, L. 31
Schatzmann, E. 442
Schild, A. 32
Schröder, E. 37
Schwinger, J. 228, 231
Shaviv, G. 451, 453, 454
Shipman, H. L. 456, 457
Sitte, K. 31
Snyder, H. S. 32
Stone, M. 227
Strittmatte, P. A. 454, 455, 456

Takano, Y. 32
Tamm, Ig. 116
Tekumalla, A. R. 227
Toledo, B. P. 30
Treder, H. 46
Tulczyjew, W. M. 164, 181

Van Horn, H. M. 442, 454, 455, 456
Van Vleck, J. H. 169
Veltman, M. 365, 366, 374
Vila, S. C. 452, 454
Vlasov, A. A. 139

Watoghin, G. 31
Wehrse, R. 457
Weidmann, V. 443
Weinstein, A. 165
Weyl, H. 149, 150, 227, 229, 236
Wheeler, J. A. 46
Wickramasinghe, D. T. 456
Wigner, E. P. 33, 149, 157

Subject Index

accidental degeneracy 399, 423
additive quantities 136
affine 271, 274
  function 271, 277
  phase space 150
alpha-clustering 485
  model 494
alpha-decay 495
alpha-transfer reactions 511
angular momentum operators 227
anharmonic oscillator 285
annihilation operators 355
anti-commutator 402, 415
arbitrary ensemble 141
associative algebra 164, 175, 183
asymptotic completeness relation 341
automorphism 266, 268
axiomatic elements 248
axiomatic quantum field theory 352
axiomatic quantum theory 312
azimuthal 228

Banach algebra 257
base normed space 275
base problem 278, 279, 281
bending of light 374
binary mixing operation 270
Bohr–Sommerfeld quantum conditions 147, 148, 171
Boolean sigma-algebra 263
bootstrap approach 325
bounded 230
Breit operator, Breit interaction 430, 432, 435
buckling 277

C^*-algebra 257, 258, 259
canonical basis 334, 335
canonical commutation relations 253, 352
canonical generators 227
canonical mapping 186
canonical transformation 183
causal automorphisms 327
causal ordering 327, 330
causality 7, 8, 9, 10, 310, 319, 322, 327
central element 334
centre of G_D 331, 332
certainty 9
Chandrasekhar limit 447
characteristic foliation 181

characteristic functional 355
charge conservation, condition 402, 407, 410, 416, 418, 421, 423, 424, 429, 432, 436, 437
charge density, operator 401, 408, 409, 418, 421, 422, 435
charge transfer salts 461, 466
classical compatibility conditions 163, 183
classical eigenstates 162
classical electromagnetic interaction 409
classical mechanics 277
classical physical system 206
classical propagators 175
Clifford algebra 235
cluster-transfer reactions 509
coexistent demixture 212
coexistent effects 209, 313
comeasurability 313
commutation relations 546
commutator 397, 415, 423, 429
compactification 327
comparison operators 286
compatibility 147, 157
compatible 256, 263, 264
  eigenequations 164
complementarity 6, 52
complementary 52, 223
complete integral 169
completeness, condition 399, 410, 416, 418, 421, 422, 423, 424, 429
completeness relation 340
complete orthonormal basis 399
complex algebra 257
conductive opacity 450, 451
conductivity 479
cone 271, 277
conformal partial wave expansions 326
consistency conditions 325
constructive quantum field theory 350, 353, 363
contact manifold 186
continuity, equation of 402, 403, 404, 405, 406, 425
convection 454
converges strongly 249
convex mixtures 268
convex prestructure 270, 271, 272
convex set 272, 277
convex structure 270, 271, 277
convex substructure 271
convexity 251

568 Subject Index

convolution kernels 339
correspondence principle 118, 119, 147, 158
cosmology 51
Coulomb interactions 447
Coulomb operator 422
coupling, spin–orbit 414
covariance 266
covariant components 338
covariant differential operator 337, 342
covariant kernels 338, 340
covariant POV measure 305, 312, 314, 315, 316, 317, 318
covariant pre-Wilson expansion 345
covariant spectral measure 300, 301, 305, 312, 313, 314, 315, 317
CPT theorem 322
creation operators 355
crossing relations 322
crossing symmetry 326
crystallization 453
current density operator 401, 402, 404, 405, 407, 408, 409, 418, 421, 422
cutoffs 356
cyclic permutation matrix 231
cyclic vector 258

Darwin term 414
de Broglie wave packets 16
Debye cooling 454
Debye temperature 454
decay fragments 19
decision effects 204, 312, 319
demixture 195, 211, 222, 223
density operator 249
detectors 19, 162
deterministic dynamics 100
dilations 327
dimension 328
Dirac-diagonal operator 403, 414, 435, 436
Dirac energy 408, 409
Dirac Hamiltonian 404, 408, 411, 422
Dirac matrices 404, 407, 411, 430
Dirac measures 171
Dirac particle 406
Dirac theory for a single relativistic particle 403
discrete analogue of CCR 227
discrete bounded spectrum 227
discrete series 328, 339, 342
discrete variable, discrete index 403, 411
dispersion (uncertainty) of x with respect to $f(x)$ 16
distinguished choice 281
distinguishing 270
doubt 11
dressing transformations 359

effect 312, 313, 319
effectiveness 266
eigenequation 153, 157, 183

eigenfunction 398, 412, 417, 421, 423
eigensembles 161
eigenvalue 53, 161, 398, 411
  lower bounds to 278
  problems 277
Einstein–Podolski–Rosen paradox 220
elastic scattering of alpha-particles by nuclei 499
electric dipole transition 555
electric monopole 548, 554
  moment 548
electric quadrupole transition 552, 554
electromagnetic field theory 365
electron degeneracy pressure 446
elementary particle 294, 298, 310, 319
energy 294, 295, 296, 298
  levels 277
ensemble 200, 204
entropy 151, 176, 177
envelope 173
EPR 473
equation of state 446, 455
ergodic 170
  Hamiltonian 181, 184
Euclidean conformal transformations 331
Euclidean group 299, 314, 315
Euclidean quantum field theory 360
Euclidean transformation 302
expectation functionals 248
expectation value 153, 176, 299, 301, 303, 306, 309
experimental propositions 260
extended relativity 322
extreme points 251, 271

face 271
fallibility 11
falsifiability 11
family of operators 345
Fermi–Dirac distribution function 450
field equations of Thirring model 336
finite dimensional vector space 227
first-class submanifold 164
flight-times 21
Fock representation 355
Fock space 253
form factor 543, 547, 548, 549, 550, 551, 552, 553, 555, 556
fourth equation of motion 15
free massless field 329, 335
free massless operator field 328
full 255
fuzzy measurement 312

gauge invariance 385
gauge transformations 368, 369
Gaussian shape 16
generalized Clifford algebra 228
generalized coordinate canonically conjugate to the Hamiltonian 14

Subject Index 569

generalized observable 312, 313
generalized special choice 285
generate(d) 267, 271
Gerlech and Stern experiment 129, 130, 131
giant multiple state 546
giant resonance 555, 556
Gleason's theorem 250, 264
GNS construction 258, 259, 265
good quantum numbers 399
gravitational field theory 365
gravitational red shift 444
Greeks 9

Haag's theorem 351, 352, 356, 357, 362
Hamiltonian 250
  Dirac 404, 411, 422
  non-relativistic 402, 403
  operator 398, 399, 400, 402, 404, 405, 406, 407, 408, 410, 411, 413, 421, 422
  quantum field theory 351, 354
  vector field 180
Hamiltonian function 397
Hamilton–Jacobi equation 166, 167
Hamilton–Jacobi theory 158, 166
Heisenberg Hamiltonian 468
Heisenberg model 462
Heisenberg picture 249
Heisenberg uncertainty principle/relation 207, 219, 249, 293, 294, 307, 318, 349, 441, 561
helicity 295, 296, 298, 303, 316, 317
helium atom 280
hidden parameters 171
hidden variables 97
high temperature expansions 363
Hilbert space 4, 399, 421, 431
hole parity, operator 417, 420, 423
Hubbard model 467
Huyghens–Freshnel interference rule 173
Huyghens–Freshnel superpositions 186
Huyghens interference rule 173
hydrodynamical model of the nucleus 543, 544, 550, 552, 553, 554, 556
hydrostatic equilibrium 446
hypermaximal 27
hyperplane 271

I-class constraints 163
I-class submanifolds 181
imprimitivity theorem 231
individual microsystems 191, 214, 216
inelastic electron scattering 553
inelastic scattering form factors 543
inelastic scattering of electrons by nuclei 543
infinite Fock space 55
information 149, 158, 161, 162, 176, 177
instantaneous measurability 117
interaction, Breit 435
  Coulomb 395, 428

interaction, Breit—*contd.*
  electromagnetic 395, 396, 408, 410, 423, 431
  magnetic 428
  orbit–orbit 431
  radiative 428, 432
  spin–orbit 414, 429, 430
  spin–other orbit 431
  spin–spin 431
interaction durations 21
interaction energy 409
interference 172
intermediate problems 278
  of the first type 279
intertwining operator 328, 342
involution 257
  algebra 257
ion–radical salts 461
irrationals 9
Ising model 363
isomorphic 271, 272
isotropic manifold 165

Jacobi identity 178, 180, 183

kernels of the second kind 341
knock-out reactions 521

Lagrangian manifold 165
Lagrangian submanifold 171, 181
latent heat of crystallization 454
lattice 204, 447, 453, 454
Lie algebra 164, 178
Lie derivative 169
light-cone expansion 338, 342
light-like data 352
lithium atom 287
localizable 265
logic 214, 217
Lorentz transformation 297, 298

macroscopically distinct states 56
macrosystem 98
magnetic monopoles 321
Markoffian theories 100
material waves 147
matrix elements, time-dependent 400, 401, 406, 409
  time-independent 400, 401, 406, 409
maximal 265
Maxwell equations 322
mean lifetimes 21
  of unstable particles 15
mean square deviation 306, 309, 310
measured values, attributing 127
measurement 53, 112
  angular momentum 129
  information side of 114
  informative side of 112

570 Subject Index

measurement—*contd.*
  momentum 121, 122, 124, 125
  particle position 120
  perturbing side of 112
  process 112, 115, 116, 118, 208, 211, 220
  state preparing side 114
  statistical action of 115
  two sides of 112
metastable states 26
method of stationary phase 159, 160
microsystem 205
microcanonical ensemble 153, 170
mixed boundary-initial conditions 18
mixed operation 269, 270
mixture(s) 96, 195, 251
model atmospheres 456
momentum 295, 296, 298, 302, 316, 318
monopole transitions 549
Mott–Hubbard gap 477
multiplier 329
multiply periodic systems 171

neutrino 293, 298, 319
  energy loss 452
Newton–Wigner position observable 318
Newton–Wigner position operator 311, 319
NMP 475
non-interacting electron gas 445, 449
non-relativistic Hamiltonian system of particles 397
non-relativistic notion of state 134
non-separability 96
non-separable 98
nuclear matter 489
number operator 240, 242

observable 102, 210, 248, 263, 294, 298, 299, 312, 313, 319
old quantum theory 171
one-parameter unitary group 161, 183
operator 27
  eigenequation 161, 179
operator-momentum 117
operator-position 117
order determining 261
orthocomplementation 261
orthocomplemented poset 261, 262
orthogonal 261, 267
  complement 261
  projections 250
orthomodular 261
orthonormality, condition 399, 410, 416, 418, 419, 421, 423, 424, 428, 429, 433
Osterwalder–Schrader axioms 362

paradox 190
parallel axiom 10
parity 418, 548
partially degenerate electrons 449
particle creation 52, 57

Pauli approximation, perturbation expansion 397, 410, 412, 418, 432
Pauli approximation for a single relativistic particle 410
Pauli exclusion principle 445
Pauli–Fierz equation 368, 369
Pauli perturbation expansion 412
persistent eigenvalue 279
perturbation theory 280, 286
phase canonically conjugate 240
phase operator 242
phase space 149, 179
phase transitions 363, 454
phases 227
photon 293, 294, 295, 296, 297, 298, 299, 300, 301, 302, 303, 304, 306, 307, 309, 310, 312, 313, 316, 317, 319
  position observable 294, 299, 302, 313, 318
  position operator 299, 301, 303
  velocity operator 309, 318
physical masses 363
physical quantities 152, 175, 182
physical space 265
physical system 191, 192, 199
pion and kaon reactions 528
Planck constant 158, 159, 171
Poincaré group 294, 297, 298, 299
Poisson bracket 151
polarization 293, 295, 299
position observable 265, 293, 294, 301, 306, 310, 312, 318, 319
position operator 301, 302, 303, 305, 306, 315, 317, 318
potential, scalar 409
  vector 409, 415
potential scattering 25
POV measure 305, 306, 308, 310, 312, 313, 314, 316, 317, 318, 319
preparing procedure 192, 196
primitive submanifold 184
principal series 339
probabilistic ideal 183, 184
probabilistic predictions 17
probability 300, 304, 305, 312
  measure 250, 261, 267
  of detecting 153
projection-valued measure 251
projective representations 154, 266
projector 177, 185
proper ensemble 153
proper equations 157
property 214
proposition system 260
propositions 214, 250, 259, 260, 261, 264, 268
proton charge density 553
pseudo-one-dimensionality 462
pseudo-properties 215

Subject Index 571

pure ensemble 138, 139, 140, 141
pure quantum states 166
pure state(s) 139, 149, 157, 164, 176, 177, 249, 251, 258

quantization of a system of electromagnetic particles 405
quantization of phase space 441, 449, 453
quantization of space-time 52
quantum electrodynamics 395
quantum field theory 15, 319
quantum logic 214, 261, 262, 263, 265
quantum mechanics 349, 399
  and construction of a new statistical theory having a constant probability character 65
  and the refection of the existence of a theory providing a more complete description of physical reality 63
  experimental test of 118
  main statements of 110
  not a theory of the consistent probability nature 62
  on discrete space 227
  statistical interpretation 109
  validity of 63
quantum statistical mechanics 441
quantum statistics 176, 444
quantum system 266, 268
quasiclassical compatibility 175
quasiclassical projection 186
quasiclassical solution 169
quasiclassical wave function 186
quasi-invariant 267
questions 8, 10, 202

radial lithium 288
radial Schrödinger equation 283
random variables 152, 175
range 264
Rayleigh-Ritz method 278
recipe 269
reciprocal time-operator 14
recording procedure 192, 197
reduced phase space 181
reflections 548
regularized products 337
relativistic electromagnetic interaction 396
relativistic mass correction 414, 429
relativistic motions of state 134
relativistic quantum theory 350, 358
relativistic statistics 133, 143
religion 10
renormalization 351, 353, 359
representation 258
rest mass 411, 412, 417, 429
retardation/advancement effect 410, 428, 432
Rosen-Podolsky paradox 89

scattering phenomena in a finite domain 18
Schrödinger equation 166, 250, 280, 398
  time-dependent 398
  time-independent 398
Schrödinger operator 278
Schrödinger picture 249
Schroer model 332
Schwinger functions 361, 362
scientific truth 10
second-type of intermediate problems 279
Segal algebra 255, 256, 258, 259
self-consistent 164, 183
  ideal 165, 170
self-energy, self-action 395, 396, 405, 432
semicovariant kernels 341, 345
semicovariant Wilson expansion 342
separation of variables 287
sheets of Minkowski space 330
short distance expansion 337
sigma-field 336
sigma-orthocomplete 261
simple 181
simplex 269
simultaneous measurability 190
simultaneous measurement 208, 248
simultaneous preparation 211
singular foliation 181
Sirius B 442, 443, 444, 446, 456
sources 162
space reflection 298
spatially equivalent 258
special choice 281
special conformal transformations 327
special relativity 349
specific heat 443, 453
spectral measure 300, 301, 303, 304, 305, 306, 308, 309, 312, 313, 314, 319
spectral theorem 251
spin 295, 315
spin and statistics 329
spin one field theory 266
spin one theory 385
spinor, wave functions 403, 404, 406, 417, 421, 433
spin two field theory 266
spin two theory 385
spinar representations 332
stability subgroup 332
state(s) 88, 98, 200, 204, 248, 251, 258
stationary state 158, 159
  wave function 398, 399, 406, 431, 432
statistical action 113, 115
  dispersion 153, 177
  ensemble 134, 135, 152, 175, 176
  principle 137
  spread 161
Stone's theorem 250
subspace 271
sum rules 543, 546, 547

572 Subject Index

superposition principle 172, 174
superrenormalizable 359
support 269
Sylvester matrix 231
symmetrized product 255
symmetry 149, 158, 161, 162, 265, 266
  group 266
  species 399
symplectic geometry 179
symplectic manifold 183, 186

tachyon monopoles 323
tachyons 322
TCNQ,I 462
tensor product decomposition theorem 338
tensor products 287
Teple's formula 282
thermodynamics 11
thermodynamic state 449
Thirring energy, momentum tensor of 336
Thirring field 335
  equations 336
Thirring model 332, 335
time–energy uncertainty relation/principle 13, 117, 132
time operator 23
time reversal 297, 318, 547, 548
TMPD 475
total 271, 272
trace norm 252
transformation group 266
transformation law 333
transition probabilities 249
transitivity 266
truncation 282
TTF 465
twistor theory 57
two-point function 328, 338
unbounded 227
  alpha-particle states 501
uncertainty correlation 25
uncertainty principle 147, 149
unit beams 275
unitarily equivalent 258
unitary representations 331

unitary transformations in Hilbert space 4
universal covering space 330

van Hove's phenomenon 358
Van Vleck determinant 170
Van Vleck solution 170
variation principle 410, 421, 423, 428, 432
vector state 258
vibrations 277
viscosity 10
von Neumann's theorem 253

wave function 166
  reduction of 113, 114
  time-dependent 398
  time-independent 398
wave packets 21, 310, 311
Weinstein–Aronszajn determinant 280
Weinstein determinant 179
  modified 280
Weinstein's methods of intermediate problems 277
Weyl commutation relations 227
Weyl form 227
Weyl group 227, 327
Weyl prescription 156, 157
Weyl–Moyal product 155
Weyl–Wigner–Moyal formulation 150
Wheeler–Everett theory 53
white dwarfs 441, 442, 443, 446, 449, 451, 452, 454, 455, 456, 457
  evolution of 450, 452, 453
Wick ordering 355
Wigner function 157, 166
Wigner–Seitz sphere 447
Wilson expansion 337
WKB 158
WKB-expansion 167, 185
WKB solution 169

yes–no measurement 202
yes–no observable 300, 312, 313

zenithal 228
Zitterbewegung 414